MicroMechatronics

Second Edition

MicroMechatronics

Second Edition

Kenji Uchino

CRC Press is an imprint of the
Taylor & Francis Group, an **informa** business

CRC Press
Taylor & Francis Group
6000 Broken Sound Parkway NW, Suite 300
Boca Raton, FL 33487-2742

First issued in paperback 2021

ISBN 13: 978-1-03-224069-5 (pbk)
ISBN 13: 978-0-367-20231-6 (hbk)

Publisher's Note
The publisher has gone to great lengths to ensure the quality of this reprint but points out that some imperfections in the original copies may be apparent.

Library of Congress Cataloging-in-Publication Data

Names: Uchino, Kenji, 1950- author.
Title: Micromechatronics / Kenji Uchino.
Description: Second edition. | Boca Raton, FL : CRC Press/Taylor & Francis Group, 2019. | Includes biblographical references and index.
Identifiers: LCCN 2019008996 | ISBN 9780367202316 (hardback : acid-free paper)
Subjects: LCSH: Mechatronics. | Microelectromechanical systems.
Classification: LCC TJ163.12 .U24 2019 | DDC 621--dc23
LC record available at https://lccn.loc.gov/2019008996

Visit the Taylor & Francis Web site at
http://www.taylorandfrancis.com

and the CRC Press Web site at
http://www.crcpress.com

Contents

Second Edition Preface

Micromechatronics, First Edition, was published in 2002. In the 15 years after the publication of this book, we have experienced various changes in the personal, academic, and industrial areas.

First of all, coauthor of the first edition, Prof. Jayne R. Giniewicz (formerly my student and a professor at Indiana University of Pennsylvania) passed away in 2003, and we lost many of our former friends and collaborators listed in the first edition Preface. On the other hand, we welcomed more younger-generation researchers who learned micromechatronics even from their undergraduate/graduate periods.

Second, after my introduction of new terminology, "micromechatronics," in 1979 for describing the application area of piezoelectric actuators, which was widely publicized in my first textbook, *Piezoelectric Actuators*, in 1983, rapid advances in semiconductor chip technology have led to a new term, micro-electro-mechanical-system (MEMS) or even nano-electro-mechanical-system (NEMS) to describe mainly thin-film sensor/actuator devices, a narrower area of my micromechatronics coverage.

Third, regarding industrial commercialization, we can point out at least the following million-selling products in the last 15 years, to which my efforts contributed:

- Inkjet printer (piezoelectric) by Epson
- Diesel injection valve (multilayer) by Siemens, Bosch, Denso (Peugeot, Toyota)
- Camera module for mobile phones (micro ultrasonic motor) by Samsung Electromechanics (Galaxy series)
- Backlight inverter (piezoelectric transformer) for laptop computers by Apple (17" MacBook)
- Piezoelectric energy harvesting device for programmable air-burst munitions by Micromechatronics Inc. (U.S. Army)

The motivations for a new edition of *Micromechatronics* are follows:

- New technologies, product developments, and commercialization provide a requirement to update the book's contents in parallel to the deletion of old contents.
- In particular, progress in high-power transducers, loss mechanisms in smart materials, energy harvesting, and computer simulations has been significant, and these topics should be revised drastically.
- We found multiple typos in the mathematical derivations/formulas because of the old computers' capabilities.
- From these 15 years of classes, various educational/instructional example problems have been accumulated, which were integrated into the new edition in order to facilitate self-learning for students and quiz/problem creation for instructors.
- As one of the pioneers in piezoelectric energy harvesting, I feel a sort of frustration on many of the recent research papers regarding the following points:
  a. Though the electromechanical coupling factor k is the smallest (i.e., the energy conversion rate from the input mechanical to electric energy is the lowest) among various device configurations, the majority of researchers primarily use the "unimorph" design in piezoelectric energy harvesting. Why?
  b. Though the typical noise vibration is in a much lower frequency range, researchers measure the amplified resonance response (even at a frequency higher than 1 kHz) and report this unrealistically harvested electric energy. Why?
  c. Though the harvested energy is lower than 1 mW, which is lower than the required electric energy to operate a typical energy-harvesting electric circuit with a DC/

DC converter (typically around 2–3 mW), researchers report the result as an energy "harvesting" system. Does this situation mean actually energy "loss"? Why?

d. Few papers have reported complete energy flow or exact efficiency from the input mechanical noise energy to the final electric energy in a rechargeable battery via the piezoelectric transducer step by step. Why?

Interestingly, the unanimous answer from these researchers to my question "Why?" is "Because the previous researchers did so," because of such easy adoption of Google search. Researchers are asked to forget the current "biased" knowledge or not to use a strategy just "because the previous researchers did so."

Otherwise, innovative breakthroughs may not happen. I present in the text a theoretical description of solid-state actuators and an overview of practical materials, device designs, drive/control techniques, and typical applications, and we consider current and future trends in the field of micromechatronics. I intend to provide the readers most of the basic, fundamental knowledge, which might not create any problematic "biased" thought in their flexible minds. An overview of the field and recent developments in microactuator technology are presented in Chapter 1, along with an overview of the materials to be presented in the text. Theoretical basics, such as origin of ferroelectricity, piezoelectric tensor treatment, and phenomenology, are discussed in Chapter 2. Practical materials are introduced in Chapter 3, including loss mechanisms and heat generation in high-power transducer materials. Chapter 4 describes transducer designs and manufacturing processes, and drive/control techniques are described in Chapter 5.

A high-power characterization system (HiPoCS) is introduced first, followed by serve control systems and pulse drive and resonance operation methods. Chapter 6 is dedicated entirely to the modeling of these effects by means of finite-element analysis (ATILA) and circuit analysis with PSpice. ATILA demonstration software is available from Micromechatronics Inc., State College, PA. OrCAD PSpice for Windows is available via internet shop easily, or most electrical engineering departments at universities provide it to students free of charge. A new topic on energy harvesting was added as Chapter 7 in this edition. Three distinct classes of applications are presented in the following chapters. Servo displacement transducer applications are presented in Chapter 8, while pulse drive motor and ultrasonic motor applications are described in Chapter 9 and Chapter 10, respectively. Finally, in Chapter 11, I consider the future of micromechatronics in terms of the technological and economic impact these new systems will have on industry and society.

This textbook is intended for graduate students and industrial engineers studying or working in the fields of electronic materials, control system engineering, optical communications, precision machinery, and robotics. This textbook is designed primarily for a graduate course with the equivalent of thirty 75-minute lectures; however, it is also suitable for self-study by individuals wishing to extend their knowledge in the field. Because the development of actuator devices for micropositioning, motors, and energy harvesting systems is highly interdisciplinary, it would be difficult for junior engineers with narrow expert areas. Hence, I have selected what I feel are the most important and fundamental topics to provide an adequate understanding of the design and application of actuator devices for micromechatronic systems in order to escape from the current "biased" knowledge, or adopting a strategy "because the previous researchers did so."

Finally, a special expression of gratitude is extended to the late Dr. Robert E. Newnham and the late Dr. L. Eric Cross of Pennsylvania State University for their initial recruitment of me to Penn State University from Japan and continuous encouragement and support to keep my directorship at the ICAT Research Center. I also express my appreciation for my wife, Michiko, and her kind daily support throughout the project of this edition.

Kenji Uchino, PhD, MS, MBA
State College, PA

First Edition Preface

Remarkable developments have taken place in the field of mechatronics in recent years. As reflected in the "Janglish" (Japanese English) word *mechatronics*, the concept of integrating electronic and mechanical devices in a single structure is one that is already well accepted in Japan. Currently, robots are employed to manufacture many products in Japanese factories, and automated systems for sample displays and sales are utilized routinely in the supermarkets of Japan. Further, the rapid advances in semiconductor chip technology have led to the need for micro (or nano in these days) displacement positioning devices. Actuators that function through the piezoelectric, magnetostriction, and shape memory effects are expected to be important components in this new age of *micromechatronic* technology. Kenji Uchino introduced the terminology, "Micromechatronics" in 1979 for describing the application area of "piezoelectric actuators," which was widely publicized in his first textbook "Piezoelectric Actuators" in 1983.

This book is the latest in a series of texts concerned with "piezoelectric actuators" written by Kenji Uchino. The first entitled *Essentials for Development and Applications of Piezoelectric Actuators* was published in 1984 through the Japan Industrial Technology Center. The second, *Piezoelectric/ Electrostrictive Actuators*, published in Japanese by the Morikita Publishing Company (Tokyo), became one of the best-sellers of that company in 1986, and was subsequently translated into Korean. The problem-based text, *Piezoelectric Actuators -Problem Solving*, was also published through Morikita and was sold with a 60 minute instructive video tape. The English translation of that text was completed and published in 1996 through Kluwer Academic Publishers. This sixth text in the series, written in collaboration with Jayne Giniewicz, offers updated chapters on topics presented in the earlier texts as well as new material on current and future applications of actuator devices in micromechatronic systems. The contents were significantly expanded to include magnetostrictive, shape memory, MEMS actuators, and rheological dampers.

We present in the text a theoretical description of solid-state actuators, an overview of practical materials, device designs, drive/control techniques, and typical applications, and we consider current and future trends in the field of micromechatronics. An overview of the field and recent developments in micro-actuator technology are presented in Chapter 1 along with an overview of the materials to be presented in the text. Theoretical basics and practical materials, design and manufacturing considerations are covered in Chapters 2 through 4 and drive/control techniques are described in Chapter 5. The unique challenges of modeling and practically regulating electromechanical losses and heat generation in actuator devices are addressed in Chapter 6. Chapter 7 is dedicated entirely to the modeling of these effects by means of finite element analysis and a demonstration of these methods is included on the CD-ROM supplement that accompanies the text. Three distinct classes of applications are presented in Chapters 8 through 10. Servo displacement transducer applications are presented in Chapter 8, while pulse drive motor and ultrasonic motor applications are described in Chapter 9 and Chapter 10, respectively. Finally, in Chapter 11 we consider the future of micromechatronics in terms of the technological and economic impact these new systems will have on industry and society.

This textbook is intended for graduate students and industrial engineers studying or working in the fields of electronic materials, control system engineering, optical communications, precision machinery, and robotics. The text is designed primarily for a graduate course with the equivalent of thirty 75-minute lectures; however, it is also suitable for self-study by individuals wishing to extend their knowledge in the field. The CD-ROM supplement that accompanies the text includes a 57-minute video in which some commercially produced actuators are presented and some examples of finite-element modeling are demonstrated. We are indebted to Dr. Philippe Bouchilloux, of Magsoft Corporation, NY, for creating this supplement and for writing the introduction to the finite element method presented in Chapter 7 of the text (He is now in Samsung Electromechanics, Korea). Because the development of actuator devices for micropositioning and motors is a new, dynamic, and highly interdisciplinary field, it would be impossible to cover the full range of related material in a single text. Hence, we have selected what we feel are the most important and fundamental topics to provide an adequate understanding of the design and application of actuator devices for micromechatronic systems.

We would like to acknowledge in particular the following individuals for permitting the citation of their studies in the application chapters of the text (the then-affiliation is shown):

Takuso Sato, Tokyo Institute of Technology
Katsunori Yokoyama and **Chiaki Tanuma**, Toshiba R&D Laboratory
Shigeo Moriyama and **Fumihiko Uchida**, Hitachi Central Research Laboratory
Teru Hayashi and **Iwao Hayashi**, Tokyo Institute of Technology, Precision Engineering Laboratory
K. Nakano and **H. Ohuchi**, Tokyo Institute of Technology, Precision Engineering Laboratory
Sadayuki Takahashi, NEC Central Research Laboratory
Ken Yano, NEC Communication Transfer Division
Michihisa Suga, NEC Microelectronics Laboratory
Toshiiku Sashida, Shinsei Industry
E. Mori, S. Ueha, and **M. Kuribayashi,** Tokyo Institute of Technology, Precision Engineering Laboratory
Akio Kumada, Hitachi Maxel
Kazumasa Onishi, ALPS Electric
Yoshiro Tomikawa and **Seiji Hirose**, Yamagata University
Kazuhiro Otsuka, Smart Structure Center, MISc
Yasubumi Furuya, Hirosaki University
Toshiro Higuchi, University of Tokyo
Ben K. Wada, Jet Propulsion Laboratory
J. L. Fanson and **M. A. Ealey**, Jet Propulsion Laboratory and Xinξtics
H. B. Strock, Strock Technology Associates
A. B. Flatau, Iowa State University
A. E. Clark, Clark Associates
Nesbit Hagood, Masachussets Institute of Technology
Mark R. Jolly and **J. David Carlson**, Lord Corporation

For the reader who seeks detailed information on various actuator technology, we recommend *The Handbook on New Actuators for Precision Control* (1087 pages) edited by the Solid State Actuator Study Committee, JTTAS (editor-in-chief, K. Uchino) and published by Fuji Technosystem, Tokyo in 1995. A thorough introduction to ferroelectric materials and applications is presented in the text entitled *Ferroelectric Devices* by Kenji Uchino (published by Dekker/CRC Press in 2000) and is strongly recommended for readers interested in learning more about ferroelectric materials.

Finally, we express our gratitude to our colleagues at the Pennsylvania State University and the Indiana University of Pennsylvania. A special expression of gratitude is extended to Dr. Robert E. Newnham and Dr. L. Eric Cross of the Pennsylvania State University for their continuous encouragement and support during the preparation of this text. Kenji Uchino expresses his appreciation of his wife, Michiko, and her kind daily support throughout the project. Jayne Giniewicz wishes to thank her colleagues and friends in the department of physics and Dean John Eck of the Indiana University of Pennsylvania for their enthusiastic support of her participation in this project. Any good that she might produce in this or any task at hand is, as always, dedicated with the fullest heart to John and Theo Giniewicz.

Kenji Uchino
Jayne Giniewicz
State College and Indiana, PA

Author

Kenji Uchino, a pioneer in piezoelectric actuators, is the founding director of the International Center for Actuators and Transducers (ICAT) and professor of Electrical Engineering and Materials Science & Engineering at the Pennsylvania State University. He was associate director ("Navy Ambassador to Japan") at The U.S. Office of Naval Research—Global Tokyo Office as IPA from 2010 to 2014. He was also the founder and senior vice president & CTO of Micromechatronics Inc., State College, PA. He is currently teaching three graduate courses entitled "Micromechatronics," "Ferroelectric Devices," and "Application of Finite Element Method" at the Penn State, using this textbook and another, *Ferroelectric Devices, Second Edition*, published by CRC Press in 2010.

He also occasionally teaches "Entrepreneurship for Engineers" at the Business School, using a text with the same name published by CRC Press in 2010. Uchino earned his MS and PhD from the Tokyo Institute of Technology, Japan. His MBA degree was awarded from St. Francis University, PA (2008), in parallel with the Micromechatronics Inc. operation. He became research associate/ assistant professor (1976) in the Physical Electronics department at that university, then joined Sophia University (Tokyo, Japan) as associate professor in the Physics Department in 1985. He was recruited from Penn State in 1991. He was also involved with the Space Shuttle Utilizing Committee in NASDA, Japan, from 1986–1988, and was vice president of NF Electronic Instruments, United States, from 1992–94. So far, in total, Uchino has been a university professor for 44 years (18 years in Japan, 28 years at Penn State), a company executive (president, vice president) for 21 years at five Japanese and U.S. firms, and a government officer for 7 years (NASDA and the U.S. Office of Naval Research).

He was the founding chair of the Smart Actuators/Sensors Committee, Japan Technology Transfer Association sponsored by the Ministry of Economics, Trading, and Industries, Japan, from 1987 to 2014, and has been a long-term chair of the International Conference on New Actuators, Messe Bremen, Germany, since 1997. He was also the associate editor for *Journal of Advanced Performance Materials*, *Journal of Intelligent Materials Systems and Structures*, and *Japanese Journal of Applied Physics*. Uchino served as Administrative Committee Member (Elected) of IEEE Ultrasonics, Ferroelectrics, and Frequency Control (1998–2000) and as Secretary of the American Ceramic Society, Electronics Division (2002–2003).

Professor Uchino's research interest is in solid state physics, especially in ferroelectrics and piezoelectrics, including basic research on theory, materials, and device designing and fabrication processes, as well as application development of solid state actuators/sensors for precision positioners, microrobotics, ultrasonic motors, smart structures, piezoelectric transformers, and energy harvesting. Uchino is known as the discoverer/inventor of the following famous topics: (1) lead magnesium niobate (PMN)-based electrostricive materials, (2) cofired multilayer piezoelectric actuators (MLAs), (3) superior piezoelectricity in relaxor-lead titanate-based piezoelectric single crystals (PZN-PT), (4) photostrictive phenomenon, (5) shape memory ceramics, (6) magnetoelectric composite sensors, (7) transient response control scheme of piezoelectric actuators (pulse-drive technique), (8) micro ultrasonic motors, (9) multilayer disk piezoelectric transformers, and (10) piezoelectric loss characterization methodology. His ongoing research projects are also in the above areas, especially in the last three items, (8), (9), and (10), recently. He has authored 582 papers and 77 books, and holds 33 patents in the ceramic actuator area. Forty-nine papers/books among his publications have been cited more than 100 times, leading to his average h-index of 71. His total citation number, 24,200, and annual average citation number, 560, are very high in the College of Engineering.

He is a fellow of the American Ceramic Society (1997), a fellow of IEEE (2012), and also a recipient of 29 awards, including Distinguished Lecturer of the IEEE UFFC Society (2018);

International Ceramic Award from Global Academy of Ceramics (2016); IEEE-UFFC Ferroelectrics Recognition Award (2013); Inventor Award from Center for Energy Harvesting Materials and Systems, Virginia Tech (2011); Premier Research Award from the Penn State Engineering Alumni Society (2011); the Japanese Society of Applied Electromagnetics and Mechanics Award on Outstanding Academic Book (2008); Society of Photo-Optical Instrumentation Engineers (SPIE), Smart Product Implementation Award (2007); R&D 100 Award (2007); American Society of Mechanical Engineers (ASME) Adaptive Structures Prize (2005); Outstanding Research Award from Penn State Engineering Society (1996); Academic Scholarship from Nissan Motors Scientific Foundation (1990); Best Movie Memorial Award at Japan Scientific Movie Festival (1989); and the Best Paper Award from the Japanese Society of Oil/Air Pressure Control (1987). He is also one of the founding members of the Worldwide University Network, which has encouraged linking between multiple U.K. and U.S. universities since 2001.

List of Symbols

D	Electric displacement
E	Electric field
P	Dielectric polarization
P_s	Spontaneous polarization
α	Ionic polarizability
γ	Lorentz factor
μ	Dipole moment
ε_o	Dielectric permittivity of free space
ε	Dielectric permittivity
ε (ε_r)	Relative permittivity or dielectric constant (assumed for ferroelectrics: $\varepsilon \approx \chi = \varepsilon - 1$)
κ	Inverse dielectric constant
χ	Dielectric susceptibility
C	Curie-Weiss constant
T_0	Curie-Weiss temperature
T_C	Curie temperature (phase transition temperature)
G	Gibbs free energy
A	Helmholtz free energy
F	Landau free energy density
x	Strain
x_s	Spontaneous strain
X	Stress
s	Elastic compliance
c	Elastic stiffness
v	Sound velocity
d	Piezoelectric charge coefficient
h	Inverse piezoelectric charge coefficient
g	Piezoelectric voltage coefficient
M, Q	Electrostrictive coefficients
k	Electromechanical coupling factor
η	Energy transmission coefficient
Y	Young's modulus
$\tan \delta$ ($\tan \delta'$)	Extensive (intensive) dielectric loss
$\tan \phi$ ($\tan \phi'$)	Extensive (intensive) elastic loss
$\tan \theta$ ($\tan \theta'$)	Extensive (intensive) piezoelectric loss

Suggested Teaching Schedule

[Thirty 75-minute sessions per semester]

	Topics	Sessions
0	Course Overview and Prerequisite Knowledge Check	1 time
1	Current Trends for Actuators and Micromechatronics	5 times
2	Theoretical Description of Piezoelectricity	4 times
3	Actuator Materials	2 times
4	Fabrication Processes and Actuator Structures	2 times
5	Drive/Control Techniques	4 times
6	Finite-Element Analysis and PSpice Simulation	2 times
	Laboratory Demonstrations	
7	Piezoelectric Energy Harvesting Systems	2 times
8	Servo Displacement Transducer Applications	1 times
9	Pulse Drive Motor Applications	1 times
10	Ultrasonic Motor Applications	4 times
11	Future of Micromechatronics	1 time
	Review/Q&A	1 time

Prerequisite Knowledge Check

The study of micromechatronics assumes certain basic knowledge. Answer the following questions by yourself prior to referring to the answers on the next page.

Q1 Provide definitions for the *elastic stiffness*, c, and *elastic compliance*, s, using stress (X) – strain (x) equations.

Q2 Sketch a *shear stress* (X_4) with arrows and the corresponding *shear strain* (x_4)/*deformation* on the square material depicted below.

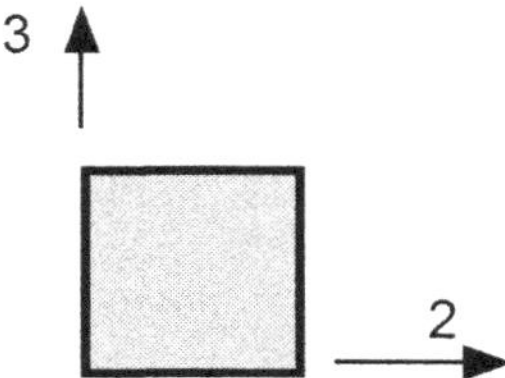

Q3 Describe an equation for the *velocity of sound*, v, in a material with mass density ρ and elastic compliance s.

Q4 Given a rod of length L, made of a material through which sound travels with a velocity v, describe an equation for the *fundamental extensional resonance frequency*, f_R.

Q5 Fill in the blank for the cosine function on the right-hand side of the following equation: $\cos(kx)\cos(\omega t) + \cos(kx - \pi/2)\cos(\omega t - \pi/2) = \cos[?]$.

Q6 Provide the capacitance, C, of a capacitor with area A and electrode gap t, filled with a material of *relative permittivity*, ε_r.

Q7 Describe an equation for the *resonance frequency* of the circuit pictured below:

Q8 Describe the *Laplace transform* for the impulse function [$\delta(t)$].

Q9 Given a power supply with an internal impedance, Z_0, what is the optimum circuit impedance, Z_1, required for maximum power transfer?

Q10 Provide the polarization, P, induced in a *piezoelectric* material with a piezoelectric strain coefficient, d, when it is subjected to an external stress, X.

Answers

[60% or better score is expected.]

Q1 $X = c\,x, x = s\,X$

Q2 $x_4 = 2\,x_{23} = 2\,\phi$

[*Note*: Radian measure is generally preferred. This shear stress is not equivalent to the diagonal extensional stress.]

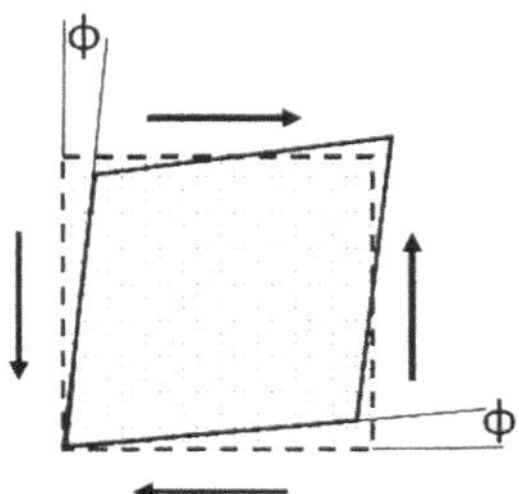

Q3 $v = 1/\sqrt{\rho s^E}$

Q4 $f = v/2L$

Q5 $(kx - \omega t)$ [A traveling wave is obtained by superimposing two standing waves.]

Q6 $C = \varepsilon_0 \varepsilon_r (S/t)$

Q7 $f = 1/2\pi\sqrt{LC}$

Q8 1

[*Note*: The impulse function is occasionally used to obtain the transfer function of the system.]

Q9 $Z_1 = Z_0$

[*Note*: The current and voltage associated with Z_1 are $V/(Z_0 + Z_1)$ and $[Z_1/(Z_0 + Z_1)]V$, respectively, the product of which yields the power. The maximum power transfer occurs when $Z_0/\sqrt{Z_1} = \sqrt{Z_1}$ (when impedance is resistive).]

Q10 $P = d\,X$

[This is called the direct piezoelectric effect.]

1 Current Trends for Actuators and Micromechatronics

1.1 THE NEED FOR NEW ACTUATORS

An *actuator* is a transducer that transforms drive energy into a mechanical displacement or force. The demand for new actuators increased significantly after the 1980s, especially for *positioner*, *mechanical damper*, and *miniature motor* applications. We started so-called *micromechatronics* to cover two different areas:

- The system is not "micro," but can manipulate the object with micro- to nanometer accuracy.
- The system/device itself is actually micro scale.

Submicrometer fabrication, common in the production of electronic chip elements, also became important in mechanical engineering. Though sensors utilizing lasers can easily detect nanometer-scale displacements, the fabrication of such precise accuracy requires submicron-meter machining equipment. In an actual machining apparatus composed of translational components (the joints) and rotating components (the gears and motor), error due to backlash will occur. Machine vibration will also lead to unavoidable position fluctuations. Furthermore, the deformations due to machining stress and thermal expansion cannot be ignored, either. The need for submicron displacement positioners to improve cutting accuracy is apparent. One example is a lathe machine that uses a ceramic multilayer actuator to adjust the cutting edge position and can achieve cutting accuracy of 2 nm, shown in Figure 1.1.[1] We will handle this product in detail in Chapter 5.

The concept of *adaptive optics* has been applied in the development of sophisticated optical systems. Earlier systems were generally designed such that parameters like position, angle, or the focal lengths of mirror and lens components remained essentially fixed during operation. Newer systems incorporating adaptive optical elements respond to a variety of conditions to essentially adjust the system parameters to maintain optimum operation. The original "lidar" system (a radar system utilizing light waves) was designed to be used on the NASA space shuttle for monitoring the shape of the galaxy.[2] Even from the space shuttle, some distribution of thin air (i.e., wind) disturbs a sharp image of the galaxy, in addition to the vibration noise and temperature fluctuation of the shuttle, and the use of a responsive positioner was considered to compensate for the detrimental effects, as shown in Figure 1.2. This product will be discussed in detail in Chapter 8.

Active and passive vibration suppression by means of solid-state devices is also a promising technology for use in space structures and military and commercial vehicles. Mechanical vibration in a structure traveling through the vacuum of space is not readily damped, and a 10-m-long array of solar panels can be severely damaged simply by the repeated impact of space dust with the structure. Active dampers using shape memory alloys or piezoelectric ceramics are currently under investigation to remedy this type of problem. After the Big Earthquake incident in the northern part of Japan in 2011, research and development on antiearthquake building structures and structure health monitoring has become accelerated. Though the size of the whole structure is huge, the superelastic behavior of shape memory alloys (SMAs) is related to micron-scale atomic phase transformation, which has attracted the attention of civil engineers.[3] A real-scale example of a superelastic SMA device is the earthquake-resistant retrofit of the Basilica San Francesco at Assisi, Italy (M. G. Castellano). The historic gable was connected with the main structure by devices using

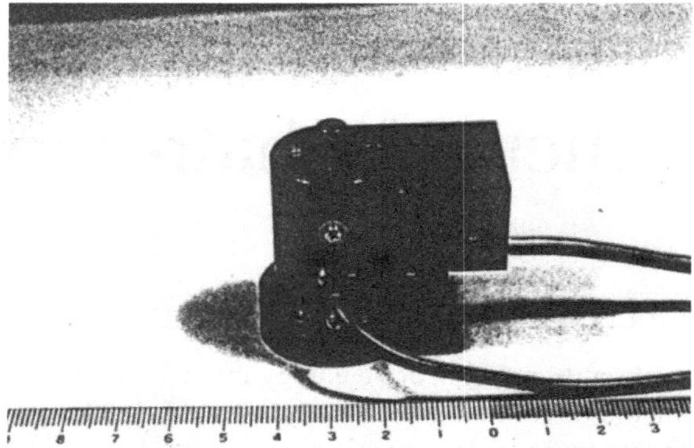

FIGURE 1.1 Ceramic multilayer actuator for a precision lathe application with 2-nm cutting accuracy. (From K. Uchino: *J. Industrial Education Soc. Jpn.* 40(a), 28, 1992.)

SMA rods, as shown in Figure 1.3. Regardless of the structure deformation, the SMA alloy keeps the same compressive force on the concrete and the gable.

The demand for other applications in the field of mechanical engineering is also increasing rapidly. One important class of devices that meets these demands is the *solid-state motor.* Market research focused on office equipment needs, such as printers and camera modules, indicates that tiny motors smaller than 10 mm will be in increasing demand (market research conducted by Uchino's group in the late 1980s). Conventional electromagnetic (EM) motor designs, however, do not provide sufficient energy efficiency for these applications. We searched motor characteristics for hundreds of commercial electromagnetic motors and plot their specific power ($\propto$ efficiency) vs. power ($\propto$ actuator size) in Figure 1.4 (data in the year 1988, unpublished). Piezoelectric motor performance is also indicated in the same figure. In general, the total power is proportional to the volume (L^3). On the contrary, the specific power ($\propto$ efficiency) is almost constant in piezoelectric motors, while that of the EM motors seems to follow the L^2, leading to a 2/3 slope between log (specific power) and log (power) (dashed line on the figure). The main reason for this trend originates from the Joule heat in the thin coil wire (usually Cu wire). Since the coil resistance is given by $\rho \times$ (area/length) $\propto L$, the heat generation is proportional to L^{-1}, leading to the 2/3 slope in Figure 1.4. The micromotor in a wrist watch has an efficiency less than 1%, in fact. Thus, piezoelectric *ultrasonic motors* (USMs), whose efficiency is insensitive to size (around 30%), are superior to conventional devices when

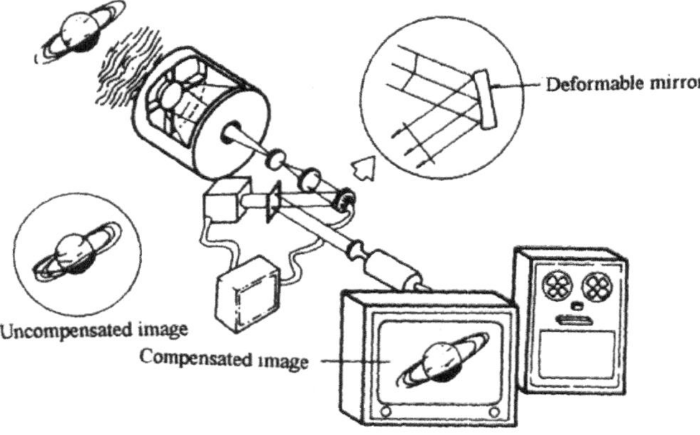

FIGURE 1.2 A telescope image correction system using a monolithic piezoelectric deformable mirror. (From J.W. Hardy, J.E. Lefebre and C.L. Koliopoulos: *J. Opt. Soc. Amer.* 67, 360, 1977.)

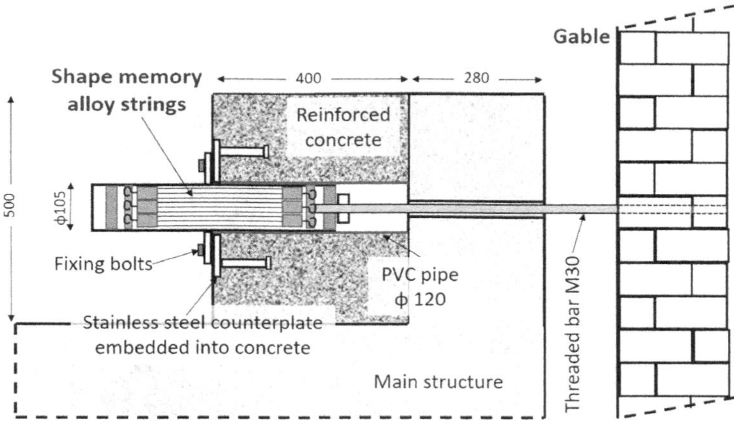

FIGURE 1.3 Shape memory alloy device for earthquake suitable connection of the historic gable and the main structure of the Basilica San Francesco in Assisi, Italy. (From S.R. Debbarma and S. Saha: *Int'l J. Civil & Structural Eng.* 2, 924, 2012.)

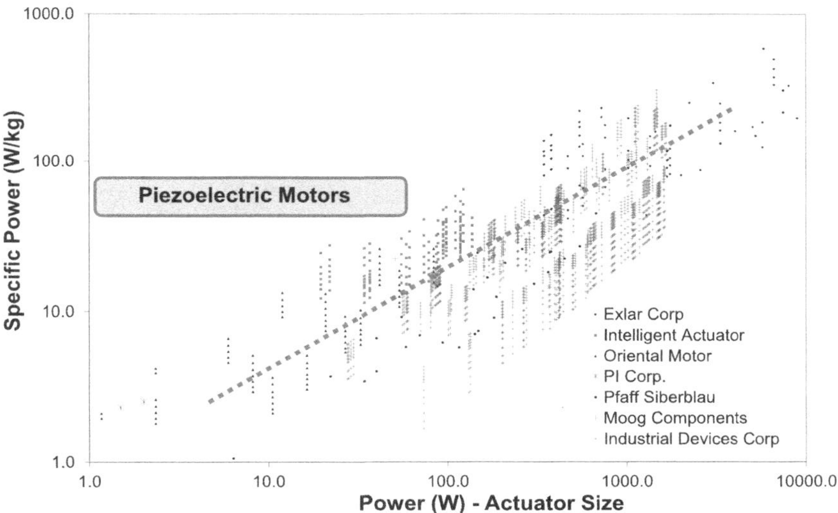

FIGURE 1.4 Specific power (∝ efficiency) vs. power (∝ actuator size) for various electromagnetic motors. Piezoelectric motor performance is also inserted [Unpublished].

motors of millimeter size are required. Note that the 30 W-level EM motor shows an efficiency of ~30%. Thus, piezomotors are superior to EM motors in the micromotor category less than 30 W. Figure 1.5 shows a 2-mm piezoultrasonic motor in the camera module of the Samsung Galaxy 6S. Details of microultrasonic motors are discussed in Chapter 10.

When we started the *micromechanism* field in the late 1980s, we searched the moving principles of various insects (this may be the original "bio-mimetic" engineering period). Figure 1.6 summarizes moving principle change with insect size. A large insect like a cockroach (~cm) moves with its six legs. With reducing the body size down to ~mm like a flea, it does not use six legs, but only two legs for jumping. For a much smaller body size, ~100 μm like paramecium and amoeba, the insect uses the body cilia or surface movement. The key factor is a ratio (surface area $\propto L^2$)/(volume or weight $\propto L^3$) to change the moving fashion. With a reduction in body weight, the friction force

FIGURE 1.5 Piezoultrasonic motor in a camera module of Samsung Galaxy 6S. (Courtesy of Samsung EM.)

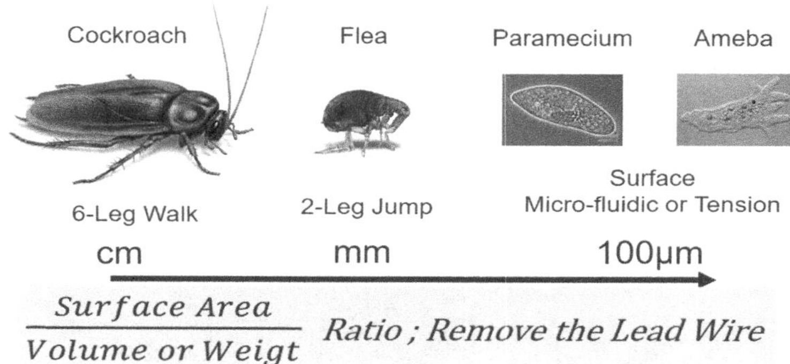

FIGURE 1.6 Moving principle change with the insect size. The key factor is a ratio (surface area)/(volume) to change the moving fashion.

obtained by the small leg contact area is not sufficient for body propulsion. Thus, full surface friction is used for a micromechanism or microrobot. Piezoelectric USMs using surface friction have been developed with this development strategy in general. One of the smallest four-wheel-drive vehicles was developed in the International Center for Actuators and Transducers at Penn State University in the early 2000s, as shown Figure 1.7, and it could climb up and down on Uchino's index finger (unpublished). Photostrictive actuators have been developed in accordance with the same direction; since the electric leadwire is too heavy to drag by a micromachine, elimination of the leadwire, that is, remote control, is definitely required. This is introduced in Section 1.3.8 in this chapter.

1.2 CONVENTIONAL METHODS FOR POSITIONING

A classification of actuators is presented in Table 1.1. This classification is based on the features of the actuator that relate to micropropulsion and positioning controllability. Electrically controlled types are generally preferred for applications where miniature devices are needed. A few of the relevant specifications for solid-state actuators included in this classification also appear in the table. Compared with conventional devices, the new-principle actuators provide much quicker response, smaller size, higher resolution, and a higher power-to-weight ratio.

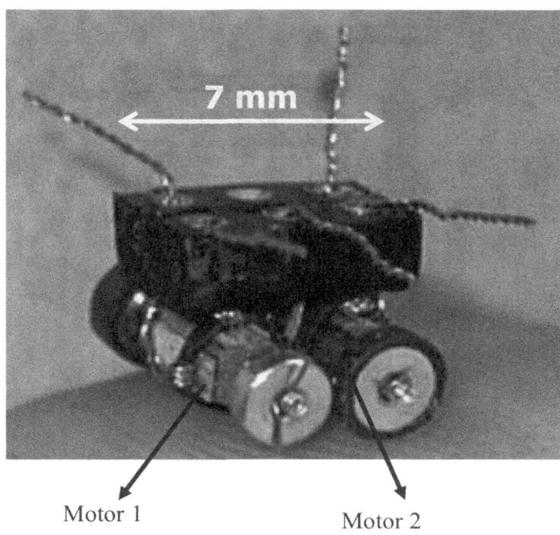

FIGURE 1.7 One of the smallest four-wheel-drive vehicles with piezomotors.

Conventional methods for micropositioning usually include displacement reduction mechanisms to suppress mechanical backlash, which are categorized into three groups: (1) oil/air pressure displacement reduction, (2) electromagnetic rotary motor with a gear, and (3) voice coil motor.[4] A brief description of each follows.

1.2.1 Oil Pressure Type Displacement Reduction Mechanism

Changing the diameter of an oil-filled cylinder (like a syringe), as illustrated in Figure 1.8, effectively reduces the resulting displacement at the output. Devices utilizing this oil pressure mechanism are generally large and have slow responses. Transducers of this type are sometimes used in reverse mode for amplifying the displacement produced by a solid-state actuator such as a piezoelectric multilayer actuator.

TABLE 1.1
Displacement Characteristics of Various Types of Actuators

Drive	Device	Displacement	Accuracy	Torque/ Generative Force	Response Time
Air pressure	Motor Cylinder	Rotation 100 mm	Degrees 100 μm	50 Nm 10^{-1} N/mm^2	10 sec 10 sec
Oil pressure	Motor Cylinder	Rotation 1000 mm	Degrees 10 μm	100 Nm 100 N/mm^2	1 sec 1 sec
Electricity	AC Servo Motor	Rotation	Minutes	30 Nm	100 msec
	DC Servo Motor	Rotation	Minutes	200 Nm	10 msec
	Stepper Motor	1000 mm	10 μm	300 N	100 msec
	Voice Coil Motor	1 mm	0.1 μm	300 N	1 msec
	Piezoelectric	100 μm	0.01 μm	30 N/mm^2	0.1 msec
	Magnetostrictor	100 μm	0.01 μm	100 N/mm^2	0.1 msec
	Ultrasonic Motor (piezoelectric)	Rotation	Minutes	1 Nm	1 msec

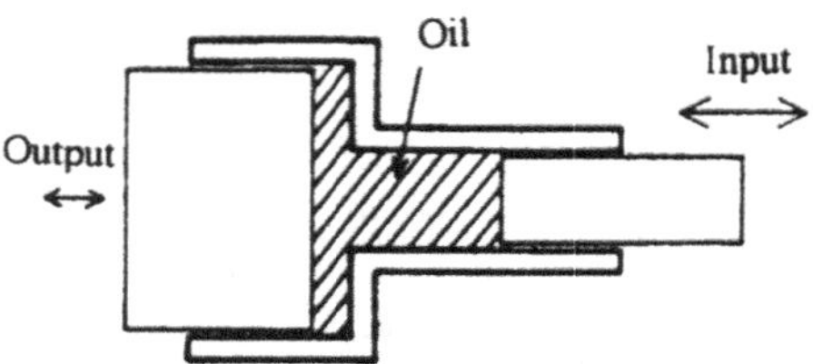

FIGURE 1.8 Schematic representation of an oil pressure displacement reduction mechanism.

1.2.2 Pure Mechanical Displacement Reduction Mechanism

There are several commonly used displacement reduction mechanisms. The mechanisms highlighted in Figure 1.9 illustrate the action of screw transfer mechanisms, a wedge, a spring constant difference, and a hinge lever.

Screw transfer mechanisms are typically used when the moving distance is long (see Figure 1.9a). Using a very precise ball screw, positioning accuracy of less than 5 μm can be obtained for a 100-mm motion. When higher accuracy is required, additional displacement reduction mechanisms are necessary. The combination of a motor with a ball screw or a displacement reduction mechanism has the advantages of quick response, a substantial generative force, and good controllability, but is generally difficult to fabricate in miniature form due to its structural complexity. In addition, the manufacturing tolerances of a typical transfer screw tend to promote backlash in positioning even when displacement reduction mechanisms are implemented.

Figure 1.9b and c show a wedge type and a spring constant difference type, respectively, for which the reader can understand the principles easily. A hinge lever type displacement reduction as in Figure 1.9d is the most useful design, which is used as a displacement amplification mechanism with a piezoelectric actuator in conjunction with the MEMS fabrication technologies these days.

1.3 AN OVERVIEW OF SOLID-STATE ACTUATORS

We overview solid-state actuators in this section after reviewing advanced microelectromagnetic motors, pure electric micro-electro-mechanical-systems, and elastomer artificial muscles. The

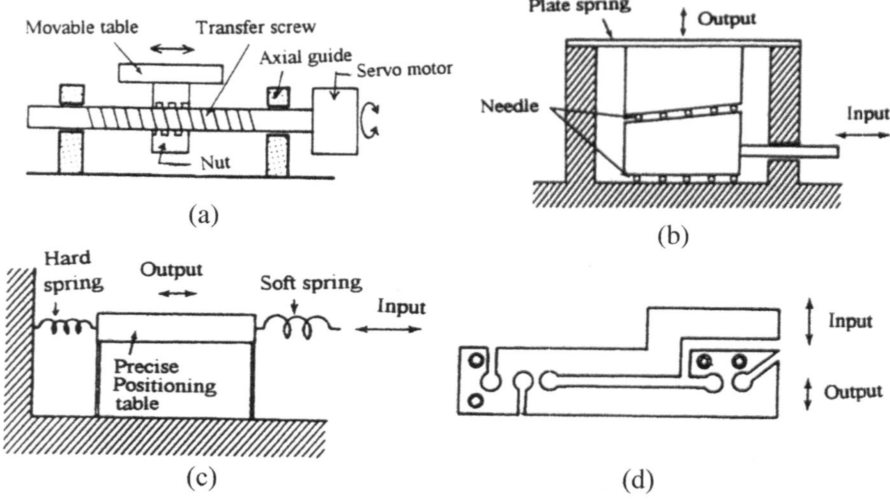

FIGURE 1.9 Displacement reduction mechanisms utilizing the action of: (a) a screw positioning mechanism, (b) a wedge, (c) a spring constant difference, and (d) a hinge lever.

operation principle is purely classic: electromagnetic, Lorentz, Coulomb, and Maxwell forces in the conventional types, while new material performances are used for solid-state actuators, such as shape memory function, magnetostrictive, piezoelectric, photostrictive effects, and electro/magneto-rheological fluid.

1.3.1 Microelectromagnetic Motors

Let us discuss the status of advanced EM motors first. The minimum size of an EM motor is generally limited to about 1 cm^3, as motors smaller than this will not provide adequate torque or efficiency. The principle of the EM motor is well known: the combination of permanent magnet and an alternating magnetic field generated by a coil. In the late 1990s, Institute of Microtechnique, Mainz GmbH, developed a micromotor with a diameter of 1.9 mm generating a torque of 7.5 μNm and rotational speed of 100,000 rpm.[5] An optional microgearbox with a reduction ratio of 47 can be used with this motor so that the drive can deliver an enhanced torque of up to 300 μNm. Figure 1.10 shows a more

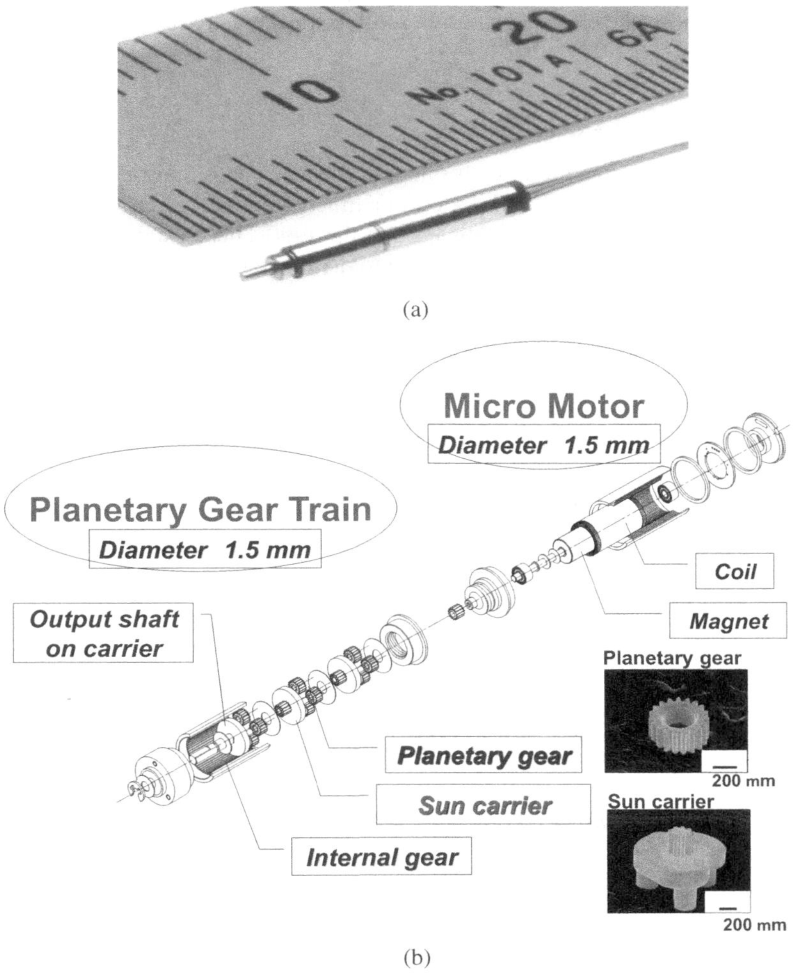

FIGURE 1.10 An EM micromotor with a diameter of 1.5 mm, developed by Namiki Precision Jewel Co., Ltd. (a) Product picture, and (b) motor structure with a planetary gear train. (http://www.namiki.net/index.htm) (Courtesy Namiki.)

advanced micromotor produced by Namiki Precision Jewel Co., Japan (Namiki Company Catalogue 2007).[6] A motor with 1.5 mm $\phi \times 9.95$ mm L was coupled with a planetary gear train (three-stage gear ratio = 254:1). Under a nominal voltage 3 V_{DC} and current 30 mA, no load speed 500 rpm and stall torque 439 μNm were realized. Use of this gearbox, however, reduces the efficiency of the motor significantly less than 1%. We need to point out again that with size reduction of the EM motor, the winding wire thickness must also be reduced, which leads to a significant increase in the electrical resistance and Joule heat. Since the microelectromagnetic motor becomes a "heater," we should not operate it continuously, but drive intermittently, as the wristwatch hand moves every second.

When we use a rotary type motor for translational positioning, we usually couple it with a screw mechanism, as introduced in Figure 1.9a. A voice coil motor is a direct linear-drive EM motor, the structure of which is shown schematically in Figure 1.11. The electromagnetic Lorentz force and the spring force balance to stabilize the movable coil position, which is controlled by the current. The voice coil motor achieves rather precise positioning and thus is popularly utilized for laser disk positioning, and so on. However, it requires relatively large input electrical energy (i.e., low efficiency), and provides a slow response and rather low generative forces because of a small spring constant. As you may know, the elastic springs of audible speakers are popularly made of a folded paper.

1.3.2 Micro-Electro-Mechanical-Systems

Silicon has become almost synonymous with integrated electronic circuitry. Due to its favorable mechanical properties, silicon can also be micromachined to create micro-electro-mechanical-systems.[7] The terminology *MEMS* started to be used in the 1990s, because *micromechatronics* was already taken in the 1980s by the author's group, and to distinguish their category merely related to silicon technologies. The techniques for micromachining silicon have developed since the 1980s in a variety of industries for a wide range of applications. Pressure and acceleration sensors are produced for application in medical instrumentation and automobiles. A successful example is the acceleration sensor used to trigger an air bag in an automobile crash.

In *bulk micromachining*, mechanical structures are fabricated directly on a silicon wafer by selectively removing wafer material. Figure 1.12 illustrates a *piezoresistive* sensor (i.e., resistance of a semiconductor highly depends on the mechanical stress) fabricated with bulk micromachining. Etching with KOH or tetramethyl-ammonium hydroxide (TMAH) is the primary technique for bulk micromachining, and is either isotropic, anisotropic, or a combination of the two states. The

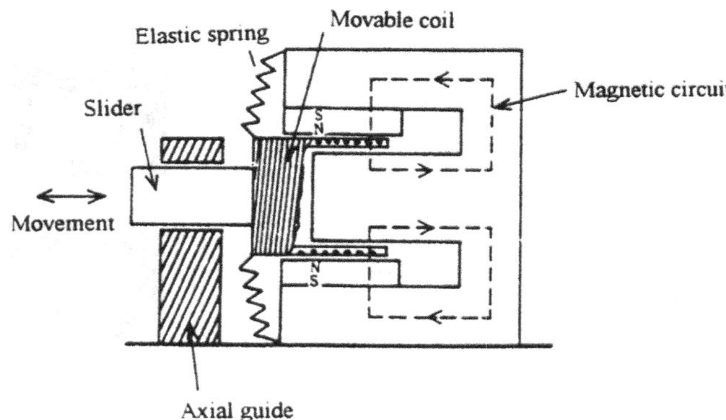

FIGURE 1.11 Schematic representation of a voice coil motor.

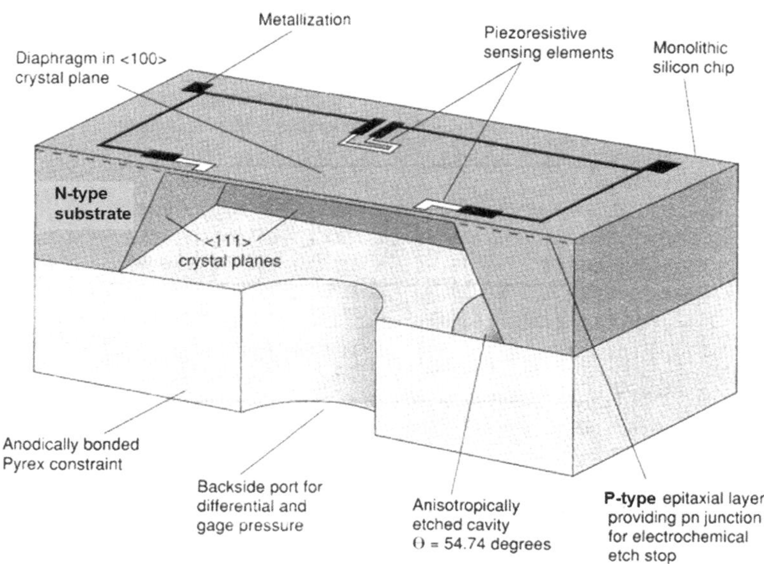

FIGURE 1.12 Piezoresistive sensor fabricated with bulk micromachining for monitoring air pressure.

n-type substrate below the active p-type layer in Figure 1.12 is to create a p-n junction, which works for stopping the etching. The etch rate for anisotropic etching also depends on the crystallographic orientation; for example, an anisotropy ratio of 100:1 is possible in the <100> direction relative to the <111> direction, which creates a cavity with a 55-degree wall angle. Etch processes can be made selective by using dopants (heavily doped regions etch slowly), or may be halted electrochemically (etching stops in a region of different polarity in a biased p-n junction). After the etching process is complete, the silicon wafer is anodically bonded to Pyrex and finally diced into individual devices. This has been a standard technique for fabricating silicon pressure sensors and micropumps.

Surface micromachining, *silicon fusion bonding*, and a process called *LIGA* (Lithographie, Galvanoformung, Abformung) have also emerged as major fabrication techniques. Surface micromachining of a wafer involves selectively applying or removing thin film layers. Thin film deposition and wet and dry etching techniques are the primary tools for this. Thin films of polysilicon, silicon oxide, and silicon nitride are used to produce sensing elements; electrical interconnections; and the structural, mask, and sacrificial layers. Silicon epitaxial layers, grown and deposited silicon oxide layers, and photoresist are used as sacrificial materials.

A typical surface micromachining process is depicted in Figure 1.13.[7] A sacrificial layer is applied (grown or deposited) in an appropriate pattern on the wafer and then removed from the areas where a mechanical structure will be attached to the substrate. Then, the mechanical layer is applied and patterned. Finally, the underlying sacrificial layer is etched away to release the mechanical structure. This process can produce structures on a scale of a few hundred micrometers, such as the microgripper appearing in Figure 1.14a. This microgripper was fabricated by Berkeley Sensor and Actuator Center, CA, using released-polysilicon surface micromachining and is activated by electrostatic forces.[7] Note the comb-teeth-like silicon structure at the bottom of the grippers, on which zero volts are applied on the top and bottom comb teeth, while 1 kV is applied on the center comb teeth, so that the distance of the comb teeth will be reduced to balance the electrostatic force and silicon beam spring force.

The first step of the LIGA process involves generating a photoresist pattern on a conductive substrate using deep x-ray lithography. The gaps between the resist patterns can be fully electroplated,

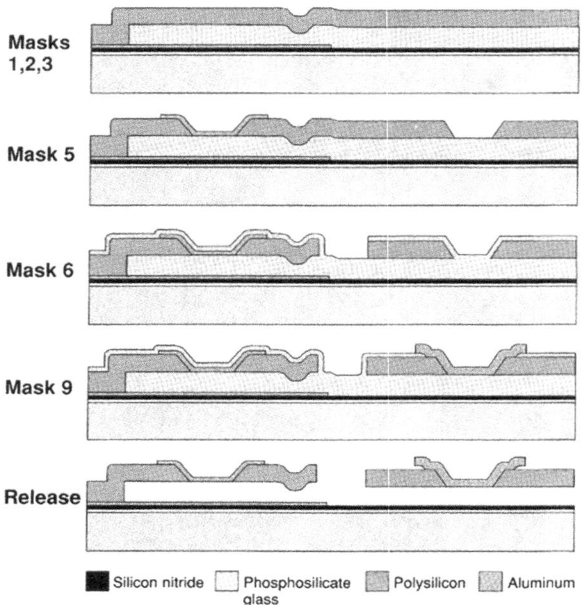

FIGURE 1.13 A typical surface micromachining process. The phosphosilicate glass layer is sacrificial in this process. (From J. Bryzek et al.: *IEEE Spectr*, (5), 20, 1994.)

yielding a highly accurate negative replica of the original resist pattern. This can be used as a mold for plastic resins such as polyimide and polymethyl methacrylate or for ceramic slurries. Once the material has been cured, the mold is removed, leaving behind microreplicas of the original pattern. An epicyclic micro-gear-train appears in Figure 1.14b that is to be fitted onto an EM micromotor, similar to the case shown in Figure 1.10. It contains 17 micro-injection-molded components made from the polymer polyoxymethylene (POM).[5] Note that the chief disadvantage of this process is the need for a short-wavelength, highly collimated x-ray source, ideally a synchrotron. Few MEMS manufacturers can afford their own synchrotron.

One of the major problems is the generative force/displacement levels produced by MEMS devices, which are generally too small to be useful for many actuator applications. Thus, we started to couple with piezoelectric thin/thick films on silicon MEMS devices, the details of which will be discussed in Chapter 4.

1.3.3 Artificial Muscle

Elastically soft actuators (i.e., artificial muscle) are composed of polymer materials, because large strains can be generated in polymer materials without causing mechanical damage because of their high elastic compliance. So-called *electroactive polymers* (EAPs) can be classified as in Table 1.2. The principle of *ionic polymer metal composite* (IPMC) can be explained as follows in a case of Pt-electroded Nafion film: cations and solvent migrate toward the cathode, resulting in a bending toward the anode side (the bending occurs during changes in potential). An AC field can be used to cause the membrane to bend from site to site with up to 50 Hz. Conducting polymers with conjugated polymers can also be used as actuators. A composite structure results when pyrrole is electropolymerized in the presence of a bulky anion like sodium dodecyl-benzene-sulphonate (DBS). In the as-prepared (oxidized) form, the polypyrrole (PPy) is positively charged and electronically conducting. The charges are balanced by DBS counter ions. When the material

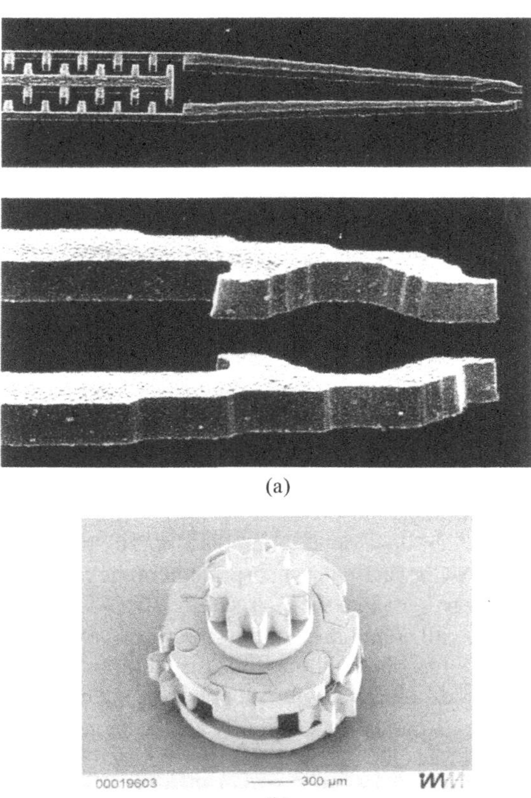

FIGURE 1.14 (a) A microgripper fabricated by Berkeley Sensor and Actuator Center, CA, using released-polysilicon surface micromachining. (From J. Bryzek et al.: *IEEE Spectr*, (5), 20, 1994.) (b) A microgearbox made of polymer POM using a LIGA technique. (From U. Berg et al.: *Proc. 6th Int'l Conf. New Actuators*, p. 552, 1998.)

is reduced, PPy chains are neutralized, and cations and solvent molecules are dragged into the structure in order to compensate for the charges of the DBS. In other words, an electrically activated swelling of the material results.[8] Since "ionic"-type actuators are slow in response (lower than 50 Hz) and environmentally sensitive (e.g., humidity), we will not detail them further in this textbook.

TABLE 1.2
Displacement Characteristics of Various Types of Polymeric Actuators

Materials	Operation Principle
Responsive Gel	Ionic
Ionic Polymer Metal Composite (IPMC)	Ionic
Conducting Polymer	Ionic
Carbon Nanotube	Ionic/Electric
Elastomer	Electric
Piezoelectric/Electrostrictive Polymer	Electric

Elastomer actuators operate through the *Maxwell force* that occurs in an electrostatic capacitor. Considering a capacitor with an area, S, and electrode gap, t, filled with a dielectric material with a relative dielectric constant, ε, the capacitance, C, and the stored energy, U, are given by:

$$C = \varepsilon_0 \varepsilon (S/t) \tag{1.1}$$

$$U = (1/2)CV^2. \tag{1.2}$$

The attractive force between the two electrodes can be obtained by

$$\begin{aligned} F &= \left(\frac{\partial U}{\partial t}\right) \\ &= -(1/2)\varepsilon_0 \varepsilon S (V/t)^2, \end{aligned} \tag{1.3}$$

the sign of which leads to a decrease in the interelectrode distance with an increase in voltage (Figure 1.15a). Hence, we see that a larger displacement can be obtained by increasing the elastic compliance and effective permittivity (and therefore the stored electric charge) of the material. A hybrid elastomer structure has been developed whereby a porous polytetrafluoroethylene (PTFE) and a stiff perfluorocyclobutyl (PFCB) phase are incorporated in the configuration depicted in Figure 1.15b.[9] The stiffer PFCB phase serves to store the charge (effectively larger ε), while the porous PTFE phase effectively increases the overall compliance of the structure. An effective piezoelectric d_{33} strain coefficient of around −600 pC/N has been obtained for this composite structure, which is 20 times larger than that obtained for a piezoelectric polymer poly-vinylidene-difluoride (PVDF). Polymer piezoelectrics such as PVDF are discussed in Section 1.3.7. In theory, one can reasonably predict a strain level as high as 100% (a twofold enhancement) for this material when a sufficiently soft polymer is used as the porous phase. SRI International, CA, developed thin polymer film actuators made of acrylic elastomer that undergoes 300% strain.[10] Rolling up the polymer film into a cigarlike rod, they produced a 90-mm polymer rod actuator, as shown in Figure 1.16a, and ±10 mm displacement is generated under an electric field. As expected from Equations 1.1 and 1.3, the capacitance of the

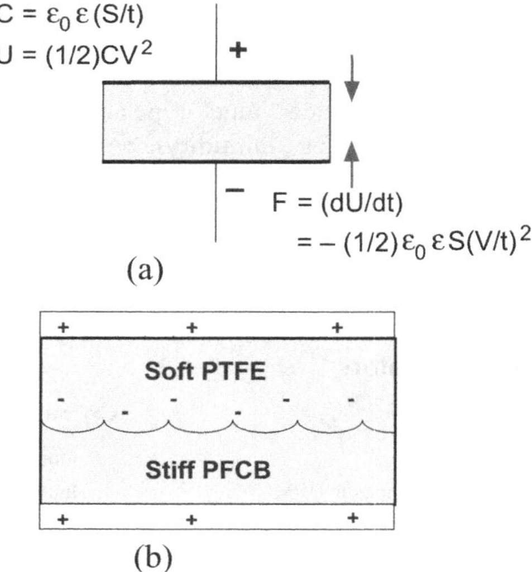

FIGURE 1.15 (a) Maxwell force in a capacitor. (b) Schematic representation of a hybrid elastomer structure incorporating a porous PTFE and a stiff PFCB phase. (From S. Bauer: *Proc. Int'l Symp. Smart Actuators*, ICAT/Penn State, PA, April, 1999.)

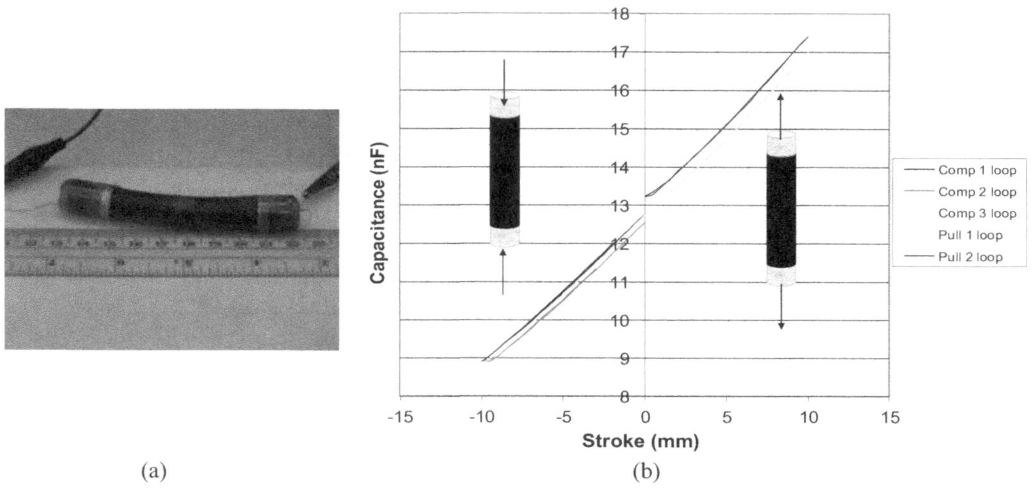

FIGURE 1.16 (a) A rolled-up elastomer film actuator. (b) ±10 mm displacement is generated under an electric field. The capacitance of the elastomer is almost proportional to the displacement. (Courtesy of SRI.)

elastomer is almost proportional to the displacement, as demonstrated in Figure 1.16b.[10] It is important to note that even though one might feasibly anticipate large displacements for a polymer, they will be realized only at the expense of the generative force and responsivity of the device.

In a manner similar to silicon MEMS devices, the operation of polymer film actuators is based on electrostatic principles. The artificial muscle described here, developed at the University of Tokyo, makes use of this type of polymer film actuator.[11] The basic design and operation of the polymer film actuator is depicted in Figure 1.17. Two polymer films with embedded electrodes are placed adjacent to each other. When three-phase voltages ($+V$, $-V$, 0) are applied in succession to every three embedded electrodes in the stator film [Figure 1.17(1)], charges of $-Q$, $+Q$, and 0 are induced on the opposing slider film [Figure 1.17(2)]. Then, when the three voltages are switched to $-V$, $+V$, and $-V$ [Figure 1.17(3)], a repulsive force is generated between the stator and slider films, and an attractive force is generated between the adjacent electrodes on the two films. This produces an electrode pitch displacement (in this case, a shift of the slider to the right) [Figure 1.17(4)]. The electrostatic force generated increases significantly as the electrode gap is reduced.

The construction of the electrostatic polymer artificial muscle is illustrated in Figure 1.18. A thin film of polyethylene terephthalate (PET), 12 μm in thickness, is used for the slider, to which a polyimide film with surface line electrodes is laminated, which serves as the stator. Spacers are included in the structure as shown to maintain the proper separation between electroded surfaces. Five stator/slider pairs, 34 mm in width and 80 mm in length, are stacked together with a 0.35-mm gap between them. The prototype structure, which weighs 43 g, is then dipped into Fluorinat (3M). When three-phase voltage at 10 Hz is applied with a pitch range of 0.1–1.0 mm, a speed of 1 m/sec is achieved without load. A propulsive force (thrust) of 1–3 N is possible with an applied root-mean-square voltage of 2.5 kV. The relatively high voltage required for operation is a major drawback of this artificial muscle.

A robot arm driven by the electrostatic artificial muscles, shown in Figure 1.19, is composed of a 40-layer electrostatic actuator (generating 320 N) in conjunction with a 20-layer actuator (generating 160 N) via a pulley mechanism.[12]

1.3.4 Smart Actuators

Let us now consider the *smartness* of a material. The various properties relating the input parameters of electric field, magnetic field, stress, heat, and light with the output parameters, charge/current,

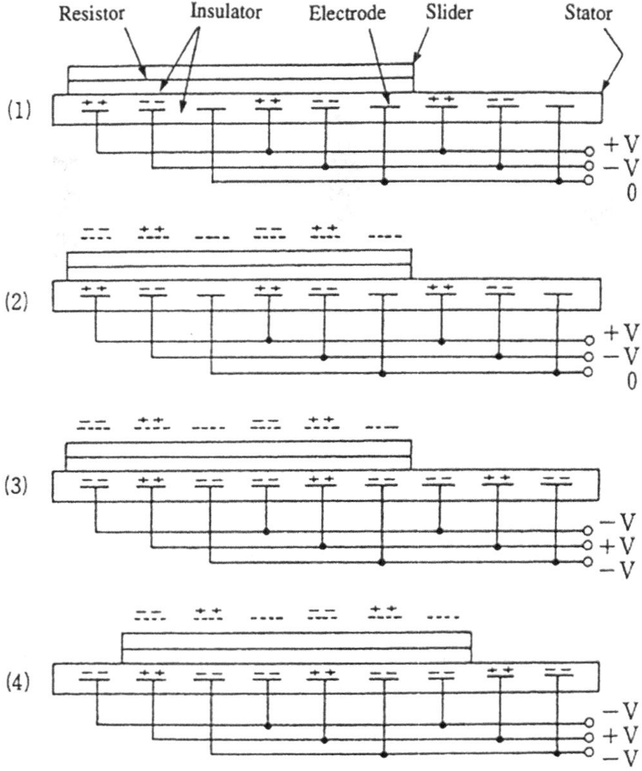

FIGURE 1.17 The basic design and operation of a polymer film actuator. (From S. Egawa et al.: *Proc. 1991 IEEE Workshop on Micro Electro Mechanical Systems*, p. 9, 1991.)

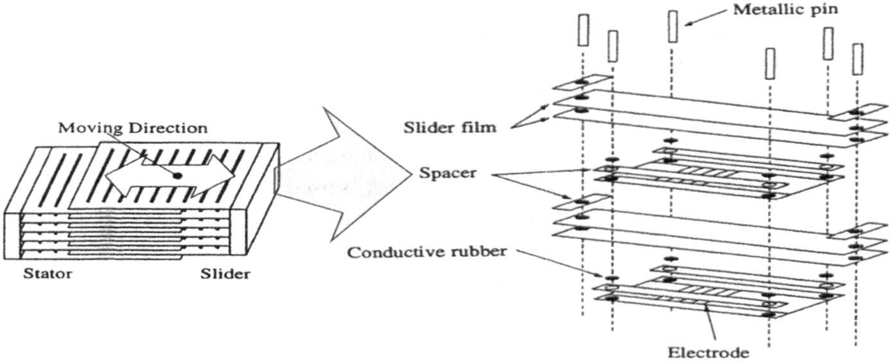

FIGURE 1.18 Construction of the electrostatic polymer artificial muscle. (From S. Egawa et al.: *Proc. 1991 IEEE Workshop on Micro Electro Mechanical Systems*, p. 9, 1991.)

magnetization, strain, temperature, and light are listed in Table 1.3. Conducting and elastic materials, which generate current and strain outputs, respectively, with input, voltage, or stress (diagonal couplings), are sometimes referred to as *trivial materials*. High-temperature superconducting ceramics are also considered trivial materials in this sense, but the figure of merit (electrical conductivity) exhibited by some new compositions has been exceptionally high in recent years, making them especially newsworthy.

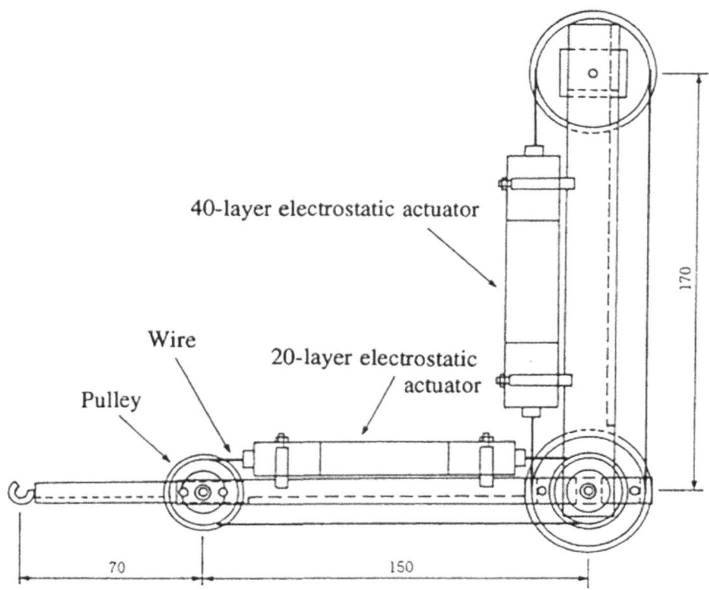

FIGURE 1.19 Robot arm incorporating two artificial muscles. (From T. Higuchi: *New Actuator Handbook for Precision Control*, p. 182, Fuji Techno System, Tokyo, 1994.)

On the other hand, pyroelectric and piezoelectric materials, which generate an electric field with the input of heat and stress, respectively, are called *smart materials*. These off-diagonal couplings have corresponding converse effects, the *electrocaloric* and *converse piezoelectric effects*, so that both "sensing" and "actuating" functions can be realized in the same material. Another example is a tooth brace made of a shape memory (superelastic) alloy. A temperature-dependent phase transition in the material responds to variations in the oral temperature, a sort of "thermal expansion," thereby generating a constant stress on the tooth.

TABLE 1.3
Various Basic and Cross-Coupled Properties of Materials

INPUT → MATERIAL DEVICE → OUTPUT

Input \ Output	CHARGE CURRENT	MAGNET-IZATION	STRAIN	TEMPERTURE	LIGHT
ELEC. FIELD	Permittivity Conductivity	Elect.-mag. Effect	Converse piezo-effect	Elec. caloric effect	Elec.-optic effect
MAG. FIELD	Mag.-elect. effect	Permeability	Magneto-striction	Mag. caloric effect	Mag.-optic effect
STRESS	Piezoelectric effect	Piezomag. effect	Elastic constant	Mechanotherm al effect	Photoelastic effect
HEAT	Pyroelectric effect	Pyromagnetic effect	Thermal expansion	Specific heat	—
LIGHT	Photovoltaic effect	—	Photostriction	Photothermal effect	Refractive index

Diagonal Coupling = Trivial material ☐ ☐ Sensor
Off-diagonal Coupling = **Smart material** ☐ Actuator

Intelligent materials must possess a "drive/control" or "processing" function, which is adaptive to changes in environmental conditions in addition to their actuator and sensing functions. Photostrictive actuators belong to this category. Some ferroelectrics generate a high voltage when illuminated (the *photovoltaic effect*). Since the ferroelectric is mostly piezoelectric, the photovoltage produced will induce a strain in the crystal. Hence, this type of material generates a drive voltage dependent on the intensity of the incident light, which actuates a mechanical response. The self-repairing nature of partially stabilized zirconia can also be considered an intelligent response. Here, the material responds to the stress concentrations produced with the initial development of the microcrack (sensing) by undergoing a local phase transformation in order to reduce the concentrated stress (control) and stop the propagation of the crack (actuation).

If one could incorporate a somewhat more sophisticated mechanism for making complex decisions into its "intelligence," a *wise material* might be created. Such a material might be designed to determine that "this response may cause harm" or "this action will lead to environmental destruction," and respond accordingly. It would be desirable to incorporate such fail-safe mechanisms in actuator devices and systems. A system so equipped would be able to monitor for and detect the symptoms of wear or damage so as to shut itself down safely before serious damage or an accident occurred.

Actuators that operate by means of a mechanism different from those found in the conventional AC/DC electromagnetic motors and oil/air pressure actuators are generally classified as *new actuators*. Some recently developed new actuators are classified in Table 1.4 in terms of input parameter. Note that most of the new actuators are made from some type of smart material with properties specifically tailored to optimize the desired actuating function. We focus on some of the popular and useful smart materials utilized for actuators in this textbook.

The displacement of an actuator element must be controllable by changes in an external parameter such as heat, magnetic field, electric field, or light irradiation. Actuators activated by changes in temperature generally operate through the *thermal expansion* or dilatation associated with a phase transition, such as ferroelectric and martensitic transformations. *Shape memory alloys*, such as Nitinol, are of this type. *Magnetostrictive* materials, such as Terfenol-D, respond to changes in an applied magnetic field. *Piezoelectric* and *electrostrictive* materials are typically used in electric field-controlled actuators. In addition to these, we will consider light-activated actuators (for which

TABLE 1.4
New Actuators Classified in Terms of Input Parameter

Input Parameter	Actuator Type/Device
Electric Field	*Piezoelectric/Electrostrictive*
	Electrostatic (Silicon MEMS)
	Electrorheological Fluid
Magnetic Field	*Magnetostrictive*
	Magnetorheological Fluid
Stress	Rubbertuator
Heat	*Shape Memory Alloy*
	Bubble Jet
Light	*Photostrictive*
	Laser Light Manipulator
Chemical	Mechanochemical
	Metal-Hydrite

Note: Italicized are smart material–based actuators

the displacements occur through the *photostrictive effect* or a photo-induced phase transformation) and electro/magnetorheological fluids.

The desired general features for a smart actuator element include:

1. Large displacement (sensitivity = displacement/driving power)
2. Good positioning reproducibility (low hysteresis)
3. Quick response
4. Stable temperature characteristics
5. Low driving energy
6. Large generative force and failure strength
7. Small size and light weight
8. Low degradation/aging in usage
9. Minimal detrimental environmental effects (mechanical noise, electromagnetic noise, heat generation, etc.)

1.3.5 Shape Memory Alloys

Many metallic, polymeric, and ceramic materials exhibit large mechanical deformations when undergoing a structural phase transition. This phase transition may be induced by temperature, stress, or electric field. In some materials, once the mechanical deformation is induced, some deformation may be retained with the release of the load and applied stress, but the original form may be restored with the application of heat. This type of action is called *shape memory.*

The stress versus strain curves for typical shape memory and superelastic alloys and a normal metal are shown in Figure 1.20.[13] When the stress applied to a normal metal exceeds the elastic limit, irreversible (nonrecoverable) plastic deformation results. A superelastic alloy subjected to the same level of stress, on the other hand, will become elastically soft at a level beyond the elastic deformation limit, due to a stress-induced phase transformation. The deformation that occurs in this case, however, is reversible and the original form is recovered as the load is removed and the stress is released. Finally, we see in the case of the shape memory alloy a response quite similar to the normal metal except that for these materials, the original form may be recovered after the load has been removed by heating the alloy at the appropriate temperature.

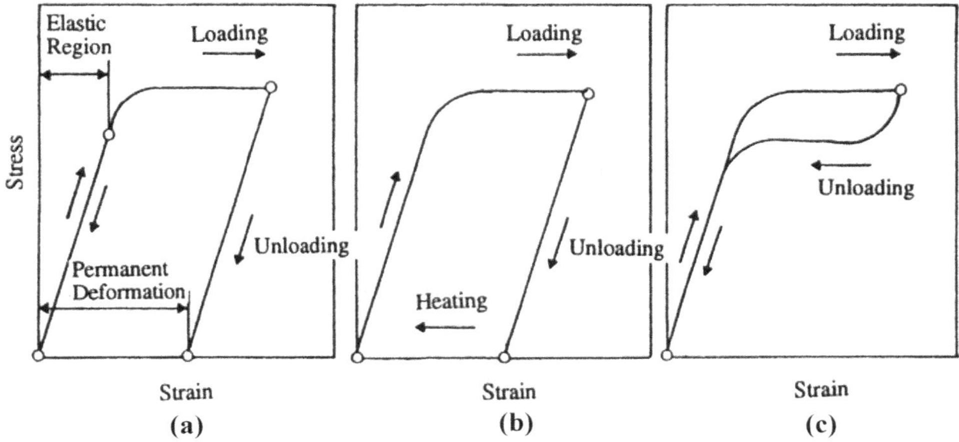

FIGURE 1.20 Typical stress versus strain curves for: (a) normal, (b) shape memory, and (c) superelastic materials. (From H. Horikawa and K. Otsuka: *New Actuator Handbook for Precision Control*, p. 454, Fuji Techno System, Tokyo, 1994.)

Depending on the temperature at which it is deformed, a shape memory alloy may exhibit one of two different types of mechanical behavior: *superelastic* or *pseudoplastic*. When Nitinol (a Ni-Ti alloy; *-nol* comes from the discovery place, the U.S. Naval Ordnance Laboratory) is behaving as a *superelastic* material, reversible deformation of up to 10% may be obtained with a very small effective modulus, which is several orders of magnitude smaller than the modulus of the parent phase. When the material is *pseudoplastic*, the deformation occurs with some slight hardening accompanied by very large strains. The pseudoplastically deformed alloy can then be restored to its initial shape by heating. The generative force may be as high as 10^8 N/m^2 during the recovery process.

The shape memory and superelastic mechanisms are considered from a crystallographic viewpoint in Figure 1.21.[14] In a normal metal, atomic shifts occur above a certain critical shear stress, leading to a shear strain, which remains as a permanent residual strain even after the stress is removed (Figure 1.21a). In a shape memory alloy, a *martensitic phase* is at room temperature and the parent phase (*austenite*) occurs above the phase transition temperature. Macroscopically, the material will retain the same shape under cooling, but on a microscopic level, many twin structures will have been generated during cooling. Since the twin planes are easily moved, the martensite material is readily deformed by the external stress (Figure 1.21b). However, when the deformed material is heated to a temperature higher than the reverse phase transition temperature, A_f *(austenite-finish)*, the parent phase (austenite) is induced and the initial shape is recovered. This is the *shape memory effect*. If

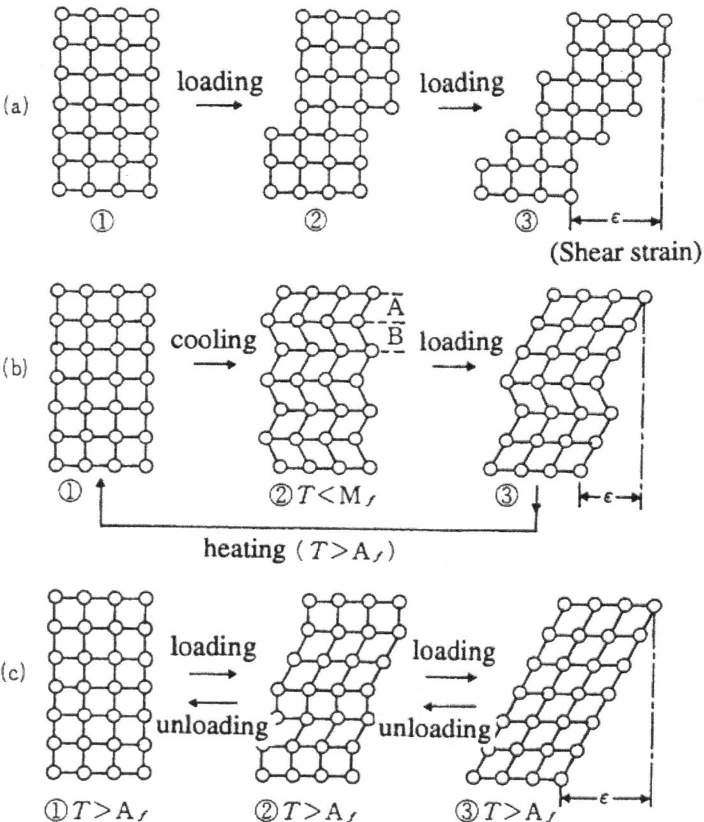

FIGURE 1.21 Microscopic lattice distortions under shear loading for: (a) normal, (b) shape memory, and (c) superelastic metals. (From K. Otsuka and C.M. Wayman: Shape Memory Materials, Chap.1, p. 1, Cambridge University Press, UK, 1998.)

stress is applied to a shape memory alloy above the transition temperature, A_f (or A_f is lower than room temperature), above a certain stress level, the martensite phase is induced by an external stress, leading to a pseudoplastic response. The material is very compliant under these conditions. As the applied stress is decreased, the reverse phase transformation (i.e., martensite to austenite) occurs and the material returns to its original elastically stiff state. This elastic phenomenon, which is similar to what is observed for rubber, is called *superelasticity*.

Interestingly, one of the first commercial successes involving the shape memory alloy was in women's lingerie. An underwire brassiere design by Wacoal, Japan, incorporating the alloy exploited the material's superelastic properties to help maintain a comfortable fit while providing adequate support. Another big market can be found in medical applications, such as stents and dental braces. Ni-Ti alloys exhibit superior adaptability properties with the human body. Figure 1.22 shows a sputtered Ni-Ti thin film cylinder for stent applications in coronary arteries, developed by TiNi Alloy Company, United States.[15] The cylinder is 20 mm long and 6 mm in diameter, with approximately 4-μm-thick Ti-Ni. Note that the film has etched pattern of 25×25 μm^2, which is essential for the body cells to take oxygen from blood. In the surgery, the dry-ice-cooled Ni-Ti cylinder (martensite) is inserted (after shaping it into a thin needle) into an artery. During the process of its temperature rising to body temperature, the needle recovers the original (austenite) cylinder shape and props open the blood vessel with constant pressure. Even if a large force on the body generates large deformation on the stent, the shape can be recovered due to the superelastic properties. The application for dental braces is popular and likely better known to readers. Since the Ni-Ti alloy can generate a similar force (stress) level over a wide deformation (strain) range, the superelastic brace is useful for a long period during the tooth alignment procedure. The medical Nitinol should have A_f temperature below the body temperature. Another popular application is for pipe couplers and electrical connectors.[16] The couplings are made by machining a cylinder of the alloy while it is in the austenite phase, usually with circumferential sealing bands on the inner diameter. A second, slightly wider cylinder is then inserted into the shape memory cylinder as it cools (with dry ice) in order to force it to expand as the material transforms to the martensite phase. When the coupling is brought back to room temperature, the outer cylinder contracts as it reverts to the austenite phase, tightly clamping the inner cylinder.

An actuator incorporating a shape memory spring is shown in Figure 1.23a. A normal steel spring and a shape memory alloy (Ni-Ti or Nitinol) spring, which will "remember" its fully extended form, are included in the design. At a low operating temperature (room temperature), the shaft will be pushed to the right as the shape memory spring becomes soft (i.e., martensite). When the temperature is raised (e.g., by means of an electric current in the shape memory wire), the spring constant of the shape memory actuator increases significantly due to the transformation into austenite, causing the shaft to be pushed to the left. A similar two-way mechanism is utilized in the flapper control unit for an air conditioning system. The temperature of the air passing from the air conditioner is constantly changing and can sometimes become uncomfortably cold. One solution to this problem is to adjust

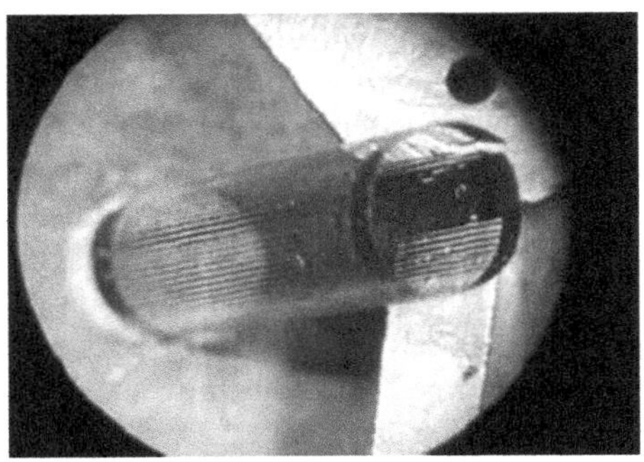

FIGURE 1.22 Sputtered TiNi thin film cylinder for stent application. (Courtesy of TiNi Alloy Company.)

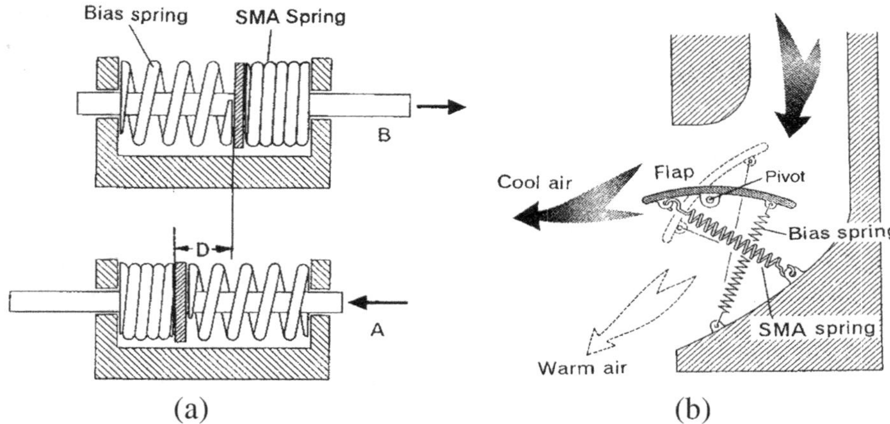

FIGURE 1.23 (a) A two-way actuator incorporating shape memory alloy (SMA) and normal metal (bias) springs. (b) An air conditioning system with a shape memory flapper control mechanism. (From T. Todoroki et al.: *Indust Mat*, 32(7), 85, 1984.)

the air flow direction using a flap mounted on the front of the air conditioner that is actuated by a shape memory alloy, as shown in Figure 1.23b.[17] When the flowing air is cold, the flap will move so as to direct the air upward, and when the air temperature exceeds body temperature, it will move so as to direct the air downward. The shape memory spring and bias spring act on the ends of the flapper to swing it like a seesaw about the pivot, thus redirecting the air flow as the temperature changes.

1.3.6 Magnetostrictive Actuators

Magnetostrictive materials convert magnetic energy into mechanical energy and vice versa. In the electronically degenerate state, the orbitals are asymmetrically occupied and get more energy. Thus, in order to reduce this extra energy, the material tries to lower the overall symmetry of the lattice, that is, undergoing distortion, which is known as *Jahn-Teller distortion*. In case of octahedral d-orbital configuration, the octahedron suffers elongation of bonds on the z-axis, thus lowering the symmetry. Therefore, a magnetostrictive material becomes strained according to the magnetization direction. Conversely, when either an applied force or torque produces a strain in a magnetostrictive material, the material's magnetic state (magnetization and permeability) will change. Magnetostriction is an inherent (intrinsic effect) material property that depends on electron spin, the orientation and interaction of spin orbitals, and the molecular lattice configuration, originating from the Jahn-Teller effect. It is also affected by domain wall motion and rotation of the magnetization (extrinsic effect) under the influence of an applied magnetic field or stress.[18]

Research on giant magnetostriction began with studies on Terfenol-D (a Tb-Fe-Dy alloy; *-nol* came again from the Naval Ordnance Laboratory) conducted by Clark et al.[19] Longitudinally and transversely induced strain curves at various temperatures for Terfenol-D appear in Figure 1.24. The domain wall motion induced with the application of a small magnetic field occurs due to the growth of domains with magnetization aligned with the applied field, at the expense of domains with magnetization opposing the field. At moderate field strengths, the magnetic moments within unfavorably oriented domains overcome the anisotropy energy and suddenly rotate such that one of their crystallographic easy axes is more closely aligned with the external field direction. This sudden rotation is generally accompanied by a large change in strain. As the field is increased further, the magnetic moments undergo coherent rotation until they are completely aligned with the applied field. At this point, the material is single domain and the strain curve becomes saturated.

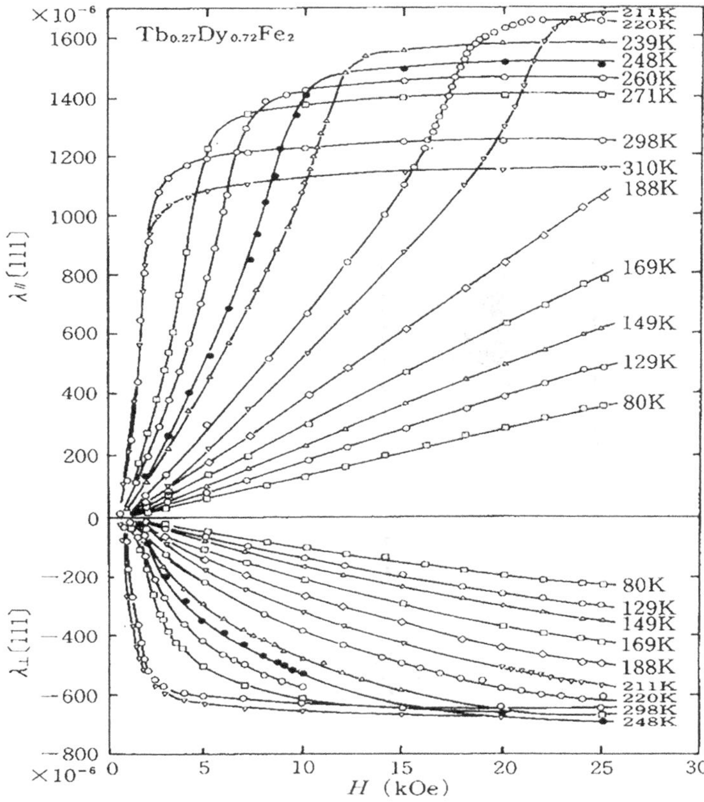

FIGURE 1.24 Longitudinally and transversely induced strains in Terfenol-D at various temperatures. (From A.E. Clark and H.S. Belson: U.S. Patent 4,378,258, 1983.)

Although one can attain strains exceeding 0.17% and sufficiently large generative stresses with magnetostrictive materials of this type, problems similar to those encountered with electromagnetic motors arise. These are related to the need for a magnetic coil and high field strengths to drive the devices. A typical design for a giant magnetostrictive actuator is depicted in Figure 1.25.[20] Two noteworthy features of this actuator are the prestressing mechanism and the magnetic coil and shield, which significantly increase the overall volume and weight of the system. The extensional prestress is important for optimum performance of the magnetostrictive alloy, and in some cases, a bias magnetic field must be maintained because the induced strain tends to be rather insensitive to the external field at low field strengths. The sonar device depicted in Figure 1.26 includes a square array of four magnetostrictive (Terfenol-D) rods mounted in a metal ring.[21] The magnetostrictive rods are 6 mm in diameter and 50 mm in length and have a free resonance frequency of 7.4 kHz. The resonance frequency of the entire ring device is 2.0 kHz.

In general, magnetostrictive actuators such as these are bulky due to the magnetic coil and shield required for their operation and hence are difficult to miniaturize. On the other hand, since they can generate relatively large forces and their efficiency increases with increasing size, they tend to be especially suitable for applications often reserved for electromagnetic motors, such as in construction/demolition machines and for vibration control in large structures. Another intriguing application can be found in surgery, where miniature magnetostrictive actuators, controlled with external magnetic fields provided by means of technology similar to what is currently used in MRI machines, can be used for specialized procedures.

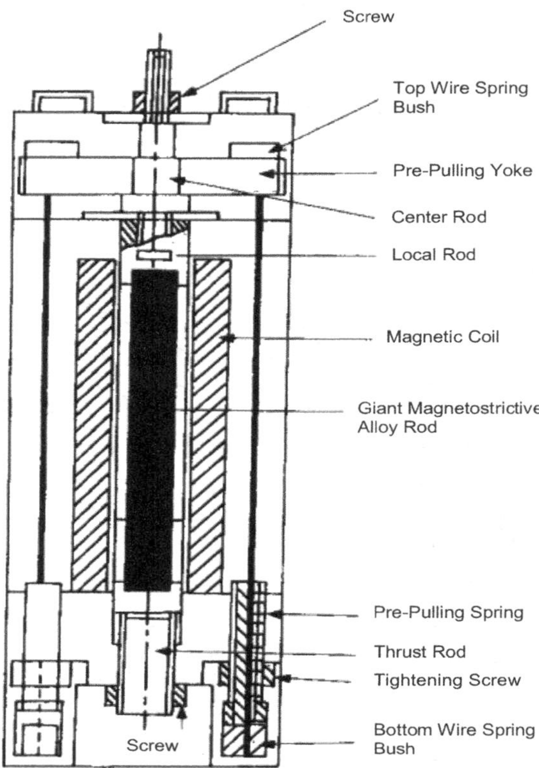

FIGURE 1.25 A design for a magnetostrictive actuator. (From H. Eda: *Giant Magnetostrictive Actuators, New Actuator Handbook for Precision Control*, p. 90, Fuji Techno System, Tokyo, 1994.)

1.3.7 Piezoelectric/Electrostrictive Actuators

Quartz is the most famous piezoelectric material, and it was discovered by the brothers Jacque and Pierre Curie in 1880. Figure 1.27 illustrates schematically a quartz crystal structure (SiO_2). Without any stress, the six dipole moments generated by Si^{4+} and O^{2-} ions cancel out originally in the hexagon (Figure 1.27a). However, when an external stress is applied to a quartz crystal, we

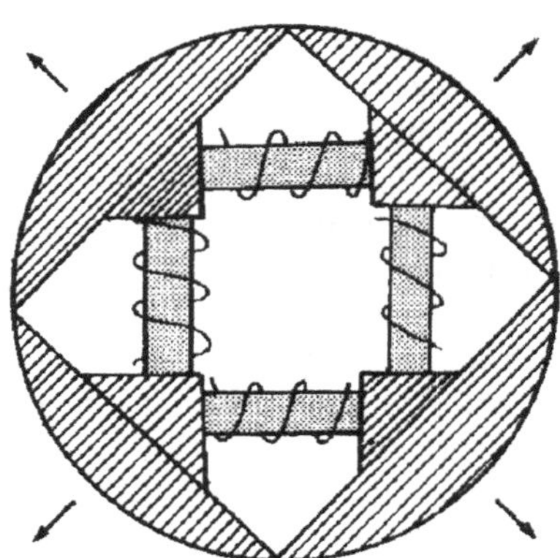

FIGURE 1.26 A high-power acoustic transducer incorporating four magnetostrictive actuators. (From J.L. Butler: *Edge Technologies*, Ames, IA, 1988.)

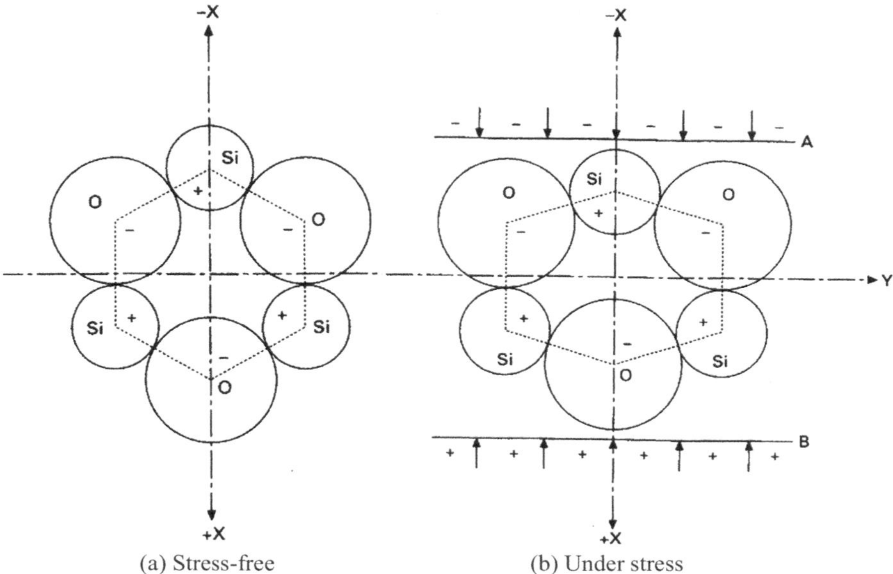

(a) Stress-free (b) Under stress

FIGURE 1.27 Quartz crystal structure distortion under compressive stress, leading to charge generation.

expect ionic shifts of Si and O ions, as in Figure 1.27b, leading to uncompensated polarization and charge generation. This effect is called the *direct piezoelectric effect*. Gabriel Lippmann discovered the *converse piezoelectric effect* one year later. That is, when an electric field is applied on quartz, the deformation (strain) is induced.[22]

When an electric field is applied to an insulating material, in general, strain will be induced in the material either through the piezoelectric effect, electrostriction, or a combination of the two effects. The *converse piezoelectric effect* is a primary electromechanical effect, where the induced strain is proportional to the applied electric field, while the electrostrictive effect is a secondary phenomenon, whereby the induced strain is proportional to the square of the applied field. A brief introduction to this class of devices is given here. A more thorough description of the piezoelectric effect and electrostriction will be presented in Chapter 2.

Electric field–induced strain curves are shown for piezoelectric lanthanum-doped lead zirconate titanate (PLZT) and electrostrictive lead magnesium niobate–based ceramics in Figure 1.28.[22] The piezoelectric response shows the characteristic linear strain versus field relation with a noticeable hysteresis, while the electrostrictive response exhibits no hysteresis and a nonlinear relation between the induced strain and the applied electric field is apparent. Due to this nonlinear behavior, a sophisticated drive circuit is generally needed for electrostrictive actuators. Note that the maximum strain and stress levels for the piezoelectric ceramics are around 0.1% and 4×10^7 N/m^2, respectively.

Among all the solid-state varieties, piezoelectric actuators have undergone the most advanced development and remain the most commonly employed type for various applications at this time. The dot matrix printer head pictured in Figure 1.29a, which operates by means of a multilayer piezoelectric actuator with sophisticated displacement amplification mechanisms, is one earlier application that was commercialized by NEC of Japan in 1987.[23] Piezoelectric ultrasonic motors have also been developed intensively. Electronic component industries have currently focused their attention on the development of these devices, mainly because piezoelectric motors offer distinct advantages of superior efficiency, miniature size (5–8 mm), and ease of manufacturing over conventional electromagnetic motors. Microultrasonic motors developed and commercialized by SEIKO Instruments, Japan, for wristwatch applications appear in Figure 1.29b.[24] The 8-mm-diameter motor is used as a silent alarm, and the 4-mm-diameter motor is part of a date change mechanism.

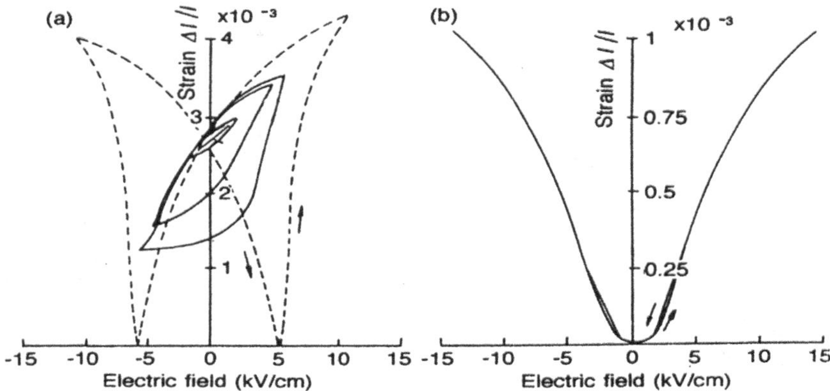

FIGURE 1.28 Electric field-induced strains in (a) piezoelectric La-doped PZT, and (b) electrostrictive lead magnesium niobate-based ceramics. (From K. Uchino: Ferroelectric Devices, Marcel Dekker, New York, NY, 2000.)

Large strains can be generated in soft polymer materials without causing mechanical damage because of their high elastic compliance. Polymer actuator materials can be classified according to two general types: *polyvinylidene difluoride–based piezoelectric polymers* and *electret/elastomer* types, the latter of which was introduced already in Section 1.3.3. PVDF or PVF2 becomes piezoelectric when it is stretched during the curing process. The molecular structure of this material is depicted in Figure 1.30a. Thin sheets of the cast polymer are drawn and stretched in the plane of the sheet in at least one direction, and frequently also in the perpendicular direction, to transform the material to its microscopically polar phase. Crystallization from the melt produces the nonpolar α-phase, which can be converted into the polar β-phase by this uniaxial or biaxial drawing operation. The resulting dipoles are then aligned by electrically poling the material. Large sheets can be manufactured and thermally formed into complex shapes. The copolymerization of PVDF with trifluoroethylene (TrFE) results in a random copolymer (PVDF-TrFE) with a stable polar β-phase. This polymer need not be stretched; the microscopically polar regions are formed during the copolymerization process so that the as-formed material can be immediately poled. The poled material has a thickness-mode coupling coefficient, k_t, of 0.30. Note that d_{33} of PVDF is negative; that is, the thickness of the film shrinks with a positive electric field application (i.e., parallel to the polarization direction), which is opposite from piezoelectric ceramics.

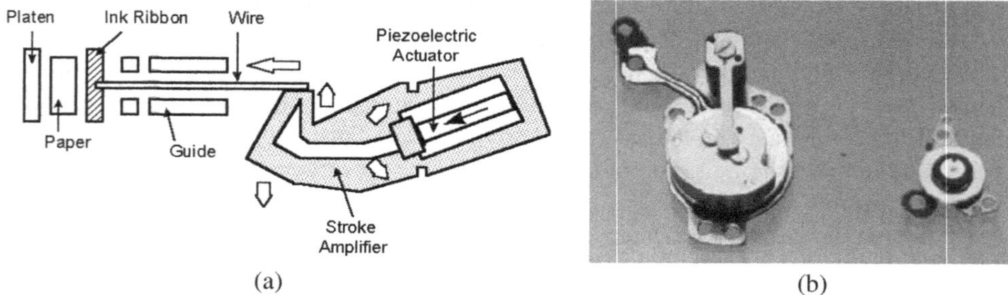

FIGURE 1.29 (a) Dot matrix printer head incorporating a multilayer piezoelectric actuator and a hinge-lever type displacement amplification mechanism (NEC, Japan). (From K. Yano et al.: *Proc. Annual Mtg. EE Japan*, p. 1–157, Spring, 1984.) (b) Microultrasonic motors used for silent alarm (8 mm diameter, left side) and date change (4 mm diameter, right side) mechanisms in a wristwatch (SEIKO, Japan). (From M. Kasuga et al.: *J. Soc. Precision Eng.*, 57, 63, 1991.)

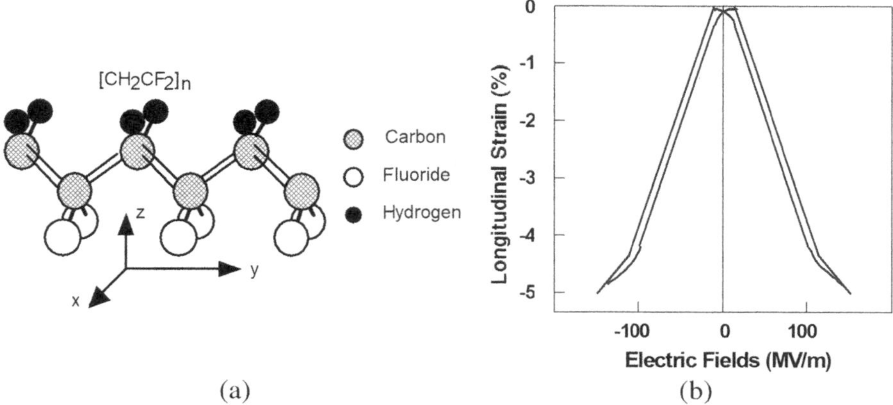

FIGURE 1.30 (a) Molecular structure of polyvinylidene difluoride (PVDF), (b) field-induced strain for an irradiated PVDF-based copolymer. (From V. Bharti et al.: *J. Appl. Phys.* 87, 452, 2000.)

Piezoelectric polymers have the following characteristics, in comparison with piezoceramics: (a) they have small piezoelectric strain coefficients, d (for actuators), but large g constants (for sensors); (b) they are lightweight and elastically soft, allowing for good acoustic impedance matching with water and human tissue; (c) they have a low mechanical quality factor, Q_m, allowing for a broad resonance band width. Piezoelectric polymers are used for directional microphones and ultrasonic hydrophones. The major disadvantages of these materials for actuator applications are the relatively small generative stress and considerable heat generation that originate from high loss or a low mechanical quality factor.

Copolymers from the system polyvinylidene difluoride-trifluoroethylene [P(VDF-TrFE)] are also known as piezoelectric materials. The strain induced in these materials is not very large, however, due to a very high coercive field. An electron irradiation treatment has been applied to materials from this system by Zhang et al. that significantly enhances the magnitude of the induced strain.[25] A 68/32 mole percent P(VDF-TrFE) copolymer film is irradiated by a 1.0-MeV electron beam at 105°C. The 70 Mrad exposure results in a diffuse phase transition and a decrease in the transition temperature, as compared with the nonirradiated material. It is believed that the observed changes in the phase transition are due to the development of a microdomain state, similar to that associated with relaxor ferroelectrics, which effectively interrupts the long-range coupling of ferroelectric domains. The strain curve pictured in Figure 1.30b for an irradiated P(VDF-TrFE) specimen demonstrates that induced strains as high as 5% are possible with an applied field strength of 150 MV/m. Note again that the longitudinal strain is negative.

1.3.8 Photo-Driven Actuators

The ongoing emphasis on miniaturization and the integration of microrobotics and microelectronics has resulted in significant development of new remote-control actuators in order to remove heavy lead wires from the actuator material. Photo-driven actuators have three categories: (1) laser manipulators, (2) photothermal actuators, and (3) photostrictive actuators.

Light rays that are reflected, refracted, or absorbed by small objects change the direction and magnitude of the propagating light wave and thereby the flux of momentum associated with the photons constituting the wave. Based on Newton's second law, it can be understood that the rate of change of the photon momentum is associated with the force acting on the photons. Consequently, if the micro-scale object bends the light rays and changes the momentum of the light, then the object undergoes an equal and opposite change in momentum. The exchange of momentum between the incident photons and the irradiated object will result in an optical force acting on the object. An all-optical trap can be created by using either two opposing beams or a single highly focused beam, called an *optical tweezer*, that creates a sufficiently high axial intensity gradient.[26]

Photothermal actuators are initiated by thermal expansion and/or thermally induced phase transformation. Shape memory alloys such as 50/50 nickel-titanium (NiTi) and 50/50 gold-cadmium (AuCd) are a group of metal alloys that do experience a significant discontinuous change in their physical structure near their crystalline phase transformation temperature. The dimensional change arising from this phenomenon is significantly greater than the linear volume change that occurs under normal thermal expansion. Furthermore, the phase transformation temperature can often be modified by varying the alloy composition. For example, a change in the composition of NiTi by 0.1% shifts the transformation temperature by 10°C.[27] Thus, light irradiation generates a significant deformation on SMAs.

Photostrictive materials, which convert photonic energy into mechanical motion directly, are of interest for their potential use in microactuation and microsensing applications. Optical actuators are also attractive for use as the driving component in optically controlled electromagnetic noise-free systems. In principle, the *photostrictive effect* arises from a superposition of the photovoltaic effect (the generation of a potential difference in response to illumination) and the converse piezoelectric effect (strain induce by an applied electric field).[28] The photostrictive effect has been studied intensively in ferroelectric polycrystalline materials by Uchino's group. Lanthanum-modified lead zirconate titanate ceramic is one of the most promising photostrictive materials due to its relatively high piezoelectric coefficient and ease of fabrication. The origin of the photovoltaic effect in PLZT is not yet clear, although several models for possible mechanisms have been proposed. Key issues in understanding the mechanisms behind the effect are both impurity doping and crystal asymmetry. One model has been proposed that describes the effect in terms of the electron energy band structure for PLZT ceramics.[29] According to this model, the donor impurity level associated with lanthanum doping will occur slightly above the valence band, as depicted in Figure 1.31a. The asymmetric potential due to crystallographic anisotropy is expected to facilitate the transition of the electron between these levels by providing it with *preferred momentum*. An asymmetric crystal exhibiting a photovoltaic response should also possess piezoelectricity and, therefore, a photostrictive response is also expected through the coupling of the two effects.

Application of the photostriction effect has been demonstrated with the PLZT ceramic photo-driven relay and micro walking devices (Figure 1.31b) developed by Uchino et al. using bimorph structures.[30] These devices are activated entirely by incident light and require no drive circuitry. Figure 1.32 shows a new class of small exploratory land vehicles for future space missions using highly efficient photostrictive PLZT films on flexible substrates.[31] The original π-shaped design of the micro walking device can be modified to assume the arch shape pictured in Figure 1.32a. It is composed of a photoactuating composite film, similar to that conventionally used for unimorphs, fabricated in the

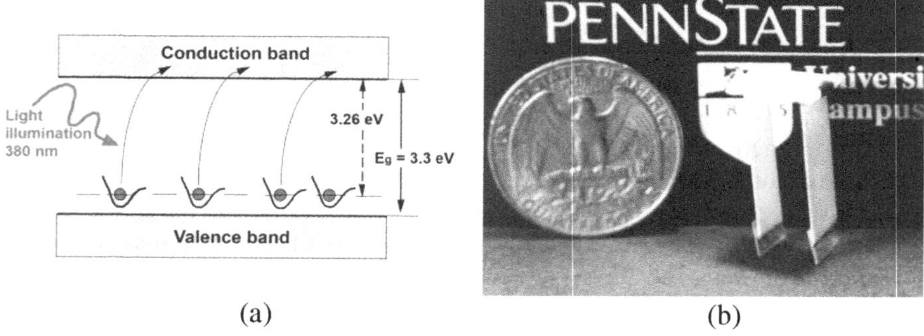

FIGURE 1.31 (a) A current source model for the photovoltaic effect in PLZT. Excited electron may receive preferred momentum. (b) Photo-driven micro walker made of photostrictive PLZT bimorph legs.

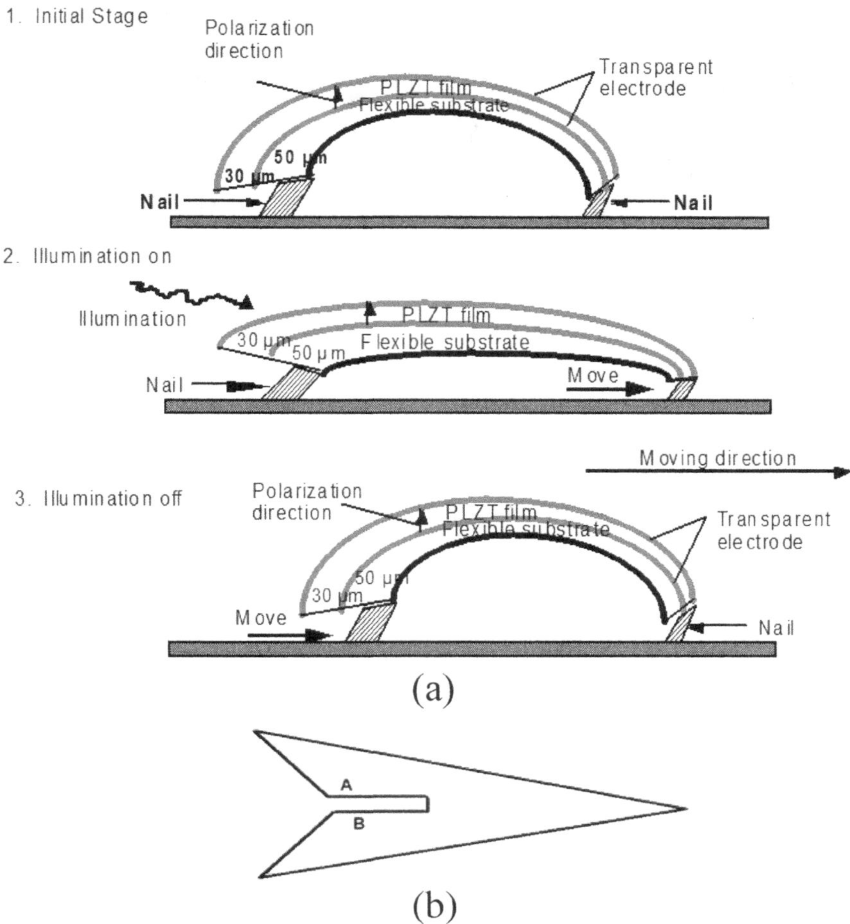

FIGURE 1.32 Schematic representation of an arch-shaped photoactuating film device: (a) action sequence and (b) the triangular top piece. (From S. Thakoor et al.: *Conf. Proc. 10th IEEE Int'l Symp. on Appl. Ferroelectrics*, 1, 205, 1996.)

form of an arch on which the triangular top piece is attached. The device executes the motion depicted in the figure when illuminated. Optimum photostrictive response has been observed for devices incorporating PLZT films approximately 30 μm in thickness. The device is driven at resonance with chopped illumination. Photomechanical resonance has been demonstrated in a PLZT bimorph.[32] Photoactuating films have been produced from PLZT solutions and applied to one side of a suitable flexible substrate designed to assume a curvature of 1 cm^{-1}. The walking device is designed to have a small difference in length between the right and left legs in order to establish a slight difference between their resonance frequencies (see Figure 1.32b). A chopped light source operating near these resonance frequencies is used to illuminate the device in order to induce the optimum vibration of the bimorph. Rotation of the walker in either the clockwise or counterclockwise direction is achieved by tuning the source to match the resonance frequency of one or the other leg.

1.3.9 Electro/Magnetorheological Fluids

Composite materials whose rheological properties can be changed with the application of an electric or a magnetic field are referred to as *electrorheological* (ER) or *magnetorheological* (MR) materials,

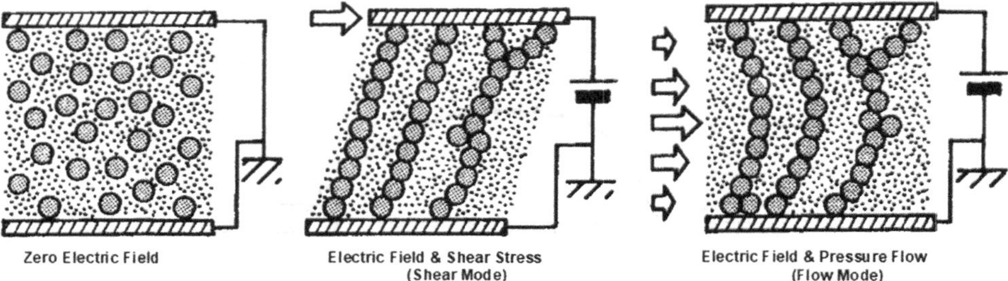

FIGURE 1.33 Schematic illustration of the response of an electrorheological fluid to applied electric field and stress.

respectively. ER and MR materials are generally in the liquid state, which does not generate displacement directly, but force in actuator systems. The rheological state of such fluids is affected when dielectric or magnetic dipoles are induced in the suspension by the applied field. Figure 1.33 illustrates the ER case, where the dielectric dipole moment is induced on resistive particles suspended in liquid. These polarized particles interact to form columnar structures parallel to the applied field, as depicted in Figure 1.33, center. These chainlike structures restrict the flow of the fluid, thereby increasing the viscosity of the suspension (Figure 1.33, right). When we use magnetic particles, magnetic dipole alignment is generated under magnetic field.

Electrorheological fluids are composed of electrically polarizable particles suspended in an insulating medium. Ferroelectric particles, such as barium titanate ($BaTiO_3$) and strontium titanate ($SrTiO_3$), are typically used in ER fluids because they have relatively high dielectric constants. On the other hand, magnetorheological fluids have a high concentration of magnetizable particles in a nonmagnetic medium. Spherical iron particles obtained from the thermal decomposition of iron pentacarbonyl are commonly used.[33] The properties of typical ER and MR fluids are summarized in Table 1.5. In general, ER and MR fluids are almost identical in terms of their rheological characteristics. However, from a production point of view, the MR fluid is preferred over the ER variety, because its properties are less affected by impurities. Once the liquid is contaminated by worn metallic (conducting) particles inside the chamber, the high electric field may cause electrical breakdown.

The yield stress and apparent viscosity of the fluids increase with the applied field because the mechanical energy required to induce and to maintain these chainlike structures increases with increasing field strength. The storage modulus is plotted as a function of applied electric field for

TABLE 1.5
Properties of Typical ER and MR Fluids

Property	ER Fluid	MR Fluid
Maximum Yield Strength	2–5 (kPa)	50–100 (kPa)
Maximum Field	4 (kV/mm)	250 (kA/m)
Plastic Viscosity	0.1–1.0 (Pa-s)	0.1–1.0 (Pa-s)
Temperature Range	+10–90 (°C) [dc] −10–125 (°C) [ac]	−40–150 (°C)
Response Time	msec	msec
Density	1–2 (g/cm^3)	3–4 (g/cm^3)
Maximum Energy Density	0.001 (J/cm^3)	0.1 (J/cm^3)
Typical Power Supply	2–5 (kV)/1–10 (mA)	2–25 (kV)/1–2 (mA)
Impurity Sensitivity	Intolerant	Unaffected

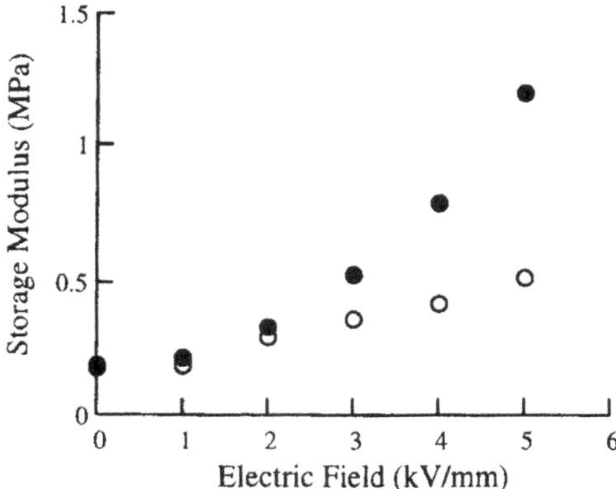

FIGURE 1.34 Storage modulus vs. applied electric field for silicone elastomers with iron. (From T. Shiga et al.: *Macromolecules*, 26, 6958, 1993.) [Random (o) and aligned (•) particle orientation states.]

silicone elastomers containing 20%–30% iron in Figure 1.34.[34] These elastomers exhibit significant variations in modulus with increasing electric field. The shear storage modulus was also found to be dependent on the alignment of the particles in the elastomer.

The basic modes of operation for ER and MR fluids are illustrated in Figure 1.35: (a) valve mode, (b) direct shear mode, and (c) squeeze mode. Magnetorheological fluid foam dampers are effective and exhibit long life. Little wear of the foam matrix occurs as the stresses are carried by the field-induced chain structure of iron particles in the MR fluid. A caliper-type MR fluid brake design is depicted in Figure 1.36a. Rather than a fully enclosed housing, the absorbent foam filled with MR fluid is attached to the pole faces of the steel yoke. These magnetorheological fluid-based devices have been successfully commercialized for use in exercise equipment[35] and in vehicle seat vibration control (Figure 1.36b).[36]

1.3.10 Bio Actuators

"Muscle" of animals is a sort of actuator with very high specific energy density (power/weight). Yamato, Tokyo Women's Medical University, Japan, demonstrated that demucosalized stomach flap can keep pulsatory motion for longer than two weeks only in isotonic sodium chloride solution with glucose (private communication 2006). Figure 1.37a shows a stomach flap (4 × 2 cm), the center of which continuously moves laterally by 3 mm distance at around 1 Hz. This can be used as a bio actuator.

Sato et al. at Nanyang Technological University, Singapore, reported that they are able to control insects and make them fly on demand.[37] A giant flower beetle can be used for spying or search and rescue missions as an insect cyborg. By strapping tiny computers and wireless radios onto the backs

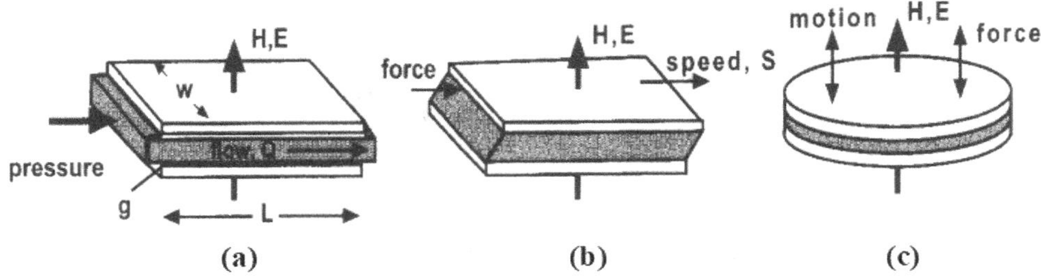

FIGURE 1.35 Basic modes of operation for ER and MR fluids: (a) valve mode, (b) direct shear mode, and (c) squeeze mode.

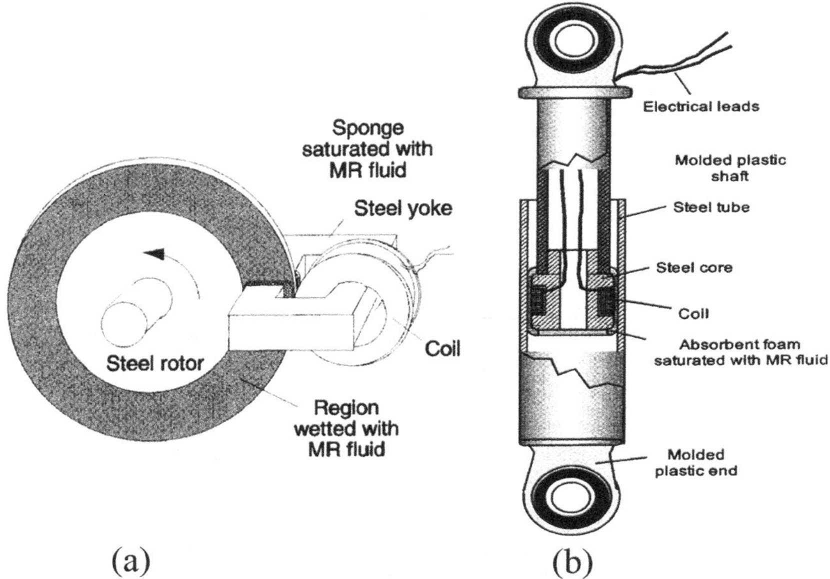

FIGURE 1.36 Construction of two simple, low-cost magnetorheological (MR) foam devices: (b) a rotary caliper brake, (From V.D. Chase: *Appliance Manufacturer*, May, p. 6, 1996.) (a) a vibration damper. (From J.D. Carlson and K.D. Weiss: U.S. Patent 5,382,373, 1995.) (Lord Corporation).

of the beetles and recording neuro-muscular data as the bugs flew untethered, they determined that a muscle known for controlling the folding of wings was also critical to steering.

1.3.11 Comparison among Solid-State Actuators

The specifications for shape memory, magnetostrictive, piezoelectric, and electrostrictive actuators are listed in Table 1.6. Certain characteristic features of each type become evident upon examining these data. *Shape memory actuators*, which operate in response to a temperature change, require a relatively large amount of input energy and typically have rather slow response speeds. A flat strip of Ni-Ti alloy, which alternates in form between an arc shape and an unbent strip to transfer energy to a spring, however, can be deformed repeatedly by heating and cooling via an electric current.

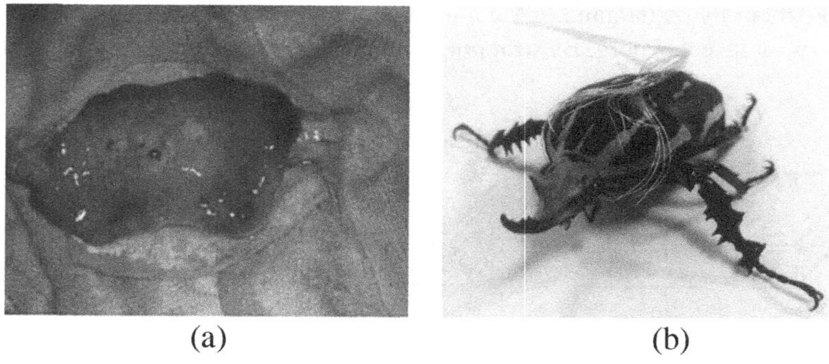

FIGURE 1.37 (a) Demucosalized stomach flap, moving pulsatorily. (b) Insect cyborg (giant flower beetle) with a tiny computer on its back. (https://www.livescience.com/54233-scientists-turn-beetles-into-cyborg-insects.html.)

TABLE 1.6
Specifications for Shape Memory, Magnetostrictive, Piezoelectric, and Electrostrictive Actuators

	Shape Memory	Magnetostrictive	Piezoelectric	Electrostrictive
Strain ($\Delta l/l$)	10^{-3}–10^{-2}	10^{-4}–10^{-3}	10^{-3}–10^{-2}	10^{-4}–10^{-3}
Hysteresis	Large	Large	Large	Small
Aging	Large	Small	Large	Small
Response	sec	μsec	msec	μsec
Drive Source	Heat	Magnetic Coil	Electric Field	Electric Field

The magnetostrictive actuator likewise demonstrates some advantages and disadvantages. Although some magnetic alloys (such as Terfenol-D) may exhibit relatively large induced strains, more commonly, the strain induced in this class of actuator is small. An additional drawback to this variety is the need for a driving coil, which in many applications can be troublesome; Joule heating is inevitable under high drive currents, and magnetic field leakage prevents it from being used adjacent to an operational amplifier or some other type of integrated circuit.

On the other hand, piezoelectric strain and electrostriction are induced by an electric field, and relatively large strains can be obtained in a variety of materials. Hence, *piezoelectric* and *electrostrictive actuators* are considered the most promising. Lead zirconate titanate (PZT)-based piezoelectric ceramics are most commonly used due to their availability, linear characteristics, low driving energy (low permittivity), and temperature stability at room temperature, as compared to electrostrictive devices. Electrostrictive actuators are preferred for applications where there may be significant temperature variations which may cause the "depoling" in a piezoelectric material (such as experienced by components used on the space shuttle) or high stress conditions (as occur for cutting machine devices). In terms of their reliability, electrostrictive materials are considered better than piezoelectrics in these cases, because they exhibit significantly less degradation and aging under severe conditions. However, the electrostrictor requires a high current power supply (due to a large capacitance) for high-speed applications.

1.4 STRUCTURE OF THE TEXT

The framework for the themes presented in this text is based on certain design concepts that are important in the development of new actuators and modern micromechatronic systems. The performance of smart actuators is dependent on complex factors, which can be divided into three major categories: (1) material properties, (2) device design, and (3) drive/control technique.

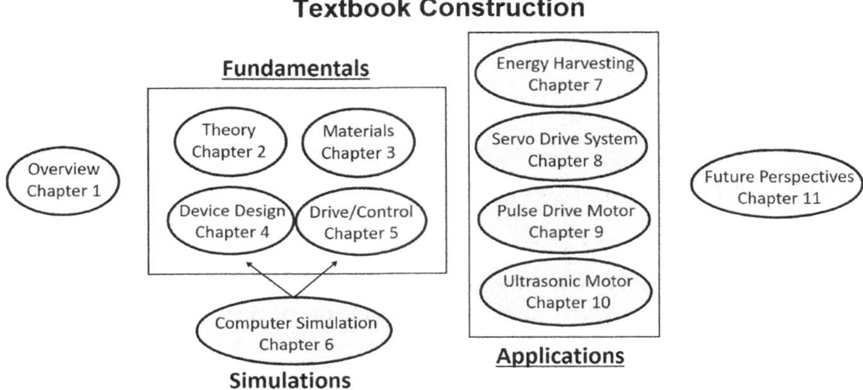

You have just read through Chapter 1 to obtain an overview of the micromechatronics area. If you are a technical engineer, you are requested to strengthen *fundamental knowledge*: physical theoretical background in Chapter 2, materials development in Chapter 3, device design principles in Chapter 4, and drive/control circuitry and systems in Chapter 5. We focus on piezoelectric and magnetostrictive devices in this text. One material-related concern of primary importance will be the optimization of the piezoceramic or single crystal composition, which may involve the incorporation of dopants. The orientation of a single crystal material is also of importance. Control of material parameters such as these is necessary in order to optimize the strains induced under high stroke level drive and to stabilize temperature and external stress dependences. Strain hysteresis and loss mechanisms, which generate significant heat in smart transducers, are also discussed in the materials chapter. The design determines to a large extent the performance, durability, and lifetime of the device. The inclusion of some failure detection or "health monitoring" mechanisms in the actuator is expected to increase its reliability significantly. When considering drive/control techniques, the pulse drive and AC drive modes require special attention. The vibration overshoot that occurs after applying an abrupt step/pulse voltage to the actuator will cause a large tensile force, and a sustained applied AC voltage will generate considerable heat within the device. The product of such a design will be one for which all of these parameters have been optimized. Chapter 6 is dedicated particularly to a device and a system engineer who needs to computer-simulate the optimum design. Transient piezoelectric displacement behavior under an arbitrary step/pulse voltage can be simulated easily with updated finite-element analysis software. Other engineers may skip this chapter.

If you are a product planning or marketing engineer, you may start reading the applicational developments with smart actuators. Chapter 7 is newly added in the second edition to provide detailed guidelines for piezoelectric energy harvesting system development. The author proposes three major phases/steps associated with piezoelectric energy harvesting: (a) mechanical-mechanical energy transfer, including mechanical stability of the piezoelectric transducer under large stresses and mechanical impedance matching; (b) mechanical-electrical energy transduction, relating to the electromechanical coupling factor in the composite transducer structure; and (c) electrical-electrical energy transfer, including electrical impedance matching, such as a DC/DC converter to accumulate the energy into a rechargeable battery. The reader can learn a comprehensive development strategy for this sort of interdisciplinary smart device. Chapters 8, 9, and 10 introduce serve drive positioning systems, pulse drive motors, and ultrasonic motors, respectively, suitable to learn each individual development procedure. Finally, Chapter 11 describes the author's perspectives for the future direction of micromechatronics. I will discuss five key trends in this chapter for future research: "Performance to Reliability," "Hard to Soft," "Macro to Nano," "Homo to Hetero," and "Single to Multifunctional." The reader may read the final chapter first, prior to learning the fundamentals or obtaining knowledge on advanced technologies. Because of the reader's sake, I dare not to hesitate minimum redundancy of physical formulas and explanations in each chapter, to escape from reading the previous chapters for refreshing the memories.

CHAPTER ESSENTIALS

1. Applications for new actuators: (a) positioner, (b) micromotor, (c) vibration suppressor, energy harvesting device
2. Disadvantages of conventional positioning devices:
   a. Oil/air pressure displacement reduction mechanism: Slow response
   b. Electromagnetic motor with transfer screw mechanism: Backlash

   c. Voice coil: Slow response, small generative force
3. Desired features for an actuator element:
   a. Large displacement (sensitivity = displacement/driving power)
   b. Good positioning reproducibility (low hysteresis)
   c. Quick response
   d. Stable temperature characteristics
   e. Low driving energy
   f. Large generative force and failure strength
   g. Small size and light weight
   h. Low degradation/aging in usage
   i. Minimal detrimental environmental effects (such as mechanical noise, electromagnetic noise, heat generation)
4. Comparison among solid-state actuators:

	Shape Memory	Magnetostrictive	Piezoelectric	Electrostrictive
Strain ($\Delta l/l$)	10^{-3}–10^{-2}	10^{-4}–10^{-3}	10^{-3}–10^{-2}	10^{-4}–10^{-3}
Hysteresis	Large	Large	Large	Small
Aging	Large	Small	Large	Small
Response	sec	μsec	msec	μsec
Drive Source	Heat	Magnetic Coil	Electric Field	Electric Field

5. Design considerations in the development of smart actuators:
   a. Materials design
   b. Device design
   c. Drive/control techniques, in terms of improved performance and reliability

CHECK POINT

1. There is a copper wire (resistivity = 17 n Ωm) 50 μm in diameter and 50 cm in length to make an electromagnetic motor coil. What is the resistance of this wire, which will generate heat via Joule heat when voltage is increased? 5 mΩ, 0.5 Ω, 5 Ω, or 50 Ω.
2. (T/F) The efficiency of a microelectromagnetic motor in a wrist watch is less than 1%. True or False?
3. What is the major role of the p-n junction between the n-type substrate below the active p-type layer during the bulk micromachining process? Answer simply.
4. Provide a popular etchant of silicon during the bulk micromachining process.
5. Name a famous shape memory alloy.
6. (T/F) Shape memory and superelastic phenomena are the same in principle, only the phase transition temperature (austenite-martensite) is lower than room temperature for a shape memory material. True or False?
7. Which is elastically stiff, the austenite or martensite phase?
8. How do we change the temperature for controlling a shape memory device, in addition to an external heater such as a hair dryer? Answer simply.
9. What effect/performance is expected in Terfenol-D?
10. Fill in the blank: In order to enhance the performance of a magnetostrictive actuator, we usually install a coil and a ( ) in the device.
11. Provide a name of a representative polymer piezoelectric.

12. Provide a full expression of "PZT."
13. The maximum longitudinal strain ($\Delta L/L$) level of typical piezoelectric and magnetostrictive materials is about 0.1%; evaluate an expected displacement of a 10-mm rod under the maximum electric or magnetic field application. 1 μm, 10 μm, 100 μm, or 1000 μm.
14. (T/F) Displacement vs. electric field hysteresis is one of the origins of heat generation in piezoelectric actuators.
15. What dielectric powder is used for preparing an electrorheological fluid?
16. What magnetic powder is used for preparing a magnetorheological fluid?
17. Which are more widely commercialized, electrorheological or magnetorheological fluid devices?
18. Provide a key reason for the commercialization superiority in the above question 17 in simple words.

CHAPTER PROBLEMS

1.1 We wish to design a compact electromagnetic motor with a coil 1 cm in diameter. We will use 100 turns of a thin wire that is 100 μm in diameter. If the conductivity of Cu is $\sigma = 6.0 \times 10^7$ $(\Omega\ \mathrm{m})^{-1}$, calculate the resistance of this coil. If a DC current of 0.5 A flows through this coil, calculate the Joule heat generated per second (neglect the heat dissipation in this case). Calculate also the loss percentage in comparison with the input electric power.

1.2 Consider an "elastomer" actuator, which consists of one polymer film (10 mm × 10 mm × 10 μm t) with thin and flexible electrodes on the top and bottom surfaces. Relative permittivity ε^X and elastic compliance s^E of this isotropic polymer are given by 20 and 800×10^{-12} m^2/N, with a Poisson's ratio (ratio of the transverse strain over the longitudinal strain) $\sigma = 0.4$. As shown in Figure 1.38, a high voltage (10 kV) is applied on the polymer film (electric field normal to the film plate [i.e., z-axis]). We will calculate the displacement along the x-axis (or y-axis). $\varepsilon_0 = 8.854 \times 10^{-12}$ [F/m].

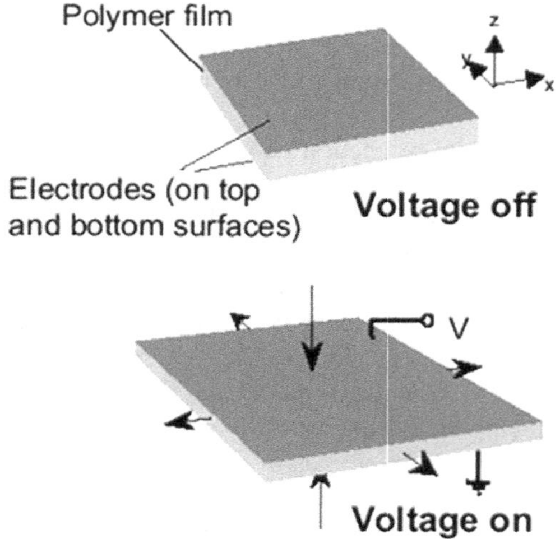

FIGURE 1.38 Elastomer actuator.

a. Calculate the Maxwell stress induced in the polymer film by the external voltage (10 kV). You may start from the equation for the electrostatic energy, $U=(1/2)CV^2$, to derive the Maxwell force and stress formulas. Suppose that the permittivity will not change by the electric field.
b. Calculate the strain induced in this polymer along the z-axis.
c. Calculate the strain induced in this polymer along the x- (or y) axis, and the actual displacement (deformation μm) along the x- (or y) axis, supposing that the Poisson's ratio σ will not change by the deformation level.
d. Choose a suitable displacement (along the x-axis) curve as a function of electric field.

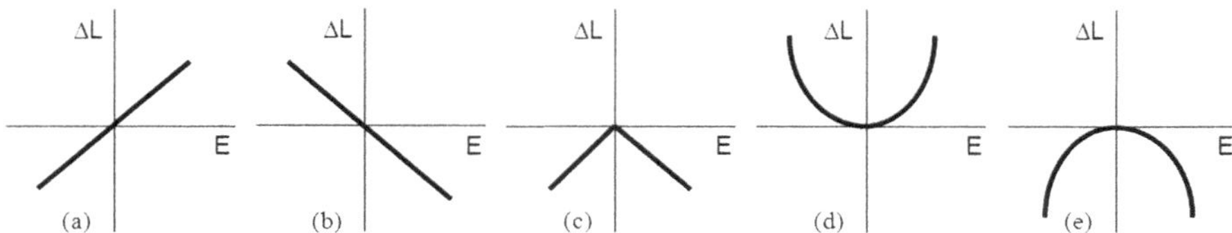

REFERENCES

1. K. Uchino: *J. Industrial Education Soc. Jpn*. 40(a), 28, 1992.
2. J.W. Hardy, J.E. Lefebre and C.L. Koliopoulos: *J. Opt. Soc. Amer.* 67, 360, 1977.
3. S.R. Debbarma and S. Saha: *Int'l J. Civil & Structural Eng.* 2, 924, 2012.
4. S. Moriyama: *Mechanical Design* 27(1), 32, 1983.
5. U. Berg, M. Begemann, B. Hagemann, K.-P. Kamper, F. Michel, C. Thurigen, L. Weber and Th. Wittig: *Proc. 6th Int'l Conf. New Actuators*, p. 552, 1998.
6. http://www.namiki.net/index.html
7. J. Bryzek, K. Petersen and W. McCulley: *IEEE Spectr.*, 5 (May issue), 20, 1994.
8. E. Smela: "Conjugated Polymer Actuators for Biomedical Applications," *Adv. Mater.*, 15(6), 481–494, 2003.
9. S. Bauer: *Proc. Int'l Symp. Smart Actuators*, ICAT/Penn State, PA, April, 1999.
10. P. Sommer-Larsen and R. Kornbluh: "Polymer Actuators,", *Proc. 8th Int'l Conf. New Actuators, B7.0*, Bremen, Germany, June 10–12, 2002.
11. S. Egawa, T. Niino and T. Higuchi: *Proc. 1991 IEEE Workshop on Micro Electro Mechanical Systems*, p. 9, 1991.
12. T. Higuchi: *New Actuator Handbook for Precision Control*, p. 182, Fuji Techno System, Tokyo, 1994.
13. H. Horikawa and K. Otsuka: *New Actuator Handbook for Precision Control*, p. 454, Fuji Techno System, Tokyo, 1994.
14. K. Otsuka and C.M. Wayman: *Shape Memory Materials*, Chap.1, p. 1, Cambridge University Press, UK, 1998.
15. V. Gupta, V. Martynov, and A.D. Johnson: *Proc. 8th Int'l Conf. New Actuators, B6.1*, Bremen, Germany, June 10–12, 2002.
16. K.N. Nelton: *Shape Memory Materials*, Chap.10, Cambridge University Press, UK, 1998.
17. T. Todoroki, K. Fukuda and T. Hayakumo: *Indust Mat*, 32(7), 85, 1984.
18. A.B. Flatau, M.J. Dapino and F.T. Calkins: *Magnetostrictive Composites, Comprehensive Composite Materials*, Chapter 5.26, p. 563, 2000.
19. A.E. Clark and H.S. Belson: U.S. Patent 4,378,258, 1983.
20. H. Eda: Giant Magnetostrictive Actuators, *New Actuator Handbook for Precision Control*, p. 90, Fuji Techno System, Tokyo, 1994.
21. J.L. Butler: *Edge Technologies*, Ames, IA, 1988.
22. K. Uchino: *Ferroelectric Devices*, Marcel Dekker, New York, NY, 2000.
23. K. Yano, T. Hamatsuki, I. Fukui and E. Sato: *Proc. Annual Mtg. EE Japan*, p. 1–157, Spring, 1984.

24. M. Kasuga, T. Satoh, N. Tsukada, T. Yamazaki, F. Ogawa, M. Suzuki, I. Horikoshi and T. Itoh: *J. Soc. Precision Eng.*, 57, 63, 1991.
25. V. Bharti, H.S. Xu, G. Shanti, Q.M. Zhang and K. Liang: *J. Appl. Phys.* 87, 452, 2000.
26. A. Ashkin: *IEEE J. Sel. Top, Quantum Electron.* 6, 841, 2000.
27. Y. Suzuki: *Keijokiokugokin no Hanashi*, Nikkan Kogyo, Tokyo, 1988. (in Japanese)
28. K. Uchino: *Mat. Res. Innovat.*, 1, 163, 1997.
29. M. Tanimura and K. Uchino: *Sensors and Mater.*, 1, 47, 1988.
30. K. Uchino: *J. Rob. Mech.*, 1(2), 124, 1989.
31. S. Thakoor, J.M. Morookian, and J.A. Cutts: *Conf. Proc. 10th IEEE Int'l Symp. on Appl. Ferroelectrics*, 1, 205, 1996.
32. P. Poosanaas, K. Tonooka, and K. Uchino: *Mechatronics* 10, 467, 2000.
33. M.R. Jolly and J.D. Carlson: Composites with Field-Responsive Rheology, *Comprehensive Composite Materials*, Chapter 5.27, p. 575, 2000.
34. T. Shiga, A. Okada and T. Kurauchi: *Macromolecules*, 26, 6958, 1993.
35. V.D. Chase: *Appliance Manufacturer*, May, p. 6, 1996.
36. J.D. Carlson and K.D. Weiss: U.S. Patent 5,382,373, 1995.
37. https://www.livescience.com/54233-scientists-turn-beetles-into-cyborg-insects.html

The following books are recommended for further study

38. K. Uchino: *Ferroelectric Devices 2nd Edition*, CRC Press, Boca Raton, FL, 2009.
39. K. Uchino: *Piezoelectric Actuators & Ultrasonic Motors*, Kluwer Academic, Boston, MA, 1996.
40. G.K. Knopf and K. Uchino: *Light Driven Micromachines*, CRC Press, Boca Raton, FL, 2018. ISBN-13: 978-1-4987-5769-0.
41. The answers for check point & chapter problems are available in K. Uchino, *Ferroelectric Devices & Piezoelectric Actuators*, DEStech Pub., Lancaster, PA, 2016. ISBN: 978-1-60595-312-0

2 A Theoretical Description of Piezoelectricity

Generally speaking, the word *electrostriction* is used to describe electric field-induced strain, and hence frequently also implies the "converse piezoelectric effect." According to solid-state theory, however, the converse piezoelectric effect is defined as a primary electromechanical coupling effect where the induced strain is directly proportional to the applied electric field, while electrostriction is a second-order coupling in which the strain is proportional to the square of the electric field. Thus, strictly speaking, they should be distinguished. However, because the piezoelectric effect for a ferroelectric, which possesses a centrosymmetric high-temperature prototype phase, originates from the electrostrictive coupling in the phenomenological theory, these two effects are closely related. In a piezoelectric ceramic, the additional strains that develop with the reorientation of ferroelectric domains are also important. A description of ferroelectricity in terms of crystal structure is presented first in this chapter; then, a tensor/matrix description of piezoelectrics is introduced. We review a phenomenological description of the piezoelectric/electrostrictive and magnetostrictive effects in detail. Domain reorientation and loss mechanisms are finally treated in this chapter.

2.1 FERROELECTRICITY

2.1.1 Crystal Structure and Ferroelectricity

In the so-called *dielectric materials*, the constituent atoms are typically ionized to a certain degree. When an electric field is applied to such ionic crystals, centers of positive charge (cation) are drawn along the field direction and centers of negative charge (anion) to the opposite direction due to electrostatic attraction, thus inducing electric dipoles within the material. Note that the dipole vector direction is defined from anion to cation, so that the dipole is induced in parallel to the electric field. This phenomenon is known as *electric polarization* (C/m^2), which is characterized by the number of electric dipoles induced (Cm) per unit volume (m^{-3}). The mechanisms of electric polarization are represented in Figure 2.1. There are three fundamental mechanisms that give rise to the net polarization of a dielectric: (1) *electronic polarization*, (2) *ionic polarization*, and (3) *dipole orientation polarization*.

A capacitor with a dielectric between its electrodes can store more electric charge than a device of similar dimensions filled with air due to polarization of the dielectric, as depicted in Figure 2.2. The physical quantity corresponding to the stored electric charge per unit area is called the *electric displacement*, ***D***, and is related to the electric field according to the equation:

$$D = \varepsilon_0 E + P = \varepsilon_0 \varepsilon E. \tag{2.1}$$

Here, ε_0 is the permittivity of free space (8.854×10^{-12} F/m), and ε is the *relative permittivity*, also referred to as the *dielectric constant*, which has a second-rank tensor property, as discussed in Section 2.3. Electric displacement is a sum of charge induced in vacuum by the externally applied electric field and charge generated in the material.

There are certain crystal structures for which the centers of positive and negative charge do not coincide, even when no external electric field is applied. Such crystals are said to possess *spontaneous polarization*. When the spontaneous polarization of the dielectric can be reversed by a practically applied electric field, the material is called a *ferroelectric*.

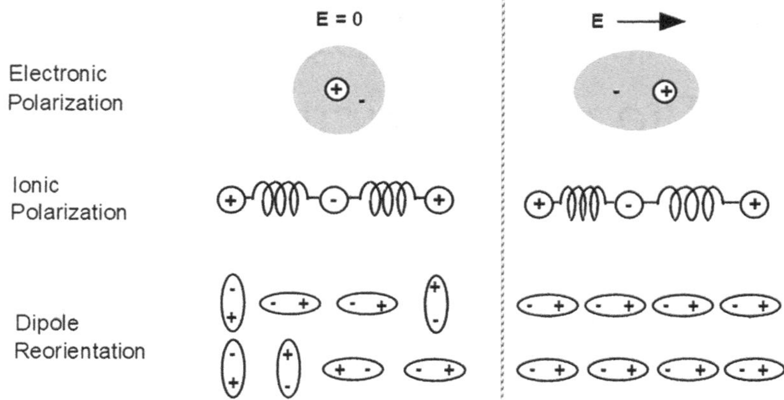

FIGURE 2.1 Microscopic origins of electric polarization.

Not every dielectric is a ferroelectric. Crystals can be classified into 32 *point groups* according to their crystallographic symmetry, and these point groups can be divided into two classes, one with a center of symmetry and the other without one. This classification of the point groups is presented in Table 2.1. There are 21 point groups that do not have a center of symmetry. Among these, 20 groups [point group (432) being the sole exception] are *piezoelectric*; that is, positive and negative charges are generated on their surfaces when stress is applied. *Pyroelectricity* is the phenomenon whereby, due to the temperature dependence of the spontaneous polarization, electric charges are generated on the surface of the crystal when the temperature of the crystal is changed. Among pyroelectric crystals, those whose spontaneous polarization can be reversed by an electric field (not exceeding the breakdown limit of the crystal) are called *ferroelectrics*. One necessary test for ferroelectricity, therefore, is to experimentally observe the polarization reversal when an electric field of the appropriate magnitude is applied.

2.1.2 Origin of Spontaneous Polarization

Why is it that crystals that, on the basis of their elastic energy alone should be stable in a nonpolar state, still experience a shifting of ions and become spontaneously polarized? For simplicity,

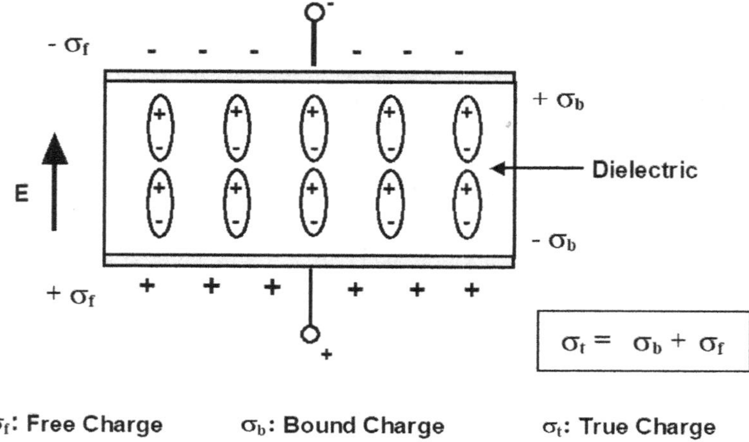

FIGURE 2.2 Charge accumulation in a dielectric capacitor under electric field.

TABLE 2.1
Crystallographic classification according to crystal centro-symmetry and polarity

Polarity	Symmetry	Crystal System										
		Cubic		Hexagonal		Tetragonal		Rhombohedral		Orthorhombic	Mono-clinic	Tri-clinic
Nonpolar (22)	Centro (11)	m3m	m3	6/mmm	6/m	4/mmm	4/m	$\bar{3}$m	$\bar{3}$	mmm	2/m	$\bar{1}$
Nonpolar (22)	Noncentro (21)	432 $\bar{4}$3m	23	622 $\bar{6}$m2	$\bar{6}$	422 $\bar{4}$2m	$\bar{4}$	32		222		
Polar (Pyroelectric) (10)	Noncentro (21)			6mm	6	4mm	4	3m	3	mm2	2m	1

Note: Inside the bold line are piezoelectrics.

let us assume that dipole moments result from the displacement of one kind of ion, A (with an electric charge q), relative to the undistorted crystal lattice. We will consider the case in which the polarization is characterized by all the A ions being displaced in a synchronous way. This kind of ionic displacement could occur with the normal thermal vibration of the lattice (i.e., at a temperature higher than 0 K). Some possible ion shift configurations are depicted for a perovskite crystal in Figure 2.3. If a particular lattice vibration effectively lowers the crystal energy, the ions will shift accordingly, thereby stabilizing the crystal structure and minimizing the energy of the system. Starting with the original cubic structure pictured in Figure 2.3a, if the configuration in Figure 2.3b is established, only the oxygen octahedra are distorted without generating dipole moments (*acoustic mode*). On the other hand, when the configurations in Figures 2.3c or d are established, dipole moments are generated (*optical mode*). The states depicted in Figures 2.3c and d correspond to ferroelectric and antiferroelectric states, respectively. If either of these modes exists, a decrease in the vibration frequency will occur with decreasing temperature (*soft phonon mode*) until finally, at a certain critical temperature corresponding to a phase transition (i.e., Curie or Néel temperature), the vibration frequency becomes zero.

Let us initially discuss the *local field* concept, which promotes ionic shifts for stabilizing the dipole arrangement. At any individual A ion site, there exists a local field associated with the

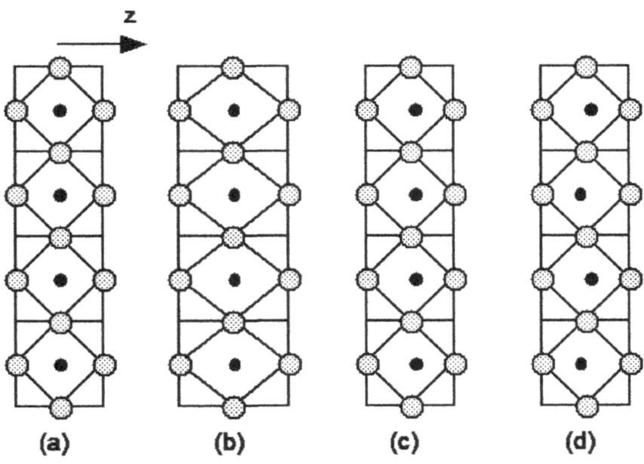

FIGURE 2.3 Some possible ion configurations in a perovskite crystal corresponding to: (a) the initial cubic structure, (b) a symmetrically elongated structure, (c) a structure with coherently shifted center cations, and (d) a structure with an antipolar shift of the center cations. (Dark: center cations; shaded: oxygen octahedra ions.)

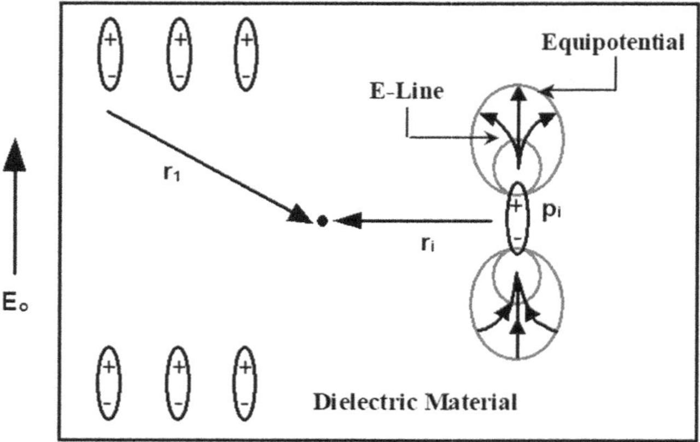

FIGURE 2.4 Schematic representation of the parameters used to define the local field, E_{loc}.

surrounding dipole moments ($\boldsymbol{p}_i$) or polarization $\boldsymbol{P}$ ($\boldsymbol{P} = \sum_i \boldsymbol{p}_i$ in unit volume 1 m^3), even if there is no external field, E_0, applied. Remember that the electric field generated by a charge q $\propto$ q/r^2, while the electric field by a dipole $p \propto p/r^3$, which is more quickly decayed with distance than a separate charge q, but a huge number of dipoles exist very closely (in the lattice parameter ~Å) in a solid crystal. It can be described as:

$$\begin{aligned} \boldsymbol{E}^{loc} &= \boldsymbol{E}_0 + \sum_i \left[3(\boldsymbol{p}_i \cdot \boldsymbol{r}_i)\boldsymbol{r}_i - r_i^2 \boldsymbol{p}_i\right] / 4\pi e_0 r_i^5 \\ &= (\gamma/3e_0)\boldsymbol{P}. \end{aligned} \tag{2.2}$$

where $\boldsymbol{r}_i$ is a position vector relating the location of a given dipole with dipole moment $\boldsymbol{p}_i$ to a particular location in space as depicted in Figure 2.4, and γ is called the *Lorentz factor.* In the case of an isotropic system, such as cubic, γ it is just unity,[1] and it is determined in general by a crystal structure. The local field is the driving force for the ion shift. Supposing that the *ionic polarizability* of ion A is α, the dipole moment, $\boldsymbol{\mu}$, of a unit cell of this crystal is given by:

$$\boldsymbol{\mu} = (\alpha\gamma/3\varepsilon_0)\boldsymbol{P} \tag{2.3}$$

The energy of this dipole moment (dipole-dipole coupling) is defined as:

$$w_{dip} = -\boldsymbol{\mu} \cdot \boldsymbol{E}^{loc} = -(\alpha\gamma^2/9\varepsilon_0^2)\boldsymbol{P}^2 \tag{2.4}$$

If there are N atoms per unit volume, the total dipole coupling energy is expressed as:

$$W_{dip} = N w_{dip} = -(N\alpha\gamma^2 / 9\varepsilon_0^2)\boldsymbol{P}^2 \tag{2.5}$$

Furthermore, when the A ions are displaced from their nonpolar equilibrium positions, the elastic energy should increase.

On the other hand, the elastic energy per unit volume can be expressed as:

$$W_{elas} = N[(k/2)\boldsymbol{u}^2 + (k'/4)\boldsymbol{u}^4] \tag{2.6}$$

where $\boldsymbol{u}$ is the magnitude of the displacement, and k and k' are harmonic and higher-order anharmonic force constants. Since the occurrence of ferro-electricity originates from nonlinear

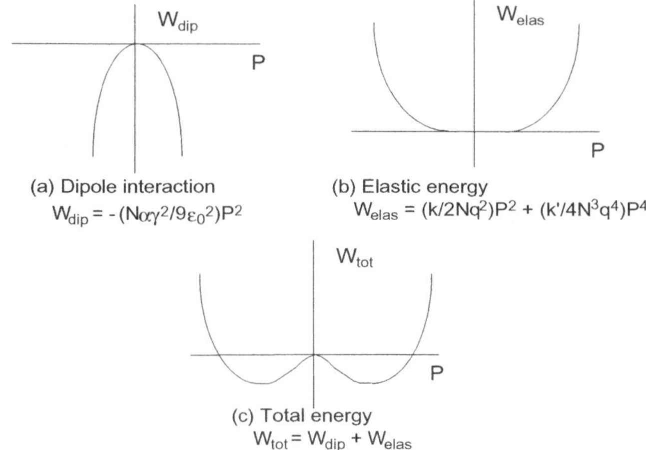

FIGURE 2.5 Energy description of the microscopic mechanism for spontaneous polarization, shown as a function of polarization: (a) dipole energy, W_{dip}; (b) elastic energy, W_{elas}; and (c) the total energy per unit volume, W_{tot}, for a perovskite-like structure.

phenomenology, considering the *nonlinear elasticity* of the ionic vibration seems to be reasonable. The higher-order force constant, k', plays an important role in determining the magnitude of the dipole moment in pyroelectric (i.e., polar) crystals, as you will recognize below. Making use of the relationship:

$$P = Nqu \tag{2.7}$$

in Equation 2.5 and combining with Equation 2.6, we may express the total energy as:

$$\begin{aligned} W_{tot} &= W_{dip} + W_{elas} \\ &= \left[(k/2Nq^2) - \left(N\alpha\gamma^2/9\varepsilon_0^2\right)\right]\boldsymbol{P}^2 + [k'/4N^3q^4]\boldsymbol{P}^4 \end{aligned} \tag{2.8}$$

The shapes of the total and individual energy functions are depicted in Figure 2.5 as a function of polarization $\boldsymbol{P}$ ($\propto$ displacement $\boldsymbol{u}$). It is apparent from this analysis that when the coefficient associated with the harmonic term (first term of Equation 2.8) of the elastic energy is equal to or greater than the coefficient associated with the dipole-dipole coupling, the polarization, $\boldsymbol{P}$, should be zero, because a single energy minimum appears; that is, the A ions are stable and will remain at their nonpolar equilibrium positions. Otherwise, a shift from the equilibrium positions will occur to establish a stable configuration and a net spontaneous polarization will result. Spontaneous polarization tends to develop more readily in perovskite crystal structures (such as barium titanate), because they generally have a higher value of the Lorentz factor ($\gamma = 10$) than most other crystal structures.[2] A slight change in the polarizability α with respect to temperature (i.e., α increases with a decrease in temperature) exhibits a change in the sign of the $\boldsymbol{P}^2$ coefficient in Equation 2.8. This particular temperature that changes the sign of the $\boldsymbol{P}^2$ coefficient is called the phase transition temperature (i.e., *Curie temperature*).

2.1.3 Physical Properties of Ferroelectrics

An important ferroelectric material is barium titanate, $BaTiO_3$ (BT). It is often presented as a classic example of a ferroelectric that exhibits so-called first-order behavior. BT has the perovskite crystal

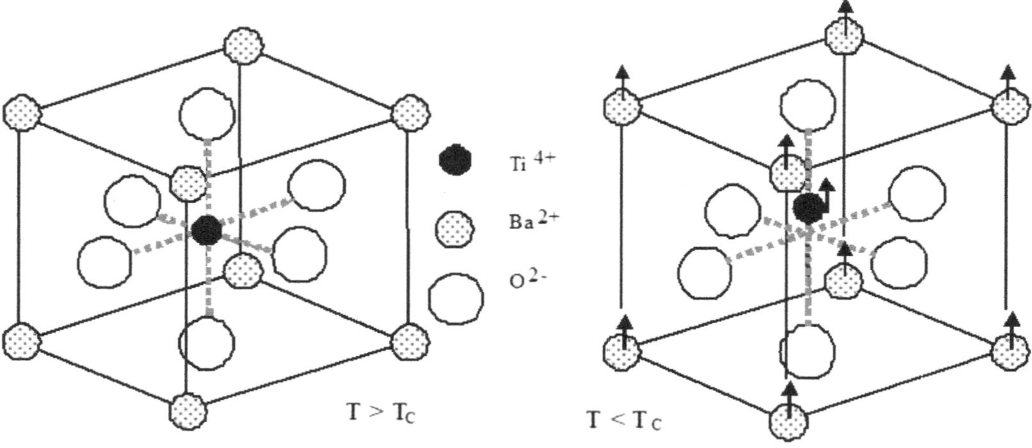

FIGURE 2.6 The crystal structure of a perovskite barium titanate, $BaTiO_3$.

structure depicted in Figure 2.6. In its high-temperature, paraelectric (nonpolar) phase, there is no spontaneous polarization, and a cubic symmetry (O_h-$m3m$) exists. Below the transition temperature, which is designated by T_C and called the *Curie temperature* (~130°C for $BaTiO_3$), spontaneous polarization develops, and the crystal structure becomes slightly elongated, assuming a tetragonal symmetry (C_{4v} – 4 mm). There are also two lower temperature phase transitions for $BaTiO_3$, one from a tetragonal to an orthorhombic phase (~0°C) and the other from an orthorhombic to a rhombohedral phase (~ −90°C). We will focus for the moment on the higher-temperature paraelectric to ferroelectric transition.

A general depiction of the temperature dependences of the spontaneous polarization, $\boldsymbol{P}_S$, and the dielectric constant, ε, for BT is shown in Figure 2.7. The general trends depicted in the figure illustrate how the spontaneous polarization, $\boldsymbol{P}_S$, decreases with increasing temperature and vanishes at the Curie point, while permittivity, ε, tends to diverge near T_C. The inverse dielectric constant,

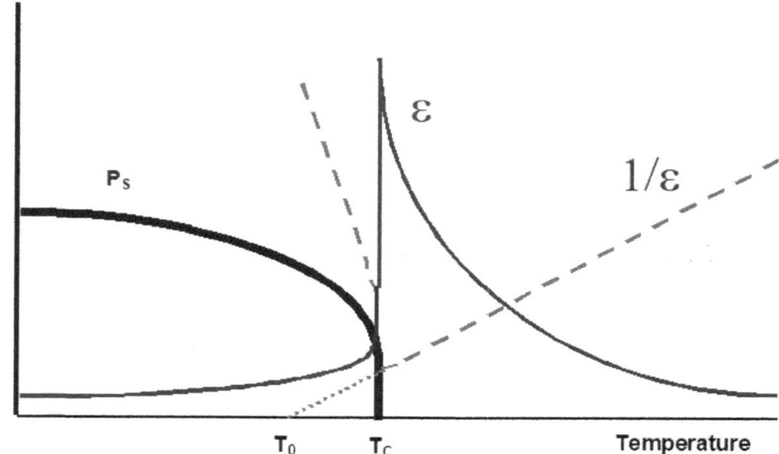

FIGURE 2.7 Temperature dependences of spontaneous polarization, P_S; dielectric constant, ε; and inverse permittivity, $1/\varepsilon$ in a ferroelectric.

$1/\varepsilon$, also shown in Figure 2.7, has a linear dependence on temperature over a broad range in the paraelectric region. This behavior is described by the so-called *Curie-Weiss law*:

$$\varepsilon = C/(T - T_0), \tag{2.9}$$

where C is the *Curie-Weiss constant* and T_0 is the *Curie-Weiss temperature*. T_0 is determined as the intersect with the horizontal (temperature) axis of $1/\varepsilon$, and usually lower than the actual phase transition temperature (i.e., Curie temperature) T_C. The inverse dielectric constant is also discontinuous at T_C for the first-order transition, as we shall see in Section 2.4. The phenomenology also derives the relation between spontaneous polarization, $\boldsymbol{P}_S$, and the spontaneous strain, x_S, by

$$x_S = Q\boldsymbol{P}_S^2 \tag{2.10}$$

The spontaneous strain, x_S, associated with barium titanate typically decreases with increasing temperature in an almost linear manner, and disappears at T_C. BT is piezoelectric in its ferroelectric phase, but becomes nonpiezoelectric in its paralectric phase. It exhibits only an electrostrictive response in this nonpolar phase.

EXAMPLE PROBLEM 2.1

Barium titanate, $BaTiO_3$, exhibits ionic displacements at room temperature, as illustrated in Figure 2.8. The lattice constants are $c = 4.036$ (Å) and $a = 3.992$ (Å). Calculate the magnitude of the spontaneous polarization for barium titanate in this tetragonal form.

Hint

$P = N\,\mu$ (N: number of the dipole moments included in a unit volume). After calculating the dipole moment in a unit cell, divide it by unit volume.

Solution

The dipole moment is defined to be the product of the magnitude of the ion charge and its displacement. The total dipole moment in a unit cell is calculated by summing the contributions of all the Ba^{2+}, Ti^{4+}, and O^{2-}-related dipoles. Each corner ion Ba contributes 1/8, each face ion O contributes 1/2, and the center Ti contributes 1. Note that four O^{2-} ions of the oxygen octahedron do not shift (this position is taken as the origin), leading to zero contribution to the dipole moment.

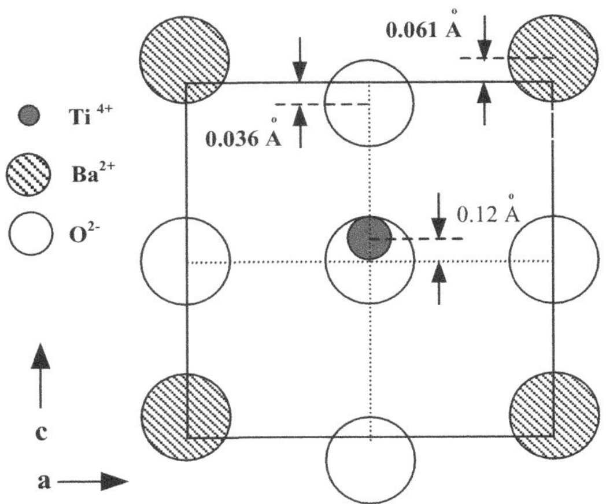

FIGURE 2.8 Ionic shifts in $BaTiO_3$ at room temperature.

$$p = 8\,[2e/8]\,[0.061\times10^{-10}(\mathrm{m})] + [4e]\,[0.12\times10^{-10}(\mathrm{m})] + 2\,[-2e/2]\,[-0.036\times10^{-10}(\mathrm{m})]$$
$$= e\,[0.674\times10^{-10}(\mathrm{m})]$$
$$= 1.08\times10^{-29}(\mathrm{C\cdot m}) \quad \text{(P2.1.1)}$$

where e is the fundamental charge: 1.602×10^{-19} (C).

Next, the unit cell volume is given by

$$v = a^2c = (3.992)^2(4.036)\times10^{-30}(\mathrm{m}^3) = 64.3\times10^{-30}(\mathrm{m}^3) \quad \text{(P2.1.2)}$$

The spontaneous polarization represents the number of (spontaneous) electric dipoles, p, per unit volume:

$$P_S = P/v = 1.08\times10^{-29}(\mathrm{C\cdot m})/64.3\times10^{-30}(\mathrm{m}^3) = 0.17(\mathrm{C/m^2}) \quad \text{(P2.1.3)}$$

This theoretical value of P_S is in reasonable agreement with the experimental value of 0.25 ($\mathrm{C/m^2}$).

2.2 MICROSCOPIC ORIGINS OF ELECTRIC FIELD–INDUCED STRAINS

Solids, especially ceramics (inorganic polycrystalline materials), are relatively hard mechanically, but still expand or contract depending on the change of the state parameters. The strain is defined to be the ratio of the length extension, Δl, to the initial length, l, of the material. When the strain is induced by temperature change and stress, the effects are referred to as *thermal expansion* and *elastic deformation*, respectively. The application of an electric field can also cause deformation in insulating materials. This is called *electric field–induced strain*.

We will consider the mechanisms for electric field–induced strains.[3] For the sake of simplicity, let us consider a simple ionic crystal like rock salt. A one-dimensional rigid-ion spring model of the crystal lattice is depicted in Figure 2.9. The springs represent the cohesive force resulting from the combined action of the electrostatic Coulomb interaction and quantum mechanical repulsion. The centrosymmetric case is shown in Figure 2.9b, and the more general noncentrosymmetric case is pictured in Figure 2.9a. The springs joining the ions are all considered identical for the centrosymmetric case of Figure 2.9. The springs joining the ions for the noncentrosymmetric case shown in Figure 2.9, however, are different for the longer and shorter ionic distances, where stiffer (higher spring constant k) springs are associated with shorter distances, while softer ones are associated with long distances. When the noncentrosymmetric crystal is subjected to an applied electric field, the cations are displaced in the direction of the electric field and the anions in the opposite direction, leading to a change in the interionic distance. Accordingly, when the direction of the electric field is rightward, as in Figure 2.9a, because of the force equality on one ion, the soft springs extend more than the stiff spring shrinkage, thus producing lattice expansion, or a positive strain, x, in the structure and a change in the unit cell length that is directly proportional to the applied electric field, E. This is the known as the *converse piezoelectric effect* and is expressed as

$$x = dE \quad (2.11)$$

where the proportionality constant d is called the *piezoelectric strain coefficient*, which is proportional to the spring constant difference between stiff and soft springs.

As for the centrosymmetric case represented in Figure 2.9b, the extension and shrinkage displacements of the springs are nearly the same because all the springs are the same. Hence, there is no induced strain if we consider the spring merely harmonic (Hooke's law). In reality, ions are not connected by such ideal springs, but the springs are *anharmonic* ($F = k_1\Delta - k_2\Delta^2$); that is, they are more easily extended than compressed. The subtle differences in ion displacement resulting

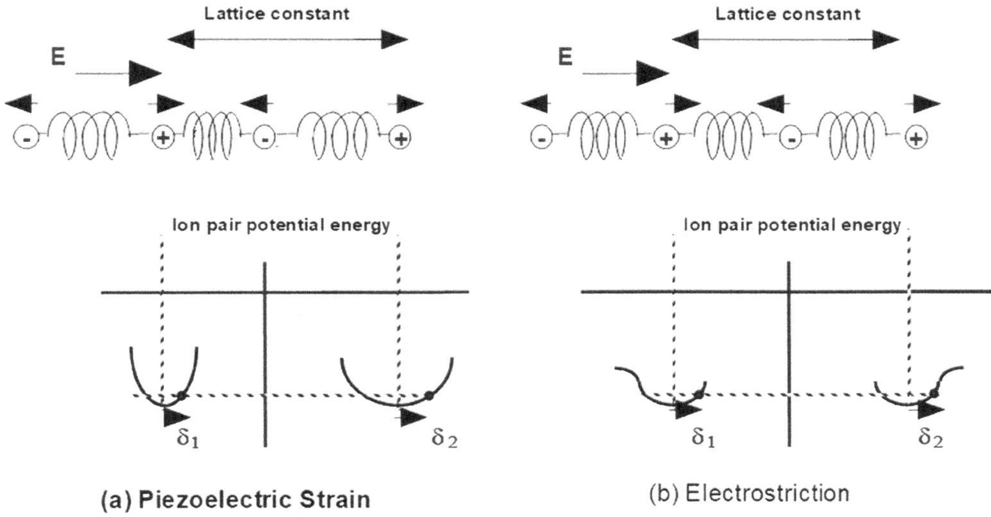

FIGURE 2.9 One-dimensional rigid-ion spring model of a simple NaCl-type crystalline lattice: (a) a noncentrosymmetric case (piezoelectric strain), and (b) a centrosymmetric case (electrostriction). (From Uchino and S. Nomura: *Bull. Jpn. Appl. Phys.*, 52, 575, 1983.)

from this anharmonicity lead to a change in the lattice parameter and an associated strain, which is independent of the direction of the applied electric field, and hence is an even function of the electric field. This is called the *electrostriction effect*:

$$x = ME^2 \tag{2.12}$$

where M is the electrostriction coefficient, which is related to the anharmonic spring constant k_2.

The one-dimensional crystal represented in Figure 2.9a also possesses a spontaneous dipole moment. The total dipole moment per unit volume is called spontaneous polarization. When a large reverse-bias electric field is applied to a crystal that has a spontaneous polarization in a particular polar direction, a transition "phase" is formed that is another stable crystal state in which the relative positions of the ions are reversed (in terms of an untwinned single crystal, this is equivalent to rotating the crystal 180° about an axis perpendicular to its polar axis). This transition, referred to as *polarization reversal*, also causes a significant change in strain, and materials that undergo such a transition are referred to as *ferroelectrics*. Generally speaking, what is actually observed as a field-induced strain is a complicated combination of the three basic effects: piezoelectric strain, electrostriction, and polarization reversal.

2.3 TENSOR/MATRIX DESCRIPTION OF PIEZOELECTRICITY

2.3.1 Tensor Representation

In the solid-state theoretical treatment of the phenomenon of piezoelectricity or electrostriction, the strain, x_{kl}, is expressed in terms of the electric field, E_i, or electric polarization, P_i, as follows:

$$x_{kl} = \sum_{i} d_{ikl}\mathbf{E}_{\mathrm{i}} + \sum_{i,j} M_{ijkl}\mathbf{E}_{\mathrm{i}}\mathbf{E}_{\mathrm{j}} = \sum_{i} g_{ikl}\mathbf{P}_{i} + \sum_{i,j} Q_{ijkl}\mathbf{P}_{\mathrm{i}}\mathbf{P}_{\mathrm{j}} \tag{2.13}$$

where d_{ikl} and g_{ikl} are the piezoelectric coefficients and M_{ijkl} and Q_{ijkl} are the electrostrictive coefficients. Represented in this way, we regard the quantities $\boldsymbol{E}_{\mathrm{i}}$ and x_{kl} as first- (vector) and second-rank tensors (matrix), respectively, while d_{ikl} and M_{ijkl} are considered third- and fourth-rank tensors, respectively. Generally speaking, if two physical properties are represented by tensors of p-rank and q-rank, *the quantity that combines the two properties in a linear relation is represented by a tensor of (p + q)-rank.*

The d_{ijk} tensor can be viewed as a three-dimensional array of coefficients composed of three "layers" of the following form:

$$\begin{array}{ll} \text{1st layer}(i=1) & \begin{pmatrix} d_{111} & d_{112} & d_{113} \\ d_{121} & d_{122} & d_{123} \\ d_{131} & d_{132} & d_{133} \end{pmatrix} \\ \text{2 nd layer}(i=2) & \begin{pmatrix} d_{211} & d_{212} & d_{213} \\ d_{221} & d_{222} & d_{223} \\ d_{231} & d_{232} & d_{233} \end{pmatrix} \\ \text{3rd layer}(i=3) & \begin{pmatrix} d_{311} & d_{312} & d_{313} \\ d_{321} & d_{322} & d_{323} \\ d_{331} & d_{332} & d_{333} \end{pmatrix} \end{array} \tag{2.14}$$

2.3.2 Crystal Symmetry and Tensor Form

A physical property measured along two different directions must have the same value if these two directions are crystallographically equivalent. This consideration sometimes reduces the number of independent tensor coefficients representing a given physical property. Let us consider the third-rank piezoelectricity tensor as an example. The converse piezoelectric effect is expressed in tensor notation as:

$$x_{kl} = d_{ikl}\, E_i \tag{2.15}$$

An electric field, E, initially defined in terms of an (x, y, z) coordinate system, is redefined as E' in terms of another system rotated with respect to the first with coordinates (x', y', z') by means of a transformation matrix (a_{ij}) such that:

$$E'_i = a_{ij} E_j \tag{2.16}$$

or

$$\begin{bmatrix} E'_1 \\ E'_2 \\ E'_3 \end{bmatrix} = \begin{pmatrix} a_{11} & a_{12} & a_{13} \\ a_{21} & a_{22} & a_{23} \\ a_{31} & a_{32} & a_{33} \end{pmatrix} \begin{bmatrix} E_1 \\ E_2 \\ E_3 \end{bmatrix} \tag{2.17}$$

The matrix (a_{ij}) is thus seen to be simply the array of direction cosines that allows us to transform the components of vector $\boldsymbol{E}$ referred to the original coordinate axes to components referred to the axes of the new coordinate system. The second-rank strain tensor is thus transformed in the following manner:

$$x'_{ij} = a_{ik}\, a_{jl}\, x_{kl} \tag{2.18}$$

or

$$\begin{bmatrix} x'_{11} & x'_{12} & x'_{13} \\ x'_{21} & x'_{22} & x'_{23} \\ x'_{31} & x'_{32} & x'_{33} \end{bmatrix} = \begin{pmatrix} a_{11} & a_{12} & a_{13} \\ a_{21} & a_{22} & a_{23} \\ a_{31} & a_{32} & a_{33} \end{pmatrix} \begin{bmatrix} x_{11} & x_{12} & x_{13} \\ x_{21} & x_{22} & x_{23} \\ x_{31} & x_{32} & x_{33} \end{bmatrix} \begin{pmatrix} a_{11} & a_{21} & a_{31} \\ a_{12} & a_{22} & a_{32} \\ a_{13} & a_{23} & a_{33} \end{pmatrix} \tag{2.19}$$

while the transformation of the third-rank piezoelectric tensor is expressed as:

$$d'_{ijk} = a_{il}a_{jm}a_{kn}\, d_{lmn} \tag{2.20}$$

Because of the strain symmetric relation: $x_{ij} = x_{ji}$, from Equation 2.15, we can derive that the d_{lmn} tensor coefficients are symmetric with respect to m and n such that $d_{lmn} = d_{lnm}$, and the following equivalences can be established:

$$\begin{array}{lll} d_{112} = d_{121} & d_{113} = d_{131} & d_{123} = d_{132} \\ d_{221} = d_{212} & d_{213} = d_{231} & d_{223} = d_{232} \\ d_{321} = d_{312} & d_{313} = d_{331} & d_{323} = d_{332} \end{array}$$

The number of independent coefficients is thus reduced from an original 27 ($= 3^3$) to only 18, and the d_{lmn} tensor may then be represented by layers of the following form. It is noteworthy that the matrix for each layer is a symmetric matrix, leading to six independent components for each layer (6 components × 3 layers = 18 components).

$$\begin{array}{ll} \text{1st layer} & \begin{pmatrix} d_{111} & d_{112} & d_{131} \\ d_{112} & d_{122} & d_{123} \\ d_{131} & d_{123} & d_{133} \end{pmatrix} \\ \text{2 nd layer} & \begin{pmatrix} d_{211} & d_{212} & d_{231} \\ d_{212} & d_{222} & d_{223} \\ d_{231} & d_{123} & d_{233} \end{pmatrix} \\ \text{3rd layer} & \begin{pmatrix} d_{311} & d_{312} & d_{331} \\ d_{312} & d_{322} & d_{323} \\ d_{331} & d_{323} & d_{333} \end{pmatrix} \end{array} \tag{2.21}$$

2.3.3 Matrix Notation

The reduction in number of tensor components just carried out for the tensor quantity, d_{ijk}, makes it possible to render the three-dimensional array of coefficients in a more tractable two-dimensional matrix form. This is accomplished by abbreviating the suffix notation used to designate the tensor coefficients according to the following scheme:

Tensor notation	11	22	33	23,32	31,13	12,21
Matrix notation	1	2	3	4	5	6

The layers of tensor coefficients represented by Equation 2.21 may now be rewritten as:

$$\begin{array}{ll} \text{1st layer} & \begin{pmatrix} d_{11} & (1/2)d_{16} & (1/2)d_{15} \\ (1/2)d_{16} & d_{12} & (1/2)d_{14} \\ (1/2)d_{15} & (1/2)d_{14} & d_{13} \end{pmatrix} \\ \text{2nd layer} & \begin{pmatrix} d_{21} & (1/2)d_{26} & (1/2)d_{25} \\ (1/2)d_{26} & d_{22} & (1/2)d_{24} \\ (1/2)d_{25} & (1/2)d_{24} & (1/2)d_{34} \end{pmatrix} \\ \text{3rd layer} & \begin{pmatrix} d_{31} & (1/2)d_{36} & (1/2)d_{35} \\ (1/2)d_{36} & d_{32} & (1/2)d_{34} \\ (1/2)d_{35} & (1/2)d_{34} & d_{33} \end{pmatrix} \end{array} \tag{2.22}$$

The last two suffixes in the tensor notation correspond to those of the strain components; therefore, for the sake of consistency, we will also make similar substitutions in the notation for the strain components.

$$\begin{bmatrix} x_{11} & x_{12} & x_{13} \\ x_{21} & x_{22} & x_{23} \\ x_{31} & x_{32} & x_{33} \end{bmatrix} \rightarrow \begin{bmatrix} x_1 & (1/2)x_6 & (1/2)x_5 \\ (1/2)x_6 & x_2 & (1/2)x_4 \\ (1/2)x_5 & (1/2)x_4 & x_3 \end{bmatrix} \tag{2.23}$$

Note here that, again, the number of independent coefficients for this second-rank tensor may also be reduced from 9 ($= 3^2$) to 6, because it is a symmetric tensor and $x_{ij} = x_{ji}$. The factors of (1/2) in this and the piezoelectric layers of Equation 2.22 are included in order to retain the general form that, like the corresponding tensor equation expressed by Equation 2.15, includes no factors of 2, so that we may write:

$$x_j = d_{ij} E_i \quad (i = 1,2,3; j = 1,2,\ldots,6) \tag{2.24a}$$

or

$$\begin{bmatrix} x_1 \\ x_2 \\ x_3 \\ x_4 \\ x_5 \\ x_6 \end{bmatrix} = \begin{pmatrix} d_{11} & d_{21} & d_{31} \\ d_{12} & d_{22} & d_{32} \\ d_{13} & d_{23} & d_{33} \\ d_{14} & d_{24} & d_{34} \\ d_{15} & d_{25} & d_{35} \\ d_{16} & d_{26} & d_{36} \end{pmatrix} \begin{bmatrix} E_1 \\ E_2 \\ E_3 \end{bmatrix} \tag{2.24b}$$

When deriving a matrix expression for the direct piezoelectric effect in terms of the matrix form of the stress X_{ij}, the factors of (1/2) are not necessary and the matrix may be represented as:

$$\begin{bmatrix} X_{11} & X_{12} & X_{13} \\ X_{21} & X_{22} & X_{23} \\ X_{31} & X_{32} & X_{33} \end{bmatrix} \rightarrow \begin{bmatrix} X_1 & X_6 & X_5 \\ X_6 & X_2 & X_4 \\ X_5 & X_4 & X_3 \end{bmatrix} \tag{2.25}$$

so that:

$$P_j = d_{ij} X_j \quad (i = 1,2,3; j = 1,2,\ldots,6) \tag{2.26a}$$

or

$$\begin{bmatrix} P_1 \\ P_2 \\ P_3 \end{bmatrix} = \begin{bmatrix} d_{11} & d_{12} & d_{13} & d_{14} & d_{15} & d_{16} \\ d_{21} & d_{22} & d_{23} & d_{24} & d_{25} & d_{26} \\ d_{31} & d_{32} & d_{33} & d_{34} & d_{35} & d_{36} \end{bmatrix} \begin{bmatrix} X_1 \\ X_2 \\ X_3 \\ X_4 \\ X_5 \\ X_6 \end{bmatrix} \tag{2.26b}$$

Although the matrix notation has the advantage of being more compact and tractable than the tensor notation, one must remember that the matrix coefficients, d_{ij}, do not transform like the tensor components, d_{ijk}. Applying the matrix notation in a similar manner to the electrostrictive coefficients, M_{ijkl}, we obtain the following equation corresponding to Equation 2.13:

$$\begin{pmatrix} x_1 \\ x_2 \\ x_3 \\ x_4 \\ x_5 \\ x_6 \end{pmatrix} = \begin{pmatrix} d_{11} & d_{21} & d_{31} \\ d_{12} & d_{22} & d_{32} \\ d_{13} & d_{23} & d_{33} \\ d_{14} & d_{24} & d_{34} \\ d_{15} & d_{25} & d_{35} \\ d_{16} & d_{26} & d_{36} \end{pmatrix} \begin{bmatrix} E_1 \\ E_2 \\ E_3 \end{bmatrix} + \begin{pmatrix} M_{11} & M_{21} & M_{31} & M_{41} & M_{51} & M_{61} \\ M_{12} & M_{22} & M_{32} & M_{42} & M_{52} & M_{62} \\ M_{13} & M_{23} & M_{33} & M_{43} & M_{53} & M_{63} \\ M_{14} & M_{24} & M_{34} & M_{44} & M_{54} & M_{64} \\ M_{15} & M_{25} & M_{35} & M_{45} & M_{55} & M_{65} \\ M_{16} & M_{26} & M_{36} & M_{46} & M_{56} & M_{66} \end{pmatrix} \begin{pmatrix} E_1^2 \\ E_2^2 \\ E_3^2 \\ E_2E_3 \\ E_3E_1 \\ E_1E_2 \end{pmatrix} \qquad (2.27)$$

Tables 2.2 and 2.3 summarize the matrices d_{ij} (or g_{ij}) and Q_{ij} (or M_{ij}) for all crystallographic point groups.[4]

EXAMPLE PROBLEM 2.2

Suppose that a shear stress, X_{31}, is applied on a crystal cube with a square cross-section such that it is deformed in a rhombus by a 1° angle, as illustrated in Figure 2.10. Calculate the induced strain x_5 $(=2x_{31})$.

HINT

Shear strain is directly related to the deformed angle of the square in the unit (radian).

SOLUTION

Since $x_5 = 2x_{31} = \tan\theta \approx \theta$ and $1° = [\pi/180]$ (rad), $x_5 = 0.017$. Remember that the angle of the rhombus less than 90° is taken as positive strain. On the contrary, when we apply 1 kV/mm on a typical PZT ceramic with $d_{15} = 500$ pC/N, we can obtain $x_5 = d_{15}\,E_3 = 5 \times 10^{-4}$ (radian), which corresponds to 0.029° or 1.7 min, a very small angle change!

EXAMPLE PROBLEM 2.3

For a cube-shaped specimen, tensile stress X and compressive stress $-X$ are applied simultaneously along the (1 0 1) and ($\bar{1}$ 0 1) axes, respectively (Figure 2.11). When we take the prime-coordinates (1′ and 3′) as illustrated in Figure 2.11, the stress tensor is represented as

$$\begin{pmatrix} X & 0 & 0 \\ 0 & 0 & 0 \\ 0 & 0 & -X \end{pmatrix}$$

Using the transformation matrix A (i.e., $\theta = -45°$ rotation along 2′ axis in Figure 2.11)

$$\begin{pmatrix} \cos\theta & 0 & \sin\theta \\ 0 & 1 & 0 \\ -\sin\theta & 0 & \cos\theta \end{pmatrix},$$

calculate $A.X.A^{-1}$, and verify that the above stress is equivalent to a pure shear stress in the original (nonprime) coordinates.

TABLE 2.2
Piezoelectric strain coefficient (d) matrices

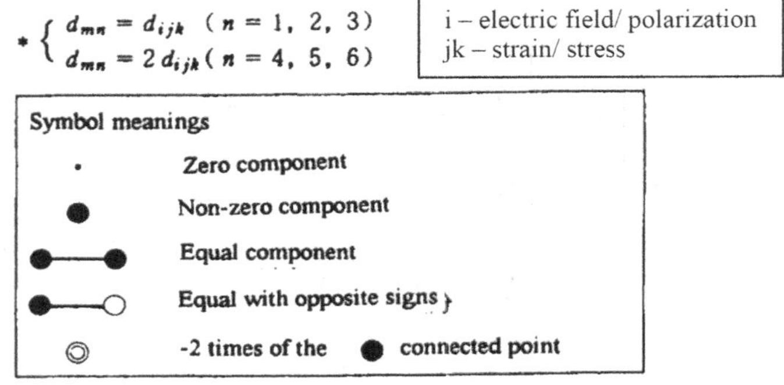

I Centro symmetric point group

Point group $\bar{1}$, $2/m$, mmm, $4/m$, $4/mmm$, $m3$, $m3m$, $\bar{3}$, $\bar{3}m$, $6/m$, $6/mmm$ All components are zero

II Non- centro symmetric point group

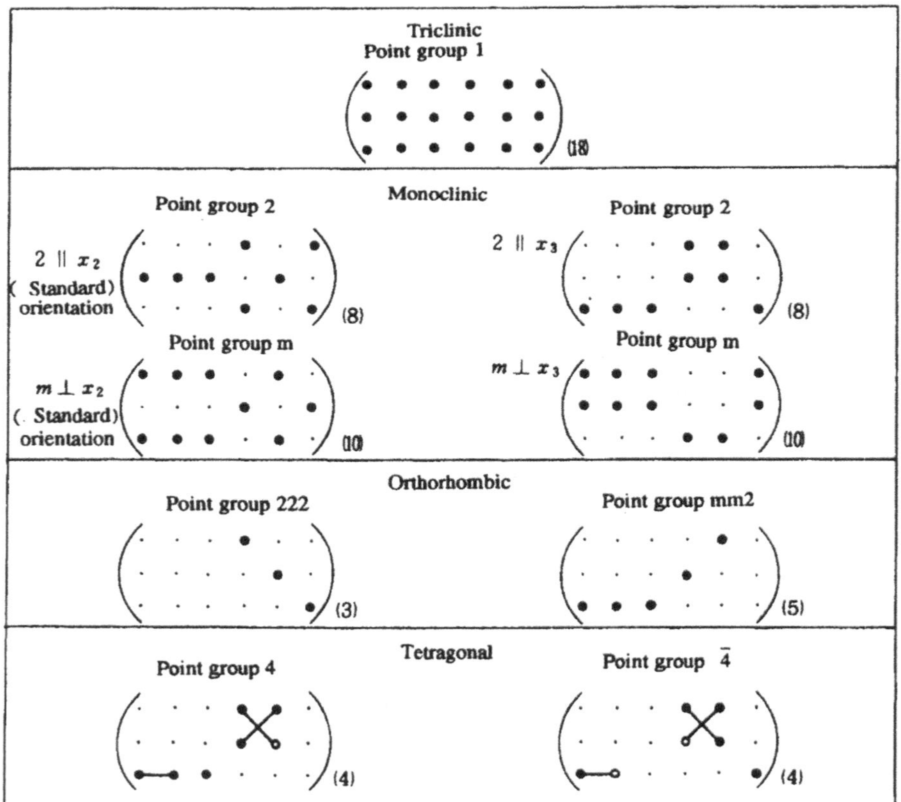

(Continued)

TABLE 2.2 (Continued)
Piezoelectric strain coefficient (d) matrices

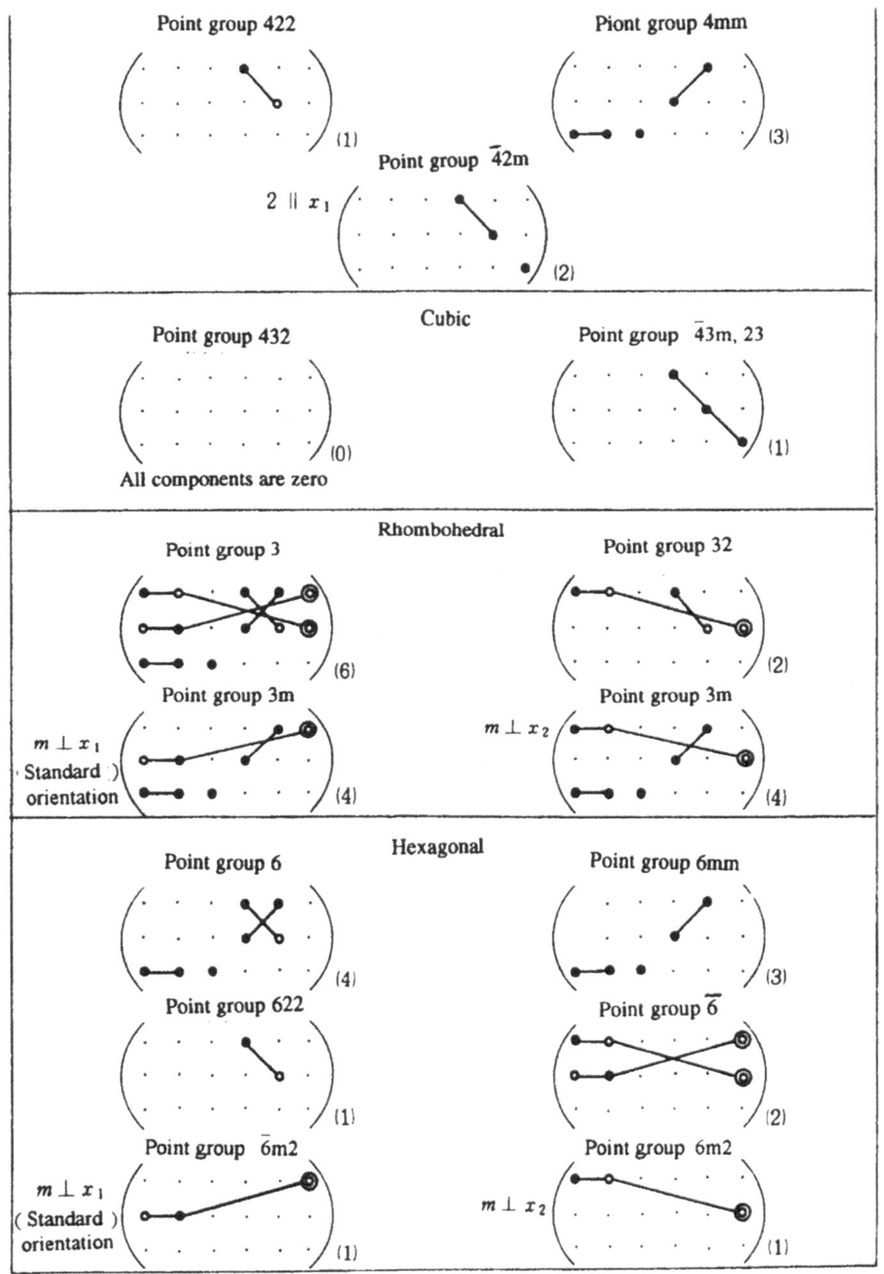

Source: J. F. Nye: *Physical Properties of Crystals, Oxford University Press*, London, p. 123, 140, 1972.

TABLE 2.3
Electrostriction Q coefficient matrices

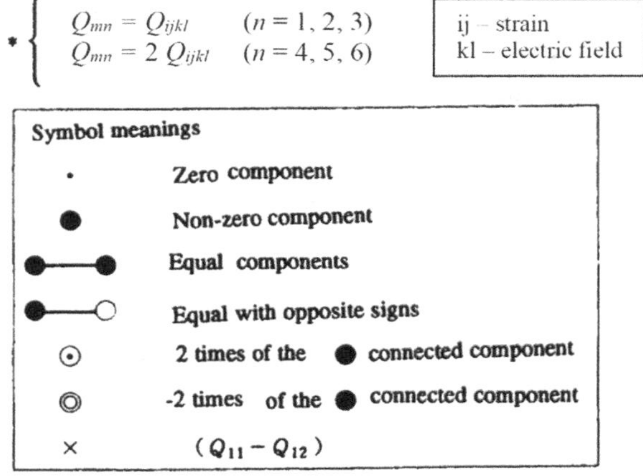

$Q_{mn} = Q_{ijkl}$ $(n = 1, 2, 3)$
$Q_{mn} = 2\,Q_{ijkl}$ $(n = 4, 5, 6)$
ij – strain
kl – electric field
Symbol meanings
Zero component
Non-zero component
Equal components
Equal with opposite signs
2 times of the connected component
-2 times of the connected component
$(Q_{11} - Q_{12})$

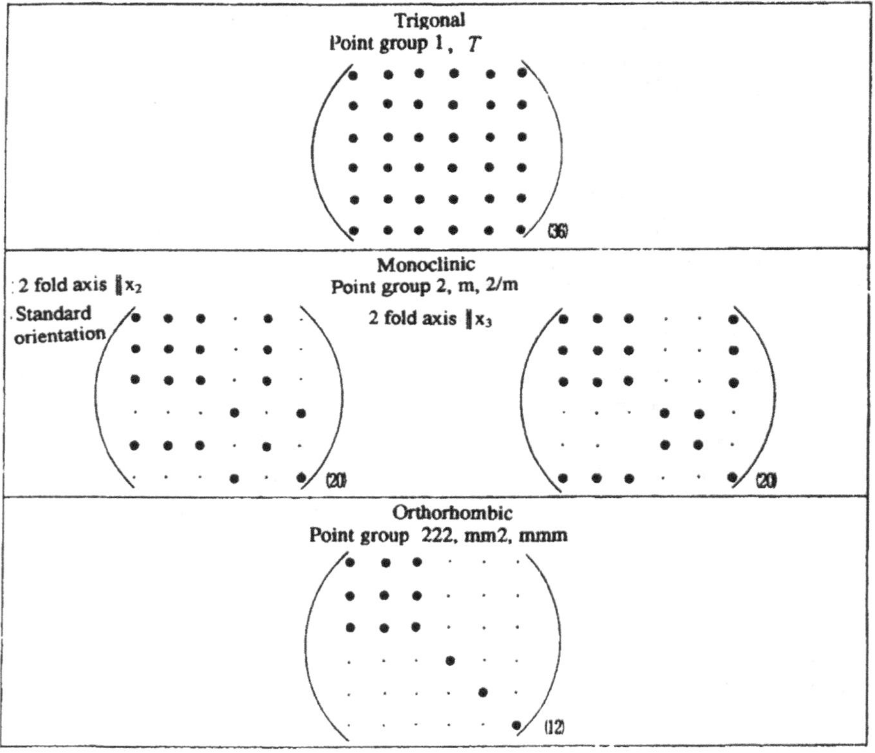

(Continued)

TABLE 2.3 (Continued)
Electrostriction *Q* coefficient matrices

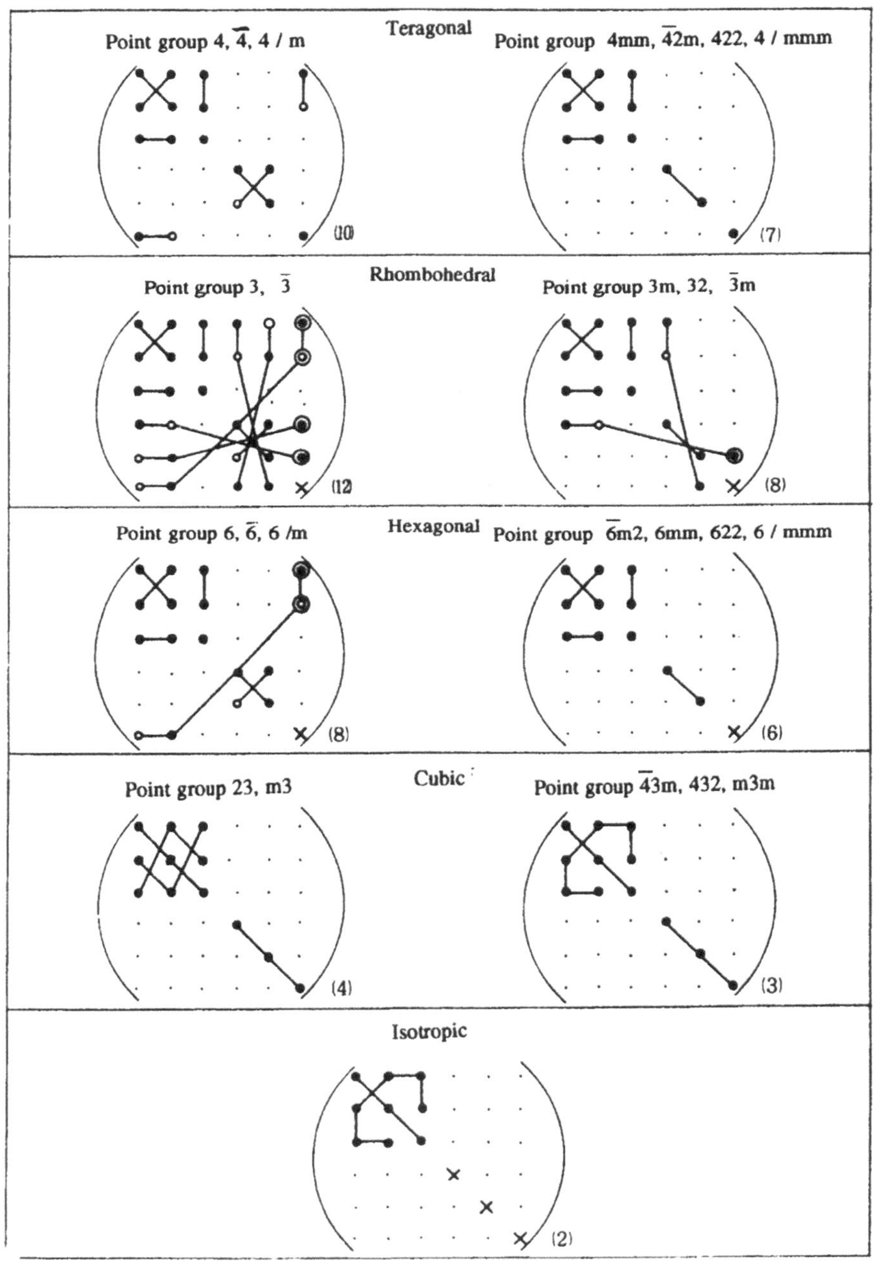

Source: J. F. Nye: *Physical Properties of Crystals, Oxford University Press*, London, p. 123, 140, 1972.

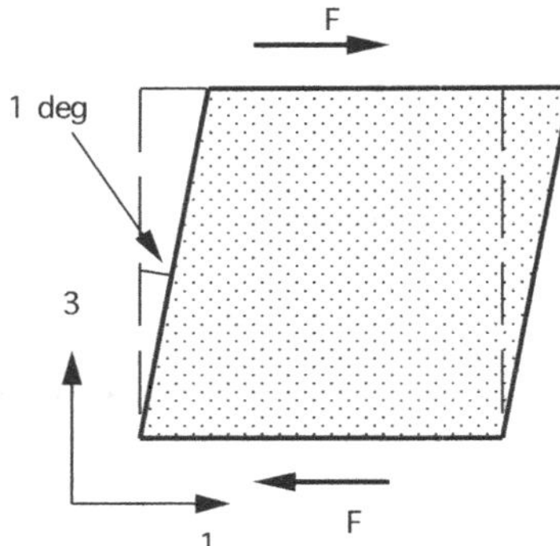

FIGURE 2.10 Strain under a shear stress.

SOLUTION

Using $\theta = -45°$, we can obtain the transformed stress representation:

$$A.X.A^{-1} = \begin{pmatrix} 1/\sqrt{2} & 0 & -1/\sqrt{2} \\ 0 & 1 & 0 \\ 1/\sqrt{2} & 0 & 1/\sqrt{2} \end{pmatrix} \begin{pmatrix} X & 0 & 0 \\ 0 & 0 & 0 \\ 0 & 0 & -X \end{pmatrix} \begin{pmatrix} 1/\sqrt{2} & 0 & -1/\sqrt{2} \\ 0 & 1 & 0 \\ -1/\sqrt{2} & 0 & 1/\sqrt{2} \end{pmatrix}$$
$$= \begin{pmatrix} 0 & 0 & X \\ 0 & 0 & 0 \\ X & 0 & 0 \end{pmatrix} \quad \text{(P2.3.1)}$$

The off-diagonal components, X_{13} and X_{31}, have the same magnitude, X, and represent a pure shear stress. Note that a shear stress is equivalent to a combination of extension and contraction stresses. Only an extensional stress applied along a diagonal direction 1′ may exhibit an apparently similar diagonal distortion (rhombus) of the crystal. However, precisely speaking, without the contraction along the 3′ direction, this is not exactly equivalent to the pure shear deformation with a volume expansion. The contraction occurs only from the Poisson's ratio of the extension.

EXAMPLE PROBLEM 2.4

Barium titanate ($BaTiO_3$) has a tetragonal crystal symmetry (point group *4* mm) at room temperature. The appropriate piezoelectric strain coefficient matrix is therefore of the form:

$$d_{ij} = \begin{pmatrix} 0 & 0 & 0 & 0 & d_{15} & 0 \\ 0 & 0 & 0 & d_{15} & 0 & 0 \\ d_{31} & d_{31} & d_{33} & 0 & 0 & 0 \end{pmatrix}$$

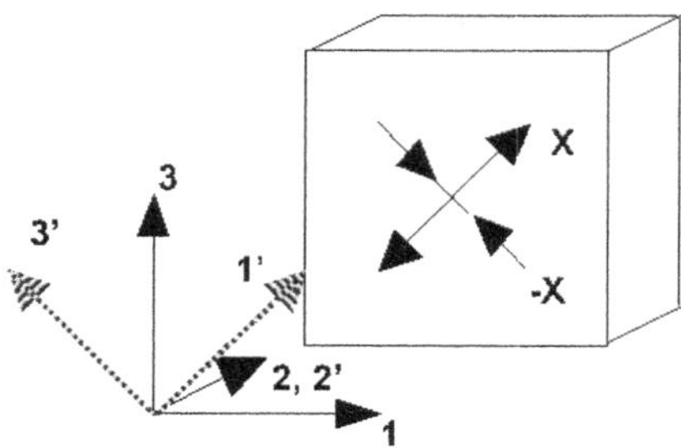

FIGURE 2.11 Application of a pair of stresses X and $-X$ to a cube of material.

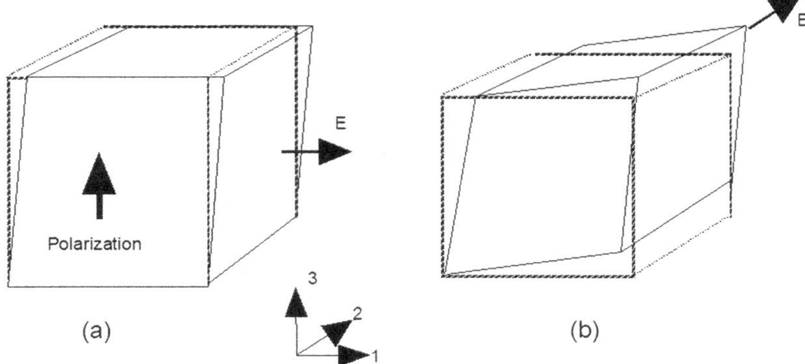

FIGURE 2.12 Induced strains considered for Problems 2.4 and 2.5: (a) shear strain induced in a BT single crystal with *tetragonal* symmetry *4*mm under E_1, and (b) electrostrictive strain induced along [111] in a PMN single crystal with *cubic* symmetry *m3m.*

a. Determine the strain induced in a piezoelectric material when an electric field is applied along the crystallographic ***c*** axis.
b. Determine the strain induced in a piezoelectric material when an electric field is applied along the crystallographic ***a*** axis.

Hint

The matrix equation that applies in this case is:

$$\begin{bmatrix} x_1 \\ x_2 \\ x_3 \\ x_4 \\ x_5 \\ x_6 \end{bmatrix} = \begin{pmatrix} 0 & 0 & d_{31} \\ 0 & 0 & d_{31} \\ 0 & 0 & d_{33} \\ 0 & d_{15} & 0 \\ d_{15} & 0 & 0 \\ 0 & 0 & 0 \end{pmatrix} \begin{bmatrix} E_1 \\ E_2 \\ E_3 \end{bmatrix} \tag{P2.4.1}$$

Solution

We can derive expressions for the induced strains:

$$x_1 = x_2 = d_{31}E_3,\ x_3 = d_{33}E_3, x_4 = d_{15}E_2, x_5 = d_{15}E_1, x_6 = 0$$

so that the following determinations can be made:

a. When E_3 is applied, elongation in the *c* direction ($x_3 = d_{33}E_3$, $d_{33} > 0$) and contraction in the *a* and *b* directions ($x_1 = x_2 = d_{31}E_3$, $d_{31} < 0$) occurs. Note that PZTs and other oxide piezoelectrics exhibit positive d_{33}, that is, extension along the electric field direction, while polymer PVDF shows negative d_{33}.
b. When E_1 is applied, a shear strain x_5 ($=2x_{31}$) $= d_{15}E_1$ is induced. The case where $d_{15} > 0$ and $x_5 > 0$ is illustrated in Figure 2.12a.

EXAMPLE PROBLEM 2.5

Lead magnesium niobate ($Pb(Mg_{1/3}Nb_{2/3})O_3$) has a cubic crystal symmetry (point group ***m3m***) at room temperature and does not exhibit piezoelectricity; however, a large electrostrictive response is induced with the application of an electric field. The relationship between the induced strain and the applied electric field is given in matrix form by:

$$\begin{bmatrix} X_1 \\ X_2 \\ X_3 \\ X_4 \\ X_5 \\ X_6 \end{bmatrix} = \begin{pmatrix} M_{11} & M_{12} & M_{12} & 0 & 0 & 0 \\ M_{12} & M_{11} & M_{12} & 0 & 0 & 0 \\ M_{12} & M_{12} & M_{11} & 0 & 0 & 0 \\ 0 & 0 & 0 & M_{44} & 0 & 0 \\ 0 & 0 & 0 & 0 & M_{44} & 0 \\ 0 & 0 & 0 & 0 & 0 & M_{44} \end{pmatrix} \begin{bmatrix} E_1^2 \\ E_2^2 \\ E_3^2 \\ E_2E_3 \\ E_3E_1 \\ E_1E_2 \end{bmatrix} \tag{P2.5.1}$$

Calculate the strain induced in the material when an electric field is applied along the [111] direction.

Hint

E_1E_2 and E_3E_1 are not independent of E_1^2, but this matrix formation handles as though independent.

Solution

Substituting the electric field applied in this case:

$$E_1 = E_2 = E_3 = E_{[111]}/\sqrt{3}$$

into Equation P2.4.1, we obtain:

$$x_1 = x_2 = x_3 = (M_{11} + 2M_{12})\frac{E_{[111]}^2}{3} (= x_{11} = x_{22} = x_{33}) \tag{P2.5.2a}$$

$$x_4 = x_5 = x_6 = M_{44}\frac{E_{[111]}^2}{3} (= 2x_{23} = 2x_{31} = 2x_{12}) \tag{P2.5.2b}$$

The resulting distortion is illustrated in Figure 2.12b. The strain, x, induced along an arbitrary direction is given by

$$x = \Sigma x_{ij} l_i l_j \tag{P2.5.3}$$

where l_i is a direction cosine defined with respect to the i-axis. The strain induced along the [111] direction, $x_{[111]\|}$, is therefore given by:

$$\begin{aligned} x_{[111]\|} &= \sum x_{ij}(1/\sqrt{3})(1/\sqrt{3}) \\ &= x_1 + x_2 + x_3 + \frac{2[(x_4/2) + (x_5/2) + (x_6/2)]}{3} \\ &= (M_{11} + 2M_{12} + M_{44})\frac{E_{[111]}^2}{3} \end{aligned} \tag{P2.5.4}$$

The strain induced perpendicular to the [111] direction, $x_{[111]\perp}$, is calculated in a similar fashion.

$$x_{[111]\perp} = [M_{11} + 2M_{12} - (M_{44}/2)]\frac{E_{[111]}^2}{3} \tag{P2.5.5}$$

The distortion associated with this strain is depicted in Figure 2.12b.

It is important to note here that the volumetric strain ($\Delta V/V$) will be given by:

$$\frac{\Delta V}{V} = x_{[111]\parallel} + 2x_{[111]\perp} = (M_{11} + 2M_{12})E^2_{[111]} \quad \text{(P2.5.6)}$$

and is independent of the applied field direction.

EXAMPLE PROBLEM 2.6

For a third-rank tensor such as the piezoelectric tensor, the transformation due to a change in the coordinate system is represented by

$$d'_{ijk} = \sum_{l,m,n} a_{il}a_{jm}a_{kn}d_{lmn} \quad \text{(P2.6.1)}$$

The transformation matrix for rotation (angle θ) about a principal z-axis is provided by:

$$\begin{pmatrix} a_{11} & a_{12} & a_{13} \\ a_{21} & a_{22} & a_{23} \\ a_{31} & a_{32} & a_{33} \end{pmatrix} = \begin{pmatrix} \cos\theta & \sin\theta & 0 \\ -\sin\theta & \cos\theta & 0 \\ 0 & 0 & 1 \end{pmatrix} \quad \text{(P2.6.2)}$$

a. When the crystal has a 4-fold axis along the z-axis (i.e., $\theta = 90°$), provide the transformation matrix explicitly.
b. Calculate d'_{122} and d'_{211}, and express them with nonprime d tensor components. The step-by-step derivation process should be described.
c. Considering that $d'_{122} = d_{122}$ and $d'_{211} = d_{211}$ for a crystal with a 4-fold symmetry, derive the relationship for the piezoelectric tensor components, d_{122} and d_{211}, with a crystallographic reason.

HINT

- Piezoelectric third-rank tensor: the transformation due to a change in coordinate system is represented by

$$d'_{ijk} = \sum_{l,m,n} a_{il}a_{jm}a_{kn}d_{lmn}$$

- When the crystal has a 4-fold axis symmetry along the z-axis, the 90° transformed d'_{ijk} should be identical to the original d_{ijk}.
- General tensor symmetry with m and n: such that $d_{123} = d_{132}$ and $d_{213} = d_{231}$.

SOLUTION

a. Taking $\cos 90° = 0$, and $\sin 90° = 1$, the rotation matrix becomes

$$\begin{pmatrix} 0 & 1 & 0 \\ -1 & 0 & 0 \\ 0 & 0 & 1 \end{pmatrix}.$$

b. $d'_{122} = a_{12}a_{21}a_{21}d_{211} = (+1)(-1)(-1)d_{211} = d_{211}$ (P2.6.3)

(*Note*: When you see "1" for the first suffix of rotation matrix component a, the second suffix is determined as "2" immediately from the 90 degree rotation matrix form.)

$d'_{211} = a_{21}a_{12}a_{12}d_{211} = (-1)(+1)(+1)d_{211} = -d_{211}$ (P2.6.4)

Reason: Because the crystal symmetry is 4-fold, the 90° tensor rotation should not identify the difference between the prime and nonprime tensor components, leading to the relations: $d'_{122} = d_{122}$ and $d'_{211} = d_{211}$.

Accordingly,

$$d'_{122} = d_{122} = d'_{211} = d_{211} = -d_{211} \tag{P2.6.5}$$

Thus, $d_{122} = d_{211} = 0$ – Required relationship

If you continue a similar reduction process of the tensor components, you will get

$$\begin{pmatrix} 0 & 0 & 0 & d_{14} & d_{15} & 0 \\ 0 & 0 & 0 & d_{15} & -d_{14} & 0 \\ d_{31} & d_{31} & d_{31} & 0 & 0 & 0 \end{pmatrix} \tag{P2.6.6}$$

for the point group **4**.

2.4 THEORY OF FERROELECTRIC PHENOMENOLOGY

2.4.1 Background of Phenomenology

A thermodynamic phenomenological theory is discussed basically in the form of an expansion series of the free energy as a function of the physical properties; polarization, $\boldsymbol{P}$, and electric field, $\boldsymbol{E}$, temperature, T, and entropy, S, stress, X, and strain, x (and, if applicable, magnetic field, $\boldsymbol{H}$, and magnetization, $\boldsymbol{M}$, in a magnetic or magnetostrictive material). In our ferroelectric discussion, the last parameters will be neglected. This derivation process is the most frequently asked question from readers.

2.4.1.1 Polarization Expansion

The free energy can be expanded in general as follows, using a so-called *order parameter*, P:

$$F(P) = a_1P + a_2P^2 + a_3P^3 + a_4P^4 + a_5P^5 + a_6P^6 + \cdots$$

We assume that the free energy of the crystal should not change with polarization reversal ($\boldsymbol{P} \rightarrow -\boldsymbol{P}$). Otherwise, the charge or permittivity in the capacitance would be changed according to the capacitor orientation/upside down, which may cause serious practical problems in electronic equipment. From $F(\boldsymbol{P}) = F(-\boldsymbol{P})$, the energy expansion series should not contain terms in odd powers of a vector component, $\boldsymbol{P}$. Thus, the expansion series includes only even powers of P:

$$F(P) = a_2P^2 + a_4P^4 + a_6P^6 + \cdots$$

2.4.1.2 Temperature Expansion

Next, we take into account the expansion series in terms of P and temperature:

$$\begin{aligned} F(P,T) = {} & a_2P^2 + a_4P^4 + a_6P^6 + \cdots \\ & + b_1T + b_2T^2 + \cdots + c_1TP^2 + \cdots \end{aligned}$$

From $S = -(\partial F/\partial T) = -b_1$, and given that a constant entropy is meaningless, we may take $b_1 = 0$. The term b_2T^2 is a higher-order term to be neglected. Thus, we adopt only c_1TP^2. Note that a possible term, TP, is omitted for the same reason, $F(\boldsymbol{P}) = F(-\boldsymbol{P})$, again. It is important to understand that the product TP^2 of the two parameters (P^2 and T) explains the *coupling effect*; that is, T change causes a change in P to keep the same free energy (this effect is called *pyroelectricity*), or E application causes a T change (this is called the *electro-caloric* effect). By combining a_2P^2 and c_1TP^2, we introduce

$$(1/2)\alpha P^2 = a_2P^2 + c_1TP^2 = (1/2)\left(\frac{T - T_0}{\varepsilon_0 C}\right)P^2$$

We also introduce the following notations by neglecting temperature dependence of the higher-order coefficients:

$$(1/4)\beta = a_4$$

$$(1/6)\gamma = a_6$$

2.4.1.3 Stress Expansion

Now we introduce the stress expansion series:

$$F(P,T,X) = (1/2)\alpha(T)P^2 + (1/4)\beta P^4 + (1/6)\gamma P^6 + \cdots$$
$$+d_1X + d_2X^2 + \cdots + e_1P^2X + \cdots \qquad [\alpha(T)(T - T_0)/\varepsilon_0 C]$$

From $x = -(\partial F/\partial X) = -d_1$, and given that constant strain is meaningless, we take $d_1 = 0$ (as the strain origin). P^2X is the fundamental electromechanical coupling (i.e., electrostrictive coupling), which explains the polarization generation under stress or strain generation under electric field. This argument is valid only when spontaneous polarization exists in the piezoelectric phase and the polarization disappears in the paraelectric phase. It is not valid in quartz, where the piezoelectric phase does not have spontaneous polarization (actually, quartz is not a ferroelectric). The odd power of stress, X, can remain in the free energy, because the crystal orientation/upside-down does not change the sign of X (X is a tensor, not a vector; negative X means compressive stress). Introducing new notations $d_2 = -(1/2)s$ (elastic stiffness), and $e_1 = -Q$ (electrostrictive coefficient), we finally obtain

$$G_1 = (1/2)[(T - T_0)/\varepsilon_0 C]P^2 + (1/4)\beta P^4 + (1/6)\gamma P^6$$
$$- (1/2)sX^2 - QP^2X$$

The above free energy, G_1, is particularly called *elastic Gibbs energy*, which is the starting formula of most conventional textbooks.

2.4.2 Landau Theory of the Ferroelectric Phase Transition

We assume that the Landau free energy, F, in one dimension is represented in terms of polarization, P (excluding stress terms) as:

$$F(P,T) = (1/2)\alpha P^2 + (1/4)\beta P^4 + (1/6)\gamma P^6 \qquad (2.28)$$

The coefficient α is temperature dependent, and β and γ are supposed to be constant. The phenomenological formulation is applied for the entire temperature range over which the crystal is in its paraelectric and ferroelectric states. It is noteworthy that the temperature dependence of α above originates from the microscopic origin, temperature dependence of polarizability α in Equation 2.8.

Because the spontaneous polarization should be zero in the paraelectric state, the free energy should be zero in the paraelectric phase at any temperatures above its Curie temperature (or the phase transition temperature). To stabilize the ferroelectric state, the free energy for a certain polarization, P, should be lower than zero. Otherwise, the paraelectric state should be realized. Thus, at least, the coefficient α of the P^2 term must be negative for the polarized state to be stable, while in the paraelectric state it must be positive, passing through zero at some temperature T_0 (Curie-Weiss temperature). From this concept, as a linear relation:

$$\alpha = (T - T_0)/\varepsilon_0 C \tag{2.29}$$

where C is taken as a positive constant (the Curie-Weiss constant) and T_0 is equal to or lower than the actual transition temperature T_C (Curie temperature). The temperature dependence of α is related on a microscopic level to the temperature dependence of the ionic polarizability α in Equation 2.8, coupled with thermal expansion and other effects of anharmonic lattice interactions. Refer to the discussion in Section 2.1.2.

The equilibrium polarization, P, established with the application of an electric field, E, satisfies the condition:

$$(\partial F/\partial P) = E = \alpha P + \beta P^3 + \gamma P^5 \tag{2.30}$$

With no electric field applied, Equation 2.30 provides two cases:

$$P(\alpha + \beta P^2 + \gamma P^4) = 0 \tag{2.31}$$

1. $P = 0 \rightarrow$ This trivial solution corresponds to a paraelectric state.
2. $\alpha + \beta P^2 + \gamma P^4 = 0 \rightarrow$ This finite polarization solution corresponds to a ferroelectric state.

2.4.2.1 The Second-Order Transition

When β is positive, γ term is often neglected because nothing special is added by this term. There are not many examples that show this second-order transition, but this description provides intuitive ideas on the phase transition because of its mathematical simplicity. Triglycine sulfate (TGS) is a close example of a ferroelectric exhibiting the second-order transition.

The polarization for zero applied field is obtained from Equation 2.31 as

$$\alpha P_S + \beta P_S^3 = 0 \quad [\alpha = (T - T_0)/\varepsilon_0 C]$$

so that either $P_S = 0$ or

$$P_S^2 = -\alpha/\beta = (T_0 - T)/\beta\varepsilon_0 C. \tag{2.32a}$$

For $T > T_0$, the unique solution $P_S = 0$ is obtained. For $T < T_0$, the minimum of the Landau free energy is obtained at:

$$P_S = \pm\sqrt{(T_0 - T)/(\beta\varepsilon_0 C)}. \tag{2.32b}$$

The phase transition occurs at $T_C = T_0$ and the polarization goes continuously to zero at this temperature; this is called a *second-order transition*.

The relative permittivity ε is calculated as:

$$1/\varepsilon = \varepsilon_0/(\partial P/\partial E) = \varepsilon_0(\alpha + 3\beta P^2) \tag{2.33}$$

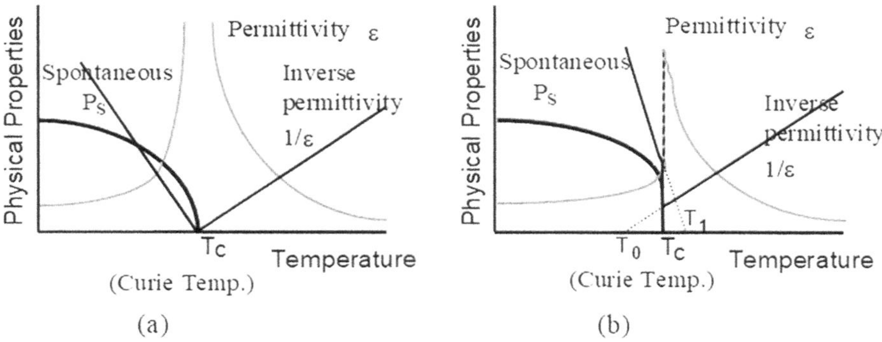

FIGURE 2.13 Phase transitions in a ferroelectric: (a) second-order and (b) first-order phase transition.

Then, for $T > T_0$, $P = 0$:

$$\varepsilon = 1/\varepsilon_0\alpha = C/(T - T_0) \quad (T > T_0) \tag{2.34}$$

For $T < T_0$, $P_S^2 = -\alpha/\beta = (T_0 - T)/\beta\varepsilon_0 C$:

$$\begin{aligned} 1/\varepsilon &= \varepsilon_0(\alpha + 3\beta P^2) = = \varepsilon_0[\alpha + 3\beta(-\alpha/\beta)] = -2\varepsilon_0\alpha \\ \varepsilon &= -1/2\varepsilon_0\alpha = C/[2(T_0 - T)] \end{aligned} \tag{2.35}$$

Figure 2.13a shows the variations of P_S and ε with temperature. It is notable that the permittivity ε becomes infinite at the transition temperature, and that the slope of the inverse permittivity $1/\varepsilon$ in the ferroelectric phase is twice that in the paraelectric phase.

2.4.2.2 The First-Order Transition

When β is negative in Equation 2.31 and is taken to be positive, a first-order transition is described. The equilibrium condition for $E = 0$ in this case is expressed by:

$$[(T - T_o)/\varepsilon_0 C]\, P_S + \beta P_S^3 + \gamma P_S^5 = 0 \tag{2.36}$$

and leads to either $P_S = 0$ or to a spontaneous polarization that is a root for:

$$[(T - T_o)/\varepsilon_0 C] + \beta P_S^2 + \gamma P_S^4 = 0 \tag{2.37}$$

$$P_S^2 = \left[-\beta + \sqrt{\beta^2 - 4\gamma(T - T_0)/\varepsilon_0 C}\right] \Big/ 2\gamma \tag{2.38}$$

The transition temperature, T_C, is obtained when the condition that the free energies of the paraelectric and ferroelectric phases are equal is applied such that $F = 0$ and:

$$[(T - T_0)/\varepsilon_0 C] + (1/2)\beta P_S^2 + (1/3)\gamma P_S^4 = 0 \tag{2.39}$$

which allows us to write (refer to Example Problem 2.7):

$$T_C = T_0 + (3/16)(\beta^2 C/\gamma) \tag{2.40}$$

According to this equation, the Curie temperature, T_C, is slightly higher than the Curie-Weiss temperature, T_0, and a discontinuity in the spontaneous polarization, P_S, as a function of temperature occurs at T_C. This is represented in Figure 2.13b, where we also see the dielectric constant as a function of temperature peak at T_C. The inverse dielectric constant is also discontinuous at T_C, and when one extends the linear paraelectric portion of this curve back across the temperature axis it intersects at the Curie-Weiss temperature, T_0. These are characteristic behaviors for the *first-order transition*. $BaTiO_3$ is an example of a ferroelectric that shows a first-order transition.

EXAMPLE PROBLEM 2.7

Landau free energy for a *first-order phase transition* is expanded in terms of the order parameter P as

$$F(P,T) = (1/2)\alpha P^2 + (1/4)\beta P^4 + (1/6)\gamma P^6 \qquad [\alpha = (T - T_0)/\varepsilon_0 C].$$

a. Is $P_S^2 = [-\beta - \sqrt{\beta^2 - 4\gamma(T - T_0)/\varepsilon_0 C}]/2\gamma$ not another root?
b. Verify the difference between the Curie (T_C) and Curie-Weiss (T_0) temperatures as expressed by:

$$T_C = T_0 + (3/16)(\beta^2 \varepsilon_0 C/\gamma).$$

HINT

The potential minima are obtained from

$$(\partial F/\partial P) = E = \alpha P + \beta P^3 + \gamma P^5 = 0. \qquad \text{(P2.7.1)}$$

This equation is valid for any temperature below and above Curie temperature. There are generally three minima, including $P = 0$ ($F = 0$). Other minima can be solved from $\alpha + \beta P^2 + \gamma P^4 = 0$. At the Curie temperature, the free energy at the nonzero polarization must be equal to the free energy of the paraelectric state; that is, zero ($F = 0$). Thus, we obtain another condition:

$$F = (1/2)\alpha P^2 + (1/4)\beta P^4 + (1/6)\gamma P^6 = 0. \qquad \text{(P2.7.2)}$$

This equation is only valid at the phase transition temperature, T_C.

SOLUTION

a. Equation P2.7.1 provides two roots, $P_S^2 = [-\beta \pm \sqrt{\beta^2 - 4\gamma(T - T_0)/\varepsilon_0 C}]/2\gamma$, in general. Since $(T - T_0) < 0$ and $\gamma > 0$, $\sqrt{\beta^2 - 4\gamma(T - T_0)/\varepsilon_0 C} > \sqrt{\beta^2} = -\beta$ (or $|\beta|$). Note that $\beta < 0$. Thus, $[-\beta - \sqrt{\beta^2 - 4\gamma(T - T_0)/\varepsilon_0 C}]/2\gamma < 0$, which is contradictory with a root of P_S^2. The spontaneous polarization should be positive or negative, but still be a real number, never an imaginary one.
b. Equations P2.7.1 and P2.7.2 are reduced for nonzero polarizations to

$$\alpha + \beta P^2 + \gamma P^4 = 0, \qquad \text{(P2.7.3)}$$

$$\alpha + (1/2)\beta P^2 + (1/3)\gamma P^4 = 0. \qquad \text{(P2.7.4)}$$

Note that Equation P2.7.3 is valid for all temperatures below T_C, but Equation P2.7.4 is only valid at $T = T_C$. Eliminating the higher-order P^4 term from these two equations (3 × Equations P2.7.4 and P2.7.3),

$$[(3/2) - 1]\beta P^2 + [3 - 1]\alpha = 0 \rightarrow P^2 = -4\alpha/\beta. \qquad \text{(P2.7.5)}$$

Then, we obtain the following equation from Equation P2.7.3:

$$\alpha + \beta(-4\alpha/\beta) + \gamma(-4\alpha/\beta)^2 = 0.$$

Or,

$$-3\alpha + \gamma \times 16\alpha^2/\beta^2 = 0.$$

Taking into account $T = T_C$, $\alpha_{T=Tc} = (T_C - T_0)/\varepsilon_0 C$, and $16\gamma\alpha/\beta^2 = 3$, the Curie temperature is calculated as

$$T_C = T_0 + (3/16)(\beta^2 \varepsilon_0 C/\gamma). \qquad \text{(P2.7.6)}$$

Since $\beta^2 > 0$, $\gamma > 0$, the Curie temperature, T_C, is higher than the Curie-Weiss temperature, T_0.

EXAMPLE PROBLEM 2.8

Draw the polarization vs. electric field hysteresis curve at a certain temperature for a ferroelectric with the second-order phase transition as

$$F(P,T) = (1/2)\alpha P^2 + (1/4)\beta P^4 \qquad [\alpha = (T - T_0)/\varepsilon_0 C].$$

a. Obtain the spontaneous polarization, P_S, at a temperature T ($<T_C$) at $E = 0$.
b. Obtain the permittivity ε at a temperature T ($<T_C$) at $E = 0$.
c. Supposing that all positive polarization, P, reverses the direction at once (without making any small domain structures), obtain the macroscopic coercive field at a temperature T ($<T_C$).
d. Draw the polarization vs. electric field hysteresis curve at a temperature T.

Solution

The potential minima are obtained from

$$(\partial F/\partial P) = E = \alpha P + \beta P^3. \qquad \text{(P2.8.1)}$$

By setting $y_1 = \alpha P + \beta P^3$ and $y_2 = E$, a visual geometrical solution technique can be used, as illustrated in Figure 2.14; that is, the intersects of these two curves ($y_1 = y_2$) provide the solution points: only one intersect exists for $T > T_C$, while for $T < T_C$, there are three intersects.

a. Spontaneous polarization is obtained from Equation P2.8.1:

$$P + \beta P^3 = E = 0.$$

Thus,

$$P_S^2 = -\alpha/\beta, \text{ or}$$
$$P_S = \sqrt{-\alpha/\beta} \qquad \text{(P2.8.2)}$$

b. Relative permittivity, ε, is obtained from $(1/\varepsilon_0)$ $(\partial P/\partial E)$, and from Equation P2.8.1,

$$\varepsilon = 1/\varepsilon_0 \left(\frac{\partial E}{\partial P}\right) = 1/\varepsilon_0(\alpha + 3\beta P^2)$$

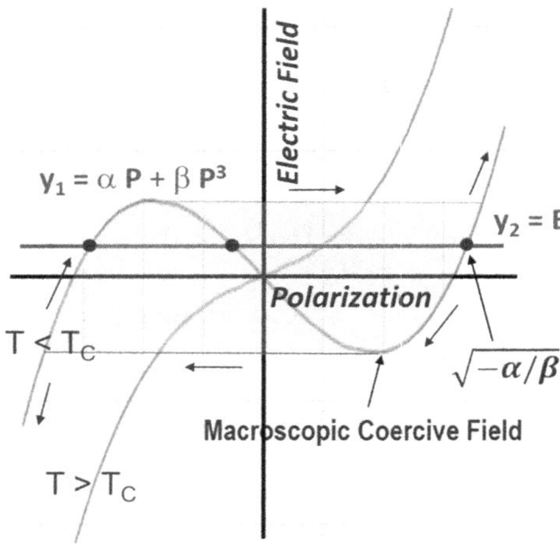

FIGURE 2.14 Polarization vs. electric field hysteresis curve, obtained from a graphic technique.

From Equation P2.8.2, relative permittivity is given by

$$\varepsilon = 1/\varepsilon_0(-2\alpha) = C/2(T_0 - T) \tag{P2.8.3}$$

c. The macroscopic coercive field is obtained from the maximum/minimum point of the y_1 curve:

$$\frac{\partial y_1}{\partial P} = 0 \rightarrow \alpha + 3\beta P^2 = 0 \rightarrow P = \sqrt{-\alpha/3\beta}.$$

Since the coercive field is obtained from y_1(max point),

$$\begin{aligned} y_1 = \alpha P + \beta P^3 &= \sqrt{-\alpha/3\beta}[\alpha + \beta(-\alpha/3\beta)] \\ &= \sqrt{-4\alpha^3/27\beta} \end{aligned} \tag{P2.8.4}$$

d. The shadowed area in Figure 2.14 shows the polarization vs. electric field hysteresis loop, when we assume the macroscopic all polarization one-time reversal.

2.4.3 Phenomenological Description of Electrostriction

In a ferroelectric whose prototype phase (high-temperature paraelectric phase) is centrosymmetric and nonpiezoelectric, the piezoelectric coupling term $\boldsymbol{P}X$ is omitted and only the electrostrictive coupling term $\boldsymbol{P}^2X$ is introduced. The theories for electrostriction in ferroelectrics were formulated in the 1950s by Devonshire[5] and Kay.[6] Let us assume that the elastic Gibbs energy should be expanded in a one-dimensional form:

$$\begin{aligned} G_1(\boldsymbol{P},X,T) &= (1/2)\alpha\boldsymbol{P}^2 + (1/4)\beta\boldsymbol{P}^4 + (1/6)\gamma\boldsymbol{P}^6 \\ &-(1/2)s\,X^2 - Q\boldsymbol{P}^2X \qquad (\alpha = (T-T_0)/\varepsilon_0 C) \end{aligned} \tag{2.41}$$

where $\boldsymbol{P}$, X, and T are the polarization, stress, and temperature, respectively, and s and Q are called the elastic compliance and electrostrictive coefficient, respectively. This leads to Equations 2.42 and 2.43.

$$\boldsymbol{E} = (\partial G_1 / \partial \boldsymbol{P}) = \alpha \boldsymbol{P} + \beta \boldsymbol{P}^3 + \gamma \boldsymbol{P}^5 - 2Q\boldsymbol{P}X \quad (2.42)$$

$$x = -(\partial G_1 / \partial X) = sX + Q\boldsymbol{P}^2 \quad (2.43)$$

Case I: X = 0

When the external stress is zero, the following equations are derived:

$$E = \alpha P + \beta P^3 + \gamma P^5 \quad (2.44)$$

$$x = QP^2 \quad (2.45)$$

$$1/\varepsilon_0\varepsilon = \alpha + 3\beta \mathrm{P}^2 + 5\gamma P^4 \quad (2.46)$$

If the external electric field is equal to zero ($E = 0$), two different states are derived:

$$P = 0 \text{ and } P^2 = (-\beta + \sqrt{\beta^2 - 4\alpha\gamma})/2\gamma.$$

1. Paraelectric phase: $P_S = 0$ or $P = \varepsilon_0\, \varepsilon\, E$ (under small E)

$$\text{Permittivity}: \varepsilon = C/(T - T_0) \quad \text{(Curie-Weiss law)} \quad (2.47)$$

$$\text{Electrostriction}: x = Q\varepsilon_0^2\varepsilon^2 E^2 \quad (2.48)$$

The previously mentioned electrostrictive coefficient M in Equation 2.27 is related to the electrostrictive Q coefficient through

$$M = Q\varepsilon_0^2\varepsilon^2 \quad (2.49)$$

2. Ferroelectric phase: $P_S^2 = (-\beta + \sqrt{\beta^2 - 4\alpha\gamma})/2\gamma$ or $P = P_S + \varepsilon_0\varepsilon\, E$ (under small E)

$$x = Q(P_S + \varepsilon_0\varepsilon E)^2 = QP_S^2 + 2\varepsilon_0\varepsilon\, QP_S E + Q\varepsilon_0^2\varepsilon^2 E^2 \quad (2.50)$$

where we define the spontaneous strain, x_S, and the piezoelectric constant, d, as:

$$\text{Spontaneous strain}: x_S = QP_S^2 \quad (2.51)$$

$$\text{Piezoelectric constant}: d = 2\varepsilon_0\varepsilon QP_S \quad (2.52)$$

We see from Equation 2.52 that piezoelectricity is equivalent to the *electrostrictive phenomenon biased by the spontaneous polarization*. The temperature dependences of the spontaneous strain and the piezoelectric constant are plotted in Figure 2.15. (Refer to Example Problem 2.10.)

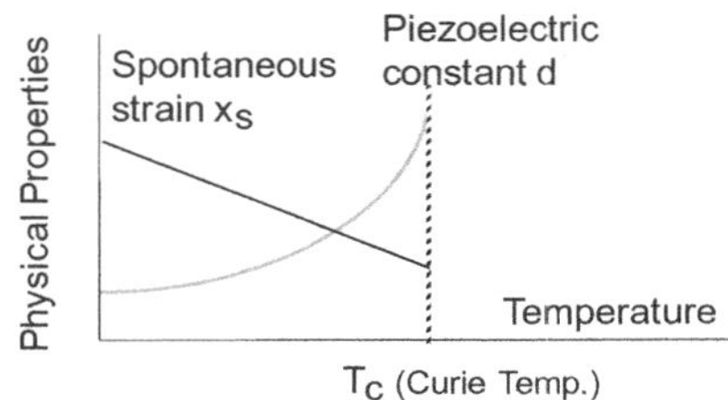

FIGURE 2.15 Temperature dependence of the spontaneous strain and the piezoelectric constant.

Case II: X ≠ 0

When a hydrostatic pressure p ($X = -p$) is applied, the inverse permittivity is changed in proportion to p:

$$\begin{aligned} 1/\varepsilon_0\varepsilon &= \alpha + 3\beta P_S^2 + 5\gamma P_S^4 + 2Qp \quad \text{(Ferroelectric state)} \\ \alpha + 2Qp &= (T - T_0 + 2Q\varepsilon_0 Cp)/(\varepsilon_0 C) \quad \text{(Paraelectric state)} \end{aligned} \tag{2.53}$$

Here, electrostrictive constant Q should be Q_h, which is defined by $(Q_{33} + 2Q_{31})$. Therefore, the pressure dependence of the Curie-Weiss temperature, T_0, or the transition temperature, T_C, is derived as follows:

$$(\partial T_0/\partial p) = (\partial T_C/\partial p) = -2Q\varepsilon_0 C \tag{2.54}$$

In general, the ferroelectric Curie temperature decreases with increasing hydrostatic pressure (i.e., $Q_h > 0$). More precisely, there is a 50°C temperature decrease per 1 GPa hydrostatic pressure increase in perovskite ferroelectrics.

EXAMPLE PROBLEM 2.9

Barium titanate has $d_{33} = 320 \times 10^{-12}$ C/N, ε_c $(= \varepsilon_3) = 800$ and $Q_{33} = 0.11$ m^4C^{-2} at room temperature. Estimate the spontaneous polarization, P_S.

Solution

Let us use the relationship:

$$d_{33} = 2\varepsilon_0\varepsilon_3 Q_{33} P_S. \tag{P2.9.1}$$

$$\begin{aligned} P_S &= d_{33}/2\varepsilon_0\varepsilon_3 Q_{33} \\ &= 320\times10^{-12}[C/N]/\{2\times8.854\times10^{-12}[F/m]\times800\times0.11[m^4C^{-2}]\} \\ &= 0.21[C/m^2] \end{aligned} \tag{P2.9.2}$$

EXAMPLE PROBLEM 2.10

In the case of a second-order phase transition, the elastic Gibbs energy is expanded in a one-dimensional form as follows:

$$\begin{aligned} G_1(P,X,T) &= (1/2)\alpha P^2 + (1/4)\beta P^4 \\ &\quad -(1/2)sX^2 - QP^2X, \end{aligned} \tag{P2.10.1}$$

where only the coefficient α is dependent on temperature, $\alpha = (T - T_0)/\varepsilon_0 C$. Obtain the dielectric constant, spontaneous polarization, spontaneous strain, and piezoelectric constant as a function of temperature (under the stress-free condition).

SOLUTION

$$E = (\partial G_1/\partial P) = \alpha P + \beta P^3 - 2QPX \quad \text{(P2.10.2)}$$

$$x = -(\partial G_1/\partial X) = sX + QP^2 \quad \text{(P2.10.3)}$$

When an external stress is zero, we can deduce the three characteristic equations:

$$E = \alpha P + \beta P^3 \quad \text{(P2.10.4)}$$

$$x = QP^2 \quad \text{(P2.10.5)}$$

$$1/\varepsilon_0\varepsilon = (\partial E/\partial P) = \alpha + 3\beta P^2 \quad \text{(P2.10.6)}$$

By setting $E = 0$ initially, we obtain the following two stable states: $P_S = 0$ or $- \alpha/\beta$.

1. Paraelectric phase $- T > T_0 - P_S = 0$

$$1/\varepsilon_0\varepsilon = \alpha, \text{ then } \varepsilon = C/(T - T_0) \quad \text{(Curie-Weiss Law)} \quad \text{(P2.10.7)}$$

2. Ferroelectric phase $- T < T_0 - P_S = \pm\sqrt{(T < T_0)/\varepsilon_0 C\beta}$ (P2.10.8)

$$1/\varepsilon_0\varepsilon = \alpha + 3\beta P^2 = -2\alpha, \text{ then } \varepsilon = C/2(T_0 - T) \quad \text{(P2.10.9)}$$

$$x_S = QP_S^2 = Q(T_0 - T)/\varepsilon_0 C\beta \quad \text{(P2.10.10)}$$

From Equations P2.10.8 and P2.10.9, the piezoelectric constant is obtained as

$$\begin{aligned} d &= 2\varepsilon_0\varepsilon QP_S \\ &= Q\sqrt{\varepsilon_0 C/\beta}(T_0 - T)^{-1/2} \end{aligned} \quad \text{(P2.10.11)}$$

Refer to Figure 2.15, which shows the case of the first-order transition.

2.4.4 Converse Effects of Electrostriction

So far we have discussed electric field–induced strains, i.e., piezoelectric strain (*converse piezoelectric effect*, $x = d\,E$) and electrostriction (*electrostrictive effect*, $x = ME^2$). Let us consider here the converse effect, that is, the material's response to an external stress, which is applicable to sensors. Actually, for piezoelectricity, this is the *direct piezoelectric effect*, that is, the *increase of the spontaneous polarization by an external stress*, expressed as

$$\Delta P = dX. \quad (2.55)$$

On the contrary, since an electrostrictive material does not have spontaneous polarization, it does not generate any charge under stress, but does exhibit a change in permittivity (first derivative of polarization, P, in terms of field, E) (see Equation 2.56):

$$\Delta(1/\varepsilon_0\varepsilon) = 2QX \quad (2.56)$$

This is the *converse electrostrictive effect.* The converse electrostrictive effect, the stress dependence of the permittivity, is used in stress sensors.[7] A bimorph structure that subtracts the static capacitances of two dielectric ceramic plates can provide superior stress sensitivity and temperature stability. The capacitance changes of the top and bottom plates have opposite signs for uniaxial stress and the same sign for temperature change. The response speed is limited by the capacitance measuring frequency up to about 1 kHz. Unlike piezoelectric sensors, electrostrictive sensors are effective in the low frequency range, especially pseudo-DC.

2.4.5 Temperature Dependence of Electrostriction

We have treated the electrostrictive coefficient as a temperature-independent constant in the last section. What is the actual situation? Several expressions for the electrostrictive coefficient Q have been proposed so far. From the data obtained by independent experimental methods such as

1. Electric field-induced strain in the paraelectric phase
2. Spontaneous polarization and spontaneous strain (x-ray diffraction) in the ferroelectric phase
3. d constants from the field-induced strain in the ferroelectric phase or from piezoelectric resonance, with permittivity, spontaneous polarization
4. Pressure dependence of permittivity in the paraelectric phase

nearly equal values of Q were obtained. Figure 2.16 shows the temperature dependence of the electrostrictive coefficients, Q_{33} and Q_{31}, observed for a complex perovskite $Pb(Mg_{1/3}Nb_{2/3})O_3$ single crystal sample, whose Curie temperature is near 0°C.[8] It is seen that there is no significant anomaly in the electrostrictive coefficient Q through the temperature range in which the paraelectric to ferroelectric phase transition occurs and piezoelectricity appears. Q is verified to be almost temperature independent. Note, however, that since electrostriction under an applied field E is obtained by $Q\varepsilon_0^2\varepsilon^2E^2$, rather significant temperature dependence is observed from the permittivity change with temperature.

2.4.6 Electromechanical Coupling Factor

2.4.6.1 Piezoelectric Constitutive Equations

We introduce the concept of the electromechanical coupling factor k in this section. We start from the fundamental piezoelectric constitutive equations (1D expression) without including losses:

$$D = \varepsilon_0\varepsilon^X E + dX \tag{2.57}$$

$$x = dE + s^E X, \tag{2.58}$$

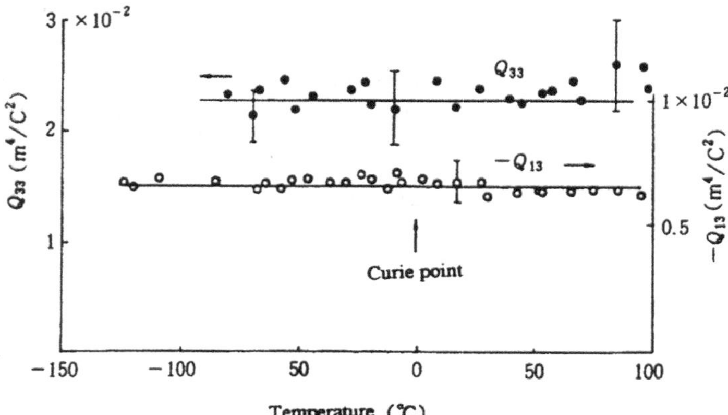

FIGURE 2.16 Temperature dependence of the electrostrictive constants, Q_{33} and Q_{31}, measured in a single crystal $Pb(Mg_{1/3}Nb_{2/3})O_3$.

where D is electric displacement (which is almost equal to polarization, P, as long as the permittivity, ε, is large), x strain, and these extensive parameters (material properties) are controlled by the intensive parameters (externally controllable parameters), electric field, E, and stress, X. The proportional constant ε^X is the relative permittivity under the stress-free condition (ε_0 is the vacuum permittivity $8.854 \times 10^{-12}\ F \cdot m^{-1}$), s^E elastic compliance under the electric field–free (or short-circuit) condition, and d the piezoelectric constant. It is worth noting that the ds appearing in Equations 2.57 and 2.58 should be thermodynamically the same. The verification process is beyond this textbook's level.

2.4.6.2 Electromechanical Coupling Factor

The term *electromechanical coupling factor k* is defined as the square value k^2 being the ratio of the converted energy over the input energy: when electric to mechanical

$$k^2 = (\text{Stored mechanical energy/input electrical energy}) \tag{2.59}$$

and when mechanical to electric:

$$k^2 = (\text{Stored electrical energy/input mechanical energy}) \tag{2.60}$$

Let us calculate Equation 2.59 first, when an external electric field E_3 is applied to a piezoelectric material. See Figure 2.17a first, when we apply electric field on the top and bottom electrodes under the stress-free condition ($X = 0$). Input electric energy must be equal to $(1/2)\varepsilon_0\varepsilon_3^X E_3^2$ from Equation 2.57, and the strain generated by E_3 should be $d_{33}E_3$ from Equation 2.58. Since the converted and stored mechanical energy is obtained as $(1/2s_{33}^E)x_3^2$, we obtain:

$$\begin{aligned} k_{33}^2 &= [(1/2)(d_{33}E_3)^2 / s_{33}^E] / [(1/2)\varepsilon_0\varepsilon_3^X E_3^2] \\ &= d_{33}^2 / \varepsilon_0\varepsilon_3^X \cdot s_{33}^E. \end{aligned} \tag{2.61a}$$

Let us now consider Equation 2.60, when an external stress, X_3, is applied to a piezoelectric material. Refer to Figure 2.17b. Under the short-circuit condition ($E_3 = 0$), the input mechanical energy must be equal to $(1/2)s_{33}^E X_3^2$ from Equation 2.58, and the electric displacement D_3 (or polarization P_3) generated by X_3 should be equal to $d_{33}X_3$ from Equation 2.57. This D_3 can be obtained by integrating the short-circuit current in terms of time through the electric lead. Since the converted and stored electric energy is obtained as $(1/2\varepsilon_0\varepsilon_3^X)D_3^2$, we obtain:

$$\begin{aligned} k_{33}^2 &= [(1/2\varepsilon_0\varepsilon_3^X)(d_{33}X_3)^2][(1/2)s_3^E X_3^2] \\ &= d_{33}^2 / \varepsilon_0\varepsilon_3^X \cdot s_{33}^E. \end{aligned} \tag{2.61b}$$

It is essential to understand that the electromechanical coupling factor, k (or k^2, which has a physical meaning of energy transduction/conversion rate), can be exactly the same for both converse

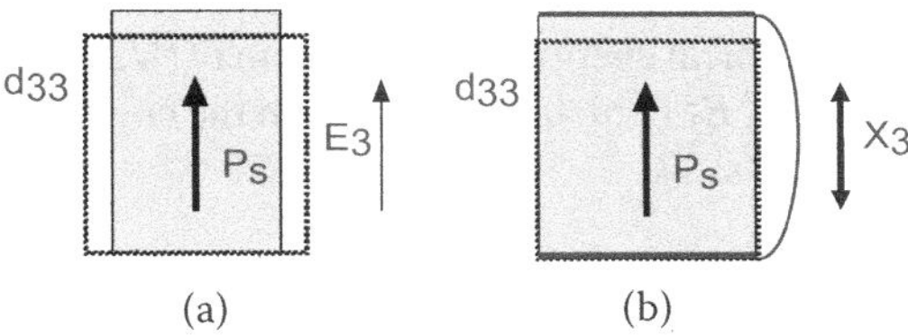

FIGURE 2.17 Calculation models of electromechanical coupling factor, k, for (a) electric input and (b) stress input.

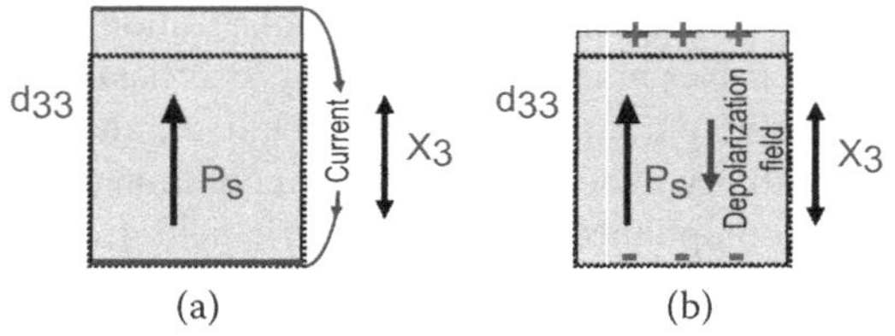

FIGURE 2.18 Calculation models of electromechanical coupling factor, k, for (a) short-circuit and (b) open-circuit conditions.

(Equation 2.61a) and direct (Equation 2.61b) piezoelectric effects. The conditions under constant X (free stress) or constant E (short-circuit) are considered nonconstrained.

Figure 2.18 illustrates further discussion on two models of the electromechanical coupling factor, k, for (a) short-circuit and (b) open-circuit conditions, by including the concept of the *depolarization field*. The former is the same as Figure 2.17b, and the result was calculated in Equation 2.65. Under the open-circuit condition, because of the electric constraint (or boundary) condition difference, the elastic compliance may be different. Thus, let us denote it as s^D (electric displacement constant condition, or $D = \text{constant}$ and $E \neq 0$). Under the stress X_3, the input mechanical energy is now given by the sum of the elastic energy $(1/2)s_{33}^D X_3^2$ per unit volume and the converted (transduced or stored) electrical energy per unit volume under open circuit is given by $(1/2)\varepsilon_0\varepsilon_3^X E_3^2$ (D_3 is supposed to be zero). When the top and bottom electrodes are open-circuited in Figure 2.18b, the *depolarization field*, E, is induced because of the piezoelectrically induced charge $P = d_{33}X_3$ in order to satisfy $D_3 = \text{constant}$ [from Gauss's Law, div $(D) = 0$ (no free charge) in a highly resistive material] in Equation 2.57:

$$E_3 = -(d_{33}/\varepsilon_0\varepsilon_3^X)X_3. \tag{2.62}$$

Note that this depolarization field is valid only for the short time period by neglecting the charge drift in the sample or air-floating free charge on a long time scale (typically much longer than 1 minute). When quickly movable free charge exists, a sort of *field screening* occurs. Thus, from Equation 2.58, the strain induced under the open-circuit condition should be smaller than that under the short circuit, as calculated:

$$x_3 = s_{33}^E X_3 + d_{33}E_3 = s_{33}^E X_3 - (d_{33}^2/\varepsilon_0\varepsilon_3^X)X_3 = s_{33}^E[1-(d_{33}^2/\varepsilon_0\varepsilon_3^X s_{33}^E)]X_3. \tag{2.63}$$

If we denote $x_3 = s_{33}^D X_3$ under the open circuit, we can obtain the following important relation between E-constant and D-constant elastic compliances through k:

$$s_{33}^D = s_{33}^E(1-k_{33}^2). \quad (k: \text{electromechanical coupling factor}) \tag{2.64}$$

The elasticity under the open-circuit condition is stiffer than that under the short-circuit condition. Finally, let us reconfirm the electromechanical coupling factor definition Equation 2.60 under the open-circuit condition: input mechanical energy is given by sum of $(1/2)s_{33}^D X_3{}^2$ and the stored electrical energy as charges given by $(1/2)\varepsilon_0\varepsilon_3^X E_3^2$ (which coincides with the converted electric energy). Taking into account Equations 2.64 and 2.62,

$$\begin{aligned} k_{33}^2 &= (1/2)\varepsilon_0\varepsilon_3^X E_3^2/[(1/2)\, s_{33}^D X_3^2 + (1/2)\varepsilon_0\varepsilon_3^X E_3^2] \\ &= (1/2)\varepsilon_0\varepsilon_3^X[-(d_{33}/\varepsilon_0\varepsilon_3^X)\, X_3]^2/(1/2)[s_{33}^E(1-k_{33}^2)]\, X_3^2 + (1/2)\varepsilon_0\varepsilon_3^X[-(d_{33}/\varepsilon_0\varepsilon_3^X)X_3]^2 \\ &= d_{33}^2/\varepsilon_0\varepsilon_3^X \cdot s_{33}^E. \end{aligned}$$

The electromechanical coupling factor, k, can be calculated exactly the same way, irrelevant of the short-circuit or open-circuit condition virtual models.

EXAMPLE PROBLEM 2.11

One of the lead zirconate titanate (PZT) ceramics has a piezoelectric coefficient, $d_{33} = 590 \times 10^{-12}$ (C/N), a dielectric constant, $\varepsilon_3 = 3400$, and an elastic compliance, $s_{33}^E = 20\times10^{-12}(m^2/N)$.

a. Calculate the induced strain under an applied electric field of $E_3 = 10 \times 10^5$ (V/m). Then, calculate the generative stress under a completely clamped (mechanically constrained) condition.
b. Calculate the induced electric field under an applied stress of $X_3 = 3 \times 10^7$ (N/m²), which corresponds to the generative stress in (a). The magnitude of the induced electric field does not correspond to the magnitude of the applied electric field in (a) [that is, 10×10^5 (V/m)]. Explain why.

SOLUTION

Part (a):

$$x_3 = d_{33}E_3 = [590\times10^{-12}(C/N)][10\times10^5(V/m)] \rightarrow x_3 = 5.9\times10^{-4} \quad \text{(P2.11.1)}$$

Under completely clamped (mechanically constrained) conditions,

$$X_3 = x_{33}/s_{33} = 5.9\times10^{-4}/20\times10^{-12}(N/m^2)] \rightarrow X_3 = 3.0\times10^7(N/m^2) \quad \text{(P2.11.2)}$$

Part (b):

$$P_3 = d_{33}X_3 = [590\times10^{-12}(C/N)][3\times10^7(N/m^2)] \rightarrow P_3 = 1.77\times10^{-2}(C/m^2) \quad \text{(P2.11.3)}$$

$$E_3 = P_3/\varepsilon_o\varepsilon = [1.77\times10^{-2}(C/m^2)]/[3400]\cdot[8.854\times10^{-12}(F/m)] \rightarrow E_3 = 5.9\times10^5(V/m) \quad \text{(P2.11.4)}$$

The induced field is only 59% of the electric field applied for the case given in Part (a). This is most readily explained in terms of the electromechanical coupling factor. Considering the mechanical energy produced through the electromechanical response of the piezoelectric with respect to the electrical energy supplied to the material, the so-called electromechanical coupling factor, k, is defined as follows:

$$k^2 = \frac{[\text{Stored Mechanical Energy}]}{[\text{Input Electrical Energy}]} = \frac{\left(\frac{1}{2}\right)\left[\frac{x^2}{s}\right]}{\left(\frac{1}{2}\right)\left[\varepsilon_0\varepsilon E^2\right]} = \frac{\left(\frac{1}{2}\right)\left[\frac{(dE)^2}{s}\right]}{\left(\frac{1}{2}\right)\left[\varepsilon_0\varepsilon E^2\right]} = \frac{d^2}{s\varepsilon_0\varepsilon} \quad \text{(P2.11.5a)}$$

An alternative way of defining the electromechanical coupling factor is by considering the electrical energy produced through the electromechanical response of the piezoelectric with respect to the mechanical energy supplied to the material, which leads to the same outcome:

$$k^2 = \frac{\text{[Stored Electrical Energy]}}{\text{[Input Mechanical Energy]}}$$

$$= \frac{\left(\frac{1}{2}\right)\left[\frac{P^2}{\varepsilon_0\varepsilon}\right]}{\left(\frac{1}{2}\right)[sX^2]} = \frac{\left(\frac{1}{2}\right)\left[(dX)^2/\varepsilon_0\varepsilon\right]}{\left(\frac{1}{2}\right)[sX^2]} = \frac{d^2}{s\varepsilon_0\varepsilon} \tag{P2.11.5b}$$

In this case, we may evaluate the electromechanical coupling factor associated with the piezoelectric response taking place through d_{33}, and find that:

$$k_{33}^2 = \frac{d_{33}^2}{s_{33}\varepsilon_0\varepsilon_3} = \frac{\left[590\times10^{-12}\left(\frac{C}{N}\right)\right]^2}{\left[20\times10^{-12}\left(\frac{m^2}{N}\right)\right]\left[8.854\times10^{-12}\left(\frac{F}{m}\right)\right][3400]}$$

$$\rightarrow k_{33}^2 = 0.58$$

So, we see why the induced electric field determined in Part (b) has a magnitude that is only a fraction of the field applied in Part (a) of 10×10^5 (V/m). Here we see that this fraction corresponds to about k^2; that is, each energy transduction ratio accompanying the $E\rightarrow M$ and $M\rightarrow E$ processes is k^2. Thus (last electrical energy)/(initial electrical energy) $= k^4$, leading to (last induced field)/(initial applied field) $= k^2$. In this sense, we may regard the quantity k^2 to be the transduction ratio associated with a particular electrical-to-mechanical or mechanical-to-electrical piezoelectric transduction event. In the case of $k = 100\%$, the magnitude of the induced electric field (b) should correspond to the same field in (a) [that is, 10×10^5(V/m)].

EXAMPLE PROBLEM 2.12

Derive the electromechanical coupling factor, k_{ij}, of piezoelectric ceramic vibrators for the following vibration modes (refer to Figure 2.19):

a. Longitudinal plate extension mode ($\perp E$): k_{31}
b. Longitudinal length extension mode ($//E$): k_{33}
c. Planar extension mode of the circular plate: k_p

Hint

Using the expression:

$$U = U_{MM} + 2U_{ME} + U_{EE}$$
$$= (1/2)\sum_{i,j} s_{ij}^E X_j X_i + 2\cdot(1/2)\sum_{m,i} d_{mi}E_m X_i + (1/2)\sum_{k,m} \varepsilon_0\varepsilon_{mk}^X E_k E_m, \tag{P2.12.1}$$

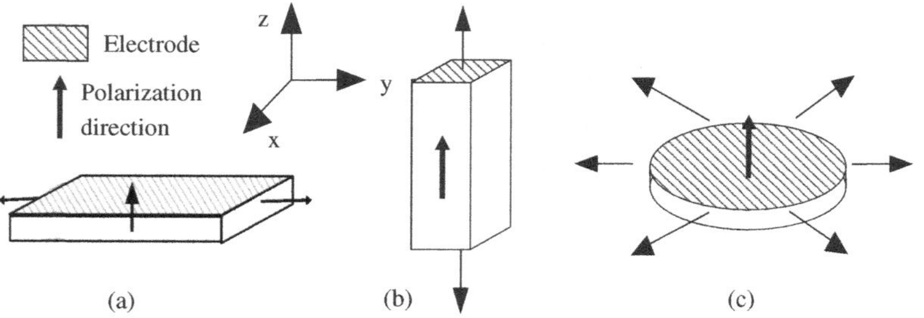

FIGURE 2.19 (a) Longitudinal plate extension via d_{31}, (b) longitudinal rod extension via d_{33}, and (c) planar extensional vibration modes of piezoelectric devices.

the electromechanical coupling factor is given by

$$k = U_{ME}/\sqrt{U_{MM} \cdot U_{EE}}. \quad \text{(P2.12.2)}$$

Solution

a. The relating equations for the k_{31} mode are:

$$x_1 = s_{11}^E X_1 + d_{31} E_3$$

$$D_3 = d_{31} X_1 + \varepsilon_0 \varepsilon_{33}^X E_3$$

$\rightarrow U_{ME}$ comes from the d_{31} term as $(1/2)d_{31}E_3X_1$, U_{MM} comes from the s_{11}^E term as $(1/2)s_{11}^E X_1^2$, and U_{EE} comes from the ε_3^X term as $(1/2)\varepsilon_{33}^X E_3^2$.
Thus,

$$\begin{aligned} k_{31} &= (1/2)d_{31}E_3X_1/\sqrt{(1/2)s_{11}^E X_1^2 \cdot (1/2)\varepsilon_0\varepsilon_{33}^X E_3^2} \\ d_{31} &= /\sqrt{s_{11}^E \cdot \varepsilon_0 \varepsilon_{33}^X} \end{aligned} \quad \text{(P2.12.3)}$$

Since $d_{31} < 0$, k_{31} should have a negative value. However, the spec data usually show the absolute value.

b. Similarly, the relating equations for the k_{33} mode are:

$$x_3 = s_{33}^E X_3 + d_{33} E_3$$

$$D_3 = d_{33} X_3 + \varepsilon_0 \varepsilon_{33}^X E_3$$

$\rightarrow U_{ME}$ comes from the d_{33} term as $(1/2)d_{33}E_3X_3$, U_{MM} comes from the s_{33}^E term as $(1/2)s_{33}^E X_3^2$, and U_{EE} comes from the ε_{33}^X term as $(1/2)\ \varepsilon_{33}^X E_3^2$.
Thus,

$$\begin{aligned} k_{33} &= (1/2)d_{33}E_3X_3/\sqrt{(1/2)s_{33}^E X_3^2 \cdot (1/2)\varepsilon_0\varepsilon_{33}^X E_3^2} \\ &= d_{33}/\sqrt{s_{33}^E \cdot \varepsilon_0 \varepsilon_{33}^X} \end{aligned} \quad \text{(P2.12.4)}$$

c. The relating equations for the k_p mode follow, including 2D x_1 and x_2 equations:

$$x_1 = s_{11}^E X_1 + s_{12}^E X_2 + d_{31} E_3$$

$$x_2 = s_{12}^E X_1 + s_{22}^E X_2 + d_{32} E_3$$

$$D_3 = d_{31} X_1 + d_{32} X_2 + \varepsilon_0 \varepsilon_{33}^X E_3$$

Assuming axial symmetry, $s_{11}^E = s_{22}^E, d_{31} = d_{32}$ and $X_1 = X_2$ $(= X_p)$ in a random ceramic specimen, the above equations are transformed to the following two equations:

$$x_1 + x_2 = 2\left(s_{11}^E + s_{12}^E\right) X_p + 2d_{31} E_3$$

$$D_3 = 2d_{31} X_p + \varepsilon_0 \varepsilon_{33}^X E_3$$

$\rightarrow U_{ME}$ comes from the d_{33} term as $(1/2)\cdot 2d_{31}E_3X_p$, U_{MM} comes from the s_{33}^E term as $(1/2)\cdot 2\left(s_{11}^E + s_{12}^E\right)X_p^2$, and U_{EE} comes from the $\varepsilon_0\varepsilon_{33}^X$ term as $(1/2)\varepsilon_0\varepsilon_{33}^X E_3^2$. Thus,

$$k_p = 2d_{31}/\sqrt{2\left(s_{11}^E + s_{12}^E\right)\cdot \varepsilon_0\varepsilon_{33}^X} = \left[d_{31}/\sqrt{s_{11}^E \cdot \varepsilon_{33}^X}\right]\cdot\sqrt{2/(1-\sigma)} = k_{31}\sqrt{2/(1-\sigma)} \tag{P2.12.5}$$

where σ is Poisson's ratio, given by

$$\sigma = -s_{12}^E/s_{11}^E. \tag{P2.12.6}$$

When we compare the values k_{31} and k_{33}, since $|d_{31}|/d_{33} \approx 1/3$, a similar difference is expected for k_{31}/k_{33}, under a supposition that the elastic compliances s_{11}^E and s_{33}^E are not different by more than 20%. On the other hand, since $\sigma \approx 1/3, k_p \approx \sqrt{3}k_{31}$ is expected. The reader needs to understand that the electromechanical coupling factor or energy transduction rate is significantly different according to the device structure/mode. Example data for PZT 4 are: $k_{31} = 33\%$, $k_{33} = 70\%$ and $k_p = 58\%$.

2.4.6.3 Intensive and Extensive Parameters

According to the International Union of Pure and Applied Chemistry (IUPAC), the extensive parameter depends on the volume of the material, while the intensive parameter is the ratio of two extensive ones and therefore is independent of the volume of the material.[9] Consequently, stress (X) and electric field (E) are intensive parameters, which are externally controllable, while strain (x) and dielectric displacement (D) are extensive parameters, which are internally determined in the material. We start with the Gibbs free energy, G, which in this case is expressed as:

$$G = -(1/2)s^E X^2 - d\,X\,E - (1/2)\varepsilon_0\varepsilon^X E^2 \tag{2.65}$$

and in general differential form as:

$$dG = -x\,dX - D\,dE - S\,dT \tag{2.66}$$

where x is the strain, X is the stress, D is the electric displacement, E is the electric field, S is the entropy, and T is the temperature. Equation 2.66 is the energy expression in terms of the externally controllable (denoted as *intensive*) physical parameters X and E. The temperature dependence of the function is associated with the elastic compliance, s^E; the dielectric constant, ε^X; and the piezoelectric charge coefficient, d. We obtain from the Gibbs energy function the following two piezoelectric equations under a constant temperature $dT = 0$:

$$x = -\frac{\partial G}{\partial X} = s^E X + dE \tag{2.67}$$

$$D = -\frac{\partial G}{\partial E} = dX + \varepsilon_0\varepsilon^X E \tag{2.68}$$

Note that the Gibbs energy function provides intensive physical parameters: E-constant elastic compliance, s^E, and X-constant permittivity, ε^X.

When we consider the free energy in terms of the extensive (that is, material-related) parameters of strain, x, and electric displacement, D, we start from the differential form of the Helmholtz free energy designated by A, such that:

$$dA = X\,dx + E\,dD - S\,dT, \tag{2.69}$$

We obtain from this energy function the following two piezoelectric equations:

$$X = \frac{\partial A}{\partial x} = c^D x - hD \tag{2.70}$$

$$E = \frac{\partial A}{\partial D} = -hx + \left(\frac{1}{\varepsilon_0}\right)\kappa^x D \tag{2.71}$$

where c^D is the elastic stiffness at constant electric displacement (open-circuit conditions), h is the inverse piezoelectric charge coefficient, and κ^x is the inverse dielectric constant at constant strain (mechanically clamped conditions).

EXAMPLE PROBLEM 2.13

Verify the relationship:

$$\frac{d^2}{s^E \varepsilon^X \varepsilon_0} = \frac{h^2}{c^D(\kappa^x/\varepsilon_0)} \tag{P2.13.1}$$

This value is defined as the square of an electromechanical coupling factor (k^2), which should be the same even for different energy description systems.

Solution

When Equation 2.67 and 2.68 are combined with Equation 2.70 and 2.71, we obtain:

$$X = c^D(s^E X + dE) - h(dX + \varepsilon_0\varepsilon^X E) \tag{P2.13.2}$$

$$E = -h(s^E X + dE) + (\kappa^x/\varepsilon_o)(dX + \varepsilon_0\varepsilon^X E) \tag{P2.13.3}$$

or upon rearranging:

$$(1 - c^D s^E + hd)X + (h\varepsilon_0\varepsilon^X - c^D d)E = 0 \tag{P2.13.4}$$

$$[hs^E - (\kappa^x/\varepsilon_o)d]\,X + [1 - (\kappa^x/\varepsilon_o)\varepsilon_0\varepsilon^X + hd)E = 0 \tag{P2.13.5}$$

Combining the latter two equations yields:

$$(1 - c^D s^E + hd)[1 - (\kappa^x/\varepsilon_o)\varepsilon_0\varepsilon^X + hd] - (h\varepsilon_0\varepsilon^X - c^D d)[hs^E - (\kappa^x/\varepsilon_o)d] = 0 \tag{P2.13.6}$$

which, when simplified, produces the desired relationship:

$$\frac{d^2}{s^E \varepsilon^X \varepsilon_0} = \frac{h^2}{c^D(\kappa^x/\varepsilon_0)} \tag{P2.13.7}$$

It is important to consider the conditions under which a material will be operated when characterizing the dielectric constant and elastic compliance of that material. When a constant electric field is applied to a piezoelectric sample, as illustrated in Figure 2.20, top, the total input electric energy (left) should be equal to a combination of the energies associated with two distinct mechanical conditions that may be applied to the material: (1) stored electric energy under the *mechanically clamped state*, where a constant strain (*zero strain*) is maintained and the specimen cannot deform, and (2) converted mechanical energy under the *mechanically free state*, in which the material is not constrained and is free to deform. This situation can be expressed by:

$$\left(\frac{1}{2}\right)\varepsilon^X\varepsilon_0 E_0^2 = \left(\frac{1}{2}\right)\varepsilon^x\varepsilon_0 E_0^2 + \left(\frac{1}{2s^E}\right)x^2 = \left(\frac{1}{2}\right)\varepsilon^x\varepsilon_0 E_0^2 + \left(\frac{1}{2s^E}\right)(dE_0)^2$$

such that:

$$\varepsilon^X\varepsilon_0 = \varepsilon^x\varepsilon_0 + \left(\frac{d^2}{s^E}\right) \tag{2.72a}$$

$$\varepsilon^x = \varepsilon^X(1-k^2) \qquad \left[k^2 = \frac{d^2}{\varepsilon^X\varepsilon_0 s^E}\right] \tag{2.72b}$$

When a constant stress is applied to the piezoelectric as illustrated in Figure 2.20, bottom, in the same manner as in Section 2.4.6.2, the total input mechanical energy will be a combination of the energies associated with two distinct electrical conditions that may be applied to the material:

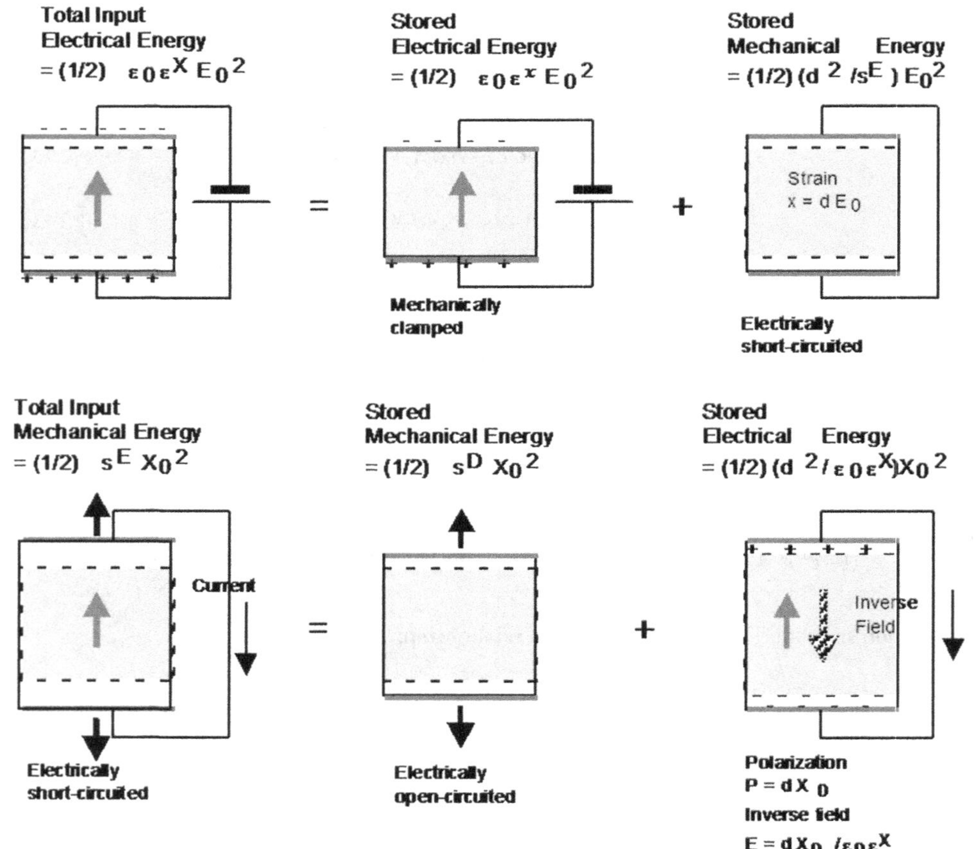

FIGURE 2.20 Schematic representation of the response of a piezoelectric material under: (a) constant applied electric field and (b) constant applied stress conditions.

(1) stored mechanical energy under the open-circuit state, where a constant electric displacement is maintained, and (2) converted electric energy under the short-circuit condition, in which the material is subject to a constant electric field. This can be expressed as:

$$\left(\frac{1}{2}\right)s^E X_0^2 = \left(\frac{1}{2}\right)s^D X_0^2 + \left(\frac{1}{2}\right)\varepsilon^X \varepsilon_0 E^2 = \left(\frac{1}{2}\right)s^D X_0^2 + \left(\frac{1}{2}\right)\varepsilon^X \varepsilon_0 (d/\varepsilon_0 \varepsilon^X) 2 X_0^2$$

which leads to:

$$s^E = s^D + \left(\frac{d^2}{\varepsilon^X \varepsilon_0}\right) \quad (2.73a)$$

$$s^D = s^E(1-k^2) \qquad \left[k^2 = \frac{d^2}{\varepsilon^X \varepsilon_0 s^E}\right] \quad (2.73b)$$

Hence, we obtain the following equations for the permittivity and elastic compliance under constrained and unconstrained conditions:

$$\varepsilon^x / \varepsilon^X = (1-k^2) \quad (2.74)$$

$$s^D / s^E = (1-k^2) \quad (2.75)$$

where

$$k^2 = \frac{d^2}{s^E \varepsilon^X \varepsilon_0} \quad (2.76)$$

We may also write equations of similar form for the corresponding reciprocal quantities:

$$\kappa^X / \kappa^x = (1-k^2) \quad (2.77)$$

$$c^E / c^D = (1-k^2) \quad (2.78)$$

where, in this context,

$$k^2 = \frac{h^2}{c^D(\kappa^x / \varepsilon_0)} \quad (2.79)$$

This new parameter k is also the electromechanical coupling factor in the extensive parameter description, and identical to the k in Equation 2.76. It will be regarded as a real quantity for the cases we examine in this text (see Example Problem 2.13).

The following three relationships exist between the intensive and extensive parameters: permittivity under constant stress ($\varepsilon^X \varepsilon_0$), elastic compliance under constant electric field (s^E), and the piezoelectric charge coefficient (d) in terms of their corresponding reciprocal quantities: inverse permittivity under constant strain (κ^x/ε_0), elastic stiffness under constant electric displacement (c^D), and the inverse piezoelectric coefficient (h). Note $k^2 = (h^2/(c^D(\kappa^x/\varepsilon_0)))$.

$$\varepsilon^X \varepsilon_0 = \frac{1}{\left(\frac{\kappa^x}{\varepsilon_0}\right)\left[1 - \frac{h^2}{c^D(\kappa^x/\varepsilon_0)}\right]} = \frac{1}{(\kappa^x/\varepsilon_0)(1-k^2)} \quad (2.80)$$

$$s^E = \frac{1}{c^D\left[1 - \dfrac{h^2}{c^D(\kappa^x/\varepsilon_0)}\right]} = \frac{1}{c^D(1-k^2)} \tag{2.81}$$

$$d = \frac{\dfrac{h^2}{c^D(\kappa^x/\varepsilon_0)}}{h\left[1 - \dfrac{h^2}{c^D(\kappa^x/\varepsilon_0)}\right]} = \frac{k^2}{h(1-k^2)} \tag{2.82}$$

2.4.7 Crystal Orientation Dependence of Piezoelectricity

Although rather thorough studies of the piezoelectric effect for several piezoelectrics in single crystal forms have been made in the past, the actual use of single crystal piezoelectrics has not been actively pursued. This is mainly because the most popular lead zirconate titanate–based materials are generally synthesized in polycrystalline form. There is currently a renewed interest in single crystal studies, however, spurred by recent developments in the production and application of epitaxially grown thin films (pseudo–single crystals) and high-quality single crystals of various compositions. Piezoelectric single crystals generally have higher strain coefficients, *d*, and generate larger strains than their polycrystalline counterparts and therefore are especially attractive for actuator/sensor applications. The crystal orientation dependence of piezoelectricity is explained in this section, and the comparison with their polycrystalline forms is discussed in Section 2.6.

A series of theoretical calculations made on perovskite ferroelectric crystals suggests that large effective piezoelectric constants and electromechanical coupling factors can be obtained when the applied electric field direction is canted by 50°–60° from the spontaneous polarization direction. This principle is visualized in Figure 2.21. When the electric field is applied in this direction, the parallel component, E_1, will generate longitudinal piezoelectric strain, and the perpendicular component, E_2, will generate shear strain through the d_{15} strain coefficient. Knowing $d_{15} > d_{33} > |d_{31}|$ in perovskite piezoelectrics, maximum strain is thus expected at a certain cant angle since the shear coupling enhances the apparent deformation.

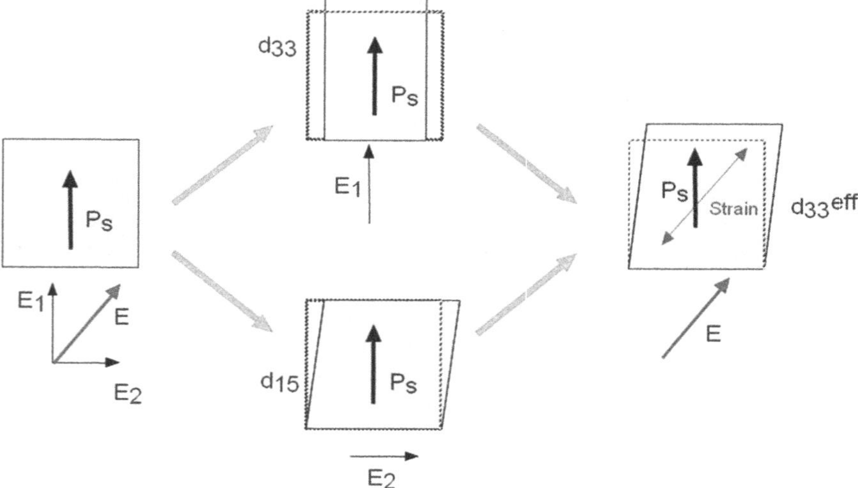

FIGURE 2.21 Deformation of a ferroelectric single crystal generated by an electric field, *E*, applied cant from the spontaneous polarization, P_S, direction.

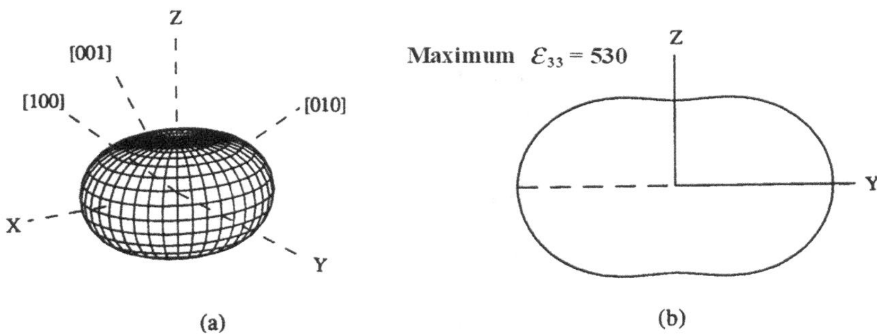

FIGURE 2.22 The dielectric constant of rhombohedral PZT 60/40: (a) the quadric surface and (b) the *y-z* cross section of the surface. (From X. -H. Du et al.: *Jpn. J. Appl. Phys.*, 36(Part 1, 9A), 5580, 1997.)

The dependence of the piezoelectric properties on crystallographic orientation has been modeled phenomenologically for PZT compositions in the vicinity of the morphotropic phase boundary.[10,11] The results summarized here have been determined by simply transforming the dielectric constant, piezoelectric coefficients, and elastic compliances from one orientation to another via the transformation matrices, using original data collected for the compositions of interest.[12] A surface representing the variation of the dielectric constant with crystallographic orientation for the rhombohedral 60/40 PZT composition appears in Figure 2.22. The dielectric constant increases monotonically as the sample cut direction is varied from the spontaneous polarization (P_S) direction along the *z*-axis (*z*-cut plate) to the *y* direction (*y*-cut plate). This trend is observed for both tetragonal and rhombohedral phases, as shown in Figure 2.23.

In contrast, electromechanical coupling exhibits an interesting crystal anisotropy. A pronounced variation in the coupling is observed with orientation, as shown in Figure 2.24, in which the quadric surface representing the effective d_{33} for the tetragonal 48/52 PZT composition and the (010) cross-section of this surface are depicted. We see from this surface that the d_{33}^{eff} attains its maximum in the direction of the spontaneous polarization, [001]. A quite different trend is observed for the rhombohedral 52/48 PZT composition, which is depicted in Figure 2.25. Here we see d_{33}^{eff} reach its maximum in the *y-z* plane at an angle 59.4° from the *z*-axis. This is very close to the

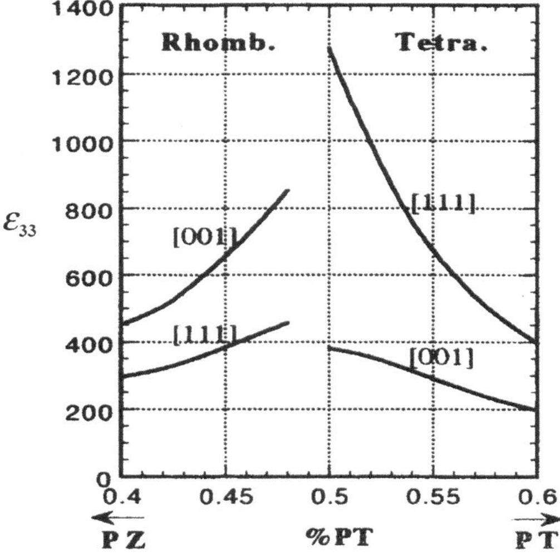

FIGURE 2.23 Dielectric constant of PZT vs. composition. (From X. -H. Du et al.: *Appl. Phys. Lett.*, 72(19), 2421, 1998.)

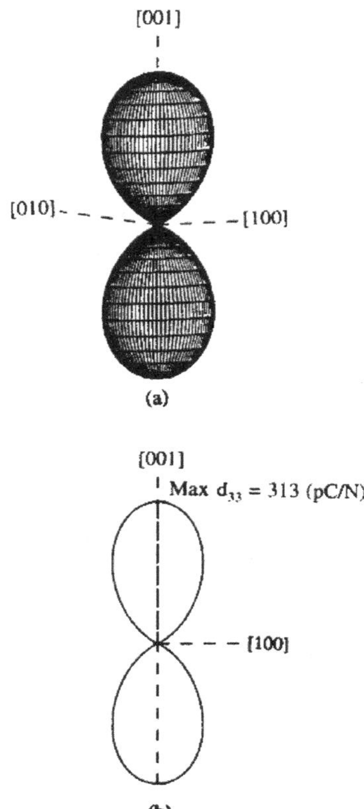

FIGURE 2.24 Effective d_{33} for (tetragonal) 48/52 PZT: (a) quadric surface and (b) (010) cross-section of the surface.

[001] directions of the paraelectric (cubic perovskite) phase. Theoretically determined effective piezoelectric strain coefficients, d_{33}^{eff}, along the [001] and [111] directions in PZT are plotted as a function of composition in Figure 2.26. A dramatic increase in the d_{33}^{eff} along [001] is apparent on approaching the morphotropic phase boundary from the rhombohedral side. This suggests that the optimum response of actuators and sensors made from PZT epitaxial thin films will be attained for those with rhombohedral compositions in the vicinity of the morphotropic phase boundary that have a [001] orientation. These results also indicate that a d_{33}^{eff} four times greater than what has been experimentally observed for PZT films so far is possible for specimens of this composition and orientation.

The exceptionally high electromechanical coupling observed for $Pb(Zn_{1/3}Nb_{2/3})O_3$-$PbTiO_3$ and $Pb(Mg_{1/3}Nb_{2/3})O_3$-$PbTiO_3$ single crystals at their morphotropic phase boundaries can be described in similar terms. We will look more closely at experimental data characterizing the response of these materials in Chapter 3.

EXAMPLE PROBLEM 2.14

The variations of the dielectric constant, ε_{33}, and the piezoelectric strain coefficient, d_{33}, with crystallographic orientation for a tetragonal PZT are shown in Figure 2.27. Let us consider a randomly oriented polycrystalline sample. Discuss the change in the ε_{33} and d_{33} before and after poling.

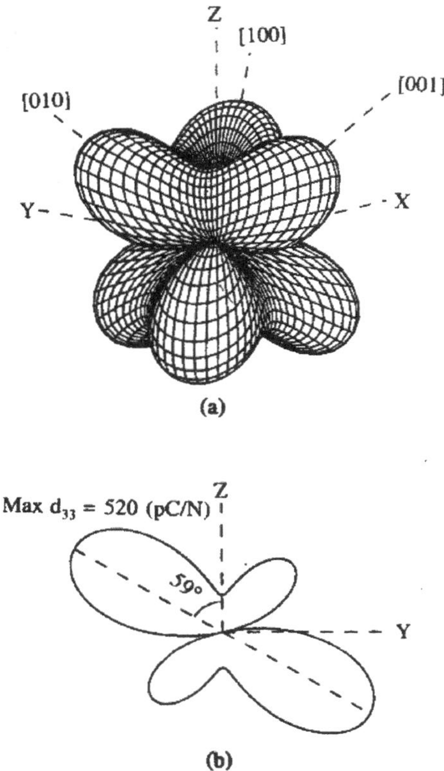

FIGURE 2.25 Effective d_{33} for (rhombohedral) 52/48 PZT: (a) quadric surface and (b) *y-z* cross-section of the quadric surface.

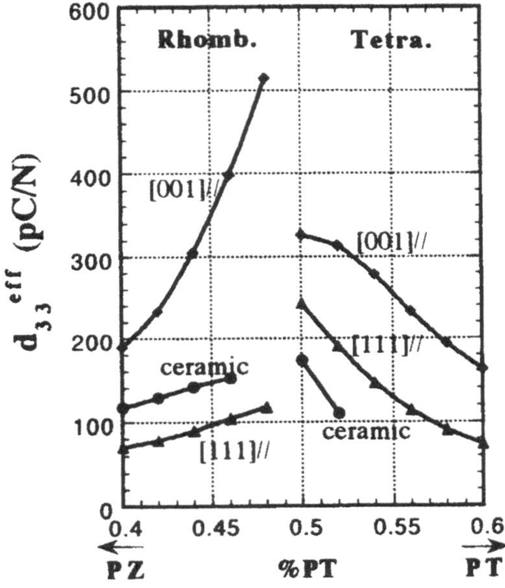

FIGURE 2.26 Effective d_{33} of PZT vs. composition. (From X. -H. Du et al.: *Appl. Phys. Lett.*, 72(19), 2421, 1998.)

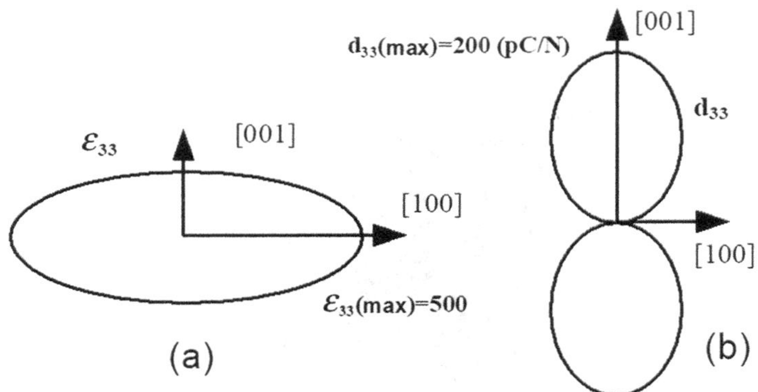

FIGURE 2.27 Quadric surfaces representing (a) the dielectric constant, ε_{33}, and (b) the piezoelectric strain coefficient, d_{33}, for a tetragonal PZT.

SOLUTION

Before the specimen is poled, the polarization directions of the individual crystallites are randomly oriented, so that the dielectric constant has a value intermediate between ε_{33}(min) and ε_{33}(max), and the piezoelectric coefficient, d_{33}, is zero. When the sample is poled, the polarization vectors of the crystallites become aligned in the poling direction (the z or [001] direction), resulting in an overall decrease in the dielectric constant. The piezoelectric coefficient, d_{33}, will increase monotonically with increasing poling field strength, saturating when the field reaches a level close to the coercive field. However, in commercially available practical PZT samples such as PZT5AH, some exceptions may exhibit the highest permittivity along the z-axis.

2.4.8 Phenomenology of Antiferroelectrics

The previous sections dealt with the case in which the directions of the spontaneous dipoles are parallel to each other in a crystal (*polar crystal*). There are cases in which antiparallel orientation lowers the *dipole-dipole interaction* energy (refer to Figure 2.3d). Such crystals are called *antipolar crystals*. Figure 2.28 shows the orientation of the spontaneous electric dipoles in an antipolar state in comparison with a nonpolar and a polar state. In an antipolar crystal, where the free energy of an antipolar state does not differ appreciatively from that of a polar state, the application of an external electric field or mechanical stress may cause a transition of the dipole orientation to a parallel state. Such crystals are called *antiferroelectrics*.

Figure 2.29 shows the relationship between $\boldsymbol{E}$ (applied electric field) and $\boldsymbol{P}$ (induced polarization) in paraelectric, ferroelectric, and antiferroelectric phases. In a paraelectric phase, the $\boldsymbol{P}$-$\boldsymbol{E}$ relation is linear; in a ferroelectric phase, there appears a hysteresis caused by the transition of the spontaneous polarization between the positive and negative directions, that is, coercive field, $\boldsymbol{E}_C$ (refer to Figure 2.14); in an antiferroelectric phase, at low electric field, the induced polarization is proportional to

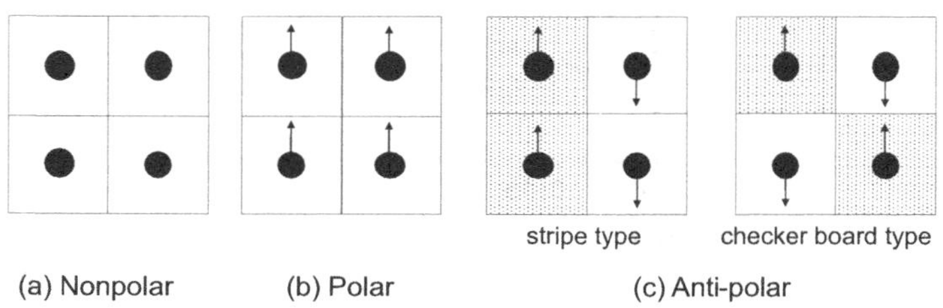

FIGURE 2.28 Schematic arrangement of the spontaneous dipoles in nonpolar, polar, and antipolar materials.

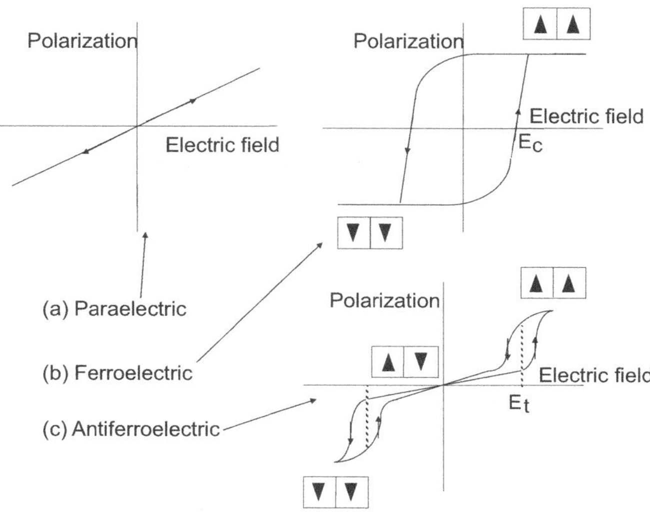

FIGURE 2.29 Polarization vs. electric field hysteresis curves in paraelectric, ferro-electric, and antiferroelectric materials.

$\boldsymbol{E}$, and when $\boldsymbol{E}$ exceeds a certain value $\boldsymbol{E}_t$, the crystal becomes ferroelectric (electric field–induced phase transition) and the polarization shows hysteresis with respect to $\boldsymbol{E}$. After removal of the electric field, the crystal returns to its anti-polar state and hence no spontaneous polarization can be observed as a whole. This is called a *double hysteresis curve*.

The simplest model for antiferroelectric materials is the so-called *one-dimensional two-sublattice* model, in which the electrostrictive coupling is described in terms of Kittel's free energy expression for antiferroelectrics.[13,14] As the name suggests, the model treats the lattice as a one-dimensional system, in which a superlattice (with dimensions twice that of the unit cell) is formed between two neighboring sublattices with sublattice polarizations, P_a and P_b. The state defined by $P_a = P_b$ represents the ferroelectric phase, while that defined by $P_a = (-P_b)$ represents the antiferroelectric phase. When considering the electrostriction effect, the coupling between the two sublattices is generally ignored and the strains from the two sublattices are QP_a^2 and QP_b^2, respectively (assuming equal electrostriction coefficients, Q, for both sublattices). The total strain induced in the crystal thus becomes:

$$x = Q\left(P_a^2 + P_b^2\right)/2, \tag{2.83}$$

where antiferroelectricity arises from a coupling between sublattices; however, it is more appropriate also to consider the sublattice coupling in our description of the electrostrictive effect. We include the coupling term, Ω, in the antiferroelectric free energy equation in order to arrive at this more complete description of the electrostriction. The free energy may then be expressed in the following form:

$$\begin{aligned} G_1 = {} & (1/4)\alpha\left(P_a^2 + P_b^2\right) + (1/8)\beta\left(P_a^4 + P_b^4\right) + (1/12)\gamma\left(P_a^6 + P_b^6\right) \\ & + (1/2)\eta P_a P_b - (1/2)\chi_T p^2 + (1/2)Q_h\left(P_a^2 + P_b^2 + 2\Omega P_a P_b\right)p \end{aligned} \tag{2.84}$$

where p is the hydrostatic pressure, χ_T is the isothermal compressibility, and Q_h is the hydrostatic electrostriction coefficient ($Q_h = Q_{33} + 2Q_{13}$). Making use of the transformations for the ferroelectric polarization, $P_F = (P_a + P_b)/2$, and the antiferroelectric polarization, $P_A = (P_a - P_b)/2$, in this equation yields:

$$\begin{aligned} G_1 &= (1/2)\alpha\left[P_F^2 + P_A^2\right] + (1/4)\beta\left[P_F^4 + P_A^2 + 6P_F^2P_F^2\right] \\ &+ (1/6)\gamma\left[P_F^6 + P_A^6 + 15P_F^4P_A^2 + 15P_F^2P_A^4\right] + (1/2)\eta\left[P_F^2 - P_A^2\right] \\ &- (1/2)\chi_T p^2 + Q_h\left[P_F^2 + P_A^2 + \Omega\left(P_F^2 - P_A^2\right)\right]p \end{aligned} \tag{2.85}$$

The dielectric and elastic equations of state may then be written as follows:

$$\begin{aligned} E &= (\partial G_1/\partial P_F) \\ &= P_F\left[\alpha + \eta + 2Q_h(1+\Omega)p + \beta P_F^2 + 3\beta P_A^2 + \gamma P_F^4 + 10\gamma P_F^2P_A^2 + 5\gamma P_A^4\right] \end{aligned} \tag{2.86}$$

$$\begin{aligned} 0 &= (\partial G_1/\partial P_A) \\ &= P_A\left[\alpha - \eta + 2Q_h(1-\Omega)p + \beta P_A^2 + 3\beta P_F^2 + \gamma P_A^4 + 10\gamma P_F^2P_A^2 + 5\gamma P_F^4\right] \end{aligned} \tag{2.87}$$

$$\begin{aligned} \Delta V/V &= (\partial G_1/\partial p) \\ &= -\chi_T p + Q_h(1+\Omega)P_F^2 + Q_h(1-\Omega)P_A^2 \end{aligned} \tag{2.88}$$

We see from Equation 2.88 how the contribution to the volume change related to the ferroelectric polarization, P_F, is given by:

$$(\Delta V/V)_{ind} = Q_h(1+\Omega)P_{F(ind)}^2 \tag{2.89}$$

Below the phase transition temperature, which is called the *Néel temperature* for an antiferroelectric, the spontaneous volume strain, $(V/V)_S$, and the spontaneous antiferroelectric polarization, $P_{A(S)}$, are related according to:

$$(\Delta V/V)_S = Q_h(1-\Omega)P_{A(S)}^2 \tag{2.90}$$

We see from this relationship that even when a given perovskite crystal has a positive hydrostatic electrostriction coefficient, Q_h, the spontaneous volume strain can be positive or negative depending on the value of Ω ($\Omega < 1$ or $\Omega > 1$). If the intersublattice coupling is stronger than the intrasublattice coupling, a volume contraction is observed at the Néel temperature. This is in stark contrast to what is observed for ferroelectrics, which always undergo a volume expansion at the Curie point.

The spontaneous strain induced in a case where $\Omega > 1$ is illustrated in Figure 2.30. When P_a and P_b are parallel, as occurs for the ferroelectric phase, the Ω term will act to effectively increase the spontaneous strain, x_S. When they are antiparallel, as occurs for the antiferroelectric phase (Figure 2.30, bottom), the Ω term acts to decrease the strain. This phenomenological treatment accurately describes what is observed experimentally for the antiferroelectric perovskite $PbZrO_3$, among others.[15] The induced strain curves for antiferroelectric $Pb_{0.99}Nb_{0.02}[(Zr_{0.6}Sn_{0.4})_{1-y}Ti_y]_{0.98}O_3$ ceramics are plotted as a function of applied electric field in Figure 2.31.[16] Figure 2.31a shows a composition $y = 0.06$, in which the induced ferro-state returns to the original antiferro-state with reducing the electric field, while Figure 2.31c shows $y = 0.065$, in which once the ferro-state is induced by the electric field, the antiferro-state will never recover during the field cycle. The isotropic volumetric change associated with the field-induced transition from the antiferro to ferroelectric phase is estimated by (from the subtraction between Equation 2.89 and 2.90):

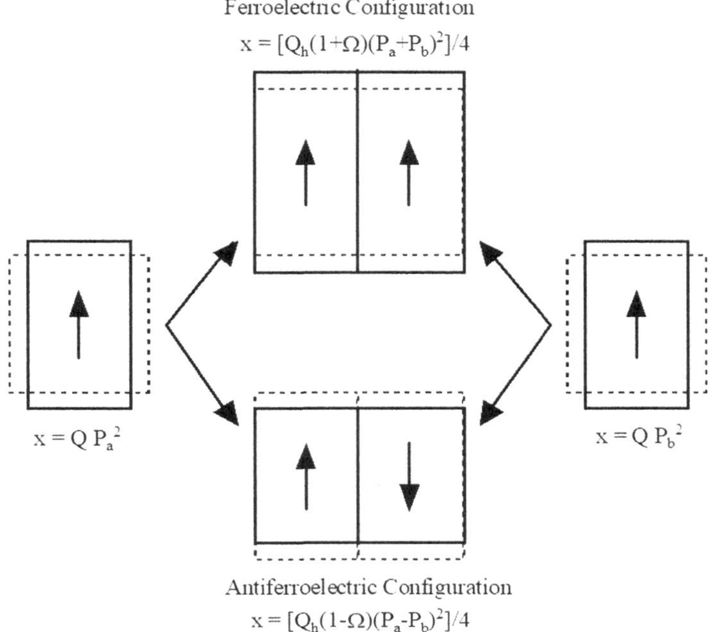

FIGURE 2.30 Schematic illustration of the sublattice coupling accompanying the electrostrictive strain response (for $\Omega > 0$).

$$\begin{aligned}(\Delta V/V)_S &= Q_h(1+\Omega)P_{F(S)}^2 \\ &\quad -Q_h(1-\Omega)P_{A(S)}^2 = 2Q_h\Omega P_{F(S)}^2\end{aligned} \tag{2.91}$$

Here, we assume that the magnitudes of P_a and P_b do not change drastically through the phase transition. It is noteworthy that the linear strain jumps associated with the phase transition experimentally observed in Figure 2.31 show almost the same value (7×10^{-4}) for all three compositions, which is almost 1/3 of the ($\Delta V/V$).

The ferroelectric strain state is characterized by both the significant strain induced by the electric field and the stability of the state, which can be optimized by carefully adjusting the Ti content. The most interesting composition is $y = 0.063$ in Figure 2.31b, in which the ferro-state strain, once induced, can be well maintained even when the applied field is removed, which is an important feature of a *shape memory function*. In order to recover the original antiferroelectric state, a small negative electric field can be applied. In comparison with shape memory alloys, where slow thermal treatment (heat generation) is required, this new antiferroelectric shape memory ceramic needs just negative electric field application for shape recovery. Application devices include a latching relay[17] and a mechanical clamper suitable for microscope sample holders.[18]

2.5 PHENOMENOLOGY OF MAGNETOSTRICTION

Similar to piezoelectrics, certain magnetic materials can also exhibit spontaneous strains along the magnetization direction and magnetostriction when a magnetic field is applied. In the electronically degenerate state, the orbitals are asymmetrically occupied and get more energy. Thus, in order to reduce this extra energy, the material tries to lower the overall symmetry of the lattice, that is, undergoing distortion, which is known as *Jahn-Teller distortion*. In the case of octahedral d-orbital configuration, the octahedron suffers elongation of bonds on the z-axis, thus lowering the symmetry.

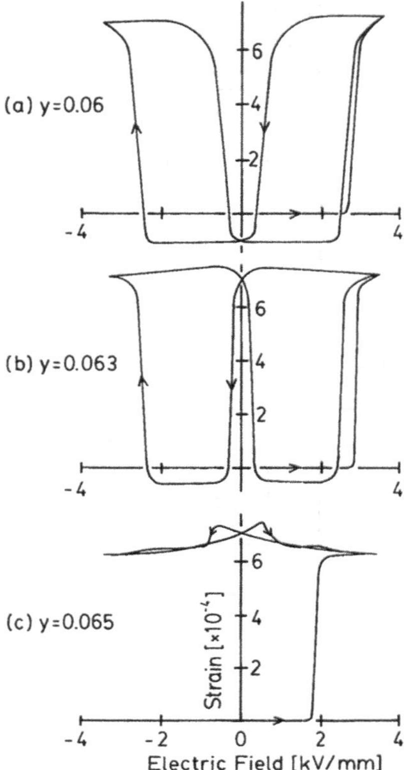

FIGURE 2.31 Induced strain for antiferroelectric $Pb_{0.99}Nb_{0.02}[(Zr_{0.6}Sn_{0.4})_{1-y} - Ti_y]_{0.98}O_3$ ceramics.

Therefore, a magnetostrictive material becomes strained according to the magnetization direction. Conversely, when either an applied force or torque produces a strain in a magnetostrictive material, the material's magnetic state (magnetization and permeability) will change. Magnetostriction is an inherent (intrinsic effect) material property that depends on electron spin, the orientation and interaction of spin orbitals, and the molecular lattice configuration, originating from the Jahn-Teller effect. It is also affected by domain wall motion and rotation of the magnetization (extrinsic effect) under the influence of an applied magnetic field or stress.

Magnetostriction occurs in these materials due to the reorientation of the spontaneous magnetization in response to an applied magnetic field. Typical strain ($\Delta l/l$) curves developed in such a magnetic material parallel to ($\lambda_{//}$) and perpendicular to ($\lambda_\perp$) the magnetic field (H) are shown in Figure 2.32. Depending on the material, the parallel strain may be positive (an extension) or

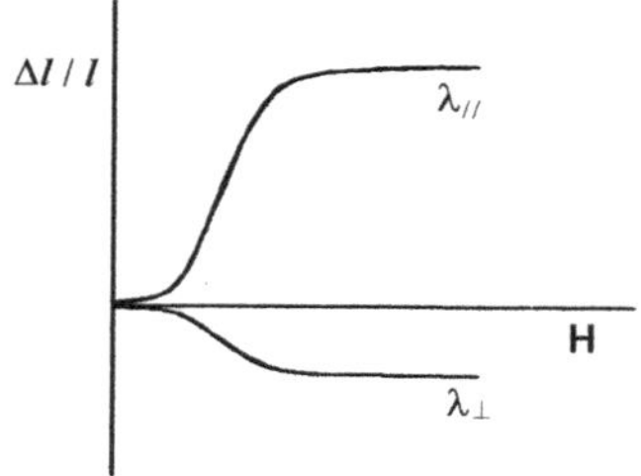

FIGURE 2.32 Strains ($\Delta l/l$) in a magnetostrictor parallel to ($\lambda_{//}$) and perpendicular to ($\lambda_\perp$) the magnetic field H.

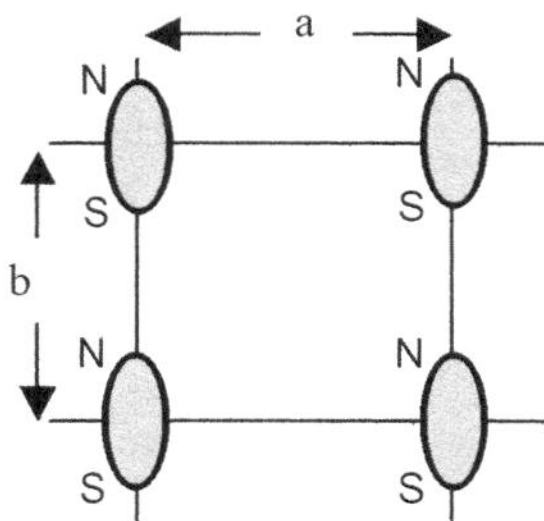

FIGURE 2.33 A simple model representing the spontaneous strain associated with the magnetization in a ferromagnetic material.

negative (a contraction). The perpendicular strain will be negative or positive depending on the value of Poisson's ratio for the material.

A simple model representing the spontaneous strain associated with the magnetization in a ferromagnetic material is pictured in Figure 2.33. If we assume that each magnet depicted in this model corresponds to the spin associated with an iron (Fe) atom in a metallic crystal, the opposite poles of adjacent magnets will attract, resulting in contraction of the lattice in the b-direction. At the same time, adjacent atoms with parallel spins repulse each other, resulting in expansion of the lattice in the a-direction. The magnetic energy associated with the resulting shift of lattice atoms is ultimately balanced by the elastic energy of the lattice, and an equilibrium level of spontaneous strain is established.

In magnetic crystals with a cubic structure, such as Fe and Ni, the magnitude of the induced strain, $\Delta l/l$, in the direction of the applied magnetic field can be expressed as:

$$\begin{aligned}\Delta l/l = (3/2)\lambda_{100}\left[\alpha_1^2\beta_1^2+\alpha_2^2\beta_2^2+\alpha_3^2\beta_3^2-1/3\right]\\ +(3)\lambda_{111}[\alpha_1\alpha_2\beta_1\beta_2+\alpha_2\alpha_3\beta_2\beta_3+\alpha_3\alpha_1\beta_3\beta_1]\end{aligned} \tag{2.92}$$

Here, λ_{100} and λ_{111} are the magnitudes of the magnetostriction when the magnetization is completely saturated along [100] and [111], respectively. Higher-order perturbation is neglected.[19] When the magnetic material is in a polycrystalline or amorphous form with isotropic characteristics, $\lambda_{100} = \lambda_{111} = \lambda_s$, so that Equation 2.92 may be simplified to:

$$\Delta l/l = (3/2)\lambda_s[\cos^2(\theta) - 1/3] \tag{2.93}$$

where θ is the angle between the magnetization and the strain directions, and λ_s is the saturated magnetostriction. When the sample is polycrystalline, the following relationship is satisfied:

$$\lambda_s = 0.4\lambda_{100} + 0.6\lambda_{111} \tag{2.94}$$

The quantities λ_{100} and λ_{111} are determined at the saturated strain levels in a single crystal, and λ_s can be determined from the saturated strain in a polycrystalline sample.

2.6 FERROELECTRIC DOMAIN REORIENTATION

2.6.1 Domain Formation

It has been assumed in the phenomenological descriptions presented so far that the materials in question are *monodomain single crystals* and the application of the field does not change their state of polarization or magnetization so easily. We provided the macroscopic coercive field in Example Problem 2.8, which gives a huge value higher than 10 kV/mm. When considering piezoelectric

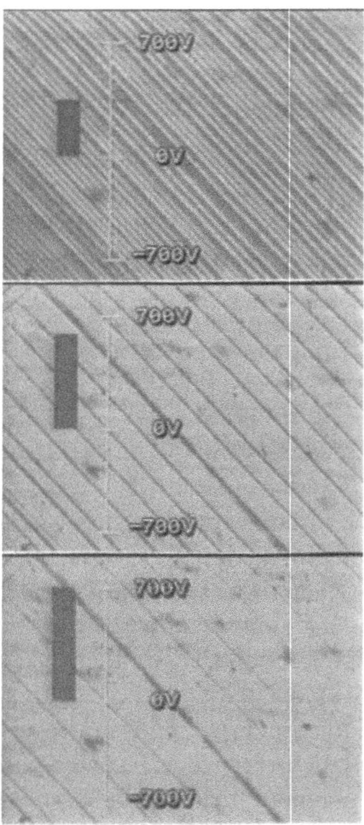

FIGURE 2.34 Domain reorientation in a $BaTiO_3$ single crystal. E direction: ↑.

ceramics, however, this assumption does not strictly hold. The material typically used in an actual device may have a multiple domain structure even in a single crystal form, and a much more complicated domain structure will exist in a polycrystalline ceramic. In practice, the apparent coercive field (to switch the polarization) is 1 kV/mm or less, 1/10 smaller than the value calculated above. The polarization reversal is realized through the domain wall movement, as shown in Figure 2.34, under a relatively small electric field. The domain pattern was observed during the polarization reorientation process in a $BaTiO_3$ single crystal (*4mm* tetragonal crystal symmetry at room temperature). Electric field was applied in the up/down ↕ directions on this page. Except under the maximum field, multiple 90° domain walls are observed. Why do multiple domains appear? The simplest answer is "because the multiple-domain state exhibits lower energy than the monodomain state."

Let us start from a multidomain model on a ferroelectric plate, as shown in Figure 2.35, where only the up/down spontaneous polarization (180° domain walls) arrangement is considered. The two key energies to be discussed are (1) *depolarization energy* and (2) *domain wall energy*. Taking into account Gauss's Law: div $\boldsymbol{D} = \rho$, where $\boldsymbol{D} = \varepsilon_0\boldsymbol{E} + \boldsymbol{P} = \varepsilon\varepsilon_0\boldsymbol{E} + \boldsymbol{P}_S$, we obtain:

$$div\,\boldsymbol{E} = \frac{1}{\varepsilon\varepsilon_0}\,(\rho - div\boldsymbol{P}_S). \tag{2.95}$$

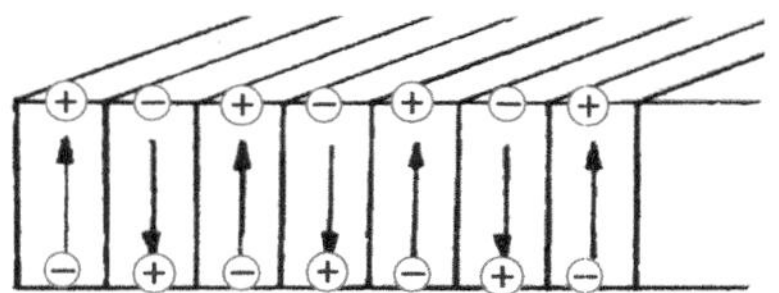

FIGURE 2.35 Multidomain model on a ferroelectric plate.

If free charge ρ is provided on the surface via surface electrode, crystal conductivity (like a p- or n-type semiconductor) or surrounding medium (wet air), the internal field $\boldsymbol{E}$ can be zero as an equilibrium status. However, if $\rho = 0$ due to a highly resistive crystal status, as in Figure 2.35, the so-called *depolarization field*:

$$\boldsymbol{E} = -(\boldsymbol{P}_S / \varepsilon\varepsilon_0) \tag{2.96}$$

should be generated in a monodomain crystal (so as to keep $\boldsymbol{D} = 0$), originating from the surface *bound charge*. Thus, the depolarization energy is calculated as

$$W_e = (1/2)\int dV\varepsilon\varepsilon_0 \mathrm{E}^2 = (1/2\varepsilon\varepsilon_0)\int dV \mathrm{P_S^2} \tag{2.97}$$

If the crystal is structured by twin up and down polarization domains with equal surface areas, the plus and minus charges can be compensated on a surface electrode, as in Figure 2.35, so that the depolarization energy can be diminished significantly.

The above discussion should be modified by considering the domain shape as below. Figure 2.36 shows an ellipsoid dielectric material for calculating the depolarization field.[20] Macroscopic field in a body can be obtained by the sum of $\boldsymbol{E} = \boldsymbol{E}_0 + \boldsymbol{E}_1$, where E_0 is the external field and E_1 is the depolarization field. Then, the depolarization field is expressed by:

$$E_{1x} = -\frac{L_x P_x}{\varepsilon_0}, E_{1y} = -\frac{L_y P_y}{\varepsilon_0}, E_{1z} = -\frac{L_z P_z}{\varepsilon_0}, \tag{2.98}$$

where L_x, L_y, and L_z are *depolarization factors*, which depend on the ratios of the ellipsoid principal axes, and $L_x + L_y + L_z = 1$. The depolarization field can be calculated as a field generated by *fictitious charges* on the ellipsoid surface (i.e., inside ellipsoid is cavity), as $L = (abc/2)A$, where A is an elliptical integral $(K(k) = \int_0^{\pi/2} d\theta/(\sqrt{1-k^2\sin^2\theta}))$. $L = 1/3$ for sphere, $L = 1$ for thin slab normal direction, and $L = 0$ for thin slab in-plane direction.

$$W_e = (1/2)\int dV\varepsilon\varepsilon_0 \mathrm{E}^2 = (\varepsilon/2\varepsilon_0)\int dV\, \mathrm{L^2 P_S^2} \tag{2.99}$$

Mitsui and Furuichi provided a simpler description of the depolarization energy by taking an aspect ratio of the domain shape (d/t_{crys}) instead of L, where d is the domain width, and t_{crys} is the ferroelectric plate thickness:[21]

$$W_e = \frac{\varepsilon^* d P^2 V}{t_{crys}} \tag{2.100}$$

where V is the crystal volume. In short summary, the depolarization energy promotes the multidomain status: the thinner domain width and the thicker crystal plate, the more preferable to reduce the depolarization energy.

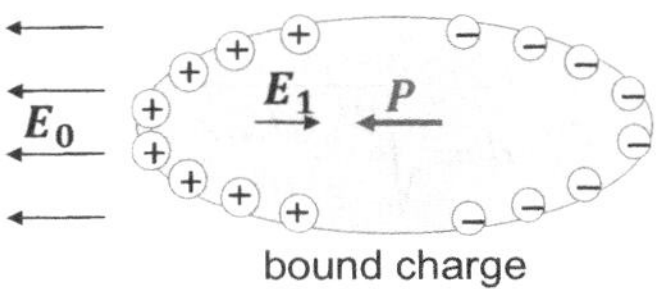

FIGURE 2.36 Ellipsoid dielectric model for calculating depolarization factor.

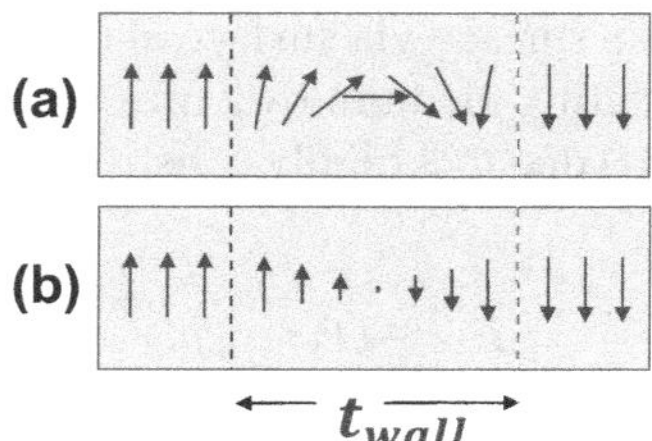

FIGURE 2.37 Ferroelectric domain wall models: (a) Néel wall, and (b) Bloch wall.

Now, we will consider the domain wall energy. The 180° *domain wall* in Figure 2.35 is the interface between upward and downward polarization domains. Though it is rather thin (similar to the unit cell dimension $\ll d$), it should have some width. In order to switch the polarization vector in an opposite direction, two possible domain wall models are proposed in Figure 2.37: (a) a Néel wall, in which, keeping the polarization magnitude constant, the vector arrow rotates, and (b) a Bloch wall, in which without changing the direction, the vector arrow length changes and the direction reverses (i.e., zero polarization at the wall center). In a ferromagnetic domain wall's case, the spin exchange energy suppresses a drastic magnetization change, leading to an increase in the domain wall thickness, while the anisotropy (elastic) energy reduces the domain wall thickness. Since the exchange energy exceeds the anisotropy energy in magnetic materials, wall thickness is rather thick (wider than hundreds of unit cells). On the contrary, in a ferroelectric domain wall's case, compared with the dipole interaction energy, the anisotropy elastic energy contributes significantly, leading to a narrow domain wall thickness, t_{wall}, around a couple of unit cell size.[22] The domain wall energy can be expressed as

$$W_{wall} = (A / t_{wall}) + Bt_{wall} \tag{2.101}$$

where A and B are *dipole interaction energy* and *anisotropy elastic energy*, respectively, and $B \gg A$ in most ferroelectric materials. Taking the energy minimization condition, $(\partial W_{wall}/\partial t_{wall}) = 0$, we can obtain the optimum t_{wall} and the wall energy W_{wall}:

$$t_{wall} = \sqrt{A/B} \tag{2.102}$$

$$W_{wall} = 2\sqrt{AB} \tag{2.103}$$

Now, we can finally obtain the stabilized domain size d on the ferroelectric crystal plate with volume V and thickness t_{crys}. We consider the sum of the depolarization energy (Equation 2.100) and the domain wall energy (Equation 2.103) × (number of walls V/d):

$$W_e + W_{wall}\left(\frac{V}{d}\right) = \frac{\varepsilon^* dP^2 V}{t_{crys}} + W_{wall}\left(\frac{V}{d}\right) \tag{2.104}$$

Minimizing again the total energy, the optimized domain width, d, can be related to the crystal plate thickness, t_{crys}, as follows:

$$d = \sqrt{\frac{W_{wall} t_{crys}}{\varepsilon^* P^2}}. \tag{2.105}$$

In summary, the depolarization energy reduces the domain width, while the domain wall energy increases the domain width, leading to a compromised domain width related to the square root of the crystal thickness. Accordingly, ferroelectric thin films (even in a single crystal-like epitaxial) may exhibit very small domain width, while we need some techniques to obtain a monodomain crystals, such as surface screening and semiconductors (charge carriers).

2.6.2 Strains Accompanied by Ferroelectric Domain Reorientation

The strains induced parallel ($\Delta l/l = x_3$) and perpendicular ($\Delta l/l = x_1$) to the applied electric field in a 7/62/38 (Pb,La)(Zr,Ti)O_3 (PLZT) ceramic are shown in Figure 2.38. For a cycle with a small maximum electric field (curve **a** in the figure), the field-induced strain curve appears nearly linear and reflects primarily the *converse piezoelectric effect*. As the amplitude of the applied electric field is increased, however, the strain curve becomes more hysteretic (curve **c**), and finally transforms into the characteristic symmetric butterfly shape (curve **e**) when the electric field exceeds a certain critical value known as the *coercive field*. This is caused by the switching of ferroelectric domains under the applied electric field, resulting in a different state of polarization. Strictly speaking, this composition of PLZT undergoes this change in polarization state in two stages:

1. Individual domain reorientation in each grain
2. Overall multidomain reorientation and domain wall movement within the entire polycrystalline structure (which may be regarded as just an assembly of randomly oriented tiny crystallites)

Different from the case in a single crystal $BaTiO_3$ in Figure 2.34, in a polycrystalline specimen, however, domain wall motion tends to be suppressed by grain boundaries, and a purely monodomain state cannot be achieved. A schematic depiction of the domain reorientation that occurs in a polycrystalline body under an applied electric field appears in Figure 2.39. Suppose an electric field is applied to a sample that has been poled in the opposite direction. The crystal should shrink as the applied field is initially increased, because the field direction opposes that of the *remnant polarization*. The shrinkage will proceed until it reaches a certain minimum that corresponds to an applied field strength equal

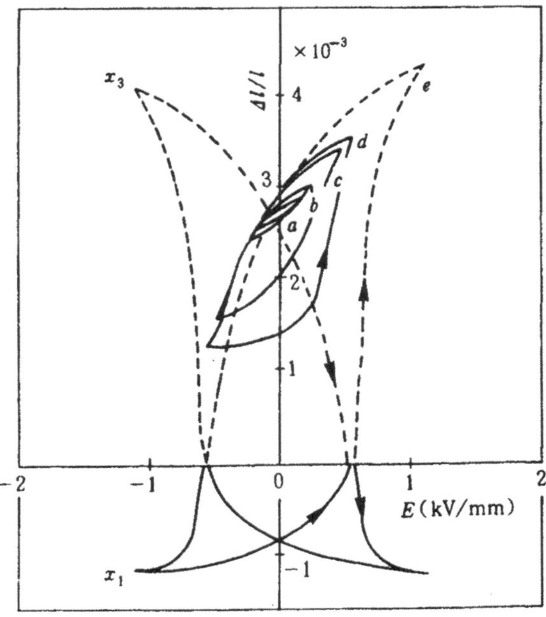

FIGURE 2.38 The electric field-induced strain in a piezoelectric PLZT 7/62/38 ceramic.

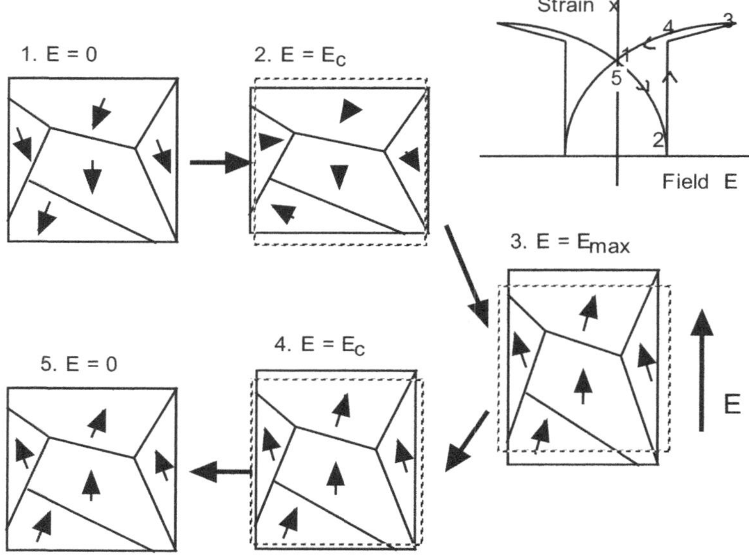

FIGURE 2.39 Schematic depiction of the reorientation of ferroelectric domains in a polycrystalline material under an applied electric field.

to the *coercive field*, E_c, when polarization reversal will begin to occur in each grain. Above E_c, the crystal will start to expand and continue to deform in this manner until the applied field strength reaches a certain saturation level, E_{max}. At E_{max}, all the reversible polarization vectors associated with the individual domains have been reversed, and the crystal displays "piezoelectric" behavior once again with a small hysteresis. As the electric field is decreased from this level, no further reorientation of the polarization should occur except for a few unstable domains that might still remain. The strain decreases monotonically as the field is reduced to zero. The polarization state at $E = 0$ is equivalent to the initial state but with all the polarization directions reversed. The crystal is now essentially poled in the positive direction. Poisson's ratio, σ, which is the ratio of the transverse strain (in this case, a contraction) to the longitudinal strain (here an expansion) is similar for all perovskite piezoelectric ceramics, about 0.3. It is interesting to note that Poisson's ratio in the electric field–induced strain $|x_1/x_3|$ is almost the same as that in pure elastically induced strain ratio $|x_1/x_3|$ in most of PZT compositions.

EXAMPLE PROBLEM 2.15

Barium titanate ($BaTiO_3$) has a rhombohedral crystal symmetry at liquid nitrogen temperature (−196°C) and the distortion angle from the cubic structure is not very large (= 1°). Determine all possible angles between the two non-180° domain walls.

Solution

The polarization of barium titanate at this temperature is oriented along the $<111>$ directions of the perovskite cell. Let us consider the three shown in Figure 2.40*a*: $[111]$, $[11\bar{1}]$, and $[1\bar{1}\bar{1}]$. Assuming a head-to-tail alignment of the spontaneous polarization across the domain wall (i.e., the normal component of the P_S should be continuous to satisfy Gauss's Law, div $\boldsymbol{D} = 0$ in highly resistive piezoelectrics), as depicted in Figure 2.40b, we expect the plane of the domain wall plane to be normal to one of the following directions or their equivalent directions:

$$[111] + [11\bar{1}] = [220]$$

$$\text{and } [111] + [1\bar{1}\bar{1}] = [200]$$

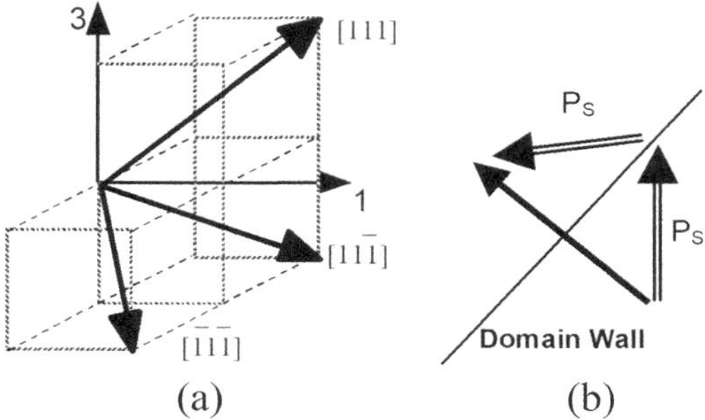

FIGURE 2.40 Rhombohedral $BaTiO_3$: (a) polarization directions and (b) the orientation of the spontaneous polarization vectors across a domain wall.

The angle between two of the non-180° domain walls is thus calculated as follows:

1. (002)/(200);(022)/0$\bar{2}$2);(002)/(220)
   $(002)\cdot(200) = 2^2[\cos(\theta)] = 0 \quad 0 \rightarrow 90°$
2. (022)/(220);(022)/2$\bar{2}$0)
   $(022)\cdot(220) = \left(2\sqrt{2}\right)^2 [\cos(\theta)] = 4 \text{ or } -4 \rightarrow \theta = 60° \text{ or } 120°$
3. (002)/(022);(002)/0$\bar{2}$2)
   $(022)\cdot(220) = 2\left(2\sqrt{2}\right)[\cos(\theta)] = 4 \; or \; -4 \rightarrow \theta = 45° \text{ or } 135°$

From the domain observation under a microscope, we can speculate the polarization direction of the domain (except for the + or – information) and the crystal symmetry (tetragonal, rhombohedral, etc.).

2.6.3 The Uchida-Ikeda Model

In this section, we discuss the *Uchida-Ikeda theory*, by which the polarization and strain curves for piezoelectric polycrystalline specimens are described and predicted in terms of domain reorientation, the crystal structure, and the coercive field.

Let us take as an example a barium titanate ($BaTiO_3$) single crystal, which has a tetragonal symmetry at room temperature. X-ray diffraction of the crystal reveals a slight elongation along the [001] direction of the perovskite unit cell with $c/a = 1.01$. Therefore, if an electric field is applied on an *a* plane single crystal (electrode on the top and bottom of the BT plane), a 90° domain reorientation from an *a* to a *c* domain is induced, resulting in a strain of 1% in the field direction. However, the situation is much more complicated in the case of a polycrystalline specimen. Uchida and Ikeda treated this problem statistically, assuming the grains (or small crystallites) are randomly oriented.[23,24]

There will be no remnant polarization in a homogeneous unpoled polycrystalline sample. Let this state be the basis for zero strain. If an electric field, E_3, is applied to this sample, a net polarization, P_3, will be induced. The strains x_1, x_2, and x_3 will also be generated, where $x_1 = x_2 = -\sigma x_3$ and σ is Poisson's ratio. Let the spontaneous polarization and *principal strain* of the individual crystallites (i.e., single crystal values) be designated by P_S and S_S, respectively. For uniaxial crystal symmetries, such as the tetragonal and rhombohedral, the principal strain, S_S, is in the direction of P_3 and is defined for each symmetry as follows:

$$S_S = [(c/a) - 1](\text{tetragonal crystal}) \tag{2.106}$$

$$S_S = (3/2)[\pi/2 - \alpha] = (3/2)\delta(\text{rhombohedral crystal}) \tag{2.107}$$

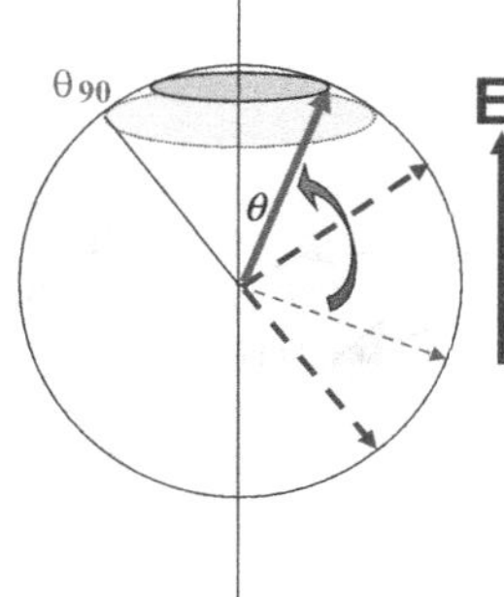

FIGURE 2.41 Polarization reorientation model for non-180° domain wall case.

The distinction between the principal and spontaneous strains should be noted here. They are *not* interchangeable terms, but in fact define two very different strains. Using $BaTiO_3$ as an example, we see that the principal strain is $S_S = 0.01$, but the spontaneous strains are defined as:

$$x_{3(S)} = [c/a_0 - 1]$$

$$x_{1(S)} = x_{2(S)} = [(a/a_0) - 1] \tag{2.108}$$

where a_0 is the lattice parameter of the paraelectric phase. When the appropriate lattice parameters are used in Equation 2.108, the spontaneous strains $x_{3(S)}$ and $x_{1(S)}$ are 0.0075 and −0.0025, respectively.

First, assuming an angle, θ, between the direction of the spontaneous polarization, P_S, of a microscopic volume, dv, in a ceramic and the direction of the electric field, E_3 (Figure 2.41), then the polarization contribution to the electric field direction, P_3, is given by $P_S{\cdot}\cos\theta$. By integrating on all volume:

$$P_3 = \frac{\int P_S \cos\theta \, dv}{\int dv} = P_S\overline{\cos\theta} \tag{2.109}$$

where $\overline{\cos\theta}$ is the average value of $\cos\theta$ in all the volume elements of the ceramic. The average strain is determined from the orientation of the strain ellipsoid:

$$x_3 = S_S\left[\frac{\int \cos^2\theta \, dv}{\int dv} - \frac{1}{3}\right] = S_S\left[\overline{\cos^2\theta} - (1/3)\right] \tag{2.110}$$

Since the strain is the second-rank tensor, we need to multiply the direction cosine ($\cos\theta$) twice to calculate the contribution along the z-axis. It is assumed in the Uchida-Ikeda model that the spontaneous strains associated with the microscopic regions change only in their orientation with no change in volume, hence Poisson's ratio is apparently $\sigma = 0.5$. This also implies that:

$$x_1 = x_2 = (-x_3/2) \tag{2.111}$$

However, this assumption does not agree well with the experimental data in Figure 2.38 ($|x_1/x_3| \approx 1/3$).

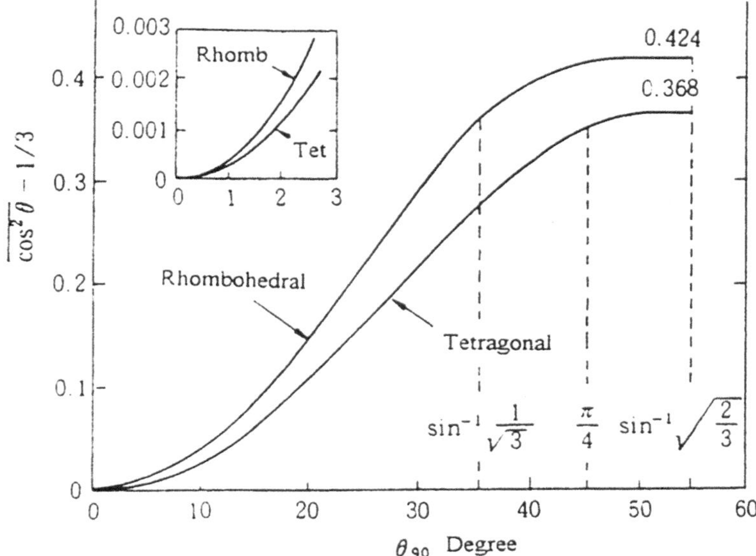

FIGURE 2.42 Strain-related quantity $\left(\overline{\cos^2\theta} - 1/3\right)$ as a function of the characteristic angle, θ_{90}, where θ_{90} is a critical angle related to the non-180° domain reorientation. (From N. Uchida and T. Ikeda: *Jpn. J. Appl. Phys.*, 6, 1079, 1967; N. Uchida: *Rev. Elect. Commun. Lab.*, 16, 403, 1968.)

In order to arrive at an expression for the induced strain as a function of the applied electric field, the relationship between θ and E_3 must be determined. This is accomplished in the context of this model by introducing a characteristic angle, θ_{90}, for non-180° domain reorientations. In tetragonal crystals, this corresponds to a 90° reorientation, and in rhombohedral crystals, 71° and 109° reorientations will occur, but in order to simplify the analysis, all reorientations are represented by the former. Suppose a 90° domain rotation occurs in a small volume element, dv, in a ceramic, and as a result the domain orientation within dv becomes θ (see Figure 2.42). It is assumed that there exists a characteristic angle, θ_{90}, such that if $\theta < \theta_{90}$, a 90° rotation will occur, whereas if $\theta > \theta_{90}$, no rotation will occur, and the region will remain in its initial state (i.e., the simplest on-off model). Given a specific angle, θ_{90}, which corresponds to a certain applied field strength, E_3, Equation 2.110 can be integrated over the range of volume for which $\theta < \theta_{90}$ is satisfied, to obtain the induced strains x_3, x_2, and x_1 as a function of θ_{90}. The quantity $\left(\overline{\cos^2\theta} - 1/3\right)$ is plotted as a function of the characteristic angle, θ_{90}, in Figure 2.42, and Figure 2.43a shows the measured values of induced strain for a rhombohedral PZT ceramic. When the equations defining these trends are combined, the curve representing the relationship between θ_{90} and E_3 is generated as depicted in Figure 2.43b. A pronounced hysteresis is apparent in this curve. The saturation values of the polarization and the strain of a tetragonal and rhombohedral ceramic under high electric field are summarized as follows:

$$\text{Tetragonal}: \quad P_3 \rightarrow (0.831)P_S \quad x_3 \rightarrow (0.368)S_S$$
$$\text{Rhombohedral}: P_3 \rightarrow (0.861)P_S \quad x_3 \rightarrow (0.424)S_S$$

It is important to note that the saturated polarization reaches higher than 85% of the single crystal value, but the strain will reach only 1/3 of the single crystal value. Further, the critical θ_{90} is around only 20° (Figure 2.43b) in practice, which corresponds to only $\left(\overline{\cos^2\theta} - 1/3\right) \sim 0.1 - 0.5$ of the single crystal spontaneous strain. In conclusion, we can evaluate the single crystal spontaneous polarization roughly from the polycrystalline experimental value just by multiplying 1.15 or so (the order is almost

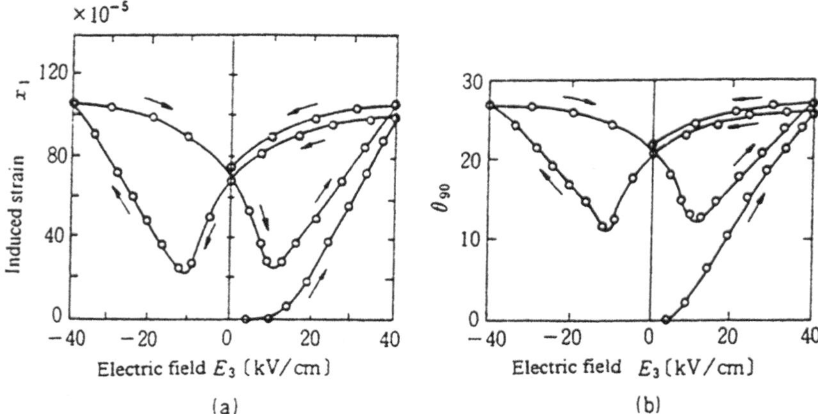

FIGURE 2.43 Field-induced transverse strain, x_1, in a $Pb(Zr_{0.57}Ti_{0.43})O_3$ ceramic sample: (a) x_1 vs. applied electric field, E_3, and (b) calculated characteristic angle, θ_{90}, vs. E_3. Measurement at 30°C. (From N. Uchida and T. Ikeda: *Jpn. J. Appl. Phys.*, 6, 1079, 1967; N. Uchida: *Rev. Elect. Commun. Lab.*, 16, 403, 1968.)

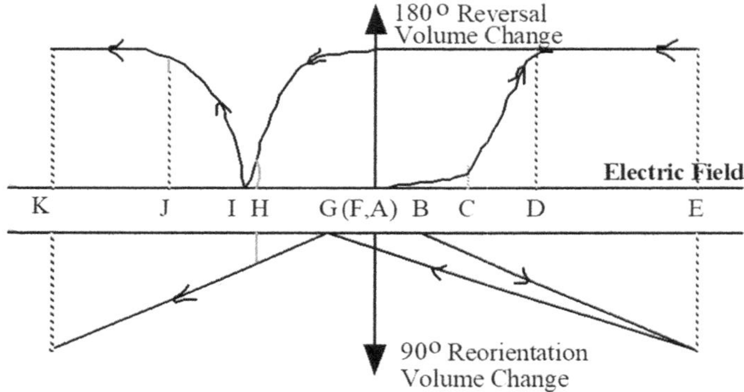

FIGURE 2.44 Electric field dependence of the volume fraction of domains that have undergone: (a) 180° reversal and (b) 90° reorientation. Notice the noncoincidence of the zero fraction points I and G (coercive field), which correspond to the 180° and 90° cases, respectively. (From N. Uchida and T. Ikeda: *Jpn. J. Appl. Phys.*, 4, 867, 1965.)

the same), while the spontaneous strain evaluation is very difficult, because the ceramic strain value is only 1/10–1/7 (one order of magnitude smaller) of the single crystal value.

Furthermore, by finding the polarization, P_3, and field-induced strain x_3 (or x_1) as a function of the electric field, E_3, it is possible to estimate the volume in which a 180° reversal or a 90° reorientation occurs. This is because only the 90° reorientation contributes to the induced strain, whereas 180° domain reversal contributes mainly to the polarization. Curves representing the volume fraction of 180° domains that have undergone reversal and 90° domains that have rotated by 90° as a function of an applied electric field are shown in Figure 2.44. We see from these curves that the 180° reversal occurs quite rapidly as compared to the slower process of 90° rotation.[25] It is notable that at G on the curve there remains some polarization and the induced strain is zero, while at H the strain is not at a minimum, but contributions to the polarization from the 180° and 90° reorientations cancelling each other so that the net polarization becomes zero. A plot of the induced strain, x_3, as a function of polarization for a PLZT (6.25/50/50) ceramic, in which 180° reversal is dominant, is shown in Figure 2.45. We see that it is characterized by a rather large hysteresis in the tetragonal ceramics, probably

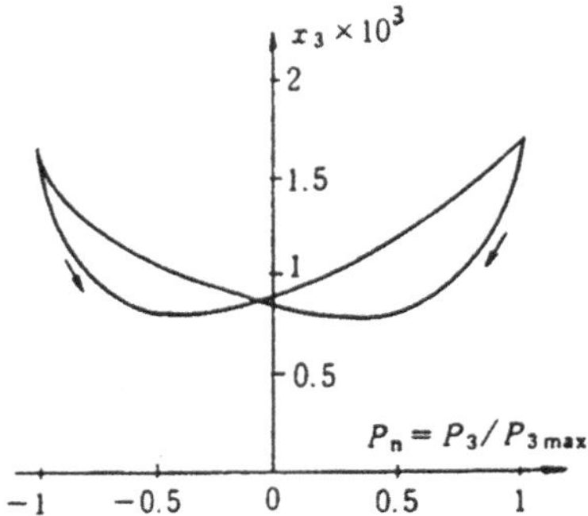

FIGURE 2.45 Induced strain, x_3, as a function of polarization for a tetragonal PLZT (6.25/50/50) ceramic. (From N. A. Schmidt: *Ferroelectrics*, 31, 105, 1981.)

due to the coercive field difference between 180° and 90° reorientation.[26] In contrast, materials whose polarization is dominated by non-180° domain rotations, such as the low-temperature rhombohedral phase of $Pb(Mg_{1/3}Nb_{2/3})O_3$, exhibit no significant hysteresis in their $x - P$ curves (see Figure 2.46b), though $P - E$ and $x - E$ curves show significant hysteresis (Figure 2.46a).[8] This $x - P$ curve fits to the electrostrictive relation $x_1 = Q_{13} P_3^2$ without hysteresis even in the ferroelectric state.

2.6.4 Crystal Structures and Coercive Fields

In the previous section, a comparison was made between tetragonal and rhombohedral ceramic systems from the viewpoint of the saturation values, P_3 and S_S, under a sufficiently large electric field. In this section, we will consider the difference between these two systems in terms of the more useful quantity, the coercive field, E_c. The principal strain, S_S; spontaneous polarization, P_S; volume percent of reoriented domains, γ_{90}; and coercive field, E_c, for a series of PLZT ceramics as reported by Schmidt[26] are summarized in Table 2.4. We see from the data presented in this table that the rhombohedral compositions generally have smaller principal strains than the tetragonal compositions. This implies that domain rotation is more readily achieved for the rhombohedral

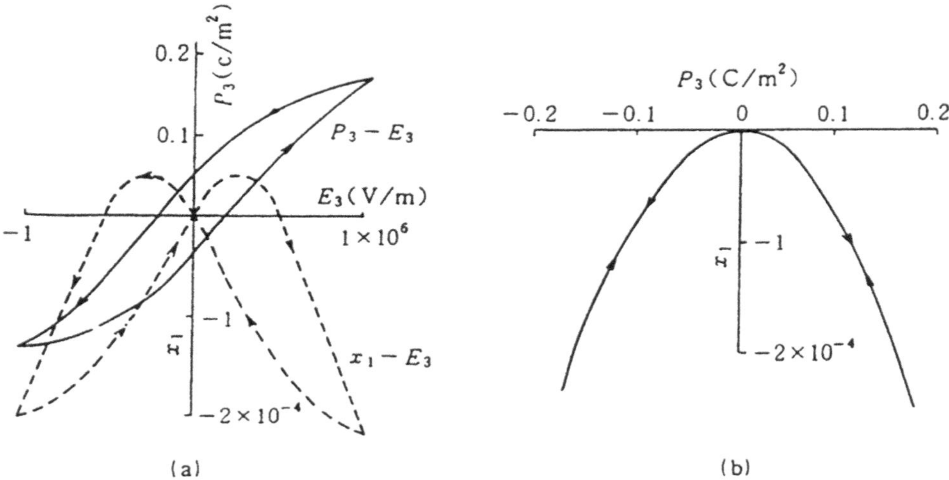

FIGURE 2.46 Rhombohedral $Pb(Mg_{1/3}Nb_{2/3})O_3$ at − 110°C: (a) polarization, P_3, and transverse strain, x_1, as a function of applied electric field, E_3, and (b) transverse strain, x_1, as a function of polarization, P_3. (From J. Kuwata et al.: *Jpn. J. Appl. Phys.*, 19, 2099, 1980.)

TABLE 2.4
Principal strain, S_S; spontaneous polarization, P_S; reoriented volume fraction; and coercive field, E_c, for some tetragonal (T) and rhombohedral (R) PLZT ceramics

Symmetry	Composition (La/Zr/Ti)	S_S (%)	P_S (μC/cm²)	Reoriented Domains (%)	E_c [measured] (kV/cm)	E_c [calculated] (kV/cm)
T	25/50/50	2.40	71.0	22.0	18.0	17.8
	25/52/48	2.20	72.0	28.0	14.7	18.8
	5//50/50	2.16	65.0	18.0	16.3	13.0
	5/52/48	1.96	64.5	23.0	14.8	13.7
	5/54/46	1.68	65.0	30.0	11.7	13.0
R	25/58/42	0.732	56.5	86.5	8.2	7.0
	25/60/40	0.740	58.5	78.5	7.6	5.4
	6/65/35	0.650	45.0	85.0	5.6	5.9
	6.25/60/40	0.610	49.0	85.0	5.7	4.8

Source: N. A. Schmidt: *Ferroelectrics*, **31**, 105, 1981.

compositions (larger γ_{90}), and the associated coercive fields are accordingly smaller. The following expression for the coercive field has been derived:

$$E_c = (\alpha Y S_S^2 \gamma_{90})/P_S \tag{2.112}$$

where Y is Young's modulus. The quantity α is a factor that takes into account the difference in domain orientation between neighboring grains, and it typically has values of 0.1 and 0.074 for tetragonal and rhombohedral compositions, respectively.[26,27]

2.7 LOSS MECHANISMS IN PIEZOELECTRICS

Heat generation is one of the significant problems in piezoelectrics for high power density applications. In this section, we review the loss phenomenology in piezoelectrics, including three losses: dielectric, elastic, and piezoelectric losses. Heat generation at off-resonance is attributed mainly to intensive dielectric loss tan δ', while heat generation at resonance is mainly originated from the intensive elastic loss tan ϕ'. The loss effect on the electromechanical resonance will be discussed in Chapter 5.

2.7.1 Loss Phenomenology in Piezoelectrics

The terminologies *intensive* and *extensive* losses are introduced here, in the relation with intensive and extensive parameters in the phenomenology by extending the concept introduced in Section 2.4.6. Assume a certain volume of material first, and imagine cutting it in half. The extensive parameter (material's internal parameters such as displacement/strain, x, and total dipole moment/electric displacement, D) depends on the volume of the material (i.e., becomes a half), while the intensive parameter (externally controllable parameters such as force/stress, X, and electric field, E) is independent of the volume of the material. These are not related to the intrinsic and extrinsic losses, which were introduced to explain the loss contribution from the monodomain single crystal state and from the others.[28] Our discussion in piezoelectrics is focused primarily on extrinsic losses, in particular losses originating from domain dynamics, in this section. However, the physical parameters of their performance such as permittivity and elastic compliance still differ in piezoelectric materials, depending on the boundary conditions: mechanically free or clamped, and electrically short circuit or open circuit. These are distinguished as intensive or extensive parameters and their associated losses.

We start from the following two *piezoelectric constitutive equations with losses* (refer to Equations 2.57 and 2.58 without losses):

$$x = s^{E} * X + d * E, \tag{2.113}$$

$$D = d * X + \varepsilon^{X} * \varepsilon_0 E, \tag{2.114}$$

where x is strain, X stress, D electric displacement, and E electric field. Equations 2.113 and 2.114 are expressed in respect to *intensive* (i.e., externally controllable) physical parameters X and E. The elastic compliance, $s^{E}*$; the dielectric constant, $\varepsilon^{X}*$; and the piezoelectric constant, $d*$, are temperature dependent in general. Note that the piezoelectric constitutive equations cannot yield a delay time–related loss, in general, without taking into account irreversible thermodynamic equations or *dissipation functions*. However, the dissipation functions are mathematically equivalent to the introduction of *complex physical constants* into the phenomenological equations, if the loss is small and can be treated as a perturbation (dissipation factor tangent $\ll 0.1$). Based on this mathematical principle, therefore, we will introduce complex parameters ε^{X*}, s^{E*}, and $d*$, using * in order to consider the small hysteresis losses in dielectric, elastic, and piezoelectric constants:

$$\varepsilon^{X*} = \varepsilon^{X}(1 - j\tan\delta'), \tag{2.115}$$

$$s^{E*} = s^{E}(1 - j\tan\phi'), \tag{2.116}$$

$$d^{*} = d\,(1 - j\tan\theta'). \tag{2.117}$$

θ' is the phase delay of the strain under an applied electric field, or the phase delay of the electric displacement under an applied stress. Both delay phases should be exactly the same if we introduce the same complex piezoelectric constant $d*$ into Equations 2.113 and 2.114. δ' is the phase delay of the electric displacement to an applied electric field under a constant stress (e.g., zero stress) condition, and ϕ' is the phase delay of the strain to an applied stress under a constant electric field (e.g., short-circuit) condition. The negative sign in front of the loss tangent comes a general consensus that the output will be slightly delayed after the input. We will consider these phase delays intensive losses.

Figures 2.47a through d correspond to the model hysteresis curves for practical experiments: D vs. E curve under a stress-free condition, x vs. X under a short-circuit condition, x vs. E under a stress-free condition, and D vs. X under a short-circuit condition for measuring current, respectively. Note that these measurements are easily conducted in practice. For example, the $D-X$ relation under a short-circuit condition can be obtained from the integration of measuring the current by changing the external stress. The average slope of the D-E hysteresis curve in Figure 2.47a corresponds to the permittivity $\varepsilon^{X}\,\varepsilon_0$, where the superscript stands for $X =$ constant (occasionally zero). Thus, tan δ' is called an intensive dielectric loss tangent. The situation of s^{E} is similar; the slope of the x-X relation is the elastic compliance under $E =$ constant condition. It is worth noting that the actual hysteresis curve exhibits rather sharp edges at the maximum and minimum external parameter (electric field or stress) points, though the complex parameter usage in Equations 2.121 through 2.123 should generate rounded edges because it is an ellipse shape, theoretically. You can easily imagine a sort of limitation of this complex parameter approach from the discrepancy with the experimental results.

Since the areas on the D-E and x-Xdomains directly exhibit the electrical and mechanical energies, respectively (see Figures 2.47a and b), the stored energies (during a quarter cycle) and hysteresis losses (during a full electric or stress cycle) for pure dielectric and elastic energies can be calculated as:

$$U_e = (1/2)\varepsilon^{X}\varepsilon_0 E_0^2, \tag{2.118}$$

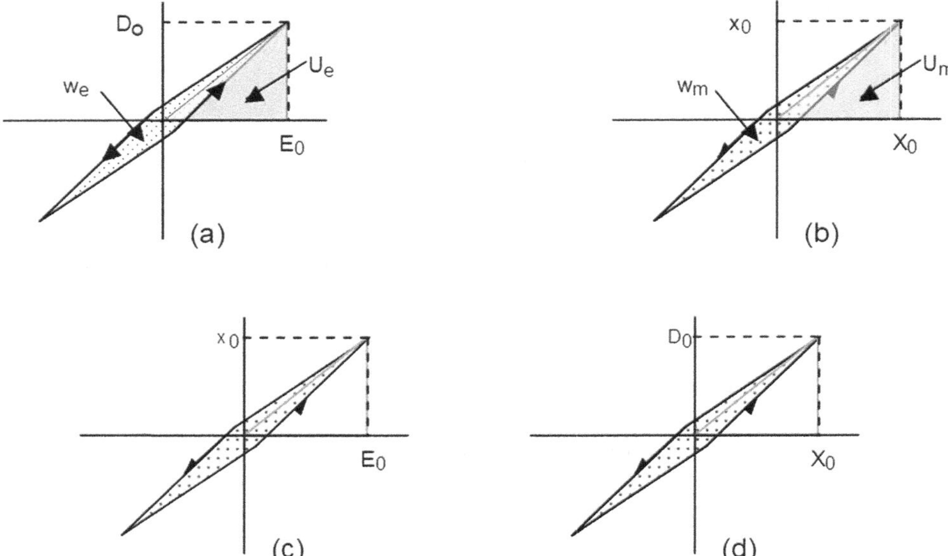

FIGURE 2.47 (a) D vs. E (stress free), (b) x vs. X (short-circuit), (c) x vs. E (stress free) and (d) D vs. X (short-circuit) curves with hysteresis.

$$w_e = \pi \varepsilon^X \varepsilon_0 E_0^2 \tan\delta', \tag{2.119}$$

and

$$U_m = (1/2) s^E X_0^2, \tag{2.120}$$

$$w_m = \pi s^E X_0^2 \tan\phi'. \tag{2.121}$$

The dissipation factors, tan δ' and tan ϕ', can experimentally be obtained by measuring the dotted hysteresis area and the stored energy area; that is, $(1/2\pi)(w_e/U_e)$ and $(1/2\pi)(w_m/U_m)$, respectively. Note that the factor (2π) comes from integral per cycle (refer to Example Problem 2.15 on the above derivation). The electromechanical hysteresis loss calculations, however, are more complicated, because the areas on the x-E and P-X domains do not directly provide energy. The areas on these domains can be calculated as follows, depending on the measuring method; when measuring the induced strain under an electric field, the electromechanical conversion energy can be calculated as follows, by converting E to stress X:

$$U_{em} = \int x\,dX = \left(\frac{1}{s^E}\right)\int x\,dx = (d^2/s^E)\int_0^{E_0} E dE = (1/2)(d^2/s^E)E_0^2, \tag{2.122}$$

where $x = dE$ was used. Then, using Equations 2.116 and 2.117, and from the imaginary part, we obtain the loss during a full cycle as

$$w_{em} = \pi(d^2/s^E)E_0^2(2\tan\theta' - \tan\phi'). \tag{2.123}$$

Note that the area ratio in the strain vs. electric field measurement should provide the combination of piezoelectric loss tan θ' and elastic loss tan ϕ' (not tan θ' directly!).

When we measure the induced charge under stress, the stored energy, U_{me}, and the hysteresis loss, w_{me}, during a quarter and a full stress cycle, respectively, are obtained similarly as

$$U_{me} = \int PdE = (1/2)(d^2/\varepsilon_0\varepsilon^X)X_0^2, \tag{2.124}$$

$$w_{me} = \pi(d^2/\varepsilon_0\varepsilon^X)X_0^2(2\tan\theta' - \tan\delta'). \tag{2.125}$$

Now, the area ratio in the charge vs. stress measurement provides the combination of piezoelectric loss, tan θ', and dielectric loss, tan δ'. Hence, from the measurements of D vs. E and x vs. X, we obtain $\tan\delta'$ and $\tan\phi'$, respectively, and either the piezoelectric (D vs. X) or converse piezoelectric measurement (x vs. E) provides tan θ' through a numerical subtraction. The above equations provide a traditional off-resonance loss measuring technique on piezoelectric actuators, an example of which is introduced in Section 2.7.2.

EXAMPLE PROBLEM 2.16

When the observed variation in electric displacement, D, can be represented as if it had a slight phase lag with respect to the applied electric field:

$$E^* = E_0 e^{j\omega t} \tag{P2.16.1}$$

$$D^* = D_0 e^{j(\omega t - \delta)} \tag{P2.16.2}$$

If we express the relationship between D^* and E^* as

$$D^* = \varepsilon^* \varepsilon_0 E^* \tag{P2.16.3}$$

where the complex dielectric constant, ε^*, is

$$\varepsilon^* = \varepsilon' - j\varepsilon'', \tag{P2.16.4}$$

$$\varepsilon''/\varepsilon' = \tan\delta \tag{P2.16.5}$$

The integrated area inside the hysteresis loop, labeled w_e in Figure 2.47a, is equivalent to the energy loss per cycle per unit volume of the dielectric. It is defined for an isotropic dielectric as:

$$w_e = -\int DdE = -\int_0^{(2\pi)/\omega} D\frac{dE}{dt}dt \tag{P2.16.6}$$

1. Substituting the real parts of the electric field, E^*, and electric displacement, D^*, into Equation P2.15.6, obtain the following equations:

$$w_e = \pi\varepsilon''\varepsilon_0 E_0^2 = \pi\varepsilon'\varepsilon_0 E_0^2 \tan\delta \tag{P2.16.7}$$

2. Verify an alternative expression for the dissipation factor:

$$\tan\delta = (1/2\pi)(w_e/U_e), \tag{P2.16.8}$$

where U_e, the integrated area so labeled in Figure 2.47a, represents the energy stored during a quarter cycle.

Solution

The integrated area inside the hysteresis loop, labeled w_e in Figure 2.47a, is equivalent to the energy loss per cycle per unit volume of the dielectric. It is defined as $w_e = -\int DdE = -\int_0^{2\pi/\omega} D(dE/dt)\,dt$.

Substituting the real parts of the electric field, E^*, and electric displacement, D^*, into Equation P2.15.6 yields:

$$w_e = \int_0^{(2\pi/\omega)} D_0\cos(\omega t-\delta)[E_0\omega\cdot\sin(\omega t)]\,dt = E_0D_0\omega\sin(\delta)\cdot\int_0^{(2\pi/\omega)}\sin^2(\omega t)dt \\ = \pi E_0D_0\sin(\delta) \tag{P2.16.9}$$

so that

$$w_e = \pi\varepsilon''\varepsilon_0E_0^2 = \pi\varepsilon'\varepsilon_0E_0^2\tan\delta \tag{P2.16.10}$$

When there is a phase lag, an energy loss (or nonzero w_e) will occur for every cycle of the applied electric field, resulting in heat generation in the dielectric material. The quantity $\tan\delta$ is referred to as the *dissipation factor*. The electrostatic energy stored during a half cycle of the applied electric field is $2U_e$, where U_e, the integrated area so labeled in Figure 2.47a, represents the energy stored during a quarter cycle.

$$2U_e = 2\,[(1/2)(E_0D_0\cos\delta)] = (E_0D_0)\cos\delta \tag{P2.16.11}$$

Knowing that $\varepsilon'\varepsilon_0 = (D_0/E_0)\cos\delta$, Equation P2.15.11 may be rewritten in the form:

$$2U_e = \varepsilon'\varepsilon_0E_0^2 \tag{P2.16.12}$$

Then, an alternative expression for the dissipation factor can be obtained:

$$\tan\delta = (1/2\pi)(w_e/U_e) \tag{P2.16.13}$$

Note that the factor 2π comes from the integration process for one cycle.

So far, we have discussed the intensive dielectric, mechanical, and piezoelectric losses (with prime notation) in terms of intensive parameters X and E. In order to consider physical meanings of the losses in the material (e.g., domain dynamics), we will introduce extensive losses[29] in respect to extensive parameters x and D. In practice, intensive losses are easily measurable, but extensive losses are not in the pseudo-DC measurement. That is, it is experimentally difficult to keep a completely open circuit without leakage current for a long period or a perfect clamped condition of the sample. We usually obtain them from the intensive losses by using the *K-matrix* introduced later. Extensive losses are essential when we consider a physical microscopic or semimacroscopic domain dynamics model. We start again from the piezoelectric constitutive equations with losses in respect to extensive parameters x and D,

$$X = c^Dx - hD, \tag{2.126}$$

$$E = -hx + \kappa^x\kappa_0D, \tag{2.127}$$

where c^D is the elastic stiffness under the D = constant condition (i.e., electrically open circuit), κ^x is the inverse dielectric constant under x = constant condition (i.e., mechanically clamped), $\kappa_0 = 1/\varepsilon_0$, and h is the inverse piezoelectric constant d. We introduce the *extensive* dielectric, elastic, and piezoelectric losses as

$$\kappa^{x}* = \kappa^{x}(1 + j\tan\delta), \tag{2.128}$$

$$c^{D}* = c^{D}(1 + j\tan\phi), \tag{2.129}$$

$$h* = h(1 + j\tan\theta). \tag{2.130}$$

The sign + here in front of the imaginary j is taken by a general induction principle, that is, polarization induced after electric field application and strain induced after stress application. However, the sign of the coupling factor (piezoelectric) loss is not very trivial. Regarding the intensive loss, tan θ', the meaning is simple and may be positive; that is, the time delay of the strain induced by the field or the delay of the polarization induced by the applied stress, but what is the physical meaning of the extensive piezo-loss, tan θ? This may be related to the piezoelectric origin. We can consider two piezoelectric origin models: ferroelectricity is primary, coupled with elasticity, or ferroelasticity is primary, coupled with polarization (phase transition order parameter difference). At present, we presume that all loss factors are positive. It is notable that the permittivity under a constant strain (e.g., zero strain or completely clamped) condition, $\varepsilon^{x}*$, and the elastic compliance under a constant electric displacement (e.g., open-circuit) condition, $s^{D}*$, can be provided as an inverse value of κ^{x} and c^{D}, respectively, in this simplest one-dimensional expression. Thus, using exactly the same losses in Equation 2.128 and 2.129,

$$\varepsilon^{x}* = \varepsilon^{x}(1 - j\tan\delta), \tag{2.131}$$

$$s^{D}* = s^{D}(1 - j\tan\phi), \tag{2.132}$$

we will consider these phase delays again as extensive losses. Care should been taken in the case of a general 3D expression, where this part must be translated as inverse *matrix components* of $\kappa^{x}*$ and $c^{D}*$ tensors.

Here, we consider the physical property difference between the boundary conditions: E constant and D constant, or X constant and x constant in a simplest 1D model. Refer to Figure 2.20, top and bottom, again, and we derive the following relations:

$$\begin{aligned} &\varepsilon^{x}/\varepsilon^{X} = (1-k^2), s^{D}/s^{E} = (1-k^2), \\ &\kappa^{X}/\kappa^{x} = (1-k^2), c^{E}/c^{D} = (1-k^2), \\ &k^2 = \frac{d^2}{s^{E}\varepsilon^{X}\varepsilon_0} = \frac{h^2}{c^{D}(\kappa^{x}/\varepsilon_0)} \end{aligned} \tag{2.133}$$

This k is called the *electromechanical coupling factor*, which is defined as a real number in this textbook.

In order to obtain the relationships between the intensive and extensive losses, the following three equations (already derived in Section 2.4.6.3) are essential:

$$\varepsilon^{X}\varepsilon_0 = \frac{1}{\left(\frac{\kappa^{x}}{\varepsilon_0}\right)\left[1 - \frac{h^2}{c^{D}(\kappa^{x}/\varepsilon_0)}\right]} \tag{2.134}$$

$$s^{E} = \frac{1}{c^{D}\left[1 - \frac{h^2}{c^{D}(\kappa^{x}/\varepsilon_0)}\right]} \tag{2.135}$$

$$d = \frac{\dfrac{h^2}{c^D(\kappa^x/\varepsilon_0)}}{h\left[1 - \dfrac{h^2}{c^D(\kappa^x/\varepsilon_0)}\right]} \tag{2.136}$$

Replacing the parameters in Equations 2.134 through 2.136 with the complex parameters in Equations 2.115 through 2.117, 2.128 through 2.136, we obtain the relationships between the intensive and extensive losses:

$$\tan\delta' = (1/(1-k^2))[\tan\delta + k^2(\tan\phi - 2\tan\theta)], \tag{2.137}$$

$$\tan\phi' = (1/(1-k^2))[\tan\phi + k^2(\tan\delta - 2\tan\theta)], \tag{2.138}$$

$$\tan\theta' = (1/(1-k^2))[\tan\delta + \tan\phi - (1+k^2)\tan\theta)], \tag{2.139}$$

where k is the electromechanical coupling factor defined by Equation 2.133, and here is a real number. It is important that the intensive dielectric, elastic, and piezoelectric losses (with prime) are mutually correlated with the extensive dielectric, elastic, and piezoelectric losses (nonprime) through the electromechanical coupling k^2, that the denominator $(1 - k^2)$ comes basically from the ratios, $\varepsilon^x/\varepsilon^X = (1 - k^2)$ and $s^D/s^E = (1 - k^2)$, and this real part reflect to the dissipation factor when the imaginary part is divided by the real part. Knowing the relationships between the intensive and extensive physical parameters and the electromechanical coupling factor k, the intensive (prime) and extensive (nonprime) loss factors have the following relationship:[30]

$$\begin{bmatrix} \tan\delta' \\ \tan\phi' \\ \tan\theta' \end{bmatrix} = [K] \begin{bmatrix} \tan\delta \\ \tan\phi \\ \tan\theta \end{bmatrix}, \tag{2.140}$$

$$[K] = \frac{1}{1-k^2}\begin{bmatrix} 1 & k^2 & -2k^2 \\ k^2 & 1 & -2k^2 \\ 1 & 1 & -1-k^2 \end{bmatrix}, \qquad k^2 = \frac{d^2}{s^E(\varepsilon^X\varepsilon_0)} = \frac{h^2}{c^D(\kappa^x/\varepsilon_0)}. \tag{2.141}$$

The matrix $[K]$ is proven to be *invertible*, that is, $K^2 = I$, or $K = K^{-1}$, where I is the identity matrix. Hence, the conversion relationship between the intensive (prime) and extensive (nonprime) exhibits full symmetry. The author emphasizes again that the extensive losses are more important for considering physical micro/macroscopic models, and can be obtained mathematically from a set of intensive losses, which can be obtained more easily from experiments (in particular, pseudo-DC measurement).

In the case of a 3D expression, a PZT polycrystalline ceramic (∞mm) holds 10 tensor components, as given below [note $s_{66} = 2(s_{11} - s_{12})$ in the ∞ symmetry]:

$$\begin{bmatrix} s_{11}^* & s_{12}^* & s_{13}^* & 0 & 0 & 0 \\ s_{12}^* & s_{11}^* & s_{13}^* & 0 & 0 & 0 \\ s_{13}^* & s_{13}^* & s_{33}^* & 0 & 0 & 0 \\ 0 & 0 & 0 & s_{44}^* & 0 & 0 \\ 0 & 0 & 0 & 0 & s_{44}^* & 0 \\ 0 & 0 & 0 & 0 & 0 & s_{66}^* \end{bmatrix}, \begin{bmatrix} 0 & 0 & 0 & 0 & d_{15}^* & 0 \\ 0 & 0 & 0 & d_{15}^* & 0 & 0 \\ d_{31}^* & d_{31}^* & d_{33}^* & 0 & 0 & 0 \end{bmatrix}, \begin{bmatrix} \varepsilon_{11}^* & 0 & 0 \\ 0 & \varepsilon_{11}^* & 0 \\ 0 & 0 & \varepsilon_{33}^* \end{bmatrix},$$

where * means a complex parameter, leading to 20 loss tensor components, taking into account both intensive and extensive losses. However, only 10 components are independent, because of the relationship between the intensive and extensive losses given by Equations 2.140 and 2.141.

2.7.2 Loss Measurement Technique—Pseudostatic Method

We can determine intensive dissipation factors, tan δ', tan ϕ', and tan θ', separately from (a) D vs. E (stress free), (b) x vs. X (short-circuit), (c) x vs. E (stress free), and (d) D vs. X (short-circuit) curves (see Figure 2.47). Using a stress-applying jig shown in Figure 2.48, Zheng et al. measured the x vs. X and D vs. X relationships.[31] Figure 2.49 summarizes intensive loss factors, tan δ', tan ϕ', and tan θ' as a function of electric field or compressive stress, measured for a "soft" PZT-based multilayer actuator. Note first that the piezoelectric loss tan θ' is not negligibly small, as believed by previous researchers, but rather large, comparable to the dielectric and elastic losses; tan θ' (=0.08) > (1/2) [tan δ' (=0.06) + tan ϕ' (=0.08)].

Then, using Equations 2.140 and 2.141, we calculated the extensive losses as shown in Figure 2.50. Note again that the magnitude of the piezoelectric loss tanθ is comparable to the dielectric and elastic losses, and increases gradually with the field or stress; now tan θ (=0.05) < (1/2)[tan δ (=0.05) + tan ϕ (=0.07)]. Also, it is noteworthy that the extensive dielectric loss tanδ increases significantly with an increase of the intensive parameter, that is, the applied electric field, while the extensive elastic loss tanϕ is rather insensitive to the intensive parameter, that is, the applied compressive stress. When similar measurements are conducted under constrained conditions; that is, D vs. E under a completely clamped state and x vs. X under an open-circuit state, respectively, we can expect smaller hystereses, that is, extensive losses, tanδ and tanϕ, theoretically. However, they are rather difficult in practice because of the migrating charge compensation on the specimen surface.

2.7.3 Heat Generation at Off-Resonance

Zheng et al. reported heat generation at an off-resonance frequency from various sizes of multilayer-type piezoelectric ceramic (soft PZT) actuators.[31] The temperature change with time in the actuators was monitored when driven at 3 kV/mm (high electric field) and 300 Hz (low frequency) (Figure 2.51a), and Figure 2.51b plots the saturated temperature as a function of V_e/A, where V_e is the effective volume (electrode overlapped part) and A is the surface area. Suppose that the temperature was uniformly generated in a bulk sample (no significant stress distribution); this linear relation is reasonable because the volume V_e generates the heat and this heat is dissipated through the area A. Thus, if we need to suppress the temperature rise, a small V_e/A design is preferred. According to

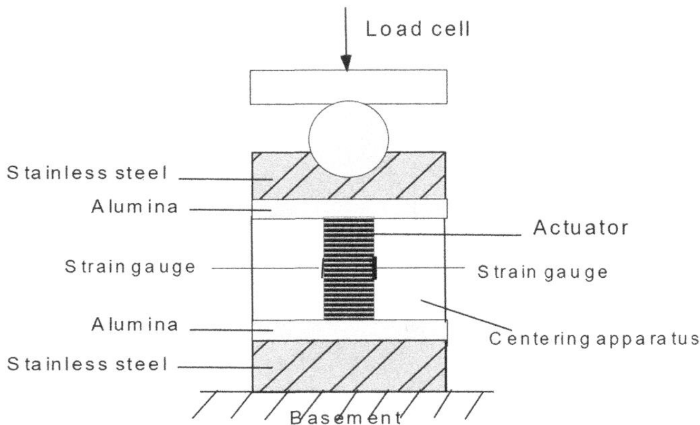

FIGURE 2.48 Stress application jig on a piezoelectric multilayer.

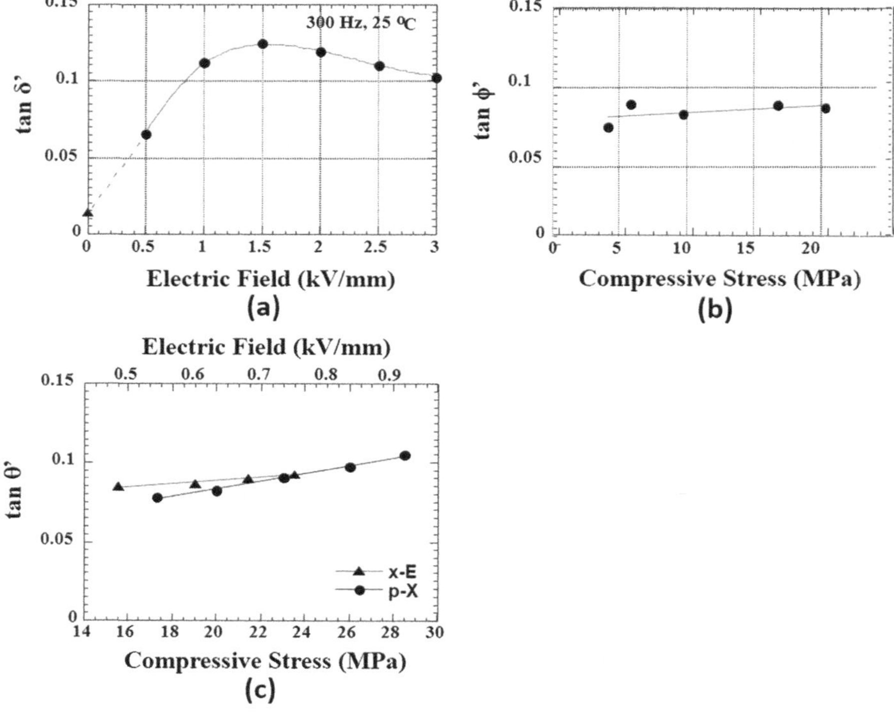

FIGURE 2.49 Intensive loss factors, *tan* δ' (a), *tan* ϕ' (b), and *tan* θ' (c) as a function of electric field or compressive stress, measured for a soft PZT actuator.

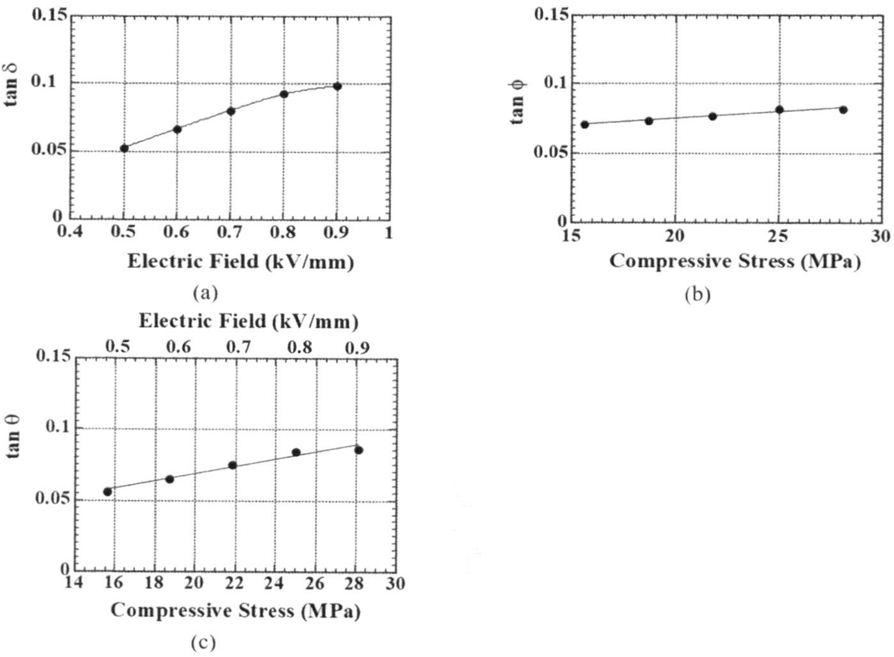

FIGURE 2.50 Extensive loss factors, *tan* δ (a), *tan* ϕ (b), and *tan* θ (c) as a function of electric field or compressive stress, for a soft PZT actuator.

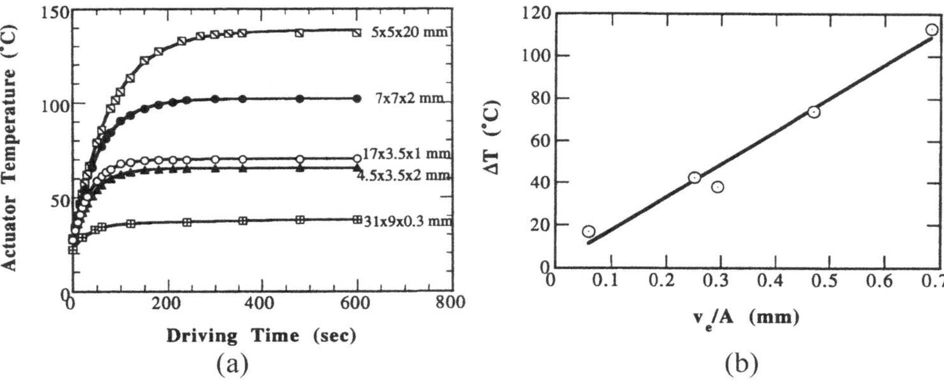

FIGURE 2.51 (a) Temperature change with driving time for PZT ML actuators with various sizes. (b) Temperature rise vs. V_e/A in ML actuators.

the law of energy conservation, the amount of heat stored in the piezoelectric, which is just the difference between the rate at which heat is generated, q_g, and that at which the heat is dissipated, q_d, can be expressed as

$$q_g - q_d = V\rho\, C(dT/dt), \tag{2.142}$$

where it is assumed a uniform temperature distribution exists throughout the sample and V is the total volume, ρ is the mass density, and C is the specific heat of the specimen. The heat generation in the piezoelectric is attributed to losses. Thus, the rate of heat generation, q_g, can be expressed as:

$$q_g = wfV_e, \tag{2.143}$$

where w is the loss per driving cycle per unit volume, f is the driving frequency, and V_e is the effective volume of active ceramic (no-electrode parts are omitted). According to the measurement conditions (off-resonance, no significant stress in the sample), this w corresponds to the dielectric hysteresis loss (i.e., P-E hysteresis), w_e, which is expressed in Equation 2.119 in terms of the intensive dielectric loss tan δ' as:

$$w = w_e = \pi\varepsilon^X\varepsilon_0 E_0^2 \tan\delta'.$$

If we neglect the transfer of heat through conduction, the rate of heat dissipation (q_d) from the sample is the sum of the heat flow by radiation (q_r) and convection (q_c):

$$q_d = q_r + q_c = eA\sigma\left(T^4 - T_0^4\right) + h_c A(T - T_0), \tag{2.144}$$

where e is the emissivity of the sample, A is the sample surface area, σ is the Stefan-Boltzmann constant, T_0 is the initial sample temperature, and h_c is the average convective heat transfer coefficient. Thus, Equation 2.142 can be written in the form:

$$wf\,V_e - A\,k(T)(T - T_o) = V\rho C(dT/dt), \tag{2.145}$$

where the quantity

$$k(T) = \sigma e\left(T^2 + T_0^2\right)(T + T_0) + h_c \tag{2.146}$$

is the overall heat transfer coefficient. If we assume that $k(T)$ is relatively insensitive to temperature change (if the temperature rise is not large), solving Equation 2.145 for the rise in temperature of the piezoelectric sample yields:

$$T - T_o = [wf\ V_e/k(T)A](1 - e^{-t/\tau}), \tag{2.147}$$

where the time constant τ is expressed as:

$$\tau = \rho\ CV/k(T)A. \tag{2.148}$$

As $t \rightarrow \infty$, the maximum temperature rise in the sample becomes:

$$\Delta T = wf\ V_e/k(T)A, \tag{2.149}$$

while, as $t \rightarrow 0$, the initial rate of temperature rise is given by

$$dT/dt = [w_T f\ V_e/\rho\ CV) = \Delta T/\tau, \tag{2.150}$$

where w_T can be regarded under these conditions as a measure of the total loss of the piezoelectric. The dependences of $k(T)$ on applied electric field and frequency are shown in Figures 2.52a and b, respectively. Note that $k(t)$ is almost constant, as long as the driving voltage or frequency is not very high ($E < 1$ kv/mm, $f < 0.3$ kHz). The total loss, w_T, as calculated from Equation 2.150 is given for three multilayer specimens in Table 2.5. The experimentally determined P-E hysteresis losses measured under stress-free conditions (Equation 2.119) are also listed in the table for comparison. It is seen that the extrinsic P-E hysteresis loss (i.e., dielectric loss) agrees well with the calculated total loss associated with the heat generated in the driven piezoelectric under off-resonance with high electric field.[31,32]

2.7.4 Microscopic Origins of Extensive Losses

So far, we have stuck to the phenomenological discussions of losses. In this section, we consider microscopic or crystallographic origins of loss mechanisms in piezoelectrics from the materials science viewpoint. Losses are considered to consist of four portions: (1) domain wall motion; (2) fundamental lattice portion, which should also occur in domain-free monocrystals; (3) microstructure

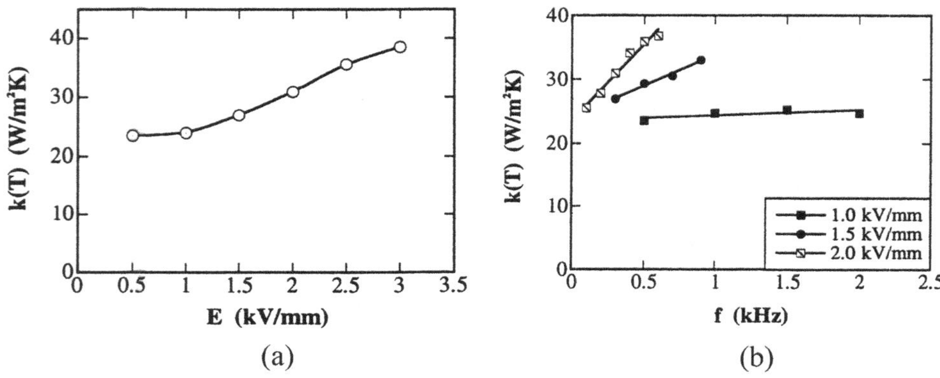

FIGURE 2.52 Overall heat transfer coefficient, $k(T)$, plotted as a function of applied electric field (a), and of frequency (b) for a PZT ML actuator with dimensions of $7 \times 7 \times 2$ mm^3 driven at 400 Hz.

TABLE 2.5
Loss and overall heat transfer coefficient for PZT multilayer samples ($E = 3$ kV/mm, $f = 300$ Hz)

Actuator	$4.5 \times 3.5 \times 2.0$ mm^3	$7.0 \times 7.0 \times 2.0$ mm^3	$17 \times 3.5 \times 1.0$ mm^3
$w_T\,(\text{kJ/m}^3) = \dfrac{\rho c v}{f v_e}\left(\dfrac{dT}{dt}\right)_{t->0}$	19.2	19.9	19.7
P-E hysteresis loss (kJ/m^3)	18.5	17.8	17.4
$k(T)$ (W/m^2 K)	38.4	39.2	34.1

Source: J. Zheng et al.: *J. Amer. Ceram. Soc.*, 79, 3193–3198, 1996.

portion, which occurs typically in polycrystalline samples; and (4) conductivity portion in highly ohmic samples. However, in the typical piezoelectric ceramic case, the loss due to the domain wall motion exceeds the other three contributions significantly.

We discuss here the relationship of the loss phenomenology with microscopic origins. To make the situation simplest, we consider here only domain wall motion-related losses. Taking into account the fact that the polarization change is primarily attributed to 180° domain wall motion, while the strain is attributed to 90° (or non-180°) domain wall motion, we suppose that the extensive dielectric and mechanical losses originate from 180° and 90° domain wall motions, respectively, as illustrated in Figure 2.53. The dielectric loss comes from the hysteresis during the 180° polarization reversal under E, while the elastic loss comes from the hysteresis during the 90° polarization reorientation under X. Regarding the piezoelectric loss, we presume that it originated from Gauss's Law, div $(D) = \sigma$ (charge). As illustrated in Figure 2.53, when we apply a tensile stress on the piezoelectric crystal, the vertically elongated cells will transform into horizontally elongated cells, so that the 90° domain wall will move rightward. However, this ferroelastic domain wall will not generate charges because the rightward and leftward

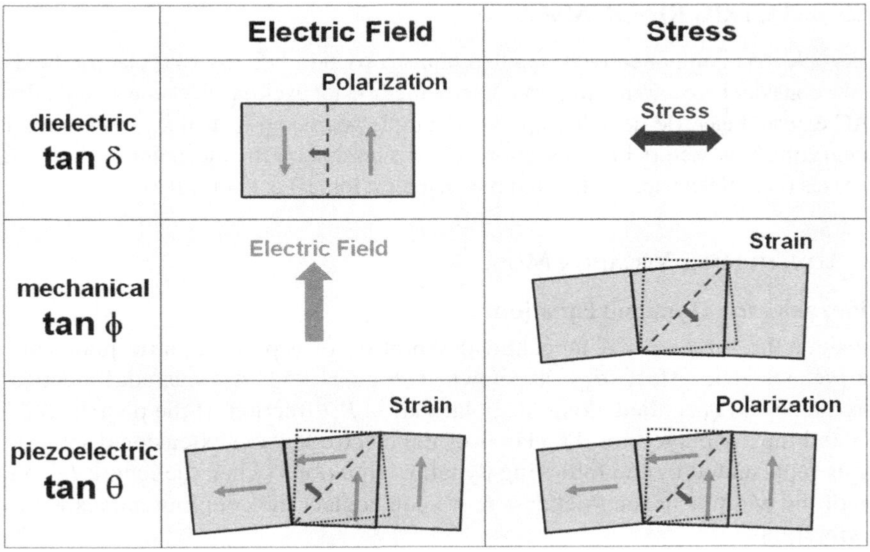

	Electric Field	Stress
dielectric tan δ	Polarization	Stress
mechanical tan ϕ	Electric Field	Strain
piezoelectric tan θ	Strain	Polarization

FIGURE 2.53 Polarization reversal/reorientation model for explaining extensive dielectric, elastic, and piezoelectric losses.

polarizations may compensate each other. Without having migrating charges σ in this crystal, div $(D) = 0$, leading to the polarization alignment head-to-tail, rather than head-to-head or tail-to-tail. After the ferroelastic transformation, this polarization alignment will need an additional time lag, which we define as the piezoelectric loss. Superposing the ferroelastic domain alignment and polarization alignment (via Gauss's Law) can generate actual charge under stress applications. In this model, the intensive (observable) piezoelectric loss is explained by the 90° polarization reorientation under E, which can be realized by superimposing the 90° polarization reorientation under X and the 180° polarization reversal under E. This is the primary reason we cannot measure the piezoelectric loss independently from the elastic or dielectric losses experimentally. Refer to Equation 2.125: $w_{me} = \pi(d^2/\varepsilon_0\varepsilon^X)X_0^2(2\tan\theta' - \tan\delta')$, the measure hysteresis area under stress cycle is a coupling between the piezoelectric and dielectric losses.

If we adopt the Uchida-Ikeda polarization reversal/reorientation model introduced in Figure 2.44,[25] we can explain the loss change with intensive parameters (externally controllable parameters). By finding the polarization, P, and the field-induced strain, x, as a function of the electric field, E, it is possible to estimate the volume in which 180° reversal or 90° rotation occurred. This is because the 180° domain reversal does not contribute to the induced strain; only the 90° rotation does, whereas the 180° domain reversal contributes mainly to polarization. The volume change of the domains with an external electric field is shown schematically in Figure 2.44. It can be seen that with the application of an electric field, 180° reversal occurs rapidly at a certain electric field, whereas 90° rotation occurs gradually from a low field. It is notable that at G in the figure, there remains some polarization while the induced strain is zero; at H the polarizations from the 180° and 90° reorientations cancel each other and become zero, but the strain is not at its minimum. Due to a sudden change in the 180° reversal above a certain electric field, we can expect a sudden increase in the polarization hysteresis and the dielectric loss (this may reflect to the extensive dielectric loss measurement in Figure 2.50a); while the slope of 90° reorientation is almost constant, we can expect a constant extensive elastic loss with changing the external parameter, E or X (extensive elastic loss in Figure 2.50b). This dramatic increase in 180° domain wall motion (i.e., extensive dielectric loss) seems to be the origin of the apparent Q_m degradation and heat generation above the maximum vibration velocity, which is discussed in Section 3.5.2.

2.8 PIEZOELECTRIC RESONANCE

So far, a piezoelectric component was handled under a pseudo-DC external electric field or stress condition. We consider here another important drive method for exciting electromechanical resonance under an AC external electric field. Two popular sample geometry k_{31} and k_{33} modes are discussed without integrating loss factors in this section. The discussion on the mechanical quality factor Q_m and three losses (i.e., dielectric, elastic, and piezoelectric losses) is not until Section 3.5.

2.8.1 k_{31} Longitudinal Vibration Mode

2.8.1.1 Piezoelectric Dynamic Equation

Let us consider the longitudinal mechanical vibration of a piezoceramic plate through the transverse piezoelectric effect (d_{31}), as shown in Figure 2.54. A sinusoidal electric field E_z (angular frequency ω) is applied along the polarization P_z direction. If the polarization is in the z-direction and the x-y planes are the planes of the electrodes, the extensional vibration in the x direction is represented by the following dynamic equations (when the length L is more than 4~6 times of the width w or the thickness b, we can neglect the coupling modes with width or thickness vibrations):

$$\rho(\partial^2 u/\partial t^2) = F = (\partial X_{11}/\partial x) + (\partial X_{12}/\partial y) + (\partial X_{13}/\partial z), \qquad (2.151)$$

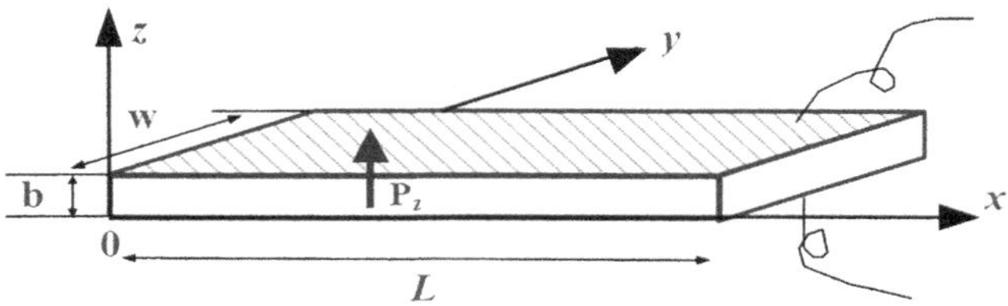

FIGURE 2.54 Longitudinal vibration through the transverse piezoelectric effect (d_{31}) in a rectangular plate ($L \gg w \gg b$).

where ρ is the density of the piezoceramic, and u is the displacement of a small volume element in the ceramic plate in the x-direction.

We integrate the piezoelectric constitutive equations (1D expression): $D = \varepsilon_0\varepsilon^X E + dX$ and $x = d E + s^E X$, where electric displacement, D, and strain, x, are controlled by the intensive parameters, electric field, E, and stress, X. The relations between stress, electric field (only E_z exists), and the induced strain in 3D expression are given by:

$$\begin{aligned} x_1 &= s_{11}^E X_1 + s_{12}^E X_2 + s_{13}^E X_3 + d_{31}E_3, \\ x_2 &= s_{12}^E X_1 + s_{11}^E X_2 + s_{13}^E X_3 + d_{31}E_3, \\ x_3 &= s_{13}^E X_1 + s_{13}^E X_2 + s_{33}^E X_3 + d_{33}E_3, \\ x_4 &= s_{44}^E X_4, \\ x_5 &= s_{44}^E X_5, \\ x_6 &= 2\left(s_{11}^E - s_{12}^E\right)X_6. \end{aligned} \tag{2.152}$$

Note $s_{66} = 2(s_{11} - s_{12})$ in the ∞mm symmetry like random ceramics (Section 2.7.1).

When the plate is very long and thin, X_2 and X_3 may be set equal to zero through the plate. Since shear stress will not be generated by the electric field E_z ($= E_3$), Equation 2.152 is reduced to only one equation:

$$\begin{aligned} x_1 &= s_{11}^E X_1 + d_{31}E_3, \text{ or} \\ X_1 &= x_1/s_{11}^E - \left(d_{31}/s_{11}^E\right)E_z. \end{aligned} \tag{2.153}$$

Introducing Equation 2.153 into Equation 2.151 and allowing for strain definition $x_1 = \partial u/\partial x$ (nonsuffix x corresponds to the Cartesian coordinate, and x_1 is the strain along the 1 [x] direction) and $\partial E_z/\partial x = 0$ (due to the equal potential on each electrode) leads to a harmonic vibration equation:

$$\begin{aligned} \rho(\partial^2 u/\partial t^2) &= \left(1/s_{11}^E\right)(\partial x_1/\partial x), \\ -\omega^2 \rho s_{11}^E u &= \partial^2 u/\partial x^2. \end{aligned} \tag{2.154}$$

Here, ω is the angular frequency of the sinusoidal drive field, E_z, and the displacement, u. Supposing the displacement u also vibrates with the frequency of ω, a general solution $u = u_1(x)e^{j\omega t} + u_2(x)e^{-j\omega t}$ is substituted into Equation 2.154, and with the boundary condition $X_1 = 0$ at $x = 0$ and L (sample length) (due to the mechanically free condition at the plate end), the following solution can be obtained (refer to Example Problem 2.16):

$$\textbf{(strain)}\ \partial u/\partial x = x_1 = d_{31}E_z[\sin\omega(L-x)/v + \sin(\omega x/v)]/\sin(\omega L/v)$$
$$= d_{31}E_z\left(\frac{\cos\left[\frac{\omega(L-2x)}{2v}\right]}{\cos\left(\frac{\omega L}{2v}\right)}\right) \tag{2.155}$$

$$\textbf{(total displacement)}\ \Delta L = \int_0^L x_1 dx = d_{31}E_z L(2v/\omega L)\tan(\omega L/2v). \tag{2.156}$$

Here, v is the *sound velocity* in the piezoceramic, which is given by

$$v = 1/\sqrt{\rho s_{11}^E}. \tag{2.157}$$

The strain distribution in Equation 2.155 is symmetrically sinusoidal with respect to the $x = L/2$ position, and the maximum strain (i.e., nodal line) exists on this line. Note that $\omega \to 0$ (i.e., pseudo-DC) makes Equation 2.155 $x_1 = d_{31}E_z$, that is, uniform strain distribution on the piezoplate.

EXAMPLE PROBLEM 2.17

Let us consider a piezoceramic plate that vibrates via the transverse piezoelectric effect (d_{31}). Substituting a general solution $u = u_1(x)e^{j\omega t} + u_2(x)e^{-j\omega t}$ into

$$-\omega^2 \rho s_{11}^E u = \partial^2 u/\partial x^2, \tag{P2.17.1}$$

then with the boundary condition $X_1 = 0$ at $x = 0$ and L, derive the strain

$$x_1 = d_{31}E_z\left(\frac{\cos\left[\frac{\omega(L-2x)}{2v}\right]}{\cos\left(\frac{\omega L}{2v}\right)}\right). \quad \left(v = 1/\sqrt{\rho s_{11}^E}\right) \tag{P2.17.2}$$

Solution

Substituting a general solution $u = u_1(x)e^{j\omega t} + u_2(x)e^{-j\omega t}$ into $-\omega^2 \rho s_{11}^E u = \partial^2 u/\partial x^2$, we obtain

$$-\left(\frac{\omega^2}{v^2}\right)(u_1(x)e^{j\omega t} + u_2(x)e^{-j\omega t}) = \left(\frac{\partial^2 u_1}{\partial x^2}\right)e^{j\omega t} + \left(\frac{\partial^2 u_2}{\partial x^2}\right)e^{-j\omega t} \tag{P2.17.3}$$

Since this equation should be satisfied for any time, u_1 and u_2 should satisfy

$$\left(\frac{\partial^2 u_1}{\partial x^2}\right) = -\left(\frac{\omega^2}{v^2}\right)u_1 \text{ and } \left(\frac{\partial^2 u_2}{\partial x^2}\right) = -\left(\frac{\omega^2}{v^2}\right)u_2,$$

respectively. If we consider only the standing wave as a solution, $u(x)$ should be a "real" parameter, leading to the relation $u_1(x) = u_2(x)$ and $u(x, t) = 2u_1(x)\cos(\omega t)$. We will neglect $\cos(\omega t)$ hereafter because this is included in $\cos(\omega t)$ of the electric field. Thus, we suppose a general solution:

$$u_1 = u_2 = A_1\cos\left(\frac{\omega}{v}x\right) + A_2\sin\left(\frac{\omega}{v}x\right) \tag{P2.17.4}$$

Now, the strain distribution on the plate can be calculated as

$$\begin{aligned} x_1 &= \partial u/\partial x = 2\left(\frac{\partial u_1}{\partial x}\right) \\ &= 2\left(\frac{\omega}{v}\right)\left[-A_1 \sin\left(\frac{\omega}{v}x\right) + A_2 \cos\left(\frac{\omega}{v}x\right)\right] \end{aligned} \quad \text{(P2.17.5)}$$

Let us consider the boundary condition. From $X_1 = 0 = x_1/s_{11}^E - \left(d_{31}/s_{11}^E\right)E_z$ at $x = 0$ and L, $x_1 = d_{31}E_z$ (i.e., $E_z = E_{max}\cos(\omega t)$) is obtained at both plate edges (without considering the time lag or loss, the strain response should be simultaneous with the electric field):

$$2\left(\frac{\omega}{v}\right)\left[-A_1 \sin\left(\frac{\omega}{v}0\right) + A_2 \cos\left(\frac{\omega}{v}0\right)\right] = 2\left(\frac{\omega}{v}\right)A_2 = d_{31}E_z$$

$$2\left(\frac{\omega}{v}\right)\left[-A_1 \sin\left(\frac{\omega}{v}L\right) + A_2 \cos\left(\frac{\omega}{v}L\right)\right] = d_{31}E_z$$

Thus, we obtain

$$A_1 = \left(\frac{1}{2}\right)\left(\frac{v}{\omega}\right)d_{31}E_z \frac{\left[\cos\left(\frac{\omega L}{v}\right) - 1\right]}{\sin\left(\frac{\omega L}{v}\right)}, \text{ and } A_2 = \left(\frac{1}{2}\right)\left(\frac{v}{\omega}\right)d_{31}E_z$$

Finally, inserting A_1 and A_2 into Equation P2.17.5, we derive

$$\begin{aligned} x_1 &= d_{31}E_z[\sin\omega(L-x)/v + \sin(\omega x/v)]/\sin(\omega L/v) \\ &= d_{31}E_z \left(\frac{\cos\left[\frac{\omega(L-2x)}{2v}\right]}{\cos\left(\frac{\omega L}{2v}\right)} \right). \end{aligned}$$

Remember that E_z is an AC field with the frequency ω. With increasing ω, the stress concentration at the nodal line ($x = L/2$) will be enhanced. The strain distribution on a rectangular plate is illustrated in Figure 2.55 for the resonance and antiresonance frequencies; cosine shape with respect to the plate center ($x = L/2$) and the amplitude depends on the drive frequency around the resonance frequency (∞ amplitude at the resonance, $\cos(\omega L/2v) = 0$).

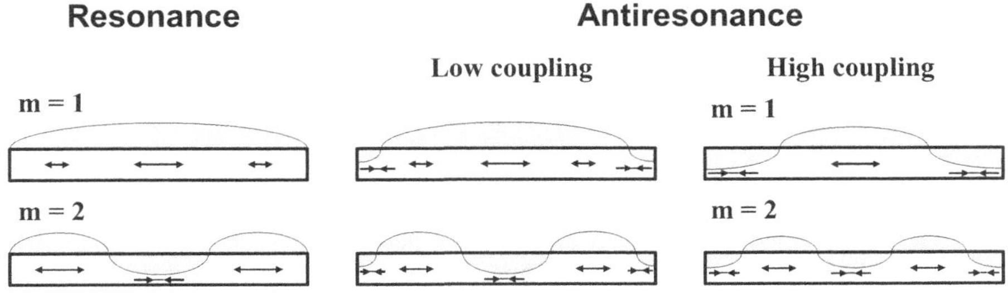

FIGURE 2.55 Strain distribution in the resonant or antiresonant state for a k_{31}-type piezoelectric plate.

2.8.1.2 Admittance around Resonance and Antiresonance

When the specimen is utilized as an electrical component such as a filter or a vibrator, the electrical impedance (applied voltage/induced current ratio) plays an important role. Now we use the first piezoelectric constitutive equation $D = \varepsilon_0 \varepsilon^X E + d X$. The current flow into the specimen is described by the surface charge increment, $\partial D_3/\partial t$. Though the electric field is uniform in the sample due to the surface electrode, the electric displacement, D, is not uniform because of the stress, X, distribution, maximum at the nodal line ($x = L/2$). The total current is given by:

$$\begin{aligned} i &= j\omega w \int_0^L D_3\, dx = j\omega w \int_0^L \left(d_{31} X_1 + \varepsilon_0 \varepsilon_{33}^X E_z\right) dx \\ &= j\omega w \int_0^L \left[d_{31} \left\{ x_1 / s_{11}^E - \left(d_{31}/s_{11}^E\right) E_z \right\} + \varepsilon_0 \varepsilon_{33}^X E_z \right] dx. \end{aligned} \tag{2.158}$$

w is the plate width. Using strain distribution in Equation 2.155, the admittance for the mechanically free sample is calculated to be:

$$\begin{aligned} (1/Z) &= (i/V) = (I/E_z \cdot t) \\ &= (j\omega wL/E_z t) \int_0^L \left[\left(d_{31}^2/s_{11}^E\right) \left(\frac{\cos\left[\frac{\omega(L-2x)}{2v}\right]}{\cos\left(\frac{\omega L}{2v}\right)} \right) E_z + \left[\varepsilon_0 \varepsilon_{33}^X - \left(d_{31}^2/s_{11}^E\right)\right] E_z \right] dx \\ &= (j\omega wL/t)\varepsilon_0 \varepsilon_{33}^{LC} [1 + \left(d_{31}^2/\varepsilon_0 \varepsilon_{33}^{LC} s_{11}^E\right)(\tan(\omega L/2v)/(\omega L/2v)], \end{aligned} \tag{2.159}$$

where w is the width, L the length, t the thickness of the rectangular piezosample, and V the applied voltage ($=E_z{\cdot}t$). ε_{33}^{LC} is the permittivity in a longitudinally clamped (LC) sample, which is given by

$$\begin{aligned} \varepsilon_0 \varepsilon_{33}^{LC} &= \varepsilon_0 \varepsilon_{33}^X - \left(d_{31}^2/s_{11}^E\right). \\ &= \varepsilon_0 \varepsilon_{33}^X \left(1 - k_{31}^2\right). \qquad \left(k_{31}^2 = d_{31}^2/\varepsilon_0 \varepsilon_{33}^X s_{11}^E\right) \end{aligned} \tag{2.160}$$

Note here that this ε_{33}^{LC} is different from the extensive permittivity ε_{33}^x introduced in Section 2.4.6.3, precisely speaking. ε_{33}^{LC} is the permittivity in the sample mechanically clamped only along the x (or 1, length) direction, free along the z (or 3, polarization direction) or y directions, while ε_{33}^x means the permittivity clamped completely in the three directions. The first term $(j\omega wL/t)\varepsilon_0\varepsilon_{33}^{LC}$ of admittance Equation 2.159 is the *damped capacitance* (longitudinally clamped), and the second term is characterized by tan(ωL/2v), which changes from 0, though $+\infty$, to $-\infty$ with ω, originating from the capacitance change with the mechanical vibration (i.e., *motional capacitance*).

Piezoelectric resonance is achieved where the admittance becomes infinite or the impedance is zero. The resonance frequency f_R is calculated from Equation 2.159 (by putting $\omega L/2v = \pi/2$ for infinite admittance $\tan(\omega L/2v) = \infty$), and the fundamental frequency is given by

$$f_R = \omega_R/2\pi = v/2L = 1/\left(2L\sqrt{\rho s_{11}^E}\right). \tag{2.161}$$

On the other hand, the antiresonance state is generated for zero admittance or infinite impedance:

$$(\omega_A L/2v)\cot(\omega_A L/2v) = -d_{31}^2/\varepsilon_{33}^{LC} s_{11}^E = -k_{31}^2/\left(1 - k_{31}^2\right). \tag{2.162}$$

The final transformation is provided by the definition,

$$k_{31} = d_{31}/\sqrt{s_{11}^E \cdot \varepsilon_{33}^X}. \tag{2.163}$$

2.8.1.3 Resonance and Antiresonance Vibration Modes

The resonance and antiresonance states are both mechanical resonance states with amplified strain/displacement states, but they are very different from the driving viewpoints. The mode difference is described by the following intuitive model. In a high electromechanical coupling material with k almost equal to 1, the resonance or antiresonance states appear for $\tan(\omega L/2v) = \infty$ or 0 [i.e., $\omega L/2v = (m - 1/2)\pi$ or $m\pi$ (m: integer)], respectively. The strain amplitude x_1 distribution for each state (calculated using Equation 2.155) is illustrated in Figure 2.55. In the resonance state, the strain distribution is basically sinusoidal with the maximum at the center of the plate ($x = L/2$) (see the numerator). When ω is close to ω_R, $(\omega_R L/2v) = \pi/2$, leading to the denominator $\cos(\omega_R L/2v) \to 0$. Significant strain magnification is obtained. It is worth noting that the stress, X_1, is zero at the plate ends ($x = 0$ and L), but the strain, x_1, is not zero, but is equal to $d_{31}E_Z$. According to this large strain amplitude, large capacitance changes (called *motional capacitance*) are induced, and under a constant applied voltage, the current can easily flow into the device (i.e., admittance, Y, is infinite). On the contrary, at the antiresonance, the strain induced in the device compensates completely (because extension and compression are compensated), resulting in no motional capacitance change, and the current cannot flow easily into the sample (i.e., admittance, Y, is zero). Thus, for a high-k material, the first antiresonance frequency, f_A, should be twice as large as the first resonance frequency, f_R.

It is notable that both resonance and antiresonance states are in the mechanical resonance, which can create large strain in the sample under minimum input electrical energy. When we use a constant voltage supply, the specimen vibration is excited only at the resonance mode, because the electrical power is very small at the antiresonance mode. This provides a common misconception among junior engineers that "antiresonance is not a mechanical resonance." In contrast, when we use a constant current supply, the vibration is excited only at the antiresonance, instead. The stress, X_1, at the plate ends ($x = 0$ and L) is supposed to be zero in both cases. However, though the strain, x_1, at the plate ends is zero/very small (precisely, $d_{31}E_z$, because of low voltage and high current drive) for the resonance, the strain, x_1, is not zero (actually the maximum) for the antiresonance (because of high voltage and low current drive). This means that there is only one vibration node at the plate center for the resonance (top left in Figure 2.55), and there are an additional two nodes at both plate ends for the first antiresonance (top right in Figure 2.55). The reason is from the antiresonance drive, i.e., high voltage/low current (minimum power) drive due to the high impedance. The converse piezo-effect strain under E directly via d_{31} (uniform strain in the sample) superposes on the mechanical resonance strain distribution (distributed strain with nodes in the sample), two strains of which have exactly the same level theoretically at the antiresonance for $k_{31} \approx 1$.

In a typical case, where $k_{31} = 0.3$, the antiresonance state varies from the previously mentioned mode and becomes closer to the resonance mode (top-center in Figure 2.55). The low-coupling material exhibits an antiresonance mode where the capacitance change due to the size change (*motional capacitance*) is compensated completely by the current required to charge up the static capacitance (called *damped capacitance*). Thus, the antiresonance frequency, f_A, will approach the resonance frequency, f_R.

2.8.2 k_{33} Longitudinal Vibration Mode

2.8.2.1 Piezoelectric Dynamic Equation

Let us consider now the longitudinal vibration k_{33} mode in comparison with the k_{31} mode. When the resonator is long in the z direction and the electrodes are deposited on each end of the rod, as shown in Figure 2.56, the following conditions are satisfied:

$$X_1 = X_2 = X_4 = X_5 = X_6 = 0 \text{ and } X_3 \neq 0.$$

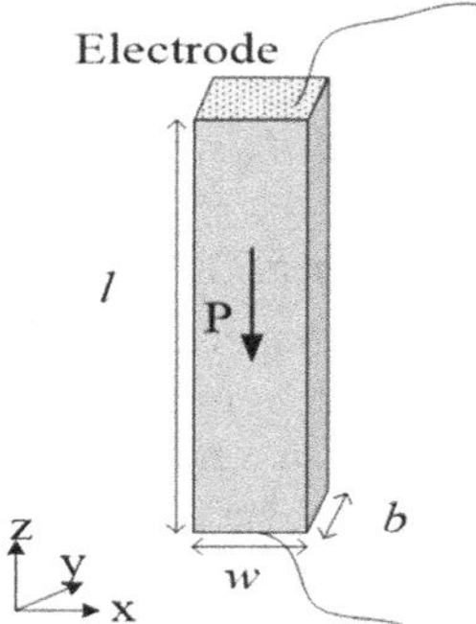

FIGURE 2.56 Longitudinal vibration through the piezoelectric effect (d_{33}) in a rod ($L \gg w \approx b$).

Thus, the constitutive equations are

$$X_3 = (x_3 - d_{33}E_z)/s_3^E \tag{2.164}$$

$$D_3 = \varepsilon_0\varepsilon_{33}^X E_z + d_{33}X_3 \tag{2.165}$$

for this configuration. Assuming a local displacement u in the z direction, from Equation 2.164 the dynamic equation is given:

$$\rho\frac{\partial^2 u}{\partial t^2} = \frac{1}{s_{33}^E}\left[\frac{\partial^2 u}{\partial z^2} - d_{33}\frac{\partial E_z}{\partial z}\right] \tag{2.166}$$

The electrical condition for the longitudinal vibration is *not* $\partial E_z/\partial z = 0$, but rather $\partial D_z/\partial z = 0$. From Equation 2.165

$$\begin{aligned} &\varepsilon_0\varepsilon_{33}^X\frac{\partial E_z}{\partial z} + \frac{d_{33}}{s_{33}^E}\left[\left(\frac{\partial^2 u}{\partial z^2}\right) - d_{33}\left(\frac{\partial E_z}{\partial z}\right)\right] = 0, \text{ or} \\ &\varepsilon_0\varepsilon_{33}^X\left(1 - k_{33}^2\right)\left(\frac{\partial E_Z}{\partial z}\right) = -\frac{d_{33}}{s_{33}^E}\left(\frac{\partial^2 u}{\partial z^2}\right) \end{aligned} \tag{2.167}$$

Thus,

$$\rho\frac{\partial^2 u}{\partial t^2} = \frac{1}{s_{33}^D}\frac{\partial^2 u}{\partial z^2} \quad \left(s_{33}^D = \left(1 - k_{33}^3\right)s_{33}^E\right) \tag{2.168}$$

Compared with the sound velocity ($v = 1/\sqrt{\rho s_{11}^E}$) in Equation 2.157 in the surface electroded (E-constant) sample along the vibration direction, the nonelectroded k_{33} (D-constant) sample exhibits $v = 1/\sqrt{\rho s_{33}^D}$, which is faster (elastically stiffened) than that in the E constant condition. In comparison with the resonance/antiresonance strain distribution status in the k_{31} mode in Figure 2.55, Figure 2.57 illustrates the strain distribution status in the k_{33} mode. Because the k_{31} and k_{33} modes possess E-constant and D-constant constraints, respectively, in k_{31}, the resonance frequency is directly related to v_{11}^E or s_{11}^E, while in k_{33}, the antiresonance frequency is directly related to v_{33}^D or s_{33}^D, c_{33}^D. The antiresonance in k_{31} and the resonance in k_{33} are subsidiary, originating from the electromechanical coupling factors. It is also worth noting that with increasing the k value toward 1, the ratio f_A/f_R

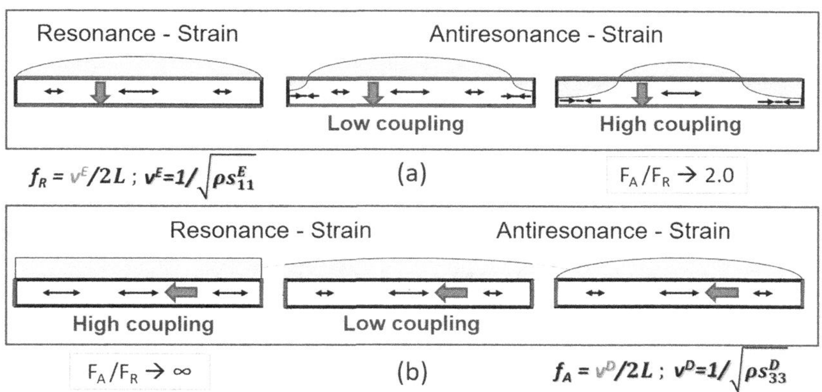

FIGURE 2.57 Strain distribution in the resonance and antiresonance states. Longitudinal vibration through the transverse d_{31} (a) and longitudinal d_{33} (b) piezo-effect in a rectangular plate.

approaches 2 in k_{31}, while it can reach ∞ in k_{33}, and that the strain distribution becomes almost flat or uniform in k_{33}, though the stress distributes sinusoidally with zero at the plate ends.

2.8.2.2 Boundary Condition: *E*-Constant vs. *D*-Constant

Both dielectric constant (permittivity), ε, and elastic compliance, s, exhibit significant difference in terms of the electromechanical coupling factor k under different boundary conditions: mechanically stress free or clamped; electrical short circuit or open circuit, as described in Equations 2.74 and 2.75. We discussed the k_{31} mode vibrator with E-constant s_{11}^E and k_{33} mode with D-constant s_{33}^D for analyzing the dynamic equations in Sections. 2.8.1 and 2.8.2. We reconsider here the relation between these status differences from the *depolarization field* viewpoint.

Let us consider boundary conditions in a piezoelectric plate with or without surface electrodes, in which a mechanical vibration wave is propagating, as shown in Figure 2.58 (one-λ standing wave-like is illustrated). The spontaneous polarization axis is perpendicular to the plate in (a) and (b) (k_{31} mode), while in parallel to the plate in (c) (k_{33} mode). The polarization fluctuation, denoted by $\boldsymbol{P}$ in the figure, is induced by the stress distribution, X, via the piezoelectric effect d_{31} or d_{33}. The bound charge inside the crystal means the end member charge (+ or –) of the crystal dipole. The key principle is from Gauss's law (static electromagnetic Maxwell relation):

$$div\,\boldsymbol{D} = \rho;\ \text{or}\ div\ \boldsymbol{E} = \frac{1}{\varepsilon_0}(\rho - div\ \boldsymbol{P}) \tag{2.169}$$

If there exist free charges, ρ, in the specimen plate surface, $div\ \boldsymbol{E}$ can be equal to zero by compensating induced $\boldsymbol{P}$ (induced by the stress as $P = dX$) with ρ, leading to E-constant with respect to space/coordinate. In contrast, if no charges, div $\boldsymbol{D} = 0$ (D-constant), by generating the so-called *depolarization electric field* $E = -(\boldsymbol{P}/\varepsilon_0)$. A mechanical/sound wave generates polarization modulation in a piezoelectric plate, as induced $P = dX$. When the surface is electroded in k_{31} mode (Figure 2.58a), charges can easily be supplied through the electrodes, as illustrated. Thus, $((\partial E_Z/\partial x) = 0)$, or E-constant is derived when the electrodes are connected to a voltage supply, or E_Z = zero in the short circuit between the top and bottom electrodes. The sound velocity along the plate is governed by the elastic compliance s_{11}^E in this electroded k_{31} plate. When the surface does not have electrodes, no charge is supplied (Figure 2.58b). Thus, the depolarization/reverse field is induced to maintain $((\partial D_Z/\partial x) = 0)$, leading to the D-constant condition. That is, as shown in Figure 2.58b, both induced polarization P and depolarization field E are positively and negatively proportional to the stress distribution and cancel each other. The resonance frequency for this no-electrode k_{31} plate

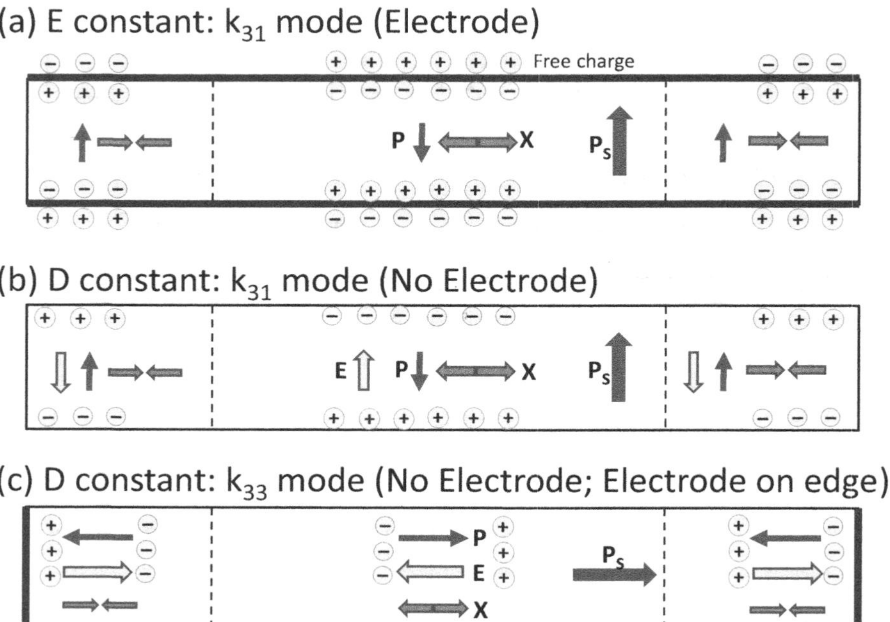

FIGURE 2.58 Boundary conditions: *E*-constant vs. *D*-constant under dynamic waves.

is determined by s_{11}^D, smaller than $s_{11}^E\left(s_{11}^D=\left(1-k_{31}^2\right)s_{11}^E\right)$, that is, in a higher frequency range than the electrode specimen. In the case of k_{33} mode (Figure 2.58c), though the edges are electroded, there is no electrode along the wave propagation z-axis (parallel to the plate). Thus, no charge is available to compensate the polarization modulation along the z direction, and the sound velocity along the polarization direction should be a *D*-constant sound velocity based on s_{33}^D. As illustrated in Figure 2.58c, the polarization fluctuation is the largest around the nodal planes (around the center and both ends in this figure). Both the induced polarization, ***P***, and depolarization field, ***E***, are positively and negatively proportional to the stress distribution, and almost cancel each other.

2.8.3 Admittance/Impedance Spectrum Characterization Method

Though we discuss a full set of piezoelectric characterization methods in Chapter 5, we introduce briefly here how to calculate the electromechanical parameters in the k_{31} mode specimen $\left(k_{31}, d_{31}, s_{11}^E, \text{and } \varepsilon_{33}^X\right)$ from the admittance/impedance spectrum measurement for the reader's convenience to get a flavor. When we measure the admittance/impedance of a piezoelectric specimen by changing the frequency, a frequency spectrum like Figure 2.59 is observed, where the first and second max/min peaks correspond to the resonance and antiresonance, respectively. The parameters used in the following procedure are described in the Figure 2.54 k_{31} mode sample.

1. The sound velocity, v, in the specimen is obtained from the resonance frequency, f_R (admittance peak frequency): $f_R = v/2L$.
2. Knowing the density, ρ, the elastic compliance, s_{11}^E, can be calculated from the sound velocity, v: $v = 1/\sqrt{\rho s_{11}^E}$.

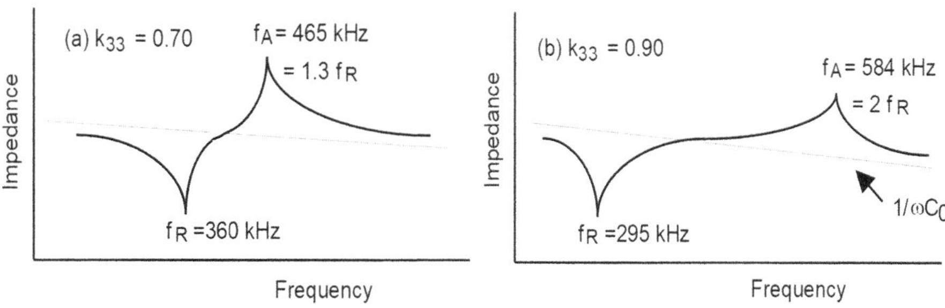

FIGURE 2.59 (a) Impedance curves for a reasonable k material (PZT 5 H, $k_{33} = 0.70$), and (b) a high k material (PZN-PT single crystal, $k_{33} = 0.90$).

3. The electromechanical coupling factor, k_{31}, is calculated from the v value and the antiresonance frequency, f_A, through Equation 2.162. Especially in low-coupling piezoelectric materials, the following approximate equation is useful:

$$k_{31}^2/\left(1-k_{31}^2\right) = (\pi^2/4)(\Delta f/f_R).\ (\Delta f = f_A - f_R) \tag{2.170}$$

4. Knowing the permittivity, ε_{33}^X, from the independent measurement (such as an LCR meter) under an off-resonance condition, the d_{31} is calculated from k_{31} through Equation 2.163: $k_{31} = d_{31}/\sqrt{s_{11}^E \cdot \varepsilon_{33}^X}$.

Figures 2.59a and b compare observed impedance curves of rod-shaped k_{33} samples for a typical k material (PZT 5H, $k_{33} = 0.70$) and a high k material (PZN-PT single crystal, $k_{33} = 0.90$). Note a large separation between the resonance and antiresonance peaks in the high k material, leading to the condition almost $f_A = 2f_R$. To the contrary, a regular PZT sample exhibits $f_A = 1.3f_R$. The *bandwidth* of the piezotransducer is defined by ($\Delta f/f_R$), leading to 100% for the PZN-PT sample and only 30% for PZT 5H. As we discuss in Chapter 5, a large mechanical vibration can be excited in a frequency range between the resonance and antiresonance frequencies, though the driving scheme is rather different according to the drive frequency: resistive (low voltage and high current, or high voltage and low current) at the resonance or antiresonance, respectively, inductive at an intermediate range.

EXAMPLE PROBLEM 2.18

Knowing the above experimental result: $f_A = 1.3\ f_R$, calculate the k_{33} of this PZT 5H.

SOLUTION

1. Similar to the most rough approximation for small k_{31}, Equation 2.170, $k_{31}^2/\left(1-k_{31}^2\right) = (\pi^2/4)(\Delta f/f_R)$, we use the following approximation for k_{33} mode:

$$k_{33}^2/\left(1-k_{33}^2\right) = (\pi^2/4)(\Delta f/f_R) = (\pi^2/4)\times 0.3 = 0.749$$

   Thus, $k_{33} = 0.654$.
2. When we use a better approximation:

$$k_{33}^2/\left(1-k_{33}^2\right) = (\pi^2/8)\left(f_A^2 - f_R^2\right)/f_R^2,$$

   we obtain $k_{33} = 0.678$, a higher value than the above approximation.

3. When we use the accurate formula:

$$(\omega_A L/2v)\cot(\omega_A L/2v) = -k_{33}^2/\left(1-k_{33}^2\right),$$

we can obtain $k_{33} = 0.677$.

The reader should understand that the k value deviates from the accurate one according to the approximation formula.

CHAPTER ESSENTIALS

1. *Functional classification*: dielectrics > piezoelectrics > pyroelectrics > ferroelectrics.
2. *Origin of spontaneous polarization*: the dipole coupling with the local field, in conjunction with nonlinear atomic elasticity.
3. *Piezoelectricity*: can be modeled with ideal springs.
   Piezostriction: can be described in terms of the difference between the harmonic terms of the two equivalent springs.
   Electrostriction: can be modeled with anharmonic equivalent springs.
4. *Origins of field-induced strains*:
   a. *Inverse Piezoelectric Effect*: x = d E
   b. *Electrostriction*: x = M E^2
   c. *Domain reorientation*: strain hysteresis
   d. *Phase Transition (antiferroelectric↔ferroelectric)*: strain "jump"
5. *Shear strain*: $x_5 = 2\,x_{31} = 2\,\phi$, taken as positive for smaller angles.
6. *The electrostriction equation*:

$$x = \underset{\text{(spontaneous strain)}}{QP_S^2} + \underset{\text{(piezostriction)}}{2Q\varepsilon_o\varepsilon P_s E} + \underset{\text{(electrostriction)}}{Q\varepsilon_O^2\varepsilon E^2}$$

7. The *piezoelectric constitutive equations*:

$$x = s^E X + dE$$

$$P = dX + \varepsilon_0 \varepsilon^X E$$

8. *Electromechanical coupling factor* (k):

$$k^2 = \frac{d^2}{s^E(\varepsilon^X \varepsilon_0)} = \frac{h^2}{c^D(\kappa^x/\varepsilon_0)}$$

9. $\varepsilon^x = \varepsilon^X(1-k^2)$; $s^D = s^E(1-k^2)$—Constraint permittivity and elastic compliance are smaller than those under free conditions.
10. Depolarization energy reduces the domain width, while domain wall energy increases the width, leading to a compromised domain width related to the square root of the crystal thickness.
11. *Polarization- and electric field-induced strain in polycrystalline materials*:

Tetragonal:	$P_3 \rightarrow (0.831)P_S$	$x_3 \rightarrow (0.368)S_S$
Rhombohedral:	$P_3 \rightarrow (0.861)P_S$	$x_3 \rightarrow (0.424)S_S$
***Coercive field*:**	Tetragonal > rhombohedral (perovskite)	

12. Three losses in piezoelectrics: dielectric, elastic, and piezoelectric losses

$$\varepsilon^{X*} = \varepsilon^X(1-j\tan\delta'), s^{E*} = s^E(1-j\tan\phi'), d^* = d\,(1-j\tan\theta')$$

13. *Microscopic origin of losses*:
    Extensive dielectric loss—180° (and 90°) domain wall motion
    Extensive mechanical loss—90° domain wall motion
    Extensive piezoelectric loss—polarization realignment due to Gauss's Law.
14. Intensive and extensive losses:

$$\begin{bmatrix}\tan\delta'\\ \tan\phi'\\ \tan\theta'\end{bmatrix}=[K]\begin{bmatrix}\tan\delta\\ \tan\phi\\ \tan\theta\end{bmatrix},\ [K]=\frac{1}{1-k^2}\begin{bmatrix}1 & k^2 & -2k^2\\ k^2 & 1 & -2k^2\\ 1 & 1 & -1-k^2\end{bmatrix},\ k^2=\frac{d^2}{s^E(\varepsilon^X\varepsilon_0)}=\frac{h^2}{c^D(\kappa^x/\varepsilon_0)}$$

15. Admittance of the k_{31} piezoelectric plate:

$$Y=(j\omega wL/t)\varepsilon_0\varepsilon_{33}^{LC}\left[1+\left(d_{31}^2/\varepsilon_{33}^{LC}s_{11}^E\right)(\tan(\omega L/2v)/(\omega L/2v)\right]$$

Resonance: Infinite $Y\rightarrow(\omega_R L/2v)=\pi/2$
Antiresonance: Zero $Y\rightarrow(\omega_A L/2v)\cot(\omega_A L/2v)=-k_{31}^2/\left(1-k_{31}^2\right)$

CHECK POINT

1. The local field is the driving force for spontaneous polarization. What is the factor γ called which enhances the applied electric field E?
2. (T/F) If the atomic spring constant is purely linear, no ferroelectricity should occur. True or False?
3. (T/F) A polycrystalline piezoelectric PZT has three independent piezoelectric d matrix components, d_{33}, d_{13}, and d_{15}. True or False?
4. (T/F) The following two force configurations are equivalent mathematically. True or False?

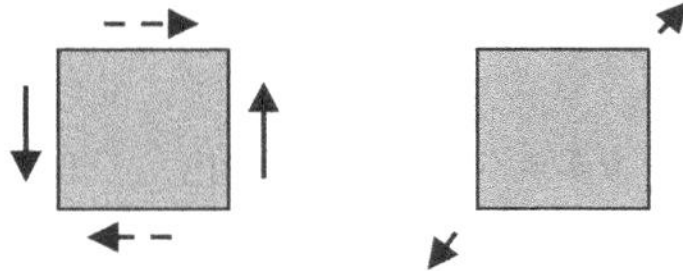

5. Elastic Gibbs energy (in 1D expression) is given by:

$$G_1=(1/2)\alpha P^2+(1/4)\beta P^4+(1/6)\gamma P^6+\cdots-(1/2)sX^2-QP^2X$$

   Why don't we include the "odd-number" power terms of polarization P? Answer simply.
6. Elastic Gibbs energy is given by:

$$G_1=(1/2)\alpha P^2+(1/4)\beta P^4+(1/6)\gamma P^6+\cdots-(1/2)sX^2-QP^2X$$

   Where can you find the primary expansion term with respect to temperature? Answer simply.
7. (T/F) The phenomenology suggests that the spontaneous polarization of a ferroelectric changes with temperature linearly. True or False?
8. (T/F) The phenomenology suggests that the spontaneous strain in a ferroelectric changes with temperature linearly. True or False?
9. (T/F) The phenomenology suggests that the permittivity of a ferroelectric material exhibits the maximum at its Curie temperature. True or False?
10. (T/F) The phenomenology suggests that the piezoelectric constant of a ferroelectric material exhibits the maximum just below its Curie temperature. True or False?
11. (T/F) The phenomenology suggests that the Curie-Weiss temperature of a ferroelectric material is always higher than (or equal to) the Curie temperature. True or False?

12. How is the piezoelectric coefficient, d, related to the electrostrictive coefficient, Q; spontaneous polarization, P_S; and relative permittivity, ε, in the Devonshire phenomenology? Using the air permittivity, ε_0, provide a simple formula.
13. In a polycrystalline PZT, we observed saturated polarization $P_S = 27\ \mu C/cm^2$. Estimate the single crystal value roughly with the Uchida-Ikeda model (consider an average calibration factor between tetragonal and rhombohedral symmetries).
14. (T/F) Piezoelectric resonance is a mechanical resonance mode, but antiresonance is not a mechanical resonance, at which the vibration amplitude is not enhanced. True or False?
15. In a k_{31} piezoceramic plate, the first antiresonance frequency, f_A, is observed higher than the resonance frequency, f_R. What value does the ratio f_A/f_R approach with increasing the electromechanical coupling factor, k_{31}?
16. (T/F) There is a highly resistive (no electric carrier/impurity in a crystal) piezoelectric single crystal (spontaneous polarization, $\mathbf{P}_S$) with a monodomain state without surface electrode. The depolarization electric field in the crystal is given by $E = -(\mathbf{P}_S / \varepsilon_0)$. True or False?

CHAPTER PROBLEMS

2.1 The room temperature form of quartz belongs to class ***32***.

a. Show that the piezoelectric matrix (d_{ij}) is given by:

$$\begin{pmatrix} d_{11} & -d_{11} & 0 & d_{14} & 0 & 0 \\ 0 & 0 & 0 & 0 & -d_{14} & -2d_{11} \\ 0 & 0 & 0 & 0 & 0 & 0 \end{pmatrix}$$

Notice that the piezoelectric tensor must be invariant for a 120° rotation around the 3-axis and for a 180° rotation around the 1-axis. The transformation matrices are, respectively:

$$\begin{pmatrix} -1/2 & \sqrt{3}/2 & 0 \\ -\sqrt{3}/2 & -1/2 & 0 \\ 0 & 0 & 1 \end{pmatrix} \text{ and } \begin{pmatrix} 1 & 0 & 0 \\ 0 & -1 & 0 \\ 0 & 0 & -1 \end{pmatrix}$$

b. The measured values of the d_{ij} for right-handed quartz are:

$$\begin{pmatrix} -2.3 & 2.3 & 0 & -0.67 & 0 & 0 \\ 0 & 0 & 0 & 0 & 0.67 & 4.6 \\ 0 & 0 & 0 & 0 & 0 & 0 \end{pmatrix} \times 10^{-12}\ (C/N)$$

i. If a compressive stress of 1 kgf/cm^2 is applied along the 1-axis of the crystal, find the induced polarization. (kgf = kilogram force = 9.8 N)
ii. If an electric field of 100 V/cm is applied along the 1-axis, find the induced strains.

2.2 In the case of a first-order phase transition, the Landau free energy is expanded as:

$$F(P,T) = (1/2)\alpha P^2 + (1/4)\beta P^4 + (1/6)\gamma P^6 \qquad (\alpha = (T - T_0)/\varepsilon_0 C)$$

Calculate the inverse permittivity near the Curie temperature, and verify that the slope $[\partial(1/\varepsilon)/\partial T]$ just below T_C is 8 times larger than the slope just above T_C.

Hint: In the first-order phase transition, P_S satisfies the following equation in the temperature range of $T < T_C$:

$$E = \alpha + \beta P_s^2 + \gamma P_s^4 = 0.$$

The permittivity is given by $1/\varepsilon_0 = \alpha + 3\beta P_s^2 + 5\gamma P_s^4$.
Thus,

$$1/\varepsilon_0\varepsilon = \alpha + 3\beta P_s^2 + 5(-\alpha - \beta P_s^2) = -4\alpha - 2\beta P_s^2$$

Since $\alpha = (T - T_0)/\varepsilon_0 C$, $P_S^2 = (\sqrt{\beta^2 - 4\alpha\gamma} - \beta)/2\gamma$ and

$$(T - T_0)/\varepsilon_0 C = (3/16)(\beta^2/\gamma) - (T_C - T)/\varepsilon_0 C,$$

we can obtain

$$1/\varepsilon_0\varepsilon = -4\alpha - 2\beta P_s^2$$

$$= -4[(3/16)(\beta^2/\gamma) - (T_C - T)/\varepsilon_0 C] + (\beta^2/\gamma) - (\beta/\gamma)\sqrt{\beta^2 - 4\gamma\left[\left(\frac{3}{16}\right)\left(\frac{\beta^2}{\gamma}\right) - \frac{T_C - T}{\varepsilon_0 C}\right]}$$

Considering $(T_C - T) \ll 1$, obtain the approximation of this equation.

2.3 Barium titanate has a tetragonal crystal symmetry at room temperature and the distortion from the cubic structure is not very large ($c/a = 1.01$). Calculate all possible angles between the two non-180° domain walls.

2.4 In calculating Equations 2.115 and 2.116, the volume element dv is given by: $dv = (2\pi r^2 \cdot dr)(\sin\theta \cdot d\theta)$. Using this dv, calculate $\int dv$, $\int \cos\theta\, dv$, and $\int \cos^2\theta\, dv$, assuming the polarization is uniformly distributed with respect to θ.

2.5 Substituting the complex parameters: $\varepsilon^{X*} = \varepsilon^X(1 - j\tan\delta')$, $s^{E*} = s^E(1 - j\tan\phi')$, $d^* = d(1 - j\tan\theta')$; $\kappa^{x*} = \kappa^x(1 + j\tan\delta)$, $c^{D*} = c^D(1 + j\tan\phi)$, $h^* = h(1 + j\tan\theta)$ into the equations:

$$\varepsilon^X\varepsilon_0 = \frac{1}{\left(\frac{\kappa^x}{\varepsilon_0}\right)\left[1 - \frac{h^2}{c^D(\kappa^x/\varepsilon_0)}\right]}, \quad s^E = \frac{1}{c^D\left[1 - \frac{h^2}{c^D(\kappa^x/\varepsilon_0)}\right]}, \quad d = \frac{\frac{h^2}{c^D(\kappa^x/\varepsilon_0)}}{h\left[1 - \frac{h^2}{c^D(\kappa^x/\varepsilon_0)}\right]},$$

verify the relationships between the intensive and extensive losses:

$$\tan\delta' = (1/(1 - k^2))[\tan\delta + k^2(\tan\phi - 2\tan\theta)],$$

$$\tan\phi' = (1/(1 - k^2))[\tan\phi + k^2(\tan\delta - 2\tan\theta)],$$

$$\tan\theta' = (1/(1 - k^2))[\tan\delta + \tan\phi - (1 + k^2)\tan\theta],$$

where k is the real electromechanical coupling factor.

2.6 Knowing the strain distribution of a k_{31} piezoplate,

$$x_1 = d_{31}E_z\left(\frac{\dfrac{\cos[\omega(L-2x)]}{2v}}{\cos\left(\dfrac{\omega L}{2v}\right)}\right)$$

illustrated in Figure 2.55, draw the displacement distribution $u(x)$ in a similar fashion for both resonance and antiresonance modes.

REFERENCES

1. C. Kittel: *Introduction to Solid State Physics*, John Wiley & Sons, New York, 1966.
2. W. Kinase, Y. Uemura and M. Kikuchi: *J. Phys. Chem. Solids*, 30, 441, 1969.
3. K. Uchino and S. Nomura: *Bull. Jpn. Appl. Phys.*, 52, 575, 1983.
4. J. F. Nye: *Physical Properties of Crystals*, Oxford University Press, London, p. 123, 140, 1972.
5. A. F. Devonshire: *Adv. Phys.*, 3, 85, 1955.
6. H. F. Kay: *Rep. Prog. Phys.*, 43, 230, 1955.
7. K. Uchino, S. Nomura, L. E. Cross, S. J. Jang and R. E. Newnham: *Jpn. J. Appl. Phys.*, 20, L367, 1981; K. Uchino: Proc. Study Committee on Barium Titanate, XXXI-171-1067, 1983.
8. J. Kuwata, K. Uchino and S. Nomura: *Jpn. J. Appl. Phys.*, 19, 2099, 1980.
9. D. R. Tobergte and S. Curtis: *IUPAC. Compendium of Chemical Terminology, (the "Gold Book")*, International Union of Pure and Applied Chemistry, p. 53, 2013.
10. X. -H. Du, U. Belegundu and K. Uchino: *Jpn. J. Appl. Phys.*, 36(Part 1, 9A), 5580, 1997.
11. X. -H. Du, J. Zheng, U. Belegundu and K. Uchino: *Appl. Phys. Lett.*, 72(19), 2421, 1998.
12. M. J. Haun, E. Furman, S. J. Jang and L. E. Cross: *Ferroelectrics,* 99, 13, 1989.
13. *C. Kittel*: *Phys. Rev.*, 82, 729, 1951.
14. *K. Uchino*: *Solid State Phys.*, 17, 371, 1982.
15. K. Uchino, L. E. Cross, R. E. Newnham and S. Nomura: *J. Appl. Phys.*, 52, 1455, 1981.
16. K. Uchino: *Jpn. J. Appl. Phys.*, 24(Suppl. 24-2), 460, 1985.
17. A. Furuta, K. Y. Oh and K. Uchino: *Sensors and Mater., 3(*4), 205, 1992.
18. A. Furuta, K. Y. Oh and K. Uchino: Proc. Int'l Symp. Appl. Ferroelectrics '90, 1991.
19. K. Ohta: *Fundamentals of Magnetic Engineering II*, Kyoritsu Publ. Co., Tokyo, 1985.
20. J. A. Osborn: *Phys. Rev.*, 67, 351, 1945; E. C. Stoner: *Phil. Mag.* 36, 803, 1945.
21. T. Mitsui and J. Furuichi: *Phys. Rev.*, 90, 193, 1953.
22. W. J. Merz: *Phys. Rev.*, 95, 690, 1954.
23. N. Uchida and T. Ikeda: *Jpn. J. Appl. Phys.*, 6, 1079, 1967.
24. N. Uchida: *Rev. Elect. Commun. Lab.*, 16, 403, 1968.
25. N. Uchida and T. Ikeda: *Jpn. J. Appl. Phys.*, 4, 867, 1965.
26. N. A. Schmidt: *Ferroelectrics*, 31, 105, 1981.
27. P. Gerthsen and G. Kruger: *Ferroelectrics*, 11, 489, 1976.
28. N. Setter (ed.): *Piezoelectric Materials in Devices*, Swiss Institute of Technology, Lausanne, Switzerland, 2002.
29. K. Uchino and S. Hirose: *IEEE-UFFC Trans.*, 48, 307–321, 2001.
30. K. Uchino, Y. Zhuang, and S. O. Ural: *J. Adv. Dielectrics*, 1(1), 17–31, 2011.
31. J. Zheng, S. Takahashi, S. Yoshikawa, K. Uchino and J. W. C. de Vries: *J. Amer. Ceram. Soc.*, 79, 3193–3198, 1996.
32. K. Uchino, J. Zheng, A. Joshi, Y. H. Chen, S. Yoshikawa, S. Hirose, S. Takahashi and J. W. C. de Vries: *J. Electroceramics*, 2, 33–40, 1998.

3 Actuator Materials

3.1 HISTORY OF ACTUATOR MATERIALS

We overview the materials history on piezoelectric, shape memory, and magnetostrictive materials in this section.

3.1.1 Piezoelectric Materials

The discovery of piezoelectricity is credited to the famous brothers Pierre and Jacques Curie, who first examined the effect in quartz crystals in 1880.[1] The discovery of ferroelectricity in 1921 in Rochelle salt, and the subsequent characterization of new ferroelectric materials, further extended the number of useful piezoelectric materials available for study and application. Prior to 1940, only two general types of ferroelectric were known, Rochelle salt and potassium dihydrogen phosphate (KDP) and its isomorphs. Extensive characterization of the new ferroelectric barium titanate ($BaTiO_3$) was independently undertaken by Wainer and Salmon[2] (U.S.), Ogawa[3] (Japan), and Vul[4] (Russia) from 1940 to 1943 (during the WWII period). Interestingly, E. W. Rath (Germany) filed a $BaTiO_3$ patent even earlier in 1936 in the United Kingdom (No. 445,495). The new material exhibited unusual dielectric properties, including an exceptionally high dielectric constant with distinctive temperature and frequency dependences. Compositional modifications of BT were made by the next generation of researchers to tailor the properties of the material for improved temperature stability and enhanced high-voltage output. The first piezoelectric transducers based on BT ceramics were also developed at this time and implemented for a variety of applications. After the discovery of PZTs, the industrial importance of barium titanate shifted from piezoelectric devices to merely capacitance applications.

In the 1950s, Jaffe and co-workers examined the lead zirconate-lead titanate solid solution system and found certain compositions exhibited an exceptionally superior piezoelectric response.[5] In particular, the maximum piezoelectric response was found for compositions near the morphotropic phase boundary between the rhombohedral and tetragonal phases. Since then, modified PZT ceramics have become the dominant piezoelectric ceramics for various applications. The development of PZT-based ternary solid solution systems has allowed for the production of even more responsive materials whose properties can be very precisely tailored for a variety of applications. PZT-based ceramics make up 90% of piezoelectric products even 70 years after their discovery.

Kawai et al. (Kureha) discovered in 1969 that certain polymers, most notably polyvinylidene difluoride, are piezoelectric when stretched during fabrication.[6] Piezoelectric polymers such as these are also useful for transducer applications. The development of piezoelectric composite materials systematically studied by Newnham et al. in 1978 allowed for the further refinement of the electromechanical properties.[7] Composite structures, incorporating a piezoelectric ceramic and a passive polymer phase, have been formed in a variety of configurations, each precisely designed to meet the specific requirements of a particular application.

The relaxor ferroelectric/electrostrictive ceramics discovered by Uchino, Cross, Newnham, and Nomura in the early 1980s are another epoch-making class of ceramic material, which has recently become important. Among the many useful compositions that have emerged is lead magnesium niobate, doped with 10% lead titanate (PT),[8] which has been used extensively for a variety of ceramic actuator applications. Advances in the growth of large, high-quality single crystals[9] have made these materials even more attractive candidates for a variety of new applications, ranging from high strain actuators to high-frequency transducers for medical ultrasound devices. The details of these materials and recent studies on thin film piezoelectrics such as zinc oxide (ZnO) and PZT developed for use in microelectromechanical devices, Pb-free piezoelectrics, and high power density transducer materials are described in Sections 3.3, 3.5, and 4.5.

3.1.2 Shape Memory Alloys

Reversible martensitic transformations are observed in many alloys; however, an especially pronounced shape memory effect is observed in alloys based on NiTi and Cu. The former is known as Nitinol (the "nol" indicates that this alloy was discovered by the U.S. Naval Ordinance Laboratory, now called the Navy Surface Warfare Center). The properties of some typical shape memory materials are summarized in Table 3.1.[10]

The induced strain (sometimes called the self-strain) for a typical martensitic transition between close-packed structures, face-centered cubic (fcc) and body-centered cubic (bcc), or hexagonal close packed (hcp) for the NiTi alloys, is as high as 20%. The R-phase in NiTi-based alloys has a smaller induced strain of about 3%, which is still an order of magnitude larger than the strains produced in piezoelectrics.[11] The large strain effect and structural reversibility of the martensitic transformation in these alloys make them effective for actuator applications. The induced strain leads to the development of stress when the transformation is impeded. The stress generated by the phase transition depends on the thermodynamics of the transformation and on the mechanical boundary conditions imposed on the specimen. The maximum value of the actuation stress, X_{max}, on cooling can be estimated according to:

$$X_{max} = q/x_o, \tag{3.1}$$

where q is the latent heat of the phase transformation, and x_o is the induced (self) strain.

3.1.3 Magnetostrictive Alloys

Magnetostriction was discovered by James Joule in the nineteenth century. Since then, the effect has been investigated in magnetic metals, such as Fe, Ni, and Co, and oxides such as the ferrites. Research on giant magnetostrictive materials, including rare earth metals, commenced in the 1960s. The magnetic structure and magnetic anisotropy of rare earth metals were under intensive study at this time. In particular, giant magnetostriction of more than 1% was reported for Tb and Dy metal single crystals at cryogenic temperatures.[12,13] This level of induced strain is two to three orders of magnitude higher than the strains typically observed in Ni and the ferrites.

TABLE 3.1
Properties of Typical Shape Memory Alloys

Property	Ni-Ti	Cu-Al-Ni	Cu-Al-Zn
Induced ("Self") Strain, x_o	20%	20%	20%
Latent Heat of Transition, q (J/kg K)	470–620	400–480	390
Maximum Recovery Stress, X_{max} (MPa)	600–800	600	700
Macroscopic Strain Effect			
Polycrystalline	4%	2%	2%
Single Crystal	10%	10%	10%

Source: K. Otsuka and S. M. Wayman: *Chap. 1 in "Shape Memory Materials"*, Cambridge University Press, Cambridge, UK, 1998.

The research of the 1970s was focused on compounds consisting of rare earth (R) and transition metal (T) elements. The RT_2 alloys, exemplified by RFe_2 (known as the cubic Laves phase), were of particular interest. Exceptionally large magnetostriction of up to 0.2% was reported for $TbFe_2$ at room temperature.[14,15] This triggered a boom in research on this enhanced effect now known as giant magnetostriction. In 1974, $Tb_{0.3}Dy_{0.7}Fe_2$, which exhibits enormous magnetostriction under relatively small magnetic field strengths at room temperature, was produced.[16]

In 1987, sophisticated crystal control technology was developed[17] and a magnetostriction jump phenomenon was discovered in [112]-oriented alloys.[18] The mechanism for the jump is a sudden realignment of the spin with one of the crystallographic easy axes closer to the external field direction. These technologies helped to stabilize the magnetostriction. Strains of 0.15% can be consistently achieved under an applied magnetic field of 500 Oe and compressive stress of 1 kgf/mm^2. The materials in the composition range $Tb_{0.27\text{-}0.3}Dy_{0.7\text{-}0.73}Fe_{1.9\text{-}2.0}$ are called Terfenol-D (Dysprosium modified terbium iron produced at "nol") and are currently commercially produced by several manufacturers.

The saturated magnetostrictive strains for some magnetic materials at 300 K are summarized in Table 3.2.[19] Note that several materials in the table have a negative magnetostriction coefficient (i.e., their length decreases under an applied magnetic field), while other materials, including Terfenol-D, have a positive magnetostriction coefficient.

There are three primary mechanical concerns in the design of Terfenol-D transducers: (1) optimization of the output strain, (2) mechanical impedance matching, and (3) strengthening against fracture.[20] A prestress in the range of 7–10 MPa is generally applied to the Terfenol-D actuator in order to enhance its magnetostrictive response. This level of stress is sufficient to align the magnetic moments parallel to an easy axis, which is almost perpendicular to an actual crystallographic axis. Actuator displacements are enhanced by a factor of three for samples subjected to the prestress treatment.[21] Mechanical impedance matching is required between a magnetostrictive actuator and its external load to ensure efficient energy transfer to the load. Terfenol-D has a low tensile strength, with a yield strength of 700 MPa in compression, but only 28 MPa in tension.[22]

Various composite structures have been investigated in recent years to more effectively address these mechanical problems and to optimize the magnetostrictive response. Terfenol-D:epoxy composites, for example, are automatically subject to prestress treatment during the curing process.[23,24] Models of the system have been developed to investigate the effects of particulate volume fraction and thermal expansion of the polymer matrix on the strain-field-load characteristics

TABLE 3.2
The Saturated Magnetostrictive Strain for Some Magnetic Materials at 300 K

Material	Induced Strain ($\Delta l/l \times 10^{-6}$)
Fe	−9
Ni	−35
Co	−60
(0.6)Co-(0.4)Fe	68
(0.6)Ni-(0.4)Fe	25
$TbFe_2$	1753
Terfenol-D	1600
$SmFe_2$	−1560
Metglass 2605SC	40

Source: J. B. Restorff: *Encyclopedia of Applied Physics*, VCH Publ., New York, 9, p.229, 1994.

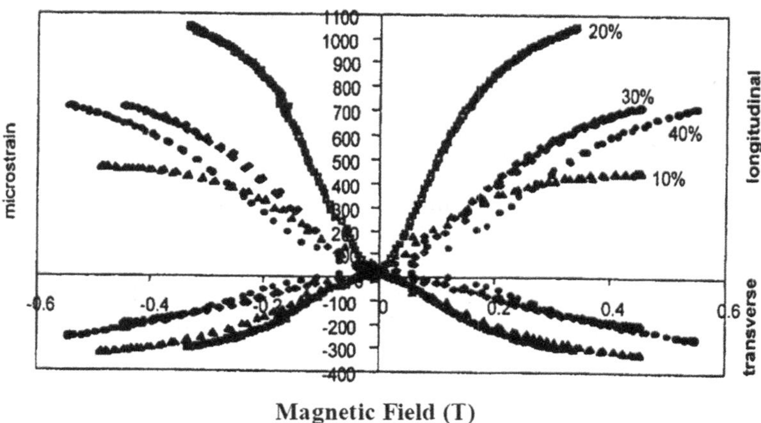

FIGURE 3.1 The longitudinal and transverse magnetostrictive strain as a function of applied magnetic field for Terfenol-D:epoxy composites with various volume fractions of Terfenol-D. (From T. A. Duenas and G. P. Carman: Proc. Adaptive Struct. & Mater. Systems, ASME, AD-57/MD-83, p. 63, 1998.)

of the composite. It was determined that a low-viscosity matrix facilitates the formation of particle chains and helps to inhibit the production of voids during the cure process. The models were verified with Terfenol-D:epoxy samples representing various particle loading conditions (10%–40%), and polymer thermal expansion coefficients ($30–50 \times 10^{-6}$°C^{-1}), Young's moduli (0.5–3 GPa), and viscosities (60–10,000 cps). The magnetostrictive strain for Terfenol-D:epoxy composites is shown in Figure 3.1.[24] We see in this response the competing influences of the magnetostrictive force needed to overcome and strain the matrix and the prestress force, which increases with decreasing volume fraction of Terfenol-D. As the volume fraction increases from 10% to 20%, an overall increase in the force and strain occurs, largely due to the higher concentration of magnetostrictive material in the composite. As the volume fraction increases beyond 20%, however, the decrease in the prestress that is applied to the magnetostrictive composite on curing ultimately limits the force and strain generated in the structure and the net response is reduced in spite of the higher concentration of magnetostrictive material.

Another major commercialized product is amorphous magnetic metals. In the 1970s, Metglas, Inc. (then Allied Signal located in Morristown, NJ), pioneered the development and production of amorphous metal, a unique alloy that exhibits a structure in which the metal atoms occur in a random pattern. Fe-Ni–based amorphous metals are commercially available and popularly used for various applications, because they exhibit medium saturation induction and high corrosion resistance, though the magnetostriction is not high.[25]

Recent research topics include $Fe_{70}Pd_{30}$ ferromagnetic shape memory alloy consisting of a face-centered tetragonal (fct) martensite phase exhibiting a large strain due to the conversion of variants by a magnetic field. Single crystals of $Fe_{70}Pd_{30}$ exhibit a magnetic field–induced strain of 0.6% at 256 K, which is triple that of Terfenol-D (0.16%).[26]

3.2 FIGURES OF MERIT FOR TRANSDUCERS

In order to compare the performance superiority of actuator and transducer materials, we usually use *figures of merit* (FOMs). There are five types of important figures of merit for transducers, in particular, piezoelectric and piezomagnetic materials: (1) the piezoelectric/piezomagnetic coefficient, d, g, and so on; (2) the electromechanical/magnetomechanical coupling factor, k; energy transmission coefficient, λ; efficiency, η; (3) the mechanical quality factor, Q_m; (4) the acoustic impedance, Z; and (5) the maximum vibration velocity, v_{max}. Each of these quantities is defined in this section.

3.2.1 The Piezoelectric/Piezomagnetic Coefficients

Let us start from the piezoelectric constitutive equations. There are four pairs of description types, depending on the intensive (E, X)/extensive (D, x) parameters:

$$x = s^E X + dE \tag{3.2}$$

$$D = dX + \varepsilon_0\, \varepsilon^X E \tag{3.3}$$

$$X = c^D x - h\, D \tag{3.4}$$

$$E = -\, h\, x + \kappa^x/\varepsilon_0\, D \tag{3.5}$$

$$x = s^D X + g\, D \tag{3.6}$$

$$E = -\, g\, X + \kappa^X/\varepsilon_0\, D \tag{3.7}$$

$$X = c^E x - e\, E \tag{3.8}$$

$$D = e\, x + \varepsilon_0\, \varepsilon^x E \tag{3.9}$$

An intensive quantity is one whose magnitude is independent of the size of the system, whereas an extensive quantity is one whose magnitude is additive for subsystems (IUPAC definition). In practice, intensive E and X are externally controllable parameters, while extensive D and x are internal material parameters. Accordingly, there are four types of piezoelectric coefficients, d, h, g, and e. The magnitude of the strain, x, induced by an applied electric field, E, is characterized by the *piezoelectric strain coefficient*, d, as:

$$x = (d)\, E \tag{3.10}$$

This quantity is an important *figure of merit for actuators* (Equation 3.2).

The induced electric field, E, is related to the applied stress, X, through the piezoelectric voltage coefficient, g, as:

$$E = (g)\, X. \tag{3.11}$$

This quantity is an important *figure of merit for sensors* (Equation 3.7). Recall that the direct piezoelectric effect is described by: $P = (d)\, X$, where P is the induced polarization (almost equal to D for a large permittivity material). When we combine this expression with Equation 3.11, we obtain an important relationship between g and d:

$$g = d/\varepsilon_0\, \varepsilon^X \tag{3.12}$$

where ε^X is the dielectric constant/relative permittivity under a free (unclamped) condition.

Equations 3.8 and 3.9 are popularly used for analyzing piezoelectric thin films, where the film strain is constrained/clamped by the substrate. The electric displacement measured via the short-circuit current by changing the strain via the substrate bending can provide the piezoelectric e constant, which is related as:

$$e = d/s^E \tag{3.13}$$

where s^E is the elastic compliance under a short-circuit condition.

Finally, h is basically an inverse component of the d tensor (like $1/d$). When we transform E and D to H and B, and permittivity, ε, to permeability, we can formulate exactly equivalent equations for piezomagnetic materials.

3.2.2 The Electromechanical Coupling Factor, *k*, and Related Quantities

The terms *electromechanical coupling factor*, k; *energy transmission coefficient*, λ_{max}; and *efficiency*, η, are sometimes misrepresented. All are related to the conversion rate between electrical energy and mechanical energy, but they are defined differently. If we convert E/D to H/B, a similar set of formulas can be obtained for the magnetomechanical coupling factor and related quantities.

3.2.2.1 The Electromechanical Coupling Factor (*k*)

Though we discussed this in Section 2.4.6, we review it again here. The electromechanical coupling factor is defined by either of the following expressions:

$$k^2 = \frac{\text{Stored Mechanical Energy}}{\text{Input Electrical Energy}} \tag{3.14a}$$

$$k^2 = \frac{\text{Stored Electrical Energy}}{\text{Input Mechanical Energy}} \tag{3.14b}$$

The input electrical energy per unit volume in Equation 3.14a is (refer to Figure 2.17):

$$\text{Input Electrical Energy} = (1/2)\ \varepsilon_0\varepsilon\ E^2$$

and the converted/stored mechanical energy per unit volume under zero external stress is given by:

$$\text{Stored Mechanical Energy} = (1/2)[x^2/s] = (1/2)[(d\ E)^2/s]$$

where s is the elastic compliance of the material. Making these substitutions into Equation 3.14a and simplifying allows us to now express k^2 as:

$$k^2 = \frac{d^2}{\varepsilon_0 \varepsilon s} \tag{3.15}$$

According to a similar derivation process, Equation 3.14b gives exactly the same expression as Equation 3.15 (refer to Figure 2.18).

3.2.2.2 The Energy Transmission Coefficient (λ_{max})

Not all the converted/stored energy can be actually used, and the actual work done depends on the mechanical load under an electric field application or on the electrical load under a mechanical force application (such as piezoelectric energy harvesting in Chapter 7). With zero mechanical load or a complete clamp (no strain), no output work is done; even a strain is generated under an electric field. Remember that the work is given by [Force] × [Displacement]. The energy transmission coefficient is defined by

$$\lambda_{max} = (\text{Output mechanical energy/Input electrical energy})_{max} \tag{3.16a}$$

or equivalently,

$$\lambda_{max} = (\text{Output electrical energy/Input mechanical energy})_{max} \tag{3.16b}$$

The difference of the above from Equations 3.14 and 3.16 is stored/converted or output/spent.

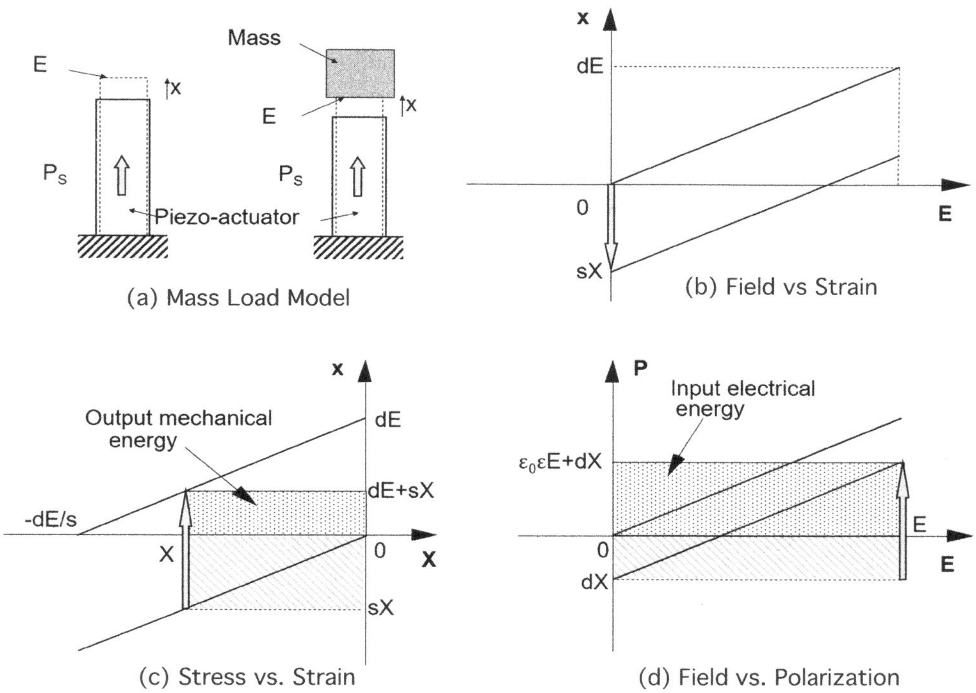

FIGURE 3.2 Calculation of the input electrical and output mechanical energy: (a) load mass model for the calculation, (b) electric field versus induced strain curve, (c) stress versus strain curve, and (d) electric field versus polarization curve.

Let us consider the case where an electric field, E, is applied to a piezoelectric under constant external stress, X (<0, because a compressive stress is necessary to work to the outside). This corresponds to the situation in which a mass is put suddenly on the actuator, as shown in Figure 3.2a. Figure 3.2b shows two electric-field versus induced-strain curves, corresponding to two conditions: under the mass load and no load. Because the area on the field-strain domain does not mean the energy, we should use the stress-strain and field-polarization domains in order to discuss the mechanical and electrical energy, respectively. Figure 3.2c illustrates how to calculate the mechanical energy. Note that the mass shrinks the actuator first by sX (s: piezo-material compliance, and $X < 0$). This mechanical energy sX^2 is a sort of "loan" of the actuator credited from the mass, which should be subtracted later (i.e., "paying back"). This energy corresponds to the bottom hatched area in Figure 3.2c. By applying the step electric field, the actuator expands by the strain level dE under a constant stress condition. This is the mechanical energy provided from the actuator to the mass, which corresponds to $|dEX|$. Like paying back the initial "loan," the output work (from the actuator to the mass) can be calculated as the area subtraction (shown by the top dotted area in Figure 3.2c)

$$\int(-X)dx = -(dE + sX)X. \tag{3.17}$$

3.2.2.2.1 Mechanical Energy Maximization

Let us initially consider merely maximizing the output mechanical energy. The maximum output energy can be obtained when the top dotted area in Figure 3.2c becomes maximum under the constraint of the rectangular corner point tracing on the line (from dE on the vertical axis to $-dE/s$ on the horizontal axis). Since the work energy Equation 3.17 can be transformed as $-s(X + dE/2s)^2 + (dE)^2/4s$, $X = -dE/2s$ provides the maximum mechanical energy of $(dE)^2/4s$. Therefore, the load should be half of the maximum generative stress when we apply the maximum electric field, E.

3.2.2.2.2 Energy Transmission Coefficient

The above mechanical energy maximization condition is not the best condition from the energy transmission coefficient viewpoint; that is, input electric energy is overspent. Figure 3.2d illustrates how to calculate the electrical energy spent. The mass load, X, generates the "loan" electrical energy by inducing $P = dX$ (see the bottom hatched area in Figure 3.2d). By applying a sudden electric field, E, the actuator (like a capacitor) receives the electrical energy of $\varepsilon_0\varepsilon\, E^2$. Thus, the total energy is given by the area subtraction (shown by the top dotted area in Figure 3.2d)

$$\int (E)dP = (\varepsilon_0\varepsilon E + dX)E. \tag{3.18}$$

We need to choose a proper load now to maximize the *energy transmission coefficient*. From the maximum condition of

$$\lambda = -(d\,E + s\,X)\,X/(\varepsilon_0\varepsilon\,E + dX)\,E, \tag{3.19}$$

we can obtain

$$\lambda_{max} = \left[(1/k) - \sqrt{(1/k^2) - 1}\right]^2 = \left[(1/k) + \sqrt{(1/k^2) - 1}\right]^{-2}. \tag{3.20}$$

Refer to Example Problem 3.2 for the derivation process. Note also that

$$k^2/4 < \lambda_{max} < k^2/2, \tag{3.21}$$

depending on the k value ($k < 0.9$). For a small k, $\lambda_{max} = k^2/4$, and for a large k, $\lambda_{max} = k^2/2$. Of course, for an extremely large $k \approx 1$, λ_{max} will approach 1.

EXAMPLE PROBLEM 3.1

We usually learn that an elastic material with strain x under stress X stores mechanical energy $(1/2)xX$ per unit volume. Similarly, a dielectric material with polarization P under electric field E stores electrical energy $(1/2)\,PE$ per unit volume. However, in the Figure 3.2a model, the mechanical and electrical energies were calculated by xX and PE, respectively, without adding (1/2). Explain the reason.

SOLUTION

When stress is gradually applied to the material, the strain is also induced gradually, following Hook's Law: $x = sX$. Thus, the total mechanical energy is calculated by

$$U_M = \int x\,dX = \int sX\,dX = (1/2)sX^2 = (1/2)xX. \tag{P3.1.1}$$

On the contrary, when constant stress is applied suddenly, for example, a mass is put on the actuator, as in this case, the mechanical energy is calculated by

$$U_M = \int X\,dx = X \cdot \int dx = X \cdot x. \tag{P3.1.2}$$

Note the difference of integration between the linear line and step function.

Similarly, when an electric field is gradually applied to the material, the polarization is induced in proportion to E, and the total electrical energy is

$$U_E = \int P\,dE = \int \varepsilon_0\varepsilon E\,dE = (1/2)\varepsilon_0\varepsilon E^2 = (1/2)PE. \tag{P3.1.3}$$

When the step field is applied, the energy is

$$U_E = \int E\,dP = E \cdot \int dP = EP. \tag{P3.1.4}$$

EXAMPLE PROBLEM 3.2

Prove that the maximum energy transmission coefficient, λ_{max}, in a piezoelectric actuator is expressed as follows using its electromechanical coupling factor, k:

$$\lambda_{max} = \left[(1/k) - \sqrt{(1/k^2) - 1}\right]^2. \tag{P3.2.1}$$

Then, verify that λ_{max} ranges $(1/4)k^2 \sim (1/2)k^2$ for a normal k value (<0.9).

Solution

The energy transmission coefficient is defined by

$$\lambda = \text{(Output mechanical energy/Input electrical energy)}.$$

Considering the case where an electric field, E, is applied to a piezoelectric under a constant external stress, X, λ can be calculated as

$$\begin{aligned} \lambda &= -x \cdot X / P \cdot E \\ &= -(dE + sX)X / (\varepsilon_0\varepsilon E + dX)E. \\ &= -[d(X/E) + s(X/E)^2] / [\varepsilon_0\varepsilon + d(X/E)] \end{aligned} \tag{P3.2.2}$$

We need to determine an appropriate stress, X, under a certain applied field, E, so as to maximize the λ value. Letting $y = X/E$, then

$$\lambda = -(sy^2 + dy)/(dy + \varepsilon_0\varepsilon). \tag{P3.2.3}$$

The maximum λ can be obtained when y satisfies

$$(d\lambda/dy) = [-(2sy + d)\cdot(dy + \varepsilon_0\varepsilon) + (sy^2 + dy)\cdot d] / (dy + \varepsilon_0\varepsilon)^2 = 0. \tag{P3.2.4}$$

Then, from $y_0^2 + 2(\varepsilon_0\varepsilon/d)y_0 + (\varepsilon_0\varepsilon/s) = 0$,

$$y_0 = (\varepsilon_0\varepsilon/d)\left[-1 + \sqrt{(1-k^2)}\right]. \tag{P3.2.5}$$

Here, $k^2 = d^2/(s\cdot\varepsilon_0\varepsilon)$. Note that only $y_0 = (\varepsilon_0\varepsilon/d)\left[-1+\sqrt{(1-k^2)}\right]$ is valid for realizing the meaningful maximum point, since $y_0 = (\varepsilon_0\varepsilon/d)\left[-1-\sqrt{(1-k^2)}\right]$ and $y_0 = (\varepsilon_0\varepsilon/d)\left[-1+\sqrt{(1-k^2)}\right]$ provide $(d^2\lambda/dy^2) > 0$ (min) and < 0 (max), respectively.

By putting $y = y_0$ into $\lambda(y)$, we can get the maximum value of λ:

$$\begin{aligned} \lambda_{max} &= -\{s[-2(\varepsilon_0\varepsilon/d)y_0 - (\varepsilon_0\varepsilon/s)] + dy_0\} / (dy_0 + \varepsilon_0\varepsilon) \\ &= [dy_0(2/k^2 - 1) + \varepsilon_0\varepsilon] / (dy_0 + \varepsilon_0\varepsilon) \\ &= \left\{\left[-1+\sqrt{(1-k^2)}\right](2/k^2 - 1) + 1\right\} / \left\{\left[-1+\sqrt{(1-k^2)}\right] + 1\right\} \\ &= \left[(1/k) - \sqrt{(1/k^2) - 1}\right]^2. \end{aligned} \tag{P3.2.6}$$

The relation between the energy transmission coefficient, λ_{max}, and electromechanical coupling factor, k, is plotted in Figure 3.3, where you can also find the ratio λ_{max}/k^2, which changes from 1/4 for small k to 1/2 for large k (<0.9). Though λ_{max}/k^2 approaches 1, when k is very close to 1, this is very unrealistic in most piezoelectrics.

It is also worth noting that the maximum condition stated above does not agree precisely with the condition that provides the maximum output mechanical energy. The maximum output energy

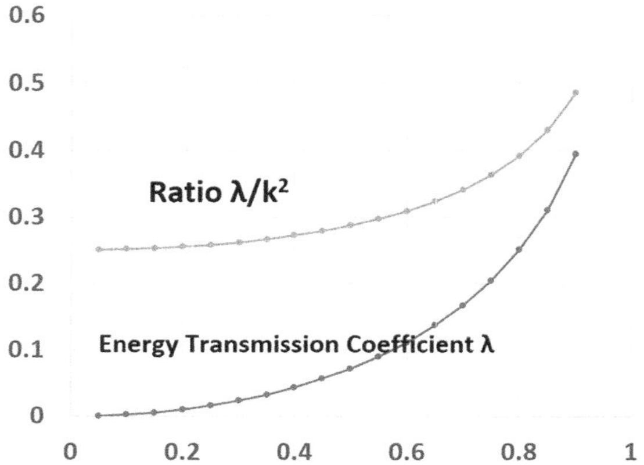

FIGURE 3.3 Energy transmission coefficient λ_{max} versus electromechanical coupling factor k.

can be obtained when the dotted area in Figure 3.2c becomes maximum under the constraint of the rectangular corner point tracing on the strain line. Therefore, the load should be half of the maximum generative stress (i.e., blocking stress), and the mechanical energy at that point: $-[dE - s(dE/2s)](-dE/2s) = (dE)^2/4s$. In this case, since the input electrical energy is given by $[\varepsilon_0\varepsilon E + d(-dE/2s)]\,E$,

$$\lambda = 1/2[(2/k^2)-1] = (1/4)\,k^2 + (1/8)\,k^4 + (4/64)\,k^6 + \ldots \quad \text{(P3.2.7)}$$

which is close to the value λ_{max} when k is small, but has a slightly different value when k is large, as predicted theoretically:

$$\lambda_{max} = \left[(1/k) - \sqrt{(1/k^2)-1}\right]^2 = (1/4)k^2 + (1/8)k^4 + (5/64)k^6 + \ldots \quad \text{(P3.2.8)}$$

3.2.2.3 Efficiency η

$$\eta = \text{(Output mechanical energy)/(Consumed electrical energy)} \quad (3.22a)$$

or

$$\eta = \text{(Output electrical energy)/(Consumed mechanical energy).} \quad (3.22b)$$

The difference between the efficiency definitions from Equations 3.16 and 3.22 is "input" energy and "consumed" energy in the denominators. In a work cycle (e.g., an electric field cycle), the input electrical energy is partially transformed into mechanical energy and the remaining is stored as electrical energy (electrostatic energy like a capacitor) in an actuator. In this way, the ineffective electrostatic energy can be returned to the power source, leading to near 100% efficiency if the loss is small. Since typical values of the total loss (dielectric, elastic, and piezoelectric losses) in PZTs are about 1%–3%, we can conclude that piezoelectric materials (and some simple components, such as piezoelectric transformers) can exhibit efficiency, η, close to 97%–99%. Piezoelectric transformer products by Micromechatronics Inc. (a spin-off from ICAT/Penn State) exhibit component efficiency around 97%–98% in practice.[27]

3.2.3 Mechanical Quality Factor, Q_M

Three losses in piezoelectrics, dielectric, elastic, and piezoelectric losses, were introduced in Section 2.7 under a pseudo-DC condition. The mechanical quality factor, Q_M, is a parameter that characterizes the sharpness of the electromechanical resonance spectrum, which originates from elastic loss under a resonance condition (i.e., $Q_M = 1/\tan\phi'$). When the motional admittance, Y_m, is plotted around the resonance frequency, ω_0, the mechanical quality factor, Q_M, is defined with respect to the full width ($2\Delta\omega$) at the admittance $Y_m/\sqrt{2}$ points as:

$$Q_M = \omega_0/2\Delta\omega. \tag{3.23}$$

Also note that Q_M^{-1} at the resonance is equal to the intensive mechanical loss ($\tan\phi'$) in a k_{31}-mode plate. When we define a complex elastic compliance, $s^E = s^{E\prime} - j\, s^{E\prime\prime}$, the mechanical loss tangent is provided by $\tan\phi' = s^{E\prime\prime}/s^{E\prime}$. The Q_M value is very important in evaluating the magnitude of the resonant displacement and strain. The vibration amplitude at an off-resonance frequency ($dE{\cdot}L$, L: length of the sample) is amplified by a factor proportional to Q_M at the resonance frequency. For example, a longitudinally vibrating rectangular plate through the transverse piezoelectric effect d_{31} generates the maximum displacement given by $(8/\pi^2)\, Q_M\, d_{31}E\, L$ at its fundamental resonance frequency. We will discuss the details in Section 5.5, including the mechanical quality factor at the antiresonance. On the other hand, for sensor applications, soft piezoelectrics with $Q_M < 50$–60 are popularly selected owing to their wide bandwidth for receiving wide frequency signals.

3.2.4 Acoustic Impedance, *Z*

The acoustic impedance, Z, is a parameter used for evaluating the acoustic energy transfer between two materials, typically between the piezoelectric transducer material and the mechanical vibration medium. It is defined, in general, by

$$Z^2 = (\text{pressure/volume velocity}). \tag{3.24}$$

In a solid material, it is expressed as

$$Z = \sqrt{\rho c}, \tag{3.25}$$

where ρ is the density and c is the elastic stiffness of the material.

In more advanced discussions, there are three kinds of mechanical impedances: (1) specific acoustic impedance (pressure/particle speed), (2) acoustic impedance (pressure/volume speed), and (3) radiation impedance (force/speed). See Reference [28] for details.

EXAMPLE PROBLEM 3.3

Why is acoustic impedance (or mechanical impedance) matching necessary for mechanical energy transfer from one material to the other? Explain the reason conceptually.

Solution

The mechanical work done by one material on the other is evaluated by the product of the applied force, F, and the displacement, ΔL:

$$W = F \times \Delta L \tag{P3.3.1}$$

Figure 3.4 shows a conceptual cartoon illustrating two extreme cases. If the material is very soft, the force, F, can be very small, leading to very small W (practically no work!). This corresponds to "pushing a curtain" (a popular Japanese proverb meaning "useless task"), exemplified by the case when the acoustic wave is generated in water directly by a hard PZT transducer. Most of the acoustic energy generated in the PZT is reflected at the interface, and only a small portion of acoustic energy transfers into water. On the other hand, if the material is very hard, the displacement will be very small, again leading to very small W. This corresponds to "pushing a wall." Polymer

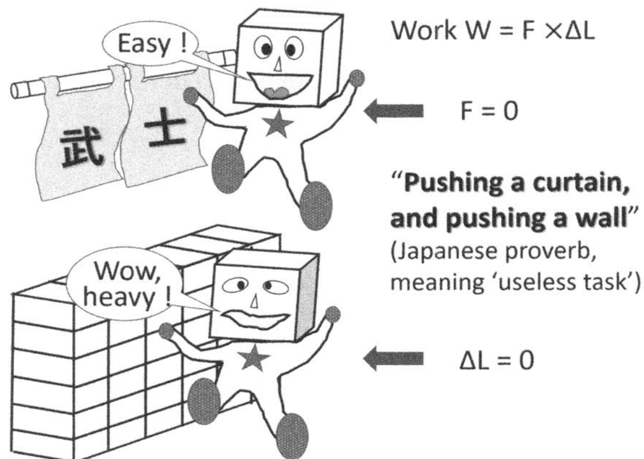
Easy !
武 士
Wow, heavy !
Work W = F ×ΔL
F = 0
"Pushing a curtain, and pushing a wall"
(Japanese proverb, meaning 'useless task')
ΔL = 0

FIGURE 3.4 Concept of mechanical impedance matching.

piezoelectric PVDF cannot drive a hard steel part efficiently. Therefore, the *acoustic impedance* must be adjusted to maximize the output mechanical power:

$$\sqrt{\rho_1 c_1} = \sqrt{\rho_2 c_2}, \quad \text{(P3.3.2)}$$

where ρ is the density, c is the elastic stiffness, and the subscripts 1 and 2 denote the two materials. In practice, acoustic impedance matching layers (elastically intermediate materials between PZT and water, such as a polymer) are fabricated on the PZT transducer surface to minimize the reflection and maximize the transfer of mechanical energy to water. More precisely, the acoustic impedance, Z, of the matching layer should be chosen as $\sqrt{Z_1 \cdot Z_2}$ (i.e., geometrical average).

3.2.5 Maximum Vibration Velocity, v_{MAX}

The power density of a piezoelectric is measured by different figures of merit for different applications:

1. Off-resonance actuator applications—positioners

$$\text{FOM} = d \text{ (piezoelectric constant)}$$

2. Resonance actuator applications—ultrasonic motors

$$\text{FOM} = v \ (\textit{vibration velocity}) \approx Q_m \cdot d \text{ (for low-level excitation)}$$

3. Resonance transducer applications—piezoelectric transformers, sonars (transmitters and receivers)

$$\text{FOM} = k \cdot v \ (k\text{: electromechanical coupling factor})$$

In order to obtain a large mechanical output power, ceramics are driven under a high vibration level, namely under a relatively large AC electric field around the electromechanical resonance frequency. Though the vibration velocity (i.e., first derivative of the vibration displacement with respect to time) is almost proportional to the applied AC electric field under a relatively small field range, with increasing the electric field, the induced vibration velocity will saturate above a certain critical field. This originates from the sudden intensive elastic loss increase or the reduction of the mechanical quality factor above this critical field. Also, heat generation becomes significant, as well as a degradation in piezoelectric properties. Therefore, a high-power device such as an ultrasonic motor requires a very hard piezoelectric with a high *mechanical quality factor*, Q_m (i.e., low elastic loss), in order to suppress heat generation. The Q_m is defined as an inverse value of the intensive elastic loss factor, tan ϕ'. It is also notable that the actual mechanical vibration velocity at the resonance frequency is directly proportional to this Q_m value (i.e., displacement amplification factor).

In order to analyze the various piezoelectric parameter changes as a function of vibration level, we occasionally use *vibration velocity* instead of the applied electric field. Though the *vibration amplitude* may be used, the vibration velocity is used in the discussion. The reason is to eliminate the size effect; that is, when the vibration amplitude is small and proportional to the applied electric field and the length, L, it should be expressed by

$$\Delta L = (8/\pi^2)\, Q_m\, d_{31} L\, E_3 \sin(\omega_R t) \tag{3.26}$$

for a d_{31}-type rectangular piezoelectric plate. Since the vibration velocity at the edge of the plate sample is the first derivative of amplitude with respect to t, we obtain

$$v = (8/\pi^2)\, Q_m\, d_{31} L\, E_3\, \omega_R \cos(\omega_R t). \tag{3.27}$$

Taking into account the fundamental resonance frequency, $f_R = \left(1/\sqrt{\rho s_{11}}\right)/2L$, the vibration velocity at the plate edge can be transformed as

$$v = (8/\pi^2) Q_m d_{31} E_3 \left(1/\sqrt{\rho s_{11}}\right) \cos(\omega_R t). \tag{3.28}$$

Note that the vibration amplitude is the sample-size L dependent, but the vibration velocity is not. Because the vibration velocity is proportional to the electric field, and the proportional constant is given primarily by $Q_m d_{31}/\sqrt{\rho s_{11}}$, which is a material constant, we use it as a measure of the vibration level. The reason of the factor $(8/\pi^2)$ is explained in Section 5.5.2.

As explained above, with increasing the electric field, the induced vibration velocity will saturate above a certain critical field, and heat generation is associated. Since the additional electric power is converted mostly to heat, rather than the vibration velocity increase, we define the *maximum vibration velocity* as the v_{max} under which the piezoelectric plate shows a 20°C temperature rise at the nodal point (i.e., the specimen center part) above room temperature. The root mean square (RMS) value of v_{max} of popular hard PZT rectangular plates ranges from 0.3 to 0.6 m/s, which is a sort of material constant, an important parameter for high power applications. When we consider the high power performance in a wide variety of piezomaterials such as Pb-free and PZT, the *maximum mechanical energy* density is suitable when taking into account the mass density, which is defined as $(1/2)\rho v_{rms}^2$ under the maximum vibration velocity condition. Current top data range from 1000 to 1500 J/m^3. By multiplying the resonance frequency, f, by the mechanical energy density, we can obtain the maximum vibration power density, the top data of which range to 30 W/cm^3.

3.3 PIEZOELECTRIC TRANSDUCER MATERIALS

Among practical piezoelectric and electrostrictive ceramics, many have the *perovskite crystal* structure ABO_3 pictured in Figure 3.5. Because of a high *Lorentz factor* in a perovskite crystal structure, perovskites tend to have large dipole-dipole coupling energy. These materials typically undergo a phase transition on cooling from a high-symmetry high temperature phase (cubic paraelectric phase) to a noncentrosymmetric ferroelectric phase. Materials with a high ferroelectric transition temperature (i.e., Curie temperature) will be piezoelectric at room temperature, whereas those with a transition temperature near or below room temperature will exhibit electrostriction. In the case of the latter, at a temperature just above the Curie temperature, the electrostriction is extraordinarily large because of the large anharmonicity of the ionic potential. Furthermore, simple compositions such as barium titanate ($BaTiO_3$), lead titanate ($PbTiO_3$), lead zirconate ($PbZrO_3$), their solid solutions

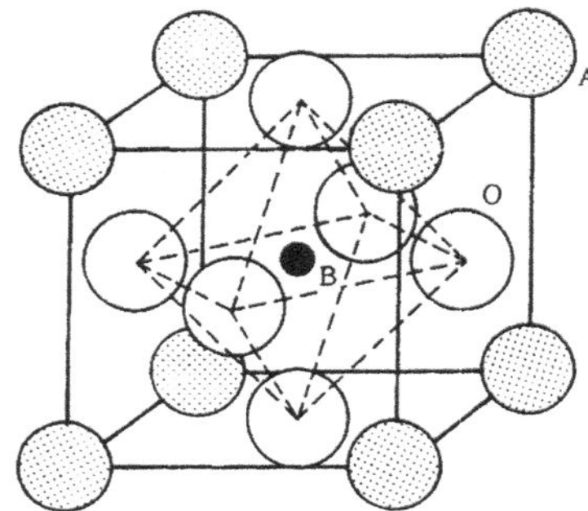

FIGURE 3.5 The perovskite crystal structure of ABO_3.

such as $A(B,B')O_3$, and complex perovskites such as $A^{2+}(B^{3+}_{1/2}B'^{5+}_{1/2})O_3$ and $A^{2+}(B^{2+}_{1/3}B'^{5+}_{2/3})O_3$ are all readily formed. Supercells of the complex perovskite structures listed above, in which ordering of the B-site cations exists, are pictured in Figures 3.6b and c. When the B and B′ cations are randomly distributed, the unit cell corresponds to the simple perovskite structure depicted in Figure 3.6a.

3.3.1 Practical Piezoelectric/Electrostrictive Materials

An overview of the current status of piezoelectric and related materials is presented in the following sections. It includes crystalline materials, piezoceramics, electrostrictive materials, shape memory ceramics, piezopolymers, and piezocomposites. The electromechanical properties for some of the most popular piezoelectric materials are summarized in Table 3.3.[29,30]

3.3.1.1 Quartz, Lithium Niobate/Tantalate Single Crystals

Although piezoelectric ceramics are used for a wide range of applications, single crystal materials are also used, especially for such applications as frequency-stabilized oscillators and surface acoustic

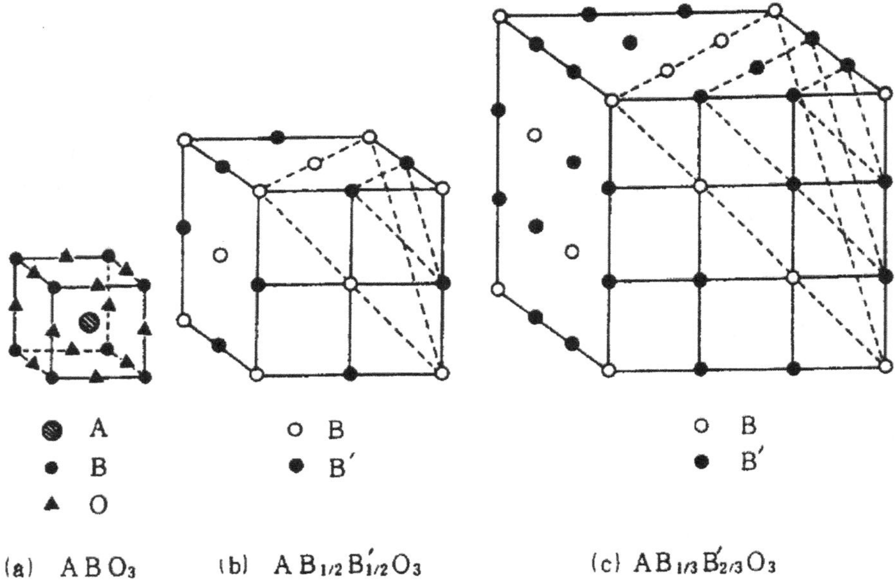

FIGURE 3.6 Complex perovskite structures with various B ion arrangements: (a) ABO_3, simple structure; (b) $A(B_{1/2}B'_{1/2})O_3$, 1:1 ordering, double cell; and (c) $A(B_{1/3}B'_{2/3})O_3$, 1:2 ordering, multiple cell.

TABLE 3.3
Piezoelectric Properties of Some Popular Piezoelectric Materials

	Quartz	$BaTiO_3$	PZT4	PZT 5H	$(Pb,Sm)TiO_3$	PVDF-TrFE
d_{33} (pC/N)	2.3	190	289	593	65	33
g_{33} (mVm/N)	57.8	12.6	26.1	19.7	42	380
k_t	0.09	0.38	0.51	0.50	0.50	0.30
k_p		0.33	0.58	0.65	0.03	
ε_{33}^X	5	1700	1300	3400	175	6
Q_m	$>10^5$		500	65	900	3–10
T_C (°C)		120	328	193	355	

wave (SAW) devices. The most popular single crystal piezoelectric materials are quartz, lithium niobate ($LiNbO_3$), and lithium tantalate ($LiTaO_3$). Piezoelectric single crystals are anisotropic, exhibiting different material properties depending on the cut.

Quartz is a well-known piezoelectric material. Alpha (α) quartz belongs to the triclinic crystal system with point group 32 and undergoes a phase transition at 537°C to its beta (β) form, which is not piezoelectric. One particular cut of quartz, the AT-cut, has a zero temperature coefficient. Quartz oscillators, made from an AT-cut crystal and operated in thickness shear mode, are used extensively for clock sources in computers, and frequency-stabilized devices are used in televisions and video recorders. An ST-cut quartz, when used as a substrate for surface acoustic waves propagating in the x direction will also have a zero temperature coefficient for such waves, thus making it suitable for frequency-stabilized surface acoustic wave devices. Another distinctive characteristic of quartz is its exceptionally high mechanical quality factor, $Q_m > 10^5$.

Lithium niobate and lithium tantalate are isomorphs and have Curie temperatures of 1210°C and 660°C, respectively. The crystal symmetry of the ferroelectric phase of these single crystals is 3 m and the polarization direction is along the c-axis. These materials have high electromechanical coupling coefficients for surface acoustic waves. Large single crystals can easily be obtained from the melt using the Czochralski technique. Thus, both materials are important for SAW device applications.

3.3.1.2 Perovskite $BaTiO_3$, Lead Zirconate Titanate Piezoelectrics

Barium titanate ($BaTiO_3$) is one of the most thoroughly studied and most widely used piezoelectric materials. Just below the Curie temperature (130°C), the material has tetragonal symmetry and the spontaneous polarization is directed along [001]. Below 5°C, the material is orthorhombic and the polarization is oriented along [011]. At −90°C, a transition to a rhombohedral phase occurs and the polarization direction is along [111]. The electromechanical properties of ceramic $BaTiO_3$ are affected by the stoichiometry and microstructure of the material, as well as the presence of dopants on both the A and B sites of the perovskite lattice. Ceramic $BaTiO_3$ has been modified with dopants such as Pb or Ca in order to stabilize the tetragonal phase over a wider temperature range. These compositions were originally designed for use in Langevin-type piezoelectric vibrators, and have currently been revived as commercial products for this and a variety of other applications because of the current trend toward Pb-free piezoelectrics.

Piezoelectric ceramics from the $Pb(Zr,Ti)O_3$ solid solution system have been widely used because of their superior piezoelectric properties. The phase diagram for the $Pb(Zr_xTi_{1-x})O_3$ system appears in Figure 3.7. The symmetry of a given composition is determined by the Zr content. Lead titanate also has a perovskite structure and a tetragonally distorted ferroelectric phase. With increasing Zr content, x, the tetragonal distortion, decreases, and at $x > 0.52$, the structure changes from tetragonal 4 mm to another ferroelectric phase with rhombohedral 3 m symmetry. The line dividing these two

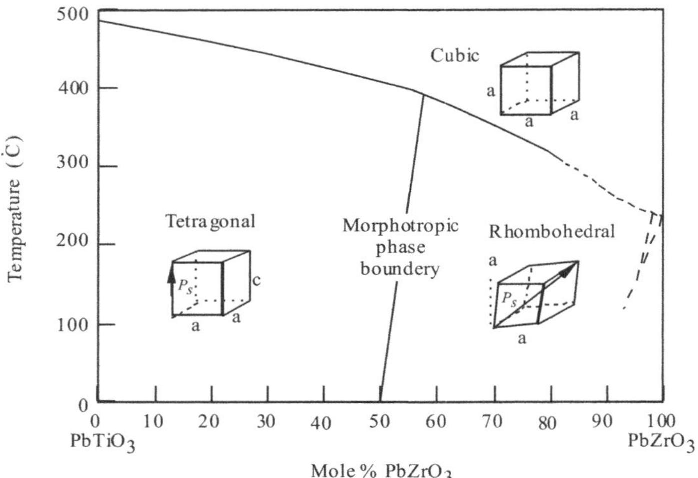

FIGURE 3.7 Phase diagram of the lead zirconate titanate $Pb(Zr_xTi_{1-x})O_3$ solid solution system.

phases in Figure 3.7 is called the *morphotropic phase boundary* (MPB). The composition at the MPB is assumed to be a mixture of the tetragonal and rhombohedral phases. The dependence of several piezoelectric strain coefficients (*d*) on composition over a narrow compositional range near the MPB is shown in Figure 3.8. All the *d* coefficients are observed to peak at the morphotropic phase boundary. The enhancement in the piezoelectric effect at the MPB has been attributed to the coexistence of the two phases, whose polarization vectors become more readily aligned by an applied electric field when mixed in this manner than may occur in either of the single-phase regions.

Doping the PZT material with donor or acceptor ions changes its properties dramatically. Materials doped with donor ions such as Nb^{5+} or Ta^{5+} are soft piezoelectrics. A soft piezoelectric has a low coercive field. The material becomes soft with donor doping largely due to the presence of Pb vacancies, which effectively facilitate domain wall motion. Materials doped with acceptor ions such as Fe^{3+} or Sc^{3+}, on the other hand, are hard piezoelectrics. A hard piezoelectric has a high coercive field. The material becomes hard with acceptor doping due to the generation of oxygen vacancies, which pin domain wall motion. We introduce these mechanisms further in Section 3.5.

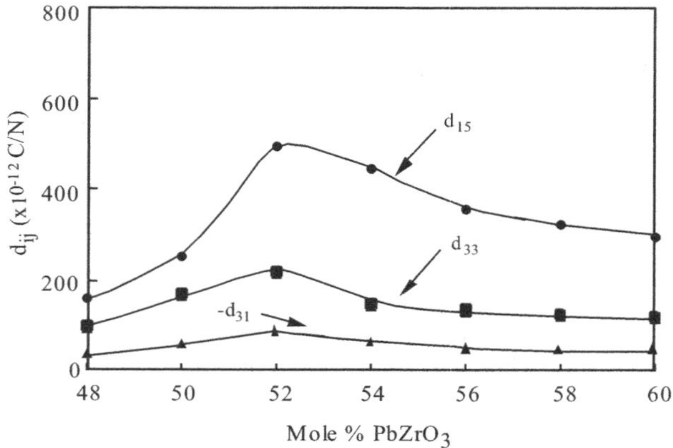

FIGURE 3.8 Dependence of several d constants on composition near the morphotropic phase boundary in the $Pb(Zr_xTi_{1-x})O_3$ solid solution system.

Various ternary solid solutions incorporating PZT and another ferroelectric perovskite have also been investigated. Some noteworthy examples utilize the following third phases: $Pb(Mg_{1/3}Nb_{2/3})O_3$, $Pb(Mn_{1/3}Sb_{2/3})O_3$, $Pb(Co_{1/3}Nb_{2/3})O_3$, $Pb(Mn_{1/3}Nb_{2/3})O_3$, $Pb(Ni_{1/3}Nb_{2/3})O_3$, $Pb(Sb_{1/2}Sn_{1/2})O_3$, $Pb(Co_{1/2}W_{1/2})O_3$, and $Pb(Mg_{1/2}W_{1/2})O_3$.[31]

A new solid solution system, $BiScO_3$-$PbTiO_3$, has been proposed from which materials with improved temperature stability can be derived for high-temperature applications.[32] The morphotropic phase boundary composition for this system has a Curie temperature of 450°C, which is about 100°C higher than that of the MPB composition in the PZT system.

Lead titanate, $PbTiO_3$, has an exceptionally large unit cell distortion. It is tetragonal at room temperature with a tetragonality ratio $c/a = 1.063$, and its Curie temperature is 490°C. Dense sintered $PbTiO_3$ ceramics are not easily obtained, because they break up into a powder when cooled through the Curie temperature due to the large spontaneous strain. Modified lead titanate ceramics incorporating A-site dopants are mechanically stronger and exhibit a high piezoelectric anisotropy. The modified compositions $(Pb,Sm)TiO_3$[33] and $(Pb,Ca)TiO_3$[34] exhibit extremely low planar coupling; that is, they have a large k_t/k_p ratio. The quantities k_t and k_p are the thickness-extensional and planar electromechanical coupling factors, respectively. Since medical imaging ultrasonic transducers made with these compositions can generate purely longitudinal waves through k_t, clear images are possible because the "ghost" image, generally associated with the transverse wave and generated through k_{31}, is not produced. Ceramics of another modified composition, $(Pb,Nd)(Ti,Mn,In)O_3$, have been developed that have a zero temperature coefficient of surface acoustic wave delay and therefore are superior substrate materials for SAW device applications.[35]

Since we investigated electric field induced strain curves on $(Pb,La)(Zr,Ti)O_3$ systems thoroughly,[36] the obtained data are introduced for the reader's reference on the actuator applications. In as-fired samples of piezoelectric ceramics, the net polarizations of individual grains are randomly oriented so as to produce a net polarization of zero for the entire sample. The net strain is likewise negligibly small under an applied electric field. Hence, as a final treatment of all piezoelectric ceramics, a relatively large electric field (>3 kV/mm) is applied in order to align the polarization directions of most of grains as much as possible. This is called *electric poling*. Refer to Figure 2.38, which illustrates the reorientation of ferroelectric domains in a polycrystalline material under an applied electric field.

The strain induced along the electric field direction (the longitudinal effect) in a poled (Pb,La)$(Zr,Ti)O_3$ (7/62/38) sample is shown in Figure 3.9.[36] When the applied field is small, the induced strain is nearly proportional to the field; this is the *converse piezoelectric effect* ($x = dE$), and the piezoelectric strain coefficient d_{33} can be determined from the slope ($\partial x/\partial E$). Note that the zero-strain should be retaken at the initial position of the strain-axis. As the field becomes larger (>100 V/mm), the strain curve deviates from this linear trend and significant *hysteresis* occurs (like a "chrysalis"

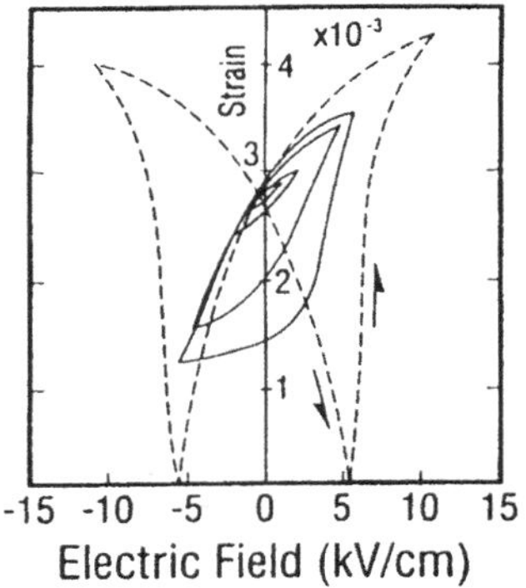

FIGURE 3.9 Field-induced strain curves for a piezoelectric PLZT (7/62/38). (From K. Furuta and K. Uchino: *Adv. Ceram. Mater.*, 1, 61, 1986.)

shape). This limits the usefulness of this material for actuator applications for positioning that requires a nonhysteretic response. Once the applied field exceeds the *coercive field* strength, E_C, the hysteresis assumes the characteristic "butterfly-shaped" curve traced with the dotted line in Figure 3.9. This strain hysteresis is caused by the reorientation of ferroelectric domains. Partial reorientation causes a pupa shape, while complete reorientation exhibits a butterfly shape.

The induced strain curves for various PLZT compositions under *bipolar* [-20 kV/cm $< E < +20$ kV/cm] and *unipolar* [$0 < E < +20$ kV/cm] drives are shown in Figure 3.10.[36] Bipolar drives are not used in general, because cracks are readily initiated with repeated polarization reversal, thus significantly shortening the lifetime of the actuator device. The coercive field in the rhombohedral phase is lower than that in the tetragonal phase; accordingly, the induced strain level also seems to be higher in the rhombohedral phase. With an increase of Lanthanum doping (donor doping), the Curie temperature decreases, and the strain increases. 8–9 molar% La on the

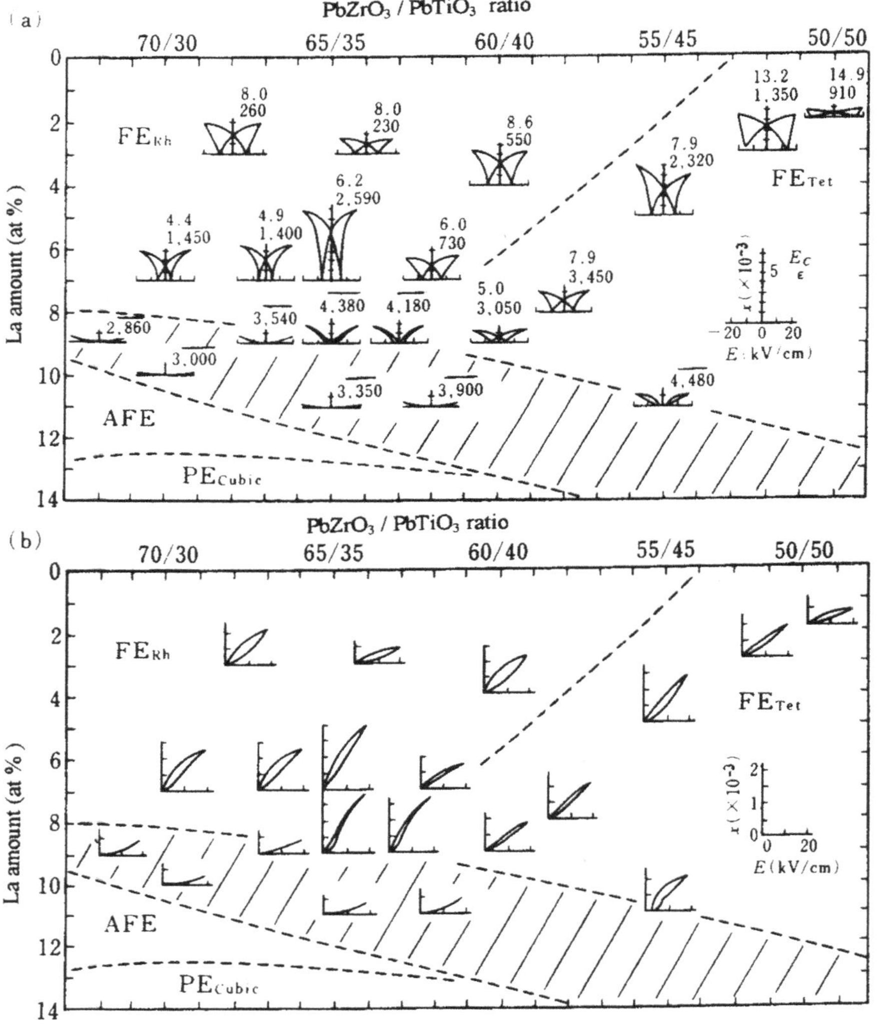

FIGURE 3.10 Field-induced strain curves for various PLZT compositions under: (a) bipolar, and (b) unipolar drive. (From K. Furuta and K. Uchino: *Adv. Ceram. Mater.*, 1, 61, 1986.) The magnitudes of the coercive field, E_C [unit: kV/cm], and the dielectric constant are written above each curve in figure (a).

morphotropic phase boundary composition provides pseudocubic symmetry and a Curie temperature around room temperature, leading to an induced strain curve close to an electrostrictive parabolic shape.

In a short summary, a typical magnitude for the piezoelectric strain coefficient d_{33} of practical piezoelectric ceramics is 100–500 × 10^{-12} m/V. If an electric field of 10^6 V/m is applied to a 10 mm sample, a strain $x_3 \approx 1 \times 10^{-3}$ is induced, which translates to a 10-μm displacement. Under such a large electric field, domain reorientation will occur in most cases, leading to a strain much larger than predicted just from the product ($d \cdot E$), where d is determined under a small electric field. The discrepancy can range from 10% to 100%. It is for this reason that piezoelectric coefficient measurements obtained at low field strengths (<10 V/mm), such as those taken at an electromechanical resonance with an impedance analyzer, are practically useless for accurately characterizing actuator response under a large electric field.

So far, we have considered mainly soft piezoelectrics (i.e., donor-doped PZT) suitable for off-resonance or pseudostatic operating conditions. Very hard, high power piezoelectrics (i.e., acceptor doped PZT) are typically used for ultrasonic resonance applications under AC conditions. Since the resonance displacement is estimated by $Q_m dE$, a material that retains a high mechanical quality factor, Q_m, and piezoelectric strain coefficient, d, at high vibration levels is essential. High power drive causes heat generation, which can lead to serious degradation of the electromechanical properties. One of the best high power piezoelectrics is a rare-earth–doped PZT-$Pb(Mn_{1/3}Sb_{2/3})O_3$ ceramic, which can generate a maximum vibration velocity of 1 m/sec.[37] We examine these high power materials and their properties in detail in Section 3.5. Note that under these high-power operating conditions, increasing the applied electric field does not increase the vibration energy. Most of the additional input electric energy is converted into heat. PZT becomes a ceramic heater, rather than an electromechanical transducer!

3.3.1.3 Complex Perovskite Relaxor Ferroelectrics

Relaxor ferroelectrics differ from the normal ferroelectrics (such as PZTs) described thus far, in that they exhibit a broad phase transition from the paraelectric to ferroelectric state (i.e., *diffuse phase transition*), a strong frequency dependence of the dielectric constant (i.e., *dielectric relaxation*), and a weak remnant polarization. Many of the lead-based relaxor materials have complex disordered perovskite structures (Figure 3.6a). They can be prepared either in polycrystalline form or as single crystals.

3.3.1.3.1 Diffuse Phase Transition

Although several compelling theories exist to describe relaxor ferroelectric phenomena, the exact mechanisms for the diffuse phase transition exhibited by these materials are yet to be fully understood. We introduce here a widely accepted description referred to as the *microscopic composition fluctuation model*, which is applicable even to macroscopically disordered structures.[38–40] The model proposes the existence of the so-called *Känzig region,* which is essentially the smallest region in which ferroelectricity can occur. The range of size for these regions is on the order of 100–1000 Å. It is assumed that there will be local fluctuations in the distribution of Mg^{2+} and Nb^{5+}, for example, on the B-site of the disordered perovskite cell for the relaxor composition $Pb(Mg_{1/3}Nb_{2/3})O_3$. A computer simulation of the composition fluctuation for a perovskite material with the general formula $A({}^{I}B_{1/2}\,{}^{II}B_{1/2})O_3$ and varying degrees of B-site short-range ordering (i.e., the possibility to find ${}^{II}B$ adjacent to ${}^{I}B$) is shown in Figure 3.11.[40] The fluctuation of the ${}^{I}B/{}^{II}B$ ratio is well represented by a Gaussian distribution function. It has been proposed in the context of this theory that local variations in composition will produce local variations in Curie temperatures, thus leading to the diffuse phase transition that is observed in the dielectric data. Short-range B-site ordering (the higher possibility of finding an Mg^{2+} ion adjacent to an Nb^{5+} ion for the ionic neutralization viewpoint) in $Pb(Mg_{1/3}Nb_{2/3})O_3$ has been observed by electron microscopy.[41] The high-resolution image revealed ordered regions with sizes in the range of 20–50 Å, as schematically depicted in the center diagram (i.e., partially ordered) of Figure 3.11.

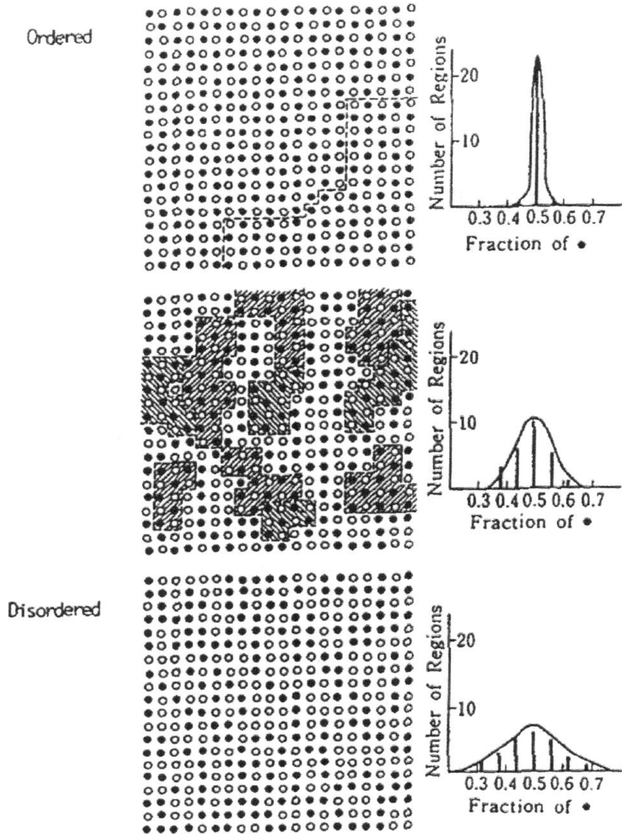

FIGURE 3.11 Computer simulation of the compositional fluctuation in a $A({}^{I}B_{1/2}{}^{II}B_{1/2})O_3$ perovskite for various degrees of B-site ordering.

3.3.1.3.2 Dielectric Relaxation

Another significant characteristic of a relaxor ferroelectric is its *dielectric relaxation* (the frequency dependence of the dielectric constant) from which their name is derived. The dielectric constant in $Pb(Mg_{1/3}Nb_{2/3})O_3$ (PMN) is plotted as a function of temperature in Figure 3.12 at various measurement frequencies.[42] We see from these data that as the measurement frequency is increased, the low temperature (ferroelectric phase) dielectric constant decreases and the peak in the curve shifts to higher temperature. This is in contrast to the trends observed for a normal ferroelectric, such as $BaTiO_3$, where the temperature of the dielectric constant peak varies very little with frequency. The origin of this effect has been examined in studies focused on $Pb(Zn_{1/3}Nb_{2/3})O_3$ single crystals.[43] The dielectric constant and loss (tan δ') are plotted as a function of temperature for an unpoled and a poled PZN sample in Figure 3.13. Domain configurations corresponding to each are also shown on the right-hand side of the figure. We see from these data that macroscopic domains are not present in the unpoled sample even at room temperature, and large dielectric relaxation and loss are observed below the Curie temperature range. Once macrodomains are induced by the external electric field, the dielectric dispersion disappears and the loss becomes very small (i.e., the dielectric behavior becomes rather normal!) below 100°C. Dielectric relaxation is thus associated with the microdomains present in this material. A sort of phase transition from macrodomain to microdomain status occurs at 100°C, rather sharply, of which further theoretical explanation is required.

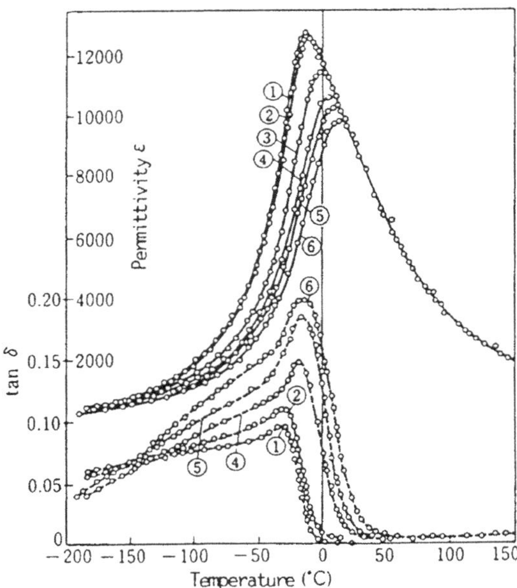

FIGURE 3.12 Temperature dependence of the dielectric permittivity in $Pb(Mg_{1/3}\text{-}Nb_{2/3})O_3$ at various frequencies (kHz): (1) 0.4, (2) 1, (3) 45, (4) 450, (5) 1500, (6) 4500.

3.3.1.3.3 Giant Electrostriction

Ceramics of PMN are easily poled when the poling field is applied near the transition temperature, but they are depoled completely when the field is removed, as the macrodomain structure reverts to microdomains (with sizes on the order of several 100 Å). This microdomain structure is believed to be the source of the exceptionally large electrostriction exhibited in these materials. The usefulness of the material is thus further enhanced when the transition temperature is adjusted to near room temperature. The longitudinal induced strain at room temperature as a function of applied electric field for 0.9PMN-0.1$PbTiO_3$ ceramic is shown in Figure 3.14.[8] Notice that the order of magnitude of the electrostrictive strain (10^{-3}) is similar to that induced under unipolar drive in PLZT (7/62/38) through the piezoelectric effect (Figure 3.9). An attractive feature of this material is the near absence of hysteresis, while the nonlinear strain behavior ($x = ME^2$) requires a sophisticated drive circuit.

Other electrostrictive materials include (Pb,Ba)(Zr,Ti)O_3[44] and various compositions from the PLZT system.[36,45] The large electrostriction effect in all these materials appears to be associated with the microdomain structure. In order to promote microscopic inhomogeneity throughout a given sample and the development of microdomain regions, dopant ions with a valence or ionic radius different from those of the ion they replace are introduced into the structure.

3.3.1.3.4 Giant Electromechanical Coupling Factor

A relaxor ferroelectric does exhibit an induced piezoelectric effect, and the electromechanical coupling factor, k_t, is observed to vary with the DC bias field strength. As the DC bias is increased, the coupling factor increases and eventually saturates. This behavior is highly reproducible and makes it possible to employ relaxor ferroelectrics as field-tunable ultrasonic transducers.[46]

Highly responsive single crystal relaxor ferroelectrics from solid solution systems with a morphotropic phase boundary are one of the epoch-making discoveries in piezoelectrics for applications as ultrasonic transducers and electromechanical actuators. Compositions very near the morphotropic phase boundary tend to show the most promise for these applications. Extremely high values for the electromechanical coupling factor ($k_{33} = 92\%–95\%$) and piezoelectric strain coefficient ($d_{33} = 1500$ pC/N) were first reported for single crystals at the MPB of the $Pb(Zn_{1/3}Nb_{2/3})$

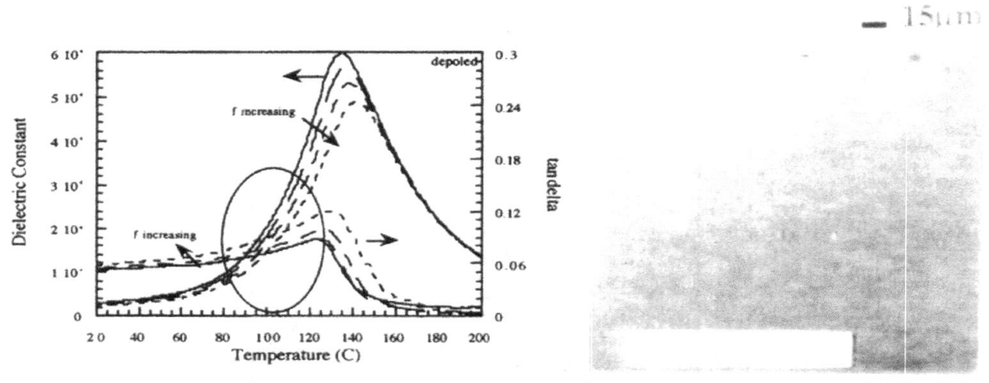

(a)

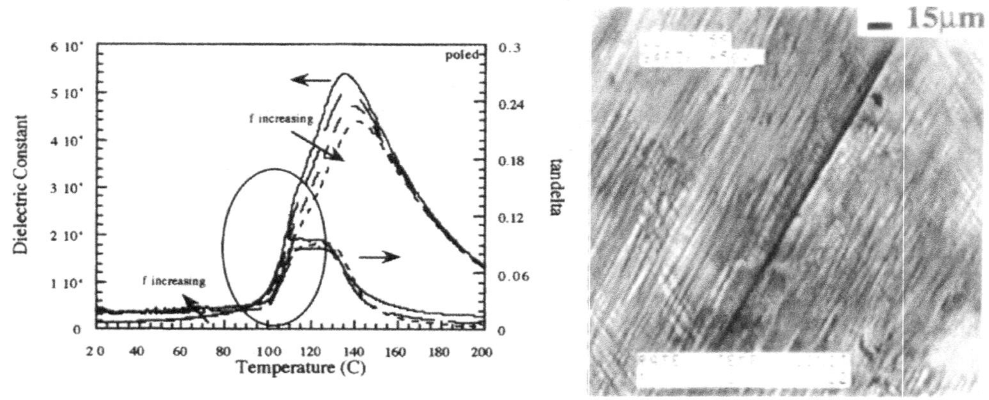

(b)

FIGURE 3.13 Dielectric constant and loss (tan δ') as a function of temperature measured along <111> (the P_S direction) in a $Pb(Zn_{1/3}Nb_{2/3})O_3$ single crystal: (a) a depoled sample (b) and a poled sample. (From H. Takeuchi et al: *Proc. IEEE 1990 Ultrasonics Symposium*, 697, 1990.) Domain configurations corresponding to each are shown on the right-hand side of the figure.

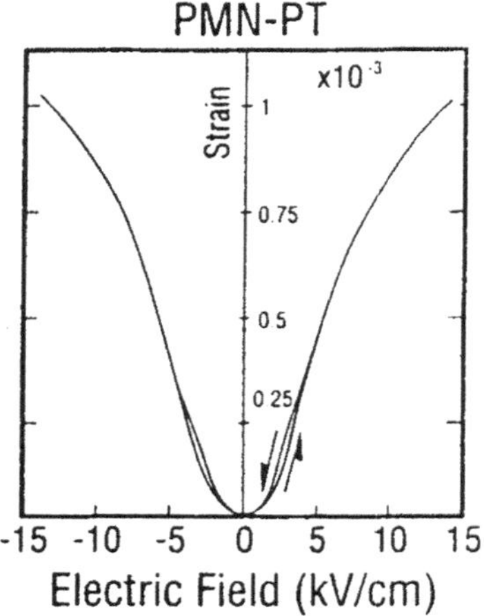

FIGURE 3.14 Electrostriction in $0.9Pb(Mg_{1/3}Nb_{2/3})O_3$-$0.1PbTiO_3$ (room temp.). (From L. E. Cross et al: *Ferroelectrics*, 23, 187, 1980.)

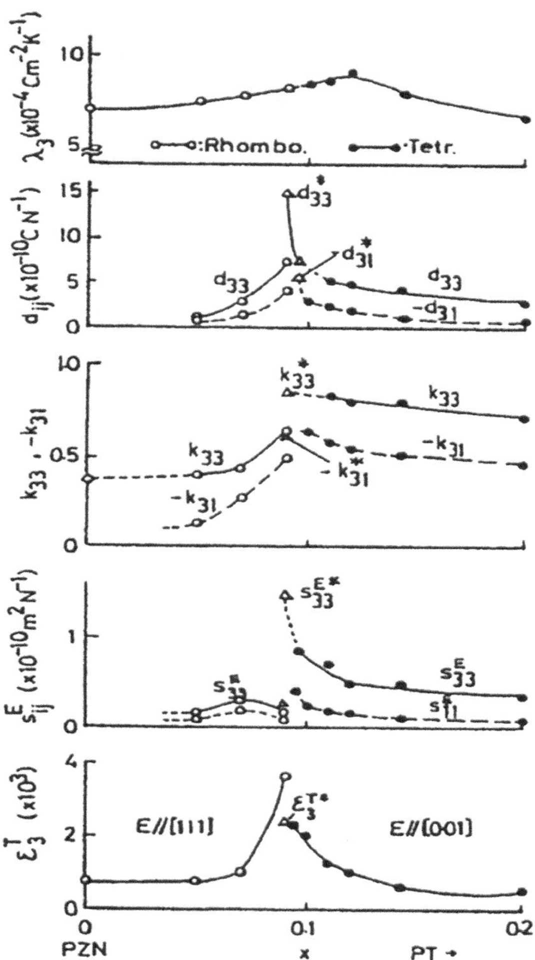

FIGURE 3.15 Electromechanical parameters in the (1-x)PZN-xPT system as a function of PT content, x. All peak at the MPB composition, $x = 0.9$. (From J. Kuwata et al: *Ferroelectrics*, 37, 579, 1981; J. Kuwata et al: *Jpn. J. Appl. Phys.*, 21(9), 1298, 1982.)

O_3-$PbTiO_3$ (PZN-PT) system in 1981 by Kuwata, Uchino, and Nomura.[47,48] Various electromechanical parameters (at room temperature) are plotted as a function of composition in Figure 3.15. Notice how two different values are plotted for the morphotropic phase boundary composition, $0.91Pb(Zn_{1/3}Nb_{2/3})O_3$-$0.09PbTiO_3$. The highest values for the piezoelectric coefficients and electromechanical coupling factors are observed for the rhombohedral composition only when the single crystal is poled along the perovskite [001] direction, not along [111], which is the direction of the spontaneous polarization.

Approximately 10 years after the discovery, Yamashita et al. (Toshiba)[49] and Shrout et al. (Penn State)[50] independently reproduced these findings, and refined data were collected in order to characterize the material for medical acoustic applications. Strains as large as 1.7% can be induced in single crystals from the morphotropic phase boundary composition of this system. The field-induced strain curve for a [001]-oriented (0.92)PZN-(0.08)PT crystal is shown in Figure 3.16.[50]

The mechanism for the enhanced electromechanical coupling in this case is basically similar to that for PZT, as described in Chapter 2. The shear coupling that occurs through d_{15} is dominant for perovskite piezoelectrics. The applied electric field should therefore be applied such that its direction is somewhat (~50°) canted from the spontaneous polarization direction in order to produce the optimum

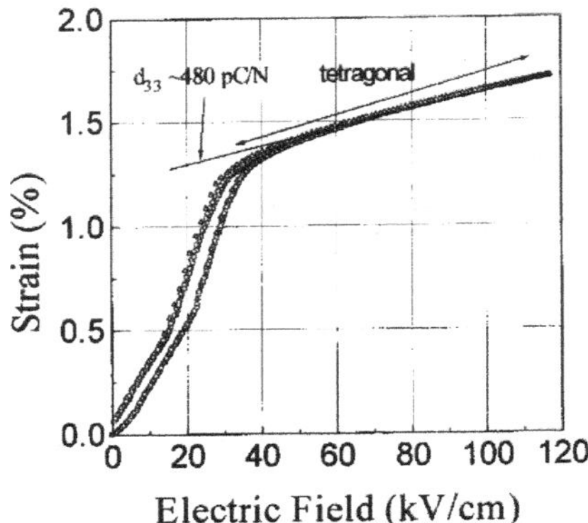

FIGURE 3.16 Field-induced strain for a [001]-oriented 0.92PZN-0.08PT single crystal. (From S. E. Park and T. R. Shrout: *Mat. Res. Innovt.*, 1, 20, 1997.)

piezoelectric response. The exceptionally high strain generated in materials with compositions near the morpotropic phase boundary (up to 1.7%) is associated additionally with the field-induced phase transition from the rhombohedral to the tetragonal phase. The applied electric field is generally directed along the perovskite [001] direction so that the electrostrictive response occurs in conjunction with a piezoelectric one, thereby enhancing the magnitude of the field-induced strain. Because of the significant cant angle between the spontaneous polarization and electric field directions, complicated multidomain structures are inevitably created in a single crystal. Accordingly, the so-called "domain engineering" concept came up to further explain the electromechanical coupling enhancement via the domain-domain interaction and domain interface/wall.[51]

It is convenient to classify electromechanical materials according to their coercive fields. A material with a coercive field much larger than 1 kV/mm is called a *hard piezoelectric*, and has a broad linear range, but exhibits relatively small induced strains. A material with a coercive field around 0.1–0.5 kV/mm is called a *soft piezoelectric*, and exhibits large induced strain, but also relatively large hysteresis. When the coercive field is between hard and soft, we sometimes call it "semi-hard" or "semi-soft," with some ambiguities. A material with a very small coercive field, much less than 0.1 kV/mm, is called an *electrostrictor*. It exhibits a strain that is approximately quadratic with respect to the applied electric field. Each type of material may also be characterized by its Curie point. The Curie point of the hard PZT piezoelectric is generally much higher (>250°C) than its usual operating temperature (room temperature), in contrast to an electrostrictive material, which has a transition temperature around or somewhat lower than room temperature. The Curie point of soft piezoelectrics ranges from 150°C to 250°C.

We now summarize and compare the practical characteristics of piezoelectric and electrostrictive ceramics:

1. The electrostrictive strain is about the same order of magnitude as the piezoelectric (unipolar) strain (0.1%). There is practically no hysteresis in the electrostrictive strain.
2. Piezoelectric materials require an electrical poling process, which can lead to significant aging. Electrostrictive materials do not need such pretreatment, but require a DC bias field for some applications due to the nonlinear response (insensitive just around $E \approx 0$).
3. Compared with piezoelectrics, the electromechanical response of electrostrictive ceramics tends not to deteriorate under severe operating conditions such as high temperature and large mechanical load, which is related to the no-poling process.
4. Piezoelectrics are superior to electrostrictive materials with regard to their temperature characteristics.
5. Piezoelectrics have smaller dielectric constants (lower capacitance) than electrostrictive materials and thus exhibit a quicker response.

3.3.1.4 Pb-Free Piezoelectrics

In 2006, the European Community started restrictions on the use of certain hazardous substances (RoHS), which explicitly limits the usage of lead (Pb) in electronic equipment. Pb-free piezoceramics started to be developed after 1999, and are basically classified into three groups; $(Bi,Na)TiO_3$ (BNT), $(Na,K)NbO_3$ (NKN) and tungsten bronze (TB), most of which were revival materials after the 1970s.

The share of the patents for bismuth compounds [bismuth layered type and $(Bi,Na)TiO_3$ type] exceeds 61%. This is because bismuth compounds are easily fabricated in comparison with other compounds. Honda Electronics, Japan, developed Langevin transducers using BNT-based ceramics for ultrasonic cleaner applications.[52] Their composition $0.82(Bi_{1/2}Na_{1/2})TiO_3$-$0.15BaTiO_3$-$0.03(Bi_{1/2}Na_{1/2})(Mn_{1/3}Nb_{2/3})O_3$ exhibits $d_{33} = 110 \times 10^{-12}$C/N, which is only 1/3 of that of a hard PZT, but the electromechanical coupling factor $k_t = 0.41$ is larger because of much smaller permittivity ($\varepsilon = 500$) than that of the PZT. Furthermore, the maximum vibration velocity of a rectangular plate (k_{31} mode) is close to 1 m/s (rms value), which is higher than that of hard PZTs.

$(Na,K)NbO_3$ systems exhibit the highest performance among the present Pb-free materials because of the morphotropic phase boundary usage. Figure 3.17 shows the current best data reported by Toyota Central Research Lab, where strain curves for oriented and unoriented (K,Na,Li) $(Nb,Ta,Sb)O_3$ ceramics are shown.[53] Note that the maximum strain reaches up to 1500×10^{-6}, which is equivalent to the PZT strain. Drawbacks include their sintering difficulty and the necessity of a sophisticated preparation technique (topochemical method for preparing flaky raw powder).

Tungsten-bronze types are an alternative choice for resonance applications because of their high Curie temperature and low loss. Taking into account the general consumer attitude on the disposability of portable equipment, Taiyo Yuden, Japan, developed microultrasonic motors using non-Pb multilayer piezoactuators.[54] Its composition is based on TB $[(Sr,Ca)_2NaNb_5O_{15}]$ without heavy metal. The basic

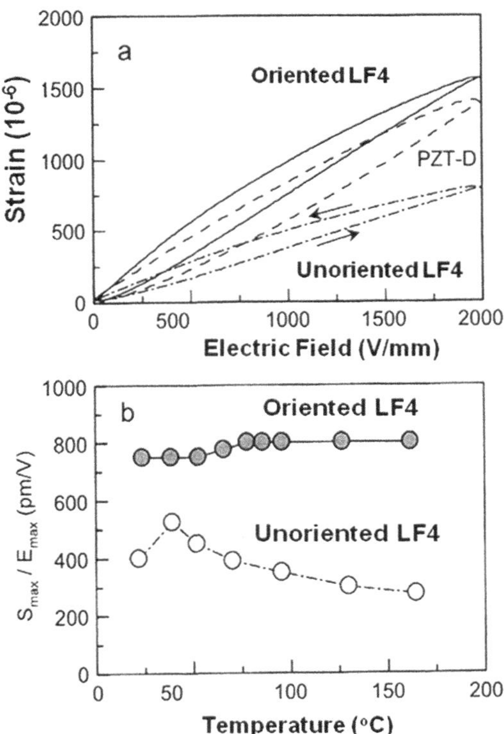

FIGURE 3.17 Strain curves for oriented and unoriented (K,Na,Li) $(Nb,Ta,Sb)O_3$ ceramics. (From Y. Saito, *Jpn. J. Appl. Phys.*, 35, 5168–73, 1996.)

piezoelectric parameters in TB ($d_{33} = 55 \sim 80$ pC/N, $T_C = 300°C$) are not very attractive. However, once *c*-axis–oriented ceramics are prepared, d_{33} is dramatically enhanced, up to 240 pC/N. Further, since the Young's modulus $Y_{33}^E = 140\,\text{GPa}$ is more than twice of that of PZT, higher generative stress is expected, which is suitable to ultrasonic motor applications. Taiyo Yuden developed a sophisticated preparation technology for oriented ceramics with a multilayer configuration, that is, preparation under a strong magnetic field, much simpler than the flaky powder preparation.

EXAMPLE PROBLEM 3.4

Table 3.4 summarizes piezoelectric, dielectric, and elastic properties of typical PZTs: soft PZT-5H, semi-hard PZT-4, and hard PZT-8. Using these data, answer the following questions and learn the interrelations between these parameters.

a. From the values of the piezoelectric *d* constant and relative permittivity, ε, calculate the piezoelectric *g* constants, and compare these calculations with the values provided in Table 3.4.
b. From the values of the piezoelectric *d* constant; permittivity, ε; and elastic compliance, s^E, calculate the electromechanical coupling factors, *k*, and compare these calculations with the values provided in Table 3.4.

TABLE 3.4
Piezoelectric, Dielectric, and Elastic Properties of Typical PZTs

	Soft PZT-5H	Semi-Hard PZT-4	Hard PZT-8
EM Coupling Factor			
k_p	0.65	0.58	0.51
k_{31}	0.39	0.33	0.30
k_{33}	0.75	0.70	0.64
k_{15}	0.68	0.71	0.55
Piezoelectric Coefficient			
d_{31} (10^{-12}m/V)	−274	−122	−97
d_{33}	593	285	225
d_{15}	741	495	330
g_{31} (10^{-3}Vm/N)	−9.1	−10.6	−11.0
g_{33}	19.7	24.9	25.4
g_{15}	26.8	38.0	28.9
Permittivity			
$\varepsilon_{33}^X / \varepsilon_0$	3400	1300	1000
$\varepsilon_{11}^X / \varepsilon_0$	3130	1475	1290
Dielectric Loss (tan δ') (%)	2.00	0.40	0.40
Elastic Compliance			
$s_{11}^E (10^{-12}\text{m}^2/\text{N})$	16.4	12.2	11.5
s_{12}^E	−4.7	−4.1	−3.7
s_{13}^E	−7.2	−5.3	−4.8
s_{33}^E	20.8	15.2	13.5
s_{44}^E	43.5	38.5	32.3
Mechanical Q_M	65	500	1000
Density ρ (10^3 kg/m^3)	7.5	7.5	7.6
Curie Temp (°C)	193	325	300

c. Calculate the elastic Poisson's ratio, $\left|s_{13}^E / s_{33}^E\right|$, and piezoelectric Poisson's ratio, $|d_{31}/d_{33}|$, for the above three PZTs, then compare the similarity.
d. From the Q_M values, calculate the elastic loss, tan ϕ', for these PZTs. Then, compare with the dielectric loss, tan δ', for these three PZTs.
e. We apply 100 W electric power on a k_{33}-type PZT-4 rod. How much will that electric energy convert to mechanical energy stored in the PZT rod? Of that stored mechanical energy, how much can we spend, maximum, for outside work? In this procedure, how much energy will be lost as heat?

Solution

a. The example calculation is made for PZT-4 with a k_{33}-type rod:

$$g_{33} = d_{33} / \varepsilon_0 \varepsilon_{33}^E = 285\times10^{-12} / 8.854\times10^{-12} \times 1300 = 24.8\times10^{-3} [\text{Vm}/\text{N}]$$

Calculate similarly for g_{31} and g_{15}, and for other PZTs.
b. The example calculation is made for PZT-4 with a k_{33}-type rod:

$$k_{33} = d_{33} / \sqrt{s_{33}^E \varepsilon_0 \varepsilon_{33}^X} = 285\times10^{-12} / \sqrt{15.2\times10^{-12} \times 8.854\times10^{-12} \times 1300}$$
$$= 0.68$$

Calculate similarly for k_{31} and k_{15}, and for other PZTs.
c. PZT-5H: $\left|s_{13}^E / s_{33}^E\right| = 7.2\times10^{-12} / 20.8\times10^{-12} = 0.35$
$|d_{31}/d_{33}| = 274 \times 10^{-12}/593 \times 10^{-12} = 0.46$
PZT-4: $\left|s_{13}^E / s_{33}^E\right| = 5.3\times10^{-12} / 15.2\times10^{-12} = 0.35$
$|d_{31}/d_{33}| = 122 \times 10^{-12}/285 \times 10^{-12} = 0.42$
PZT-8: $\left|s_{13}^E / s_{33}^E\right| = 4.8\times10^{-12} / 13.5\times10^{-12} = 0.36$
$|d_{31}/d_{33}| = 97 \times 10^{-12}/225 \times 10^{-12} = 0.43$
Both Poisson's ratios are close in number, but $|d_{31}/d_{33}|$ seems to be a little larger than $\left|s_{13}^E / s_{33}^E\right|$.
d. PZT-5H: tan $\phi' = 1/Q_M = 1/65 = 0.015 \leftrightarrow$ tan $\delta' = 0.02$
PZT-4: tan $\phi' = 1/Q_M = 1/500 = 0.002 \leftrightarrow$ tan $\delta' = 0.004$
PZT-5H: tan $\phi' = 1/Q_M = 1/1000 = 0.001 \leftrightarrow$ tan $\delta' = 0.004$
The dielectric loss, tan δ', seems to be a little larger than the elastic loss, tan ϕ'.
e. Mechanically converted and stored energy= $100\,\text{W} \times k_{33}^2 = 49\,\text{W}$. Of the stored mechanical energy, 1/4 $\sim$ 1/2 is usually spent out, depending on the k value at the λ_{max} condition. Just for the maximum output mechanical energy condition, 1/2 of the stored energy = 25 W is usable under the mechanical impedance matching condition. When this procedure is made quasistatically (off-resonance), the loss is primarily dielectric loss (0.4%) = 0.4 W. When this procedure is made at this PZT rod's resonance, the heat originates from the elastic loss (0.2%) = 0.2 W, because the applied field is small, but the strain/stress excited in the sample is large. Learn the situation in Figure 3.18, taking

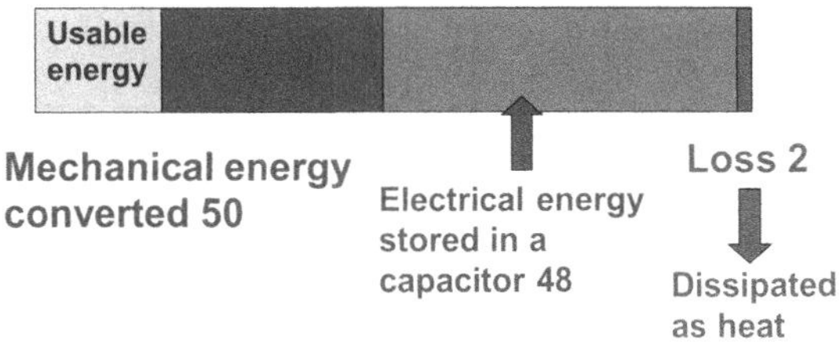

FIGURE 3.18 Energy conversion rate in a typical piezoelectric ceramic (PZT-4).

into account the difference between the electromechanical coupling factor, energy transmission coefficient, and efficiency. If we can recover the electrical energy stored in a damped capacitance, the efficiency of the piezoelectric device is as high as 97%–99%, exemplified by piezoelectric transformers.

3.3.2 Antiferro- to Ferroelectric Phase-Change Ceramics

Shape memory/superelastic alloys utilize the thermally or stress-induced phase transformation between the martensite and austenite phases, while the phase-change–induced strains employed for electric field–driven actuators typically involve a transition from an antiferroelectric to a ferroelectric phase, which is accompanied by an induced polarization.[55] The change in the spontaneous strain that occurs with this type of field-induced phase transition is theoretically much larger than that associated with a paraelectric to ferroelectric phase change or the non-180° domain reorientation that occurs for certain ferroelectrics under an applied field (see Section 2.4.8). The field-induced strain for an antiferroelectric lead zirconate stannate-based composition from the $Pb_{0.99}Nb_{0.02}[(Zr_xSn_{1-x})_{1-y}Ti_y]_{0.98}O_3$ (PZST) system is shown in Figure 3.19.[55] We see from these data that a maximum induced strain of about 0.4% is possible for this material, which is much larger than what is typically observed for conventional piezoelectrics or electrostrictors. A more rectangular-shaped hysteresis curve is observed for some compositions from this system (Figure 3.19a), because the change in strain occurring at the phase transition is much larger than the strain change induced in either the antiferroelectric or the ferroelectric phases. This mode of operation is referred to as *digital displacement* because the material effectively switches between on and off strain states.

A *shape memory effect* is observed for certain compositions (Figure 3.19b). Once the ferroelectric phase has been induced, the material will retain or "memorize" its ferroelectric state, even if the field is completely removed. A relatively small reverse bias field restores the material to its original state, as shown in the figure. One advantage to this type of shape memory ceramic is that it does not require a sustained applied electric field; instead, only a pulse drive is required, thereby saving considerable energy. The response speed of the phase transition is quick enough to generate a resonant mechanical vibration.[56]

3.3.3 Piezoelectric Polymers

Polyvinylidene difluoride, PVDF or PVF2, becomes piezoelectric when it is stretched during the curing process. The basic structure of this material is depicted in Figure 1.30. Thin sheets of the cast polymer are drawn and stretched in the plane of the sheet, in at least one direction, and frequently

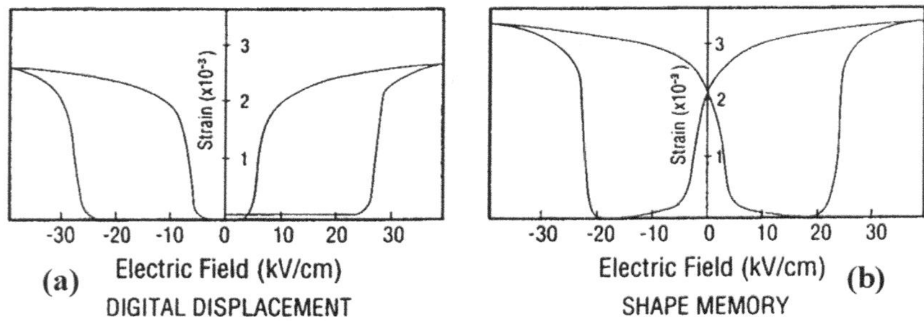

FIGURE 3.19 Field-induced strain in an antiferroelectric composition from the $Pb_{0.99}Nb_{0.02}[(Zr_{0.6}Sn_{0.4})_{1-y}Ti_y]_{0.98}O_3$ (PZST) system (longitudinal effect): (a) $y = 0.06$, digital displacement response and (b) $y = 0.063$, shape memory response. (From K. Uchino and S. Nomura: *Ferroelectrics*, 50, 191, 1983.)

also in the perpendicular direction, to transform the material to its microscopically polar phase. Crystallization from the melt produces the nonpolar α-phase, which can be converted into the polar β-phase by this uniaxial or biaxial drawing operation. The resulting dipoles are then aligned by electrically poling the material. Large sheets can be manufactured and thermally formed into complex shapes.

The copolymerization of polyvinylidene difluoride with trifluoroethylene (TrFE) results in a random copolymer (PVDF-TrFE) with a stable polar β-phase. This polymer need not be stretched; the microscopically polar regions are formed during the copolymerization process so that the as-formed material can be immediately poled. The poled material has a thickness-mode coupling coefficient, k_t, of 0.30.

Though polymer piezoelectrics do not include Pb (i.e., they are nontoxic), they cannot be disposed of because (1) they will remain permanently in the ground, or (2) fluorine gas is generated when burned. Murata Manufacturing Co. is seeking biodegradable devices using L-type poly-lactic acid (PLLA). PLLA has a vegetable corn-based composition.[57] Because it exhibits a pure piezoelectric without pyroelectric effect, the stress sensitivity is sufficient for leaf-grip remote controllers, which do not need a very long lifetime.

Piezoelectric polymers have the following characteristics: (a) they have small piezoelectric strain coefficients, d, (for actuators), but large g constants (for sensors); (b) they are lightweight and elastically soft, allowing for good *acoustic impedance matching* with water and human tissue; and (c) they have a low mechanical quality factor, Q_m, allowing for a broad resonance bandwidth. Piezoelectric polymers are used for directional microphones, ultrasonic hydrophones (i.e., sensors), and slow and large deformation actuators, similar to electroactive polymer/elastomer applications. The major disadvantages of these materials for actuator applications are the relatively small generative stress and the considerable heat generation that is generally associated with a low mechanical quality factor. PZT:polymer composites are equivalently used to piezoelectric polymers from an application viewpoint, but the composites are discussed in Section 4.3.6.

3.4 RELIABILITY ISSUES OF ACTUATOR MATERIALS

In addition to the primary large displacement/driving power performance, the following reliability issues should be satisfied in practical actuator materials.

1. Temperature stability in displacement
2. Quick response, low degradation/aging in usage
3. Mechanical stress dependence of displacement, fracture toughness
4. High power characteristic, heat generation (Section 3.5)

We consider the above reliability issues in this section and Section 3.5.

3.4.1 Temperature Stability in Field-Induced Strain

The temperature dependence of the field-induced strain in an electrostrictive $0.9Pb(Mg_{1/3}Nb_{2/3})O_3$-$0.1PbTiO_3$ ceramic under an applied field of 3.3 kV/cm is shown in Figure 3.20. Researchers strived to reduce the temperature dependence of the strain response when developing new electrostrictive materials. Two methods in particular have been effective in improving the temperature stability of new materials. The first is a macroscopic composite approach, where two materials with different temperature characteristics are combined in a suitable configuration to mutually compensate for and effectively cancel their individual temperature dependences. The other method involves the design of new solid solutions with intermediate compositions that provide optimum temperature characteristics.

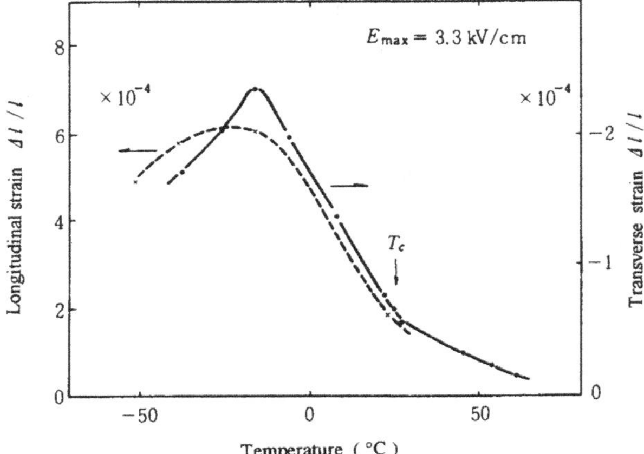

FIGURE 3.20 Temperature dependence of the electrostriction for $0.9Pb(Mg_{1/3}\text{-}Nb_{2/3})O_3\text{-}0.1PbTiO_3$ ceramics.

3.4.1.1 Macroscopic Composite Method

A composite structure incorporating $Pb(Mg_{1/3}Nb_{2/3})O_3$ and $Pb(Zr,Ti)O_3$ calcined powders has been developed as a new material possessing dielectric and electromechanical properties that are very stable over a broad range of temperatures.[58] The dielectric permittivity is shown as a function of temperature for diphasic mixtures of PMN and PZT (molar ratio 0.9:0.1) sintered at various temperatures in Figure 3.21a. The induced piezoelectric strain coefficient, d_{31}, is shown as a function of temperature for some diphasic PMN-PZT composites, 0.9PMN-0.1PT, and pure PMN under a bias of 1 kV/mm in Figure 3.21b. We see from these data that the composite materials have electromechanical responses that are far more stable with temperature (over the temperature range of −10°C to 70°C) than those exhibited by the pure PMN and 0.9PMN-0.1PT ceramics.

3.4.1.2 Microscopic Approach

Another approach for improving the temperature dependence of electrostriction involves microscale modifications of the material. A simple relationship between the temperature coefficient of

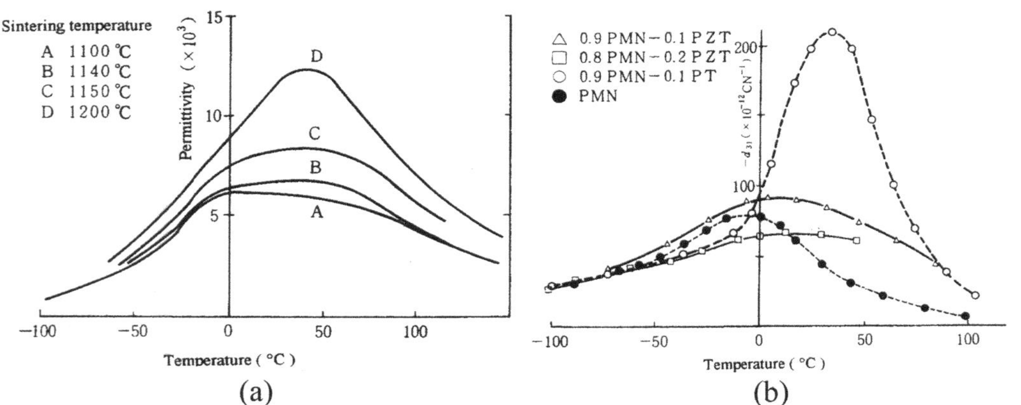

FIGURE 3.21 (a) Dielectric permittivity vs. temperature for diphasic mixtures of $Pb(Mg_{1/3}Nb_{2/3})O_3$ (PMN) and $Pb(Zr,Ti)O_3$ (PZT) powders (molar ratio 0.9:0.1) sintered at various temperatures. (b) Induced piezoelectric strain coefficient, d_{31}, vs. temperature for some diphasic PMN-PZT composites under a bias of 1 kV/mm. (From S. Nomura and K. Uchino: *Ferroelectrics*, 41, 117, 1982.)

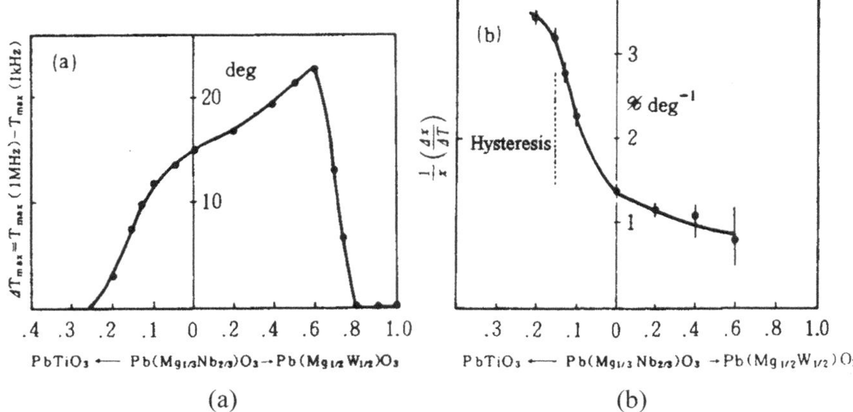

FIGURE 3.22 Dielectric and electrostriction data for materials from the $Pb(Mg_{1/3}Nb_{2/3})O_3$-$Pb(Mg_{1/2}W_{1/2})O_3$ and $Pb(Mg_{1/3}Nb_{2/3})O_3$-$PbTiO_3$ systems: (a) difference in the temperatures of the dielectric constant maximum for measurements made at 1 MHz and at 1 kHz, and (b) temperature coefficient of electrostriction (TCE), both as a function of composition. (From K. Uchino et al: *Jpn. J. Appl. Phys.*, 20(Suppl. 20–4), 171, 1981.)

electrostriction (TCE) and the diffuseness of the phase transition has been discovered for the solid solution systems $Pb(Mg_{1/3}Nb_{2/3})O_3$-$Pb(Mg_{1/2}W_{1/2})O_3$ (PMN-PMW) and $Pb(Mg_{1/3}Nb_{2/3})O_3$-$PbTiO_3$ (PMN-PT).[59] The diffuse phase transition (DPT) observed for these materials has been associated with the *short-range ordering* of cations on the B-site of the perovskite lattice. Long-range ordering of B-site cations tends to occur for certain annealed $Pb(^{I}B_{1/2}{}^{II}B_{1/2})O_3$ perovskites, while only limited short-range ordering has been observed for $Pb(^{I}B_{1/3}{}^{II}B_{2/3})O_3$-type compounds. A clear correlation between the degree of long-range ordering of B-site cations and the diffuseness of the phase transition has been observed for complex perovskites of this type. The difference between the temperatures of the dielectric constant maximum for measurements made at 1 MHz and 1 kHz, ΔT_{max}, is plotted as a function of composition for these systems in Figure 3.22a. This quantity, ΔT_{max} [$=T_{max}$(1 MHz) $- T_{max}$(1 kHz)], is one measure of the diffuseness of the phase transition. The noteworthy feature of this plot is the maximum that occurs at the $0.4Pb(Mg_{1/3}Nb_{2/3})O_3$-$0.6Pb(Mg_{1/2}W_{1/2})O_3$ composition. This is an indication that the diffuseness of the phase transition is strongly affected by the degree of cation order on the B-site.

The temperature coefficient of electrostriction is defined as follows:

$$\frac{\partial[\ln(x)]}{\partial T} = \frac{[x(0°C) - x(50°C)]}{50°C[x(25°C)]} \tag{3.29}$$

The TCE at room temperature for materials from the PMN-PMW and PMN-PT systems is plotted as a function of composition in Figure 3.22b. We see from these data that the TCE decreases to values as low as 0.8% ($°C^{-1}$) (at the 0.4PMN-0.6PMW composition) as ordering at the B-site increases (which occurs with increasing PMW content). These data were collected under low bias fields (1–3 kV/cm). A simple empirical equation that relates the ΔT_{max} for a given material to its TCE has been derived from the results depicted in Figures 3.22a and b:

$$\Delta T_{max}[\text{TCE}] = \Delta T_{max}\left[\frac{\Delta x}{x(\Delta T)}\right] = 0.22 \pm 0.03 \tag{3.30}$$

We see from this expression that the product of these parameters should be nearly constant for diffuse phase transition materials.

The electrostrictive materials developed for practical transducers can be classified into three categories based on the structural features giving rise to the DPT:

Category I: Perovskite structures with disordered B-site cations:

Type I(a): Materials produced from the combination of ferroelectric and nonferroelectric compounds [such as the solid solution system (Pb,Ba)(Zr,Ti)O_3 in which $BaZrO_3$ is the nonferroelectric component].

Type I(b): Materials in which A-site lattice vacancies also occur, which further promotes diffuse phase transition [such as the solid solution system (Pb,La,□)(Zr,Ti)O_3].

Category II: Perovskite structures with some degree of short-range ordering of B-site cations [such as the solid solution systems $Pb(Mg_{1/3}Nb_{2/3},Ti)O_3$ and $Pb(Mg_{1/2}W_{1/2},Ti)O_3$].

This classification also represents general techniques for improving the TCE. Improvement by means of the mechanism identified with Case (II) has been demonstrated with materials from solid solutions of PMN-PT with PMW or $Ba(Zn_{1/3}Nb_{2/3})O_3$ (BZN). The addition of PMW tends to generate microregions with 1:1 B-cation ordering, while incorporation of PZN tends to promote the formation of microregions with 1:2 ordering. Unfortunately, these modifications also tend to cause a decrease in the electrostrictive coefficient, M. The search continues for new solid solution systems such as these with compositions for which the TCE is optimized and acceptably high electrostriction coefficients are maintained.

3.4.2 Response Speed of Actuators

The response speed of piezoelectric/electrostrictive actuators depends not only on the material properties, but also on the mechanical resonance frequency of the device and on the specifications of the drive power supply.

3.4.2.1 Material Restrictions

Since the speed of domain wall motion is limited and very much dependent on the magnitude of the applied electric field, the polarization and induced strain will also exhibit pronounced frequency dependences. This is the source of the observed *hysteresis* in the polarization and induced strain with respect to the applied electric field. It is also the cause of *zero point drift* (i.e., the gradual shift in the strain level at zero field following the application of the unipolar drive voltage), *creep*, and *aging* (i.e., long-term strain degradation due to *depoling*). The creep characteristic of PZT-4 is shown in Figure 3.23, where we see the transverse strain as measured after an electric field pulse of 1 kV/cm is applied to the sample.[60] The induced strain, x, can be represented as the superposition of a fast-response part (x_0) and a slow-response part (Δx). The time dependence of Δx is approximated by an exponential function. The ratio of $\Delta x_\infty : x_o$(where Δx_∞ represents Δx at saturation, or t equal to infinity) depends not only on temperature, but also on the applied electric field. When the field is small, as is the case for the data appearing in Figure 3.23, this ratio can be as high as 20% or 30%.

The frequency dependence of the field-induced strain in simple disk samples of PLZT is shown in Figure 3.24.[36] We see from these data that the coercive field and hysteresis generally increase with increasing drive frequency. This is attributed to a decrease in the slow-response domain contribution at higher frequencies. The response in the field-induced strain (Figure 3.19a) in a phase-transition related material, antiferroelectric lead zirconate stannate-based PNZST composition $Pb_{0.99}Nb_{0.02}[(Zr_{0.6}Sn_{0.4})_{0.94}Ti_{0.06}]_{0.98}O_3$ system is much slower than that of piezoelectric materials because of the latent heat/energy associated with the transition.[56] The electric field required for the antiferro- to ferroelectric phase transition is increased with increasing the drive frequency, as shown in Figure 3.25. We see that a maximum induced longitudinal strain of about 0.4% is possible for this material at low frequencies (<5 Hz). However, above 10–30 Hz, the phase transition does not complete anymore

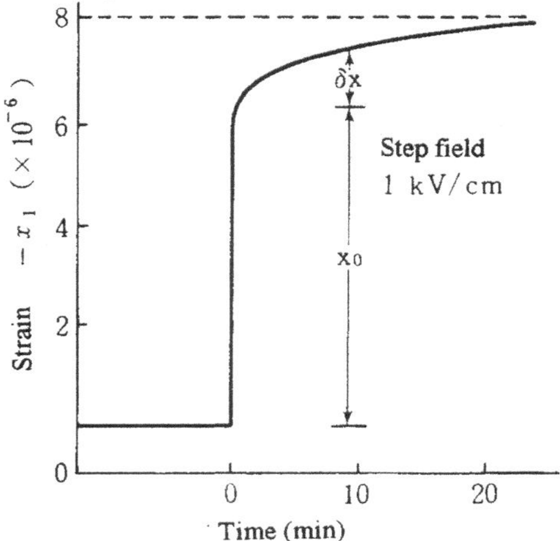

FIGURE 3.23 Strain creep characteristic in PZT-4. (From R. W. Basedow and T. D. Cocks: *J. Phys. E: Sci. Inst.*, 13,840, 1980.)

with this voltage supply. We can conclude that the maximum response is slower than 1 msec as a *digital displacement* actuator, which effectively switches between on and off strain states.

3.4.2.2 Device Restriction

There are three key factors in designing the actuator device: capacitance, device size, and vibration mode. The electric energy charging-up speed is restricted by the device capacitance under a certain

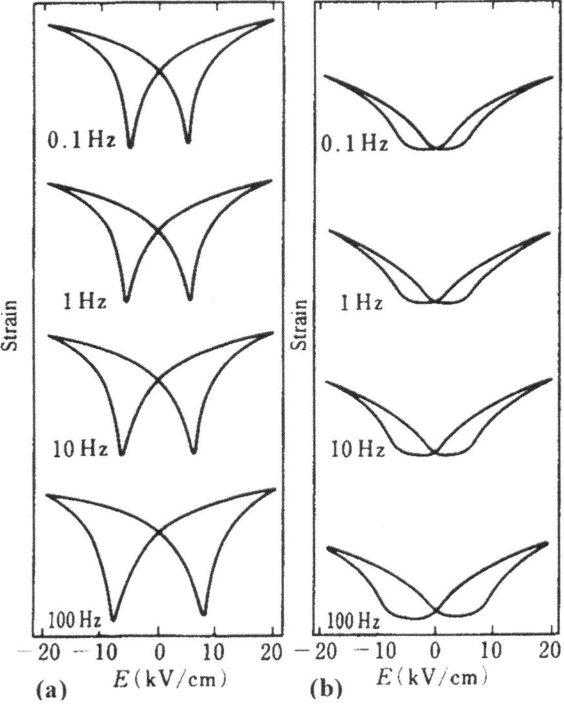

FIGURE 3.24 Frequency dependence of the field-induced strain in simple disk samples of PLZT: (a) PLZT (7/65/35) and (b) PLZT (9/63/37). (From K. Furuta and K. Uchino: Adv. Ceram. Mater., 1, 61, 1986.)

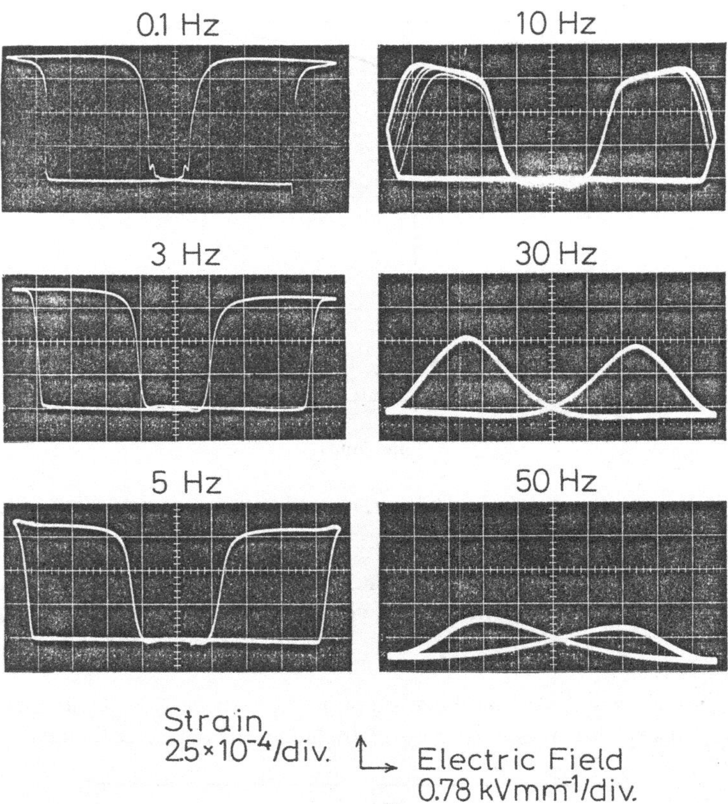

FIGURE 3.25 Frequency dependence of the field-induced strain in an antiferroelectric PNZST digital displacement response.

limited power supply. Remember the fundamental formula $Q = CV$ (Q: charge, V: voltage, C: capacitance). Supposing that 100 V is applied on a multilayer piezoactuator with capacitance 1 μF; we need to accumulate 10^{-4} C. If the power supply can provide 10 Amp, 10 μsec is enough for charging up. But if only 0.1 Amp is supplied from the driver (such as a standard impedance analyzer), 1 msec is required for charging up; that is, a standard impedance analyzer cannot measure the resonance response in this multilayer (ML) actuator with 150 kHz of the resonance frequency. We need to reduce the measuring voltage less than 0.1 V to obtain the impedance spectrum, though lowering the voltage increases the noise/error of the experimental results. This is the major reason electrostrictive actuators (with permittivity >10,000) are used only for slow pseudo-DC actuators, and piezoelectrics (with permittivity ~1000) are chosen for pulse drive and high frequency drive applications.

At frequencies in the vicinity of the mechanical resonance frequency of the piezoelectric device, the amplitude of the induced strain is enhanced considerably. Well above the resonance frequency, however, the strain is completely suppressed. The quickest response is thus achieved for operation frequencies near resonance. For a rod-shaped multilayer ceramic actuator, the fundamental resonance frequency is given by

$$f_R = v/2L, \tag{3.31}$$

where v is the sound velocity of the actuator material, and L is the rod length. Since a typical $v = 3000$ m/s in PZTs, a 10-mm-long ML rod exhibits 150 kHz, or the maximum response speed around 6.7 μsec.

Note that the resonance frequency of an actuator device also depends on the vibration mode, in addition to the size itself. A 10-mm bimorph bending actuator (flextensional type) whose displacement amplification factor is large (×1000 times), is not high (around 100 Hz–1 kHz, depending on the thickness and length), which ultimately limits the response speed of actuators of this type to this range. On the other hand, moonies/cymbals, which are composed of a sandwich structure of a PZT disk and two metal endcaps, exhibit intermediate displacement amplification (10–100 times) and response speed (~10 kHz) between the MLs and bimorphs. We will consider the operation of these devices further in Section 4.3.

3.4.2.3 Drive Circuit Limitation

Driving a ceramic actuator is essentially a process of charging a large capacitor (such as 1 μF in a multilayer actuator). Assuming a conventional power supply is used with an output impedance of 100 Ω, the *time constant*, which is determined by the product of the resistance and the capacitance (RC), is just 10^{-4} s. This corresponds to a response speed of 10 kHz, which is the practical limit for a ML device with a capacitance of 1 μF. If we use a power supply with an output impedance of 10 Ω, the response speed would be increased to 100 kHz. Thus, we need to choose a power supply with the output impedance lower than 1 Ω if we work up to the 1-MHz range. Another consideration is the maximum current of the power supply, which must be sufficient to charge the actuator. When 100 V is applied to the 1 μF actuator for 10^{-5} sec (corresponding to 100 kHz), the transient current attained is 10 A. In order to reduce the current drawn from the power supply, the capacitance of the device, or more specifically the dielectric constant of the material, must be reduced. This is the primary reason a soft piezoelectric is preferred over an electrostrictive ceramic for pulse drive applications. We examine the specifications of a ceramic actuator drive system again in Example Problem 5.11.

3.4.3 Mechanical Properties of Actuators

Solid-state actuators are used in mechanical systems, occasionally under certain prestress conditions, in order to optimize efficiency as well as to improve mechanical reliability. We will consider in this section the stress dependence of piezoelectricity and the mechanical strength of materials. The use of acoustic emission (AE) from piezoelectrics as a means of monitoring for device failure is also introduced.

3.4.3.1 Uniaxial Stress Dependence of Piezoelectric/Electrostrictive Strains

Even elastically stiff ceramic actuators will deform under an applied stress. Electric field–induced strain is affected by a bias stress. The uniaxial compressive stress dependence of the longitudinal field-induced strain in $BaTiO_3$-based, PZT, and lead magnesium niobate-lead titanate (0.65PMN-0.35PT) ceramics is shown in Figure 3.26a.[61] PZT- and PMN-based ceramics exhibit *maximum field-induced strains* (which is indicated on the plot where the individual lines intersect with the strain axis) much larger than the BT-based material. The *maximum generative stress* (indicated where the lines intersect the stress axis and sometimes referred to as the *blocking force*), on the other hand, is very nearly the same (about 3.5×10^7 N/m^2) for all the samples. This is because the elastic compliance of lead-based ceramics tends to be relatively large (i.e., they are elastically soft).

The strain as a function of stress is also plotted for electrostrictive PMN-PT actuators in Figure 3.26b.[62] We see from these data that the 0.9PMN-0.1PT devices exhibit the larger maximum field-induced strains of the two compositions, but the maximum generative forces for each composition at similar field strengths are nearly the same (≈46 MPa). It is worth noting here that (1) the electrostriction exhibits significant nonlinear behavior against the stress (NB: electrostriction has a quadratic relation with the electric field), (2) the elastic compliance of the 0.9PMN-0.1PT electrostrictor decreases abruptly with increasing bias stress, while (3) the piezoelectric 0.65PMN-0.35PT specimens are practically linear/straight at each applied electric field strength. Precisely

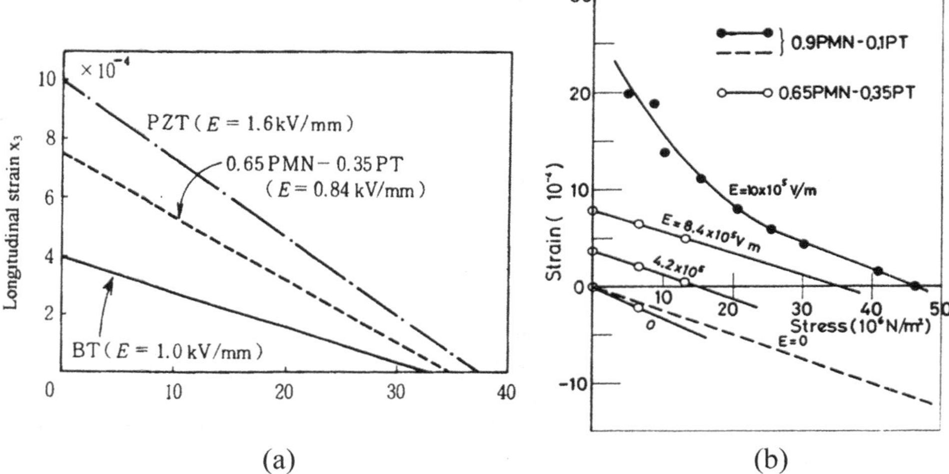

FIGURE 3.26 (a) Uniaxial stress dependence of the electric field-induced strain in $BaTiO_3$ (BT)-, $Pb(Mg_{1/3}Nb_{2/3})O_3$ (PMN)-, and $Pb(Zr,Ti)O_3$ (PZT)-based piezoelectric ceramics.(From K. Uchino, *Piezoelectric/ Electrostrictive Actuators*, Morikita Pub. Co., Tokyo, 1985.) (b) Uniaxial stress dependence of the longitudinal strain for PMN-PT actuators. (From Y. Nakajima et al: *Jpn. J. Appl. Phys*., 24(2), 235, 1985.) 0.9PMN-0.1PT: electrostrictor, 0.65PMN-0.35PT: piezoelectric.

speaking, note also that (4) with increasing the applied electric field, the stress-strain curve slope decreases; that is, the elastic compliance decreases (elastically hardening).

The variation of the large field "effective" piezoelectric strain coefficient, d_{33} (which corresponds to the maximum value of $\partial x/\partial E$ for electrostrictive materials), with applied compressive stress for various electrostrictive, soft piezoelectric, and hard piezoelectric ceramics is shown in Figure 3.27.[63] All the samples show a dramatic decrease in the piezoelectric coefficient above a certain critical level of applied stress. This change is reversible later in the stress cycle for electrostrictive materials, but in hard piezoelectric ceramics it is not, and the lower d_{33} is retained even after the compressive stress is removed. This degradation has been related to the reorientation of domains induced by the applied stress.[64] Note that this critical level of stress tends to be highest for hard piezoelectrics, somewhat less for soft piezoelectrics, and lowest for electrostrictors. We can deduce from these data that for conditions where the maximum stress is less than 10^8 N/m^2 (1 ton/cm^2), piezoelectric ceramics are useful, while under larger stresses, electrostrictive ceramics seem to be more reliable as long as the ceramic can endure such high stress levels (100 MPa).

The variation of the weak field effective piezoelectric coefficient with applied compressive stress is quite different from the large field response. The piezoelectric strain coefficients, d_{33} and d_{31}, for various PZT compositions (from soft [PZT-5H] to hard [PZT-8]) are plotted as a function of applied compressive stress in Figure 3.28.[65] Reduced units are defined with respect to the actual d values measured at $X_3 = 0$. We see in all cases the piezoelectric coefficients increase up to a certain critical stress, then dramatically decrease beyond this level of applied stress. The degradation in this case is associated with a stress-induced depoling effect. It is irreversible for soft piezoelectric materials.

The most interesting feature of the data plotted in Figure 3.28 is the increase in the piezoelectric coefficients when subjected to the lower applied stresses. This is related to the poling direction with respect to the applied stress direction. The dependence of the piezoelectric d_{33} coefficient on specimen orientation with respect to the poling direction is shown in Figure 3.29.[66] After the ceramic PZT specimens were poled, they were cut as illustrated in Figure 3.29a, and the effective d_{33} coefficient (normalized by $d_{33//Ps}$, the value of d_{33} along the poling direction) was measured for each.

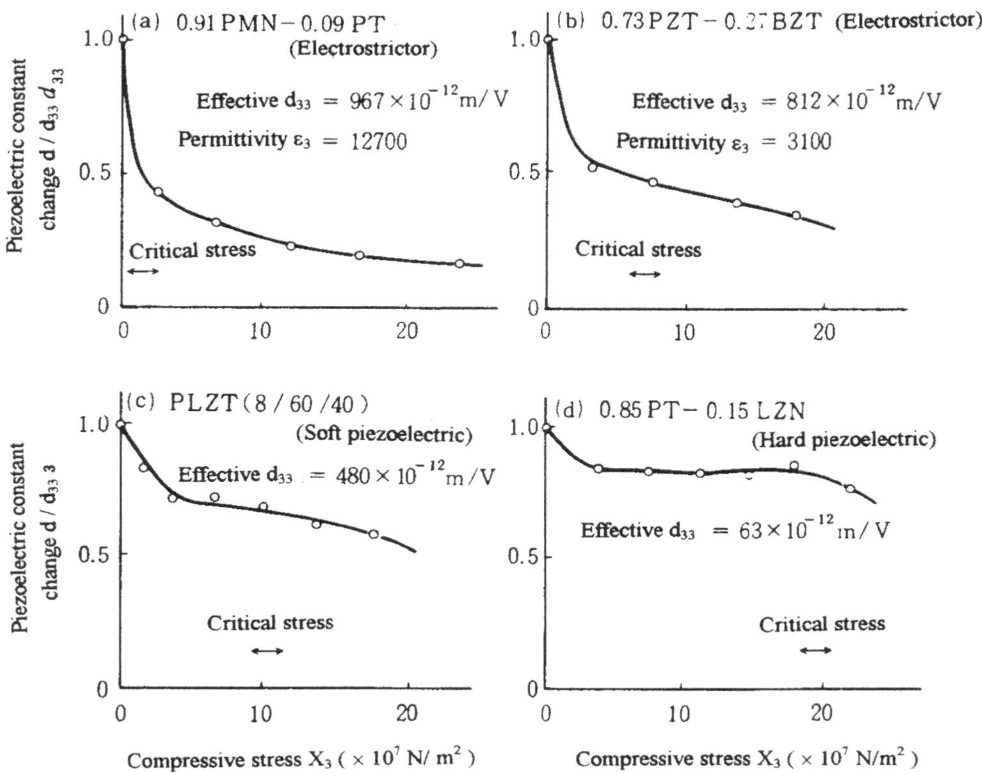

FIGURE 3.27 Variation of the large field effective piezoelectric strain coefficient, d_{33}, with applied compressive stress for various electrostrictive, soft piezoelectric, and hard piezoelectric ceramics. (From S. Nomura et al: *Abstract Jpn. Appl. Phys.*, 764, Spring, 1982.)

The effective d_{33} values for the specimens from this series are plotted as a function of orientation with respect to the poling direction in Figure 3.29b. The actual magnitudes of the effective strain coefficients of specimens oriented parallel to the poling direction, $d_{33//}$, for the tetragonal PZT (48/52) and rhombohedral PZT (54/46) compositions are 98 pC/N and 148 pC/N, respectively. Note that for rhombohedral PZT, d_{33} increases as the cutting angle increases from 0° to 45°, attaining at 45° an enhancement of 1.25 times the response observed for the specimen with the parallel alignment. This is directly related to the crystal orientation effects, d_{33}, discussed in Section 2.4.7. Similar to what occurs for the samples previously described, the spontaneous (or remnant) polarization for these ceramic specimens subjected to a large compressive uniaxial stress (greater than the critical stress) will be somewhat canted and the samples will become partially depoled. Thus, for applied stresses below the critical stress, the piezoelectric coefficient increases, but beyond this, permanent degradation occurs. It is due to this stress-enhanced piezoelectric response that so-called "bolt-clamped design," through which the appropriate level of stress is maintained on the device during operation, was developed and commonly applied to hard PZT Langevin tansducers.[67]

Recall also the argument conducted in Section 3.2.2.2 on the energy transmission coefficient: the load should be half of the maximum generative stress when we apply the maximum electric field, E. Actual output work can most efficiently be taken under this level of compressive stress condition.

3.4.3.2 Mechanical Strength

The mechanical properties and mechanical strength of the ferroelectric ceramics used in actuator devices are as important as the electrical and electromechanical properties. We now consider some

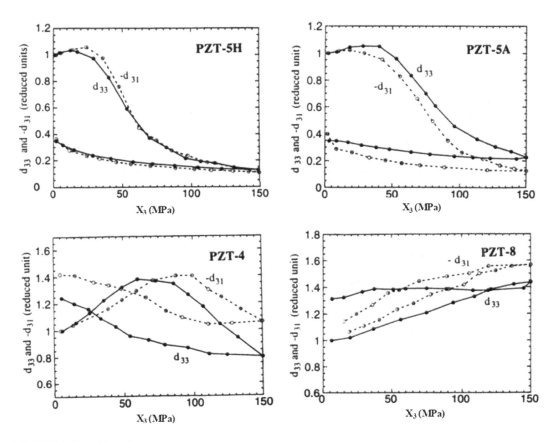

FIGURE 3.28 Weak field piezoelectric strain coefficients, d_{33} (solid line) and d_{31} (dashed line), for various PZT compositions (from soft [PZT-5H] to hard [PZT-8]) as a function of applied compressive stress. (From Q. M. Zhang et al: *J. Mater. Res.*, 12, 226, 1996.)

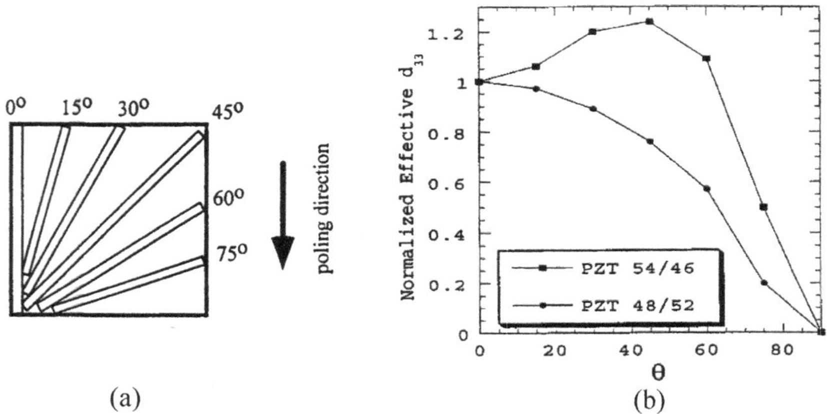

FIGURE 3.29 Dependence of the effective piezoelectric d_{33} coefficient on the specimen orientation: (a) sample orientation with respect to the poling direction, and (b) effective d_{33} as a function of orientation with respect to the poling direction, θ. (From X. –H. Du et al: *J. Ceram. Soc. Jpn.*, 107(2), 190, 1999.) (Data are normalized by $d_{33//Ps}$.)

general principles of fracture mechanics, the fracture toughness of piezoelectric ceramic as it relates to grain size, and the mechanical strength of the material as it is affected by field-induced strains. The use of acoustic emission from piezoelectrics as a means of monitoring for device failure will also be presented.

3.4.3.2.1 Fracture Mechanics

Ceramics are generally *brittle*, and fracture of these bodies tends to occur suddenly and catastrophically.[68] Improvement of the *fracture toughness* (which is essentially the material's resistance to the development and propagation of cracks within it) is thus a key issue in the design of new ceramics for actuator applications. The brittle nature of ceramic materials is directly related to their crystal structure. The atoms in ceramic crystallites are typically *ionically* or *covalently bonded* and their displacement under external influences, such as an applied stress, is limited. The mechanisms for relieving stress in such structures are few and, therefore, even a small stress can cause fracture. In polycrystalline samples, there are two types of crack propagation: (1) *transgranular*, in which the cracks pass through grains, and (2) *intergranular*, in which the cracks propagate along grain boundaries.

Three fundamental stress modes that may act on a propagating crack are depicted in Figure 3.30. Mode I involves a tensile stress, as shown in Figure 3.30a. Mode II leads to a sliding displacement of the material on either side of the crack in opposite directions, as shown in Figure 3.30b. The stress distribution associated with Mode III type produces a tearing of the medium. The fracture of most brittle ceramics corresponds primarily to Mode I; thus, this is the mode of principal concern to designers of new ceramic materials for actuator devices.[69]

Theoretically, the fracture strength of a material can be evaluated in terms of the cohesive strength between its constituent atoms, and is approximately one-tenth of the Young's modulus, Y. The experimentally determined values do not support this premise, however, as they are typically three orders of magnitude smaller than the theoretically predicted values. This discrepancy has been associated with the presence of microscopic cracks, which are sometimes called *stress raisers*,[70] in the material before any stress is applied to it. It has been shown that for a crack with an elliptical shape, oriented with its major axis perpendicular to the applied stress, σ_0, (Mode I), the local stress concentration, σ_m, at the crack tip increases significantly as the ellipticity of the defect increases. The relationship between the applied and local stresses is given by:

$$\sigma_m = K_t \sigma_o \tag{3.32}$$

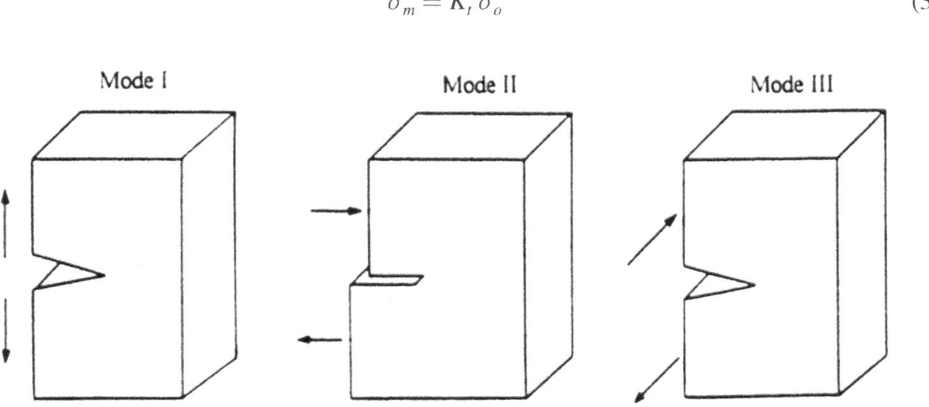

FIGURE 3.30 Three fundamental stress modes that may act on a propagating crack: (a) *Mode I*: tensile mode in which a tensile stress acts normal to the crack plane, (b) *Mode II*: sliding mode in which a shear stress acts normal to the crack edge plane, (c) *Mode III*: tearing mode in which a shear stress acts parallel to crack edge plane.

where K_t is the stress concentration factor. Fracture is expected to occur when the applied stress level exceeds some critical value, σ_c. The *fracture toughness*, K_{Ic}, is defined as follows:

$$K_{Ic} = F\sigma_c\sqrt{\pi a} \tag{3.33}$$

where F is a dimensionless parameter that depends on both the sample and crack geometries, σ_c is the critical stress, and a is the microcrack size. The fracture toughness is a fundamental material property that depends on temperature, strain rate, and microstructure.

3.4.3.2.2 Measurement of Fracture Toughness

There are four common methods for measuring the fracture toughness of a ceramic: (1) *indentation microfracture* (IM), (2) controlled surface flow (CSF), (3) chevron notch (CN), and (4) single-edge notched beam (SENB). We will focus most of our attention on the first of these methods, as it is of the four perhaps one of the most commonly employed. The test is initiated by artificially generating cracks on a polished surface of the sample with a Vickers pyramidal diamond indenter. The fracture toughness, K_{Ic}, is then determined from the indentation size and the crack length.

Three types of cracks are produced by indentation: (1) the *Palmqvist crack*, (2) the *median crack*, and (3) the *lateral crack*, as illustrated in Figure 3.31. A *Palmqvist crack* is generated at the initial stage of loading. It has the shape of a half ellipse, and occurs around the very shallow region of plastic deformation near the surface. As the load is increased, a crack in the shape of a half circle starts to form at the boundary between the plastic and elastic deformation regions. This is called a *median crack*. Above a certain critical indenter load, the median crack will reach the surface. The indentation also produces a residual stress around the indented portion of the sample, in which are generated more Palmqvist and median cracks, as well as *lateral cracks* in the plastic deformation region, which will not reach the surface.

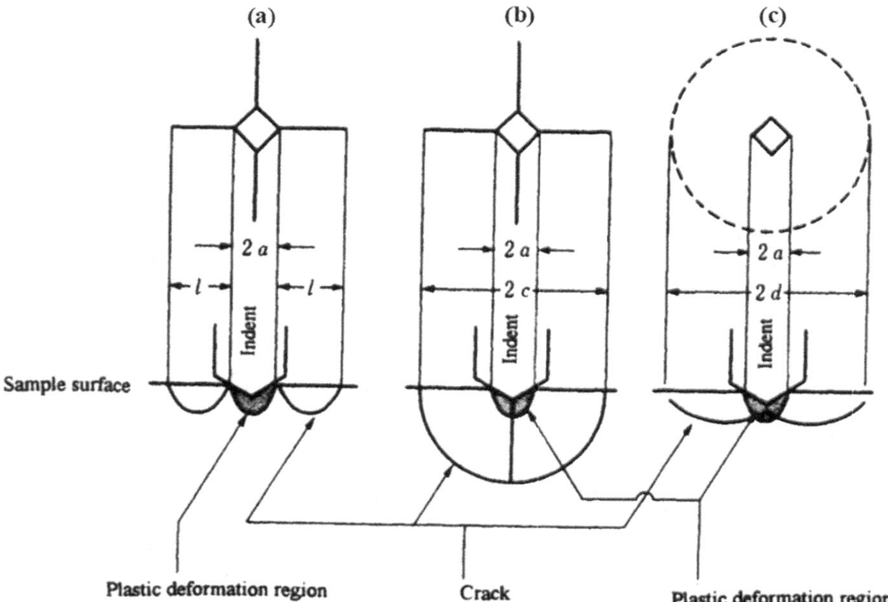

FIGURE 3.31 Crack shapes generated by the Vickers indentation: (a) the Palmqvist crack, (b) the median crack, and (c) the lateral crack. (Top and sectional views are shown.)

The indentation microfracture method is used to characterize Palmqvist and/or median cracks in order to determine K_{Ic}. A theoretical equation descriptive of median cracks has been derived, which has the following form:

$$\left[\frac{K_{Ic}\phi}{H\sqrt{a}}\right]\left(\frac{Y\phi}{H}\right)^{2/5} = (0.055)\log\left(\frac{8.4a}{c}\right) \tag{3.34}$$

where ϕ is a restriction constant, H is the hardness, Y is the Young's modulus, and a and c are half of the indentation diagonal and the crack length, respectively.[71] The theory that produced this equation was further refined, and modified expressions[72] that individually characterize the Palmqvist and median cracks were derived as follows:

$$\left[\frac{K_{Ic}\phi}{H\sqrt{a}}\right]\left[\frac{H}{Y}\right]^{2/5} = (0.018)\left[\frac{(c-a)}{a}\right]^{-1/2} \left[\text{Palmqvist crack}\right] \tag{3.35}$$

$$\left[\frac{K_{Ic}\phi}{H\sqrt{a}}\right] = (0.203)\left[\frac{c}{a}\right]^{-3/2} \left[\text{Median crack}\right] \tag{3.36}$$

The fracture toughness, K_{Ic}, can be calculated using the crack length determined from the IM method. Other methods require precracking a sample prior to failure testing, and then K_{Ic} is calculated using the fracture stress determined from the test. The controlled surface flow method, for example, uses the median crack generated by a Vickers indenter as a precrack to determine the relationship between the fracture stress, σ_f, and the crack length, $2c$, under three- or four-point bending tests. The fracture toughness, K_{Ic}, is then calculated from:

$$K_{Ic} = (1.03)\sigma_f\sqrt{\frac{\pi b}{q}} \tag{3.37}$$

The parameter q is defined to be:

$$q = \Phi^2 - 0.212\left[\frac{\sigma_f}{\sigma_y}\right]^2 \tag{3.38}$$

where σ_y is the tensile yield stress and Φ is the second perfect elliptic integration given by:

$$\Phi = \int_0^{\pi/2} [\cos^2\theta + (b/c)^2\sin^2\theta]^{1/2} d\theta \tag{3.39}$$

Here, b is half of the minor axis of the elliptically shaped crack (referred to as the crack depth) and c is half of the major axis of the elliptical crack (referred to as the crack length).

The probability of fracture, $p(\sigma_f)$, for a sample subjected to a stress, σ_f, in a three- or four-point bend test is described by the following equation:

$$1 - p(\sigma_f) = \exp[-(\sigma_f/\sigma_o)^m] \tag{3.40}$$

where σ_0 is the average bending strength and m is referred to as the *Weibull coefficient*. The Weibull stress is a well-known means of predicting the likelihood of weakest link brittle fracture. Imagine a necklace chain composed of many small gold rings connected/linked each other. When we pull this chain with a strong force, the chain will be torn into two pieces. The torn part should be at the mechanically weakest gold ring. Weibull stress is based on its use of a two-parameter Weibull distribution, a commonly used distribution in probabilistic engineering. The distribution is defined by a shape parameter, the Weibull modulus, and a scaling parameter. The parameters σ_0 and m in Equation 3.40 are the average fracture stress and distribution level of the fracture stress, respectively. When this relationship is rendered in terms of the natural logarithm (see Example Problem 3.5), we should find that the quantity ln(ln[1/(1 − p)]) is linearly related to ln(σ_f) (σ_f: measured fracture stress). A graph of this function is called a *Weibull plot.*

EXAMPLE PROBLEM 3.5

Three-point bend tests were carried out on a series of barium titanate-based multilayer actuators (laminated along the *l* direction). The sample configuration is illustrated in Figure 3.32. The results of the tests are summarized in the following table:

Sample	Length, *l* (cm)	Width, *b* (cm)	Height, *h* (cm)	Load, *M* (kg)
1	0.38	0.420	0.106	4.60
2	0.48	0.415	0.121	4.00
3	0.49	0.420	0.093	2.00
4	0.40	0.420	0.112	5.50
5	0.31	0.415	0.102	2.90
6	0.55	0.420	0.099	2.34

The samples were collapsed under the load 2.34–4.60 kgf. Calculate the average bending strength, σ_0, and the Weibull coefficient, m.

SOLUTION

The fracture stress, σ_f, can be obtained from:

$$\sigma_f = 3\ M\ l/2\ b\ h^2 \tag{P3.5.1}$$

which yields, for each of the samples tested, the following table:

Sample	1	2	3	4	5	6
σ_f [MN/m^2]	54.4	46.5	39.7	61.4	30.6	46.0

The sample sequence is changed according to the fracture stress (small to large) in the Excel table in Figure 3.33, top. The nonfracture probability, $1 - p(\sigma_f)$, of the sample under the applied stress, σ_f, of a three-point bend test is described by:

$$1 - p(\sigma_f) = \exp[-(\sigma_f\ /\ \sigma_o)^m] \tag{P3.5.2}$$

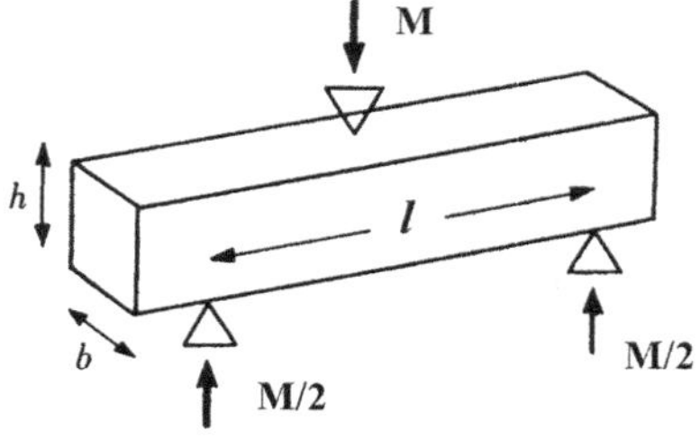

FIGURE 3.32 Sample configuration for the three-point bend test.

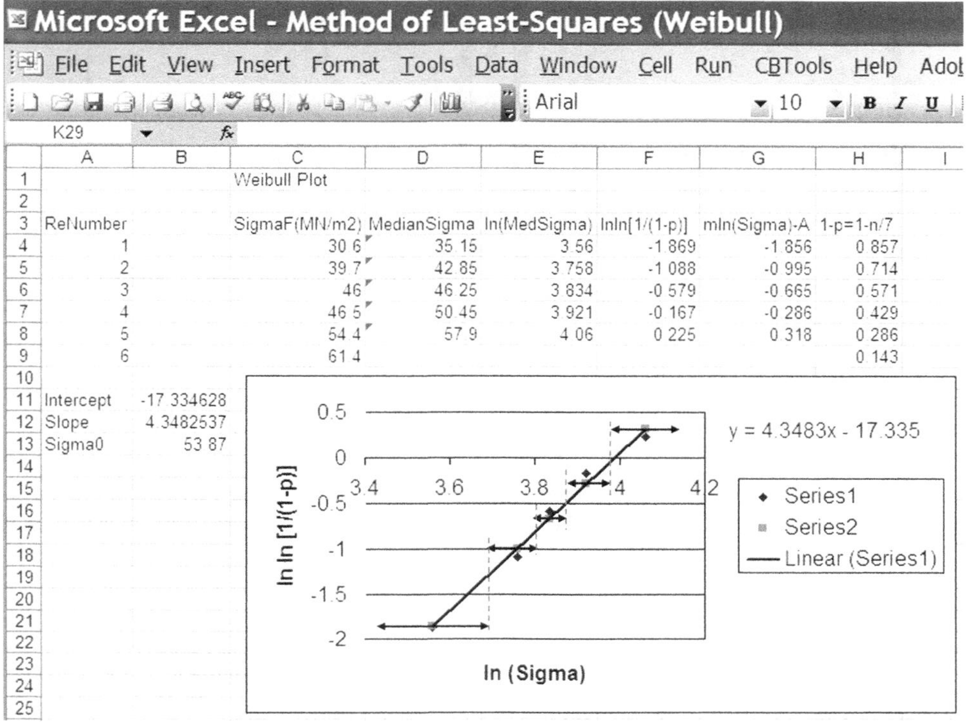

Microsoft Excel - Method of Least-Squares (Weibull)

	A	B	C	D	E	F	G	H
1			Weibull Plot					
2								
3	ReNumber		SigmaF(MN/m2)	MedianSigma	ln(MedSigma)	lnln[1/(1-p)]	mln(Sigma)-A	1-p=1-n/7
4		1	30.6	35.15	3.56	-1.869	-1.856	0.857
5		2	39.7	42.85	3.758	-1.088	-0.995	0.714
6		3	46	46.25	3.834	-0.579	-0.665	0.571
7		4	46.5	50.45	3.921	-0.167	-0.286	0.429
8		5	54.4	57.9	4.06	0.225	0.318	0.286
9		6	61.4					0.143
10								
11	Intercept	-17.334628						
12	Slope	4.3482537						
13	Sigma0	53.87						

FIGURE 3.33 Weibull plot for the data collected from the three-point bend tests conducted on a series of BT-based multilayer actuators.

where $p(\sigma_f)$ is the fracture probability, σ_0 is the average bending strength, and m is the Weibull coefficient. This equation can be rewritten in the following form:

$$\ln(\ln[1/(1-p(\sigma_f))]) = m\ [\ln(\sigma_0) - \ln(\sigma_f)] \quad \text{(P3.5.3)}$$

Designating the total number of samples as N, the nonfracture probability can be evaluated by:

$$[1-p(\sigma)] = 1-[n/(N+1)] \quad \text{(P3.5.4)}$$

where $(\sigma_n < \sigma < \sigma_{n+1})$.

When the sample number is small (only six in this case), one certain nonfracture probability has a wide range, which is shown in the plot in Figure 3.33: from one fracture strength to the next fracture strength. Thus, in such a case, the intermediate fracture strength (average) may be used to enhance the calculation accuracy. Using just the obtained fracture stresses calculated directly from the experimental data without considering this wide range increases the calculation error in the Weibull constant and average sigma zero increase. The Weibull plot for this set of fracture tests appears in Figure 3.33, with the regression line with Excel software. We are able to determine the Weibull coefficient to be $m = 4.3$ and the average bending strength to be $\sigma_0 = 54$ [MN/m^2] from this plot. The Weibull coefficient is a measure of the fracture strength distribution and directly related to the product quality control: the smaller, the better in the production line!

3.4.3.2.3 Grain Size and Fracture Toughness

The mechanical strength of ceramics is strongly dependent on the grain size. Micrographs of indentations made on two PLZT 9/65/35 samples with different grain sizes are shown in Figure 3.34.[73] Notice that the crack length for the sample with the smaller grain size (1.1 μm) is much shorter than that of the sample with the larger grain size (2.4 μm), while the center indentation size is almost the

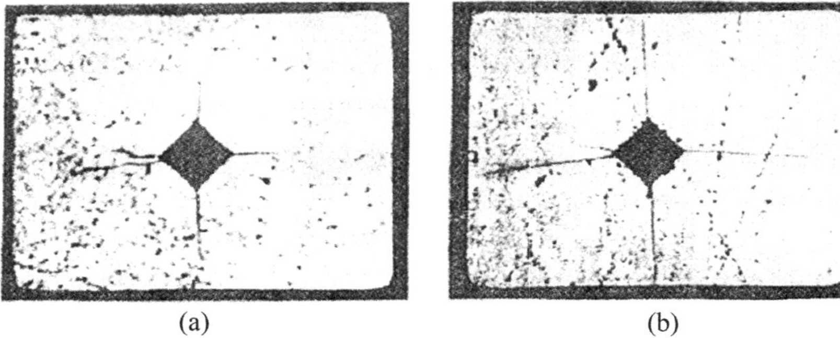

FIGURE 3.34 The indentation mark and cracks generated on a (9/65/35) PLZT surface: (a) grain size 1.1 μm, crack length $c = 208$ μm, and (b) grain size 2.4 μm, crack length $c = 275$ μm. (From K. Uchino and T. Takasu: *Inspec.*, 10, 29, 1986.)

same. The grain size dependence of the hardness, H_v, and the fracture toughness, K_{Ic}, are shown in Figure 3.35.[73] We see from these data that the hardness, H_v, is insensitive to the grain size (i.e., similar center indentation size), while the fracture toughness, K_{Ic}, increases dramatically for specimens with grain sizes below 1.7 μm (i.e., shorter crack length). A multidomain grain model has been proposed to explain this increase in K_{Ic}. The ferroelastic domain structure in ferroelectric and antiferroelectric ceramics has a multidomain state if the grain size is large and assumes a single-domain state with decreasing grain size. In general, ceramic tends to retain a residual compressive stress, which is generated during sintering. This residual stress is relieved by a multidomain structure. The residual stress is not relieved in a single domain state. It is this residual stress that increases the apparent fracture toughness, K_{Ic}, in a small grain sample. Smaller grains possess smaller domain size in general.

3.4.3.2.4 Poling and Mechanical Strength

Piezoelectric ceramics require an electric poling process, which induces an anisotropy to the mechanical strength.[74] A schematic representation of the microindentation and cracks generated in a poled PLZT 2/50/50 (tetragonal) specimen appears in Figure 3.36. This study confirmed that a crack that is oriented perpendicular to the poling direction propagates much faster than one oriented parallel to it. Figure 3.37 shows the bending test results in terms of the poling direction. The test setting is illustrated in Figure 3.37a, and Weibull plots for samples bending perpendicular and parallel to the poling direction in a three-point bending test are shown in Figure 3.37b. These data

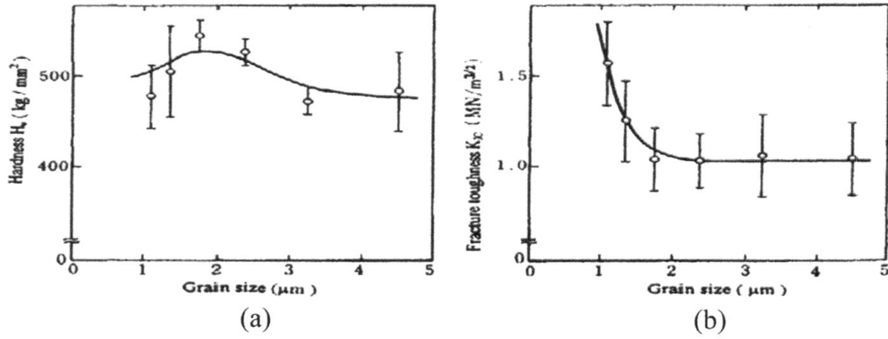

FIGURE 3.35 Mechanical strength data for (9/65/35) PLZT ceramics showing the grain size dependence of: (a) the hardness, H_v, and (b) the fracture toughness, K_{Ic}. (From K. Uchino and T. Takasu: *Inspec.*, 10, 29, 1986.)

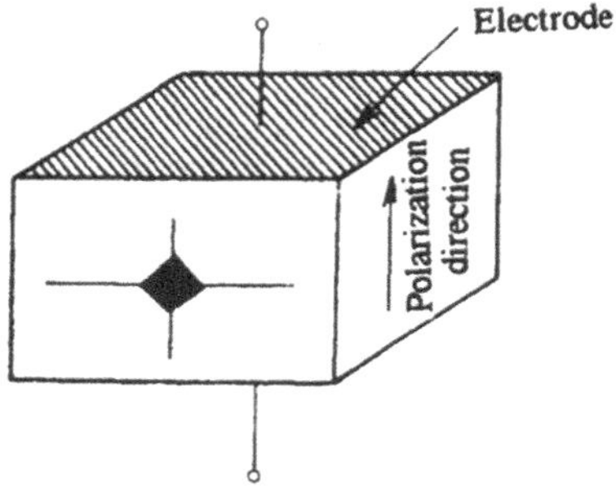

FIGURE 3.36 Microindentation and the cracks generated in a poled piezoceramic.

indicate that cracks propagate more easily in the direction perpendicular to the poling direction.[74] This has been attributed to the anisotropic internal stress caused by the strain induced during the poling process.

Crack propagation in unpoled PZT and PMN ceramics subjected to an applied electric field has also been investigated.[75] In this investigation, it was found that the applied electric field promotes crack propagation perpendicular to the electric field and inhibits it parallel to the field. It was also observed that no mechanical stress is generated in a mechanically constrained, defect-free electrostrictive ceramic subjected to a uniform electric field, even when an electrostrictive response is induced. This has been attributed to microstructural inhomogeneities. The microcrack is regarded as a source of the internal stress, which is concentrated at the crack tip. Calculations based on this model have accurately predicted these experimentally observed trends.

3.4.3.3 Acoustic Emission in Piezoelectrics

Acoustic emission is utilized popularly for monitoring crack propagation in infrastructures and apparatuses. In a piezoelectric material, an AE accompanies the reorientation of domains, a phase transformation, in addition to crack propagation in a ceramic. It is in general an inaudible (to human), high-frequency acoustic burst signal caused by mechanical vibration in the specimen.[76,78] Thus, we need to distinguish the two types of AE signals, crack-related and normal domain-reorientation AE, during monitoring.

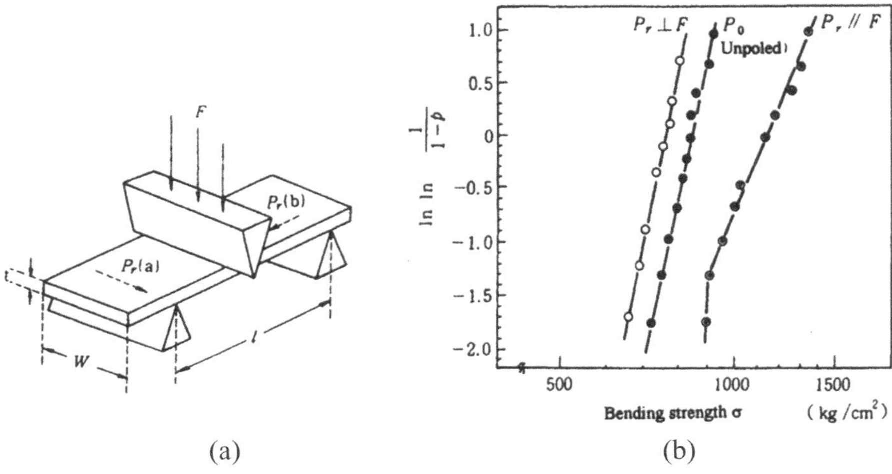

FIGURE 3.37 (a) Three-point bend test configuration, and (b) Weibull plots for samples bending perpendicular and parallel to the poling direction.

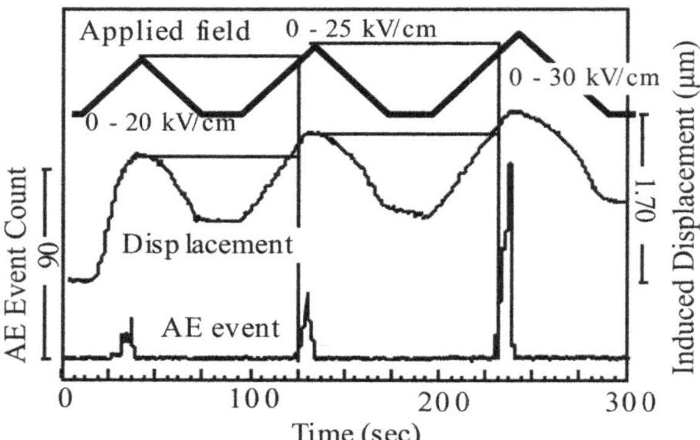

FIGURE 3.38 Acoustic emission (AE) from the cyclically induced displacement, started from an unpoled state in a soft PZT ceramic disk. (From H. Aburatani and K. Uchino: *Jpn. J. Appl. Phys.*, 35, L516, 1996.)

3.4.3.3.1 The Kaiser Effect Associated with Field-Induced Acoustic Emission

The acoustic emission from the induced displacement under electric field cycles, started from an unpoled/virgin soft PZT ceramic disk, is shown as a function of time in Figure 3.38.[77] The threshold level of the acoustic emission signal was 400 mV at 100 dB signal amplification. An electric field of 20 kV/cm was applied to the sample during the first cycle. It was increased to 25 kV/cm during the second cycle, and up to 30 kV/cm during the third cycle.

The sample was poled during the first cycle of the applied electric field. The induced displacement and total residual displacement (observed as a zero point shift) are observed to increase with increasing applied field. This is because the degree of poling becomes greater as the applied field strength is increased. During the first cycle, an acoustic emission occurs with the first displacement as the field is increased. No acoustic emission signal is generated as the field is decreased immediately following this first displacement. The next acoustic emission is observed during the second cycle as the applied field is increased above 20 kV/cm. Another acoustic emission occurs during the third cycle, as the applied field is this time increased above 25 kV/cm. We see from these data that acoustic emission events subsequent to the first will occur only when the applied field exceeds the maximum value of the previous cycle (i.e., memorizing the strain history). This characteristic electric field dependence of the acoustic emission, which is known as the *Kaiser effect*, involves a field-induced deformation of the PZT ceramic. It has been proposed that domain motion is the cause of the field-induced acoustic emission.[77] However, not all types of domain motion, but rather only certain domain reorientation processes, can be identified with an acoustic emission event. Another possible cause of the acoustic emission is the mechanical stress generated in a highly strained sample.

3.4.3.3.2 Fractal Dimension of the Electric Field-Induced Acoustic Emission

The acoustic emission count rate and the induced displacement as a function of applied field for a specimen subjected to ± 35 kV/cm at 1.5×10^{-3} (Hz) are shown in Figure 3.39a.[77] The threshold level of the acoustic emission signal was set as 400 mV at 100 dB amplification. The characteristic butterfly-shaped induced displacement is observed, which occurs due to domain reorientation. At a critical level of the applied electric field, where the acoustic emission is first stimulated, we see a point of inflection in the curve for the displacement as a function of the applied field. A spike is apparent at this field strength in the corresponding *d(displacement)/dE* curve, which is also shown in Figure 3.39a. Note this field is slightly higher than the strain minimum point (i.e., the so-called *coercive electric field*).

It is noteworthy that the maximum in the acoustic emission count rate does not occur at the maximum applied field strength. It actually occurs at about 27 kV/cm. Though the internal stress in a ferroelectric increases monotonously with the applied field, this AE decrease indicates that a

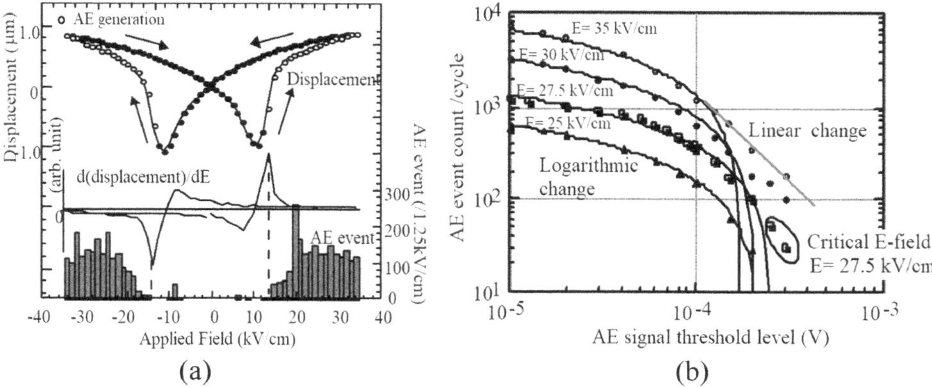

FIGURE 3.39 (a) Acoustic emission count rate and the induced displacement as a function of applied field for a PZT subjected to ± 35 kV/cm at 1.5×10^{-3} (Hz). (b) AE count rate per cycle as a function of the signal threshold level. (From H. Aburatani and K. Uchino: *Jpn. J. Appl. Phys.*, 35, L516, 1996.)

mechanism other than internal stress can also be the source of an acoustic emission. The induced displacement in ferroelectric ceramics is associated with two types of deformation: one due to domain reorientation and the other a pure piezoelectric deformation that involves no domain reorientation. It has been proposed that initially the field-induced acoustic emission in PZT ceramics subjected to a bipolar electric field is stimulated in conjunction with the domain reorientation deformation. Once the domain reorientation is complete, subsequent AE events are believed to occur in conjunction with the piezoelectric deformation and the internal stresses that are induced at high field strengths. The decrease in the AE count rate observed may, therefore, correspond to the completion of domain reorientation in the specimen and its associated deformation.

The acoustic emission count rate can be described by the following function:

$$f(x) = c\, x^{-D} \tag{3.41}$$

where x is the AE signal amplitude, c is a constant, and D is a number representing a fractal dimension. The fractal dimension, D, is used to estimate the degree of stress freedom in the materials. The integrated AE function:

$$F(x) = \int_x^\infty f(x)\cdot dx = \frac{1}{D-1} cx^{-D+1} (\mathrm{D} > 1) \tag{3.42}$$

is obtained experimentally by monitoring the change in the AE signal as the signal threshold level is varied.

The AE count rate per cycle is shown as a function of the signal threshold level in Figure 3.39b.[77] The AE count rate is observed to decrease logarithmically with signal threshold level under an applied field of ±25 kV/cm. A fractal dimension of $D = 1$ was determined from these data. At higher applied field strengths, the acoustic emission rate as a function of threshold level no longer follows a logarithmic trend. The critical electric field amplitude, above which a nonlogarithmic AE function is observed, was found to be around 27 kV/cm. These curves also followed a nearly logarithmic form at lower signal threshold levels. At higher AE signal threshold levels (typically $>10^{-4}$ mV), a linear decrease with threshold level is observed. When an electric field of ±35 kV/cm is applied, a fractal dimension of $D = 2.8$ is obtained. Since the domain reorientation should be complete at electric fields greater than the critical electric field (27 kV/cm), it is assumed here that the AE characterized by fractal dimension of 2.8 is representative of pure piezoelectric deformation.

Regarding the macrodomain reorientation in PZT under low electric field, clear plate-like 90°-domain walls move primarily in perpendicular to the plate (1D motion). If the AE comes from this motion, the fractal dimension $D = 1$ can be explained. However, when the electric field is larger than coercive field in PZT, mono-domain-like state will be realized. The deformation beyond this field level is caused by intrinsic pure piezoelectricity (elastic deformation), that is, 3D deformation (1D extension associated with 2D shrinkage). This suggests the fractal dimension $D = 2.8$, close to the lattice dimension 3.[79]

3.4.3.3.3 The Felicity Effect and Mechanical Strength

When an elastic material is damaged, the reapplied stress, where new acoustic emission starts, becomes lower than the previous maximum stress (the *Felicity effect*).[80] This decrease in the stress at the onset of AE count can be caused by a friction between free and damaged surface. The *Felicity ratio* is defined as follows:

$$\text{Felicity ratio} = \sigma_{\text{stress at onset of AE}} / \sigma_{\text{previous maximum stress}}. \tag{3.43}$$

Figure 3.40 shows the induced displacement and AE event count rate in a multilayer piezoelectric actuator (MPA 5 × 5 × 2, 40 μm × 48 layers) for a unipolar cyclic test with a peak voltage of 180 V. In the first cycle[81], a large number of AE event counts started from 160 V and could be caused by both stress and cracking (Figure 3.40a). The AE observed during the applied voltage decrease is probably due to the friction at the damaged and undamaged surfaces (i.e., delamination), which is against the Kaiser effect during domain reorientations. In the second cycle, the total AE event count per cycle decreased. This is assumed to be stress relaxation in the MPA due to cracking. A larger AE event rate than that of the first cycle was observed from 130 to 160 V (Figure 3.40b). At the 10th cycle, even the overall AE event decreased, and the AE onset voltage stayed around 130 V (Figure 3.40c).

During the cyclic test, the total AE event count decreased with the number of applied voltage cycles, but the AE onset voltage was found to be around 130 V. It is supposed that the damage in the MPA sample lowered the durability against applied voltage. This damage effect on the AE generation can indicate the Felicity effect in the MPA, and the Felicity ratio is calculated to be 130 V/180 V = 0.72. The Felicity ratio in terms of the applied voltage can be used to evaluate the existing damage in the MPA and to determine the maximum safety driving voltage for the MPA that will not cause any major cracking.

After measuring the AE event count and observing the Felicity effect in the sample, the MPA was carefully sliced and polished for microscopic observation. Figure 3.41 shows a micrograph of the crack actually generated in the sample poled at 200 V[81]. Since the crack was observed in the vicinity of the internal electrode edge at an active layer next to the inactive top layer, the stress concentration caused by the inactive layer should be the origin of the cracking. Failures that can raise the destruction probability of the MPA can be detected during the poling process using AE monitoring.

The acoustic emission counts accumulated in a single drive cycle are plotted in Figure 3.42 as a function of the total number of drive cycles for three different multilayer devices.[82] The field-induced strain generates a large stress concentration near the internal electrode edge, which can initiate a crack (see Section 4.3.2). All the samples exhibit a dramatic increase in the acoustic emission count while the cracks are propagating and level off after the crack propagation ceases. The significant differences in the durability/lifetime (measured by the number of drive cycles before failure occurs) among these three samples are attributed mainly to the magnitude of the maximum strain attainable by each device: 0.1% for the electrostrictor, 0.2% for the piezoelectric, and 0.4% for the antiferroelectric sample. Assuming the grain-grain adhesion is similar among the samples, it is reasonable to expect that the larger strain will lead to larger stress and therefore a greater likelihood of fracture.

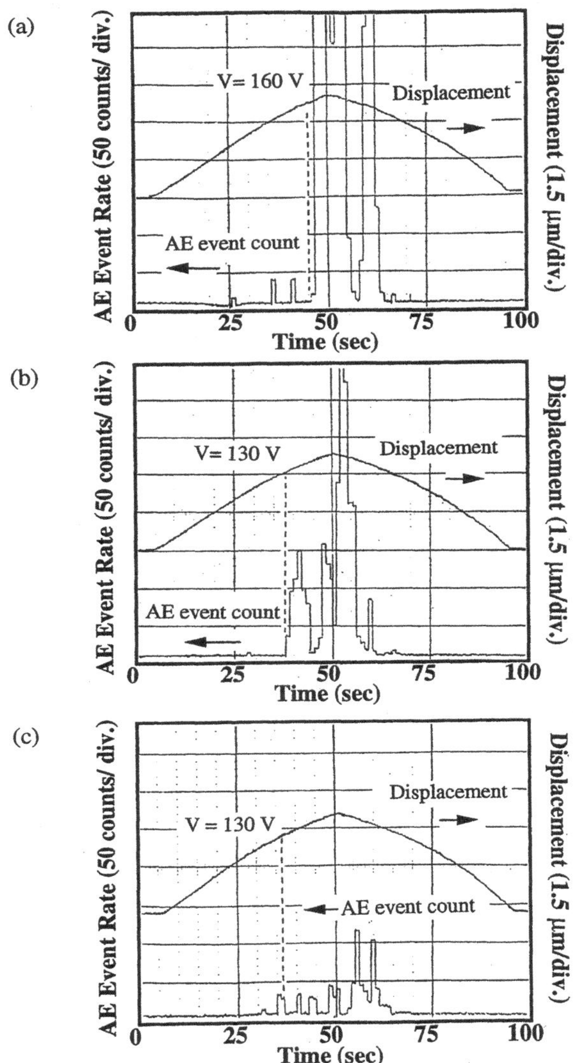

FIGURE 3.40 The AE count and induced displacement as a function of time: (a) 1st cycle, (b) 2nd cycle, and (c) 10th cycle. (From H. Aburatani and K. Uchino: *Japan. J. Appl. Phys.*, 37, Part 1, (1), 204–209, 1998.)

3.5 HIGH-POWER PIEZOELECTRICS

With the acceleration of the commercialization of resonating piezoelectric actuators and transducers for portable equipment applications, we identified the bottleneck of the piezoelectric devices; that is, significant heat generation limits the maximum power density. The current maximum handling energy of a commercialized hard $Pb(Zr,Ti)O_3$ is only around 10 W/cm^3. When we increase the input drive electric power, most of the additional energy will convert into heat, not into the vibration level; the PZT becomes a ceramic heater, not a transducer! We desire 100 W/cm^3 or higher power density for further miniaturization of devices (such as piezo MEMS actuators) without losing efficiency. The Penn State University research group introduced the concept of the "maximum vibration velocity" as a practical high-power performance measure; the maximum vibration velocity is defined by the vibration velocity at the end tip of a piezoelectric specimen

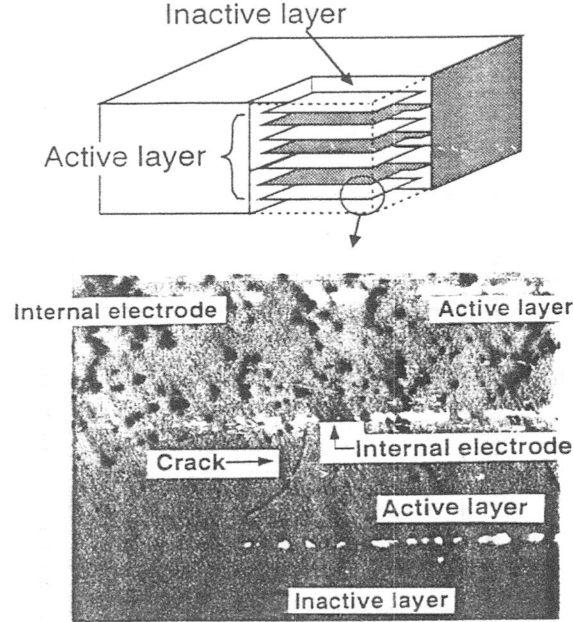

FIGURE 3.41 The crack in the vicinity of the internal electrode edge (poling voltage is 200 V). (From H. Aburatani and K. Uchino: *Japan. J. Appl. Phys.*, 37, Part 1, (1), 204–209, 1998.)

when the maximum temperature spot of the sample (typically at its node) reaches 20°C above room temperature. Thus, the main research focus seems to be shifting from the "real parameters" such as larger polarization and displacement, to the "imaginary parameters" such as polarization/displacement hysteresis, heat generation, and mechanical quality factor, which originate from three loss factors (dielectric, elastic, and piezoelectric losses). Reducing hysteresis and heat generation and increasing the mechanical quality factor to amplify resonance displacement are the primary targets. However, since resonance displacement is provided by a product of the mechanical quality factor, Q_M, and the real parameter (piezoelectric constant d), and the Q_M value is usually reduced with increasing d in the piezoceramics, the target of the material development is highly complicated.

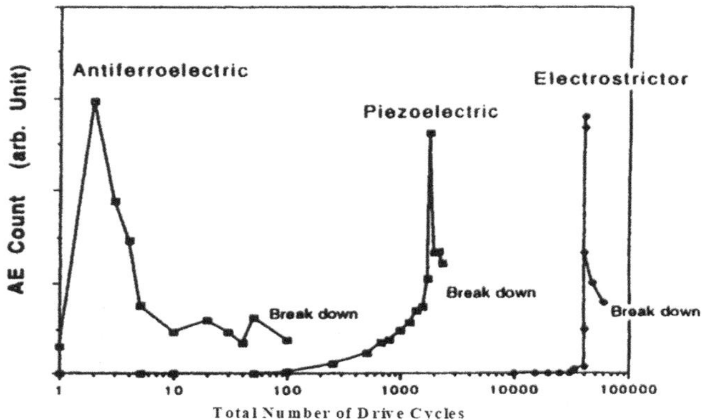

FIGURE 3.42 Acoustic emission counts as a function of total number of drive cycles for electrostrictive, piezoelectric, and phase-change materials. (H. Aburatani et al: *Jpn. J. Appl. Phys.*, 33, Pt.1, (5B), 3091, 1994.)

Loss and hysteresis in piezoelectrics exhibit both merits and demerits in general. For positioning actuator applications, hysteresis in the field-induced strain causes a serious problem, and for resonance actuation such as ultrasonic motors, loss generates significant heat in piezoelectric materials. Further, in consideration of the resonant strain amplified in proportion to a mechanical quality factor, Q_m, low (intensive) mechanical loss materials are preferred for ultrasonic motors. In contrast, for force sensors and acoustic transducers driven typically by a burst mode, a low mechanical quality factor, Q_m (which corresponds to high mechanical loss), is essential to widen a frequency range for receiving signals. In summary, this section focuses on high power density piezoelectric materials not for sensor applications.

Historically, Haerdtl wrote a review article on electrical and mechanical losses in ferroelectric ceramics in the early 1980s.[83] Losses are considered to consist of four portions: (1) domain wall motion; (2) a fundamental lattice portion, which should also occur in domain-free monocrystals; (3) a microstructure portion, which occurs typically in polycrystalline samples; and (4) a conductivity portion in highly ohmic samples. However, in the typical high-quality piezoelectric ceramic case, the loss due to the domain wall motion exceeds the other three contributions significantly. Haerdtl reported interesting experimental results on the linear relationship between dielectric (tan δ) and elastic (tan ϕ) losses (i.e., $\tan\phi/\tan\delta = 0.32$) in piezoceramics, $Pb_{0.9}La_{0.1}(Zr_{0.5}Ti_{0.5})_{1-x}Me_xO_3$, where Me represents the dopant ions Mn, Fe, or Al and x varies between 0 and 0.09 (Figure 3.43). However, he measured the mechanical losses on poled ceramic samples, while the electrical losses were on unpoled samples at 520 kHz, that is, in a different polarization state, which leads to an ambiguity in the discussion.

3.5.1 Loss and Mechanical Quality Factor Relations

Mezheritsky formulated the relations of the mechanical quality factors, Q_m, with loss parameters (including *piezoelectric loss*) for both resonance and antiresonance modes.[84,85] Since Zhuang et al. transformed them in user-friendly formulae,[86,87] we introduce the latter in the following.

3.5.1.1 Loss and Mechanical Quality Factor in k_{31} Mode

3.5.1.1.1 Resonance (A-type) Q_A

Let us start from the mathematical formulation of the mechanical quality factor by integrating three losses in lossless piezoelectric equations.[88] We introduce the complex parameters into the

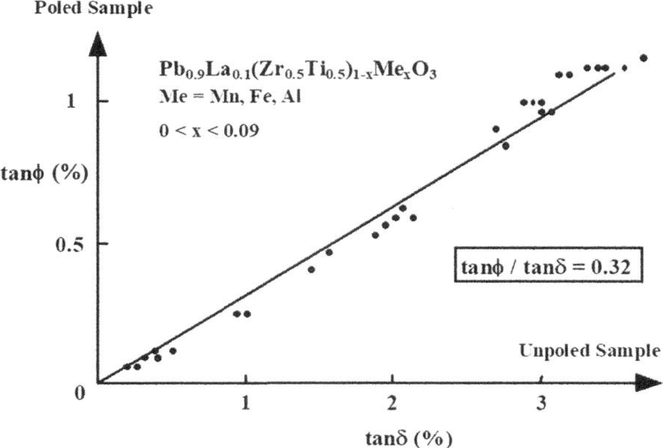

FIGURE 3.43 Correlation between mechanical loss, tanϕ, and dielectric loss, tanδ, for a series of piezoelectric ceramics with $Pb_{0.9}La_{0.1}$ $(Zr_{0.5}Ti_{0.5})_{1-x}Me_xO_3$, where Me represents the dopant ions Mn, Fe, or Al, and x covers a range between 0 and 0.09. [f = 520 kHz]. (From K. H. Haerdtl: *Ceram. Int'l.*, 8, 121–127, 1982.)

admittance formula in Equation 2.158 in k_{31} mode: $\varepsilon_{33}^{X*} = \varepsilon_{33}^{X}(1 - j\tan\delta_{33}')$, $s_{11}^{E*} = s_{11}^{E}(1 - j\tan\phi_{11}')$, and $d_{31}^{*} = d_{31}(1 - j\tan\theta_{31}')$.

$$
\begin{aligned}
Y &= Y_d + Y_m \\
&= (j\omega wL/b)\varepsilon_0\varepsilon_{33}^{LC*}[1 + (d_{31}^{*2}/\varepsilon_0\varepsilon_{33}^{LC*}s_{11}^{E*})(\tan(\omega L/2v^*)/(\omega L/2v^*)]. \\
&= j\omega C_d(1 - j\tan\delta_{33}) + j\omega C_d K_{31}^2[(1 - j(2\tan\theta_{31}' - \tan\phi_{11}')] \\
&\quad [(\tan(\omega L/2v_{11}^{E*})/(\omega L/2v_{11}^{E*})],
\end{aligned}
\tag{3.44}
$$

where

$$v = 1/\sqrt{\rho s_{11}^{E}}) \tag{3.45}$$

$$C_0 = (wL/t)\varepsilon_0\varepsilon_{33}^{X}, \text{(free electrostatic capacitance, real number)} \tag{3.46}$$

$$C_d = (1 - k_{31}^2)C_0. \text{(damped / clamped capacitance, real number)} \tag{3.47}$$

$$K_{31}{}^2 = \frac{k_{31}^2}{1 - {}_{31}^{2}} \tag{3.48}$$

Note that the loss for the first term ("damped/clamped" conductance) is represented by the semiextensive (1D clamped, not 3D clamped) dielectric loss tan δ, not by the intensive loss tan δ'. Remember that $\tan\delta = (1/(1 - k_{31}^2))[\tan\delta' + k_{31}^2(\tan\phi' - 2\tan\theta')]$. Taking into account

$$v_{11}^{E*} = \frac{1}{\sqrt{\rho s_{11}^{E}(1 - j\tan\phi_{11}')}} = v_{11}^{E}\left(1 + j\frac{\tan\phi_{11}'}{2}\right), \tag{3.49}$$

we further calculate $1/[\tan(\omega L/2v^*)]$ with an expansion-series approximation around the A-type resonance (fundamental resonance) frequency $(\omega_A L/2v) = \pi/2$, taking into account that the resonance state is defined in this case for the minimum impedance point. Using new frequency parameters,

$$\Omega_A = \omega_A L/2v_{11}^{E} = \pi/2, \Delta\Omega_A = \Omega - \pi/2(\ll 1), \tag{3.50}$$

$$\frac{1}{\tan\Omega^*} = \cot\left(\frac{\pi}{2} + \Delta\Omega_A - j\frac{\pi}{4}\tan\phi_{11}'\right) = \Delta\Omega_A - j\frac{\pi}{4}\tan\phi_{11}', \tag{3.51}$$

the *motional admittance*, Y_m, is approximated around the first resonance frequency, ω_A:

$$Y_m = j(8/\pi^2)\omega_A C_d K_{31}^2[(1 - j(2\tan\theta_{31}' - \tan\phi_{11}')]/[(4/\pi)\Delta\Omega_A - j\tan\phi_{11}']. \tag{3.52}$$

The maximum Y_m is obtained at $\Delta\Omega_A = 0$:

$$Y_m^{\max} = (8/\pi^2)\omega_A C_d K_{31}^2(\tan\phi_{11}')^{-1} = (8/\pi^2)\omega_A C_0 k_{31}^2 Q_A, \tag{3.53}$$

The mechanical quality factor for A-type resonance $Q_A = (\tan\phi'_{11})^{-1}$ can be proved as follows: Q_A is defined by $Q_A = \omega_A/2\Delta\omega$, where $2\Delta\omega$ is a full width of the 3 dB down (i.e., $1/\sqrt{2}$, because $20\log_{10}(1/\sqrt{2}) = -3.01$) of the maximum value $Y_m^{\max}$ at $\omega = \omega_A$. Since $|Y| = |Y|^{\max}/\sqrt{2}$ can be obtained when the "conductance = susceptance", $\Delta\Omega_A = (\pi/4)\tan\phi'_{11}$,

$$Q_A = \Omega_A/2\Delta\Omega_A = (\pi/2)/2(\pi/4)\tan\phi'_{11} = (\tan\phi'_{11})^{-1}. \tag{3.54}$$

Similarly, the maximum displacement u^{max} is obtained at $\Delta\Omega = 0$:

$$U^{max} = (8/\pi^2)\, d_{31}\, E_Z\, L\, Q_A. \quad (3.55)$$

The maximum displacement at the resonance frequency is $(8/\pi^2)Q_A$ times larger than that at a nonresonance frequency, $d_{31}\, E_Z\, L$.

3.5.1.1.2 Antiresonance (B-type) Q_B

On the other hand, the antiresonance corresponds to the minimum admittance.[85,86] Since the expansion series of tan Ω is convergent in this case, we can apply the following expansion.

$$\tan(\Omega^*) = \tan\left(\Omega - j\frac{\Omega \tan\phi'_{11}}{2}\right) = \tan\Omega - j\frac{\Omega\tan\phi'_{11}}{2\cos^2\Omega}. \quad (3.56)$$

The admittance expression, Equation 3.44, can be simplified as follows:

$$Y' = 1 - k_{31}^2 + k_{31}^2 \frac{\tan(\omega l/2v_{11}^E)}{\omega l/2v_{11}^E}. \quad (3.57)$$

Introducing losses for the parameters leads to

$$Y' = 1 - k_{31}^2[1 - j(2\tan\theta'_{31} - \tan\delta'_{33} - \tan\phi'_{11})] + k_{31}^2[1 - j(2\tan\theta'_{31} - \tan\delta'_{33} - \tan\phi'_{11})]\frac{\tan\Omega^*}{\Omega^*}. \quad (3.58)$$

We separate Y' into conductance G (real part) and susceptance B (imaginary part) as $Y' = G + jB$

$$G = 1 - k_{31}^2 + k_{31}^2\frac{\tan\Omega}{\Omega}. \quad (3.59)$$

$$B = \left(k_{31}^2 - k_{31}^2\frac{\tan\Omega}{\Omega}\right)(2\tan\theta'_{31} - \tan\delta'_{33} - \tan\phi'_{11}) - \frac{k_{31}^2}{2}\left(\frac{1}{\cos^2\Omega} - \frac{\tan\Omega}{\Omega}\right)\tan\phi'_{11}. \quad (3.60)$$

Using new parameters,

$$\Omega = \Omega_B + \Delta\Omega_B, \quad (3.61)$$

the antiresonance frequency, Ω_B, should satisfy

$$1 - k_{31}^2 + k_{31}^2\frac{\tan\Omega_B}{\Omega_B} = 0. \quad (3.62)$$

Similar to $\Delta\Omega_A$ for the resonance, $\Delta\Omega_B$ is also a small number and the first-order approximation can be utilized.

$$\frac{\tan\Omega}{\Omega} = \frac{\tan\Omega_B}{\Omega_B} + \frac{1}{\Omega_B}\left(\frac{1}{\cos^2\Omega_B} - \frac{\tan\Omega_B}{\Omega_B}\right)\Delta\Omega_B. \quad (3.63)$$

Neglecting high-order terms that have two or more small factors (loss factors or $\Delta\Omega_B$),

$$G = \frac{k_{31}^2}{\Omega_B}\left(\frac{1}{\cos^2\Omega_B} - \frac{\tan\Omega_B}{\Omega_B}\right)\Delta\Omega_B. \quad (3.64)$$

$$B = (2\tan\theta'_{31} - \tan\delta'_{33} - \tan\phi'_{11}) - \frac{k_{31}^2}{2}\left(\frac{1}{\cos^2\Omega_B} - \frac{\tan\Omega_B}{\Omega_B}\right)\tan\phi'_{11}. \tag{3.65}$$

Consequently, the minimum absolute value of admittance can be achieved when $\Delta\Omega_B$ is 0. The antiresonance frequency, Ω_B, is determined by Equation 3.62. In order to find the 3 dB-up point, let $G = B$:

$$\frac{k_{31}^2}{\Omega_B}\left(\frac{1}{\cos^2\Omega_B} - \frac{\tan\Omega_B}{\Omega_B}\right)\Delta\Omega_B = (2\tan\theta'_{31} - \tan\delta'_{33} - \tan\phi'_{11}) - \frac{k_{31}^2}{2}\left(\frac{1}{\cos^2\Omega_B} - \frac{\tan\Omega_B}{\Omega_B}\right)\tan\phi'_{11}. \tag{3.66}$$

Further, the antiresonance quality factor is given by

$$Q_{B,31} = \frac{\Omega_B}{2\,|\Delta\Omega_B|}. \tag{3.67}$$

Thus, Equation 3.66 can be represented as

$$\frac{k_{31}^2}{2Q_{B,31}}\left(\frac{1}{\cos^2\Omega_B} - \frac{\tan\Omega_B}{\Omega_B}\right) = -(2\tan\theta'_{31} - \tan\delta'_{33} - \tan\phi'_{11}) + \frac{k_{31}^2}{2}\left(\frac{1}{\cos^2\Omega_B} - \frac{\tan\Omega_B}{\Omega_B}\right)\tan\phi'_{11}. \tag{3.68}$$

Considering Equation 3.68, we can obtain the result as

$$\frac{1}{Q_{B,31}} = \frac{1}{Q_{A,31}} - \frac{2}{\mathrm{k}_{31}{}^2}(2\tan\theta'_{31} - \tan\delta'_{33} - \tan\phi'_{11}) \Big/ \left(\frac{1}{\cos^2\Omega_B} - \frac{\tan\Omega_B}{\Omega_B}\right). \tag{3.69}$$

Taking into account the following relation:

$$\frac{1}{\cos^2\Omega_B} - \frac{\tan\Omega_B}{\Omega_B} = \frac{(1-k_{31}^2)^2\Omega_B^2 + k_{31}^2}{k_{31}^4}, \tag{3.70}$$

The equation of the antiresonance quality factor is given by

$$\frac{1}{Q_{B,31}} = \frac{1}{Q_{A,31}} - \frac{2}{1+\left(\frac{1}{k_{31}} - k_{31}\right)^2\Omega_B^2}(2\tan\theta'_{31} - \tan\delta'_{33} - \tan\phi'_{11}) \tag{3.71}$$

3.5.1.2 Loss and Mechanical Quality Factor in Other Modes

To obtain the loss factor matrix, five vibration modes need to be characterized in PZTs, as shown in Figure 3.44 and Table 3.5. The methodology is based on the equations of quality factors, Q_A (resonance) and Q_B (antiresonance), in various modes with regard to loss factors and other properties.[86,87] We measure Q_A and Q_B for each mode by using the 3 dB-up/down method in the impedance/admittance spectra (see Figure 3.45). In addition to some derivations based on fundamental relations of the material properties, all 20 loss factors can be obtained for piezoelectric ceramics. We derived the relationships between mechanical quality factors, Q_A (resonance) and Q_B (antiresonance), in all required five modes in Table 3.5. The results are summarized below:

a. k_{31} *mode*:

$$Q_{A,31} = \frac{1}{\tan\phi'_{11}}$$

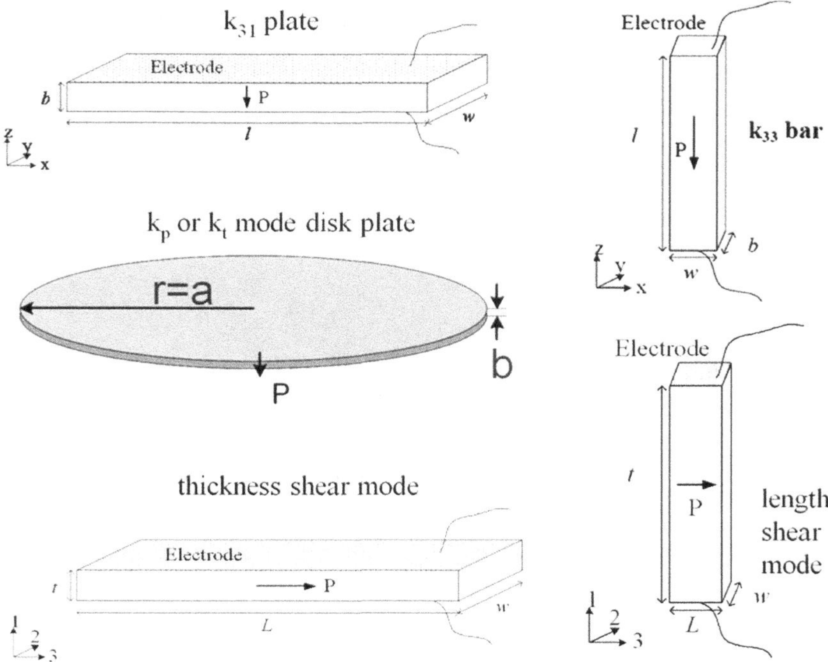

FIGURE 3.44 Sketches of the sample geometries for five required vibration modes.

TABLE 3.5
The Characteristics of Various Piezoelectric Resonators with Different Shapes and Sizes

	Factor	Boundary Conditions	Resonator Shape	Definition
a	k_{31}	$T_1 \neq 0, T_2 = T_3 = 0$ $S_1 \neq 0, S_2 \neq 0, S_3 \neq 0$		$\frac{d_{31}}{\sqrt{s_{11}^E \varepsilon_0 \varepsilon_{33}^T}}$
b	k_{33}	$T_1 = T_2 = 0, T_3 \neq 0$ $S_1 = S_2 \neq 0, S_3 \neq 0$		$\frac{d_{33}}{\sqrt{s_{33}^E \varepsilon_0 \varepsilon_{33}^T}}$
c	k_p	$T_1 = T_2 \neq 0, T_3 = 0$ $S_1 = S_2 \neq 0, S_3 \neq 0$		$k_{31}\sqrt{\frac{2}{1-\sigma}}$
d	k_t	$T_1 = T_2 \neq 0, T_3 \neq 0$ $S_1 = S_2 = 0, S_3 \neq 0$		$h_{33}\sqrt{\frac{\varepsilon_0 \varepsilon_{33}^S}{c_{33}^D}}$
e	$k_{24} = k_{15}$	$T_1 = T_2 = T_3 = 0, T_5 \neq 0$ $S_1 = S_2 = S_3 = 0, S_5 \neq 0$		$\frac{d_{15}}{\sqrt{s_{55}^E \varepsilon_0 \varepsilon_{11}^T}}$

$$\frac{1}{Q_{B,31}} = \frac{1}{Q_{A,31}} + \frac{2}{1+\left(\dfrac{1}{k_{31}} - k_{31}\right)^2 \Omega_{B,31}^2} (\tan\delta'_{33} + \tan\phi'_{11} - 2\tan\theta'_{31})$$

($\tan\delta'_{33}$, $\tan\phi'_{11}$, $\tan\theta'_{31}$: intensive loss factors for $\varepsilon_{33}^T, s_{11}^E, d_{31}$; $\Omega_{B,31}$: antiresonance frequency.)

$$\Omega_{A,31} = \frac{\omega_a l}{2v_{11}^E} = \frac{\pi}{2}\left[v_{11}^E = 1/\sqrt{\rho s_{11}^E}\right], \Omega_{B,31} = \frac{\omega_b l}{2v_{11}^E}, 1 - k_{31}^2 + k_{31}^2 \frac{\tan\Omega_B}{\Omega_B} = 0$$

b. k_t *mode*:

$$Q_{B,t} = \frac{1}{\tan\phi_{33}}$$

$$\frac{1}{Q_{A,t}} = \frac{1}{Q_{B,t}} + \frac{2}{k_t^2 - 1 + \Omega_{A,t}^2 / k_t^2} (\tan\delta_{33} + \tan\phi_{33} - 2\tan\theta_{33})$$

($\Omega_{A,33}$: resonance frequency parameter.)

$$\Omega_{B,t} = \frac{\omega_b l}{2v_{33}^D} = \frac{\pi}{2}\left[v_{33}^D = 1/\sqrt{\rho/c_{33}^D}\right], \Omega_{A,t} = k_t^2 \tan\Omega_{A,t}.$$

c. k_{33} *mode*:

$$Q_{B,33} = \frac{1-k_{33}^2}{\tan\phi'_{33} - k_{33}^2(2\tan\theta'_{33} - \tan\delta'_{33})} \quad \left[\approx \frac{1}{\tan\phi_{33}}\right]$$

$$\frac{1}{Q_{A,33}} = \frac{1}{Q_{B,33}} + \frac{2}{k_{33}^2 - 1 + \Omega_A^2 / k_{33}^2} (2\tan\theta'_{33} - \tan\delta'_{33} - \tan\phi'_{33})$$

$$\left[\approx \frac{1}{Q_{B,33}} + \frac{2}{k_{33}^2 - 1 + \Omega_A^2 / k_{33}^2} (\tan\delta_{33} + \tan\phi_{33} - 2\tan\theta_{33})\right]$$

$$\Omega_{B,33} = \frac{\omega_b l}{2v_{33}^D} = \frac{\pi}{2}\left[v_{33}^D = 1/\sqrt{\rho s_{33}^D}\right], \Omega_{A,33} = k_{33}^2 \tan\Omega_{A,33}$$

d. k_{15} *mode (constant E—length shear mode)*:

$$Q_{A,15}^E = \frac{1}{\tan\phi'_{55}}$$

$$\frac{1}{Q_{B,15}^E} = \frac{1}{Q_{A,15}^E} - \frac{2}{1+\left(\dfrac{1}{k_{15}} - k_{15}\right)^2 \Omega_B^2} (2\tan\theta'_{15} - \tan\delta'_{11} - \tan\phi'_{55})$$

$$\Omega_B = \frac{\omega_b L}{2v_{55}^E} = \frac{\omega_b L}{2v_{55}^E}\sqrt{\rho s_{55}^E}, 1 - k_{15}^2 + k_{15}^2 \frac{\tan\Omega_B}{\Omega_B} = 0$$

e. k_{15} *mode (constant D—thickness shear mode)*:

$$Q_{B,15}^D = \frac{1}{\tan\phi_{55}}$$

$$\frac{1}{Q_{A,15}^D} = \frac{1}{Q_{B,15}^D} + \frac{2}{k_{15}^2 - 1 + \Omega_A^2 / k_{15}^2}(\tan\delta_{11} + \tan\phi_{55} - 2\tan\theta_{15})$$

$$\Omega_A = \frac{\omega_a t}{2v_{55}^D} = \frac{\omega_a t}{2}\sqrt{\frac{\rho}{c_{55}^D}},\ \Omega_A = k_{15}^2 \tan\Omega_A$$

Note again that because k_{31} and k_{33}/k_t modes possess E-constant and D-constant constraints, respectively, in k_{31}, the resonance frequency is directly related to v_{11}^E or s_{11}^E as $f_A = \frac{v_{11}^E}{2L} = 1/2L\sqrt{\rho s_{11}^E}$, while in k_{33}/k_t, the antiresonance frequency is directly related to v_{33}^D or s_{33}^D, c_{33}^D as $f_B = \frac{v_{33}^D}{2L} = 1/2L\sqrt{\rho s_{33}^D}$ or $1/2b\sqrt{\rho / c_{33}^D}$. It is important to distinguish k_{33} ($X_1 = X_2 = 0$, $x_1 = x_2 \neq 0$) from k_t ($X_1 = X_2 \neq 0$, $x_1 = x_2 = 0$) from the boundary conditions. The argument is valid only when the length of a rod k_{33} is not very long ($c_{33}^D \approx 1/s_{33}^D$). The antiresonance in k_{31} and the resonance in k_{33}/k_t are subsidiary, originating from the electromechanical coupling factors. We can also derive the relation for the *electromechanical coupling factor losses* from Equations 2.140 and 2.141:

$$(2\tan\theta' - \tan\delta' - \tan\phi') = -(2\tan\theta - \tan\delta - \tan\phi). \tag{3.72}$$

Since the side is not clamped ($x_1 = x_2 \neq 0$) in k_{33} mode, 1D type Equation 3.72 is not precisely true. Thus, "$\approx$" is used in the above for roughly translating intensive losses into extensive losses.

It is important to discuss the assumption in the IEEE Standard: $Q_A = Q_B$. This situation occurs only when $(2\tan\theta' - \tan\delta' - \tan\phi') = 0$ or $\tan\theta' = (\tan\delta' + \tan\phi')/2$. The IEEE Standard discusses only the case when the piezoelectric loss is equal to the average value of the dielectric and elastic losses, which exhibits a serious contradiction to the PZT well-known experimental results $Q_A < Q_B$. As we can see in Figure 3.45 from the peak sharpness, PZTs exhibit Q_A (resonance) $< Q_B$ (antiresonance) irrelevant to the vibration mode (Figure 3.45 is an example for k_{31}). This

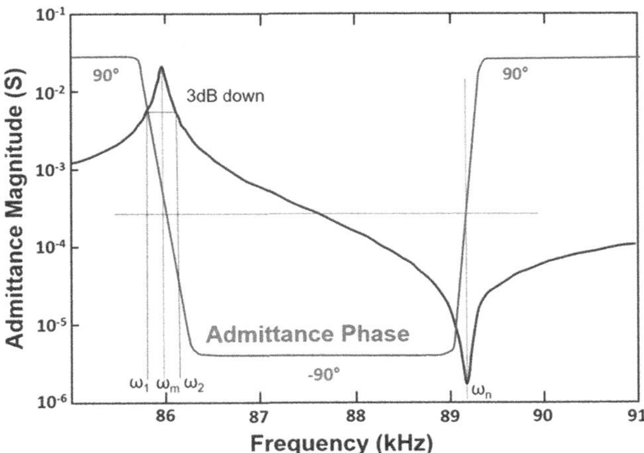

FIGURE 3.45 Admittance magnitude and phase spectra for a rectangular piezoceramic plate for a fundamental longitudinal mode (k_{31}) through the transverse d_{31} effect.

means that ($\tan \delta'_{33} + \tan \phi'_{11} - 2 \tan \theta'_{31}) < 0$, or ($\tan \delta'_{33} + \tan \phi'_{11})/2 < \tan \theta'_{31}$ for k_{31}, and ($\tan \delta_{33} + \tan \phi_{33} - 2 \tan \theta_{33}) > 0$, or ($\tan \delta_{33} + \tan \phi_{33})/2 > \tan \theta_{33}$ for k_t. It is worth noting that the intensive piezoelectric loss is larger than the average of the dielectric and elastic intensive losses in Pb-contained piezoceramics. We introduce in Section 3.5.3 that in Pb-free piezoelectric ceramics, the piezoelectric loss contribution is not significant, and ($\tan \delta'_{33} + \tan \phi'_{11})/2 > \tan \theta'_{31}$, which may suggest different loss mechanisms in different piezoceramics.

3.5.2 Heat Generation under Resonance Conditions

3.5.2.1 Vibration Velocity versus Mechanical Quality Factor, Q_m

Uchino et al. reported the degradation mechanism of the mechanical quality factor, Q_m, with increasing electric field and vibration velocity.[89] Figure 3.46 shows the vibration velocity dependence of the mechanical quality factors, Q_A and Q_B, and corresponding temperature rise for A (resonance) and B (antiresonance) type resonances of a longitudinally vibrating PZT ceramic transducer through the transverse piezoelectric effect, d_{31} (the sample size is inserted).[90] Q_m is almost constant for a small electric field/vibration velocity, but above a certain vibration level, Q_m degrades drastically, where a temperature rise starts to be observed.[90] In order to evaluate the mechanical vibration level, we introduced the vibration velocity at the rectangular plate tip, rather than the vibration displacement, because the displacement is a function of size, while the velocity is not. The *maximum vibration velocity* is defined at the velocity where a 20°C temperature rise at the highest temperature point (i.e., nodal point) from room temperature occurs. Note that even if we further increase the driving voltage/field, additional energy will convert to merely heat (i.e., PZT becomes a ceramic heater!) without increasing the vibration amplitude. Thus, the reader can understand that the maximum vibration velocity is a sort of material constant that ranks high-power performance. Note that most of the commercially available hard PZTs exhibit a maximum vibration velocity around 0.3 m/sec, which corresponds to roughly 5 W/cm^3 (i.e., 1 cm^3 PZT can generate maximum 5 W mechanical energy).

When we compare the change trends in Q_A and Q_B, Q_B is higher than Q_A in all vibration levels (this is true in Pb-contained piezoelectrics such as PZTs). The same result was already discussed for a small vibration level in Figure 3.45. Accordingly, heat generation in the B-type (antiresonance) mode is superior to the A-type (resonance) mode under the same vibration velocity level (in other words, the maximum vibration velocity is higher for Q_B than for Q_A).

Figure 3.47b depicts an important notion on heat generation from piezoelectric material, where the damped and motional resistances, R_d and R_m, in the equivalent electrical circuit of a PZT sample

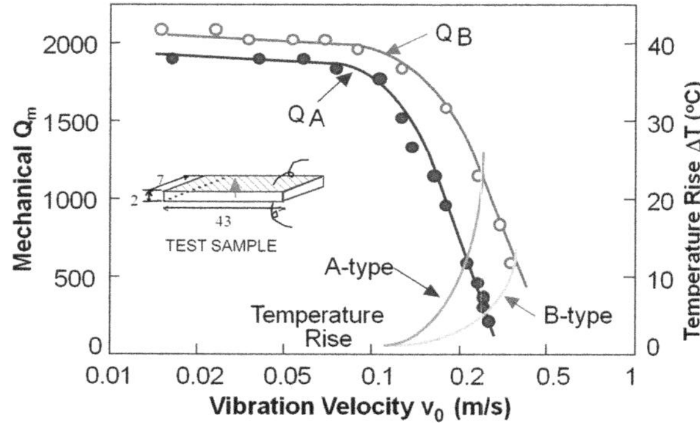

FIGURE 3.46 Vibration velocity dependence of the mechanical quality factors, Q_A and Q_B, and corresponding temperature rise for A (resonance) and B (antiresonance) type resonances of a longitudinally vibrating PZT ceramic transducer through the transverse piezoelectric effect, d_{31} (the sample size is inserted). (K. Uchino et al: *J. Electroceramics*, 2, 33–40, 1998.)

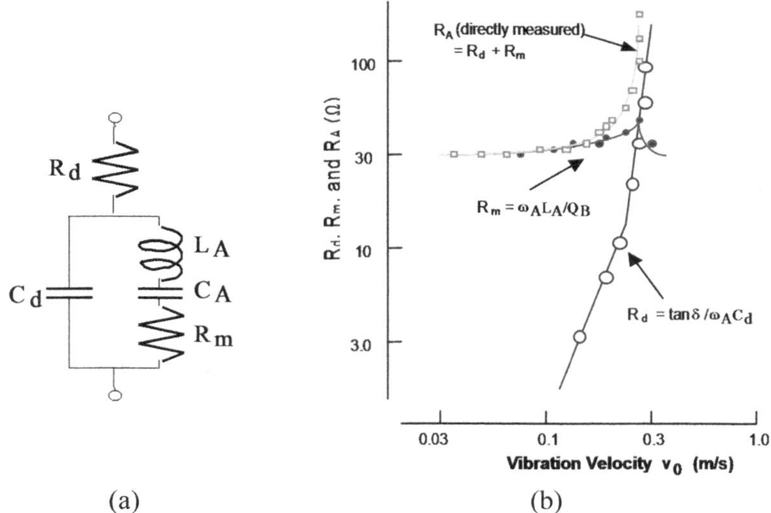

FIGURE 3.47 (a) Equivalent circuit of a piezoelectric sample for the resonance under high power drive. (b) Vibration velocity dependence of the resistances R_d and R_m in the equivalent electric circuit for a longitudinally vibrating PZT ceramic plate through the transverse piezoelectric effect, d_{31}. (From S. Hirose et al: *Proc. Ultrasonics Int'l* '95, Edinburgh, pp. 184–187, 1995.)

(Figure 3.47a) are separately plotted as a function of vibration velocity.[89] Note that R_m, which we speculate to be mainly related to the extensive mechanical loss (90° domain wall motion), is insensitive to the vibration velocity, while R_d, related to the extensive dielectric loss (180° domain wall motion), increases significantly around a certain critical vibration velocity. Thus, the resonance loss at a small vibration velocity is mainly determined by the extensive mechanical loss that provides a high mechanical quality factor, Q_m, and with increasing vibration velocity, the extensive dielectric loss contribution significantly increases. This is consistent with the discussion of Figure 2.50. After R_d exceeds R_m, we started to observe heat generation. We did not include the piezoelectric loss in the equivalent circuit previously. An updated discussion will be given in Section 5.5.2, "Equivalent Circuit."

3.5.2.2 Heat Generation under Resonance Conditions

Tashiro et al. observed heat generation in a rectangular piezoelectric plate during a resonating drive.[91] Even though the electric field is not large, considerable heat is generated due to the large induced strain/stress at the resonance. The maximum heat generation was observed in the nodal regions for the resonance vibration, which corresponds to the locations where the maximum strains/stresses are generated.

The ICAT at Penn State University also worked on heat generation comprehensively in rectangular piezoelectric k_{31} plates when driven at the resonance.[89] The temperature distribution profile in a PZT-based plate sample was observed with a pyroelectric infrared camera, as shown in Figure 3.48, where the temperature variations are shown in the sample driven at the first (28.9 kHz) (a) and second resonance (89.7 kHz) (b) modes, respectively. The highest temperature (bright spot) is evident at the nodal areas for the specimen in Figures 3.48a and b. This observation supports the idea that the heat generated in a resonating sample is associated with the intensive elastic loss, $\tan\phi'$. As we discussed in Section 3.5.1, the resonance and antiresonance modes are both mechanical resonances with the impedance equal to minimum and maximum, respectively, and Q_B at antiresonance is higher than Q_A at resonance in PZTs. Figures 3.49a and b show temperature variations in a PZT-based plate specimen driven at the antiresonance (a) and resonance frequency

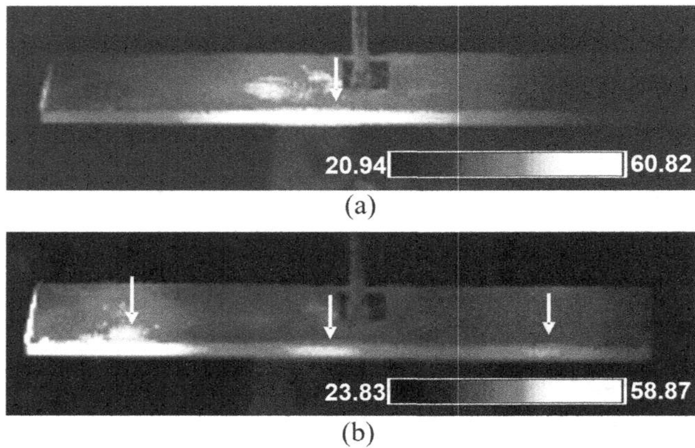

FIGURE 3.48 Temperature variations in a PZT-based plate sample. The specimens are driven at two different resonance frequencies: (a) first resonance mode (28.9 kHz) and (b) second resonance mode (89.7 kHz).

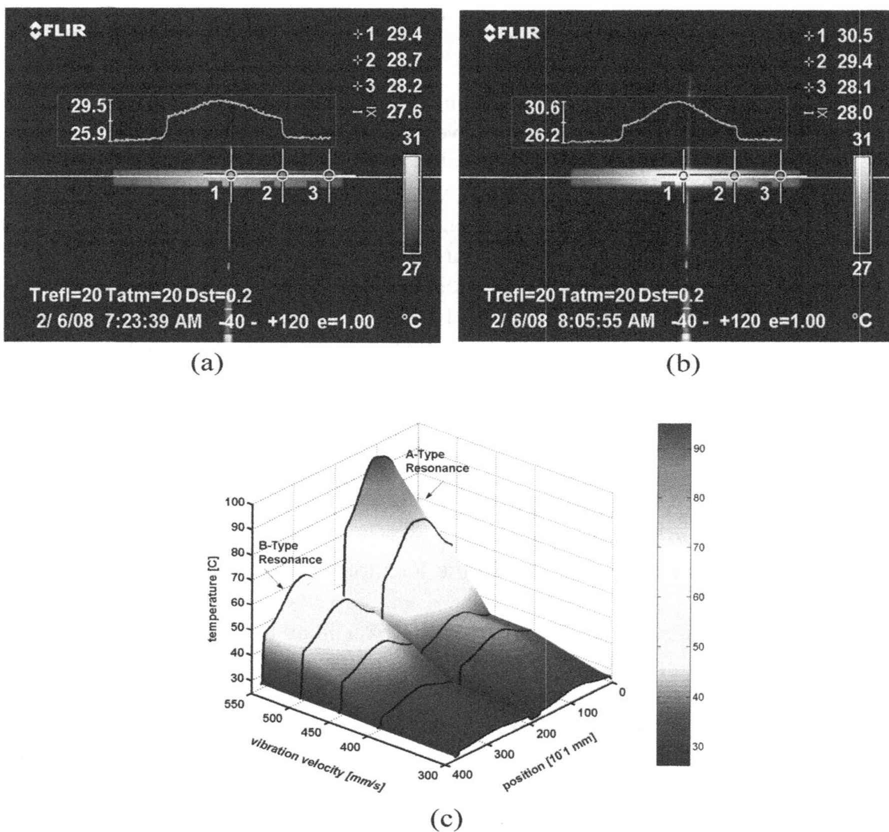

FIGURE 3.49 Temperature variations in a PZT-based plate sample observed with a pyroelectric infrared camera. The specimens are driven at the antiresonance (a) and resonance frequency (b). (c) Numerical temperature profile for the A- and B-type resonance modes.

(b) under the same vibration velocity (i.e., the same output mechanical energy), which clearly exhibits a lower temperature rise in the antiresonance than the resonance drive. Numerical profiles of the temperature distribution for the A- and B-type resonance modes are shown in Figure 3.49c and seem to be sinusoidal curves.

Extending the heat flow (Equation 2.142) for a uniform temperature profile, we need to define the coordinate-dependent energy generation profile at the resonance mode.[92] Because the strain/stress distribution is almost sinusoidal, we suppose volumetric heat generation is provided by

$$u_G(x) = u_G \cos^2\left(\frac{\pi x}{L}\right). \tag{3.73}$$

Here the coordinate x is taken based on the center of the rectangular k_{31} specimen with length L.

The mechanical quality factor, Q_m, is defined as the ratio of elastic energy of an oscillator to the power being dissipated by elastic mechanisms

$$Q_m = 2\pi f_r \frac{U_e}{P_d}, \tag{3.74}$$

where U_e is the maximum stored mechanical energy and P_d is the dissipated power, measured in this experiment by heat generation measurements through temperature, under a supposition $P_d =$ electrically spent energy. Because the compliance of a piezoelectric material has nonlinearity, the maximum kinetic energy is used to define the stored energy term. For a longitudinally vibrating rod, the kinetic energy as a function of displacement, u_x, is

$$U_e = \frac{1}{2} A \int_{-\frac{L}{2}}^{\frac{L}{2}} \rho \left(\frac{\partial u_x}{\partial t}\right)^2 dx \tag{3.75}$$

Using the geometry of the k_{31} plate in Figure 3.44, and assuming sinusoidal forcing at a frequency near the fundamental resonance and antiresonance, the spatial vibration can be described as

$$u_x(x,t) = V_{\mathrm{RMS}}\sqrt{2}\sin\left(\frac{\pi x}{L}\right)\sin(2\pi f t) \tag{3.76}$$

The maximum kinetic energy can be calculated as

$$U_e = \frac{1}{2} A \int_{-\frac{L}{2}}^{\frac{L}{2}} \rho \left(V_{\mathrm{RMS}}\sqrt{2}\sin\left(\frac{\pi x}{L}\right)\right)^2 dx = V_{\mathrm{RMS}}^2 \rho A \frac{L}{2}. \tag{3.77}$$

The dissipated power due to convection and radiation is equal to the heat generation at steady state, so P_d can be expressed as

$$P_d = A \int_{-\frac{L}{2}}^{\frac{L}{2}} u_G \cos^2\left(\frac{\pi x}{L}\right) dx = \frac{1}{2} u_G A L. \tag{3.78}$$

Thus, substituting these expressions into Equation 3.74, the quality factor in terms of heat generation and RMS vibration velocity can be formed

$$Q_m = 2\pi f_r \times \frac{\rho V_{\text{RMS}}^2}{u_G}. \tag{3.79}$$

Note that the quality factor does not depend explicitly on geometrical terms, as is expected if the mode of vibration is known. We see that if the vibration velocity is increased, the heat generation must also increase. Although this derivation is specific to the longitudinal vibration of a rod around the fundamental resonance, it can easily be extended to higher modes of vibration and also other structures such as disks. This can be accomplished by altering the shape of the heat generation distribution and the vibration velocity distribution and then reinvoking the definition of the mechanical quality factor.

Shekhani et al. measured the temperature distribution on a PZT sample with a resonance frequency at 20.04 kHz at room temperature with $Q_m = 507$ with a 3 dB down method on an admittance spectrum. The sample was excited under the vibration velocity of 400 mm/s for 30 seconds, which corresponds to a heat dissipation of 11.6 W/m^2. Figure 3.50 shows the Q_m obtained at three frequencies above the resonance frequency. An increase of Q_m with increasing the frequency is obvious. This is an alternative characterization method for determining the mechanical quality factor, Q_m, at any frequency (not only either the resonance or antiresonance frequency) just from vibration velocity and heat generation (nonelectric measurement!).

3.5.3 Composition Dependence of High-Power Performances

3.5.3.1 Lead Zirconate Titanate–Based Ceramics

"High power" in this article stands for "high power density" in mechanical output energy converted from the maximum input electrical energy under the drive condition with 20°C temperature rise at the highest temperature point (i.e., usually the nodal point). For an off-resonance drive condition, the figure of merit of piezoactuators is given by the piezoelectric d constant ($\Delta L = d \cdot EL$). Heat generation can be evaluated by the intensive dielectric loss tan δ' (i.e., P-E hysteresis) (refer to Section 2.7.3). In contrast, for a resonance drive condition, the figure of merit is primarily the vibration velocity, v_0, which is roughly proportional to $Q_m \cdot dEL$ (under small vibration levels). Q_m can be considered an amplification factor of the vibration amplitude and velocity. Heat generation originates

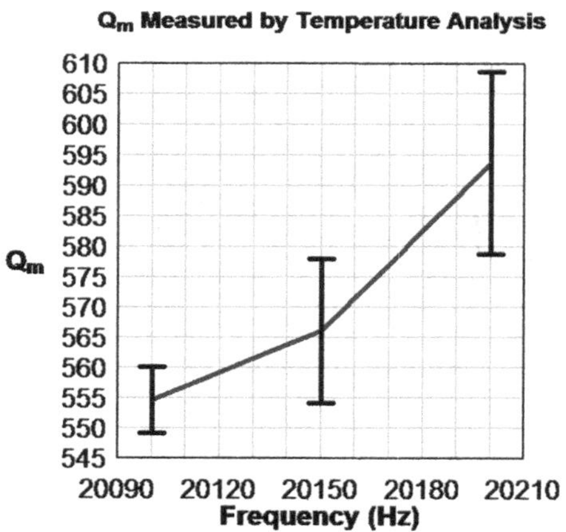

FIGURE 3.50 Change in Q_m with frequency ($f_r \approx 20.006$ kHz). (From H. N. Shekhani and K. Uchino: *J. Amer. Ceram. Soc.*, 97(9), 2810–2814, 2014.)

from the intensive elastic loss, tan ϕ' (inverse value of Q_m), in a k_{31}-type plate. The mechanical power density can be evaluated by the square of the maximum vibration velocity (v_0^2), which is a sort of material constant. Remember that there exists a maximum mechanical energy density above which level piezoelectric material becomes a mere ceramic heater. High vibration velocity piezomaterials are suitable for actuator applications such as ultrasonic motors. Our primary target for high power density materials is set around $v_0 = 0.8$ m/sec or 40 W/cm^3 (of course, the higher the better in the future), in comparison with the commercially available $v_0 = 0.3$ m/sec or 5 W/cm^3. Further, when we consider transformers and transducers, where both transmitting and receiving functions are required, the figure of merit will be the product of $v_0 \cdot k$ (k: electromechanical coupling factor).

Let us discuss high vibration velocity materials based on PZTs first. Figure 3.51 shows the mechanical Q_m versus basic composition x at two effective vibration velocities $v_0 = 0.05$ and 0.5 m/s for $Pb(Zr_xTi_{1-x})O_3$ doped with 2.1 at.% of Fe.[93] The decrease in mechanical Q_m with an increase of vibration level is minimum around the rhombohedral-tetragonal morphotropic phase boundary (MPB 52/48). In other words, the smallest Q_m material under a small vibration level becomes the highest Q_m material under a large vibration level, which is very suggestive. That is, the data obtained by a conventional impedance analyzer with a small voltage/power do not provide any information relevant to high power characteristics. The reader should note that most of the materials with $Q_m > 1200$ in a company catalog are degraded dramatically at an elevated power measurement. This is the major reason our ICAT/Penn State group has been putting significant efforts into High Power Piezoelectric Characterization System (HiPoCS™) developments in the last 30 years.

Conventional piezoceramics have a limitation in the maximum vibration velocity (v_{max}), since the additional input electrical energy is converted into heat, rather than into mechanical energy. The typical rms value of v_{max} for commercially available materials, defined by the temperature rise of 20°C from room temperature, is around 0.3 m/sec for rectangular samples operating in k_{31} mode (like a Rosen-type transformer).[90] $Pb(Mn,Sb)O_3$ (PMS)–$Pb(Zr,Ti)O_3$ ceramics with a v_{max} of 0.62 m/sec are currently used for NEC, Japan, transformers.[93] By doping PMS-PZT or $Pb(Mn,Nb)O_3$-PZT with rare-earth ions such as Yb, Eu, and Ce, we further developed high power piezoelectrics, which can operate with v_{max} up to 1.0 m/sec.[94,95] Compared with commercially available piezoelectrics, 10 times (square of 3.3 times of v_0) higher input electrical energy and output mechanical energy

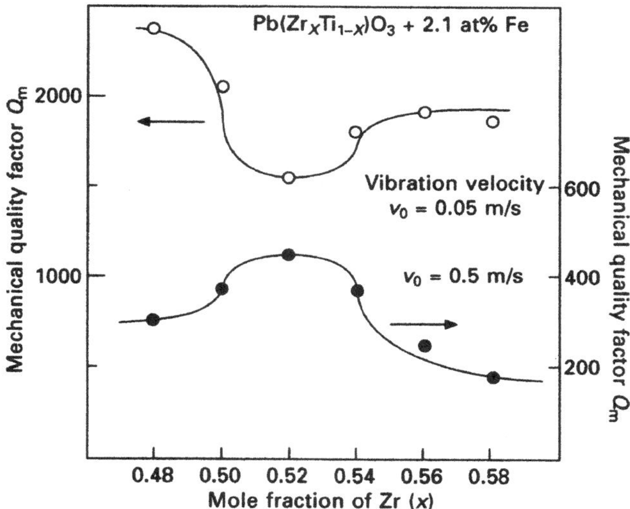

FIGURE 3.51 Mechanical Q_m versus basic composition, x, at two effective vibration velocities, $v_0 = 0.05$ m/s and 0.5 m/s for $Pb(Zr_xTi_{1-x})O_3$ doped with 2.1 at.% of Fe. (From S. Takahashi and S. Hirose, *Jpn. J. Appl. Phys.*, 32, 2422–2425, 1993.)

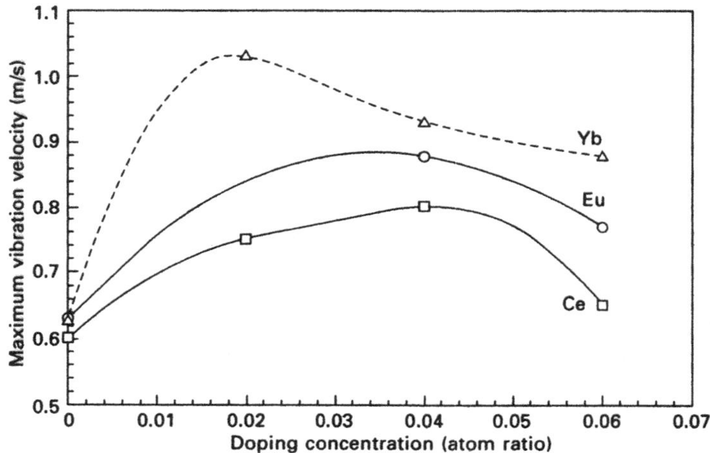

FIGURE 3.52 Dependence of the maximum vibration velocity, v_0 (20°C temperature rise), on the atomic % of rare-earth ion, Yb, Eu, or Ce in $Pb(Mn,Sb)O_3$ (PMS)–PZT based ceramics. (From J. Ryu et al: *Jpn. J. Appl. Phys.*, 42(3), 1307–1310, 2003.)

can be expected from these new materials without generating a significant temperature rise, which corresponds to 50 W/cm^2. Figure 3.52 shows the dependence of the maximum vibration velocity v_0 (20°C temperature rise) on the atomic % of rare-earth ion, Yb, Eu, or Ce in $Pb(Mn,Sb)O_3$ (PMS)–PZT-based ceramics. Enhancement in the v_0 value is significant by the addition of a small amount of the rare-earth ion.[96] The high power performance improvement by doping will be discussed in Section 3.5.4 from a mechanism viewpoint.

3.5.3.2 Pb-Free Piezoelectrics

The twenty-first century is called "The Century of Environmental Management." In 2006, the European Community started RoHS, which explicitly limits the usage of lead in electronic equipment. Basically, we may need to regulate the usage of lead zirconate titanate, most famous piezoelectric ceramics, in consumer usage. The Japanese and European communities may experience governmental regulation on PZT usage in 5 years. Pb-free piezoceramics started to be developed after 1999 and are classified into three types: $(Bi,Na)TiO_3$,[97] $(Na,K)NbO_3$,[98] and tungsten bronze.[99] NKN systems exhibit the highest performance because of their morphotropic phase boundary usage, in particular in structured ceramic manufacturing with flaky powders. TB types are an alternative choice for resonance applications, because of their high Curie temperature and low loss. A sophisticated preparation technology for fabricating oriented ceramics was developed with a multilayer configuration, that is, preparation under strong magnetic fields.[97]

Though there is a lot of research related to lead-free piezoelectric materials, there are quite a few research studies on high-power characterization of lead-free piezoelectric materials. The ICAT's study illuminated the high power characteristics of Pb-free piezoelectric ceramic compared to hard lead-zirconate-titanate. Though the RoHS regulation is the trigger for the Pb-free piezoelectric development, if we find superior performances in Pb-free piezoelectrics, this new category of material may contribute significantly to industrial and military applications. High power characteristics were investigated with our high power piezoelectric characterization system.[100,101] Samples were prepared in collaboration with Toyota R&D, Honda Electronics, Korea University, and Rutgers University, including:

1. NKN—$(Na_{0.5}K_{0.5})(Nb_{0.97}Sb_{0.03})O_3$ prepared with 1.5% mol CuO addition
2. BNT-BT-BNMN—$0.82(Bi_{0.5}Na_{0.5})TiO_3$-$0.15BaTiO_3$-$0.03(Bi_{0.5}Na_{0.5})(Mn_{1/3}Nb_{2/3})O_3$
3. BNKLT—$0.88(Bi_{0.5}Na_{0.5})TiO_3$–$0.08(Bi_{0.5}K_{0.5})TiO_3$–$0.04(Bi_{0.5}Li_{0.5})TiO_3$ with Mn-doped and -undoped

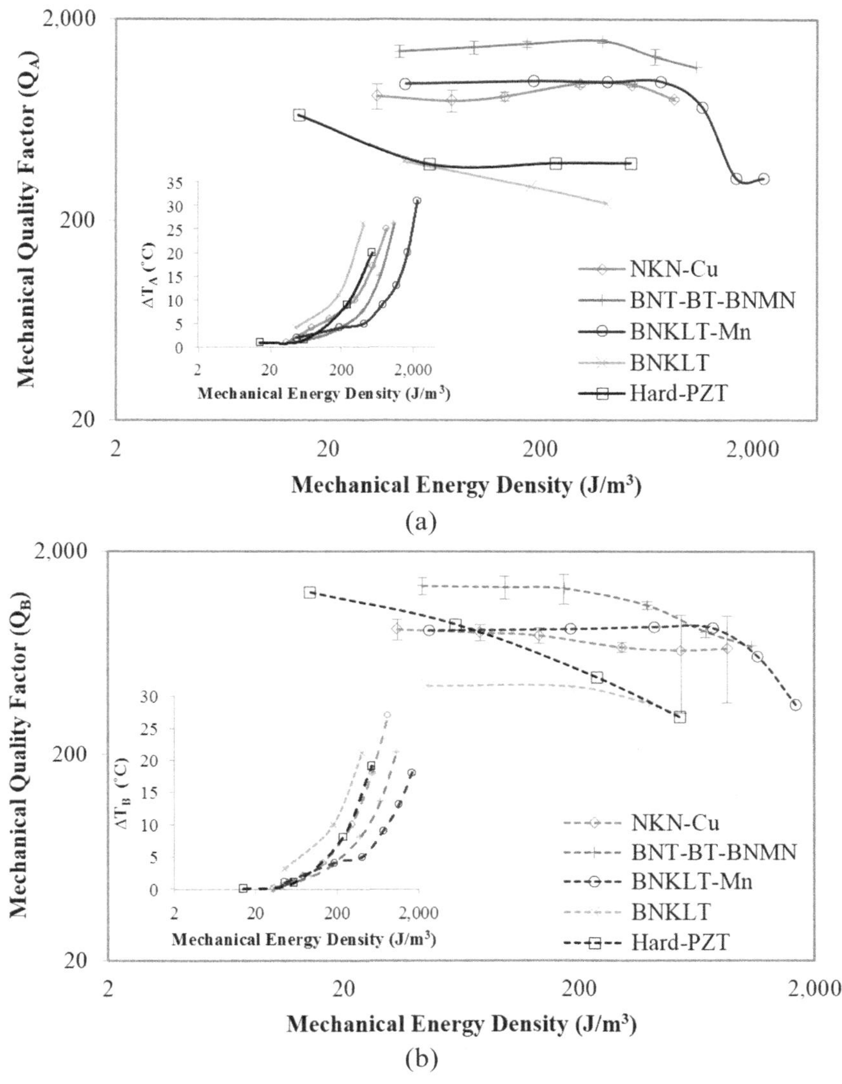

FIGURE 3.53 Temperature rise and mechanical quality factors for (a) resonance (ΔT_A and Q_A) and (b) antiresonance (ΔT_B and Q_B) as functions of mechanical energy density ($u_{\text{mech}} \sim (1/2)\rho v_{\text{rms}}^2$) for lead-free and hard-PZT piezoelectrics.

Figure 3.53 shows the high power results for a disk sample. The mechanical quality factors at resonance (Q_A) and at antiresonance (Q_B) do not decrease with increasing vibration level [*mechanical energy density* ($(1/2)\rho v_{\text{rms}}^2$)] in Pb-free piezoelectrics, in comparison with the PZT trend. It is worth noting that the *vibration velocity* (v_{rms}) reaches 0.8 m/s in all three Pb-free compositions, but that a low mass density can easily generate a large vibration velocity. Therefore, *mechanical energy density* ($1/2\rho v_{\text{rms}}^2$) should be used in order to compare the performance superiority in terms of the maximum vibration level of the materials. The maximum mechanical energy density defined with a 20°C increase of the temperature on the nodal point in NKN and BNT-BT-BNMN is superior when compared to hard PZTs with their sharp decrease in Q_A with the increasing vibration level. At antiresonance, the high power behavior trend for this material did not change. The mechanical quality factor at the antiresonance (Q_B) also remained constant up to the maximum vibration level.

Regarding the figure-of-merit (actual vibration level, $Q_m \cdot k_p$) change with the vibration level, the PZT shows better performance in the low power vibration range because of higher k_p values in PZT. However, in the high power range (>1000 J/m^3), only Pb-free piezoelectrics can practically be adopted.

The Q_A and Q_B values in Pb-free piezoelectrics are almost the same up to the maximum vibration velocity, v_{max}. This trend is also distinctly different from the hard PZT, where Q_B is always greater than Q_A, experimentally (depending on the piezoelectric loss). Comparison of mechanical quality factors at resonance (Q_A) and antiresonance (Q_B) provides an aspect to the material losses (i.e., dielectric [tan δ'], mechanical [tan ϕ'], and piezoelectric [tan θ']) behavior for each different composition. Both soft and hard PZTs result in $Q_B > Q_A$, as we have already discussed. However, regarding Pb-free piezoelectrics, hard Bi-perovskite ceramics (i.e., BNT-BKT-BLT-Mn and BNT-BT-BNMN) seem to have a minimum influence from piezoelectric loss (tan θ') when excited at high vibration levels since they show the $Q_A > Q_B$ relationship. Cu-doped NKN ceramics seem to have more influence ($Q_A \cong Q_B$ for NKN-Cu) from the piezoelectric loss (tan θ'). Softer BNT-BKT-BLT ceramics seem to have the largest influence ($Q_B > Q_A$ for BNT-BKT-BLT), similar to the hard PZT ($Q_B > Q_A$ for hard PZT) at high power levels.

3.5.4 Doping Effect on Piezoelectric Losses

3.5.4.1 Hard and Soft Lead Zirconate Titanates

Small amounts of dopants sometimes dramatically change the dielectric and electromechanical properties of ceramics. Donor doping tends to facilitate domain wall motion, leading to enhanced piezoelectric charge coefficients, d, and electromechanical coupling factors, k, producing what is referred to as a *soft piezoelectric* (i.e., compliant from both electrical and mechanical viewpoints). Acceptor doping, on the other hand, tends to "pin" domain walls and impede their motion, leading to an enhanced mechanical quality factor, Q_m, producing what is called a *hard piezoelectric*. Table 3.6 summarizes the advantages and disadvantages of soft and hard piezoelectrics and compares their characteristics with an electrostrictive material, $Pb(Mg_{1/3}Nb_{2/3})O_3$. Electrostrictive ceramics are commonly used for positioning devices where hysteresis-free performance is a primary concern. However, due to their high permittivity, electrostrictive devices are generally used only for applications that require slower response times. On the other hand, soft piezoelectric materials with their relatively low permittivity and high piezoelectric charge coefficients, d, can be used for applications requiring a quick response time, such as pulse-driven devices like inkjet printers. However, soft piezoelectrics generate a significant amount of heat when driven at the resonance, due to their small mechanical quality factor, Q_m. Thus, for ultrasonic motor and piezotransformer applications, hard piezoelectrics with a larger mechanical quality factor are preferred despite the slight sacrifices incurred with respect to their smaller piezoelectric strain coefficients, d, and the electromechanical coupling factors, k.

TABLE 3.6
Advantages (+) and Disadvantages (–) of Soft and Hard Piezoelectrics, Compared with the Features of an Electrostrictive Material, $Pb(Mg_{1/3}Nb_{2/3})O_3$ (PMN)

Material	d	k	Q_m	Off-Resonance Applications	Resonance Applications
PMN	High$^+$ (DC Bias)$^-$	High$^+$ (DC Bias)$^-$	Low$^-$ (DC Bias)$^-$	High Displacement$^+$ No Hysteresis$^+$	Broad Bandwidth$^+$
Soft PZT-5H	High$^+$	High$^+$	Low$^-$	High Displacement$^+$	Heat Generation$^-$
Hard PZT-8	Low$^-$	Low$^-$	High$^+$	Low Strain$^-$	High AC Displacement$^+$

Let us introduce first the most popular "domain wall pinning" model, taking into account the crystallographic defects produced on the perovskite lattice due to doping. There are two key factors to be considered in the "hardening/softening" mechanism: (1) Gauss's Law, div $D = \rho$, and (2) generation of the movable charge, ρ, by ion doping.

The soft and hard characteristics are reflected in the coercive field, E_c, or more precisely in the stability of the domain walls. A piezoelectric is classified as hard if it has a coercive field greater than 1 kV/mm, and soft if the coercive field is less than 0.1 kV/mm (others are called semi-hard or semi-soft). Consider the transient state of a 180° domain reversal that occurs at a domain wall associated with a configuration of head-to-head polar domains. We know from Gauss's law that:

$$\mathrm{div}\, D = \rho, \tag{3.80}$$

where D is the electric displacement and ρ is the charge density. The domain wall is very unstable in a highly insulating material and therefore readily reoriented, and the coercive field for such a material is found to be low. However, this head-to-head configuration is stabilized in a less resistive (some movable charges) material and thus a higher coercive field is required for polarization reversal and the associated domain wall movement to occur. These two cases are illustrated schematically in Figure 3.54. The difference between hard and soft is determined by the movable charges associated with defect structures in a doped PZT material.

Now we consider the ion doping effect. Acceptor ions, such as Fe^{3+}, lead to the formation of oxygen deficiencies (□) in the PZT lattice, and the resulting defect structure is described by:

$$Pb(Zr_y Ti_{1-y-x} Fe_x)(O_{3-x/2} \square_{x/2})$$

Acceptor doping allows for the easy reorientation of deficiency-related dipoles. These dipoles are composed of an Fe^{3+} ion (effectively the negative charge because it is situated in the 4+ Ti site) and an 2- oxygen vacancy (effectively the positive charge). The oxygen deficiencies are produced at high temperature (>1000°C) during sintering. The oxygen ions are still able to migrate at temperatures well below the Curie temperature (even at room temperature), because the oxygen and its associated vacancy are only 2.8 Å apart and the oxygen may readily move into the vacant site as depicted in Figure 3.55a.

In the case of donor (higher valence ion) dopant ions, such as Nb^{5+}, a Pb deficiency is produced and the resulting defect structure is designated by the following:

$$(Pb_{1-x/2} \square_{x/2})(Zr_y Ti_{1-y-x} Nb_x)O_3$$

Donor doping is *not* effective in generating movable dipoles, since the Pb ion cannot easily move to an adjacent A-site vacancy due to the proximity of the surrounding oxygen ions, as depicted in Figure 3.55b. Soft characteristics are, therefore, observed for donor-doped materials. Another factor that should be considered here is that lead-based perovskites, such as PZTs, tend to be p-type

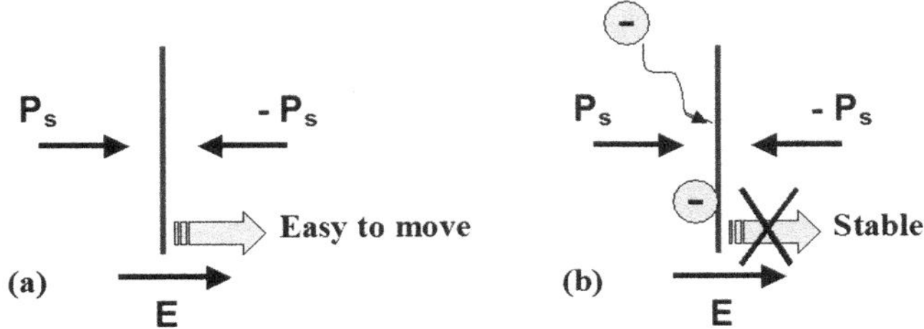

FIGURE 3.54 Stability of 180° domain wall motion in: (a) an insulating material and (b) a material with free charges.

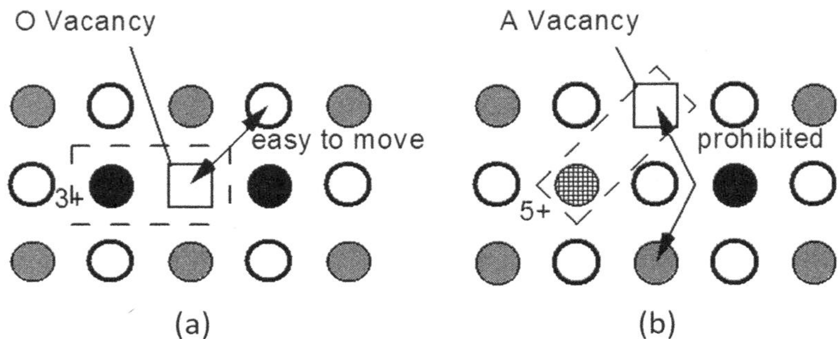

FIGURE 3.55 Lattice vacancies in PZT containing: (a) acceptor (lower valence ion) and (b) donor (higher valence ion) dopants.

semiconductors due to the evaporation of lead during sintering and are thus already hardened to some extent by the lead vacancies that are produced. Hence, donor doping will compensate the p-type, facilitate the domain wall motion, then exhibit large piezoelectric charge coefficients, *d*, but will also exhibit pronounced aging due to their soft characteristics.

3.5.4.2 Dipole Alignment Models

We introduced in the previous section just one model, domain wall pinning. There are three types of the impurity dipole (originating from a pair of the acceptor ion and oxygen vacancy) alignment possibility, as schematically visualized in Figure 3.56: (a) random alignment, (b) unidirectionally fixed alignment, and (c) unidirectionally reversible alignment. Accordingly, the expected polarization vs. electric hysteresis curve will be (a) a high coercive field P-E loop, in comparison with that of the original undoped ferroelectric; (b) a DC-bias field P-E loop; and (c) a double hysteresis loop.

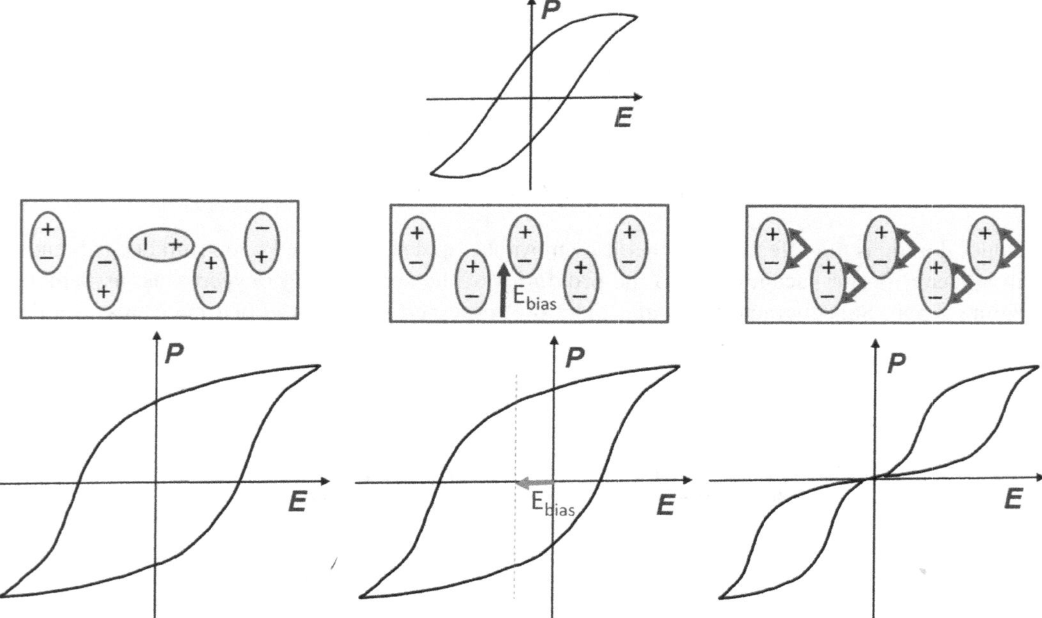

FIGURE 3.56 Impurity dipole alignment possibility and the expected P vs. E hysteresis curves: (a) random alignment, (b) unidirectionally fixed alignment, and (c) unidirectionally reversible alignment.

3.5.4.2.1 Dipole Random Alignment

This corresponds to the domain wall pinning model described in the previous section. From Gauss's law, div $D = \rho$, the domain wall (head-to-head, tail-to-tail) is stabilized in a less resistive (some movable charges) material. The presence of acceptor dopants, such as Fe^{3+}, in the perovskite structure is found to produce oxygen deficiencies, while donor dopants, such as Nb^{5+}, produce A-site deficiencies. Only acceptor doping generates movable dipoles, which correspond to ρ and can stabilize the domain walls.

These simple defect models help us to understand and explain various changes in the properties of a perovskite ferroelectric of this type that occur with doping. The effect of donor doping in PZT on the field-induced strain response of the material was examined for the soft piezoelectric composition $(Pb_{0.73}Ba_{0.27})(Zr_{0.75}Ti_{0.25})O_3$.[102] The parameters *maximum strain*, x_{max}, and the *degree of hysteresis*, $\Delta x/x_{max}$, are defined in terms of the hysteresis response depicted in Figure 3.57a. The maximum strain, x_{max}, is induced under the maximum applied electric field. The degree of hysteresis, $\Delta x/x_{max}$, is the ratio of the strain difference induced at half the maximum applied electric field to the maximum strain, x_{max}. The effect of acceptor and donor dopants (2 at% concentration) on the induced strain and degree of hysteresis is shown in Figure 3.57b. It is seen that materials incorporating high valence donor-type ions on the B-site (such as Ta^{5+}, Nb^{5+}, W^{6+}) exhibit excellent characteristics as positioning actuators, namely enhanced induced strains and reduced hysteresis. On the other hand, the low valence acceptor-type ions (+1, +2, +3) tend to suppress the strain and increase the hysteresis and the

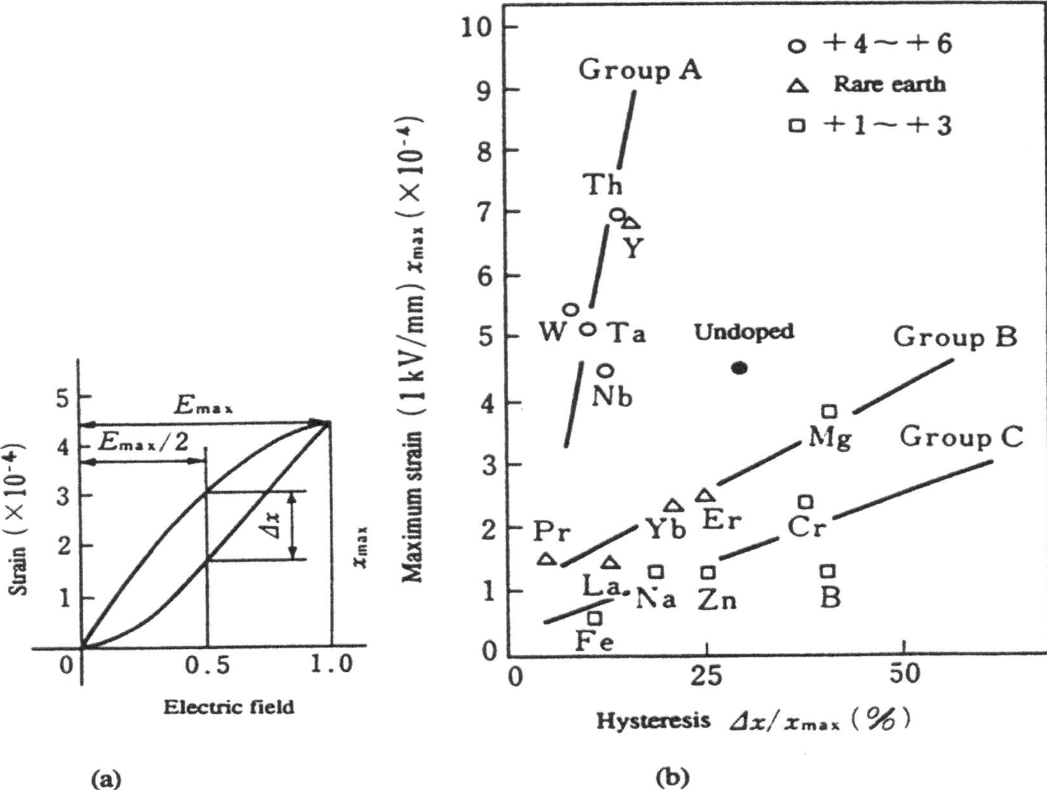

FIGURE 3.57 Maximum strain and hysteresis in $Pb_{0.73}Ba_{0.27}(Zr_{0.75}Ti_{0.25})O_3$-based ceramics: (a) hysteresis curve showing the parameters needed for defining the maximum strain, x_{max}, and the degree of hysteresis, $\Delta x/x_{max}$, and (b) the dopant effect on actuator parameters. (From A. Hagimura and K. Uchino: *Ferroelectrics*, 93, 373, 1989.)

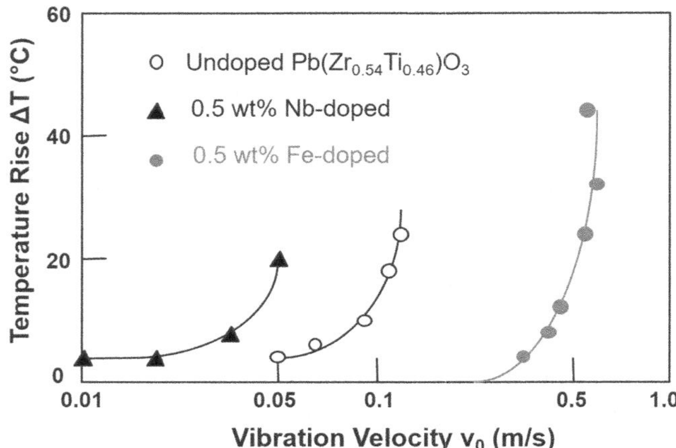

FIGURE 3.58 Temperature rise, ΔT, plotted as a function of effective vibration velocity, v_0, for undoped, Nb-doped, and Fe-doped PZT samples. (From K. Uchino et al: J. *Electroceramics*, 2, 33–40, 1998.)

coercive field. Although acceptor-type dopants are not desirable when designing actuator ceramics for positioner applications, acceptor doping is important in producing hard piezoelectric ceramics, which are preferred for ultrasonic motor applications. In this case, the acceptor dopant acts to pin domain walls, resulting in the high mechanical quality factor characteristic of a hard piezoelectric.

Let us now consider high power piezoelectric ceramics for ultrasonic (AC drive) applications. When the ceramic is driven at a high vibration rate (i.e., under a relatively large AC electric field) heat will be generated in the material, resulting in significant degradation of its piezoelectric properties. A high power device such as an ultrasonic motor therefore requires a very hard piezoelectric with a high mechanical quality factor, Q_m, to reduce the amount of heat generated. The temperature rise (at the nodal point, i.e., plate center) in undoped, Nb-doped, and Fe-doped PZT k_{31} plate samples is plotted as a function of vibration velocity (rms) value measured at the plate end in Figure 3.58.[89] A significant reduction in the generation of heat is apparent for the Fe-doped (acceptor-doped) ceramic. Commercially available hard PZT ceramic plates tend to generate the maximum vibration velocity around 0.3 m/s (rms) when operated in k_{31} mode. Even when operated under the higher applied electric field strengths, the vibration velocity will not increase for these devices; the additional energy is just converted into heat.

Higher maximum vibration velocities have been realized in PZT-based materials modified by dopants that effectively reduce the amount of heat generated in the material and thus allow for the higher rates of vibration. NEC, Japan, developed multilayer piezoelectric transformers with the $(1-z)Pb(Zr_xTi_{1-x})O_3$-$(z)Pb(Mn_{1/3}Sb_{2/3})O_3$ composition.[103] The maximum vibration velocity of 0.62 m/s occurs at the $x = 0.47$, $z = 0.05$ composition and is accompanied by a 40°C temperature rise from room temperature. The incorporation of additional rare earth dopants to this optimum base composition results in an increase in the maximum vibration velocity to 0.9 m/s at a 20°C temperature rise, already introduced in Figure 3.52.[94] This increased vibration rate represents a threefold enhancement over that typically achieved by commercially produced hard PZT devices and corresponds to an increase in the vibration energy density by an order of magnitude with minimal heat generated in the device. The mechanism of this stable high power performance is explained in the next section.

3.5.4.2.2 Unidirectionally Fixed Dipole Alignment

Hard PZT is usually used for high power piezoelectric applications, because of its high coercive field, in other words, the stability of the domain walls. Acceptor ions, such as Fe^{3+}, introduce oxygen deficiencies in the PZT crystal (in the case of donor ions, such as Nb^{5+}, Pb deficiency is

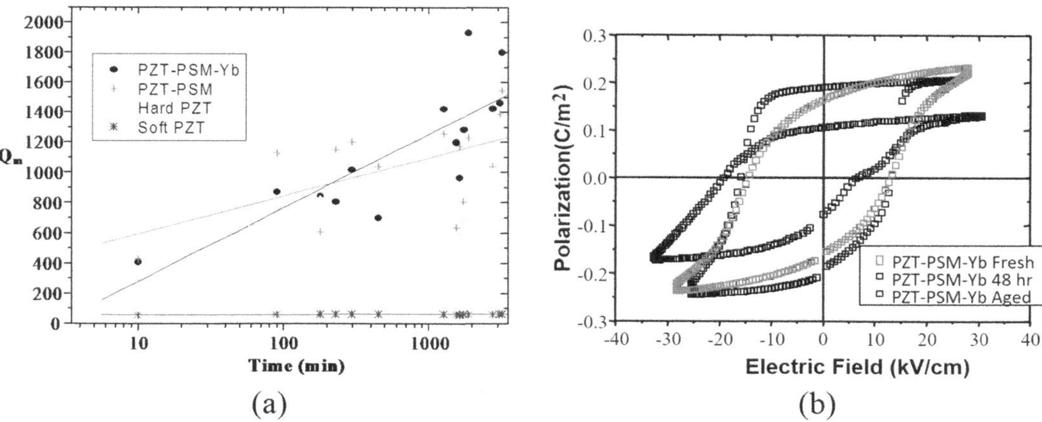

FIGURE 3.59 (a) Change in the mechanical Q_m with time lapse (minutes) just after the electric poling, measured for various commercial soft and hard PZTs, PSM-PZT, and PSM-PZT doped with Yb. (b) Polarization vs. electric field hysteresis curves measured for Yb-doped Pb(Mn,Sb)O_3-PZT sample just after poling (fresh), 48 hours after, and a week after (aged).

introduced). Thus, in the conventional model, the acceptor doping causes domain wall pinning through the easy reorientation of deficiency-related dipoles, leading to hard characteristics (domain wall pinning model [Reference 88]). ICAT/Penn State University explored the origin of our high power piezoelectric ceramics, and found that the "internal bias field model" seems to be better for explaining our material's characteristics.

High mechanical Q_m is essential in order to obtain a high power material with a large maximum vibration velocity. Figure 3.59a exhibits suggestive results in the mechanical Q_m increase with time lapse (minute) after electric poling, measured for various commercial soft and hard PZTs, PSM-PZT, and PSM-PZT doped with Yb.[96,104] It is worth noting that the Q_m values for commercial hard PZT and our high power piezoelectrics were almost the same, slightly higher than soft PZTs, and around 200~300 immediately after poling. After a couple of hours passed, the Q_m values increased more than 1000 for the hard materials, while no change was observed in the soft material. The increasing slope is the maximum for the Yb-doped PSM-PZT. We also found a contradiction that this gradual increase (in a couple of hours) in the Q_m cannot be explained by the above-mentioned domain wall pinning model, which is hypothesized that the oxygen-deficit-related dipole should move rather quickly at the millisecond scale. Figure 3.59b shows the polarization vs. electric field hysteresis curves measured for the Yb-doped Pb(Mn,Sb)O_3–PZT sample immediately after poling (fresh), 48 hours after, and a week after (aged).[105,106] Remarkable aging effects could be observed: (a) in the decrease in the magnitude of the remnant polarization, and (b) in the *positive internal bias* electric field growth (i.e., the hysteresis curve shift leftward with respect to the external electric field axis). Phenomenon (a) can be explained by the local domain wall pinning effect, but the large internal bias (close to 1 kV/mm) growth (b) seems to be the origin of the high power characteristics. Suppose that the vertical axis in Figure 2.44 (Uchida-Ikeda domain reorientation model) shifts rightward (according to 1 kV/mm positive internal bias field); the larger negative electric field is required for realizing the 180° polarization reversal, leading to resistance enhancement against generating hysteresis or heat with increasing the applied AC voltage.

Finally, let us propose the origin of this internal bias field growth. Based on the presence of oxygen deficiencies and the relatively slow (a couple of hours) growth rate, we assume here the oxygen deficiency diffusion model, which is illustrated in Figure 3.60.[105,106] Under the electric poling process, the defect dipole P_{defect} (a pair of acceptor ion and oxygen deficiency) will be arranged parallel to the external electric field. After removing the field, oxygen diffusion occurs, which can be estimated in a scale of hours at room temperature (this is a major contradiction of the so-called

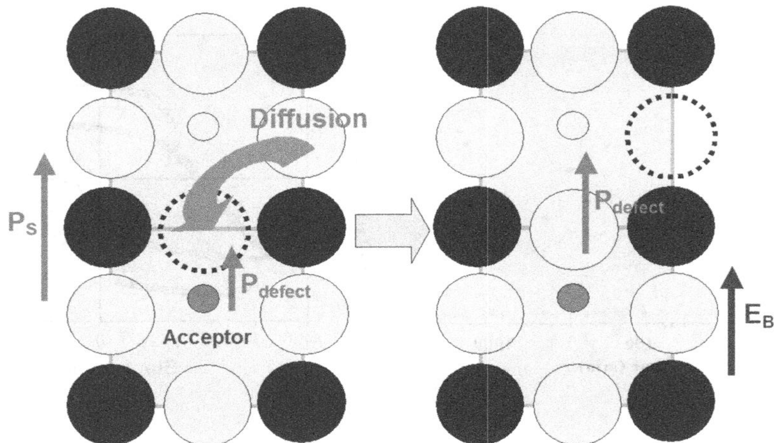

FIGURE 3.60 Oxygen deficiency diffusion model for explaining the internal bias electric field growth. (From Y. Gao et al: *J. Appl. Phys.*, 101(11) 114110, 2007; Y. Gao et al: *J. Appl. Phys.*, 101(5) 054109, 2007.)

"domain wall pinning" model). Taking into account slightly different atomic distances between the A and B ions in the perovskite crystal in a ferroelectric (asymmetric) phase, the oxygen diffusion probability will be slightly higher for the downward, as shown in the figure schematically, leading to the increase in the defect dipole or local defect polarization with time. The local polarization generates the local electric field via the Lorentz factor. This may be the origin of the internal bias electric field. You can easily understand that the dipole alignment is unidirectionally fixed along the polarization direction.

3.5.4.2.3 Unidirectionally Reversible Dipole Alignment

Tan and Viehland reported a double hysteresis in K-doped PZT ceramics (PKZT).[107] Figure 3.61a shows a P-E hysteresis curved observed in an "aged" sample of 4 at% doped $Pb(Zr_{0.65}T_{0.35})O_3$. Probably due to the unidirectionally (along the spontaneous polarization direction) switchable impurity dipole, though the hysteresis curve is symmetry (no internal DC bias field), the polarization is pinched around $E = 0$. However, with increasing the temperature, this pinching curve was released to the normal high coercive field hysteresis. Near 150°C, mobile defects are reported to begin to move into a random alignment.

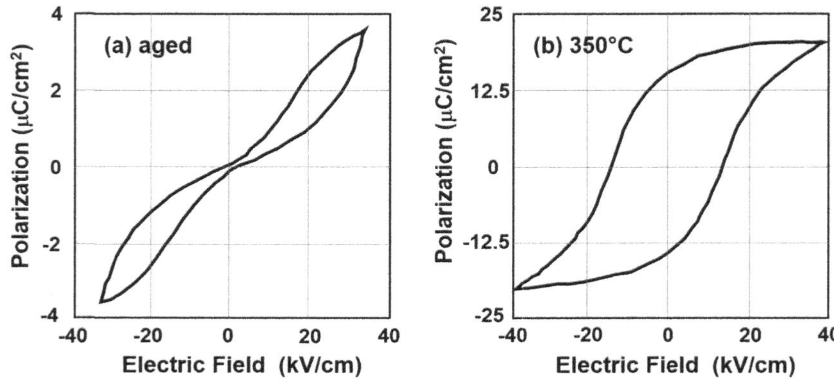

FIGURE 3.61 (a) Room temperature P-E hysteresis curve observed in an aged 4 at% doped $Pb(Zr_{0.65}T_{0.35})O_3$. (b) The P-E hysteresis at 350°C. (From Q. Tan and D. Viehland: Philosophical Magazine, 1997.)

3.5.5 Grain Size Effect on Hysteresis and Losses

To understand the grain size dependence of the piezoelectric properties, we must consider two size regions: the μm range in which a multiple domain state becomes a monodomain state, and sub-μm range in which the ferroelectricity becomes destabilized. We will primarily discuss the former region in this book. Figure 3.62a shows the transverse field-induced strains of 0.8 at.% Dy doped fine-grain ceramic $BaTiO_3$ (grain diameter around 1.5 μm) and of the undoped coarse-grain ceramic (50 μm), as reported by Yamaji et al.[108] As the grains become finer, under the same electric field, the absolute value of the strain decreases and the hysteresis becomes smaller. This is explained by the increase in coercive field for 90° domain rotation with decreasing grain size. The grain boundaries (with many dislocations on the grain boundary) pin the domain walls and do not allow them to move easily. Also the decrease of grain size seems to make the phase transition of the crystal much more diffuse. Although the effective value of d_{33} decreases in the Dy-doped sample, the temperature dependence is remarkably improved for practical applications. It should be noted that Yamaji's experiment cannot separate the effect due to intrinsic grain size from that due to dopants.

Uchino and Takasu studied the effects of grain size on PLZT.[109] We obtained PLZT (9/65/35) powders by coprecipitation. Various grain sizes were prepared by hot-pressing and by changing sintering periods, without using any dopants. PLZT (9/65/35) shows significant *dielectric relaxation* (frequency dependence of the permittivity) below the Curie point of about 80°C, and the dielectric constant tends to be higher at lower frequency. We prepared various grain size samples in the range of 1–5 μm. For grain size larger than 1.7 μm, the dielectric constant decreases with decreasing grain size. Below 1.7 μm, the dielectric constant increases rapidly. Figure 3.62b shows the dependence of the longitudinal field-induced strain on the grain size. As the grain size becomes smaller, the maximum strain decreases monotonically. However, when the grain size becomes less than 1.7 μm, the hysteresis is reduced. This behavior can be explained as follows: with decreasing grain size, (anti)ferroelectric (ferroelastic) domain walls become difficult to form in the grain, and the domain reorientation contribution to the strain becomes smaller (*multidomain–monodomain transition model*). The critical size is about 1.7 μm. However, note that the domain size is not

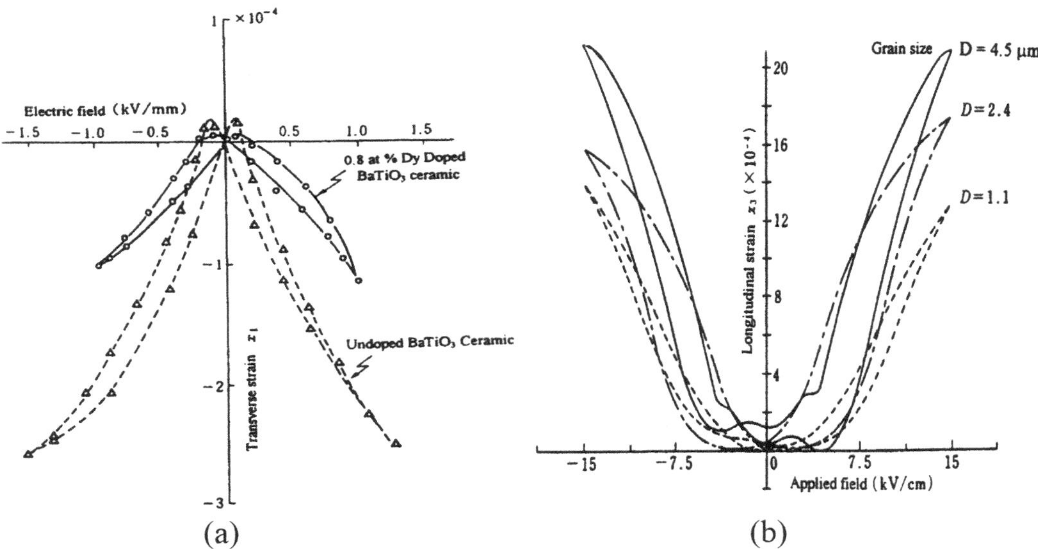

FIGURE 3.62 (a) Electric field–induced strain curves in Dy-doped and undoped $BaTiO_3$ ceramic samples. (From A. Yamaji et al: *Proc. 1st Mtg. Ferroelectr. Mater. Appl.*, Kyoto, p.269, 1977.) (b) Grain size dependence of the induced strain in PLZT ceramics. (From K. Uchino and T. Takasu, Inspection, 10, 29, 1986.)

constant, but is dependent on the grain size, and that in general the domain size decreases with decreasing grain size.

Sakaki et al. studied the grain size effect on high power performances, from a practical device application viewpoint.[110] The vibration velocity versus the temperature rise was investigated for various grain size soft Nb-doped PZTs from 0.9 to 3 μm. For two different grain sizes, 0.9 and 3.0 μm indicates that a higher maximum velocity of about 0.40 m/s is observed for the 0.9 μm-grain ceramic than that of the 3.0 μm-grain ceramic (0.30 m/s). Furthermore, the temperature rise for the fine-grain ceramic is observed to be about 40% lower than that for the coarse-grain ceramic near the maximum vibration velocity (0.30 m/s). This trend has suggested that a higher vibration velocity and lower heat generation can be achieved by reducing the grain size, which has been confirmed by investigating the heat generation phenomenon as a function of the grain size at various vibration velocities, and the observed trend is shown in Figure 3.63. A linear trend in heat generation can been found with grain size. In addition, the slope of heat generation is found to increase with an increment of the vibration velocity. This in turn suggested effective control over heat generation by lowering grain size of ceramic.

It is known that a Nb-doped PZT with a molecular formula $(Pb_{1-y/2}\square_{y/2})(Zr_xTi_{1-x-y}Nb_y)O_3$ is free of oxygen vacancies; hence, the movable space charge or impurity dipole may not be observed in this material. Thus, it is apparent that the increment of the mechanical quality factor (Q_m) with reducing the grain size is not caused by space charge effect of oxygen vacancies, but by the grain size effect. The origin of such an effect is speculated due to the following reasons:

1. The change in the configuration of the domain structure
2. The pinning effect by the grain boundaries with reduction in the grain size (i.e., grain boundaries contribute as additional pining points)

3.5.6 Loss Anisotropy: Crystal Orientation Dependence of Losses

3.5.6.1 Loss Anisotropy in Lead Zirconate Titanate

Zhuang et al. determined all 20 loss dissipation factors for a PZT ceramic using the ICAT HiPoCS admittance spectrum method introduced in Section 5.2.[87] Table 3.7 summarizes all elastic, dielectric,

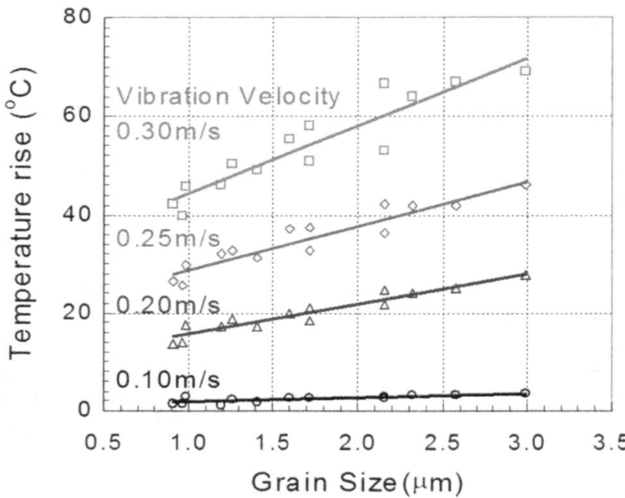

FIGURE 3.63 Grain size dependence of the temperature rise of the specimen for various vibration velocity. Note that heat generation is suppressed with reducing the grain size. (From C. Sakaki et al: *Jpn. J. Appl. Phys.*, 40, 6907–6910, 2001.)

TABLE 3.7
The Loss Factors of PZT APC 850 with Experimental Uncertainties

	$\tan\phi'_{11}$	$\tan\phi'_{12}$	$\tan\phi'_{13}$	$\tan\phi'_{33}$	$\tan\phi'_{55}$	
Result	0.01096	0.0095	0.01507	0.01325	0.0233	
Uncertainty	0.00007	0.0003	0.00034	0.00033	0.0022	
Relative	0.6%	3.2%	2.2%	2.5%	9.6%	
	$\tan\phi_{11}$	$\tan\phi_{12}$	$\tan\phi_{13}$	$\tan\phi_{33}$	$\tan\phi_{55}$	
Result	0.0105	0.0104	0.0076	0.00433	0.0149	
Uncertainty	0.0018	0.0028	0.0013	0.00008	0.0003	
Relative	17%	28%	17%	1.7%	2.1%	
	$\tan\delta'_{33}$	$\tan\delta'_{11}$	$\tan\delta_{33}$	$\tan\delta_{11}$		
Result	0.0143	0.0176	0.0058	0.0092		
Uncertainty	0.0002	0.0004	0.0011	0.0023		
Relative	1.4%	2.3%	20%	25%		
	$\tan\theta'_{31}$	$\tan\theta'_{33}$	$\tan\theta'_{15}$	$\tan\theta_{31}$	$\tan\theta_{33}$	$\tan\theta_{15}$
Result	0.0184	0.0178	0.0296	0.0133	0.0004	0.0024
Uncertainty	0.0006	0.0004	0.0026	0.0081	0.0004	0.0013
Relative	3.2%	2.1%	8.8%	61%	100%	57%

and piezoelectric losses determined on a soft PZT, APC 850 (APC International, State College, PA). Note the following general conclusions:

1. The antiresonance, Q_B, is always larger than the resonance, Q_A, in PZTs: This is a significant contradiction of the IEEE Standard assumption, leading to the necessity of the revision of the conventional "Standard" method.
2. The intensive (prime) losses are larger than the corresponding extensive (nonprime) losses: This is understood by the boundary or constraint condition difference between intensive and extensive; that is, *mechanically free* or *clamped* or *electrically open-* or *short-circuit* conditions.
3. There is apparent loss anisotropy in dielectric, elastic, and piezoelectric losses, indicating the anisotropy in domain wall mobility in the crystal.
4. The intensive piezoelectric losses are larger than the intensive dielectric or elastic losses in PZTs. That is, $\tan\theta' > (\tan\delta' + \tan\phi')/2$, while the opposite is true for Pb-free piezoelectrics.

Further specific conclusions include:

5. $\tan\ \delta_{33} < \tan\ \delta_{11}$: Polarization seems to be more stable along spontaneous polarization, similar to the permittivity trend $\left(\varepsilon_{33}^X < \varepsilon_{11}^X\right)$.
6. $\tan\ \phi_{33} < \tan\ \phi_{11}$: Elastic compliance also seems to be more stable along spontaneous polarization.
7. $\tan\ \theta_{33} < \tan\ \theta_{31}$: Piezoelectric constant also seems to be more stable along spontaneous polarization.

Choi et al. explored the loss anisotropy in piezoelectric PZT ceramics.[111] To observe the polarization angle dependence of losses, two different models for k_{31}- and k_{33}-mode vibration were prepared using three different compositions of 1% Nb-doped PZT for the tetragonal, rhombohedral, and morphotropic phase boundary structures. The polarization angle is defined by the angle of the polarization measured from the electric field direction (i.e., 0°: $P_S // E$; 90°: $P_S \perp E$). Larger dielectric loss is observed with higher polarization angle, regardless of the crystal structure. However, the capacitance showed different tendencies by compositions, indicating the tetragonal structure would have larger capacitance when depoled, while the rhombohedral structure would have smaller capacitance by depoling. The elastic compliance and mechanical loss exhibit complicated behavior by the poling angle, though the elastic loss tangent change is less significant than other two losses. The real and imaginary part seems to have a linear tendency, though further improvement of measuring accuracy is necessary to discuss in detail. The piezoelectric constant and coupling factor gradually decrease with increasing polarization angle, where the degradation of piezoelectricity can be explained by depoling. The extensive piezoelectric loss factor increases in MPB and rhombohedral structures with an increasing polarization angle, while it remains constant in the tetragonal structure. The quality factor at resonance becomes bigger than the one at antiresonance, where thê loss tangent increases.

3.5.6.2 $Pb(Mg_{1/3}Nb_{2/3})O_3$-$PbTiO_3$ Single Crystal

Extensive studies have been made of the different characteristics between ferroelectric single crystals and polycrystalline PZT. $Pb(Mg_{1/3}Nb_{2/3})O_3$-$PbTiO_3$ single crystals, for example, have significant loss anisotropy and doping dependence, the maximum vibration velocities (defined by 20°C temperature rise) of single crystals and ceramics (k_{31} mode).[112] The performance of single crystals is not as good as high-power hard PZTs, but better than soft types. Mechanical quality factors and electromechanical coupling factors exhibit significant crystallographic orientation dependence.

3.6 HIGH POWER MAGNETOSTRICTORS

3.6.1 Piezomagnetic Loss Equations

We consider loss mechanisms and high power characteristics in piezomagnetic/magnetostrictive materials, analogous to piezoelectric materials.[113,114] Three types of losses, mechanical loss ($\tan\phi'_m$), magnetic loss ($\tan\delta'_m$), and piezomagnetic loss ($\tan\theta'_m$), are introduced in the intensive parameter definitions (magnetic field, H, and stress, X):

$$\mu^{X*} = \mu^X (1 - j\tan\delta'_m), \tag{3.81}$$

$$s^{H*} = s^H (1 - j\tan\phi'_m), \tag{3.82}$$

$$d_m^* = d_m(1 - j\tan\theta'_m). \tag{3.83}$$

where μ^X is the magnetic permeability, s^H elastic compliance, and d_m the piezomagnetic constant.

A k_{33}-type magnetostrictive rod longitudinal vibrator was prepared, as shown in Figure 3.64, made of Terfenol-D. With negligible cross-sectional dimensions (length ≫ radius in Figure 3.64a), all stresses will be zero except X_3. When magnetic flux, B, and stress, X, are chosen as independent variables, piezomagnetic equations are expressed for the length expander mode as:

$$x_3 = s_{33}^B X_3 + g_{33} B_3 \tag{3.84}$$

$$H_3 = -g_{33} X_3 + \upsilon_{33}^X B_3 \tag{3.85}$$

where $g_{33} = d_m / \mu_{33}^X$, and $\upsilon_{33}^X = 1/\mu_{33}^X$ is designated as the reluctivity at constant stress. Eddy current loss generated by applying an alternating magnetic electric field is not considered here, since the laminated magnetostrictive specimen is used to constrict the part of eddy current loss,

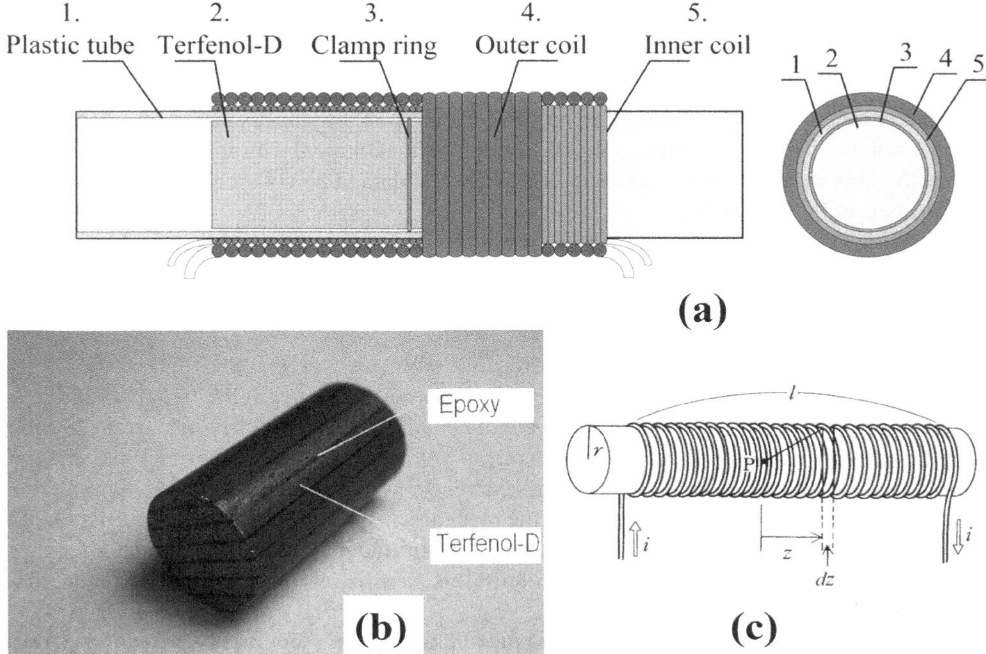

FIGURE 3.64 (a) Structure of a Terfenol-D rod (k_{33})-type piezomagnetic transducer. (b) Picture of the laminated Terfenol-D sample. (c) Coil schematic structure.

and the resonance frequencies are also controlled below the eddy current cut-off frequency range (see Figure 3.64b). Using the dynamic motion equation:

$$\rho \frac{\partial^2 u_3}{\partial t^2} = \frac{1}{s_{33}^B} \frac{\partial^2 u_3}{\partial x_3^2} \tag{3.86}$$

we can obtain the admittance expression as:

$$Y = \frac{I}{V} = \frac{1}{j\omega L_d} \left[1 - k_{33}^2 \frac{\tan(\omega l / v^B)}{(\omega l / 2v^B)} \right] \tag{3.87}$$

where $L_d = \frac{\mu_{33}^x N^2 A_r}{l}$, N total number of turns, A_r cross-section area, l length of the solenoid coil, and $v^B = 1/\sqrt{\rho s_{33}^B}$ (Figure 3.64c). Now we integrate the loss factors as the complex parameters in Equation 3.87, then derive the mechanical quality factors, Q_A and Q_B, very similarly to the piezoelectric case:

$$\boldsymbol{Q}_{B,33} = \frac{1 - \boldsymbol{k}_{33}^2}{\tan \phi_{33}' - \boldsymbol{k}_{33}^2 (2 \tan \boldsymbol{\theta}_{33}' - \tan \boldsymbol{\delta}_{33}')} \left[= \frac{1}{\tan \phi_{33}''} \right] \tag{3.88}$$

$$\frac{1}{\boldsymbol{Q}_{A,33}} = \frac{1}{\boldsymbol{Q}_{B,33}} + \frac{2}{\boldsymbol{k}_{33}^2 - 1 + \Omega_A^2 / \boldsymbol{k}_{33}^2} (2 \tan \boldsymbol{\theta}_{33}' - \tan \boldsymbol{\delta}_{33}' - \tan \phi_{33}') \tag{3.89}$$

$$\Omega_{B,33} = \frac{\omega_b l}{2 v_{33}^B} = \frac{\pi}{2}, \Omega_{A,33} = \boldsymbol{k}_{33}^2 \tan \Omega_{A,33} \tag{3.90}$$

3.6.2 Impedance Spectrum Measurement

The key differences in the characterization from the piezoelectric case include:

1. Inductive measurement—maximum impedance corresponds to the resonance and minimum impedance to the antiresonance mode. Since the mechanical vibration is excited by a coil, the coil structure also influences the measured data. The coil schematic structure is illustrated in Figure 3.64c.
2. In addition to the material losses (elastic, magnetic, and piezomagnetic losses), eddy current loss is added in experiments in general. We need to minimize this influence by using the sample lamination technique.

The impedance spectra of giant magnetostrictors were measured with Terfenol-D ($Tb_xDy_{1-x}Fe_{1.92}$) on the laminated specimen (eddy current loss seems to be negligibly small). Figure 3.65 shows (a) impedance spectra and (b) vibration velocity change with an AC driving current (magnetic field), while (c) impedance spectra and (d) vibration velocity change with an DC bias magnetic field (DC current A/m). With increasing the AC current/magnetic field, we can see a monotonous increase in vibration velocity, but a decrease in the mechanical quality factor, Q_m, in particular at the resonance. On the contrary, with increasing the DC current/magnetic bias field (under keeping the AC current), the elastic stiffness increase (hardening) is significant, in addition to the mechanical quality factor increase.

Table 3.8 summarizes elastic, magnetic, and piezomagnetic loss factors in the magnetostrictive Terfenol-D, in comparison with the piezoelectric hard PZT (APC 841). (1) The intensive piezomagnetic

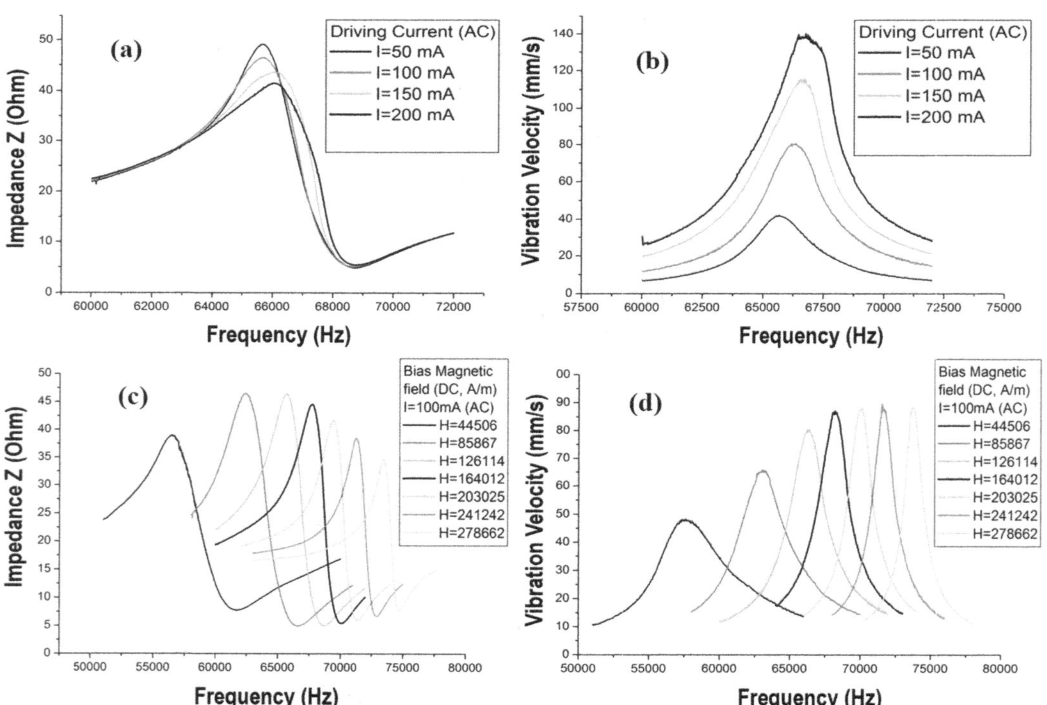

FIGURE 3.65 (a) Impedance sprectra and (b) vibration velocity change with an AC driving current (magnetic field); (c) impedance sprectra and (d) vibration velocity change with an DC bias magnetic field (DC current A/m), measured on Terfenol-D ($Tb_xDy_{1-x}Fe_{1.92}$).

TABLE 3.8
Comparison of Loss Factors among Magnetostrictive Terfenol-D and Piezoelectric Hard PZT

	Intensive loss factors		Extensive loss factors	
	Terfenol-D			
Elastic loss (s_{33})	$\tan \phi_m'$	0.0657	$\tan \phi_m$	0.0556
Magnetic loss (μ_{33})	$\tan \delta_m'$	0.0440	$\tan \delta_m$	0.0338
Piezomagnetic loss (d_{33})	$\tan \theta_m'$	0.0893	$\tan \theta_m$	0.0103
	PZT (APC 841)			
Elastic loss (s_{33})	$\tan \phi'$	0.00090	$\tan \phi$	0.00054
Dielectric loss (ε_{33})	$\tan \delta'$	0.0035	$\tan \delta$	0.0031
Piezoelectric loss (d_{33})	$\tan \theta'$	0.0025	$\tan \theta$	0.0015

loss is the largest among three losses. (2) The extensive loss factors are always smaller than the corresponding intensive losses. (3) The mechanical quality factor at the antiresonance, Q_B, is higher than Q_A at the resonance. (4) In comparison with the piezoelectric loss in a hard PZT 0.3%, the piezomagnetic loss is 9%, significantly larger. Thus, piezomagnetic materials may not be suitable for a long period of operation at their resonance frequency, due to heat generation. Burst mode should be used for their operation.

CHAPTER ESSENTIALS

1. Hard Piezoelectrics: $E_c > 1$ kV/mm
   Soft Piezoelectrics: $E_c < 0.1$ kV/mm
   Electrostrictors: $E_c \approx 0$ kV/mm
2. Comparison between *electrostriction* and *piezoelectricity*:
   a. Electrostriction produces strains of about the same order of magnitude as the piezoelectric (unipolar) strain (0.1%). An additional attractive feature of this effect is the absence of any significant hysteresis.
   b. Piezoelectrics require an electrical poling process, which makes these materials subject to significant aging caused by depoling over time. Electrostrictors do not need such a preliminary treatment, but do require a DC bias field for some applications because of their nonlinear response.
   c. In contrast to piezoelectrics, electrostrictive ceramics are durable and reliable when operated under severe conditions, such as high temperature and large mechanical load.
   d. The temperature characteristics (variation of electromechanical properties with temperature) of piezoelectrics are superior to electrostrictors.
   e. Piezoelectrics have smaller dielectric constants than electrostrictors, thus allowing for a quicker response.
3. *Materials classification* for developing compositions with an optimum temperature coefficient of electrostriction:
   Category I: Perovskite structures with disordered B-site cations:
   Type I(a): Materials produced from the combination of ferroelectric and nonferroelectric compounds [such as the solid solution system (Pb,Ba)(Zr,Ti)O_3 in which $BaZrO_3$ is the nonferroelectric component].

Type I(b): Materials in which A-site lattice vacancies also occur, which further promotes the diffuse phase transition [such as the solid solution system (Pb,La,□)(Zr,Ti)O_3].

Category II: Perovskite structures with some degree of short-range ordering of B-site cations [such as the solid solution systems Pb($Mg_{1/3}Nb_{2/3}$,Ti)O_3 and Pb($Mg_{1/2}W_{1/2}$,Ti)O_3].

4. *Figures of Merit for Transducers:*
   a. *The Electromechanical Coupling Factor, k*:

$$k^2 = \frac{\text{Stored (Mechanical or Electrical) Energy}}{\text{Input (Electrical or Mechanical) Energy}} = \frac{d}{\varepsilon_o \varepsilon s}$$

   where d is the piezoelectric strain coefficient, ε is the dielectric constant, and s is the elastic compliance.

   b. *The Energy Transmission Coefficient*, λ_{max}:

$$\lambda_{max} = \left[\frac{\text{Output (Mechanical or Electrical) Energy}}{\text{Input (Electrical or Mechanical) Energy}}\right]_{max}$$

$$\lambda_{max} = \left[(1/k) - \sqrt{(1/k^2) - 1}\right]^2 = \left[(1/k) + \sqrt{(1/k^2) - 1}\right]^{-2}$$

   where $k^2/4 < \lambda_{max} < k^2/2$.

   c. *The Efficiency*, η:

$$\eta = \frac{\text{Output (Mechanical or Electrical) Energy}}{\text{Consumed (Electrical or Mechanical) Energy}}$$

   which for many actuator designs is practically 98%–99%.

5. *Mechanical Strength:*
   a. *The differences in durability among the electromechanical and phase-change transducers* are attributed mainly to the magnitude of the maximum strain attainable by each device: 0.1% for the electrostrictor, 0.2% for the piezoelectric, and 0.4% for the antiferroelectric sample.
   b. *The grain size dependence of the Vickers indentation*: The crack length becomes shorter with decreasing grain size, while the center indentation size is virtually independent of grain size.
   c. *The electric poling process* for piezoelectric ceramics tends to induce a significant anisotropy in the mechanical strength of the material.

6. *Mechanical quality factors* at resonance and antiresonance (k_{31} piezoplate):

$$Q_{A,31} = \frac{1}{\tan\phi'_{11}}, \frac{1}{Q_{B,31}} = \frac{1}{Q_{A,31}} - \frac{2}{1+\left(\frac{1}{k_{31}} - k_{31}\right)^2 \Omega_B^2}(2\tan\theta'_{31} - \tan\delta'_{33} - \tan\phi'_{11})$$

$$\Omega_{B,31} = \frac{\omega_b l}{2v_{11}^E}. \text{ Antiresonance frequency } \Omega_B \text{ satisfies } 1 - k_{31}^2 + k_{31}^2 \frac{\tan\Omega_B}{\Omega_B} = 0$$

7. *Heat generation:*
   a. Large E drive at off-resonance—Uniform heat generation from dielectric loss tan δ'
   b. Small E drive at resonance—Sinusoidal temperature distribution from elastic loss tan ϕ'

8. *Maximum vibration velocity, v_0*: Vibration velocity that generates heat generation 20°C higher than room temperature at the sample's nodal point.
   a. PZT—0.3–0.6 m/s: $\tan\theta' > (1/2)[\tan\delta' + \tan\phi']$
   b. Pb-free piezoelectrics—0.8 m/s: $\tan\theta' \leq (1/2)[\tan\delta' + \tan\phi']$
   c. Mechanical energy density $(1/2)\rho \cdot v_0^2$ is more suitable in power comparison.
9. *Domain wall pinning* hardening/softening mechanism:
   1. Gauss's Law, $div\, D = \rho$
   2. Generation of the movable charge ρ by ion doping
10. *Impurity dipole alignment* models:
   a. Random alignment = Domain wall pinning model—High coercive field
   b. Unidirectionally fixed alignment—Internal bias DC field → Realistic model for practical high power density piezoelectrics
   c. Unidirectionally reversible alignment—Double P-E hysteresis
11. *Piezomagnetic Losses:*
   a. The intensive piezomagnetic loss is the largest among elastic, magnetic, and piezomagnetic losses.
   b. The extensive loss factors are always smaller than the corresponding intensive losses.
   c. The mechanical quality factor at the antiresonance, Q_B, is higher than Q_A at the resonance.
   d. In comparison with the piezoelectric loss in a hard PZT 0.3%, the piezomagnetic loss is 9%, significantly larger.

CHECK POINT

1. How to make a soft PZT? Provide a typical dopant ion in the PZT.
2. How to make a hard PZT? Provide a typical dopant ion in the PZT.
3. What is a typical number for k_{33} in soft PZT ceramics? 1%, 10%, 50%, or 70%?
4. There is a PZT multilayer actuator with a cross-section area 5×5 mm^2. Provide a generative (blocking force) roughly. 1 N, 10 N, 100 N, or 1 kN?
5. What is the fundamental longitudinal resonance frequency of a PZT multilayer actuator with a length 10 mm? 1 kHz, 10 kHz, 100 kHz, or 1 MHz?
6. A typical PZT shows $d_{33} = 300 \times 10^{-12}$ m/V and $d_{31} = -100 \times 10^{-12}$ m/V. Calculate its Poisson's ratio.
7. A typical PZT shows $d_{15} = 600 \times 10^{-12}$ m/V. When an electric field 1 kV/mm is applied to this PZT perpendicular (1-axis) to the polarization direction (3-axis), calculate the angle change of this ceramic sample between the 1- and 3-axes in degree angle. 0.0003, 0.003, 0.03, 0.3, or 3 degree?
8. Which is a hard PZT result, the left- or right-hand-side figure?

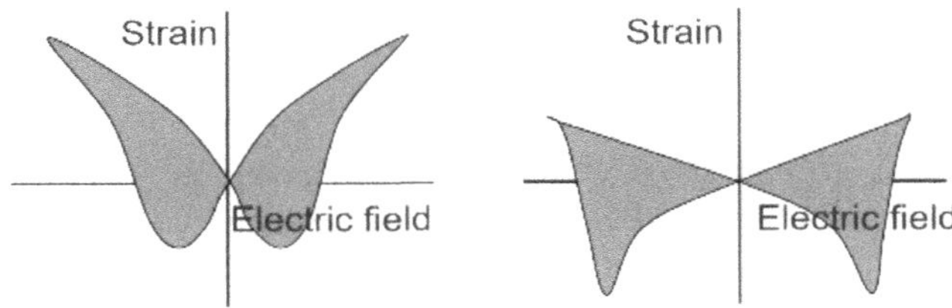

9. What is the MPB composition of the PZT system (MPB: morphotropic phase boundary) at room temperature? $PbZrO_3$:$PbTiO_3$ = (48: 52), (50:50), (52:48), or none of these?
10. (T/F) The MPB composition of the PZT system exhibits the maximum electromechanical coupling, *k*; piezoelectric coefficient, *d*; and minimum permittivity, ε. True or False?

11. Glassware is usually harder than porcelain pots. Even a human nail can mark raw porcelain. However, when we pour boiling water, glassware cracks, but porcelain does not. What is the key physical parameter to distinguish this difference? Answer in professional terminology.
12. Using an impedance analyzer, we obtained the elastic loss, tan $\phi' = 0.5\%$. What is the mechanical quality factor, Q_m, value?

CHAPTER PROBLEMS

3.1 The piezoelectric strain matrix (d) for a tetragonal PZT is:

$$\begin{pmatrix} 0 & 0 & 0 & 0 & 200 & 0 \\ 0 & 0 & 0 & 200 & 0 & 0 \\ -60 & -60 & 180 & 0 & 0 & 0 \end{pmatrix} \quad [pC/N]$$

The sample has dimensions [10 × 10 × 10 mm³]. The spontaneous polarization is along the c-axis, and an electric field of 10 kV/cm is applied along the a-axis. Supposing that polarization reorientation should not occur during the field application, calculate the deformation of the cubic specimen numerically and make a drawing illustrating the deformation.

3.2 A monochromatic acoustic wave is initially propagating through a solid with mass density, ρ_1, and elastic stiffness, c_1 (top layer in the right figure). It encounters a planar boundary with another solid with mass density, ρ_2, and elastic stiffness, c_2 (bottom layer), from which it is reflected normal to the boundary. Derive an equation for the reflectance for this wave.

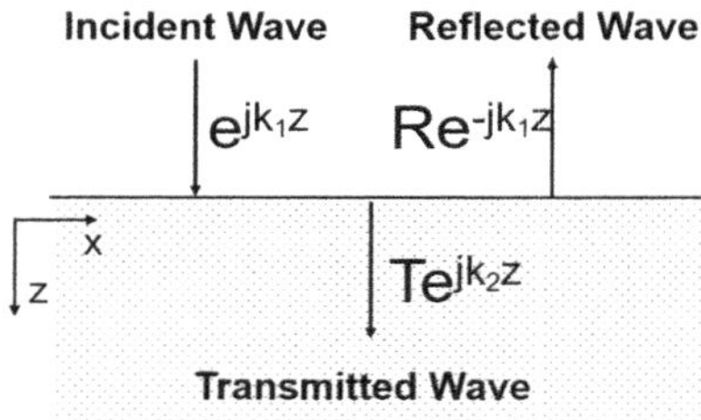

Hint: We describe the incident, reflected, and transmitted stress waves as e^{jk_1z}, Re^{-jk_1z}, and Te^{jk_2z} (normalized notation), respectively, where R and T denote the reflection and transmission coefficients. The total normalized stress in the upper (+) and lower (–) media are expressed as:

$$X^+(z) = e^{jk_1z} + Re^{-jk_1z}; \text{ and } X^-(z) = Te^{jk_2z}.$$

From the stress continuation condition at $z = 0 \rightarrow 1 + R = T$

Taking into account $\rho(\partial^2u/\partial t^2) = (\partial X/\partial z)$, and imposing continuity of particle velocity, we need the relation: $(k_1/\rho_1)(1 - R) = (k_2/\rho_2)T$. Considering wave vector k and sound velocity v, describe the final relation with density ρ_1, ρ_2, and elastic stiffness, c_1, c_2.

REFERENCES

1. K. Uchino: *Advanced Piezoelectric Materials – Science and Technology*, Ed. K. Uchino, Woodhead Pub./Elsevier, Duxford, UK, 2017. ISBN: 978-0-08-102135-4
2. E. Wainer and N. Salomon: *Trans. Electrochem. Soc.* 89, 1946.

3. T. Ogawa: *On Barium Titanate Ceramics [in Japanese]*, Busseiron Kenkyu, No.6, 1–27, 1947.
4. B. M. Vul, "High and ultrahigh dielectric constant materials" [in Russian], *Electrichestvo*, (3), 1946.
5. B. Jaffe, R. W. Cook and H. Jaffe: *Piezoelectric Ceramics*, Academic Press, New York, 1971.
6. H. Kawai: *Jpn. J. Appl. Phys.*, 8, 975, 1969.
7. R.E. Newnham, D.P. Skinner and L.E. Cross: *Materials Research Bulletin*, 13, 525, 1978.
8. L. E. Cross, S. J. Jang, R. E. Newnham, S. Nomura and K. Uchino: *Ferroelectrics*, 23, 187, 1980.
9. J. Kuwata, K. Uchino and S. Nomura: *Ferroelectrics*, 37, 579, 1981.
10. K. Otsuka and S. M. Wayman: *Chap. 1 in "Shape Memory Materials"*, Cambridge University Press, Cambridge, UK, 1998.
11. A. L. Roytburd, J. Slutsker and M. Wuttig: *Chap. 5.23, Smart Composites with Shape Memory Alloys, in "Comprehensive Composite Materials,"* Elsevier Science, Oxford, UK, 2000.
12. S. Legvold, J. Alstad and J. Rhyne: *Phys. Rev. Lett.*, 10, 509, 1963.
13. A. E. Clark, B. F. DeSavage and R. Bozorth: *Phys. Rev.*, 138, A216, 1965.
14. N. C. Koon, A. Schindler and F. Carter: *Phys. Lett.*, A37, 413, 1971.
15. A. E. Clark and H. Belson: *AIP Conf. Proc.*, 5, 1498, 1972.
16. A. E. Clark: *Ferroelectric Materials*, Vol. 1, p. 531, Eds. K. H. J. Buschow and E. P. Wohifarth, North-Holland, Amsterdam, 1980.
17. J. D. Verhoven, E. D. Gibson, O. D. McMasters and H. H. Baker: *Metal Trans. A.*, 18A(2), 223, 1987.
18. A. E. Clark, J. D. Verhoven, O. D. McMasters and E. D. Gibson: *IEEE Trans. Mag.*, MAG-22, 973, 1986.
19. J. B. Restorff: *Encyclopedia of Applied Physics*, VCH Publ., New York, 9, p.229, 1994.
20. A. B. Flatau, M. J. Dapino and F. T. Calkins: *Chap. 5.26, Magnetostrictive Composites, in "Comprehensive Composite Materials,"* Elsevier Science, Oxford, UK, 2000.
21. R. Kellogg and A. B. Flatau: *Proc. SPIE*, 3668(19), 184, 1999.
22. J. L. Butler: *Application Manual for the Design Etrema Terfenol-D Magnetostrictive Transducers*, Edge Technologies, Ames, IA, 1988.
23. G. P. Carman, K. S. Cheung and D. Wang: *J. Intelligent Mater. Systems and Structures*, 6, 691, 1995.
24. T. A. Duenas and G. P. Carman: *Proc. Adaptive Struct. & Mater. Systems*, ASME, AD-57/MD-83, p. 63, 1998.
25. https://metglas.com/magnetic-materials/
26. R. D. James and M. Wuttig: *Philos. Mag. A.*, 7, 1273, 1998.
27. http://mmech.com/transformers/pts-operation
28. L. E. Kinsler, A. R. Frey, A. B. Coppens and J. V. Sanders: *Fundamental of Acoustics*, John Wiley & Sons, New York, 1982.
29. Y. Ito and K. Uchino: *Piezoelectricity, Wiley Encyclopedia of Electrical and Electronics Engineering*, John Wiley & Sons, New York, 1999.
30. W. A. Smith: *Proc. SPIE - The International Society for Optical Engineering 1733*, 1992.
31. K. H. Hellwege et al.: *Landolt-Bornstein, Group III*, Springer-Verlag, Vol.11, New York, 1979.
32. T. R. Shrout, R. Eitel, C. A. Randall, P. Rehrig, W. Hackenberger and S. –E. Park: *Proc. 33rd. Int'l Smart Actuator Symp., State College*, April, 2001.
33. H. Takeuchi, S. Jyomura, E. Yamamoto and Y. Ito: *J. Acoust. Soc. Am.*, 74, 1114, 1982.
34. Y. Yamashita, K. Yokoyama, H. Honda and T. Takahashi: *Jpn. J. Appl. Phys.*, 20(Suppl. 20–4), 183, 1981.
35. Y. Ito, H. Takeuchi, S. Jyomura, K. Nagatsuma and S. Ashida: *Appl. Phys. Lett.*, 35, 595, 1979.
36. K. Furuta and K. Uchino: *Adv. Ceram. Mater.*, 1, 61, 1986.
37. Y. Gao, Y. H. Chen, J. Ryu, K. Uchino and D. Viehland: *Jpn. J. Appl. Phys.*, 40, 79–85, 2001.
38. B. N. Rolov: *Fiz. Tverdogo Tela*, 6, 2128, 1963.
39. V. A. Isupov: *Izv. Akad. Nauk SSSR, Ser. Fiz.*, 28, 653, 1964.
40. K. Uchino, J. Kuwata, S. Nomura, L. E. Cross and R. E. Newnham: *Jpn. J. Appl. Phys.*, 20(Suppl. 20–4), 171, 1981.
41. H. B. Krause, J. M. Cowley and J. Wheatley: *Acta Cryst.*, A35, 1015, 1979.
42. G. A. Smolensky, V. A. Isupov, A. I. Agranovskaya and S. N. Popov: *Soviet Phys.- Solid State*, 2, 2584, 1961.
43. M. L. Mulvihill, L. E. Cross and K. Uchino: *Proc. 8th European Mtg. Ferroelectricity*, Nijmegen, 1995.
44. K. M. Leung, S. T. Liu and J. Kyonka: *Ferroelectrics*, 27, 41, 1980.
45. A. Varslavans: *USSR Licenzintorg*, Peter Stuchka Latvian State University, 1980.
46. H. Takeuchi, H. Masuzawa, C. Nakaya and Y. Ito: *Proc. IEEE 1990 Ultrasonics Symposium*, 697, 1990.
47. J. Kuwata, K. Uchino and S. Nomura: *Ferroelectrics*, 37, 579, 1981.
48. J. Kuwata, K. Uchino and S. Nomura: *Jpn. J. Appl. Phys.*, 21(9), 1298, 1982.
49. K. Yanagiwara, H. Kanai and Y. Yamashita: *Jpn. J. Appl. Phys.*, 34, 536, 1995.

50. S. E. Park and T. R. Shrout: *Mat. Res. Innovt.*, 1, 20, 1997.
51. S. Wada: *Future Development of Lead-Free Piezoelectrics by Domain Wall Engineering*, Elsevier 2015/08/01, 2015.
52. T. Tou, Y. Hamaguchi, Y. Maida, H. Yamamori, K. Takahashi and Y. Terashima, *Jpn. J. Appl. Phys.*, 48, 07GM03, 2009.
53. Y. Saito, *Jpn. J. Appl. Phys.*, 35, 5168–73, 1996.
54. Y. Doshida, *Proc. 81st Smart Actuators/Sensors Study Committee*, JTTAS, Dec. 11, Tokyo, 2009.
55. K. Uchino and S. Nomura: *Ferroelectrics*, 50, 191, 1983.
56. A. Furuta, K. Y. Oh and K. Uchino: *Sensors and Mater.*, 3, 205, 1992.
57. www.murata.co.jp/corporate/ad/article/metamorphosis16/Application_note/
58. S. Nomura and K. Uchino: *Ferroelectrics*, 41, 117, 1982.
59. K. Uchino, J. Kuwata, S. Nomura, L. E. Cross and R. E. Newnham: *Jpn. J. Appl. Phys.*, 20(Suppl. 20–4), 171, 1981.
60. R. W. Basedow and T. D. Cocks: *J. Phys. E: Sci. Inst.*, 13,840, 1980.
61. K. Uchino, *Piezoelectric/Electrostrictive Actuators*, Morikita Pub. Co., Tokyo, 1985.
62. Y. Nakajima, T. Hayashi, I. Hayashi and K. Uchino: *Jpn. J. Appl. Phys.*, 24(2), 235, 1985.
63. S. Nomura, O. Osawa, K. Uchino and I. Hayashi: *Abstract Jpn. Appl. Phys.*, 764, Spring, 1982.
64. H. Cao and A. G. Evans: *J. Amer. Ceram. Soc.*, 76, 890, 1993.
65. Q. M. Zhang, J. Zhao, K. Uchino and J. Zheng: *J. Mater. Res.*, 12, 226, 1996.
66. X. –H. Du, Q. –M. Wang, U. Belegundu and K. Uchino: *J. Ceram. Soc. Jpn.*, 107(2), 190, 1999.
67. E. Mori, S. Ueha, Y. Tsuda: *Proc. Ultrasonics International*, 83, 154–159, 1983.
68. W. D. Callister, Jr.: *Materials Science and Engineering*, Wiley, p. 189, 1984.
69. T. Nishida and E. Yasuda ed.: *Evaluation of Mechanical Characteristics in Ceramics*, Nikkan –Kogyo, p. 68, 1986.
70. A. A. Griffith: *Phil. Trans. Roy. Soc. (London)*, A221, 163, 1920.
71. A. G. Evans and E. A. Charles: *J. Amer. Ceram. Soc.*, 59, 371, 1976.
72. K. Niihara, R. Morena and D. P. H. Hasselman: *Commun. Amer. Ceram. Soc.*, C-116, Jul. 1982.
73. K. Uchino and T. Takasu: *Inspec.*, 10, 29, 1986.
74. T. Yamamoto, H. Igarashi and K. Okazaki: *Ferroelectrics*, 50, 273, 1983.
75. H. Wang and R. N. Singh: *Ferroelectrics*, 168, 281, 1995.
76. H. Aburatani, S. Harada, K. Uchino, A. Furuta and Y. Fuda: *Jpn. J. Appl. Phys.*, 33, 3091, 1994.
77. H. Aburatani and K. Uchino: *Jpn. J. Appl. Phys.*, 35, L516, 1996.
78. H. Aburatani, J. P. Witham and K. Uchino: *Jpn. J. Appl. Phys.*, 37, 602, 1998.
79. K. Uchino: *J. Nanotech. Mater. Sci.*, 1(1), 1–15, 2014.
80. R. Halmashaw: *Non-Destructive Testing*, 2nd Ed., Chap. 2, Edward Amold, London, p. 273, 1991.
81. H. Aburatani and K. Uchino: *Japan. J. Appl. Phys.*, 37, Part 1, (1), 204–209, 1998.
82. H. Aburatani, S. Harada, K. Uchino, A. Furuta and Y. Fuda: *Jpn. J. Appl. Phys.*, 33, Pt.1, (5B), 3091, 1994.
83. K. H. Haerdtl: *Ceram. Int'l.*, 8, 121–127, 1982.
84. A. Mezheritsky: *Ferroelectrics*, 266, 277, 2002.
85. A. Mezheritsky: *IEEE UFFC*, 49(4), 484, 2002.
86. Y. Zhuang, S. O. Ural, A. Rajapurkar, S. Tuncdemir, A. Amin and K. Uchino: *Japan. J. Appl. Phys.*, 48, 041401, 2009.
87. Y. Zhuang, S. O. Ural, S. Tuncdemir, A. Amin and K. Uchino: *Japan. J. Appl. Phys.*, 49, 021503, 2010.
88. K. Uchino: *Ferroelectric Devices*, 2nd Ed., CRC, Boca Raton, FL, 2010.
89. K. Uchino, J. Zheng, A. Joshi, Y. H. Chen, S. Yoshikawa, S. Hirose, S. Takahashi and J. W. C. de Vries: *J. Electroceramics*, 2, 33–40, 1998.
90. S. Hirose, M. Aoyagi, Y. Tomikawa, S. Takahashi and K. Uchino: *Proc. Ultrasonics Int'l '95, Edinburgh*, pp. 184–187, 1995.
91. S. Tashiro, M. Ikehiro and H. Igarashi: *Jpn. J. Appl. Phys.*, 36, 3004–3009, 1997.
92. H. N. Shekhani and K. Uchino: *J. Amer. Ceram. Soc.*, 97(9), 2810–2814, 2014.
93. S. Takahashi and S. Hirose, *Jpn. J. Appl. Phys.*, 32, 2422–2425, 1993.
94. J. Ryu, H. W. Kim, K. Uchino and J. Lee: *Jpn. J. Appl. Phys.*, 42(3), 1307–1310, 2003.
95. Y. Gao, K. Uchino and D. Viehland: *J. Appl. Phys.*, 92, 2094–2099, 2002.
96. Y. Gao, and K. Uchino: *J. Materials Tech.*, 19(2), 90–98, 2004.
97. T. Tou, Y. Hamaguchi, Y. Maida, H. Yamamori, K. Takahashi and Y. Terashima: *Japan. J. Appl. Phys.*, 48, 07GM03, 2009.
98. Y. Saito: *Japan. J. Appl. Phys.*, 35, 5168–73, 1996.

99. Y. Doshida: *Proc. 81st Smart Actuators/Sensors Study Committee*, JTTAS, Dec. 11, Tokyo, 2009.
100. E. A. Gurdal, S. O. Ural, H. Y. Park, S. Nahm, and K. Uchino: *Sensors and Actuators A: Physical*, 200, 44, 2013.
101. M. Hejazi, E. Taghaddos, E. Gurdal, K. Uchino, and A. Safari: *J. Amer. Ceram. Soc.*, 97(10), 3192–3196, 2014.
102. A. Hagimura and K. Uchino: *Ferroelectrics*, 93, 373, 1989.
103. S. Takahashi, Y. Sasaki, S. Hirose and K. Uchino: *Proc. Mater. Res. Soc. Symp.*, 360, 305, 1995.
104. Y. Gao, K. Uchino and D. Viehlan: *Jpn. J. Appl. Phys.*, 45(12), 9119–9124, 2006.
105. Y. Gao, K. Uchino and D. Viehland: *J. Appl. Phys.*, 101(11) 114110, 2007.
106. Y. Gao, K. Uchino and D. Viehland: *J. Appl. Phys.*, 101(5) 054109, 2007.
107. Q. Tan and D. Viehland: *Philosophical Magazine*, 1997.
108. A. Yamaji, Y. Enomoto, E. Kinoshita, and T. Tanaka, *Proc. 1st Mtg. Ferroelectr. Mater. Appl.*, Kyoto, p.269, 1977.
109. K. Uchino and T. Takasu, *Inspection*, 10, 29, 1986.
110. C. Sakaki, B. L. Newarkar, S. Komarneni and K. Uchino: *Jpn. J. Appl. Phys.*, 40, 6907–6910, 2001.
111. M. Choi, K. Uchino, E. Hennig and T. Scholehwar: *J. Electroceramics* doi: 10.1007/s10832-017-0085-y, 2017.
112. A. Rajapurkar, S. O. Ural, Y. Zhuang, H.-Y. Lee, A. Amin, and K. Uchino: *Japan. J. Appl. Phys.*, 49, 071502, 2010.
113. M. Tao, Y. Zhuang, D. Chen, S. O. Ural, Q. Lu and K. Uchino: *Adv. Mater. Res., Trans Tech.*, 490–495, 922–926, 2012.
114. M. Tao, Y. Zhuang, D. Chen, S. O. Ural, Q. Lu and K. Uchino: *Adv. Mater. Res., Trans Tech.*, 490–495, 985–989, 2012.It is known that a Nb-doped PZT with a molecular formula $(Pb_{1-y/2}\square_{y/2})(Zr_xTi_{1-x-y}Nb_y)O_3$ is free of oxygen vacancies; hence, the movable space charge or impurity dipole may not be observed in this material. Thus, it is apparent that the increment of the mechanical quality factor (Q_m) with reducing the grain size is not caused by space charge effect of oxygen vacancies, but by the grain size effect. The origin of such an effect is speculated due to the following reasons:

4 Ceramic Fabrication Methods and Actuator Designs

4.1 FABRICATION PROCESSES OF CERAMICS AND SINGLE CRYSTALS

The active materials used in many actuator designs are most readily incorporated into the device structure when they are in ceramic form. The process for producing the ceramic generally occurs in two stages: (1) the preparation of the ceramic powders and (2) the sintering of the assembled structures. Single crystals are occasionally incorporated into structures designed for certain special applications.

4.1.1 Preparation of the Ceramic Powders

Particle shape, particle size distribution, and compositional uniformity are key factors to control when producing ceramic powders in order to optimize the reproducibility of the electromechanical response. A conventional method for producing powders is the *mixed-oxide method*, which involves firing a mixture of oxide powders in a process called *calcination*. The calcined material is then mechanically crushed and milled into fine powders. One major disadvantage of the mixed-oxide method is that it tends to produce materials with pronounced microscale compositional fluctuations. Wet chemical methods (such as the *co-precipitation* and *alkoxide methods*) are thus generally preferred, as they produce more compositionally homogeneous ceramics. In this section, these processes for producing barium titanate, lead zirconate titanate, and lead magnesium niobate ceramics are reviewed.[1]

4.1.1.1 Solid-State Reaction (Mixed-Oxide Method)

Let us first consider the general process involved in producing ceramic powders by the mixed-oxide method as it occurs in the preparation of $Pb(Zr_xTi_{1-x})O_3$. The oxide powders PbO, ZrO_2, and TiO_2 are weighed out in appropriate proportions, mixed, and then calcined at a temperature in the range of 800°C–900°C (depending on the composition) for 1–2 h. The calcined powder is subsequently crushed and milled into a fine powder. (Figure 4.1). The milling step often leads to certain undesirable features in the product. First, particle sizes of less than 1 μm generally cannot be produced by mechanical milling. Second, contamination of the powder by the milling media is unavoidable. The preparation of $BaTiO_3$ by this method can in principle be carried out in a similar way using equimolar quantities of the oxides BaO and TiO_2. In practice, $BaCO_3$ is generally used in place of BaO, however, because high-purity BaO is expensive and chemically less reactive. The process must be modified to some extent when it is used to prepare $Pb[(Mg_{1/3}Nb_{2/3})_{1-x}Ti_x]O_3$ ceramics from the starting oxides PbO, MgO, Nb_2O_5, and TiO_2. If the oxides are simply mixed and calcined, a second phase (*pyrochlore*) in addition to the perovskite phase is formed. One effective means of suppressing the formation of this second phase has been to add several mole percent of excess PbO during the final sintering stage.[2] An effective solution to this problem as it applies to the preparation of pure $Pb(Mg_{1/3}Nb_{2/3})O_3$ has been developed, which involves an initial two-stage calcination process whereby a *columbite* $MgNb_2O_6$ precursor is first prepared, and then reacted with PbO to form the desired perovskite phase according to the following reaction:

$$3PbO + MgNb_2O_6 \rightarrow 3Pb(Mg_{1/3}Nb_{2/3})O_3$$

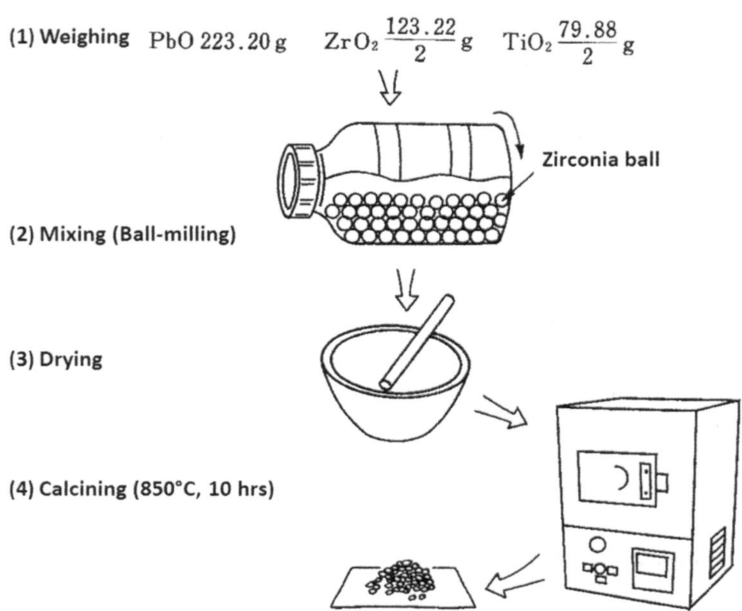

FIGURE 4.1 Powder preparation process of 1 mole $Pb(Zr_{0.5}Ti_{0.5})O_3$.

The product prepared in this manner is almost entirely perovskite phase, with only the very slightest traces of the pyrochlore phase present.[3] When this method is used for the preparation of PMN-PT, the MgO, Nb_2O_5, and TiO_2 are first mixed and fired at 1000°C to form the columbite precursor. Then, PbO is added to the columbite phase, and the mixture is calcined at 800°C–900°C. The addition of several mole percent excess MgO was found to be particularly effective in obtaining the perfect perovskite phase.

4.1.1.2 The Co-Precipitation Method

Since many of the popular piezoelectric/electrostrictive ceramics are of the complex perovskite type described in Section 3.3, compositional homogeneity has become as important an issue as phase purity in the production of these compositions. The mixed-oxide method is especially prone to problems in both these areas and thus is not generally the preferred method for preparing ceramics. The co-precipitation method has been found to produce materials with a much higher level of compositional homogeneity. The method basically involves adding a precipitant into a liquid solution of mixed metal salts to generate a homogeneous precipitate, which is then subjected to a thermal dissolution process to produce a homogeneous powder of the desired composition.

As an example, let us consider the preparation of a $BaTiO_3$ sample by this method. Oxalic acid is added to an aqueous solution of $BaCl_2$ and $TiCl_4$ to generate a precipitate of $BaTiO(C_2H_4)_2.4H_2O$ with a perfect 1:1 ratio of Ba to Ti on the atomic scale. Thermal dissolution of this precipitate produces highly stoichiometric $BaTiO_3$ powders with good sintering characteristics. When this method is used to prepare $(Pb,La)(Zr,Ti)O_3$ ceramics, $Pb(NO_3)_2$, $La(NO_3)_3.6H_2O$, $ZrO(NO_3)_2.2H_2O$, and $TiO(NO_3)_2$ are used as the starting materials.[4] First, the nitrates are mixed in the desired proportion to produce an aqueous solution; then, a half volume of ethanol is added to the mixture. Oxalic acid diluted with ethanol is then dripped slowly into the nitric solution, and a PLZT oxalate is precipitated. The thermal dissolution is carried out at 800°C.

In all the cases described so far, a final thermal dissolution of the precipitate is required to obtain the desired powder. The powder specimen can be obtained without this final step, however, for certain compositions, by what is referred to as the *direct precipitation method*. One such composition

TABLE 4.1
Some Ferroelectric Compositions That Can Be Synthesized by the Alkoxide Hydrolysis Method

Crystalline	$BaTiO_3$
	$Ba(Zr,Ti)O_3$
	$(Ba,Sr)TiO_3$
Amorphous	$Pb(Mg_{1/3}Nb_{2/3})O_3$
	$Ba(Zn_{1/3}Nb_{2/3})O_3$
	$Pb(Zr,Ti)O_3$
	$(Pb,La)(Zr,Ti)O_3$

is $BaTiO_3$. When it is prepared by the direct precipitation method, $Ti(OR)_4$ (R: propyl) is dripped into a $Ba(OH)_2$ water solution to produce high-purity, stoichiometric $BaTiO_3$ powders with an average particle size of 10 nm.[1]

4.1.1.3 Alkoxide Hydrolysis

When metal alkoxides $M(OR)_n$ (M: metal atom, R: alkyl) are mixed in alcohol in appropriate proportions and water is added, the hydrolytic reaction produces alcohol and a metal oxide or metal hydrate. This process is sometimes referred to as the *sol-gel method*. Some ferroelectric compositions that can be synthesized in this way are listed in Table 4.1. The sol-gel method can produce very fine, high-purity powders. Since metal alkoxides tend to be volatile, purification is easily accomplished through distillation. High purity can be sustained throughout the hydrolytic reaction because there is no need to introduce any ions other than those of the desired composition. The mechanisms of hydrolysis and condensation are summarized as follows:

Hydrolysis:

$$\mathrm{H-\underset{\underset{\displaystyle H}{|}}{O}} + \mathrm{M-OR} \rightarrow \mathrm{H-O-M} + \mathrm{ROH}$$

Alkoxylation: (Removal of H as an alcohol)

$$\mathrm{M-\underset{\underset{\displaystyle H}{|}}{O}} + \mathrm{M-OR} \rightarrow \mathrm{M-O-M} + \mathrm{ROH}$$

Oxolation: (Removal of H in water)

$$\mathrm{M-\underset{\underset{\displaystyle H}{|}}{O}} + \mathrm{M-OH} \rightarrow \mathrm{M-O-M} + \mathrm{H_2O}$$

When this method is employed to produce $BaTiO_3$ powders, the metal alkoxides $Ba(OC_3H_7)_2$ and $Ti(OC_5H_{11})_4$ are diluted with isopropyl alcohol (or benzene). Very fine, stoichiometric $BaTiO_3$ powders with good crystallinity and particle sizes in the range of 10–100 Å (agglomerate size = 1 μm) can be obtained by this method. The hydrolytic process produces powder with purity of more than 99.98%, which leads to a significant increase in the permittivity of the sintered ceramic as compared with samples prepared by the mixed-oxide method.[5]

When this method is employed to prepare $Pb(Zr,Ti)O_3$ powders, it is found that the lead alkoxide is relatively difficult to obtain as compared to the titanium and zirconium alkoxides. A modified two-stage approach has been developed to synthesize this more challenging composition. In the first stage of the process, $(Zr,Ti)O_2$ is prepared by the alkoxide method. The $(Zr,Ti)O_2$ is combined with PbO in the second stage, during which a solid-state reaction occurs.[6] A partial sol-gel method such as this, carried out with inexpensive ready-made nanosize powders, is an attractive cost-effective alternative for the commercial production of these powders. Another method that has presented some promise involves combining zirconium n-butoxide $Zr[O(CH_2)_3CH_3]_4$ and titanium isopropoxide $Ti[OCH(CH_3)_2]_4$ with lead acetyl acetonate $Pb(CH_3COCHCOCH_3)$ to obtain a PZT precursor phase.[5]

4.1.2 The Sintering Process

The calcined powders are generally mixed with an appropriate binder and formed into an appropriate shape by pressing, extrusion, or some other casting method. The green body is subjected to a low temperature "burn-out" process just prior to the final high-temperature firing in order to volatilize the binder from the body. This final firing process in which the ceramic attains its optimum density is called *sintering* and is typically carried out at high temperatures (but still below the melting temperature) and sometimes also at high pressure (*hot pressing*). The process promotes accelerated diffusion of the constituent atoms on the particulate surfaces due to the surface energy (surface tension), which leads to microcrystal bonding at the interface between adjacent particles, as depicted in Figure 4.2. The ceramic body may thus acquire sufficient mechanical strength while retaining its intended shape as it uniformly shrinks. The physical properties of the sintered body will depend not only on the properties of the particulates, but also on features of the microstructure such as the grain boundaries and the configuration of any remaining porosity. The mechanical strength, for example, will depend on the bonding between grains as well as the mechanical strength of the individual particulates. Mechanical failure in ceramics can occur either at the grain boundary (*intergranular fracture*) or across individual grains (*intragranular fracture*). The mechanical strength is thus enhanced for ceramic bodies with mechanically tough crystallites and strong intergranular bonding.

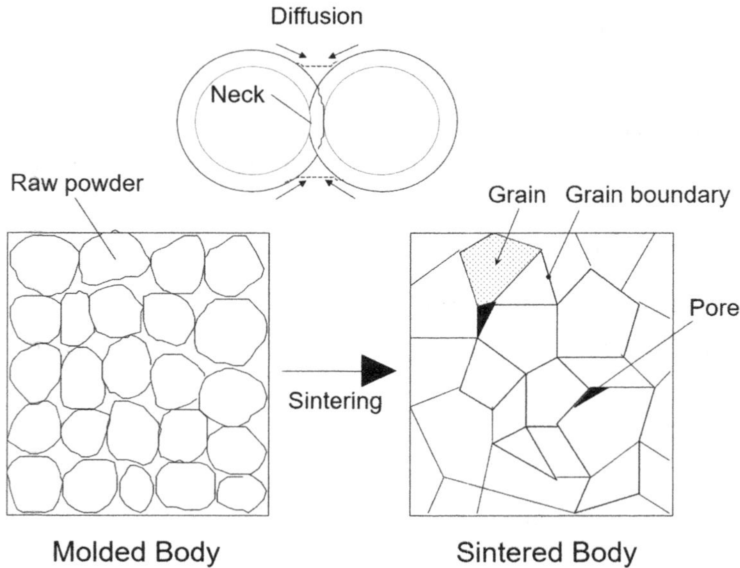

FIGURE 4.2 Schematic diagrams representing the mechanism for grain growth and the bonding of crystallites during the sintering process.

During sintering, the ceramic grains grow and their shape changes significantly. The features of the raw powder are found to strongly affect the dynamics of the sintering and the characteristics of the final product. In general, the diffusion processes that take place during sintering are accelerated as the particle size of the raw powder is decreased, because the driving force of sintering is related to the surface energy of the particles. Moreover, for fine powders, the diffusion length of the constituent atoms becomes shorter, which accelerates pore diffusion and elimination. This results in high-density ceramics.

There have been many studies on grain growth.[7] The relationship between the grain size, D, and the sintering period, t, is generally defined by:

$$D^{\beta} - D_o^{\beta} \propto t \tag{4.1}$$

where *normal grain growth* is characterized by $\beta=2$, and *abnormal grain growth* by $\beta=3$. The microstructures of PLZT (9/65/35) ceramics fabricated from powders prepared by the oxalic acid/ethanol method and sintered at 1200°C for 1 and 16 h are pictured in Figure 4.3a and b.[8] These represent cases of normal grain growth characterized by the function shown in Figure 4.3c, for which we see a good linear relation between the sintering period and the square of the grain size.

Doping is another method commonly used to regulate sintering conditions and grain growth in certain ceramic compositions. The desired effect of the dopant is to decrease the sintering temperature, and additional effects such as suppression or enhancement of grain growth are sometimes observed. Excess PbO or Bi_2O_3 added to PZT, for example, tends to inhibit grain growth. The addition of

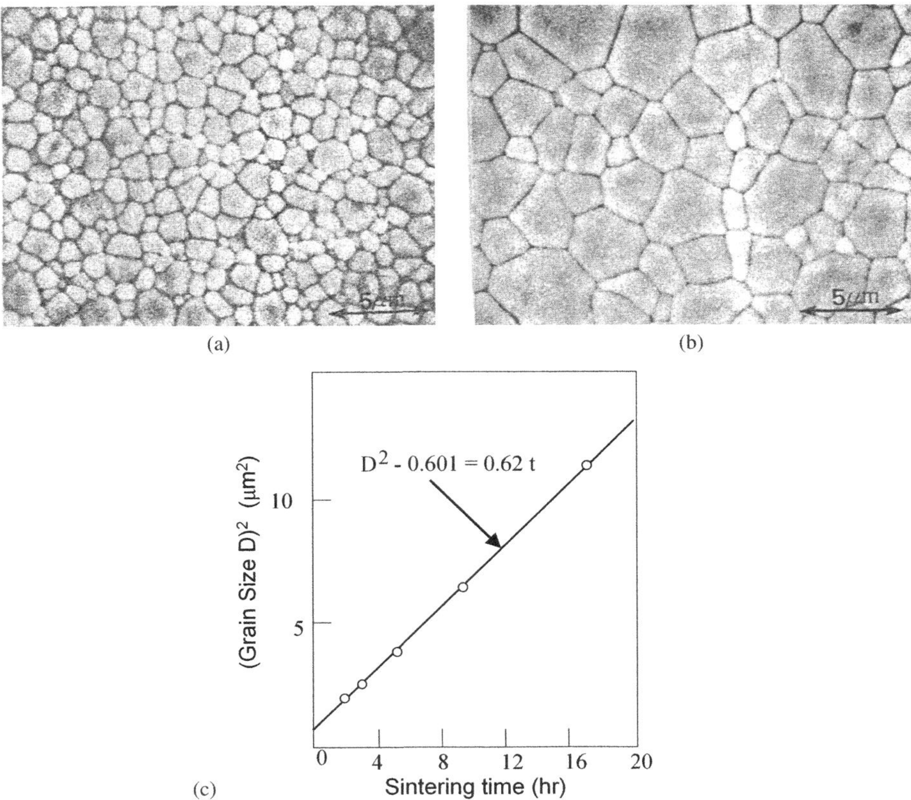

FIGURE 4.3 Grain growth in PLZT (9/65/35) ceramics sintered for (a) 1 h and (b) 16 h. (c) Grain growth in PLZT as a function of sintering time.

0.8 atom percent of Dy to $BaTiO_3$ ceramics has been found to effectively suppress the grain size to less than 1 μm,[9] as introduced in Section 3.5.5.

EXAMPLE PROBLEM 4.1

The cations K^{1+}, Bi^{3+}, Zn^{2+}, and Nb^{5+} constitute a disordered complex perovskite crystal, where the first two occupy the A-site and the last two occupy the B-site of the lattice. Determine the possible compositional formulas.

SOLUTION

The perovskite structure is generally described by the following:

$$ABO_3 = \left(K^{1+}_{1-x}Bi^{3+}_{x}\right)\left(Zn^{2+}_{1-y}Nb^{5+}_{y}\right)O^{2-}_{3}$$

We begin by accounting for charge neutrality:

$$[1(1-x)]+[3(x)]+[2(1-y)]+[5(y)]=+6 \qquad \text{(P4.1.1)}$$

which yields:

$$2(x)+3(y)=3 \qquad \text{(P4.1.2)}$$

where $0<x<1$, $1/3<y<1$.

A continuous range of compositions satisfy the conditions for x and y given by Equation P4.1.2. The composition $(K_{3/4}Bi_{1/4})\,(Zn_{1/6}Nb_{5/6})O_3$ is one possibility within this range.

4.1.3 Single Crystal Growth

Single crystals are not as commonly used in piezoelectric/electrostrictive devices as are ceramics, but there are nevertheless some promising designs under investigation which incorporate single crystals.

4.1.3.1 Quartz, LN, LT

The popular single crystals are quartz, grown by hydrothermal synthesis, and $LiNbO_3$ and $LiTaO_3$, by the Czochralski method. Nakamura et al. reported on a monomorph bending actuator fabricated from a thin plate of $LiNbO_3$ crystal, in which half of the thickness is reverse-polarized to function like a bimorph.[10] Although this device is fragile and the bending displacement is not large, linearity in the displacement curve without hysteresis is attractive for some special applications such as scanning tunneling microscopes.

4.1.3.2 $Pb(Zn_{1/3}Nb_{2/3})O_3$-$PbTiO_3$, $Pb(Mg_{1/3}Nb_{2/3})O_3$-$PbTiO_3$

Since single crystal growth of PZT was rather difficult with a composition near the morphotropic phase boundary, $Pb(Zn_{1/3}Nb_{2/3})O_3$-$PbTiO_3$ and $Pb(Mg_{1/3}Nb_{2/3})O_3$-$PbTiO_3$ have been the focus of medical acoustic transducer applications, because large single crystals of more than 1 inch^3 can easily be grown by a simple flux method, and enormously high electromechanical coupling factors (95%) and piezoelectric d constants (1570×10^{-12} C/N) can be obtained when they are poled in a special crystal direction.[11,12] A more recent result is covered in Reference 13.

4.1.3.3 $Pb(Zr_{1-x}Ti_x)O_3$

Recently, Ye has successfully grown single crystals of $Pb(Zr_{1-x}Ti_x)O_3$, which is considered one of the most important electroceramic materials.[14] The growth of PZT single crystals, which had been a long-standing challenge in the ferroelectrics field, has allowed the discovery of a series of new phenomena, including optical isotropy and monoclinic phase, unusual structural factors, structural heterogeneity, and diffuse scattering. Because of the current applicational and industrial success of PMN-PT single crystals, however, investment in PZT single crystal growth has not been enthusiastic.

4.1.4 Templated Grain Growth

Without using the rather expensive Czochralski, hydrothermal, or flux method, inexpensive "template grain growth" methods have been introduced for preparing BT, PZN-PT, PMN-PT, PZT, and Pb-free piezoelectric crystals. Figure 4.4 shows a large PMN-PT single crystal prepared from a seed $BaTiO_3$ crystal with solid-state single crystal growth (SSCG) (Ceracomp, Korea).[15] Based on the crystal orientation of the seed, PMN-PT powders are realigned during the sintering or firing process.

Crystallographic texturing of polycrystalline ferroelectric ceramics exhibits significant enhancements in the piezoelectric response. *Templated grain growth* (TGG) is a method to produce textured ceramics with single crystal-like properties. In TGG, nucleation and growth of the desired crystal on aligned single crystal template particles with heating results in an oriented material. To facilitate alignment during forming, template particles are anisometric in shape. To serve as the preferred sites for epitaxy and subsequent oriented growth of the matrix, the template particles need to be single crystal and chemically stable up to the growth temperature. Messing et al. reported that the resulting ceramics show texture levels up to 90% and significant enhancements in the piezoelectric properties in comparison with randomly oriented ceramics.[16] For example, the piezoelectric *d* coefficient of textured PMN-32.5PT piezoelectrics was about 1150 pC/N, 2~3 times higher than randomly oriented ceramics, and as high as 90% of the single crystal values.

Toyota Central Research Lab is using reactive-templated grain growth (RTGG) as a key processing technique for preparing textured Pb-free piezoelectric ceramics. Bismuth layer structured ferroelectric $ABi_4Ti_4O_{15}$ ($A = Na_{0.5}Bi_{0.5}$, Ca, Sr) ceramics with a highly preferred [001] orientation were fabricated with enhanced piezoelectric properties using $Bi_4Ti_3O_{12}$ platelets. Textured simple perovskite-type ceramics were also prepared in $Bi_{0.5}Na_{0.5}TiO_3$ (BNT), $BaTiO_3$, and $K_{0.5}Na_{0.5}NbO_3$ (KNN)-based compositions with Lotgering's factor higher than 0·8, which exhibited enhanced electromechanical coupling coefficients and piezoelectric constants when compared with their randomly oriented counterparts.[17] Figure 4.5a and b show the strain versus electric field curves and the displacement change with temperature for textured and randomly oriented piezoceramics, $(K,Na,Li)(Nb,Ta,Sb)O_3$.[18] Slurry with plate-like NKN powders was tape-casted to align the NKN

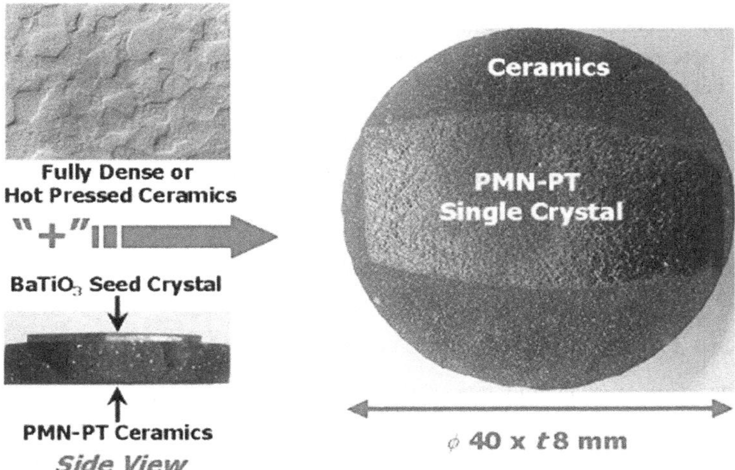

FIGURE 4.4 Growth of PMN-PT single crystals using SSCG technique. (Courtesy Ho-Yong Lee.) (From http://www.ceracomp.com/.)

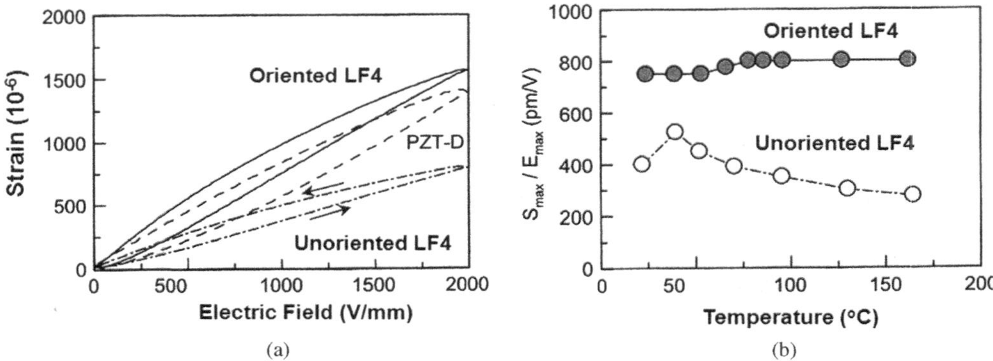

FIGURE 4.5 Strain versus electric field curves (a), and the displacement change with temperature (b), for textured and randomly oriented piezoceramics, (K,Na,Li)(Nb,Ta,Sb)O_3. Note significant enhancement in the strain level in the textured ceramic, which is almost comparable to the PZT ceramic strain.

crystal orientation, then sintered. Note that significant enhancement occurred in the strain level in the textured ceramic, which became almost comparable to the PZT ceramic strain.

4.2 SIZE EFFECT OF FERROELECTRICITY

So-called "fine ceramic" used in the 1980s has been renamed *nanotechnology* in the 2000s, that is, manufacturing ceramics that possess controlled grain size or very fine grains (note that "nano" does not mean a real nanometer range, but a micron range). In parallel, so-called "amorphous ferroelectric" was studied in 1980s. Lines theoretically suggested that significantly large permittivity might be realized when a ferroelectric ceramic is prepared in an amorphous form.[19] Note that he was not suggesting that an amorphous form of this material exhibits ferroelectric properties. All results reported during the 1980s period concluded that there is little possibility of realizing "amorphous" ferroelectrics. Different from ferromagnetics, which originate from short-range spin-exchange coupling, ferroelectrics originate from a cooperative phenomenon based on rather long-range Coulombic coupling. In conclusion, relatively large crystalline size (sub-micron meter) seems to be required to realize ferroelectricity by annealing amorphous ferroelectrics to increase the crystallinity.[20]

Small particle nanotechnology seems to eliminate ferroelectric functionality, unfortunately. However, thin film technology with nanometer thickness seems to be able to realize good ferroelectricity. There remains a future question to clarify the followings from both experimental and theoretical viewpoints:

1. 3D Problem: What is a critical particle size below which ferroelectricity will disappear?
2. 2D Problem (1D connectivity): What is a critical fiber diameter below which ferroelectricity will disappear?
3. 1D Problem (2D connectivity): What is a critical film thickness below which ferroelectricity/piezoelectricity will disappear? (Discussed in Section 4.5.1.)

4.2.1 Grain Size Effect on Ferroelectricity

To understand the grain size dependence of the dielectric properties, we must consider two size regions: the μm range in which a multiple domain state becomes a monodomain state, and the sub-μm range in which ferroelectricity becomes destabilized. Figure 4.6a shows the transverse field-induced strains of 0.8 at.% Dy-doped fine-grain ceramic $BaTiO_3$ (grain diameter around 1.5 μm) and of undoped coarse-grain ceramic (50 μm), as reported by Yamaji.[9] As the grains become finer, under the same electric field, the absolute value of the strain decreases and the hysteresis becomes smaller.

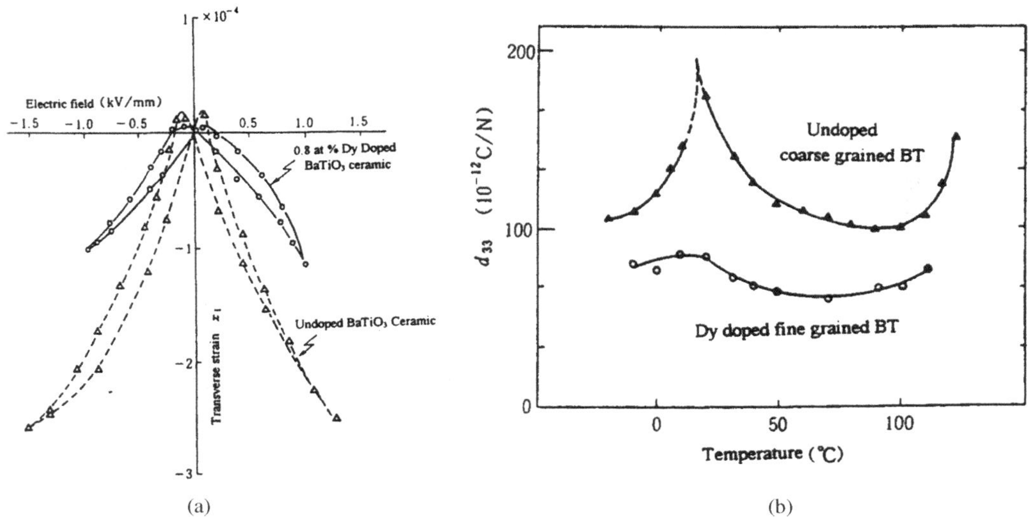

FIGURE 4.6 (a) Electric field–induced strain curves, and (b) temperature dependence of the piezoelectric d_{33} in Dy-doped and undoped $BaTiO_3$ ceramics. (From A. Yamaji Y. et al.: *Proc. 1st Mtg. Ferroelectric Mater. & Appl.*, Kyoto, p. 269, 1977.)

This is explained by the increase in coercive field for 90° domain rotation with decreasing grain size. The grain boundaries (with many dislocations on the grain boundary) pin the domain walls and do not allow them to move easily. Also, the decrease of grain size seems to make the phase transition of the crystal much more diffuse. Figure 4.6b shows the temperature dependence of the piezoelectric coefficient, d_{33}. Although the absolute value of d_{33} decreases in the Dy-doped sample, the temperature dependence is remarkably improved for practical applications. It should be noted that Yamaji's experiment cannot separate the effect due to intrinsic grain size from that due to dopants.

Takasu et al. studied the effects of grain size on PLZT.[8] They obtained PLZT (9/65/35) powders by coprecipitation. Various grain sizes were prepared by hot-pressing and by changing sintering periods, without using any dopants. PLZT (9/65/35) shows significant *dielectric relaxation* (frequency dependence of the permittivity) below the Curie point of about 80°C, and the dielectric constant tends to be higher at lower frequencies. Figure 4.7a shows the dependence of the peak dielectric constant on grain size. For grain size larger than 1.7 μm, the dielectric constant decreases with decreasing grain size. Below 1.7 μm, the dielectric constant increases rapidly. Figure 4.7b shows the dependence of the longitudinal field-induced strain on the grain size. As the grain size becomes smaller, the maximum strain decreases monotonically. However, when the grain size becomes less than 1.7 μm, hysteresis is reduced. This behavior can be explained as follows: with decreasing grain size, (anti)ferroelectric (ferroelastic) domain walls become difficult to form in the grain, and the domain rotation contribution to the strain becomes smaller (*multidomain–monodomain transition model*). The critical size is about 1.7 μm. However, note that the domain size is not constant, but is dependent on the grain size, and that in general the domain size decreases with decreasing grain size.

4.2.2 *3D* Particle Size Effect on Ferroelectricity

Regarding the much smaller grain/particle size range, Uchino et al. reported previously on a number of informative experiments. Figure 4.8a shows the most cited figure in recent nanotechnology papers. The degree of tetragonality (i.e., *c/a* ratio) is plotted as a function of particle size in pure $BaTiO_3$ at room temperature.[21] The *c/a* value decreases drastically below 0.2 μm and becomes 1 (i.e., cubic!) at 0.12 μm, defined as a *critical particle size*. Figure 4.8b shows the temperature dependence of the *c/a* ratio for various particle size powders. This demonstrates the correlation between the critical particle size and

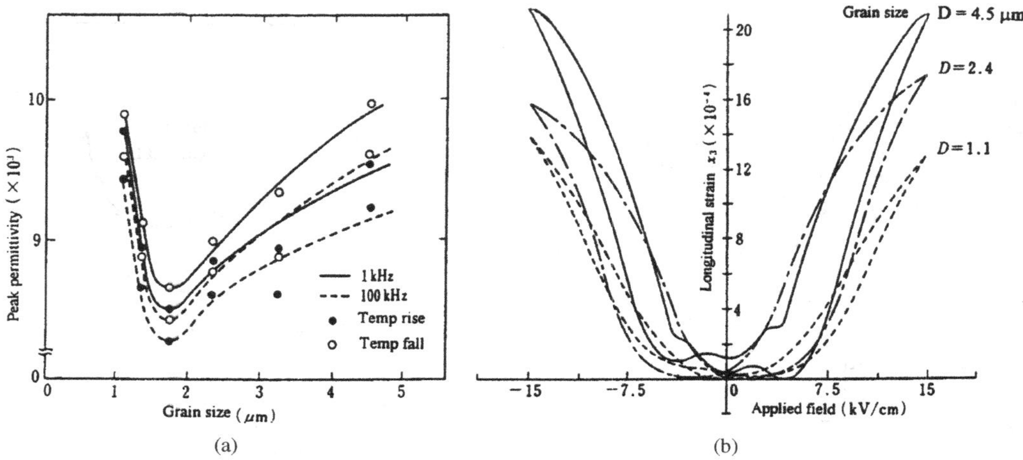

FIGURE 4.7 Grain size dependence of the peak permittivity (a), and of the induced strain in PLZT (9/65/35) ceramics. (From K. Uchino and T. Takasu: *Inspec.*, 10, 29, 1986.)

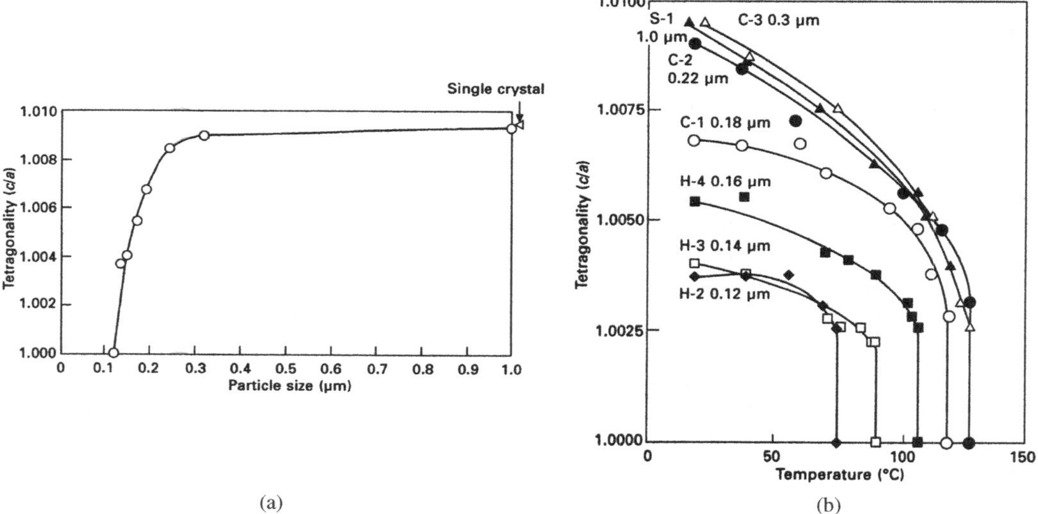

FIGURE 4.8 (a) Particle size dependence of the tetragonality in $BaTiO_3$ at room temperature. (b) Temperature dependence of the tetragonality in $BaTiO_3$ for various particle size samples. (From K. Uchino et al.: *J. Amer. Ceram. Soc.*, 72, 1555, 1989.)

the Curie temperature, which decreases with decreasing particle size. This paper clearly indicated that ferroelectricity disappears with a reduction of the three-dimensional particle size. It is interesting to note that tetragonality between 1 and 1.0025 was not observed experimentally, suggesting that the first-order phase transition seems to be sustained regardless of the crystalline size. The tetragonality change curve is simply shifted to lower temperature with reduction of the particle size.

Though multiple recent papers have discussed the critical particle size of 0.12 μm being too small or too large, experimentally, it is not the key point, because there is a large ambiguity in determining the average particle size, as Reference 21 explicitly explained. Specific surface area, x-ray diffraction, electron microscope, or optical microscope provides large deviation on the particle size of more than several times.

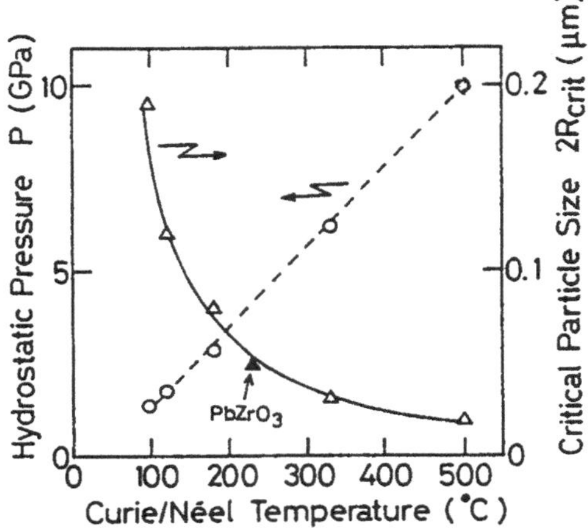

FIGURE 4.9 Relation between the critical particle size, D_{crit}, or the critical hydrostatic pressure and the phase transition temperature. (From K. Uchino et al.: *Ceramic Trans.*, 8, Ceramic Dielectrics, 107, 1990.)

A similar critical particle size was also reported in (Ba,Sr)TiO_3,[22] (Ba,Pb)TiO_3,[23] and for antiferroelectric $PbZrO_3$.[22] Figure 4.9 shows the relationship between the critical particle size, D_{crit}, and the Curie (or Néel) temperature, T_C, for these materials. An important empirical rule is obtained:

$$D_{crit} \times (T_C - \text{Room Temp.}) = \text{Const.} \tag{4.2}$$

Although there have been many reports on the critical grain/particle size, it seems to be true that there is a kind of critical size below which ferroelectricity disappears (i.e., the crystal becomes cubic).[24]

No satisfactory explanation has yet been presented. One possible explanation is based on a *hydrostatic pressure model*. In general, the ferroelectric transition temperature decreases sharply (50°C /MPa) with increasing hydrostatic pressure. Systematic data can be found in Samara's papers,[25] which used external hydrostatic oil pressure. The effective *surface tension* γ on a fine particle causes a hydrostatic pressure p intrinsically:[25]

$$p = 2\gamma/R \quad (R: \text{particle radius}) \tag{4.3}$$

From the critical particle size, D_{crit} (=2 R_{crit}), and the critical hydrostatic pressure, p_{crit}, above which the cubic structure is realized at room temperature, we can calculate the effective surface tension, γ, as listed in Table 4.2. The γ value is almost constant for all perovskite ferroelectrics. The γ value, 50 times larger than that of nonpolar oxides (such as MgO), may be due to the additional

TABLE 4.2
Critical Particle Size, Critical Hydrostatic Pressure, and Surface Tension Energy for Various Perovskites ($P = 2\ \gamma/R$)[25]

Material	Curie Temp (°C)	$2R_{crit}$ (μm)	P_{crit} (GPa)	γ (N/m)
$Ba_{0.9}Sr_{0.1}TiO_3$	95	0.19	1.2	57
$BaTiO_3$	125	0.12	1.8	54
$Ba_{0.85}Pb_{0.15}TiO_3$	180	0.08	2.9	58
$Ba_{0.5}Pb_{0.5}TiO_3$	330	0.032	6.2	50
$PbTiO_3$	500	0.02	10	50

energy from the surface charge contribution, and/or from a crystallographically different skin phase on the particle surface (i.e., *core-shell model*).

EXAMPLE PROBLEM 4.2

Why can we not expect amorphous ferroelectricity or ferroelectric nanoparticles?

Answer

The first approach is from the crystal energy stability viewpoint. To understand the reason ferroelectricity will disappear with decreasing particle size, we can consider the energy fluctuation for a nanosize ferroelectric particle as follows. Consider a 1-dimensional finite chain of two kinds of ions, $+q$ and $-q$, arranged alternately with a distance of a (see Figure 4.10a). A nanosize crystal grows gradually, starting from a single positive ion and adding a pair of negative or positive ions, thus keeping a crystal size $2na$ ($n=1, 2, 3, \ldots$). With increasing crystal size, the crystal Coulomb energy will be changed continuously as:

$$U_1 = (2/4\pi\varepsilon_0\varepsilon)[-(q^2/a)]$$
$$U_2 = (2/4\pi\varepsilon_0\varepsilon)[-(q^2/a)+(q^2/2a)]$$
$$U_3 = (2/4\pi\varepsilon_0\varepsilon)[-(q^2/a)+(q^2/2a)-(q^2/3a)]$$
$$\ldots\ldots$$
$$U_n = (2/4\pi\varepsilon_0\varepsilon)\,[-(q^2/a)+(q^2/2a)-\cdots+(q^2/na)]$$

Knowing the relation: $\ln(1+x)=x-x^2/2+x^3/3-x^4/4+\ldots$, the Madelung constant, M, which is defined as the saturated energy, $U=(-M/4\pi\varepsilon_0\varepsilon)(q^2/a)$, is calculated. The value $2\ \ln 2=1.386$ is the Madelung constant for a 1D chain. The Coulomb potential change at the center point is shown in Figure 4.10b as a function of the crystal size, n, with a final saturating value of Madelung energy. When the energy fluctuation is so large, that small energy imbalance between a paraelectric and a ferroelectric state may not cause a phase transition. Suppose that the minimum crystal size ($2na$) is required to maintain a potential energy fluctuation of less than $\pm$ 10% at the center positive ion, even when adding or subtracting a pair of ions (i.e., $n+1$ or $n-1$). If the basic crystal Coulomb energy is not stabilized to less than this degree of fluctuation, we cannot expect the ferroelectric phase transition to occur as a cooperative phenomenon. By equating $1/n=10\%$, we get $n=10$. If we use $a=4$ Å, $2na=80$ Å$=8$ nm. A minimum 10-nm-size single crystal may be required for realizing ferroelectricity. According to a study on amorphous $PbTiO_3$,[26] the soft phonon mode and the maximum permittivity that indicate ferroelectricity appearance start to be observed around

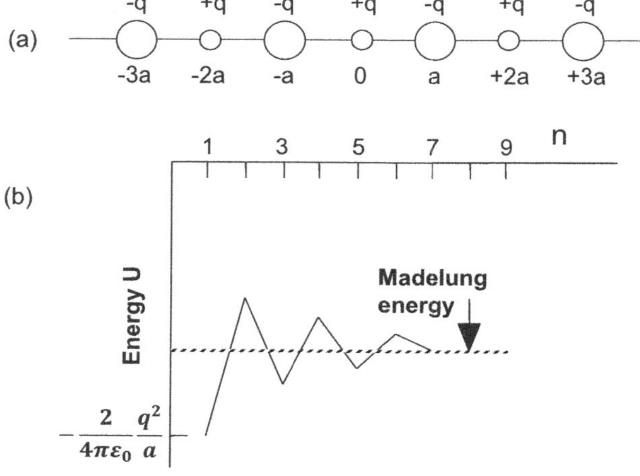

FIGURE 4.10 1D finite chain of two kinds of ions, $+q$ and $-q$.

100 Å with increasing the crystalline size during the annealing process. This crystalline size is in the same range as the above theoretically estimated crystal size. The reader is requested to extend the theory to two- and three-dimensional models.

The second explanation is from the local field strength viewpoint. As we discussed in Chapter 2, the local electric field is given by $(\gamma/3\varepsilon_0)P$, and the polarization is expressed by using the polarizability, α, as:

$$P = N\alpha\left(E + \frac{\gamma}{3\varepsilon_0}P\right) \quad \text{(P4.2.1)}$$

where N is atomic density, α polarizability, and γ the Lorentz factor. Thus, dielectric susceptibility is expressed as

$$\chi = P/E = N\alpha/\left(1 - \frac{\gamma}{3\varepsilon_0}N\alpha\right). \quad \text{(P4.2.2)}$$

The appearance of ferroelectricity requires $\chi \to \infty$ (at a certain temperature), and a large Lorentz factor is essential, like $\gamma = 10$ in $BaTiO_3$ perovskite crystal. From this consideration, a pure amorphous, that is, isotropic, material should exhibit $\gamma = 1$ theoretically, leading to a conclusion that it is difficult for amorphous material to be ferroelectric.[27]

4.3 ACTUATOR/DEVICE DESIGN

A classification of electromechanical ceramic actuators based on structure type is presented in Figure 4.11. Simple devices directly use longitudinally or transversely induced strain. The simple

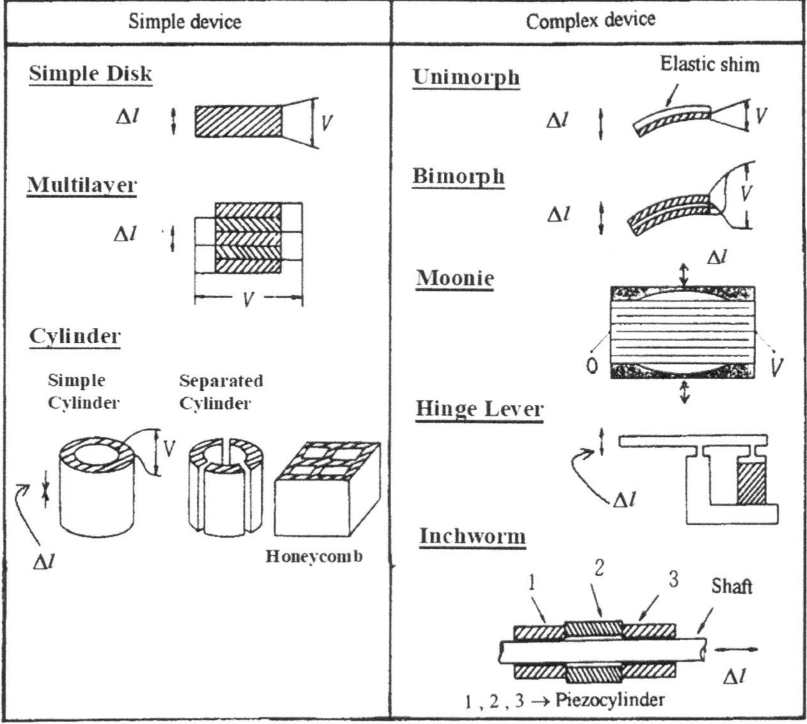

FIGURE 4.11 A classification of electromechanical ceramic actuators based on structure and displacement mechanism.

disk and multilayer types make use of the longitudinal strain (d_{33}), and the cylinder types (the simple cylinder, separated cylinder, and honeycomb designs) utilize the transverse strain (d_{31}). Complex devices do not use the induced strain directly, but rather a magnified displacement, produced through a spatial magnification mechanism (demonstrated by the unimorph, bimorph, moonie, and hinge lever designs) or through a sequential drive mechanism (inchworm). Among the designs shown in Figure 4.11, the multilayer and bimorph types are the most commonly used structures. Although the multilayer type produces only relatively modest displacements (10 μm), it offers a respectable generative force (1 kN), a quick response speed (10 μsec), long lifetime (10^{11} cycles), and high electromechanical coupling factor, k_{33} (70%). The bimorph type provides large displacements (300 μm), but can only offer a relatively low generative force (1 N), a much slower response speed (1 msec), a shorter lifetime (10^8 cycles), and a rather low electromechanical coupling factor, k_{eff} (10%). We will examine each structure more closely in the sections that follow.

4.3.1 Disk Actuators

Single-disk devices are generally not commercially popular because of their low displacement/voltage. They are still considered useful in the laboratory for preliminary tests, however, and find their widest application in this setting.

EXAMPLE PROBLEM 4.3

The apparent dielectric constant of a 1-mm barium titanate–based ceramic disk is measured and found to be 500. Due to a fabrication error, however, a thin air gap of 0.5 μm was discovered between the ceramic and the electrodes that extends over most of the electroded area on both sides of the device. Estimate what the actual dielectric constant of the ceramic should be without this defect.

Solution

The measured capacitance in this case will be the total capacitance of the ceramic and the two air gaps effectively arranged in series within the device. If we denote the capacitor area, thickness, and air gap by S, d, and δ, respectively, the total capacitance is given by:

$$\frac{1}{C} = \frac{1}{(\varepsilon_o KS/d)} + \frac{2}{(\varepsilon_o S/\delta)} = \left(\frac{1}{\varepsilon_o S}\right)\left[\frac{d}{K} + 2\delta\right]. \tag{P4.3.1}$$

The apparent dielectric constant was calculated from the measured capacitance according to the relationship:

$$\frac{Cd}{\varepsilon_o S} = 500 \tag{P4.3.2}$$

This expression combined with Equation P4.3.1 allows us to write:

$$\frac{1}{K} + \frac{2\delta}{d} = \frac{1}{500}. \tag{P4.3.3}$$

Substituting $d = 10^{-3}$ m, $\delta = 0.5 \times 10^{-6}$ m into this equation, we determine that the dielectric constant of the defect-free device should be $K = 1000$.

This kind of fabrication flaw tends to occur when alcohol is used to clean the ceramic disk after polishing and it is not dried completely on a hot plate (above 100°C). This demonstrates well why it is important to take care not to generate even a submicron air gap during the electroding process.

4.3.2 Multilayer Actuators

Ferroelectric ceramic multilayer devices have been investigated intensively for capacitor and actuator applications, because they have low driving voltages and they are highly suitable for miniaturization and integration onto hybrid structures. Miniaturization and hybridization are key concepts in the development of modern micromechatronic systems. The layer thickness of multilayer actuators ranges around 30–100 μm, which is much thicker than the current standard 3 μm for multilayer capacitors. Because the Ag-Pd electrode is mainly used for actuators to suppress Ag metal migration into the ceramic, very thin layer devices are too costly from the practical commercialization viewpoint. A multilayer structure typically exhibits a field-induced strain of 0.1% along its length, L (e.g., a 1-cm sample will exhibit a 10-μm displacement), and has a fundamental resonance frequency given by:

$$f_r = \frac{1}{2L\sqrt{\rho s_{33}^D}} \tag{4.4}$$

where ρ is the density and s_{33}^D is the elastic compliance (e.g., a 1-cm sample will have a 100–150 kHz resonance frequency). New multilayer configurations or heterostructures composed of electromechanical materials and modified electrode patterns are anticipated that will be incorporated into ever more sophisticated smart systems. There are two general methods for fabricating multilayer ceramic devices: (1) the *cut-and-bond method* and (2) the *tape-casting method.*

A schematic diagram of a commercially manufactured piezopile from NTK-NGK is shown in Figure 4.12a.[28] It is composed of 100 cut and polished PZT ceramic discs, each 1 mm thick, which are stacked and interleaved with metal foils that serve as electrodes within the device. A 100-μm displacement is generated by the pile with an applied voltage of 1.6 kV, and the maximum generative force is about 3 tons. Since the entire device is clamped by bolts and mechanically biased, delamination and mechanical fracture do not readily occur. Although devices such as this offer substantial displacements and generative forces, the cut-and-bond method in general has its disadvantages. One major drawback is the labor-intensive process itself, which is not at all well suited to mass production. The devices also tend to require rather high drive voltages because the minimum layer thickness possible for stacks prepared in this way is only about 0.2 mm.

The tape-casting method, in which ceramic green sheets with printed electrodes are laminated and *co-fired* with compatible internal electrodes, is far more conducive to mass production and produces devices with much thinner layers so that low drive voltages may be employed.

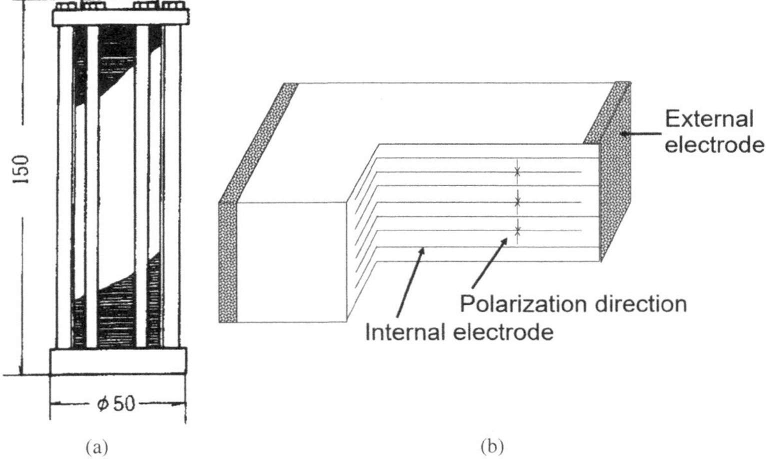

FIGURE 4.12 (a) Schematic diagram of a piezopile manufactured by NTK-NGK. (From S. Yamashita: *Jpn. J. Appl. Phys.*, 20(Suppl. 20-4), 93, 1981.) (b) The structure of a multilayer actuator.

The multilayer structure is essentially composed of alternating ferroelectric and conducting layers, which are co-fired to produce a dense, cohesive unit as shown in Figure 4.12b. A ferroelectric layer sandwiched between a pair of electrodes constitutes a single displacement element. Hundreds of these units may be connected in parallel to the potential difference supplied by the external electrodes, which are connected to the many interleaved internal electrodes of the stack, as shown in Figure 4.12b. A flowchart for the manufacturing process is shown in Figure 4.13. *Green sheets* are prepared in two steps. First, the ceramic powder is combined with an appropriate liquid solution to form a *slip*. The slip mixture generally includes the ceramic powder and a liquid composed of a *solvent*, a *deflocculant*, a *binder*, and a *plasticizer*. During the second part of the process, the slip is cast into a film under a special straight blade, called a *doctor blade*, whose distance above the carrier determines the film thickness. Once dried, the resulting film, called a green sheet, has the elastic flexibility of synthetic leather. The volume fraction of the ceramic in the now-polymerized matrix at this point is about 50%. The green sheet is cut into an appropriate size, and internal electrodes are printed using silver, palladium, or platinum ink. Several tens to hundreds of these layers are then laminated and pressed using a hot press. After the stacks are cut into small chips, the green bodies are sintered at around 1200°C in a furnace, with special care taken to control the initial binder evaporation at 500°C. The sintered chips are polished and externally electroded, lead wires are attached, and finally the chips are coated with a waterproof spray.

A cross-sectional view of a conventional *interdigital electrode* configuration is shown in Figure 4.14a.[29] The area of the internal electrode is slightly smaller than the cross-sectional area of the device. Notice that every two layers of the internal electrodes extend to one side of the device and connect with the external electrode on that side so that all active layers of the device are effectively connected in parallel. The small segments in each layer that are not addressed by the internal electrodes remain inactive, thereby restricting the overall generative displacement and leading to detrimental stress concentrations in the device. A multilayer structure is represented in Figure 4.15a, and the strain distribution measured in a test device is shown in Figure 4.15b. The derivative of the displacement distribution provides an estimate of the stress concentration in the device.[30] The internal stress distribution was also predicted using the finite-element method. The results of this analysis are summarized in Figure 4.16. The maximum tensile and compressive stresses are 1×10^8 and 1.2×10^8 N/m^2, respectively, which are very close to the critical strength of the ceramic.[30]

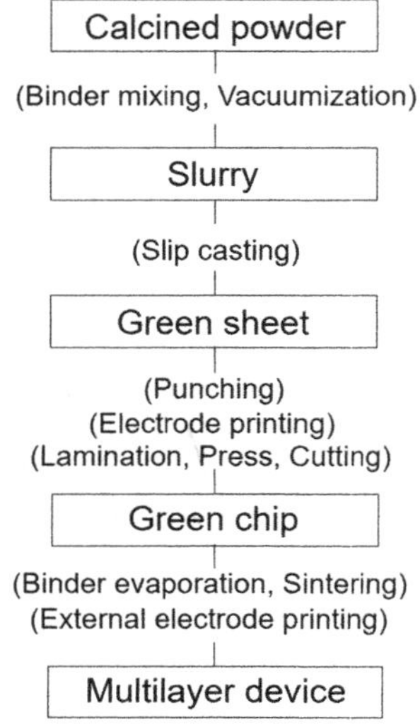

FIGURE 4.13 A flowchart for the fabrication of a multilayer ceramic actuator.

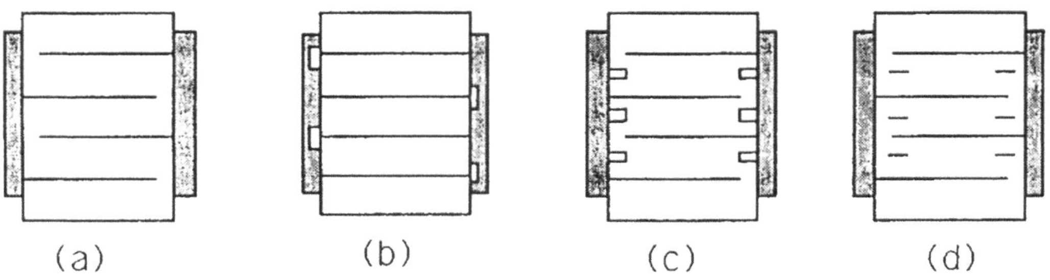

FIGURE 4.14 Various internal electrode configurations for multilayer actuators: (a) interdigital, (b) plate-through, (c) slit-insert, and (d) interdigital with float electrode.

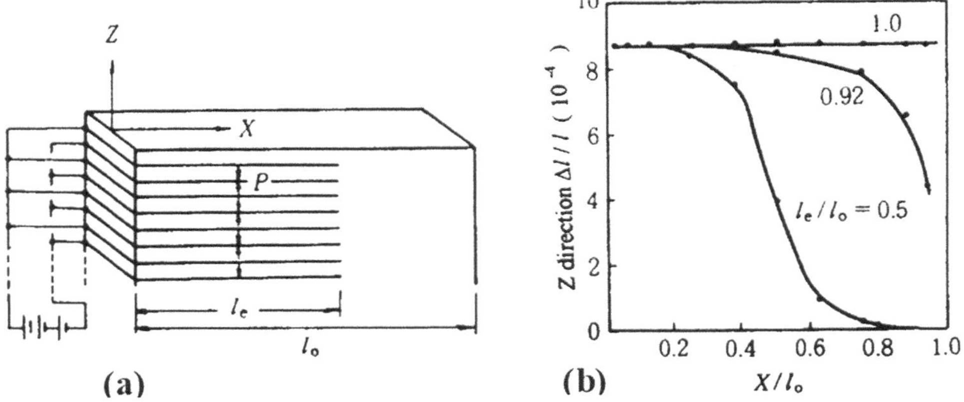

FIGURE 4.15 A test multilayer structure: (a) a schematic depiction and (b) the induced strain distribution measured in a test device. (From S. Takahashi et al.: *Jpn. J. Appl. Phys.*, 22[Suppl. 22-2], 157, 1983.)

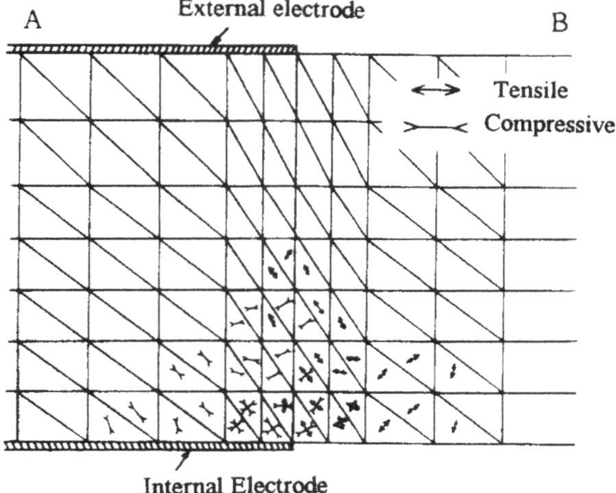

FIGURE 4.16 Internal stress distribution for a multilayer actuator as predicted by finite-element analysis. (From S. Takahashi et al.: *Jpn. J. Appl. Phys.*, 22[Suppl. 22-2], 157, 1983.)

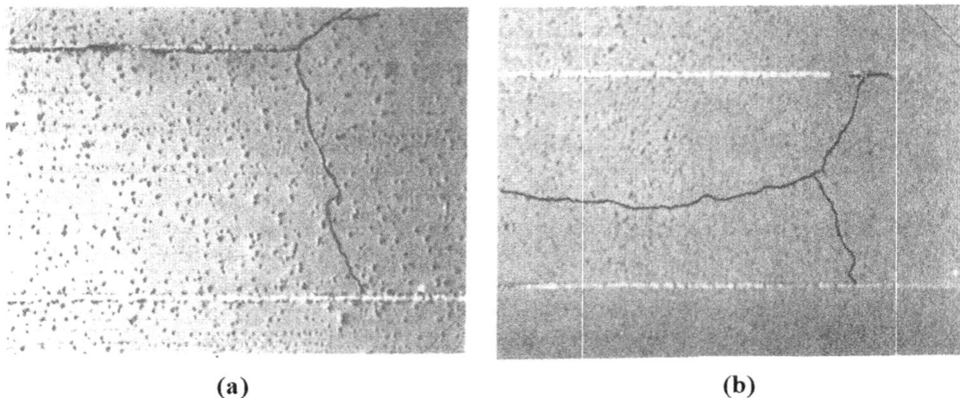

FIGURE 4.17 Crack generation in multilayer ceramic actuators under bipolar drive: (a) piezoelectric PNNZT, and (b) antiferroelectric PNZST. (From A. Furuta and K. Uchino: *J. Amer. Ceram. Soc.*, 76, 1615, 1993.)

Crack propagation has been investigated in a variety of multilayer systems.[31] A crack pattern commonly observed in a piezoelectric soft PZT, PNNZT ($Pb(Ni_{1/3}Nb_{2/3})O_3 - PZT$) actuators under bipolar drive is shown in Figure 4.17a. It occurs much as predicted in the theoretical treatment of these systems.[30] The crack originates at the edge of an internal electrode and propagates toward the adjacent electrode. Delamination between the electrode and the ceramic occurs simultaneously, leading to a lying Y-shaped crack. An interesting difference was observed in the case of the antiferroelectric PNZST system[31] shown in Figure 4.17b. Here a Y-shaped crack is again produced, but it originates in the ceramic between the electrodes.[32] This is probably due to the combination of two distinct induced strains in the ceramic: the anisotropic piezoelectric strain and the more isotropic strain (i.e., negative Poisson's ratio!) associated with the antiferroelectric response to the applied field.

A modified electrode configuration called a plate-through design (see Figure 4.14b) was developed by NEC as a solution to this particular mechanical problem.[33] The electrode in this modified configuration extends over the entire surface of the ceramic so that the stress concentration cannot develop. This modification requires that an insulating tab terminate every two electrode layers on the sides of the device where the external electrodes are painted. Key issues in producing reliable devices with this alternative electrode design are concerned with the precise application of the insulating terminations and improvement of the adhesion between the ceramic and internal electrode layers. The designers of NEC addressed the first of these issues by developing an electrophoretic technique for applying glass terminators to the device. The problem of adhesion was resolved by making use of a special electrode paste containing powders of both the Ag-Pd electrode material and the ceramic phase. The displacement curve for a (0.65)PMN-(0.35)PT multilayer actuator with 99 layers of 100-μm-thick sheets ($2 \times 3 \times 10$ mm^3) is shown in Figure 4.18a.[34] We see from these data that a 8.7-μm displacement is generated by an applied voltage of 100 V, accompanied by a slight hysteresis. This curve is consistent with the typical response of a disk device. The transient response of the induced displacement after the application of a rectangular voltage is shown in Figure 4.18b. Rising and falling responses as quick as 10 μsec are observed.

The *slit-insert design* and the *interdigital with float electrode* are shown in Figure 4.14c and d, respectively. The induced stress concentration is relieved in the slit-insert design, while electric field concentration is avoided with the interdigital with float electrode configuration.[35]

Another common crack pattern is the vertical crack, which occasionally occurs in the layer just adjacent to the top or bottom inactive layer. The inactive layer serves as an interface between the device and the object to which the actuator is attached. A crack occurring in the layer adjacent to the inactive bottom layer of a multilayer device is pictured in Figure 4.19. It originates from a transverse tensile stress produced in this layer due to clamping from the adjacent thick inactive

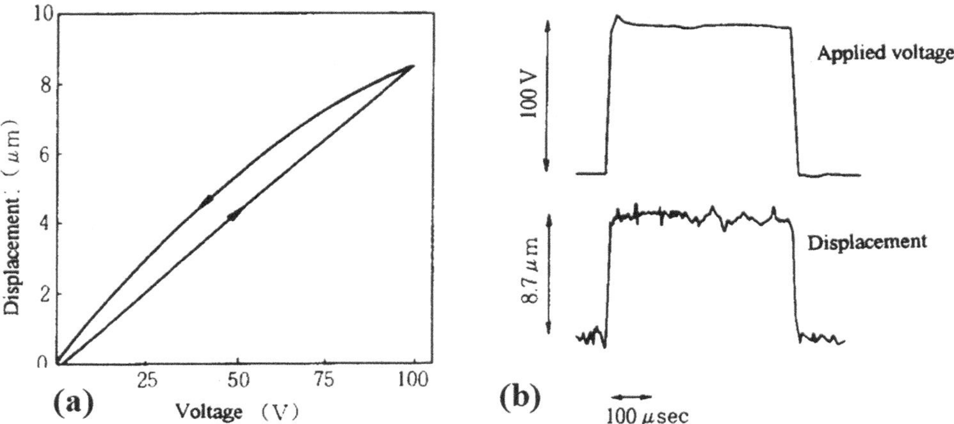

FIGURE 4.18 Response of a (0.65)PMN-(0.35)PT multilayer actuator: (a) the displacement as a function of applied voltage and (b) the displacement response to a step voltage. (From S. Takahashi: *Sensor Technol.*, 3(12), 31, 1983.)

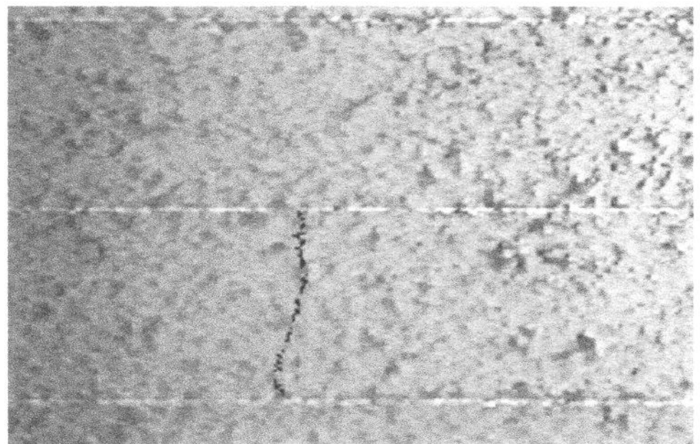

FIGURE 4.19 A vertical crack observed in the layer just adjacent to the bottom inactive layer in a multilayer actuator.

layer. One possible solution that has been proposed for this problem is to pre-pole the top and bottom layers of the device in order not to generate large residual stress remaining during the initial poling process.

Vertical cracks are sometimes initiated even inside multilayer devices, when the internal electrode metal particle generates a wedge-shaped agglomerate on the electrode layer, which enhances the electric field concentration around this defect, leading to crack initiation.

EXAMPLE PROBLEM 4.4

A piezoelectric multilayer actuator under a certain applied voltage, V, exhibits an amplified displacement as compared with the displacement generated in a single disk of the active material. An even more pronounced displacement amplification is expected from an electrostrictive device. Verify this situation, using the following equations for the piezoelectric and electrostriction effects: $x = d\,E$ and $x = M\,E^2$.

Solution

Let l, Δl, and n be the total thickness, displacement, and number of ceramic layers, respectively. The strain, x, is just equal to $\Delta l/l$. The displacement produced by each type of device is given by the following equations.

a. The Piezoelectric Actuator:

$$\Delta l = l \times d\,E = l\,d[V/(l/n)] = n\,d\,V \qquad \text{(P4.4.1)}$$

We see from this equation that the displacement is amplified in proportion to the number of layers, n.

b. The Electrostrictive Actuator:

$$\Delta l = lx = l\,M\,E^2 = l\,M[V/(l/n)]^2 = n^2(M/l)V^2 \qquad \text{(P4.4.2)}$$

We see in this case that the displacement is amplified in proportion to the square of the number of layers, n. The electrostrictive device is thus more effective than the piezoelectric one.

A multilayer actuator incorporating a new interdigital internal electrode configuration has been developed by Tokin.[36] In contrast to devices with the conventional interdigital electrode configuration, for the modified design, line electrodes are printed on piezoelectric green sheets, which are stacked so that alternate electrode lines are displaced by one-half pitch. This actuator produces displacements perpendicular to the stacking direction (i.e., in-plane extension) through the longitudinal piezoelectric effect. Long ceramic actuators up to 74 mm in length have been manufactured, which can generate longitudinal displacements up to 55 μm.

A three-dimensional positioning actuator with a stacked structure has been proposed by PI Ceramic, in which both transverse and shear strains are induced to generate displacements.[37] As shown in Figure 4.20a, this actuator consists of three parts: the top 10-mm-long Z-stack generates the displacement along the z direction, while the second and bottom 10-mm-long X and Y stacks provide the x and y displacements through shear deformation, as illustrated in Figure 4.20b. The device can produce 10-μm displacements in all three directions when 500 V is applied to the 1-mm-thick layers.

Various failure detection techniques have been proposed to implement in smart actuator devices to essentially monitor their own "health."[38] One such "intelligent" actuator system that utilizes acoustic emission detection is shown in Figure 4.21. The actuator is controlled by two feedback

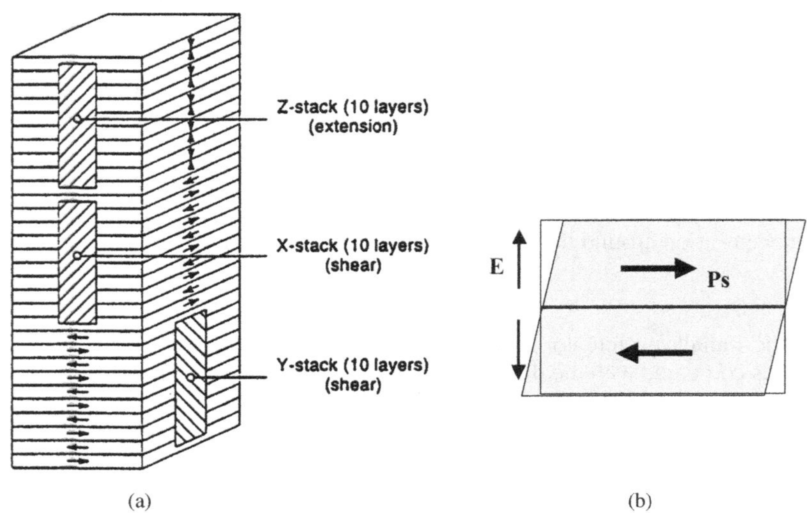

FIGURE 4.20 A 3D positioning actuator with a stacked structure proposed by PI Ceramic: (a) a schematic diagram of the structure and (b) an illustration of the shear deformation. (From A. Banner and F. Moller: *Proc. 4th. Int'l Conf. New Actuators, AXON Tech. Consult. GmbH*, p. 128, 1995.)

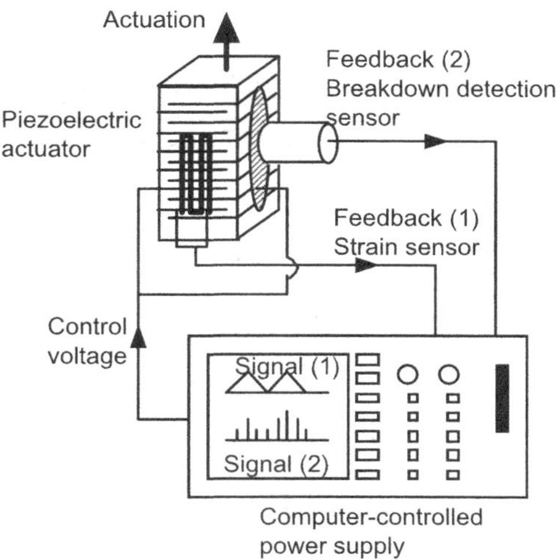

FIGURE 4.21 An intelligent actuator system with both position and breakdown detection feedback mechanisms.

mechanisms: position feedback, which can compensate for positional drift and hysteresis, and breakdown detection feedback, which can shut down the actuator system safely in the event of an imminent failure. As previously described in Section 3.4.3.3, acoustic emission from a piezoelectric actuator driven by a cyclic electric field is a good indicator of mechanical failure. The emissions are most pronounced when a crack propagates in the ceramic at the maximum speed. A portion of this smart piezoelectric actuator is therefore dedicated to sensing and responding to acoustic emissions. The AE rate in a piezoelectric device can increase by three orders of magnitude just prior to complete failure. During the operation of a typical multilayer piezoelectric actuator, the AE sensing portion of the device will monitor the emissions and respond to any dramatic increase in the emission rate by initiating a complete shutdown of the system.

Another development for device failure self-monitoring is based on a strain gauge–type internal electrode configuration, as pictured in Figure 4.22a.[39] Both the electric field-induced strain and the occurrence of cracks in the ceramic can be detected by closely monitoring the resistance of a strain gauge–shaped electrode embedded in a ceramic actuator. The resistance of such a smart device is plotted as a function of applied electric field in Figure 4.22b and c. The field-induced strain of a "healthy" status is represented by the series of curves depicted in Figure 4.22b. Each curve corresponds to a distinct number of drive cycles. A sudden decrease in the resistance, as shown in Figure 4.22c, is a typical symptom of device failure.

The effect of aging is manifested clearly by a gradual increase in the resistance with the number of drive cycles, as shown for a smart device in Figure 4.22b. Ceramic aging is an extremely important factor to consider in the design of a reliable actuator device, although there have been relatively few investigations done to better understand and control it. Aging is associated with two types of degradation: (1) depoling and (2) mechanical failure. Creep and zero-point drift in the actuator displacement are caused by depoling of the ceramic. The strain response is also seriously impaired when the device is operated under conditions of very high electric field, elevated temperature, high humidity, and high mechanical stress. The lifetime of a multilayer piezoelectric actuator operated under a DC bias voltage can be described by the empirical relationship:

$$t_{DC} = AE^{-n} \exp(W_{DC}/kT) \tag{4.5}$$

where W_{DC} is an activation energy ranging from 0.99–1.04 eV.[40]

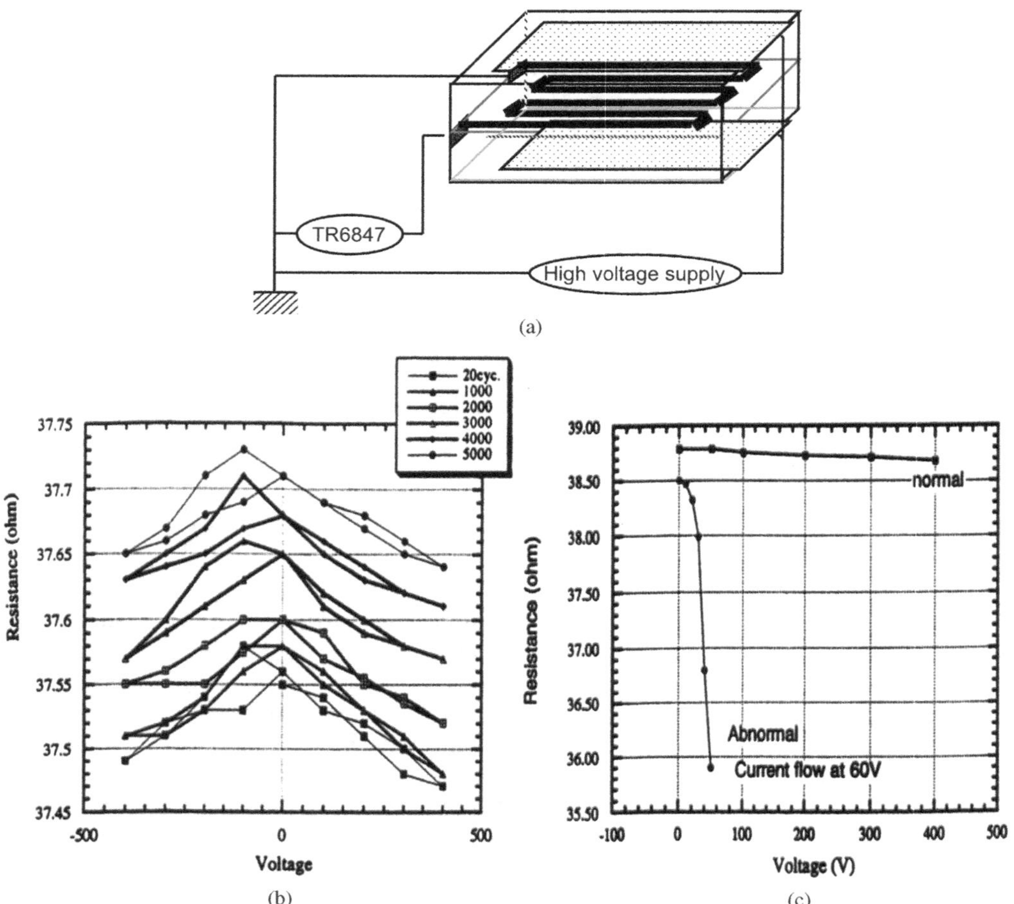

FIGURE 4.22 (a) Strain gauge configuration of the internal electrode for an intelligent health monitoring actuator. (b) Resistance as a function of applied electric field for a smart actuator: response of a healthy device and (c) the response of a failing device. (From H. Aburatani and K. Uchino: *Amer. Ceram. Soc. Annual Mtg. Proc., SXIX-37-96*, Indianapolis, April, 1996.)

Investigations have been conducted on the heat generation from multilayer piezoelectric ceramic actuators of various sizes.[41] The temperature rise, ΔT, monitored in actuators driven at 3 kV/mm and 300 Hz is plotted as a function of the quantity, V_e/A, in Figure 4.23, where V_e is the effective actuator volume (corresponding to the electroded portion of the device) and A is its surface area. The linear relationship observed is expected for a ratio of device volume, V_e, to its surface area, A. We see from this trend that a configuration with a small V_e/A will be the most conducive to suppressing heating within the device. Flat and cylindrical shapes, for example, are preferable over cube and solid rod structures, respectively.

4.3.3 Cylindrical Devices

In general, the longitudinally induced strain (parallel to the electric field) exhibits less hysteresis than the strain transversely induced in an identical piezoelectric ceramic. The magnitude of the strain induced in either mode, however, does depend on the device configuration. Two cylindrical PZT-5 actuators used for controlling the optical path length in an optical interferometer are illustrated in Figure 4.24.[42] Although both designs utilize the displacement along the cylinder axis, the simple

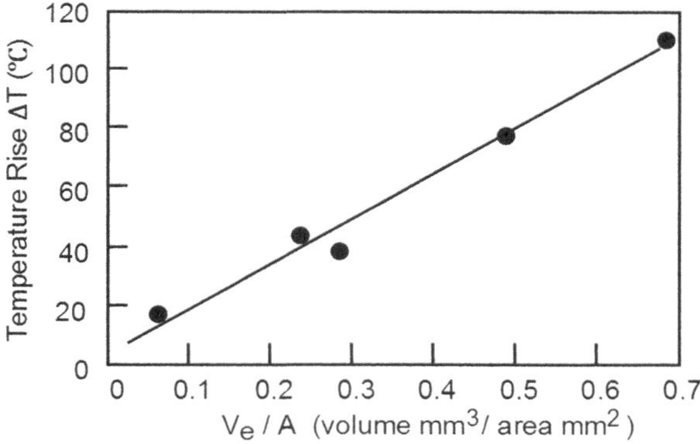

FIGURE 4.23 Temperature rise, ΔT, in actuators driven at 3 kV/mm and 300 Hz, plotted as a function of V_e/A (V_e: effective actuator volume; A: surface area). (From J. Zheng et al.: *J. Amer. Ceram. Soc.*, 79, 3193, 1996.)

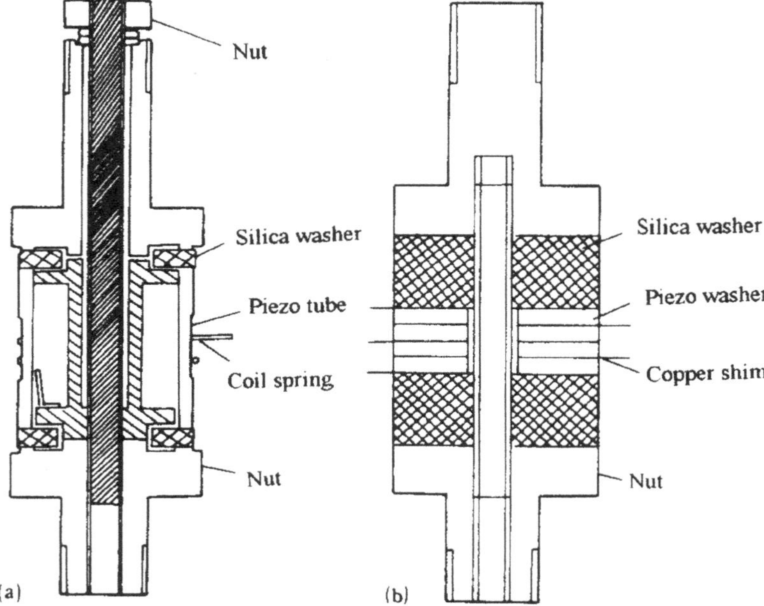

FIGURE 4.24 Two cylindrical PZT actuators used for controlling the optical path length in optical interferometers: (a) a simple cylinder type operated through the transverse piezoelectric effect and (b) a multilayer cylinder design operated through the longitudinal effect. (From R. W. Basedow and T. D. Cocks: *J. Phys. E: Sci. Inst.*, 13, 840, 1980.)

cylinder type pictured in Figure 4.24a operates through the transverse effect (d_{31}), while the multilayer cylinder design pictured in Figure 4.24b operates through the longitudinal effect (d_{33}). The simple cylinder structure tends to exhibit lower hysteresis than the multilayer structure.

4.3.4 Unimorph/Bimorph/Monomorph

4.3.4.1 Unimorph/Bimorph

Unimorph and *bimorph* devices are simple structures composed of piezoceramic and inactive elastic plates bonded surface to surface. Unimorph devices have one piezoplate, and bimorph structures have two ceramic plates bonded onto an elastic shim.

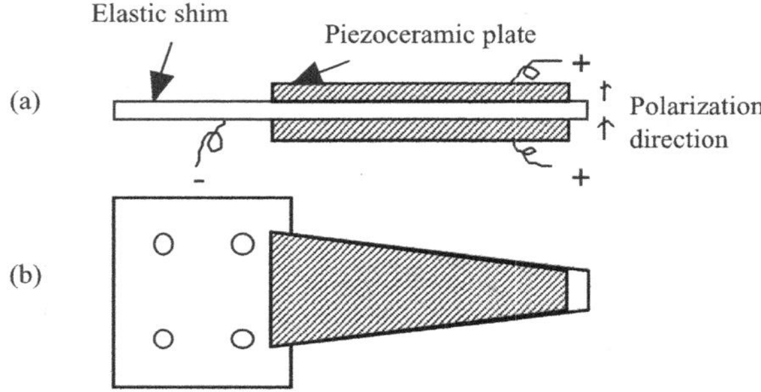

FIGURE 4.25 Basic structure of a piezoelectric bimorph: (a) side view of the device and (b) top view.

The bending deformation in a bimorph occurs because the two piezoelectric plates are bonded together and each plate produces its own extension or contraction under the applied electric field. This effect is also employed in piezoelectric speakers. The induced voltage associated with the bending deformation of a bimorph is used in accelerometers. This is a very popular and widely used structure mainly because it is easily fabricated (the two ceramic plates are just bonded with an appropriate epoxy resin) and the devices readily produce a large displacement. The drawbacks of this design include a low response speed (1 kHz) and low generative force due to the bending mode. A metallic sheet shim is occasionally used between two piezoceramic plates to increase the reliability of the bimorph structure, as illustrated in Figure 4.25. When this type of shim is used, the structure will maintain its integrity even if the ceramic fractures. The bimorph is also generally tapered in order to increase the response frequency while maintaining optimum tip displacement. Anisotropic elastic shims, made from such materials as oriented carbon fiber reinforced plastics, have been used to enhance the displacement magnification rate by a factor of 1.5 as compared to the displacement of a similar device with an isotropic shim.[43]

There have been many studies conducted on these devices, which have produced equations describing the tip displacement and the resonance frequency. Two shimless bimorph designs are illustrated in Figure 4.26. Two poled piezoceramic plates of equal thickness and length are bonded together with their polarization directions either opposing or parallel to each other. When the devices are operated under an applied voltage, V, with one end clamped (the *cantilever condition*), the tip displacement, δ, is given by:

$$\delta = (3/2)d_{31}(l^2/t^2)V \tag{4.6a}$$

$$\delta = 3d_{31}(l^2/t^2)V \tag{4.6b}$$

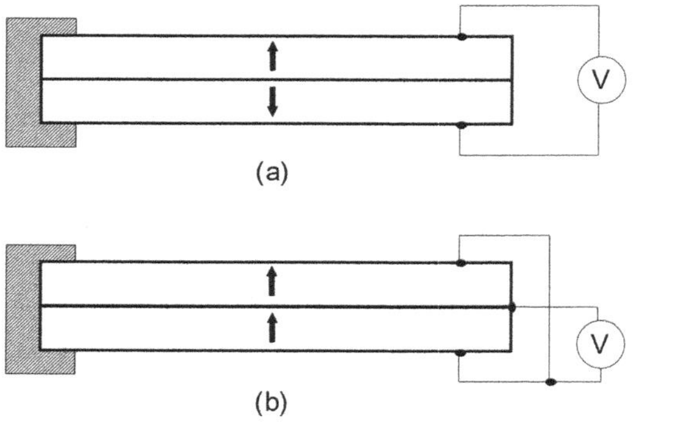

FIGURE 4.26 Two types of piezoelectric bimorphs: (a) the antiparallel polarization type and (b) the parallel polarization type.

where d_{31} is the piezoelectric strain coefficient of the ceramic, t is the combined thickness of the two ceramic plates, and l is the length of the bimorph. Equation 4.6a applies to the antiparallel polarization condition and Equation 4.6b to the parallel polarization condition. Notice that the difference between the two cases arises from the difference in electrode gap. The separation between the electrodes is equal to the combined thickness of the two plates for the antiparallel polarization case (Figure 4.26a) and half that thickness for the parallel polarization case (Figure 4.21b). The merit of type (a) is a simpler manufacturing process, leading to cheaper cost (the lead-wire connection from the center electrode in type [b] is tricky). The fundamental resonance frequency in both cases is determined by the combined thickness of the two plates, t, according to the following equation:

$$f_o = 0.161\left[\frac{t}{l^2\sqrt{\rho s_{11}^E}}\right] \tag{4.7}$$

where ρ is the mass density of the ceramic, and s_{11}^E is its elastic compliance.[44]

EXAMPLE PROBLEM 4.5

Using a PZT-based ceramic with a piezoelectric strain coefficient of $d_{31} = -300$ pC/N, design a shimless bimorph with a total length of 30 mm (where 5 mm is used for cantilever clamping) that can produce a tip displacement of 40 μm under an applied voltage of 20 V. Calculate the response speed of this bimorph. The mass density of the ceramic is $\rho = 7.9$ g/cm³ and its elastic compliance is $s_{11}^E = 16\times10^{-12}\, m^2/N$.

SOLUTION

When the device is to be operated under low voltages, the parallel polarization type of device pictured in Figure 4.26b is preferred over the antiparallel polarization type pictured in Figure 4.26a because it produces a two times larger displacement under these conditions. Substituting a length of $l = 25$ mm into Equation 4.6b, we obtain the combined piezoelectric plate thickness, t.

$$\begin{aligned} t &= l\sqrt{3(d_{31}V/\delta)} \\ &= 25\times10^{-3}(\text{m})\sqrt{3[300\times10^{-12}(C/N)\cdot 20(V)/40\times10^{-6}(\text{m})]} \end{aligned} \tag{P4.5.1}$$

$$\rightarrow\rightarrow t = 530\,\mu\text{m}.$$

The piezoceramic is sliced into plates 265 μm in thickness, 30 mm in length, and 4–6 mm in width. The plates are electroded and poled and then bonded together in pairs. The width of the bimorph is usually chosen such that $[w/l < 1/5]$ in order to optimize the magnitude of the bending displacement.

The response time is estimated by the resonance period. We can determine the fundamental resonance frequency of the structure from the following equation:

$$f_o = 0.161\left[\frac{t}{l^2\sqrt{\rho s_{11}^E}}\right] \tag{P4.5.2}$$

$$= 0.161\left[\frac{530\times10^{-6}\,\text{m}}{[25\times10^{-3}\,\text{m}]^2\sqrt{7.9\times10^3\,\text{kg/m}^3\,16\times0^{-12}\,\text{m}^2/\text{N}}}\right] = 378\text{Hz}$$

$$\rightarrow\rightarrow \text{Response Time} \approx \frac{1}{f_o} = \frac{1}{378\,s^{-1}} = 2.6\,\text{ms}$$

This bimorph was used for the camera shutter in practice in the 1980s.

EXAMPLE PROBLEM 4.6

A unimorph bending actuator can be fabricated by bonding a piezoceramic plate to a metallic shim.[45] The tip deflection, δ, of the unimorph supported in a cantilever configuration is given by:

$$\delta = \frac{d_{31}El^2Y_ct_c}{\left(Y_m[t_o^2-(t_o-t_m)^2]+Y_c[(t_o+t_c)^2-t_o^2]\right)} \quad \text{(P4.6.1)}$$

Here, E is the electric field applied to the piezoelectric ceramic, d_{31} the piezoelectric strain coefficient, l the length of the unimorph, Y the Young's modulus for the ceramic or the metal, and t the thickness of each material. The subscripts c and m denote the ceramic and the metal, respectively. The quantity t_o is the distance between the *strain-free neutral plane* and the bonding surface, and is defined according to the following:

$$t_o = \frac{t_ct_m^2(3t_c+4t_m)Y_m+t_c^4Y_c}{6t_ct_m(t_c+t_m)Y_m} \quad \text{(P4.6.2)}$$

Assuming $Y_c=Y_m$, calculate the optimum (t_m/t_c) ratio that will maximize the deflection, δ, under the following conditions:

a. A fixed ceramic thickness (you are a purchaser of a PZT plate), t_c, and
b. A fixed total thickness (you need to keep the resonance frequency), t_c+t_m.

Solution

Setting $Y_c=Y_m$, Equations P4.6.1 and P.4.6.2 become:

$$\delta = \frac{d_{31}El^2t_c}{\left([t_o^2-(t_o-t_m)^2]+[(t_o+t_c)^2-t_o^2]\right)} \quad \text{(P4.6.3)}$$

$$t_o = \frac{t_ct_m^2(3t_c+4t_m)+t_c^4}{6t_ct_m(t_c+t_m)} \quad \text{(P4.6.4)}$$

Substituting t_o as it is expressed in Equations P4.6.4 into P4.6.3 yields

$$\delta = \frac{d_{31}El^2 3t_mt_c}{(t_m+t_c)^3} \quad \text{(P4.6.5)}$$

a. The function $f(t_m)=(t_mt_c)/(t_m+t_c)^3$ must be maximized for a fixed ceramic thickness, t_c.

$$\frac{df(t_m)}{dt_m} = \frac{(t_c-2t_m)t_c}{(t_m+t_c)^4} = 0 \quad \text{(P4.6.6a)}$$

The metal plate thickness should be $t_m=t_c/2$ and $t_o=t_c/2$.

b. Equation P4.6.6a becomes under a fixed total thickness, $t_{tot}=t_c+t_m$:

$$\frac{df(t_m)}{dt_m} = \frac{(t_{tot}-2t_m)}{t_{tot}^3} = 0 \quad \text{(P4.6.6b)}$$

Thus, it is determined that both the metal and ceramic plate thickness should be $t_m=t_c=t_{tot}/2$ and $t_o=t_{tot}/3$.

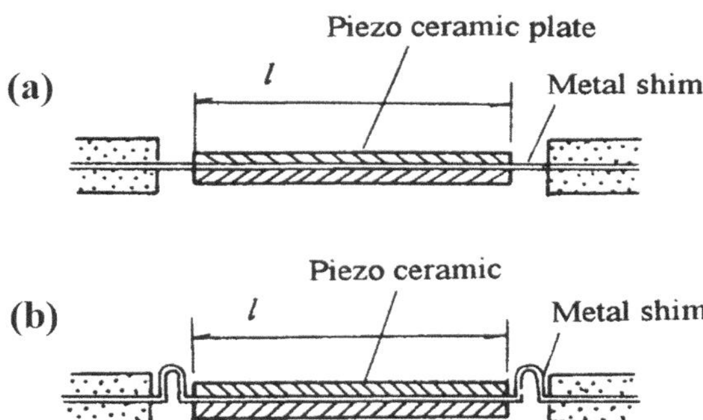

FIGURE 4.27 Double support mechanisms for a bimorph: (a) conventional design and (b) special shim design. (From C. Tanuma et al.: *Jpn. J. Appl. Phys.*, 22[Suppl. 22-2], 154, 1983.)

The generative displacement of a bimorph is decreased when it is supported at both ends, as shown in Figure 4.27a. When the special shim design pictured in Figure 4.27b is used, however, the center displacement of the bimorph is enhanced by a factor of 4 as compared to the displacement of a conventionally supported device.[46] This is because circled metal shim parts reduce the horizontal clamping force dramatically.

Bimorph displacement inevitably includes some rotational motion. A special mechanism generally must be employed to eliminate this rotational component of the induced motion. Several such mechanisms are pictured in Figure 4.28.

A twin structure developed by SONY is shown in Figure 4.28a.[47] It is employed for video tracking control. The design incorporates a flexible head support installed at the tips of two parallel bimorphs. The complex bimorph design pictured in Figure 4.28b has been proposed by Ampex. It has electrodes on the top and bottom of the device that are independently addressed such that any deviation from

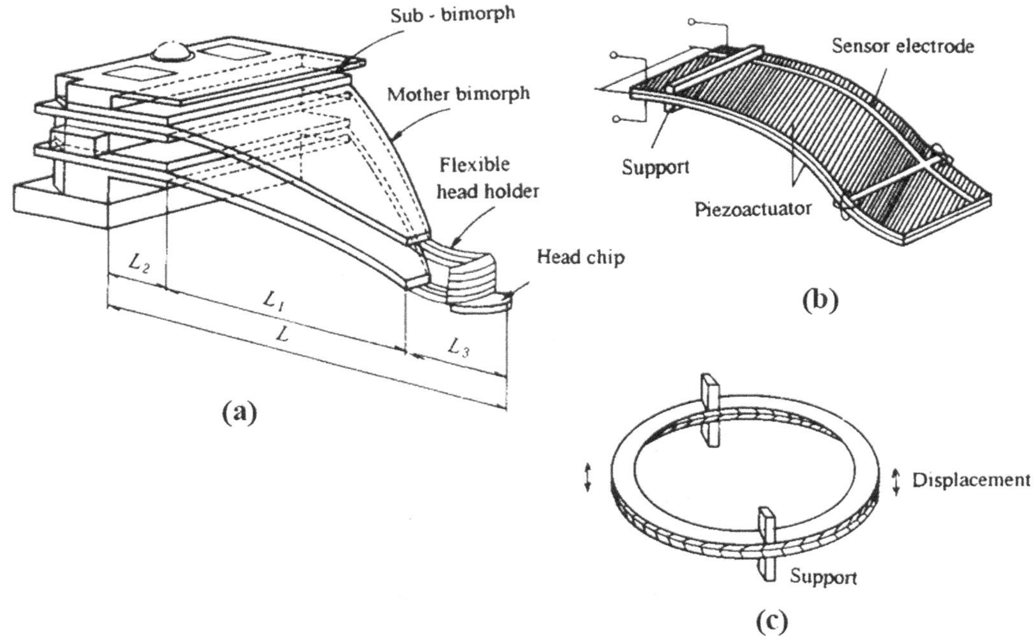

FIGURE 4.28 Bimorph structures designed to produce perfectly parallel motion: (a) parallel spring (twin), (b) complex bimorph (s-shaped), and (c) ring configurations.

perfectly parallel motion is compensated for and effectively eliminated. The proposed design also includes a sensor electrode that detects the voltage generated in proportion to the magnitude of the bending displacement. The ring-shaped bimorph design pictured in Figure 4.28c has been developed by Matsushita Electric for use in a video tracking control system. The design is compact and provides a large displacement normal to the ring. It also remains firmly supported in operation due to the dual-clamp configuration incorporated in the design.

The Thunder actuator developed by Face International is essentially a unimorph, but has a curved shape, and a tensile stress is maintained on the piezoceramic plate.[48] This actuator is very reliable when operated under both high field and high stress conditions. The device is fabricated at high temperatures, and its curved shape occurs due to the thermal expansion mismatch between the bonded PZT and metal plates, which creates significant uniaxial compressive stress on the PZT plate.

4.3.4.2 Monomorph/Rainbow

Conventional bimorph bending actuators are composed of two piezoelectric plates, or two piezoelectrics and an elastic shim, bonded together. The bonding layer, however, causes both an increase in hysteresis and a degradation of the displacement characteristics, as well as delamination problems. Furthermore, the fabrication process for such devices, which involves cutting, polishing, electroding, and bonding steps, is rather laborious and costly. Thus, a monolithic bending actuator (monomorph) that requires no bonding is a very attractive alternative structure.

Such a monomorph device has been produced from a single barium titanate ceramic plate.[49] The operating principle is based on the combined action of a semiconductor contact phenomenon and the piezoelectric or electrostrictive effect. When metal electrodes are applied to both surfaces of a semiconductor plate and a voltage is applied, as shown in Figure 4.29, the electric field is concentrated on one side, (i.e., a *Schottky barrier* is formed), thereby generating a nonuniform field within the plate. When the piezoelectric is slightly semiconducting, contraction along the surface occurs through the piezoelectric effect only on the side where the electric field is concentrated. The nonuniform field distribution generated in the ceramic causes an overall bending of the entire plate. The energy of a modified structure including a very thin insulating layer is represented in Figure 4.30a.[50] The thin insulating layer increases the breakdown voltage.

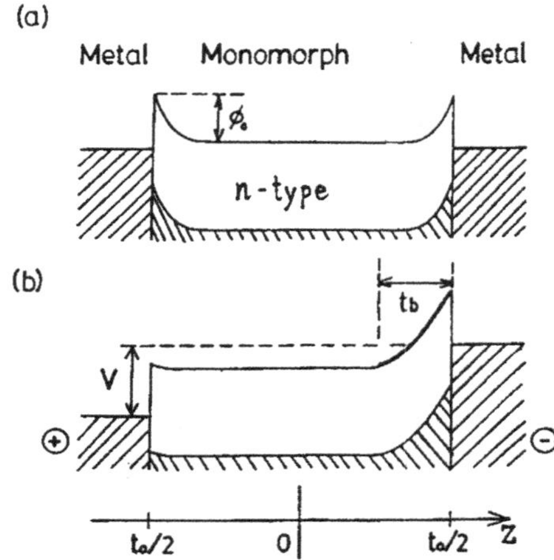

FIGURE 4.29 Schottky barrier generated at the interface between semiconductive (n-type) piezoceramic and metal electrodes: (a) before and (b) after the electric field is applied. (From K. Uchino et al.: *Japan. J. Appl. Phys.*, 26, 1046, 1987.)

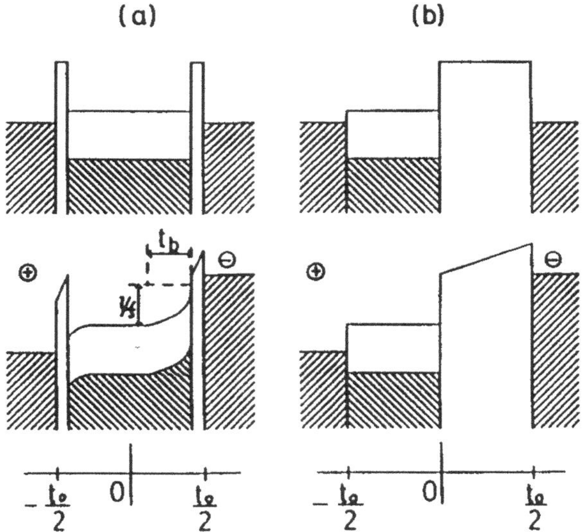

FIGURE 4.30 Energy diagrams for modified monomorph structures: (a) a device incorporating a very thin insulating layer and (b) the rainbow structure. (From K. Uchino et al.: *Jpn. J. Appl. Phys.*, 26[Suppl.26-2], 201, 1987.)

Research is underway to develop new compositions of doped barium titanate and lead zirconate titanate piezoelectric ceramics with the semiconducting properties required for use in these monomorph devices. Modified PZT ceramics prepared from a solid solution with the semiconductor perovskite compound $(K_{1/2}Bi_{1/2})ZrO_3$ (*n*-type) have the desired electrical properties. When 300 V is applied to a ceramic plate 20 mm long and 0.4 mm thick and fixed at one end, a maximum tip deflection of 200 μm can be obtained (see Figure 4.31a). This is comparable with the optimum deflection of a bimorph device.[51] Figure 4.31b shows opposite bending deformation due to the p-type BT-based piezoceramic. Note that the displacement diminishes dramatically when we adopt the Ohmic-like electrode InGa, which verifies that the bending is originated from the junction effect. The *rainbow actuator* by Aura Ceramics is a modification of the basic semiconducting piezoelectric monomorph design, where half of the piezoelectric plate is reduced so as to make

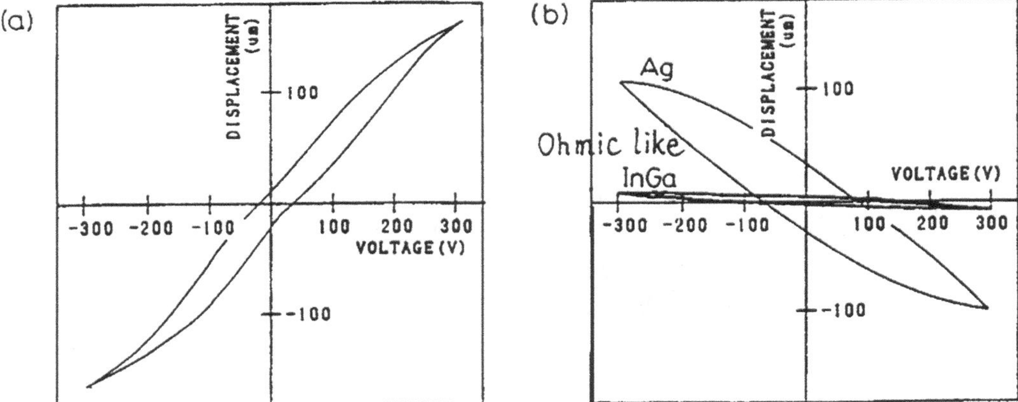

FIGURE 4.31 Tip displacement curves of (a) PZT (n-type) and (b) BT (p-type) monomorph structures. (From K. Uchino et al.: *Ferroelectrics*, 95, 161, 1989.)

a thick semiconducting electrode that enhances the bending action.[52] The energy diagram for the "rainbow" device is shown in Figure 4.30b.[50]

Another monomorph structure has been reportedly made from a $LiNbO_3$ single crystal plate, which has no bonding layer but instead uses an inversion polar layer generated in the plate.[53] The 500-μm-thick *z*-cut plates were cut from a single domain lithium niobate crystal. Then, the unelectroded samples were heat-treated at 1150°C for 5 h in flowing Ar gas containing water vapor. After rapid cooling at a rate of 50°C per minute, a two-layer domain configuration structure was created: the spontaneous polarization P_s in the original upward direction in the first layer and an opposing polarization originating from the positive side of the plate in the second. The thickness of the inversion layer depends strongly on the conditions of heat treatment, such as temperature, time, and atmosphere. As the annealing temperature or time increases, the inversion layer becomes thicker and finally reaches the median plane of the plate. After electroding both sides, this single crystal plate will execute a bending deformation with the application of a suitable electric field. Even though the magnitude of the displacement for this device is not as high as a PZT-based bimorph and the actuator is fragile due to crystal cleavage, it has the advantages of a highly linear displacement curve and zero hysteresis.

4.3.5 Flextension/Hinge Lever Amplification Mechanisms

4.3.5.1 Displacement Amplification Mechanism

Market research conducted by the author in the late 1990s suggested that the target specifications for actuators required by the market are:

a. Stroke = 100 μm
b. Force = 100 N
c. Response time = 100 μsec

Neither multilayer nor bimorph actuators can satisfy the above specifications directly. A multilayer needs a displacement amplification mechanism, while a bimorph is required to improve its force and response speed.

Hinge-lever and flextension-type displacement amplification mechanisms are adopted for multilayer actuators. The most famous hinge-type design is the one utilized in a dot-matrix printer developed by NEC, as depicted in Figure 4.32a.[54] Figure 4.32b shows a flextensional type developed by Cedrat.[55]

A *monolithic hinge structure* is made from a monolithic ceramic body by cutting indented regions in the monolith, as shown in Figure 4.33a. This effectively creates a lever mechanism that may function to either amplify or reduce the displacement. It was initially designed to reduce the

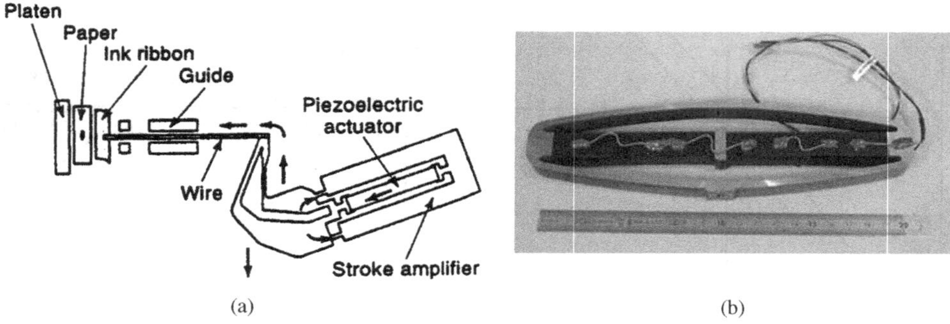

FIGURE 4.32 (a) A hinge-lever type displacement amplification mechanism (NEC) for a dot-matrix type printer head, and (b) a flextensional amplified piezoelectric actuator based on multilayer actuators (Cedrat).

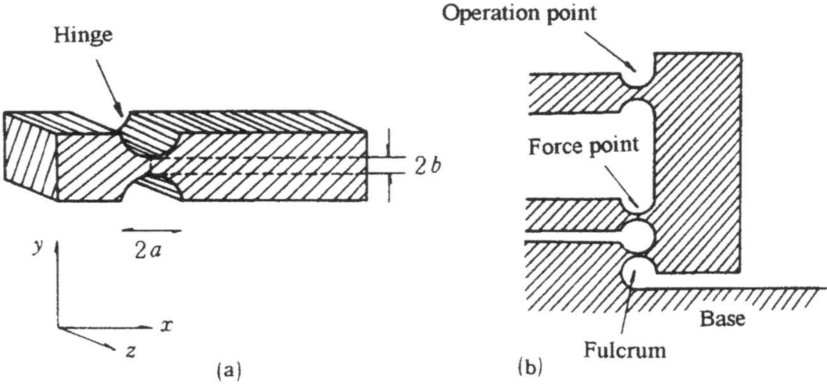

FIGURE 4.33 Monolithic hinge lever mechanisms: (a) monolithic hinge structure and (b) hinge lever mechanism.

displacement (including backlash) produced by a stepper motor and thus increase its positioning accuracy. However, monolithic hinge levers have been combined with piezoelectric actuators to amplify their displacement, or even on MEMS structures.

If the indented region of the hinge can be made sufficiently thin to promote optimum bending while maintaining extensional rigidity (ideal case!), a mechanical amplification factor for the lever mechanism close to the apparent geometric lever length ratio is expected. The actual amplification, however, is generally less than this ideal value. If we consider the dynamic response of the device, a somewhat larger hinge thickness (identified as $2b$ in Figure 4.33a) producing an actual amplification of approximately half the apparent geometric ratio is found to be optimum for achieving maximum generative force and response speed. The characteristic response (i.e., in terms of the displacement, generative force, and response speed) of an actuator incorporating a hinge lever mechanism is generally intermediate between that of multilayer and bimorph devices.

When a piezoelectric actuator is to be utilized in a large mechanical control system, the *oil pressure displacement amplification mechanism* shown in Figure 4.34a is often implemented, in particular in automobile applications. It is a mechanism especially well suited for oil pressure servo systems. The output displacement Δl_2 is effectively amplified by a Part 1 to Part 2 area ratio according to:

$$\Delta l_2 = (d_1/d_2)^2 \Delta l_1 \tag{4.8}$$

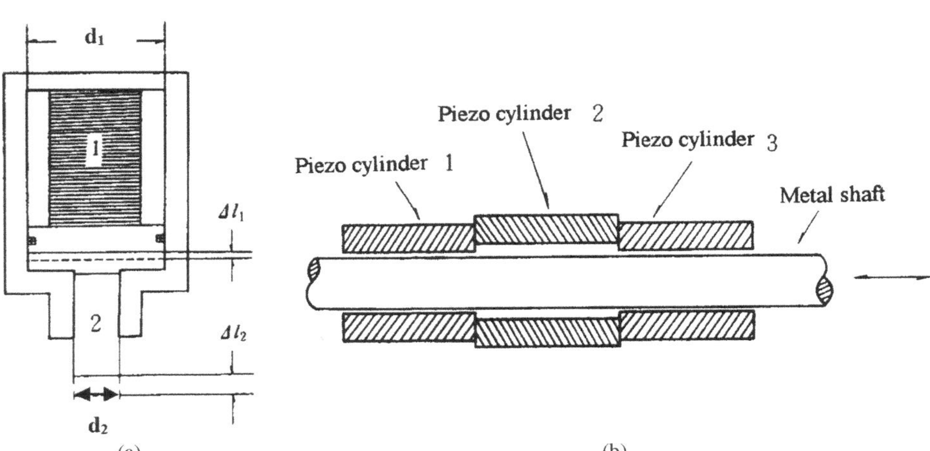

FIGURE 4.34 (a) Oil-pressure displacement amplification mechanism. (b) Inchworm structure.

Although the response is slow due to the viscosity of the oil, the overall loss associated with the amplified displacement is very small.

An *inchworm* is a linear motor that advances in small steps over time (i.e., displacement amplification in terms of time). An example of an inchworm device designed by Burleigh Instruments is pictured in Figure 4.34b.[56] It is composed of three piezoelectric tubes. Two of the tubes (labeled 1 and 3 in the figure) function to either clamp or release the metal shaft, while the remaining tube (labeled 2 in the figure) actually moves the inchworm along the shaft. When three separate drive voltages of differing phase are applied to the three tubes, the inchworm can translate forward and backward along the metal shaft. Although this motor is rather slow (0.2 mm/sec), the high resolution per step (1 nm) is a very attractive feature of the device.

4.3.5.2 Moonie/Cymbal

A composite actuator structure called the *moonie* or *cymbal* has been developed to amplify the pressure sensitivity and small displacements induced in a piezoelectric ceramic.[57] The moonie/cymbal has intermediate characteristics between the conventional multilayer and bimorph actuators; it exhibits an order of magnitude larger displacement (100 μm) than the multilayer, and much larger generative force ($10\ \mathrm{kg} \cdot f = 100\ \mathrm{N}$) with quicker response (100 μsec) than the bimorph.

The moonie device consists of a thin multilayer ceramic element and two metal plates with a narrow moon-shaped cavity bonded together (Figure 4.35a). A moonie with a size of $5 \times 5 \times 2.5\ \mathrm{mm}^3$ can generate a 20-μm displacement under 60 V, which is 8 times as large as the generative displacement of a multilayer of the same size. By modifying the end cap design (cymbal type) as shown in Figure 4.35b, twice the displacement can be obtained.[58] Also, the generative displacement is rather uniform on the top flat part, independent of the position from the center of the end cap. Another advantage of the cymbal over the moonie is its easy fabrication process. One-step punching can make endcaps from a metal plate.[59]

A "ring-morph" has been developed that is made with a metal shim and two PZT rings rather than PZT disks.[60] The bending displacement is amplified by more than 30% when an appropriate inner ring diameter is chosen.

4.3.6 Flexible Composites

When needle- or plate-shaped piezoelectric ceramic bodies are arranged and embedded in a polymer matrix, advanced composites can be fabricated, which provide enhanced sensitivity by keeping the actuation function. These composites exhibit intermediate performances between solid piezoceramics and soft piezopolymers. Figure 4.36a shows such a 1–3 composite device, where PZT rods are arranged in a polymer in a two-dimensional array. The simplest composite from a fabrication viewpoint is a 0–3 connectivity type, which is made by dispersing piezoelectric ceramic powders uniformly in a polymer matrix (Figure 4.36b). The *connectivity* "*x–y*" in a two-phase composite stands for x-dimensional connection of the primary phase (such as PZT) and y-dimensional connection of the secondary phase (such as polymer). "0" and "1" connection correspond to "powder" and "needle" shapes, respectively, "2" to plane, and "3" is a volumetric connection.

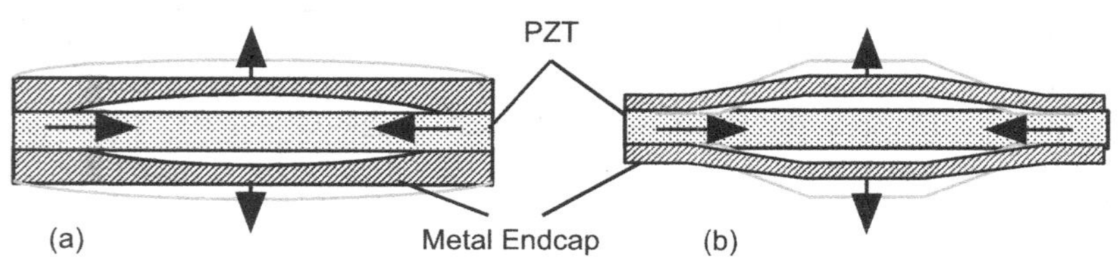

FIGURE 4.35 Flextensional structures: (a) the moonie and (b) the cymbal. (From A. Dogan: Ph. D. Thesis, Penn State University, 1994; Y. Sugawara et al.: *J. Amer. Ceram. Soc.*, 75, 996, 1992.)

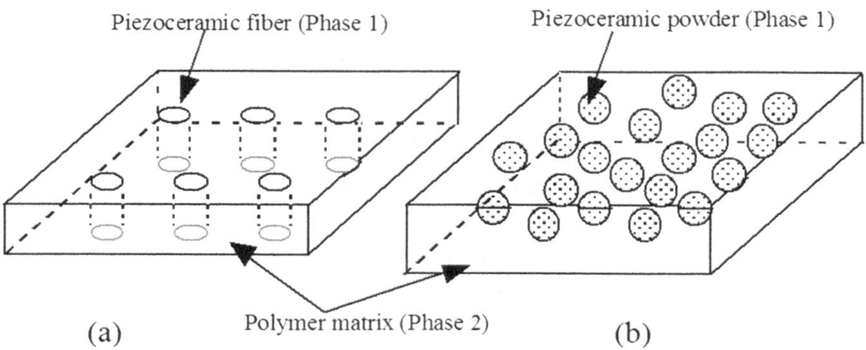

FIGURE 4.36 PZT: polymer composites: (a) 1–3 connectivity and (b) 0–3 connectivity.

The fabrication processes of the 0–3 composites are classified into a melting and a rolling method.[61] Figure 4.37 shows a flowchart for the fabrication processes. The powders are mixed with molten polymer in the first method, while the powders are rolled into a polymer using a hot roller in the second method. The original fabrication process of the 1–3 types involves the injection of epoxy resin into an array of PZT fibers assembled with a special rack.[62] After the epoxy being cured, the sample is cut, polished, electroded on the top and bottom, and finally electrically poled. The die casting technique has also been employed to make rod arrays from a PZT slurry.

A 1–3 composite deformable mirror is pictured in Figure 4.38.[63] The glass mirror is attached to one surface of the composite on which the common electrode is applied, while the individual PZT rods are addressed from the other side of the device. As each rod in the array can be individually addressed, the surface of the deformable mirror can be shaped to any desired contour. This 1–3 composite design is merely to separate the acting rods. A 0–3 composite composed of PZT powder in a polymer matrix can be used to make flexible bimorphs. Although a large curvature can be obtained with such a bimorph, excessive heat generation during AC drive is a serious problem for these devices. This is due primarily to the highly thermo-insulating nature of the polymer matrix. The heat generated through dielectric loss of the piezoelectric ceramic is not readily dissipated.

Active fiber composites (AFCs) have been developed at MIT, composed of PZT needles that are arranged in a polymer matrix, pictured in Figure 4.39.[62] One application for this composite is the helicopter blade vibration control device. Fine PZT needles, which are fabricated by an extrusion technique, are arranged in an epoxy resin, and the composite structure is sandwiched between the interdigital electrodes (Ag/Pd), which are coated on a Kapton substrate. The AFC

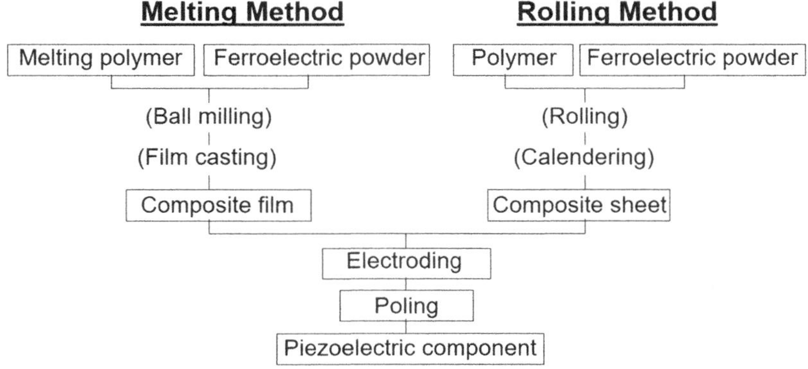

FIGURE 4.37 Fabrication process for PZT: polymer composites.

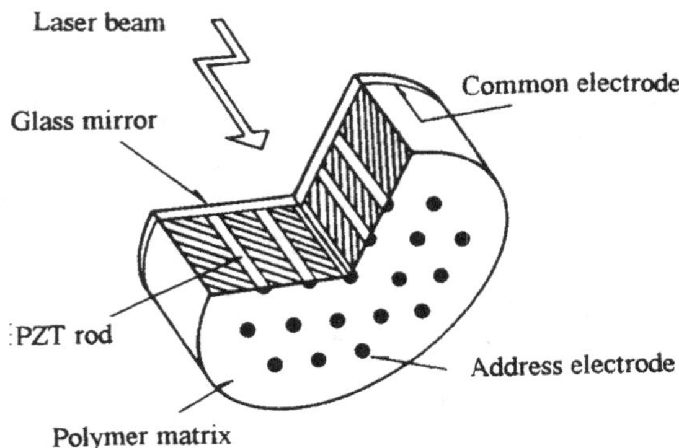

FIGURE 4.38 A PZT: polymer composite deformable mirror. (From J. W. Hardy et al.: *J. Opt. Soc. Amer.*, 67, 360, 1977.)

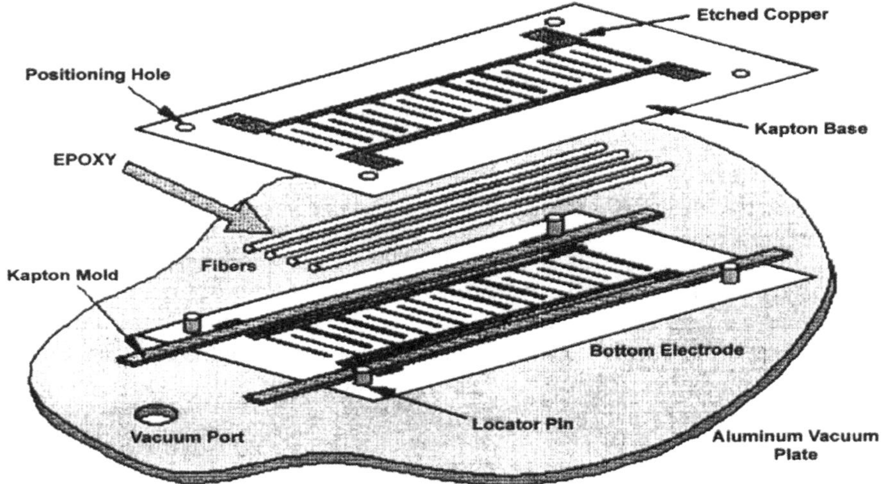

FIGURE 4.39 Active fiber composite (AFC), composed of oriented PZT fibers and epoxy. (From B. Z. Janos and N. W. Hagood: *Proc. 6th Conf. New Actuators, Actuator '98*, Bremen, Germany, p. 193, 1998.)

structures are laminated on the helicopter blade such that the PZT fiber axes are oriented 45° with respect to the blade direction in order to produce a torsional deformation of the blade. A major problem with this design is the limitation in the effective applied electric field strength due to the restricted contact area between the flat electrode surface and the rounded contour of the PZT fibers. One remedy for this problem has been to incorporate conducting particles in the epoxy matrix.

4.4 ELECTRODE MATERIALS

4.4.1 Actuator Electrodes: An Overview

Electrodes are installed on piezoelectric devices primarily to apply voltage. In addition, for actuator applications, (1) electrical conductivity, (2) mechanical strength, and (3) reliability issues should further be improved significantly. When piezoactuators are driven under an off-resonance slow electric field, the electric conductivity of the electrode does not affect the performance much, but when driven at the resonance frequency, poor electrode conductivity degrades the resonance performance dramatically due to small resonance impedance of the piezotransducer. On the other hand, the insertion of electrode layers in multilayer actuator ceramics decreases the mechanical strength and fracture toughness, leading to easy delamination and crack propagation, in particular in pulse/impact drive actuators. Regarding reliability, the major problem with the most commonly used electrode material, silver (Ag), is that it tends to migrate into ceramics under pseudo-DC high electric fields, in particular, under high temperature, high humidity, and high atmospheric pressure.

4.4.2 Ag-Based Electrodes

In order to solve the migration problem in pure Ag electrodes, alloys such as silver-palladium (Ag-Pd) (30%Pd, in particular) are currently used for actuator applications. However, precious metals such as gold (Au), platinum (Pt), and silver-palladium (Ag-Pd) are particularly expensive and thus not very suitable for mass production for automobile applications. In addition, it is worth noting that the electric conductivity of Pt and Pd is worse than pure Ag. Thus, the recent research direction is to return to pure Ag (better and cheaper!) for resonance transducer components such as piezoelectric transformers and ultrasonic motors. However, to escape from the oxidation of Ag during sintering, we need to reduce the firing temperature.

The *bonding method* developed by Ohnishi and Morohashi[64] is one such method. The structure fabricated by this method is a multilayer stack of 22 layers composed of 0.3-mm-thick sintered ceramic disks arranged alternately with 10-μm-thick silver (Ag) foils. The multilayer structure is hot-pressed at 900°C for 4 h. A device of this type with dimensions $5 \times 5 \times 9$ mm^3 demonstrates a bending strength of approximately 100 MPa. The cost of manufacturing multilayer devices by the bonding method is thus significantly cheaper than the conventional methods, because other inexpensive base metal foils may be utilized for the internal electrodes.

In order to strengthen the adhesion between the piezoceramic layer and electrode layer, we usually use a specially prepared electrode ink for printing the electrode patterns on the green sheet. We mix the piezoceramic powders into Ag/Pd electrode ink (~10 vol.%) so that these ceramic powers bridge between the ceramic layers tighter during the sintering process (unpublished information).

We sometimes use encapsulated multilayer actuators for severe atmospheric conditions, such as automobile diesel injection valve control, which is at 160°C. We initially used a completely shielded capsule, which unexpectedly degraded the lifetime dramatically at 160°C. Then, we created a pinhole on the capsule to leak N_2 high pressure owing to the temperature rise, which recovered the lifetime. This shows that high N_2 pressure (though zero humidity) enhances Ag migration (unpublished information).

Siemens group reported an intriguing grain/domain texture generated by the internal stress with decreasing the layer thickness of multilayer (ML) devices. Two types of ML PZT actuators were prepared with layer thicknesses of 280 and 17 μm, as shown in Figure 4.40a and b.[65] This PZT composition has tetragonal symmetry, and when the x-ray diffraction pattern was taken for both samples from the top surface, Figure 4.41a was obtained. The sample with 280-μm layer thickness seemed to have no texture by showing the x-ray intensity ratio roughly 1:2 for the (001) and (100) reflections, while the intensity of the (001) reflection was very high for the sample with 17-μm layer thickness. The high intensity of the (001) reflection for the sample with 17-μm layer thickness indicates the polarization direction aligned normal to the layer plane, as schematically shown in Figure 4.41b.

If this domain texture was induced by mechanical stress through the manufacturing process, the thermal expansion difference between the PZT and metal electrode may be the origin. Because the metal electrode shrinkage is larger than the PZT during the cooling-down process, the PZT layer will experience compressive stress (the thinner layer sample will experience larger compressive stress, because of rather constant electrode thickness), leading to domain alignment normal to the layer surface (Figure 4.42).

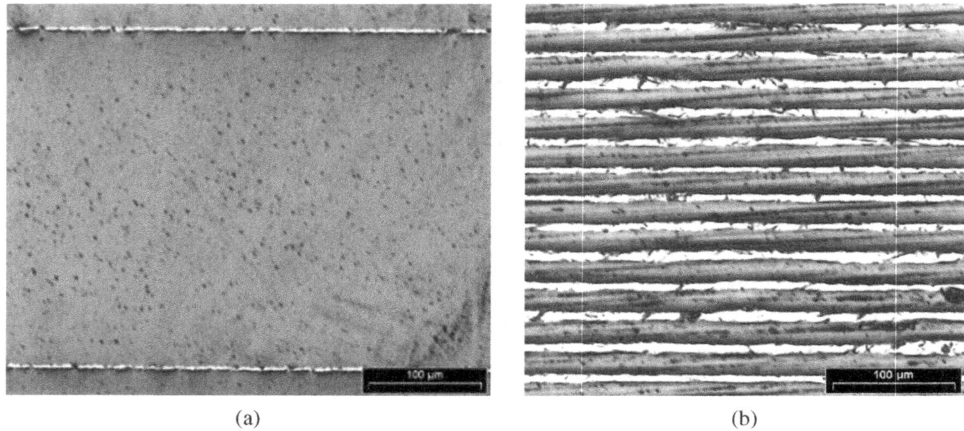

FIGURE 4.40 PZT multilayer actuators with their layer thicknesses of (a) 280 and (b) 17 μm. (From C. Schuh: *Proc. New Actuators 2004*, Bremen, 2004.)

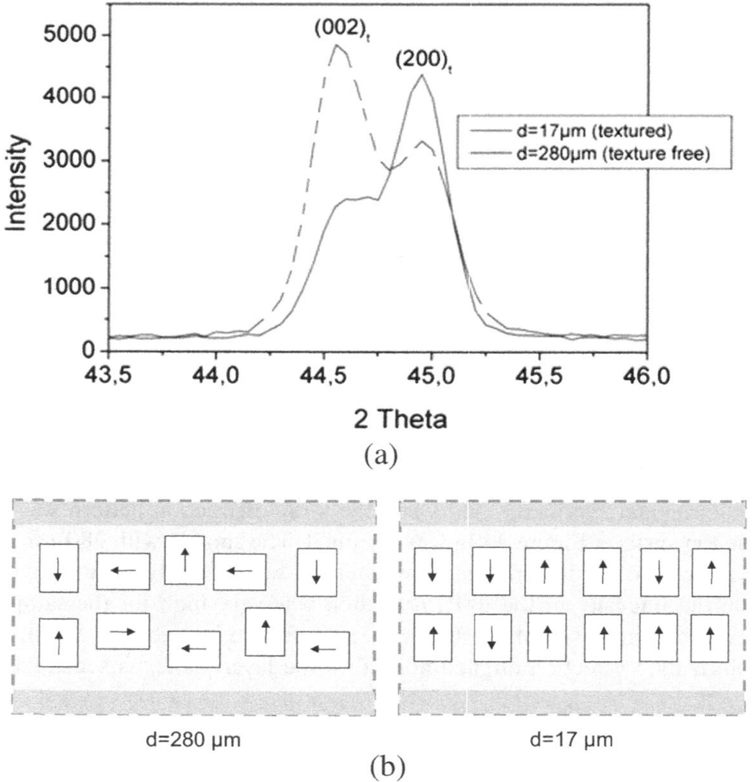

FIGURE 4.41 (a) Two x-ray diffraction patterns for ML actuators with 280- and 17-μm layer thicknesses. (b) Two domain texture models for the two ML actuators. (From C. Schuh: *Proc. New Actuators 2004*, Bremen, 2004.)

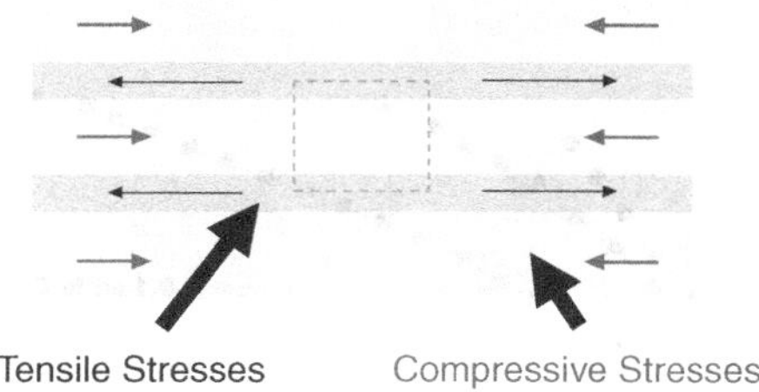

FIGURE 4.42 Domain texture creation model for the ML actuators with 17-μm layer thickness. (From C. Schuh: *Proc. New Actuators 2004*, Bremen, 2004.)

4.4.3 Base Metal Electrodes

When we use an Ag/Pd internal electrode for 200-layer piezoelectric actuators, roughly 90% of the raw material's cost is shared by the electrode. As new piezoelectric materials that can be sintered at low temperature are developed, electrode materials like copper (Cu) and nickel (Ni) became feasible inexpensive alternatives.

An innovative fabrication technique has been developed by Morgan Electroceramics, which involves incorporating a copper (Cu) internal electrode in a PZT multilayer actuator.[66] The PZT and copper electrode materials are co-fired. Ordinarily the sintering temperature of PZT is too high to allow for co-firing with a Cu electrode. The sintering temperature of the PZT used in these structures was reduced to 1015°C by adding excess PbO. Special measures are used to optimize the co-firing process. The oxygen (O_2) pressure is precisely regulated by sintering in a nitrogen (N_2) atmosphere (10^{-10} atm). The pieces are also fired in a special sintering sand, which is essentially a mixture of Cu and PZT powders. This helps to inhibit the oxidation of the Cu electrode during firing. A four-layer PZT actuator (layer thickness of 25 μm) fabricated by this method is pictured in Figure 4.43.

Based on the high-power piezoelectric ceramics, the Sb, Li, and Mn-substituted $0.8Pb(Zr_{0.5}Ti_{0.5})O_3$-$0.16Pb(Zn_{1/3}Nb_{2/3})O_3$-$0.04Pb(Ni_{1/3}Nb_{2/3})O_3$, we further modified them by adding CuO and Bi_2O_3 in order to lower the sintering temperature at 900°C of the ceramics.[67,68] Under a sintering condition of 900°C for 2 h, the properties were: $k_p=0.56$, Q_m (31-mode) = 1023, $d_{33}=294$ pC/N, $\varepsilon_{33}/\varepsilon_0=1326$, and tan $\delta=0.59\%$ when CuO and Bi_2O_3 were added 0.5 wt% each. The maximum vibration velocity of this composition was 0.41 m/s. Figure 4.44 shows the maximum vibration velocity vs. applied field change with various amounts of CuO. Note that with increasing CuO content (0.4 wt% or higher) only 6–8 V_{rms}/mm is required for obtaining $v_0=0.48$ m/s. These compositions are suitable for piezoelectric transformers and transducers.

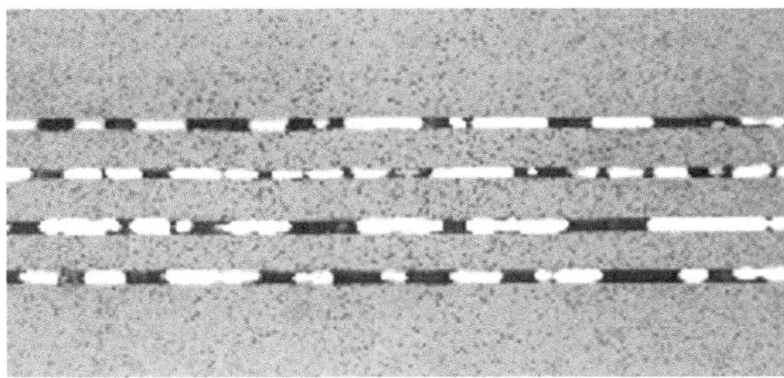

FIGURE 4.43 A PZT multilayer actuator with Cu internal electrodes co-fired at 1015°C. (From W. A. Groen et al.: *Proc. 33rd Int'l Smart Actuator Symp.*, State College, PA, April, 2001.)

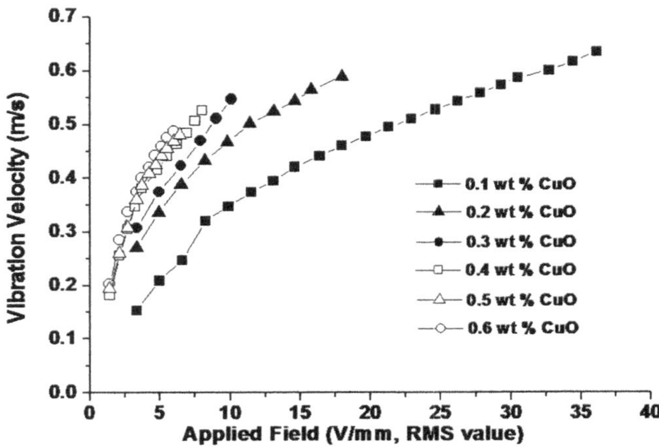

FIGURE 4.44 Vibration velocity variation with applied field in $0.8Pb(Zr,Ti)O_3$-$0.2Pb[0.7\{0.7(Zn,Ni)_{1/3}$-$(Nb,Sb)_{2/3}$-$0.3Li_{1/4}(Nb,Sb)_{3/4}\}$-$0.3Mn_{1/3}(Nb,Sb)_{2/3}]O_3$ for various x wt% of CuO added to the ceramic.

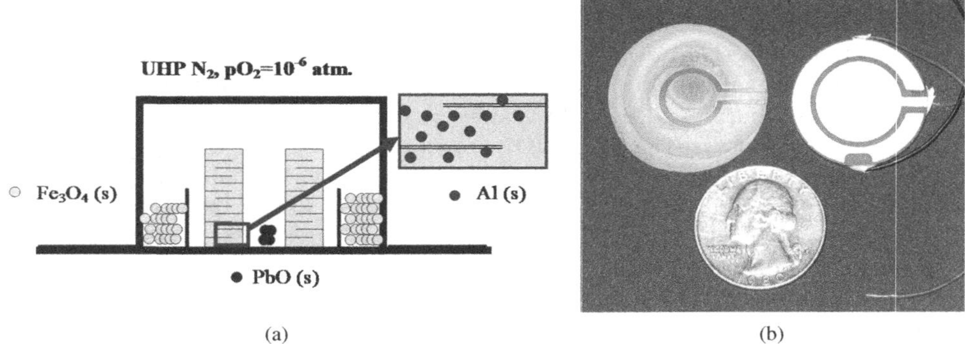

FIGURE 4.45 (a) Experimental setup for sintering Cu-electrode-embedded multilayer transformers in a reduced N_2 atmosphere. (b) Multilayer co-fired transformer with hard PZT and Cu (left) or pure Ag (right) electrode, sintered at 900°C (Penn State trial products). (From S. Ural et al.: *Proc. 10th Int'l Conf. New Actuators*, Bremen, Germany, June 14–16, 2006, P23, pp. 556–558, 2006.)

Using the above low-temperature-sinterable hard PZT, Penn State developed Cu and pure-Ag embedded multilayer piezotransformers (Figure 4.45b), which were sintered at 900°C in a reduced atmosphere with N_2, as illustrated in Figure 4.45a.[69] Ring-dot disk multilayer types (OD = 27, center dot $D = 14$ mm) with Cu and Ag/Pd (or Ag/Pt) (as references) revealed the maximum power density (at 20°C temperature rise) 42 and 30 W/cm³, respectively. This big difference comes from the poor electric conductivity of Pd or Pt compared to Cu or pure Ag. Note that the power density depends not only on the piezoceramic composition, but also on the electrode species.

EXAMPLE PROBLEM 4.7

When you prepare the electrodes on a piezoelectric ceramic specimen, you probably use either Ag paste or ink. Describe the difference between Ag paste and Ag ink.

Solution

Both silver paste and ink are for preparing silver electrodes on piezoelectric ceramics. Silver ink is silver nanoparticle suspended liquid that is used for printing internal electrodes on piezoceramic green sheets. Using a sophisticated printing screen, the electrode thickness can be controlled even below 1 μm. The adhesion strength between the Ag electrode and piezoceramic layers is usually small. Thus, for the actuator application, we occasionally mix some amount of the piezopowder into this silver ink.

On the contrary, silver paste contains glass particles in addition to silver nanoparticles, and is popularly used for making external electrodes on sintered piezoelectric components. When we use high-temperature curing-type paste, the piezocomponents need to be fired around 500°C for a half hour or so in order to melt the glass phase to solidify the electrode layer on the piezoceramic surface. The piezoceramic surface is slightly roughened in order to increase the adhesion strength. Low-temperature curing-type silver paste includes resin (such as epoxy) instead of glass particles. Though we do not need to fire the components (no disturbance to the poling status), the mechanical strength of the electrode is low. It is not recommended to misuse silver paste for the internal electrodes of multilayer actuators. First, due to the glass particles included, the printing screen will be damaged; second, because of the glass material in the internal electrode layer, the glass particle initiates vertical cracking under the electric field drive.

4.4.4 Ceramic Electrodes

Mechanical weakness at the junction between the ceramic and the electrode metal often gives rise to delamination problems. One solution to this problem has been to make a more rigid electrode by mixing ceramic powder of the composition used for the actuator material with the metal electrode paste. We examined some of the more popular conducting ceramic materials currently used as electrodes in ceramic actuators in this section.[71]

4.4.4.1 Ceramic Electrode

Attractive *ceramic electrode materials* are conducting or semiconducting perovskite oxides because of their compatibility with the crystal structure of the actuator ceramics. Among the conducting ceramics $Sr(Fe,Mo)O_3$, $(La,Ca)MnO_3$, and $Ba(Pb,Bi)O_3$ (which is actually a superconductor at low temperatures) were considered first, but their conductivity was drastically reduced when they were sandwiched between lead-based PZT ceramic layers and sintered. One of the few successful structures utilizing a ceramic conductor is a unimorph device composed of a piezoceramic and a $Ba(Pb,Bi)O_3$ fabricated by hot-pressing. Semiconducting $BaTiO_3$-based ceramics with a *positive temperature coefficient of resistivity* (PTCR) also appear to be promising alternative electrode materials for these structures.

In general, good ceramic electrode materials should possess the following characteristics:

1. High conductivity
2. Sintering temperature and shrinkage similar to those of piezoceramics
3. Good adhesion with piezoelectric ceramics
4. Slow diffusion into piezoelectric ceramics during sintering

The last characteristic may be the most critical, and the ultimate success of the device structure depends on a well-defined interface between the electrode and piezoelectric layers.

4.4.4.2 $BaTiO_3$-Based Multilayer Actuator

Barium titanate ($BaTiO_3$), which is a ferroelectric with a Curie temperature of 120°C, is known to become electrically conductive when polycrystalline samples are doped with rare-earth ions.[71,72] The resistivity of some $BaTiO_3$-based ceramics (with 5 mol% SiO_2 and 2 mol% Al_2O_3) is plotted in Figure 4.46 as a function of La_2O_3 concentration. Relatively small concentrations of La in the range of 0.1 to 0.25 atom.% lead to a change in the resistivity of more than 10 orders of magnitude. Multilayer actuators composed of alternating layers of undoped resistive $BaTiO_3$ and $BaTiO_3$-based semiconducting composition doped with 0.15 atom.% La have been

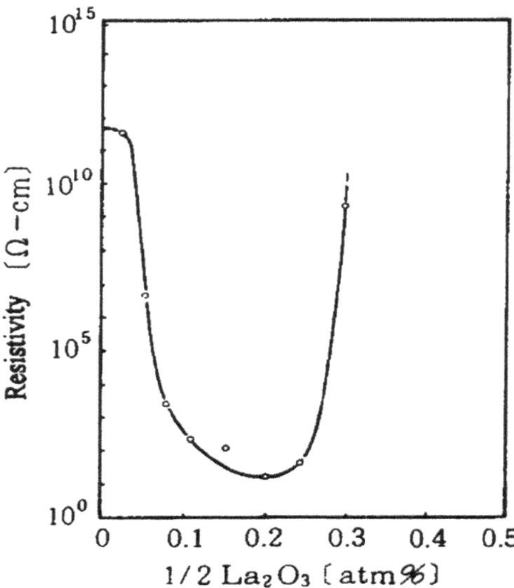

FIGURE 4.46 Resistivity of $BaTiO_3$-based ceramics (with 5 mol% SiO_2 and 2 mol% Al_2O_3) plotted as a function of La_2O_3 concentration.

fabricated by the tape-casting process. The main advantage in using these materials is that atomic diffusion across the interface between layers during the sintering process tends to be suppressed because the layers are compositionally similar, resulting in a well-defined interface between the actuator and electrode layers. Another beneficial feature of this combination is that the sintering temperature and shrinkage of both materials are almost the same, so that no residual stress is present in the sample after sintering. The fabrication of this structure is also somewhat simpler since it requires no binder burn-out process, which is a time-consuming step required in the fabrication of devices with metal electrodes. A prototype device of this type has been produced with eight 0.5-mm-thick actuator layers, sandwiched between 0.25-mm-thick electrode layers.[70]

4.4.4.3 Mechanical Strength of Ceramic Electrode Devices

The mechanical strength of multilayer samples having an overall platelike shape with their piezoelectric and ceramic electrode layers perpendicular to the plate as shown in Figure 4.47a has been tested by a *three-point bend method*.[70] Data were also collected for samples of this configuration having the same piezoelectric ceramic as the test specimens and palladium (Pd) electrode layers, which served as a reference. The *Weibull plots* for both are shown in Figure 4.47b. The mechanical strength of the ceramic electrode device is observed to be about 50 MN/m^2, which is approximately three to four times higher than that observed for the metal electrode actuator. Refer to Example Problem 3.5 for detailed analysis of this Weibull plot. It is also noteworthy that the Weibull coefficient is generally much larger when ceramic electrodes are used, indicating that the deviation in fracture strength is much smaller. Fracture was observed to occur mainly at the ceramic-electrode interface for the structures with metal electrodes, while no such tendency was observed for the ceramic electrode device.

The electric field–induced strains measured in the prototype ceramic electrode device are shown in Figure 4.48a. It was determined in the study of this structure that the longitudinal strain induced in the actuator layers was $x_3 = 5 \times 10^{-4}$ and the transverse strain $x_1 = -1.2 \times 10^{-4}$. The measured strains are 30% and 60% smaller than the predicted values for the longitudinal and transverse strains, respectively. The clamping effect associated with this strain anisotropy can be reduced by decreasing the thickness of the electrode layers.

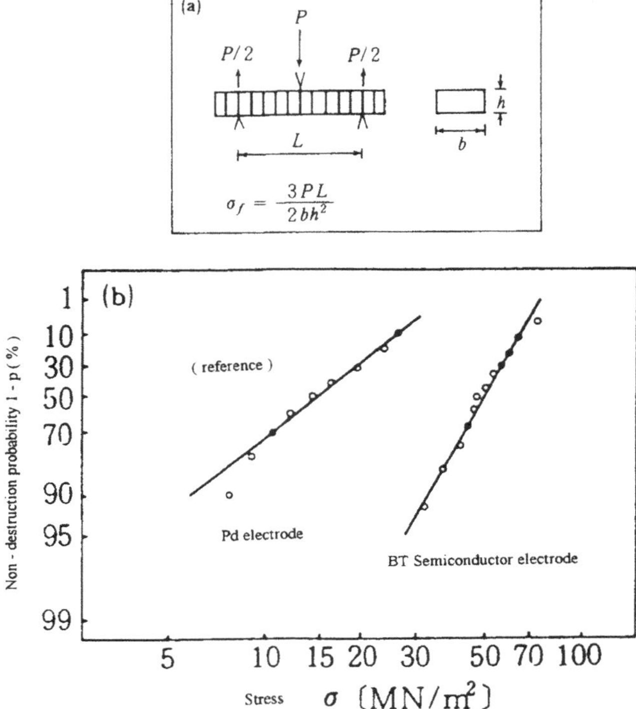

FIGURE 4.47 Three-point bend testing of multilayer actuators: (a) sample configuration and experimental setup, and (b) the Weibull plots for the ceramic electrode and metal electrode (reference) structures. (From K. Abe et al.: *Ferroelectrics*, 68, 215, 1986.)

The uniaxial stress dependence of the strain induced in $BaTiO_3$ under several applied electric fields is shown in Figure 4.48b. The typical strain responses of PZT and PMN-PT specimens are also inserted in the figure for comparison. The magnitude of the induced strain in $BaTiO_3$ is not as large as that generated in lead-based materials, but the generative stress (320 kgf/cm^2 = 32 MPa) level is comparable with those of the PZT and PMN-PT samples, mainly due to the relatively small elastic compliance [$s_{33} = 13 \times 10^{-12}$ (m^2/N)] of $BaTiO_3$.

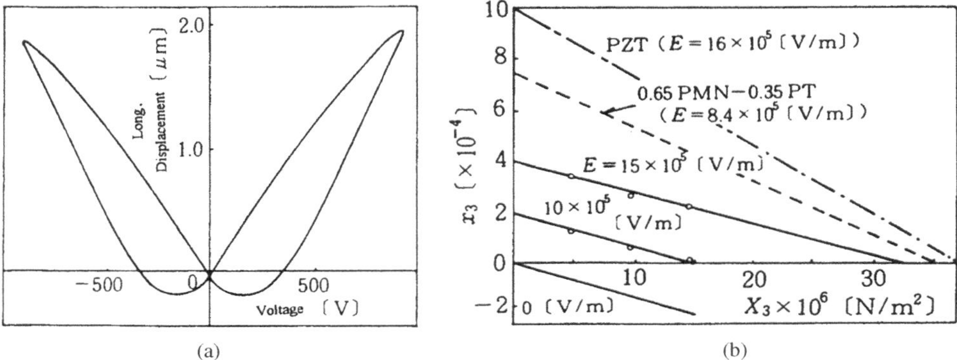

FIGURE 4.48 (a) Displacement curve for a $BaTiO_3$-based multilayer actuator with ceramic electrodes.[71] (b) Uniaxial stress dependence of the strain induced in $BaTiO_3$ under several applied electric fields. The typical strain responses of PZT and PMN-PT specimens are also shown for comparison.

4.5 PIEZOELECTRIC THIN/THICK FILMS AND MICRO-ELECTRO-MECHANICAL-SYSTEMS

Piezoelectric sensors and actuators have been ignited significantly in the MEMS or NEMS area, with new thin film technologies. Though sensor applications have been successfully commercialized even for automobile devices, actuator applications are still primitive. The key reason is the total volume of the piezoelectric materials required for practical devices. Note that the current best PZTs can handle maximum 30 W/cm^3 for the input/output energy level. Recall the maximum vibration velocity, above which additional input energy converts merely to heat generation. Since 1 mW level output mechanical energy is required even for a micro actuator, this leads to a PZT volume of 3×10^{-11} m^3, in comparison with 1 μW in sensors. If we use 10-μm-thick PZT film, we need an area of 3 mm^2 (or 1.7 mm square). If we use only 100 nm thin film, 17 mm square is required as an actuator, which is not a small device anymore! Thus, the MEMS actuator target should start from 10–30-μm-thick films from the practical application viewpoints.

The definition of "thin" and "thick" films is not clear, but we define thin film with the thickness <10 μm, while thick film has thickness >10 μm, because the performance of the films >10 μm is almost the same as that of bulk materials.

4.5.1 Piezoelectric Thin Films

4.5.1.1 Thin Film Fabrication Techniques

Techniques for fabrication of oxide thin films are classified into physical and chemical processes:

a. Physical Processes
   - Electron beam evaporation
   - RF sputtering, DC sputtering
   - Ion beam sputtering
   - Ion plating

b. Chemical Processes
   - Sol-gel method (dipping, spin coating, etc.)
   - Chemical vapor deposition (CVD)
   - Metal-oxide CVD (MOCVD)
   - Liquid phase epitaxy, melting epitaxy, capillary epitaxy, etc.

Sputtering has been most commonly used for ferroelectric thin films such as $LiNbO_3$, PLZT,[73] and $PbTiO_3$.[74] Figure 4.49 shows the principle of a magnetron sputtering apparatus. Heavy Ar plasma ions bombard the cathode (target) and eject its atoms. These atoms are deposited uniformly on the substrate in an evacuated enclosure. Choosing a suitable substrate and deposition condition, single crystal-like epitaxially deposited films can be obtained. The sol-gel technique has also been employed for processing PZT films.[75] Applications of thin film ferroelectrics include memories, surface acoustic wave devices, piezosensors, and micromechatronic or MEMS devices. As we discussed in Section 2.4.7, (001) epitaxially oriented PZT rhombohedral composition films are most suitable from the piezoelectric application viewpoint. Kalpat et al. demonstrated (001)- and (111)-oriented PZT films on the same Pt-coated Si substrate by changing the rapid thermal annealing profile.[76] Epitaxially grown rhombohedral (70/30) PZT films with (001) and (111) orientations were prepared by using a *rapid thermal annealing method*. Figure 4.50a shows optimum annealing profiles; a popular one-step process (very rapid to 700°C) exhibits (100)-oriented films, while the two-step process (very rapid to 550°C, followed by a little slower up to 750°C) gives (111)-oriented films. The x-ray diffraction patterns for these films grown according to these profiles are shown in Figure 4.50b.[76]

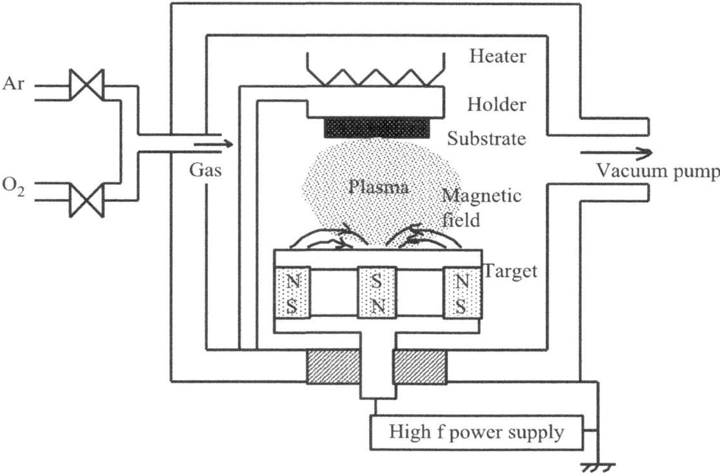

FIGURE 4.49 Principle of a magnetron sputtering apparatus.

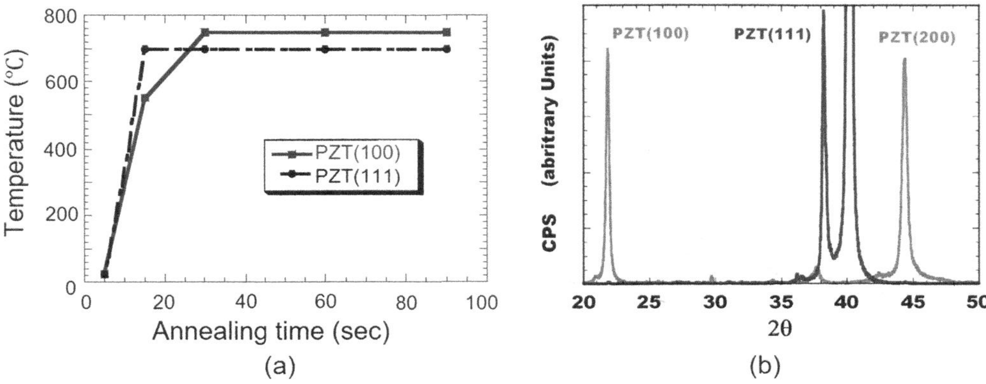

FIGURE 4.50 Epitaxially grown rhombohedral (70/30) PZT films with (001) and (111) orientations: (a) optimum rapid thermal annealing profiles, and (b) x-ray diffraction patterns for films grown according to these profiles. (From S. Kalpat et al.: *Jpn. J. Appl. Phys.*, 40, 158–162, 2001.)

4.5.1.2 Orientation Dependence of Piezoelectric Performances

Polarization, P, versus electric field, E, hysteresis loops were measured on epitaxially grown rhombohedral (70/30) PZT films with (001) and (111) orientations, as shown in Figure 4.51. It is noteworthy that the saturated polarization in PZT (100) film is $1/\sqrt{3}$ of that in PZT (111) film, which corresponds exactly to the direction cosine of the spontaneous polarization of this rhombohedral crystal. Second, the P-E hysteresis is very slanted in PZT (111) film with the coercive field $E_C = 65$ kV/cm and remnant polarization $P_r = 25\ \mu C/cm^2$, while the P-E hysteresis is rather square in PZT (100) film with smaller $E_C = 35$ kV/cm and $P_r = 20\ \mu C/cm^2$. Square hysteresis with a smaller coercive field is suitable to ferroelectric memory applications. Third, the piezoelectric properties are compared between (111) and (100) films. Figure 4.52a shows a phenomenological simulation of the effective piezoelectric constant d_{33} as a function of PZT composition for different crystal orientations, reported in our previous papers (see Section 2.4.7).[77] We have demonstrated that the (100)-oriented crystal with a rhombohedral composition should exhibit the highest piezoelectric performances. As shown in the experimental results in Figure 4.52b on our films, where the measured effective piezoelectric constant, e_{31}, is plotted as a function of PZT composition for different crystal

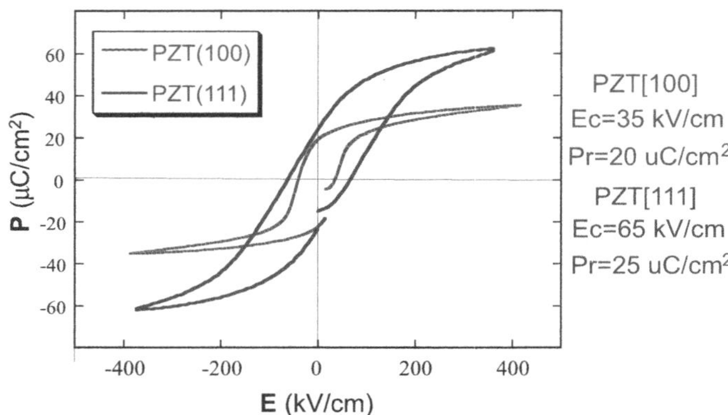

FIGURE 4.51 Polarization vs. field hysteresis loops on epitaxially grown rhombohedral (70/30) PZT films with (001) and (111) orientations. (From S. Kalpat et al.: *Jpn. J. Appl. Phys.*, 40, 158–162, 2001.)

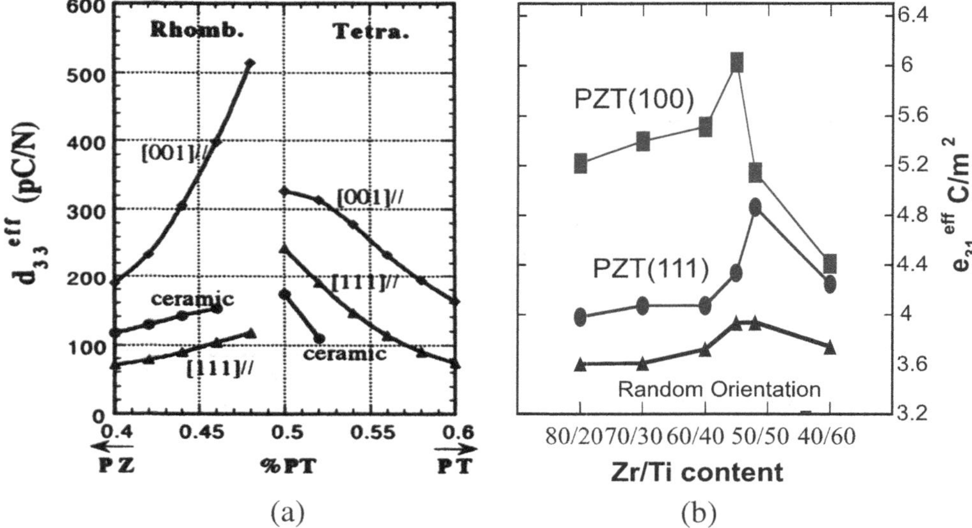

FIGURE 4.52 (a) Phenomenological simulation of the effective piezoelectric constant d_{33} as a function of PZT composition for different crystal orientations. (From X.-H. Du et al.: *Appl. Phys. Lett.*, 72(19), 2421, 1998.) (b) Measured effective piezoelectric constant e_{31} as a function of PZT composition for different crystal orientations. (From S. Kalpat et al.: *Jpn. J. Appl. Phys.*, 40, 158–162, 2001.)

orientations,[76] PZT (100) films in the rhombohedral composition exhibit the highest value in practice, though the enhancement is not as high as theoretically predicted due to the substrate constraint.

4.5.1.3 Thickness Dependence of Physical Performance

We discussed in Section 4.2.2 the 3D particle size dependence of piezoelectricity, where a critical particle size (~sub-micron meter) exists below which the ferroelectricity and piezoelectricity disappear in perovskite-type ferroelectric materials. We know the fact that PZT films exhibit ferroelectricity; even the thickness becomes thinner than 100 nm (0.1 μm), as long as the area is much larger than micron meter. The critical thickness in 2D-connected films may be much thinner than 100 nm. Figure 4.53 summarizes the experimental results reported by Chen et al.[78] and Perez de la Cruz[79] on PZT thin films. The former reported for a wide thickness range, 1–10 μm, while the latter for a thinner range, 100 nm–1 μm. We can generally conclude that dielectric permittivity,

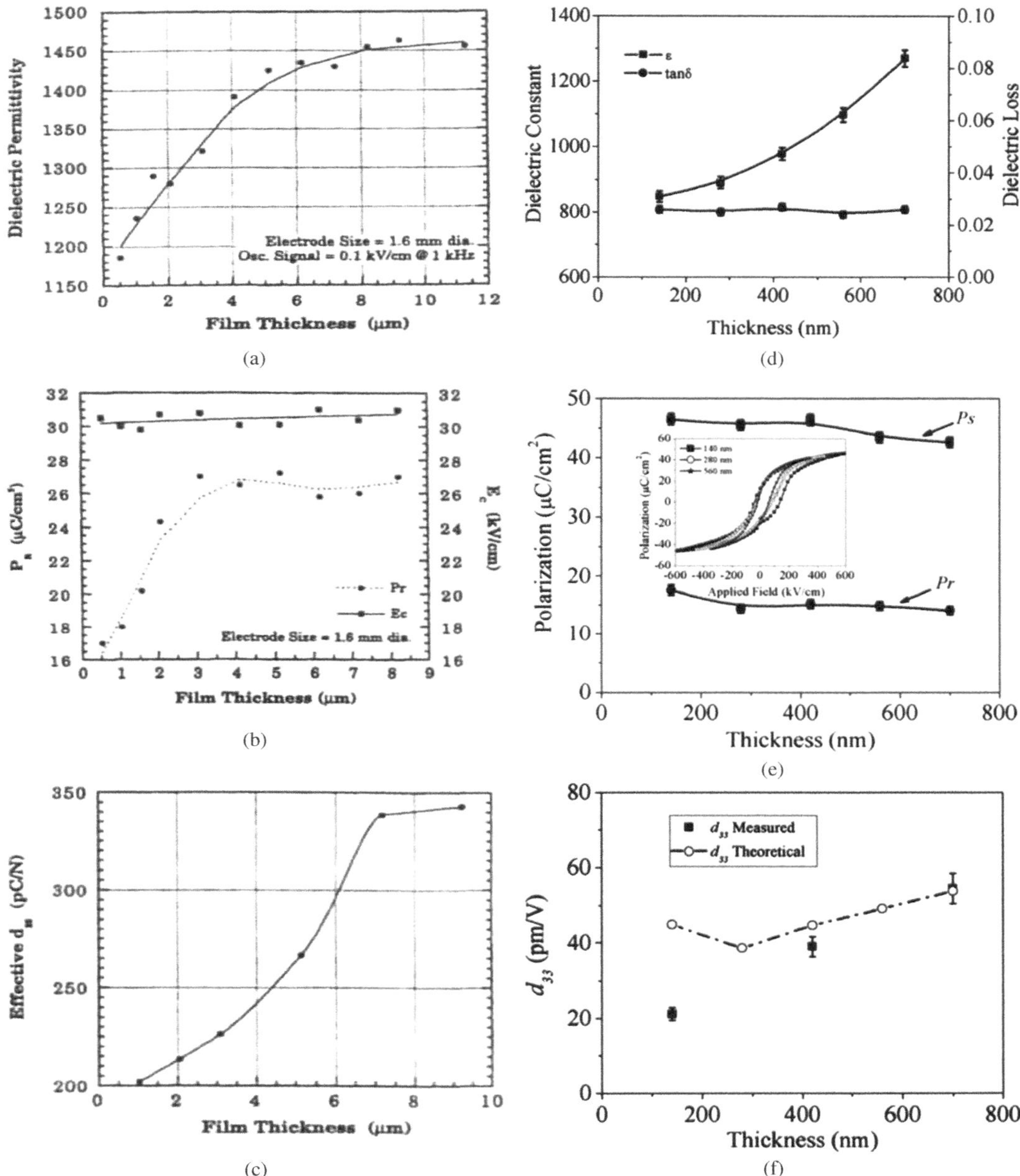

FIGURE 4.53 Film thickness dependence of dielectric constant (a,d), polarization (b,e) and piezoelectric constant (c,f) for the thickness range <10 μm (left) and <1 μm (right). (From H. D. Chen et al: *J. Amer. Ceram. Soc.*, 79, 2189, 1996; J. Pérez de la Cruz et al.: *J. Appl. Phys.*, 108, 114106, 2010.)

spontaneous polarization, and the piezoelectric constant, *d*, start to decrease significantly below 6 μm with reduction of the thickness in a wide thickness range. Films with >6 μm exhibit similar properties to bulk ceramics. In a thinner range, 100–800 nm, the variation is not very large down to 100 nm, still keeping reasonable ferroelectricity and piezoelectricity.

Recently, Gao et al. reported on the absence possibility of critical thickness in perovskite ferroelectric films.[80] Although the size effect in ferroelectric thin films has been claimed for a long time, the underlying mechanism is not yet fully understood, and whether there is a critical thickness below which ferroelectricity vanishes is still under debate. They directly measured thickness-dependent polarization in ultrathin $PbZr_{0.2}Ti_{0.8}O_3$ films via quantitative annular bright field imaging, and found that the polarization is significantly suppressed for films <10-unit cells thick (~ 4 nm). However, approximately, the polarization never vanishes. The residual polarization is about 16 μCcm^{-2} (~17%) at 1.5-unit cell (~0.6 nm)-thick film on bare $SrTiO_3$ and 22 μCcm^{-2} at 2-unit cell-thick film on $SrTiO_3$ with $SrRuO_3$ electrodes. The residual polarization in these ultrathin films is mainly attributed to the robust covalent Pb–O bond. This atomic study provides new scientific insights into mechanistic understanding of nanoscale ferroelectricity and the size effects.

4.5.1.4 Constraints in Thin Films

The thin film structure is inevitably affected by four important factors:

1. *Size constraints*: Similar to a powder sample, there may exist a critical film thickness below which ferroelectricity would disappear. Further research will be required from this viewpoint.
2. *Stress from the substrate*: Tensile or compressive stress is generated due to thermal expansion mismatch between the film and substrate, leading to sometimes a higher coercive field for domain reorientation. The Curie temperature is also modified with a rate of 50°C per 1 GPa.
3. *Epitaxial growth*: Crystal orientation dependence should also be considered, similar to the case in single crystals. An example can be found in a rhombohedral composition PZT, which is supposed to exhibit the maximum performance when the P_s direction is arranged 57° cant from the film normal direction [i.e., (001) crystallographic orientation].
4. *Preparation constraint*: Si substrate requires a low sintering temperature of the PZT film. Typically, 800° for a short period is the maximum for preparing the PZT, which may limit the crystallization of the film, leading to a reduction of the properties. A metal electrode on a Si wafer such as Pt also limits the crystallinity of the PZT film.

Wasa investigated piezoelectric performance enhancement by intentionally using the crystal lattice mismatch of PZT-based thin films on Si and sapphire.[81] Referring to Figure 4.54a, lattice mismatch statuses between the thin film and substrate are considered in the case where the lattice parameter of the piezoelectric film is larger than that of the substrate. When we deposit the film on the substrate, the film will impose compressive stress in the plane direction. However, if the misfit rate is large, dislocations may be generated in every certain unit lattice. Since the dislocation degrades the piezoelectric performance significantly, most researchers used to try to reduce the mismatch rate between the film and substrate. Wasa's group proposed an innovative preparation

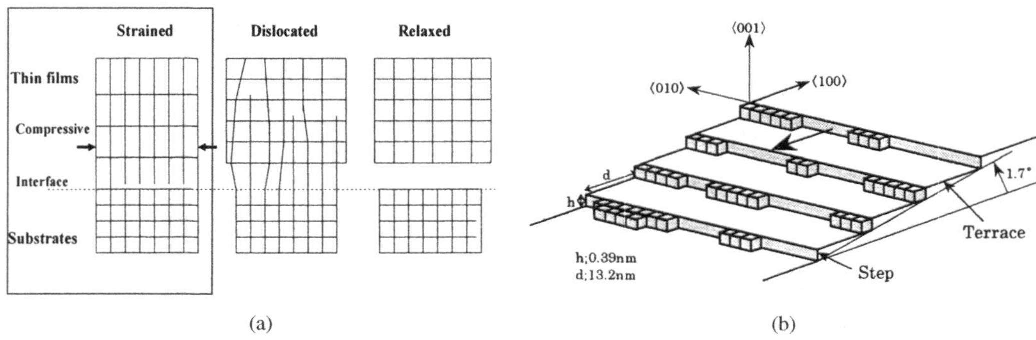

FIGURE 4.54 (a) Lattice mismatch statuses between the thin film and substrate. (b) Preparation method of the strained thin film on the substrate without dislocation.

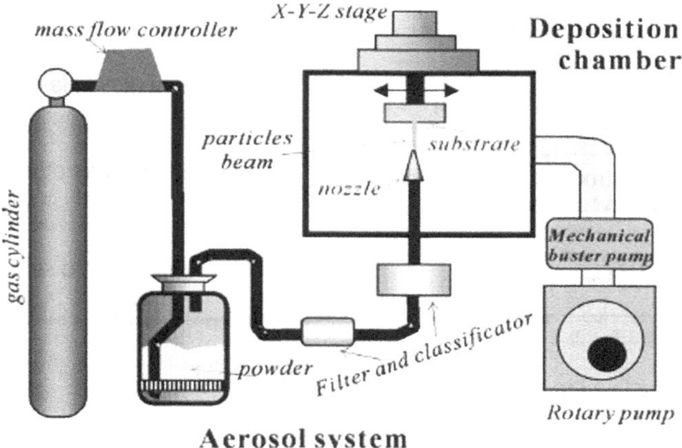

FIGURE 4.55 Schematic of aerosol deposition (AD) apparatus. (From J. Akedo: Advanced Piezoelectric Materials. Ed. K. Uchino, Chapter 14 Aerosol Techniques for Manufacturing Piezoelectric Materials, p. 493, Woodhead Pub., Cambridge, UK, 2010.)

method of the strained thin film on the substrate without dislocation. Figure 4.54b illustrates their idea with using a 1.7° angled surface of the substrate to deposit PZT film. Because the PZT interface plane changes one-step height every 30 horizontal unit-cells, the lattice misfit stress can be relaxed without generating dislocations. This angle can be different, depending on the lattice misfit rate. They demonstrated PZT films with Curie temperatures as high as 600°C, 250°C higher than the original Curie temperature under zero stress.

4.5.2 Piezoelectric Thick Films

Aerosol deposition (AD) is a recent technique for depositing thick ceramic films, where submicron ceramic particles are accelerated by gas flow, then impacted on a substrate.[82] AD can rapidly form a thick, dense, uniform, and hard ceramic layer at room temperature without additional heat treatment for solidifying ceramic powders, even in a low vacuum condition, by using an inexpensive and simple production facility, as schematically illustrated in Figure 4.55. It is expected to reduce energy and costs and to ease the fabrication of thick films with complicated material compositions. Compared to the sputtering or sol-gel spin-coating methods, the AD method is the most promising technique to realize 30-μm-thick PZT films at mass-production scale.

4.5.3 Piezoelectric Micro-Electro-Mechanical-Systems

As discussed in Section 1.3.2, MEMS started in the 1980s from the pure silicon technologies. Though sensor applications were commercialized widely already in the 1990s in vehicle usage, such as engine knock sensors, actuator applications have not been successful due to the micro power level. Since an actuation principle in pure silicon MEMS is basically electrostatic force with a comb-teeth-like silicon structure, exemplified in Figure 1.14, the energy density is not sufficient to move micro subjects mechanically (typically ~1 mW is required). Moreover, high voltage ~1 kV is required even in this micro comb-teeth structure.

With the expansion of piezofilm research in PZTs in the 1990s, the attempt to combine piezofilm with MEMS accelerated. Using epitaxially grown sophisticated PZT film [i.e., rhombohedral (100) oriented], Uchino's group prototyped MEMS micropumps in the late 1990s.[76] The micromachining process used to fabricate the PZT micropump is illustrated in Figure 4.56a.[76] The etching process for the silicon:PZT unit is shown on the left-hand side of the figure

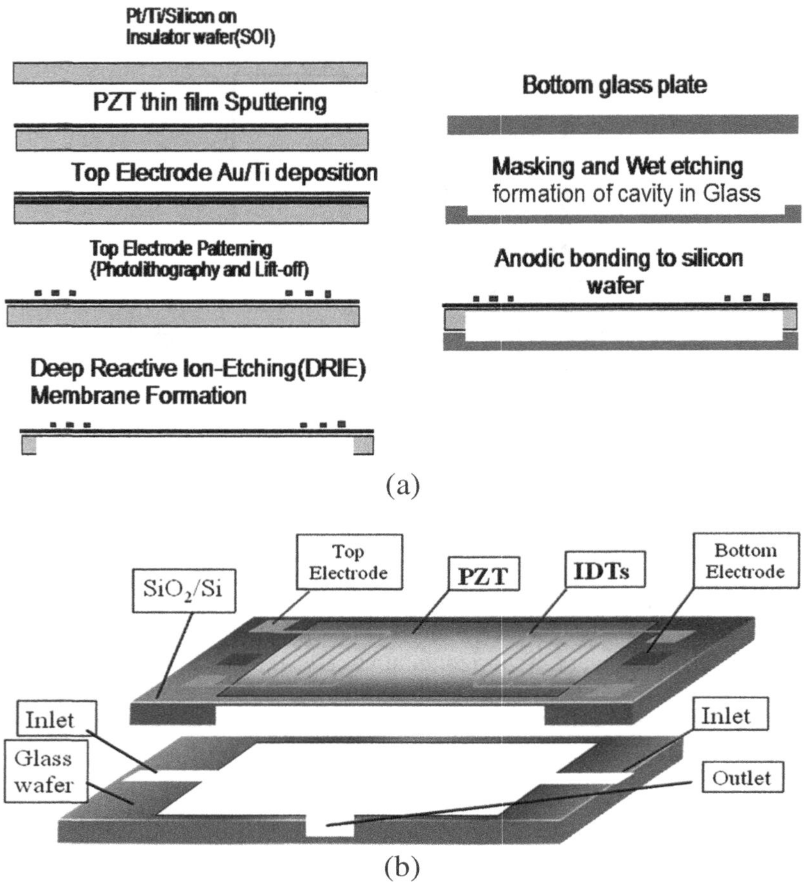

FIGURE 4.56 (a) The micromachining process used to fabricate a PZT micropump. (b) A schematic diagram of the structure of a PZT micropump. Actual size: $4.5 \times 4.5 \times 2\ \text{mm}^3$. (From S. Kalpat et al.: *Jpn. J. Appl. Phys.*, 40, 158–162, 2001.)

and that for the glass plate on the right side. A schematic of the micropump for blood tester is shown in Figure 4.56b. The blood sample and test chemicals enter the system through the two inlets, identified in Figure 4.56b, are mixed in the central cavity, and finally are passed through the outlet for analysis. The movement of the liquids through the system occurs through the bulk bending of the PZT diaphragm (analogous to the surface acoustic wave) in response to the drive potential provided by the *interdigital surface electrodes*. Figure 4.57 visualizes the fabrication process and the IDT electrode patterns of a PZT-MEMS pump for blood test applications. Using a 3-μm-thick PZT film, a minimum of this size shown in Figure 4.56 caption was required to soak/flow the blood (~1 mW). Though the operation condition was higher than the maximum vibration velocity level, the liquid could cool down the heat generated on the PZT film.

Refer to [83,84] for learning up-to-date piezo-MEMS technologies.

EXAMPLE PROBLEM 4.8

Predict on a conceptual basis the polarization and strain hysteresis curves under a large electric field operation for epitaxially grown rhombohedral (111)- and (001)-oriented rhombohedral PZT thin films, taking into account the domain configuration and reorientation mechanism on these two crystallographically different sample configurations.

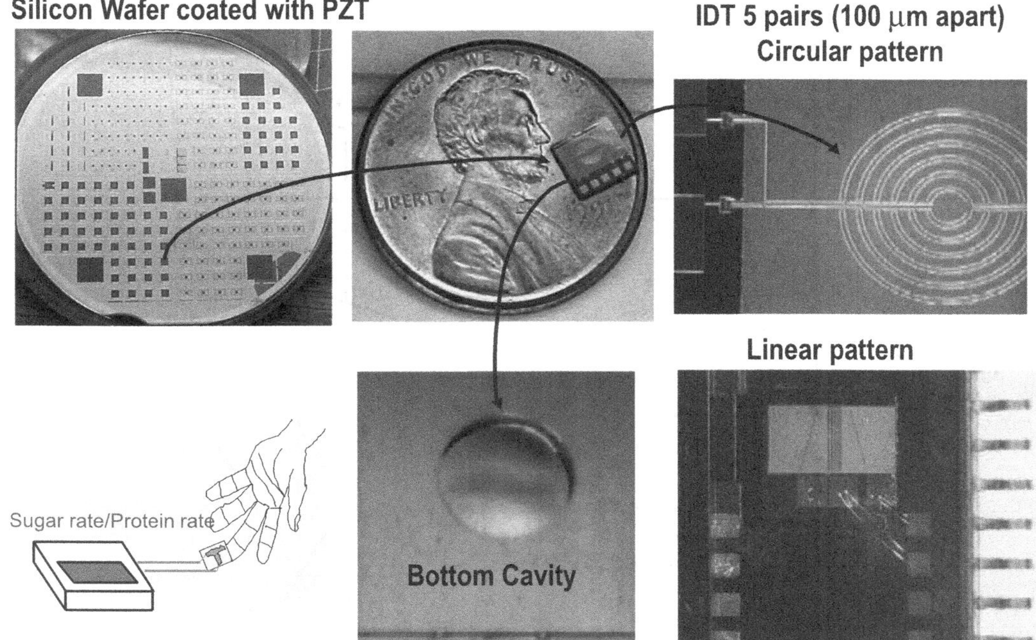

FIGURE 4.57 Fabrication process and IDT electrode patterns of a PZT micropump (actual size: $4.5 \times 4.5 \times 2\ mm^3$). 3-μm-thick PZT film requires a minimum of this area to generate ~1 mW output energy. (From S. Kalpat et al.: *Jpn. J. Appl. Phys.*, 40, 158–162, 2001.)

Solution

Since the spontaneous polarization is directed along the (111) perovskite axes, there are four equivalent polarization states for the (111) film, while there are only two states for the (001) film (top 1, 2, 3, and 4; bottom 5, 6, 7, and 8), as illustrated schematically in Figure 4.58.[83] The coercive field is, therefore, very well defined for the (001) film, leading to a square *P-E* curve (Figure 4.59, top right). Furthermore, since polarization reorientation in this film is more restricted than in the (111) film, one would expect an extended linear portion of the strain curve for field strengths greater than the coercive field, as depicted in Figure 4.59, bottom right. On the contrary, in the case of (111) film, there exist various nonequivalent routes for polarization reorientation from 5 to 1; such as $5 \rightarrow 2 \rightarrow 1$, $5 \rightarrow 6 \rightarrow 1$, $5 \rightarrow 6 \rightarrow 4 \rightarrow 1$, and so on. We expect a very slanted hysteresis, as shown in Figure 4.59, top left. Refer to the experimental results shown in Figure 4.51. Note that the remnant polarization in the (100) film should be $1/\sqrt{3}$ that of (111) film, and that the effective piezoelectric constant, d_{33}, is much larger in the (100) film, due to a significant contribution from the shear strain via d_{15}.

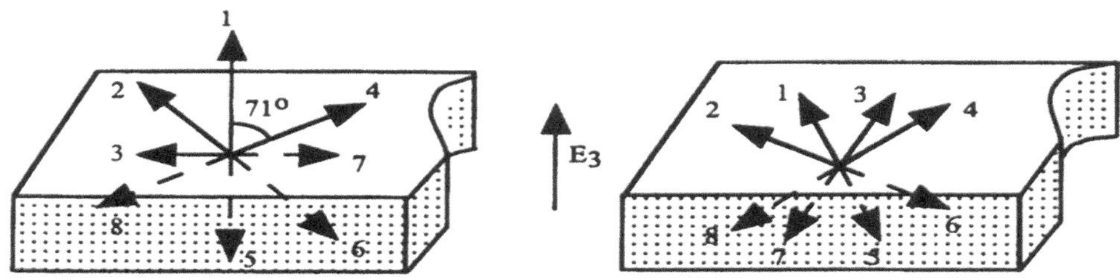

FIGURE 4.58 Equivalent polarization states for epitaxially grown rhombohedral PZT films: (a) (111)-oriented and (b) (001)-oriented. (From X.-H. Du et al.: *Japan. J. Appl. Phys.*, 36(9), 5580, 1997.)

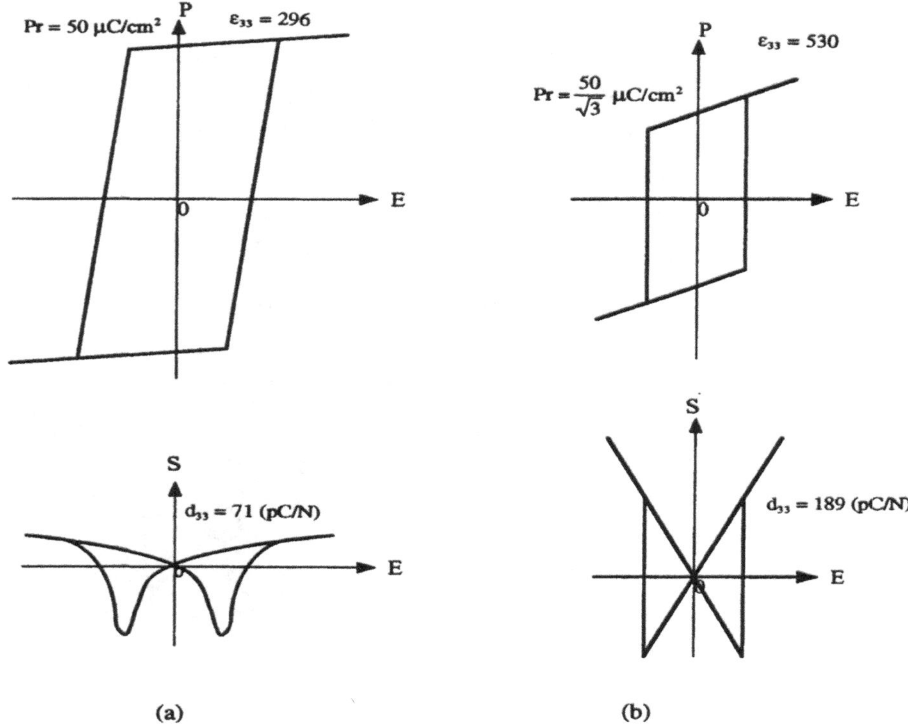

FIGURE 4.59 Anticipated polarization (top) and strain (bottom) hysteresis curves for epitaxially grown rhombohedral PZT films: (a) (111)-oriented and (b) (001)-oriented. (From X.-H. Du et al.: *Japan. J. Appl. Phys.*, 36(9), 5580, 1997.)

CHAPTER ESSENTIALS

1. *Preparation of ceramic powders*:
   a. Mixed-oxide technique
   b. Co-precipitation method
   c. Alkoxide hydrolysis
2. *Critical particle size*: There seems to exist a critical particle size (3D restriction) around 0.01–0.1 μm, below which the ferroelectricity and piezoelectricity disappear, while, in the case of thin films (1D restriction), the critical thickness below which the ferroelectricity disappears seems to be very low, less than a couple of nm.
3. *Displacement magnification mechanisms*:
   a. Multilayer
   b. Unimorph/bimorph
   c. Monolithic hinge lever
   d. Moonie/cymbal
   e. Inchworm
4. *Comparison between multilayer and bimorph actuators*:
   a. Though the multilayer type does not exhibit large displacements (10 μm), large generative force (100 kgf = 1 kN); quick response speed (10 μs); long lifetime (10^{11} cycles); and high electromechanical coupling, k_{33} (70%), are attractive.
   b. The bimorph type exhibits large displacements (300 μm), but relatively low generative force (100 gf = 1 N); slow response speed (1 ms); shorter lifetime (10^8 cycles); and low electromechanical coupling, k_{eff} (10%), are the demerits.

5. *Electrode configurations for multilayer actuators*:
   a. Interdigital type: in which internal stresses tend to develop. The following configurations are specifically designed to relieve these stresses.
   b. Plate-through
   c. Slit-insert
   d. Float-electrode types
6. *Multilayer structure* typically exhibits field-induced strain of 0.1% along its length, l (for example, a 1-cm sample will exhibit a 10-μm displacement), and has a fundamental resonance frequency given by:

$$f_r = \frac{1}{2l\sqrt{\rho s_{33}^D}}$$

where ρ is the density and s_{33}^D is the elastic compliance under open circuit (for example, a 1-cm sample will have a 100-kHz resonance frequency).
7. *Bimorph tip displacement* (δ): Clamped at one end condition (cantilever):

$$\delta = \frac{3d_{31}l^2V}{2t^2} \quad \text{or} \quad \delta = \frac{3d_{31}l^2V}{t^2}$$

depending on the configuration. The fundamental resonance frequency of the bimorph is determined by the total thickness, t, according to:

$$f_r = (0.16)\frac{t}{l^2\sqrt{\rho s_{11}^E}}$$

for both configurations.
8. *Silver paste and ink*: Silver ink is a silver nanoparticle suspended liquid that is used for printing internal electrodes on piezoceramic green sheets. Silver paste contains glass particles, in addition to silver nanoparticles, and is used for making external electrodes on sintered piezoelectric components.
9. *Ag, Cu electrodes*: Compared with Ag/Pd or Pt electrodes, Ag and Cu electrodes are essential from both (1) cost and (2) performance viewpoints. Note that poor electrical conductivity in Pd and Pt exhibit deteriorates performance in resonance piezoelectric devices such as transformers.
10. *Thin/thick films*: Piezoelectric thin film is useful for sensor applications, but it is not sufficient to generate sufficient mechanical energy for actuator applications (~mW). A 30-μm thickness is desired for practical actuator applications, because the maximum energy density handled in the PZT is ~30 W/cm^3.
11. *Four important factors in the thin film structure*:
    a. *Size constraints:* a critical film thickness below which the ferroelectricity would disappear. Further research will be required from this viewpoint.
    b. *Stress from the substrate:* tensile or compressive stress is generated due to thermal expansion mismatch. Curie temperature is modified with the stress.
    c. *Epitaxial growth:* crystal orientation dependence should also be considered, similar to the case in single crystals. A rhombohedral PZT exhibits better performance when the *Ps* direction is arranged 57° canted from the film normal direction, that is, (001) crystallographic orientation.
    d. *Preparation constraint:* Si substrate requires a low sintering temperature of the PZT film. Typically, 800°C for a short period is the maximum for preparing the PZT, which may limit the crystallization of the film.
12. *Aerosol Deposition*: Sputtering and sol-gel spin-coating methods are suitable for making thin films (<1 μm, while aerosol deposition is the best for making thick films (>10 μm) in a short period (~30 min).

CHECK POINT

1. (T/F) Piezoelectric ceramics exhibit larger electric-field induced strain with reducing the grain size less than 1 μm. True or False?
2. (T/F) The piezoelectric ceramic powder has a critical particle size below which the piezoelectricity disappears. True or False?
3. (T/F) There are two bimorph actuators with the same PZT composition: Type I: 40 mm long, 1 mm thick, 6 mm wide, and Type II: 20 mm long, 0.5 mm thick, 3 mm wide. The resonance frequencies for these two bimorphs are the same. True or False?
4. (T/F) There are two bimorph actuators with the same PZT composition: Type I: 40 mm long, 1 mm thick, 6 mm wide, and Type II: 20 mm long, 0.5 mm thick, 3 mm wide. When the same DC voltage is applied, the tip displacement under a cantilever support for these two bimorphs is the same. True or False?
5. What is the fundamental resonance frequency of a bimorph actuator with the size of 40 mm long, 1 mm thick, 6 mm wide? 300, 3, 30, or 300 kHz?
6. What is the maximum displacement induced in a multilayer piezo-actuator with length 10 mm and area 5×5 mm^2? 1, 10, 100, or 1000 μm?
7. There is a piezoelectric multilayer actuator with a length 10 mm and a cross-section area 5×5 mm^2. We coupled a monolithic hinge mechanism to amplify the displacement up to 100 μm. How large a force is expected from the tip of this hinge mechanism? 1, 10, 100 N, 1 kN, or 10 kN?
8. (T/F) The performance of a piezoelectric resonator does not depend significantly on the electrode material, as long as a rare metal is used. True or False?
9. Which metal has the highest conductivity among Ag, Ag/Pd, Pd, and Pt?
10. (T/F) Base metals Cu and Ni are cheaper than rare metals Ag/Pd and Pt, but the electrical conductivity of the base metals is not as low as that of rare metals. True or False?

CHAPTER PROBLEMS

4.1 A two-dimensional array of positive and negative ions forms a square pattern, with alternative charges at each adjacent position. Calculate the Madelung constant, which is larger than the Madelung constant for the 1D chain ($2 \ln 2 = 1.386$).

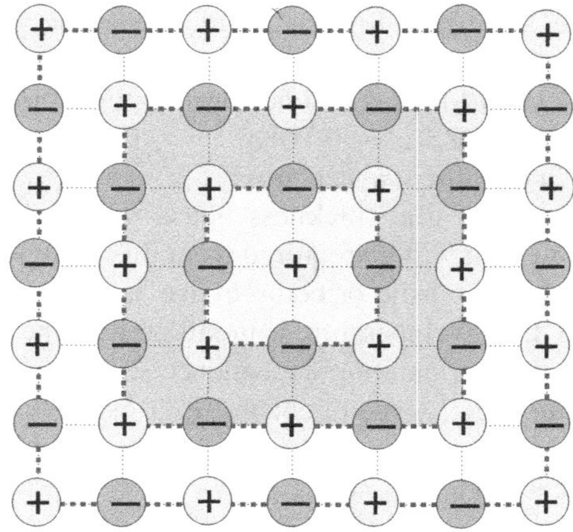

2-D square pattern.

Hint: Select an area with boundaries "through" the outermost ions. Use the fraction on these ions that is shared by this area, that is, one-half of the edge ions and one-fourth of the corner ions.

1. $-\frac{2}{1}+\frac{1}{\sqrt{2}}=-1.2929$

2. $-\frac{2}{1}+\frac{3}{\sqrt{2}}+\frac{2}{2}-\frac{4}{\sqrt{5}}+\frac{1}{2\sqrt{2}}=-0.3140$

3. $\frac{2}{2}-\frac{4}{\sqrt{5}}+\frac{3}{2\sqrt{2}}-\frac{2}{3}+\frac{4}{\sqrt{10}}-\frac{4}{\sqrt{13}}+\frac{1}{3\sqrt{2}}=-0.0036$

   [The next square is small enough.]

   $1)+2)+3)\rightarrow 1.610(5)$

4.2 Making use of a PZT-based ceramic plate with a piezoelectric strain coefficient, $d_{31}=-300$ (pC/N), dimensions of $25\times 5\times 0.5$ mm^3, and a phosphor bronze electrode (a high Q_M material!), design a *unimorph* with a total active length of 25 mm that will execute its maximum tip displacement under an applied voltage of 100 V. Determine the optimum thickness of the phosphor bronze electrode, then calculate the maximum displacement. Here, the density and elastic compliance of the ceramic are $\rho=7.9$ (g/cm^3) and $s_{11}^E=16\times 10^{-12}$ m^2/N, respectively. You will need to search for the necessary data (Young's modulus, etc.) for phosphor bronze. (Refer to Example Problem 4.5.)

4.3 Summarize the problems related to the use of ceramic electrode materials as compare to those associated with conventional metal electrode pastes.
Hint: Due to the limited range of conductivity for the ceramic electrode materials, there will be a cut-off frequency, which presents a fundamental limitation in the response speed.

4.4 The bend curvature ($1/R$) of a monolithic piezoelectric plate under an electric field E is given by:

$$\frac{1}{R}=\frac{\int_{-(t_o/2)}^{(t_o/2)} Y[d_{31}(Z)][E_3(Z)]Zw\,dZ}{\int_{-(t_o/2)}^{(t_o/2)} YZ^2w\,dZ}$$

where t_o is the plate thickness, w is the plate width, Y is the Young's modulus of the ceramic, and d_{31} is the piezoelectric strain coefficient of the piezoceramic. The reference coordinate system Z is placed such that its origin is at the center of the plate. When Y and w are constant, the equation for the bend curvature reduces to:

$$\frac{1}{R}=-\left[\frac{12}{t_o^3}\right]\int_{-(t_o/2)}^{(t_o/2)}[d_{31}(Z)][E_3(Z)]Z\,dZ$$

Given a monomorph plate for which the field distribution $E(Z)$ and the barrier thickness t_b are given by

$$E(Z)=\left[\frac{qN_d}{\varepsilon_o\varepsilon}\right][Z-(t_o/2)+t_b]$$

where $[((t_o/2) - t_b) < Z < (t_o/2)]$ and

$$t_b = \sqrt{\left[\frac{2\varepsilon_o\varepsilon}{qN_d}\right](\phi_o + V)}$$

where ε_o is the vacuum permittivity, ε is the relative permittivity of the monomorph, q is the electron charge, N_d is the donor density, and ϕ_o is the work function. Verify that the curvature is given by

$$\frac{1}{R} = \frac{-\beta d_{31} t_b^2 (3t_o - t_b)}{t_o^3}$$

or

$$\frac{1}{R} = \frac{-\beta d_{31} t_b^2 (2t_o - t_b)}{t_o^3}$$

corresponding to the completely poled case $[d_{31}(Z) = d_{31}(\text{constant})]$ or the gradually poled case $[d_{31}(Z) = d_{31}(Z - (t_o/2) + t_b)/t_b)]$.[49]

4.5 Referring to the Ellingham diagram for Ni and Cu that follows, discuss the higher stability of Cu with PZT when they are co-fired. Find the Ellingham diagram for PZT by yourself.

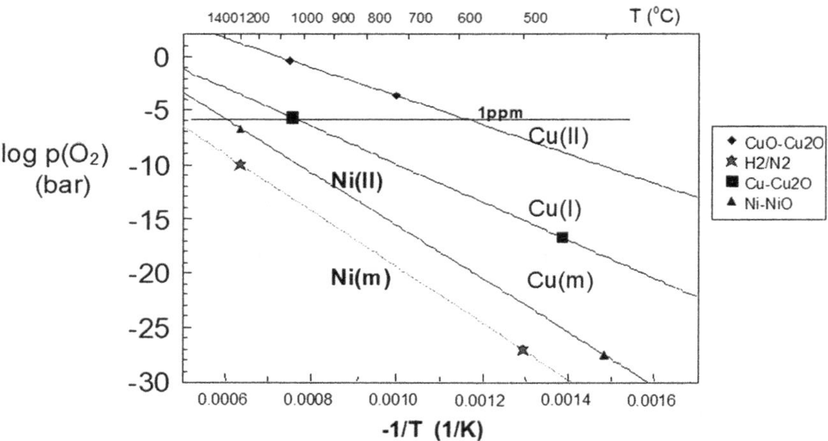

Ellingham diagram for Ni and Cu

REFERENCES

1. T. Kato: *Fine Ceramics Technology, Vol.3 Fabrication Technology of Ceramic Powder and Its Future*, Industry Research Center, Nagoya, Japan, p. 166, 1983.
2. M. Lejeune and J. P. Boilot: *Ferroelectrics*, 54, 191, 1984.
3. S. L. Swartz, T. R. Shrout, W. A. Schulze and L. E. Cross: *J. Amer. Ceram. Soc.*, 67, 311, 1984.
4. Y. Tanada and S. Yamamura: *Abstract 22nd Jpn. Ceram. Soc. Fundamental Div.*, 3B5, 81, 1984.
5. Y. Ozaki: *Electronic Ceramics*, 13(Summer), 26, 1982.
6. Kakegawa, Mohri, Imai, Shirasaki and Takahashi: *Abstract 21st Jpn. Ceram. Soc. Fundamental Div.*, 2C6, 100, 1983.
7. H. Abe: *Recrystallization*, Mater. Sci. Series 2, Kyoritsu Pub., Tokyo, 1969.
8. K. Uchino and T. Takasu: *Inspec.*, 10, 29, 1986.

9. A. Yamaji, Y. Enomoto, E. Kinoshita and T. Tanaka: *Proc. 1st Mtg. Ferroelectric Mater. & Appl.*, Kyoto, p. 269, 1977.
10. K. Nakamura, H. Ando and H. Shimizu: *Jpn. J. Appl. Phys.*, 26(Suppl.26-2), 198, 1987.
11. J. Kuwata, K. Uchino and S. Nomura: *Ferroelectrics* , 37, 579, 1981.
12. J. Kuwata, K. Uchino and S. Nomura: *Jpn. J. Appl. Phys.*, 21(9), 1298, 1982.
13. L. Luo, X. Zhao and H. Luo: *Chapter 7 of Advanced Piezoelectric Materials 2nd Ed.*, Ed. K. Uchino, Woodhead Pub./Elsevier, Cambridge, MA, 2017.
14. A. A. Bokov, X. Long, and Z.-G. Ye: *Phys. Rev. B*, 81(17), 172103/1–4, 2010.
15. http://www.ceracomp.com/
16. G. Messing, et al.: *Critical Reviews in Solid State and Mater. Sci.*, 29(2), 45, 2004.
17. T. Tani and T. Kimura: *Advances in Appl. Ceram.*, 105(1), 55, 2006.
18. Y. Saito, H. Takao, T. Tani, T. Nonoyama, K. Takatori, T. Homma, T. Nagaya and M. Nakamura: *Nature*, 432(4), 84–7, 2004.
19. M. E. Lines: *Phys. Rev.*, B15, 497, 1980.
20. T. Nakamura: *Solid State Physics*, 20(8), 660, 1985.
21. K. Uchino, E. Sadanaga and T. Hirose: *J. Amer. Ceram. Soc.*, 72, 1555, 1989.
22. T. Yamakawa and K. Uchino: *Proc. Int'l. Symp. Appl. Ferroelectrics '90*, p. 610, 1991.
23. K. Saegusa et al.: *Amer. Ceram. Soc., 91th Ann. Mtg.*, 1989.
24. K. Uchino, E. Sadanaga, K. Oonishi and H. Yamamura: *Ceramic Trans.*, 8, Ceramic Dielectrics, 107, 1990.
25. G. A. Samara: *Ferroelectrics* , 2, 277, 1971.
26. M. Takashige and T. Nakamura: *Jpn. J. Appl. Phys.*, 22(Suppl. 22-2), 29, 1983.
27. M. E. Lines: *Phys. Rev. B*, 15, 497, 1980.
28. S. Yamashita: *Jpn. J. Appl. Phys.*, 20(Suppl. 20-4), 93, 1981.
29. K. Uchino, S. Nomura, L. E. Cross, R. E. Newnham and S. J. Jang: *J. Mater. Sci.*, 16, 569, 1981.
30. S. Takahashi, A. Ochi, M. Yonezawa, T. Yano, T. Hamatsuki and I. Fukui: *Jpn. J. Appl. Phys.*, 22(Suppl. 22-2), 157, 1983.
31. A. Furuta and K. Uchino: *J. Amer. Ceram. Soc.*, 76, 1615, 1993.
32. A. Furuta and K. Uchino: *Ferroelectrics*, 160, 277, 1994.
33. S. Takahashi: Fabrication and Application of Piezoelectric Materials, Ed T. Shiosaki. *Chap. 14 Actuators*, CMC Pub., Tokyo, Japan, 1984.
34. S. Takahashi: *Sensor Technology*, 3(12), 31, 1983.
35. H. Aburatani, K. Uchino, A. Furuta and Y. Fuda: *Proc. 9th Int'l Symp. Appl. Ferroelectrics*, p. 750, 1995.
36. J. Ohashi, Y. Fuda and T. Ohno: *Jpn. J. Appl. Phys.*, 32, 2412, 1993.
37. A. Banner and F. Moller: *Proc. 4th. Int'l Conf. New Actuators, AXON Tech. Consult. GmbH*, p. 128, 1995.
38. K. Uchino and H. Aburatani: *Proc. 2nd Int'l Conf. Intelligent Mater.*, p. 1248, 1994.
39. H. Aburatani and K. Uchino: *Amer. Ceram. Soc. Annual Mtg. Proc., SXIX-37-96*, Indianapolis, April, 1996.
40. K. Nagata: *Proc. 49th Solid State Actuator Study Committee, JTTAS*, Japan, 1995.
41. J. Zheng, S. Takahashi, S. Yoshikawa, K. Uchino and J. W. C. de Vries: *J. Amer. Ceram. Soc.*, 79, 3193, 1996.
42. R. W. Basedow and T. D. Cocks: *J. Phys. E: Sci. Inst.*, 13, 840, 1980.
43. T. Kitamura, Y. Kodera, K. Miyahara and H. Tamura: *Jpn. J. Appl. Phys.*, 20(Suppl. 20–4), 97, 1981.
44. K. Nagai and T. Konno Ed.: *Electromechanical Vibrators and Their Applications*, Corona Pub., Tokyo, Japan, 1974.
45. K. Abe, K. Uchino and S. Nomura: *Jpn. J. Appl. Phys.*, 21, L408, 1982.
46. C. Tanuma, Y. Suda, S. Yoshida and K. Yokoyama: *Jpn. J. Appl. Phys.*, 22(Suppl. 22-2), 154, 1983.
47. Okamoto et al.: *Broadcast Technol.*7, 144, 1982.
48. http://www.faceco.com
49. K. Uchino, M. Yoshizaki, K. Kasai, H. Yamamura, N. Sakai and H. Asakura: *Japan. J. Appl. Phys.*, 26, 1046, 1987.
50. K. Uchino, M. Yoshizaki and A. Nagao: *Jpn. J. Appl. Phys.*, 26(Suppl.26-2), 201, 1987.
51. K. Uchino, M. Yoshizaki and A. Nagao: *Ferroelectrics*, 95, 161, 1989.
52. Aura Ceramics, Inc.: USA, Catalogue "Rainbow".
53. K. Nakamura, H. Ando and H. Shimizu: *Appl. Phys. Lett.*, 50, 1413, 1987.
54. T. Yano, E. Sato, I. Fukui and S. Hori: *Proc. Int'l Symp. Soc. Information Display*, p. 180, 1989.
55. http://www.cedrat.com/en/mechatronic-products.html
56. Burleigh Instruments Inc.: East Rochester, N.Y., Catalog.

57. A. Dogan: Ph. D. Thesis, Penn State University, 1994.
58. Y. Sugawara, K. Onitsuka, S. Yoshikawa, Q. C. Xu, R. E. Newnham and K. Uchino: *J. Amer. Ceram. Soc.*, 75, 996, 1992.
59. H. Goto, K. Imanaka and K. Uchino: *Ultrasonic Technol.*, 5, 48, 1992.
60. S. Dong and K. Uchino: *IEEE UFFC-Trans.*, 2002. [in press].
61. *Kitayama*: *Ceramics*, 14, 209, 1979.
62. B. Z. Janos and N. W. Hagood: *Proc. 6th Conf. New Actuators, Actuator '98*, Bremen, Germany, p. 193, 1998.
63. J. W. Hardy, J. E. Lefebre and C. L. Koliopoulos: *J. Opt. Soc. Amer.*, 67, 360, 1977.
64. K. Ohnishi and T. Morohashi: *J. Japan. Ceram. Soc.*, 98, 895, 1990.
65. C. Schuh: *Proc. New Actuators 2004*, Bremen, 2004.
66. W. A. Groen, D. Hennings and M. Thomas: *Proc. 33rd Int'l Smart Actuator Symp.*, State College, PA, April, 2001.
67. S.-H. Park, S. Ural, C.-W. Ahn, S. Nahm and K. Uchino: *Jpn. J. Appl. Phys.*, 45, 2667–2673, 2006.
68. S.-H. Park, Y.-D. Kim, J. Harris, S. Tuncdemir, R. Eitel, A. Baker, C. Randall and K. Uchino: *Proc. 10th Int'l Conf. New Actuators*, Bremen, Germany, June 14–16, 2006, B7.3, pp. 432–435, 2006.
69. S. Ural, S.-H. Park, S. Priya and K. Uchino: *Proc. 10th Int'l Conf. New Actuators*, Bremen, Germany, June 14–16, 2006, P23, pp. 556–558, 2006.
70. K. Abe, K. Uchino and S. Nomura: *Ferroelectrics* , 68, 215, 1986.
71. O. Saburi: *J. Phys. Soc. Jpn.*, 14, 1159, 1959.
72. Matsuo, M. Fujimura, H. Sakaki, K. Nagase and S. Hayakawa: *Amer. Ceram. Soc. Bull.*, 47, 292, 1968.
73. M. Ishida, H. Matsunami, and T. Tanaka: *Appl. Phys. Lett.*, 31, 433, 1977.
74. M. Okuyama et al.: *Ferroelectrics*, 33, 235, 1981.
75. S. K. Dey and R. Zuleeg: *Ferroelectrics*, 108, 37, 1990.
76. S. Kalpat, X. Du, I. R. Abothu, A. Akiba, H. Goto and K. Uchino: *Jpn. J. Appl. Phys.*, 40, 158–162, 2001.
77. X.-H. Du, J. Zheng, U. Belegundu and K. Uchino: *Appl. Phys. Lett.*, 72(19), 2421, 1998.
78. H. D. Chen, K. R. Udayakumar, C. J. Gaskey, L. E. Cross, J. J. Bernstein and L. C. Niles: *J. Amer. Ceram. Soc.*, 79, 2189, 1996.
79. J. Pérez de la Cruz, E. Joanni, P. M. Vilarinho, and A. L. Kholkin: *J. Appl. Phys*, 108, 114106, 2010.
80. P. Gao, et al.: *Nature Comms.*, 15549, doi: 10.1038, 2017.
81. K. Wasa: *Proc. 69th ICAT Int'l Smart Actuator Symp.*, State College, PA, October. 4–5, 2016.
82. J. Akedo: Advanced Piezoelectric Materials. Ed. K. Uchino, *Chapter 14 Aerosol Techniques for Manufacturing Piezoelectric Materials*, p. 493, Woodhead Pub., Cambridge, UK, 2010.
83. K. Wasa: Advanced Piezoelectric Materials 2nd Edition, Ed. K. Uchino, *Chapter 13 Thin Film Technologies for Manufacturing Piezoelectric Materials*, p. 481, Woodhead Pub., Cambridge, UK, 2017.
84. N. Korobova: Advanced Piezoelectric Materials 2nd Edition, Ed. K. Uchino, *Chapter 14 Piezoelectric MEMS Technologies*, p. 533, Woodhead Pub., Cambridge, UK, 2017.

5 Drive/Control Techniques for Piezoelectric Actuators

There are three general methods for actuator drive/control that are most commonly employed: DC drive, pulse drive, and AC drive. These methods are typically used for displacement transducers (Chapter 8), pulse drive motors (Chapter 9), and ultrasonic motors (Chapter 10), respectively. Displacement transducers are usually controlled in a closed-loop mode. Open-loop control can also be employed, but only when strain hysteresis is negligible and temperature fluctuation during operation is very small. Closed-loop control is a feedback method whereby the electric field–induced displacement of a ceramic actuator is monitored, deviation from the desired displacement is detected, and an electric signal proportional to this deviation is fed back to the ceramic actuator through an amplifier to effectively correct the deviation. The pulse drive motor is typically operated in an open-loop mode, but special care must be taken to suppress displacement overshoot and/or vibration ringing that can occur after the pulse voltage is applied. The AC electric power around the resonance/antiresonance frequency range applied to ultrasonic motors is not very large, but significantly amplified displacement can be excited, where the drive frequency must be precisely matched with the resonance or antiresonance frequency of the device for the optimum performance. Heat generation, which is a potentially significant problem with this design, can be effectively minimized with the proper selection of operating parameters in the range around the resonance and antiresonance.

Figure 5.1 shows an example admittance spectrum of a 20-mm-long k_{31} type piezoelectric plate specimen with a resonance frequency around 86 kHz. When the operating frequency is lower than 10 kHz, this is considered an "off-resonance" drive, and its characteristic is purely "capacitive" with the admittance phase lag of 90°. When the operating frequency is 86 or 89 kHz, the characteristic becomes "resistive" with the phase lag of 0°, which corresponds to the resonance or antiresonance frequency. In order to induce the same level of vibration velocity, low voltage and high current or high voltage and low current are required at the resonance or antiresonance drive, respectively. We also introduce an operating frequency at 88 kHz in the inductive region in order to minimize the required input drive power for obtaining the same vibration level. On the contrary, the pulse drive includes a wide range of frequencies (pseudo-DC to multiple higher-order resonance frequencies) that exhibit linear or parabolic total displacement in addition to overshoot and/or vibration ringing.

This chapter describes a *high power piezoelectric characterization system* first, followed by the operating principles of servo displacement transducers, pulse drive motors, and ultrasonic motors, including practical drive circuits.

5.1 CLASSIFICATION OF PIEZOELECTRIC ACTUATORS

Piezoelectric and electrostrictive (or magnetostrictive) actuators are classified into two major categories, based on the type of drive voltage applied to the device and the nature of the strain induced by the voltage, as depicted in Figure 5.2. They are: (1) *rigid displacement devices,* for which the strain is induced unidirectionally, aligned with the applied DC field, and (2) *resonant displacement devices,* for which an alternating strain is excited by an AC field, in particular, at the mechanical resonance frequency (*ultrasonic motors*). The first category can be further divided into two general types: *servo displacement transducers* (*positioners*), which are controlled by a feedback system through a position detection signal, and *pulse drive motors*, which are operated in a simple on/off switching mode. The drive/control techniques presented in this chapter will be discussed in terms of this classification scheme.

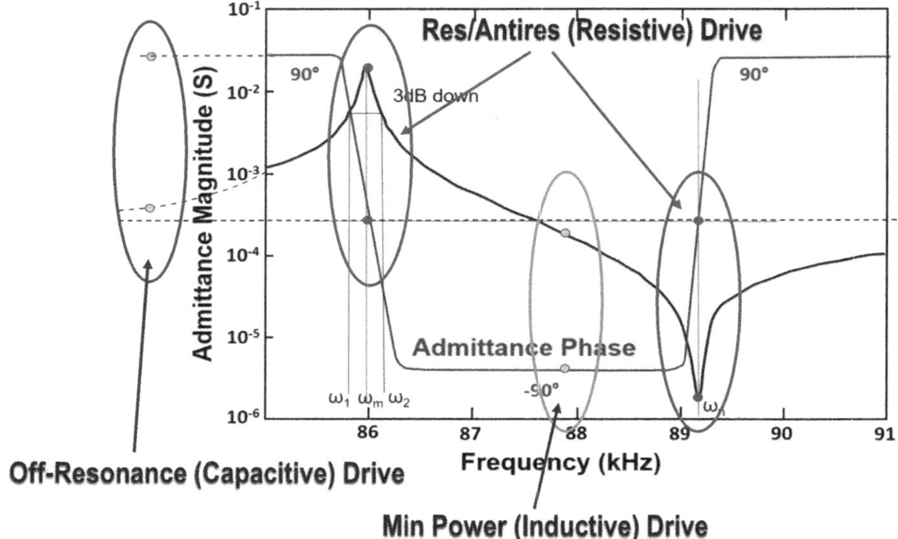

FIGURE 5.1 Application frequency ranges for displacement transducers, pulse drive motors, and ultrasonic motors.

The response of the resonant displacement device is not directly proportional to the applied voltage, but is dependent strongly on the drive frequency. Although the positioning accuracy of this class of devices is not as high as that of rigid displacement devices, ultrasonic motors are able to produce very rapid motion due to their high-frequency operation. Servo displacement transducers, which are controlled by a feedback voltage superimposed on a DC bias, are used as positioners for optical and precision machinery systems. In contrast, a pulse drive motor generates only on/off strains, suitable for the impact elements of inkjet printers or injection valves.

The material requirements for each class of devices are different, and certain compositions will be better suited for particular applications. The servo displacement transducer suffers most from

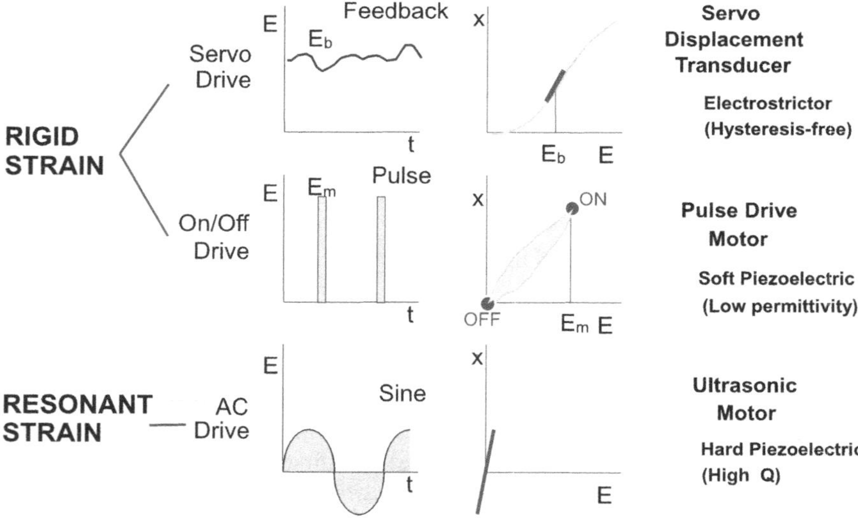

FIGURE 5.2 Classification of piezoelectric/electrostrictive actuators according to the type of drive voltage and the nature of the induced strain.

strain hysteresis and, therefore, a $Pb(Mg_{1/3}Nb_{2/3})O_3$-$PbTiO_3$ electrostrictive material is preferred for this application. It should be noted that even when a feedback system is employed, the presence of a pronounced strain hysteresis general results in a much slower response speed. The pulse drive motor, for which a quick response rather than a small hysteresis is desired, requires a low-permittivity material under a current limitation of a power supply. Soft $Pb(Zr,Ti)O_3$ piezoelectrics are preferred over the high-permittivity PMN for this application. The ultrasonic motor, on the other hand, requires a very hard piezoelectric with a high mechanical quality factor, Q_m, in order to maximize the AC strain and to minimize heat generation. Note that the figure of merit for the resonant strain is characterized by $dEL \cdot Q_m$ (d: piezoelectric strain coefficient, E: applied electric field, L: sample length, Q_m: mechanical quality factor). Although hard PZT materials have smaller d coefficients, they also have significantly larger Q_m values, thus providing the high resonant strains needed for these devices.

5.2 HIGH POWER CHARACTERIZATION SYSTEM

There are various methods for characterizing high power performances and losses in piezoelectric materials: pseudostatic, admittance/impedance spectrum, and transient/burst mode methods. The admittance/impedance spectrum method is further classified into (1) constant voltage, (2) constant current, and (3) constant vibration velocity methods. Piezoelectric resonance can be excited by either electrical or mechanical driving, as shown in Figure 5.3. In a k_{31}-mode piezoelectric plate, for example, as long as the surface is electroded, the sound velocity along the rod direction is v^E originating from s_{11}^E, while in no-electrode specimens, they are v^D and s_{11}^D. In the normal IEEE Standard measuring technique,[1] the specimen should have electrodes and be excited by electrical AC signal, while the resonance on a no-electrode specimen can be excited only by mechanical excitation. A short-circuit condition realizes the resonance, and an open-circuit condition provides the antiresonance mode under the mechanical excitation method. In order to measure the D-constant parameters (s_{11}^D and its extensive elastic loss tan ϕ) directly, we need to use a nonelectrode sample under the mechanical driving method. The improvement of high power piezoelectric characterization systems at Penn State University is introduced in this section in chronological order.

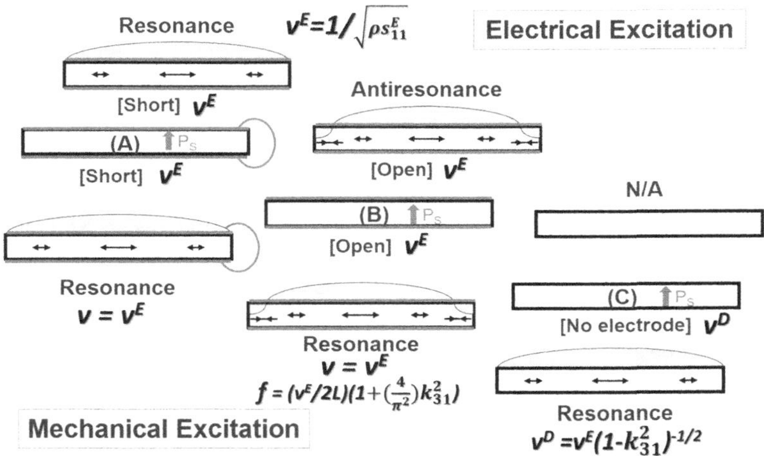

FIGURE 5.3 Resonance and antiresonance mode excitation under electrical or mechanical driving methods (visualization for k_{31} mode).

5.2.1 Loss Measuring Technique I: Pseudostatic Method

The pseudo-DC measuring method is introduced in Section 2.7.2. We can determine intensive dissipation factors, tan δ', tan ϕ', and tan θ', separately from (a) D vs. E (stress free), (b) x vs. X (short-circuit), (c) x vs. E (stress free), and (d) D vs. X (short-circuit) curves (see Figure 2.47). This pseudo-DC method is a traditional high power (i.e., high DC electric field or high DC stress) characterization technique, directly derived from the *piezoelectric constitutive equations*. Using a stress-applying jig, as shown in Figure 2.48, Zheng et al. measured the x vs. X and D vs. X relationships.[2] Figures 2.49 and 2.50 summarize intensive and extensive loss factors, tan δ', tan ϕ', and tan θ'; tan δ, tan ϕ, and tan θ as a function of electric field or compressive stress measured for a soft PZT-based multilayer actuator. Note initially that the piezoelectric losses, tan θ' and tan θ, are not negligibly small as believed by previous researchers, but rather large, comparable to dielectric and elastic losses. Also, it is noteworthy that the extensive dielectric loss, tan δ, increases significantly with an increase of the intensive parameter, that is, the applied electric field, while the extensive elastic loss, tan ϕ, is rather insensitive to the intensive parameter, that is, the applied compressive stress.[3] If similar measurements could be conducted under constrained conditions; that is, D vs. E under a completely clamped state and x vs. X under an open-circuit state, respectively, we could expect smaller hysteresis, that is, extensive losses, tan δ and tan ϕ, theoretically. However, they are rather difficult in practice because of the migrating charge compensation on the specimen surface. The charge leak is usually significant during the x vs. X measurement period ~10 min.

5.2.2 Loss Measuring Technique II: Admittance Spectrum Method

After the theoretical formulas of the mechanical quality factors, Q_A (resonance) and Q_B (antiresonance), were conducted, as discussed in Section 3.5.1, the characterization target has been shifted on how to precisely determine the Q_B, in addition to the Q_A with the admittance spectrum technique. Recall the following relations between Q_A and Q_B through the material's three loss factors in the k_{31} case:

$$Q_{A,31} = \frac{1}{\tan\phi'_{11}},$$

$$\frac{1}{Q_{B,31}} = \frac{1}{Q_{A,31}} + \frac{2}{1+((1/k_{31})-k_{31})^2\Omega_{B,31}^2}(\tan\delta'_{33} + \tan\phi'_{11} - 2\tan\theta'_{31}),$$

where $\tan\delta'_{33}$, $\tan\phi'_{11}$, $\tan\phi'_{31}$ are intensive loss factors for ε_{33}^T, ε_{11}^E, d_{31}, and $\Omega_{B,31}$ is the antiresonance frequency.

5.2.2.1 Constant Voltage Drive: Gen I (1980s)

Because commercially, the then-available Impedance Analyzer did not generate high voltage/current for measuring high power piezoelectric performance, Uchino (as an NF Corporation deputy director) commercialized the Frequency Response Analyzer (500 V, 20 A, 1 MHz maximum) from NF Corporation, Japan. Uchino reported the existence of the critical threshold voltage and maximum vibration velocity for a piezoelectric, above which the piezoelectric drastically increases the heat generation and becomes a ceramic heater. Though this measurement technique is simple yet sophisticated, there have been problems in heat generation in the sample around the resonance range, further in significant distortion of the admittance frequency spectrum when the sample is driven by a constant voltage, due to the nonlinear behavior of elastic compliance at high vibration amplitude.[4] Figure 5.4a exemplifies the problem, where the admittance spectrum is skewed with a jump around the maximum admittance point. Thus, we cannot determine the resonance frequency or the mechanical quality factor precisely from these skewed spectra. Thus, HiPoCS in this version

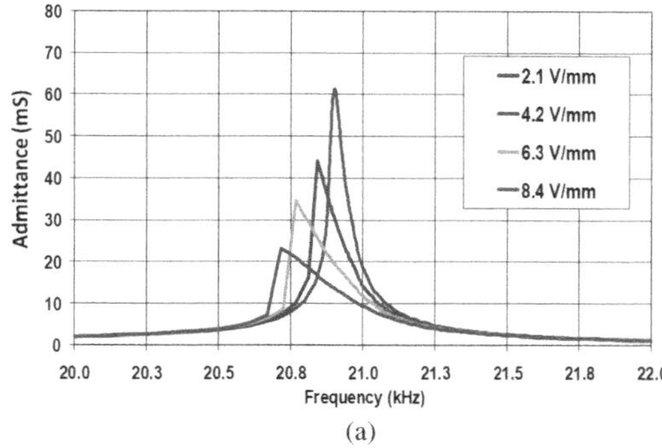

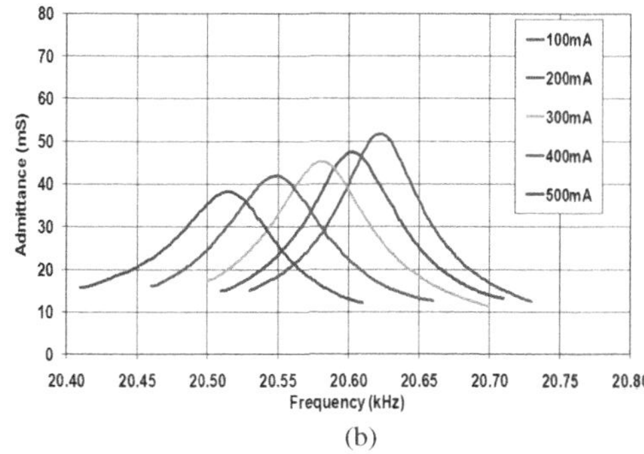

FIGURE 5.4 Experimentally obtained admittance frequency spectra under (a) constant voltage, and (b) constant current condition. Note the skew-distorted spectrum with a jump under a constant voltage condition. (Data taken by Michael R. Thibeault, The Penn State Univ.)

did not have the capability to measure the piezoelectric loss. This NF high voltage/current power supply is still a key device in the present HiPoCS.

5.2.2.2 Constant Current Drive: Gen II (1990s)

In order to escape from the problem with a constant voltage measurement, we proposed a *constant current measurement* technique.[3] Since the vibration amplitude is primarily proportional to the driving current (not the voltage) at the resonance, a constant current condition guarantees almost constant vibration amplitude through the resonance frequency region, escaping spectrum distortion due to elastic nonlinearity. As demonstrated in Figure 5.4b, the spectra exhibit symmetric curves, from which we can determine the resonance frequency and the mechanical quality factor, Q_A, precisely.

Although the traditional constant voltage measurement was improved by using a constant current measurement method, the constant current technique [HiPoCS (II)] is still limited to the vicinity of the resonance. In order to identify a full set of high power electromechanical coupling parameters and the loss factors of a piezoelectric, both resonance and antiresonance vibration performance (in particular Q_A and Q_B) should be precisely measured simultaneously. Basically, Q_A can be determined by the constant current method around the resonance (A-type), while Q_B should be determined by the constant voltage method around the antiresonance (B-type). The mechanical quality factor, Q_m (or the inverse value, loss factor, tan ϕ'), is obtained from $Q_m = \omega_R/2\Delta\omega$, where $2\Delta\omega$ is a full width of the $1/\sqrt{2}$ (i.e., 3 dB down) of the maximum admittance value at ω_R. HiPoCS (II) improved the capability for measuring the mechanical quality factor, Q_m, at the resonance region, but did not provide information on the piezoelectric loss, tan θ', easily because we need to switch the measuring systems between constant-current and constant-voltage types.

We reported the degradation mechanism of the mechanical quality factor, Q_m, with increasing electric field and vibration velocity.[3] Figure 5.5 shows the vibration velocity dependence of the mechanical quality factors, Q_A and Q_B, and corresponding temperature rise for *A* (resonance) and *B* (antiresonance) type resonances of a longitudinally vibrating PZT ceramic transducer through the transverse piezoelectric effect, d_{31} (the sample size is inserted).[5] Q_m is almost constant for a small electric field/vibration velocity, but above a certain vibration level, Q_m degrades drastically, where temperature rise starts to be observed.[5] In order to evaluate the mechanical vibration level, we introduced the vibration velocity at the rectangular plate tip, rather than the vibration displacement, because the displacement is a function of size, while the velocity is not. The *maximum vibration velocity* at the edge of the plate specimen is defined at the velocity where a 20°C temperature rise at the nodal point (the center of the plate) from room temperature occurs. Note that even if we further increase the driving voltage/field, additional energy will convert to merely heat (i.e., PZT becomes a

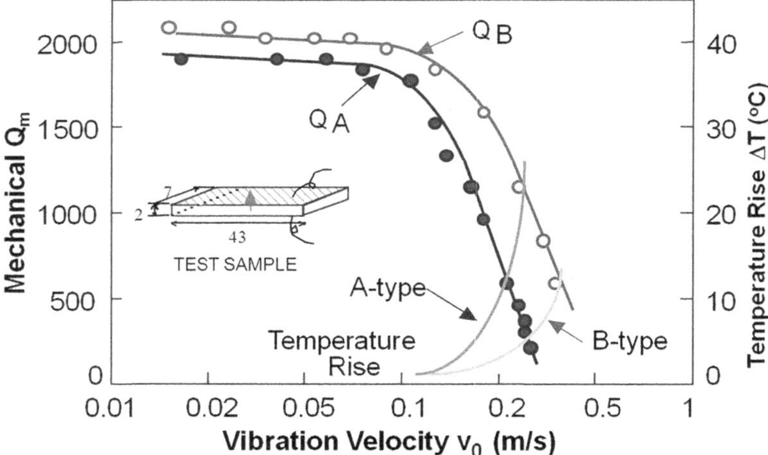

FIGURE 5.5 Vibration velocity dependence of the mechanical quality factors Q_A and Q_B, and corresponding temperature rise for A- and B-type resonances in a longitudinally vibrating PZT plate d_{31} transducer.

ceramic heater!) without increasing the vibration amplitude. Thus, the reader can understand that the maximum vibration velocity, v_0, is a sort of material constant that ranks the high power performance. More precisely, we use the vibration energy density, (1/2) ρv_0^2, for high power evaluation. Note that most of the commercially available hard PZTs exhibit a maximum vibration velocity around 0.3 m/sec, which corresponds to roughly 5 W/cm^3 (i.e., 1 cm^3 PZT can generate maximum 5 W mechanical energy).

When we compare the change trends in Q_A and Q_B, Q_B is higher than Q_A at all vibration levels (this is true in PZTs). The same result was already introduced for a small vibration level in Figure 5.1. Accordingly, the heat generation in the B-type (antiresonance) mode is superior to the A-type (resonance) mode under the same vibration velocity level (in other words, the maximum vibration velocity is higher for Q_B than for Q_A). Figure 5.6b depicts an important notion of heat generation from the piezoelectric material, where the damped and motional resistances, R_d and R_m, in the equivalent electrical circuit of a PZT sample (Figure 5.6a) are separately plotted as a function of vibration velocity.[5] Note that R_m, which we speculate to be mainly related to the extensive mechanical loss (90° domain wall motion), is insensitive to the vibration velocity, while R_d, related to the extensive dielectric loss (180° domain wall motion), increases significantly around a certain critical vibration velocity. Thus, the resonance loss at a small vibration velocity is mainly determined by the extensive mechanical loss, which provides a high mechanical quality factor, Q_m, and with increasing vibration velocity, the extensive dielectric loss contribution significantly increases. This is consistent with the discussion of Figure 2.50 under off-resonance drive. After R_d exceeds R_m, we started to observe heat generation in practice. Though we realized the difference of Q_A and Q_B, we did not include the piezoelectric loss in the equivalent circuit analysis previously. An updated discussion will be conducted in Section 5.5.2.

5.2.2.3 Constant Vibration Velocity Method: Gen III (2000s)

Zhuang and Uchino derived an expansion-series approximation of the mechanical quality factors at both resonance and antiresonance modes, as introduced in Section 3.5.1, and finally obtained a useful formula (modified from the formula by Mezheritsky):[6,7] the mechanical quality factors, Q_A and Q_B, are described in the case of a k_{31}-type rectangular piezoplate as

$$Q_{A,31} = \frac{1}{\tan\phi'_{11}},$$
$$\frac{1}{Q_{B,31}} = \frac{1}{Q_{A,31}} + \frac{2}{1+((1/k_{31})-k_{31})^2\Omega_{B,31}{}^2}(\tan\delta'_{33} + \tan\phi'_{11} - 2\tan\theta'_{31}), \tag{5.1}$$

where $\tan\delta'_{33}$, $\tan\phi'_{11}$, $\tan\theta'_{31}$ are intensive loss factors for ε^T_{33}, s^E_{11}, d_{31}, and $\Omega_{B,31}$ is the antiresonance frequency. The key is the different values of Q_A and Q_B, and if we precisely measure both values, information on the piezoelectric loss, tan θ', can be obtained. If we do not introduce the piezoelectric loss, tan θ', we derive theoretically $Q_A > Q_B$ from Equation 5.1. Since $Q_A < Q_B$ is always the case in PZT experiments, this discrepancy introduced the importance of the piezoelectric loss (as subtraction!) in the research. Thus, we proposed a simple, easy, and user-friendly method to determine the piezoelectric loss factor, tan θ', in k_{31} mode through admittance/impedance spectrum analysis.

A method for determining the piezoelectric loss is summarized for a piezoelectric k_{31} mode plate sample here[6] (refer to a review paper, Reference [7], for other modes):

1. Obtain tan δ' from an impedance analyzer or a capacitance meter at a frequency away from the resonance (usually a lower frequency, 1–10 kHz). Section 5.2.3.2 introduces an additional method.
2. Obtain the following parameters experimentally from an admittance/impedance spectrum around the resonance (A-type) and antiresonance (B-type) range (3 dB bandwidth method) or the burst mode: ω_a, ω_b, Q_A, Q_B, and the normalized frequency $\Omega_B = \omega_b l/2v$.
3. Obtain tan ϕ' from the inverse value of Q_A (quality factor at the resonance) in k_{31} mode.

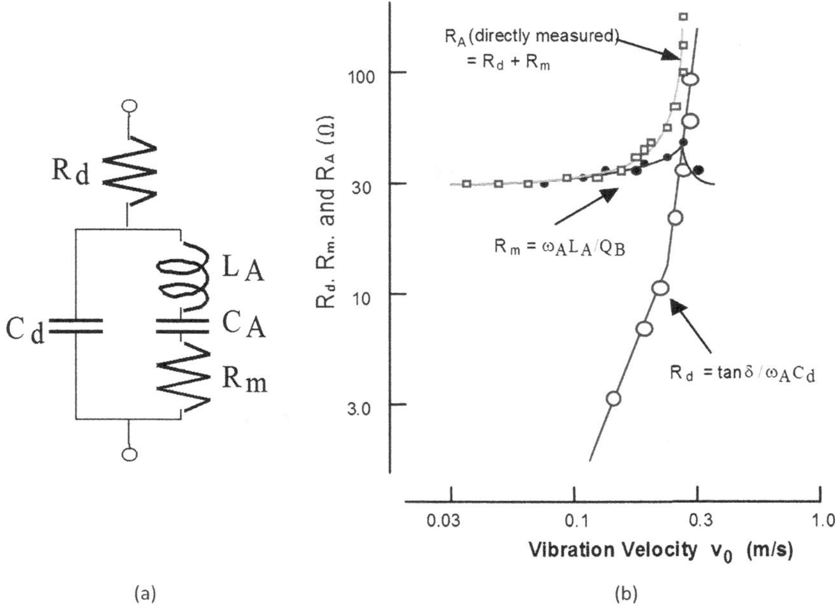

FIGURE 5.6 (a) Equivalent circuit of a piezoelectric sample for the resonance under high power drive. (b) Vibration velocity dependence of the resistances R_d and R_m in the equivalent electric circuit for a longitudinally vibrating PZT ceramic plate through d_{31}. (From S. Hirose et al.: *Proc. Ultrasonics Int'l '95*, Edinburgh, pp. 184–187, 1995.)

4. Calculate the electromechanical coupling factor, k, from ω_a and ω_b with the IEEE Standard equation in k_{31} mode:

$$\frac{k_{31}^2}{1-k_{31}^2} = \frac{\pi}{2}\frac{\omega_b}{\omega_a}\tan\left[\frac{\pi(\omega_b - \omega_a)}{2\omega_a}\right].$$

5. Finally obtain $\tan\theta'$ by the following equation in k_{31} mode:

$$\tan\theta' = \frac{\tan\delta' + \tan\phi'}{2} + \frac{1}{4}\left(\frac{1}{Q_A} - \frac{1}{Q_B}\right)\left[1 + \left(\frac{1}{k_{31}} - k_{31}\right)^2 \Omega_B^2\right].$$

In order to identify both mechanical quality factors, Q_A and Q_B, precisely for adopting the above-mentioned user-friendly methodology, both resonance and antiresonance vibration performance should be measured simultaneously. Basically, Q_A can be determined by the constant current method around the resonance (A-type), while Q_B should be determined by the constant voltage method around the antiresonance (B-type). Thus, we developed HiPoCS Version III, shown in Figure 5.7, which is capable of measuring the impedance/admittance curves by keeping the following various conditions: (1) constant voltage, (2) constant current, (3) constant vibration velocity of a piezoelectric sample, and (4) constant input power. In addition, the system is equipped with an infrared image sensor to monitor the heat generation distributed in the test sample. We demonstrated the usefulness of the new system in a rectangular piezoelectric plate in the whole frequency range, including resonance and antiresonance frequencies. Figure 5.8 shows an interface display of HiPoCS Ver. III, demonstrating a rectangular k_{31} plate measurement under a constant vibration velocity condition.[8] In order to keep the vibration velocity constant (i.e., stored/converted mechanical energy is constant), the current is almost constant and the voltage is minimized around the resonance, while the voltage

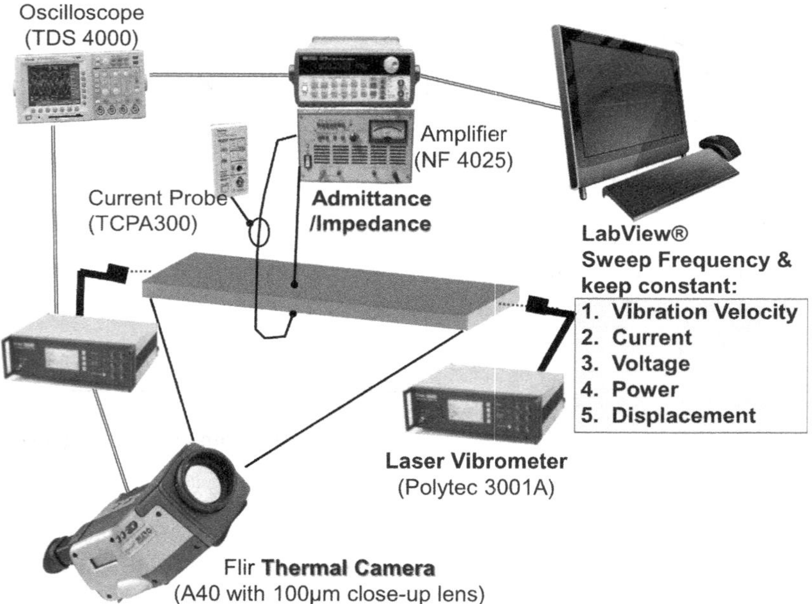

FIGURE 5.7 Setup of HiPoCS Version III.

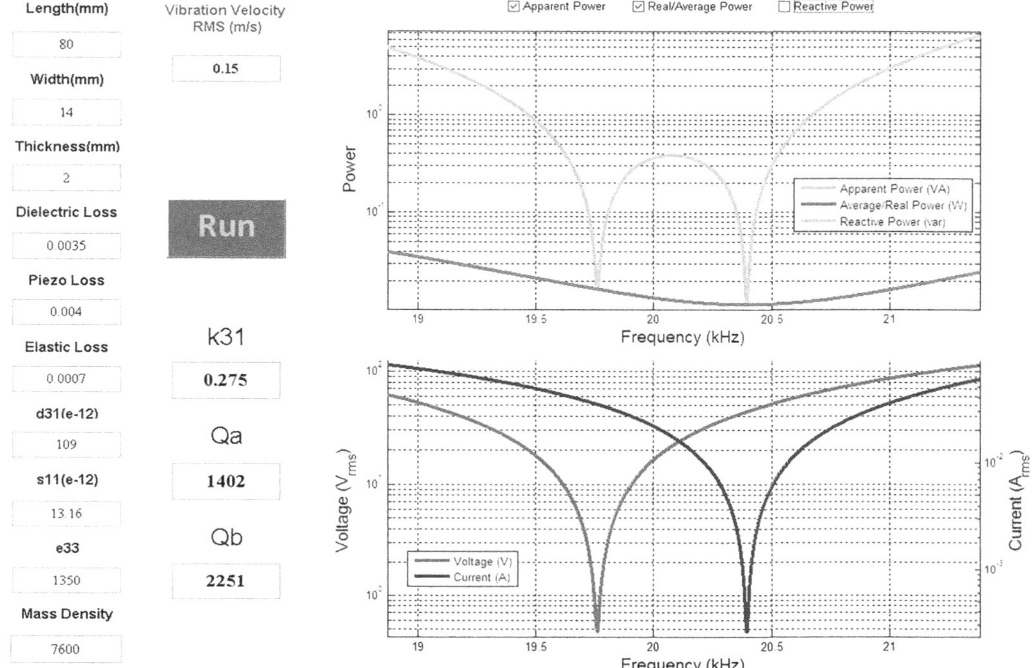

FIGURE 5.8 Voltage and current change with frequency under the constant vibration velocity condition. (From S. O. Ural et al.: *Japan. J. Appl. Phys.*, 48, 056509, 2009.)

is almost constant and the current is minimized around the antiresonance frequency. The *apparent power* (i.e., $V \cdot I$) is shown in the top of Figure 5.8, and more detailed results are shown in Figure 5.9, which clearly indicates that antiresonance operation requires less power than the resonance mode for generating the same vibration velocity, or stored mechanical energy. We can conclude that the PZT transducer should be operated at the antiresonance frequency, rather than the resonance mode, when we consider energy efficiency.

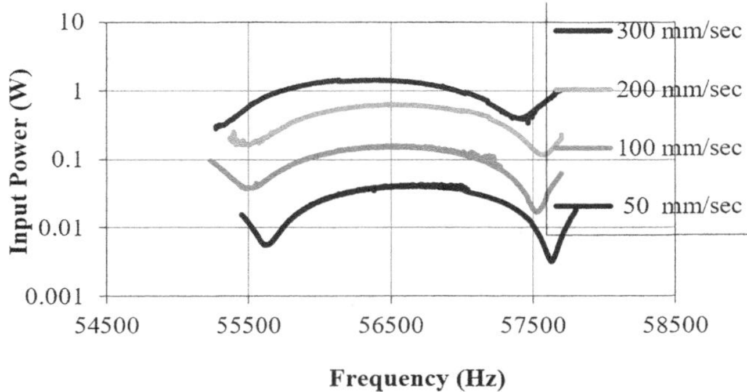

FIGURE 5.9 Frequency spectra of power under constant vibration velocity, conducted across the resonance and antiresonance frequencies.

5.2.2.4 Real Electric Power Method: Gen V (2010s)

Because the conventional admittance spectrum method can provide mechanical quality factors only at two frequency points (i.e., resonance, Q_A, and antiresonance, Q_B), we had frustration over knowing Q_m at any frequency. A unique methodology for characterizing the quality factor in piezoelectric materials has been developed in the ICAT by utilizing *real electrical power* measurements (including the phase lag), for example, $P = V \cdot I \cos \varphi$, rather than the *apparent power*, $V \cdot I$.[9] The relation between mechanical quality factor and real electrical power and mechanical vibration is based on two concepts: (1) at equilibrium, the power input is the power lost and (2) the stored mechanical energy can be predicted using the known vibration mode shape. When the electromechanical coupling factor, k_{31}, is less than 50%, the vibration modes are rather similar sinusoidal curves for both resonance and antiresonance modes, and the mechanical stored energy is evaluated from the plate tip vibration velocity. We can derive the following equation from these concepts, which allows the calculation of the mechanical quality factor at any frequency from the real electrical power (P_d) and tip RMS vibration velocity (V_{RMS}) measurements for a longitudinally vibrating piezoelectric resonator (k_t, k_{33}, k_{31}):

$$Q_{m,l} = 2\pi f \frac{(1/2)\rho V_{RMS}^2}{P_d/(Lwt)}, \tag{5.2}$$

where ρ and L, w, t are the PZT density and the plate size.[9] The change in mechanical quality factor was calculated for a hard PZT (APC 841) ceramic plate (k_{31}) under constant vibration condition of 100 mm/s RMS tip vibration velocity (i.e., stored mechanical energy constant). The experimental key in the HiPoCS Ver. V usage is to determine the phase difference φ precisely to obtain the $\cos(\varphi)$ value. The required power and mechanical quality factor, Q_m, are shown in Figure 5.10. The quality factor obtained at the resonance is within 2% agreement with results from the impedance spectrum method (3 dB-down bandwidth). This technique reveals the behavior of the mechanical quality factor at any frequency between the resonance and antiresonance frequencies, and, very interestingly, the mechanical quality factor reaches a maximum value between the resonance and antiresonance frequency, the point of which may suggest the optimum condition for the transducer operation merely from an efficiency viewpoint, and also for understanding the behavior of piezoelectric material properties under high power excitation. Refer to Section 5.7.2.3 for this application.

5.2.2.5 Determination Methods of the Mechanical Quality Factor

Let us discuss the precise determination method of the mechanical quality factor. The admittance spectrum in k_{31} mode and its *admittance circle* under the relation between the susceptance jB

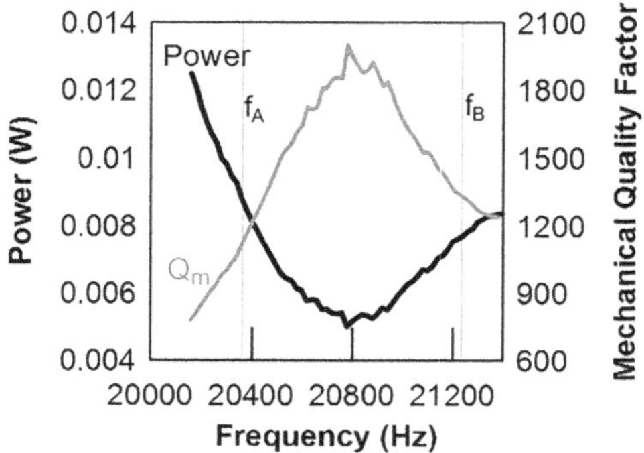

FIGURE 5.10 Mechanical quality factor measured using real electrical power (including the phase lag) for a hard PZT k_{31} plate. (From H. N. Shekhani and K. Uchino: *J. Euro. Ceram. Soc.* 35(2), 541–544, 2014.)

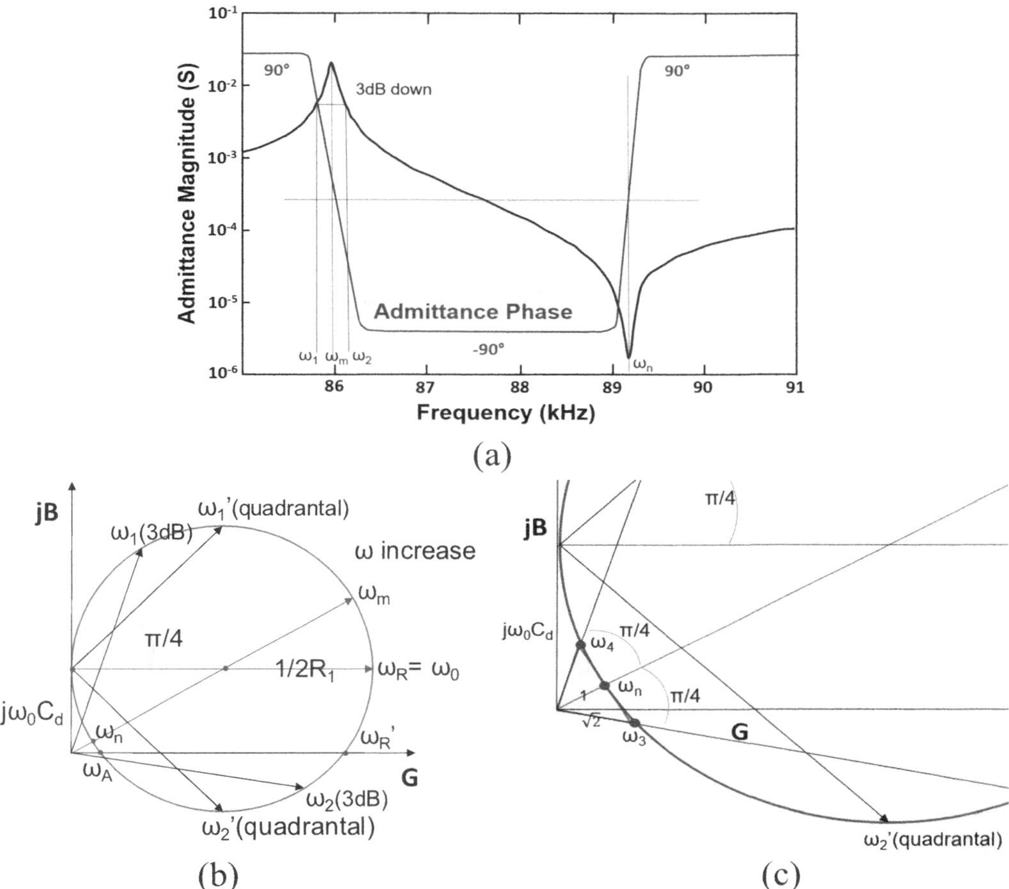

FIGURE 5.11 (a) Admittance spectrum on k_{31} mode. (b) Admittance circle. (c) Admittance circle magnified around the antiresonance frequency.

and conductance G are shown in Figures 5.11a and b, respectively. Figure 5.11c is a magnified vision around the antiresonance frequency. Note that when we plot the *impedance circle* among the reactance jX and resistance R, the antiresonance frequency range is emphasized, and the circle looks like Figure 5.11b for Q_B determination.

a. *Resonance/Antiresonance Frequencies*: The resonance frequency is defined by ω_R, the rightmost on the motional admittance circle, while the antiresonance frequency is defined by ω_A, the intersect between the admittance circle and the susceptance $B = 0$. Popularly used maximum and minimum frequencies of the admittance magnitude, ω_m and ω_n, are not exactly the resonance and antiresonance frequency. Note the following general relation: $\omega_m < \omega_R = \omega_0 < \omega_A < \omega_n$.
b. Q_A *Determination*: See Figure 5.11b.
   - 3 dB down method around ω_m

$$\left(Q_A^{-1}\right) = \frac{\omega_2 - \omega_1}{\omega_m} \tag{5.3}$$

- Quadrantal frequency method around ω_R

$$\left(Q_A^{-1}\right)' = \frac{\omega_2' - \omega_1'}{\omega_R} \tag{5.4}$$

Note ω_1(3 dB) $< \omega_1'$ (quadrantal) $< \omega_2$(3 dB) $< \omega_2'$ (quadrantal). The difference between $\left(Q_m^{-1}\right)$ and $\left(Q_m^{-1}\right)'$ can be estimated as

$$\left(Q_A^{-1}\right) / \left(Q_A^{-1}\right)' = 1 + 1/2M^2 \tag{5.5}$$

where $M = |Y_m|/|Y_d| = 1/R_1\omega_R C_d = Q_A K$ and $K = C_1/C_d$ ($1/K$: capacitance ratio). When we consider $Q_m \approx 1000$, the deviation of Q_m values among these methods is less than 1 ppm (negligibly small).

c. Q_B *Determination*: See Figure 5.11c.
   - 3 dB up method around ω_n

$$\left(Q_B^{-1}\right) = \frac{\omega_4 - \omega_3}{\omega_n} \tag{5.6}$$

   - Instead of the admittance circle, the impedance circle can provide the antiresonance frequency, ω_A, as the rightmost point, and the *quadrantal frequency method* around ω_A can be adopted.

5.2.3 Loss Measuring Technique III: Transient/Burst Drive Method

5.2.3.1 Pulse Drive Method

The *pulse drive method* is a simple method for measuring high-voltage piezoelectric characteristics, developed at ICAT/Penn State in the early 1990s. By applying a step electric field to a piezoelectric sample, the transient vibration displacement corresponding to the desired mode (extensional, bending, etc.) is measured under a short-circuit condition (NB: the output impedance of the voltage supply is small, <50 Ω). Using a rectangular piezoelectric ceramic plate (length: L, width: w, and thickness: b, poled along the thickness; see Figure 5.12), we explain how to determine the electromechanical coupling parameters k_{31}, d_{31}, and Q_m. The density, ρ; permittivity, ε_{33}^X; and size (L, w, b) of the ceramic plate must be known prior to the experiments.

The *resonance period*, *stabilized displacement*, and *damping constant* are obtained under a short-circuit condition experimentally (Figure 5.13), from which the *elastic compliance*, *piezoelectric constant*, *mechanical quality factor*, and *electromechanical coupling factor* can be calculated under a certain applied electric field in the sequence described as follows:

1. From the stabilized displacement, D_s, we obtain the piezoelectric coefficient, d_{31}:

$$D_s = d_{31} E L \tag{5.7}$$

2. From the ringing period, we obtain the elastic compliance, s_{11}^E:

$$T_0 = 2L/v = 2L\left(\rho s_{11}^E\right)^{1/2} \tag{5.8}$$

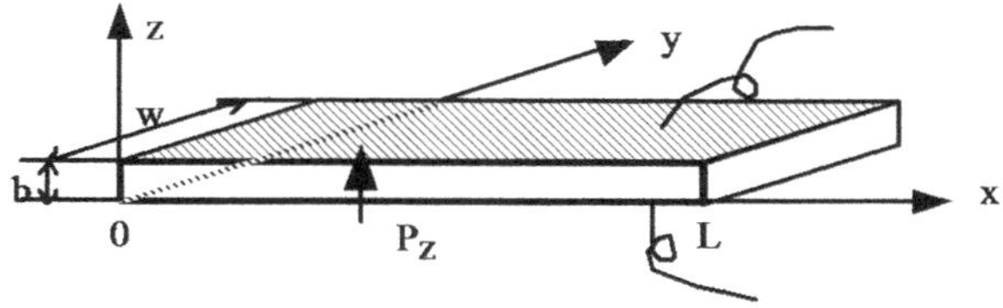

FIGURE 5.12 A rectangular piezoceramic plate ($L \gg w \gg b$) for a longitudinal mode through the transverse piezoelectric effect (d_{31}).

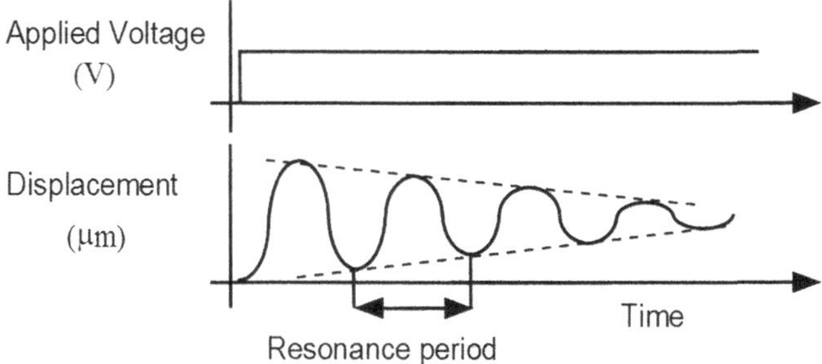

FIGURE 5.13 Pulse drive technique for measuring electromechanical parameters.

3. From the damping constant, τ, which is determined by the time interval to decrease the displacement amplitude by $1/e$, we obtain the mechanical quality factor, Q_A:

$$Q_A = (1/2)\omega_0\tau, \tag{5.9}$$

where the resonance angular frequency $\omega_0 = 2\pi/T_0$.

4. From the piezoelectric coefficient, d_{31}; elastic compliance, s_{11}^E; and permittivity, ε_3, we obtain the electromechanical coupling factor, k_{31}:

$$k_{31} = d_{31}/\left(\varepsilon_0\varepsilon_3 s_{11}^E\right)^{1/2} \tag{5.10}$$

On the other hand, the antiresonance, Q_B, can be obtained as follows: By removing a large electric field suddenly from a piezoelectric sample and keeping the open circuit, the transient vibration displacement corresponds to the antiresonance mode. Although the experimental accuracy of the pulse drive method is not very high, the simple setup is attractive, especially for its low cost. Moreover, unlike the resonance/antiresonance method, this technique requires only one voltage pulse during the measurement and thus does not generate heat (i.e., the temperature effect can be eliminated). Thus, the electric field dependence of piezoelectricity can be measured.

5.2.3.2 Burst Mode Method: Gen IV (2010s)

Because the equilibrium HiPoCS characterization (continuous method, admittance/impedance spectrum measurement) of high power piezoelectrics conventionally used is inevitably associated with a temperature rise around the resonance and antiresonance ranges, the pure vibration amplitude/velocity dependence of Q_M cannot be discussed by removing the temperature influence. In order to eliminate the heat generation effect, the *burst method* can be chosen. In this method, the internally generated heat is close to none, and observations would be a direct function of the ambient temperature on piezoelectric properties. The pulse drive method was introduced by Uchino in the

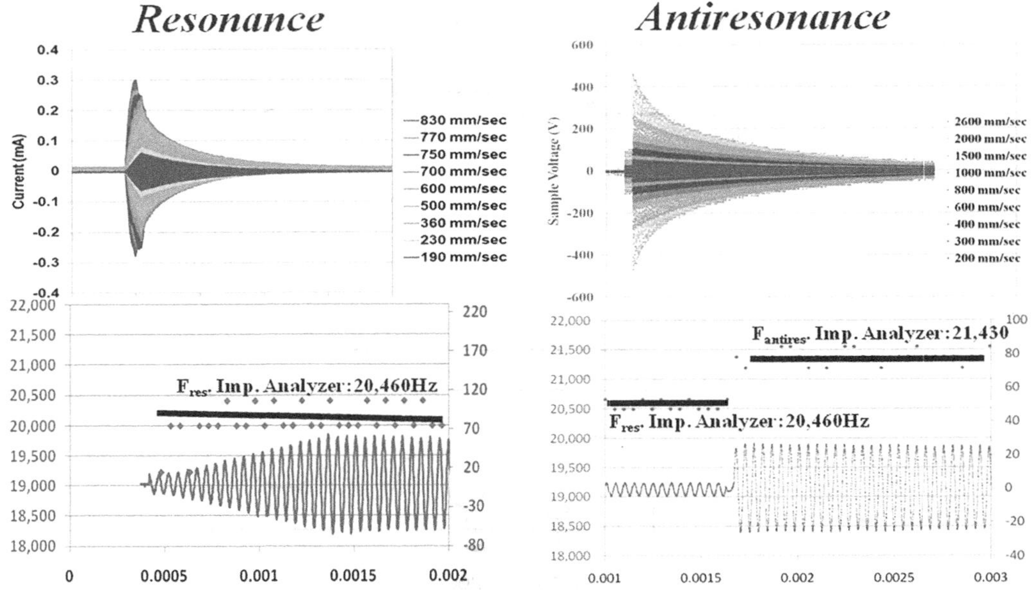

FIGURE 5.14 Vibration ring-down characteristics for the short- (left) and open-circuit condition (right). Note the sudden change of the ringing frequency for the open circuit.

early 1990s (see Section 5.2.3.1), but it was enhanced systematically by Umeda et al. as a *burst mode* for determining the equivalent circuit parameters of a piezoelectric transducer.[10] It was thereafter adapted to measure the properties of piezoelectric ceramic samples.[11,12] Both pulse and burst mode are basically the mechanical excitation technique, because the measurement is conducted after shutting out the electrical connection (i.e., short circuit or open circuit).

We introduce a short-circuit or open-circuit system that functions immediately after the resonance or antiresonance burst drive (only for 1 msec or 10–100 AC cycles) without generating heat (0.2°C maximum in practice), which generates the ring-down of vibration amplitude for the resonance and antiresonance mode, respectively. The burst drive around the resonance/antiresonance frequency can amplify the mechanical displacement significantly from just the pulse drive. Figure 5.14 illustrates the results for the short- and open-circuit (top is the wide time scale of current/voltage, and bottom is the zoomed-in view to show the mechanical resonant vibration and the vibration frequency).[13] In the short circuit, the initial time interval (0.5–1.5 msec) is for exciting the mechanical vibration at the resonance frequency, followed by the short-circuit status. The current and vibration velocity are proportional, and the decay rate provides the mechanical quality factor, Q_A (resonance), while in the open circuit, the voltage and displacement (at the plate end) are proportional, and the decay rate gives the Q_B (antiresonance). Note the resonance to antiresonance frequency jump in Figure 5.14, right, owing to a sudden electrical open circuit, from the initial resonance excitation. Regarding the modification of HiPoCS Ver. III, short-circuit and open-circuit relays should be integrated (Ver. IV). With the new blocking circuit added to our burst method characterization with the HiPoCS, we can also monitor the mechanical quality factor drop in the antiresonance (Gen IV).

5.2.3.2.1 *Force Factor and Voltage Factor*

We derive the *force factor* (A_{31}) and *voltage factor* (B_{31}) comprehensively, first, in terms of material properties and sample geometry from the constitutive equations for the k_{31} piezoelectric sample.[13] The force factor is the relationship between current and vibration in resonance, and it is related to the piezoelectric stress coefficient. The force factor analysis for k_{31} has been presented previously by Takahashi,[11] but its explicit derivation has not, while the voltage factor (B_{31}) is the relationship between open-circuit voltage and displacement in resonance, related to the converse piezoelectric coefficient (h_{31}), which has been newly applied in bulk piezoceramics in the antiresonance condition. The relationships are derived below.

The usual k_{31} mode sample geometry is shown in Figure 5.12. The derivations assume a rectangular plate with $a \ll b \ll L$, fully electroded, and poled along the thickness. Another assumption is that the vibration occurs in the length direction around the fundamental resonance frequency, traditionally corresponding to the x direction ($x = 0$ is at the plate center different from Figure 5.12). In general, the mode shape of a piezoelectric resonator with stress-free boundary conditions, undergoing vibration in 1D, with losses and finite displacement can be described as

$$u(x,t) = u_0 f(x) \sin \omega t, \tag{5.11}$$

where ω is the operating frequency (not just the resonance frequency) and $f(x)$ is a function symmetric about the origin normalized to the displacement at the ends of the piezoelectric resonator, where $f(0) = 0$ (i.e., no center of gravity motion). Then, according to the fundamental theorem of calculus in terms of strain $\partial u/\partial x$, we obtain:

$$\int_{-L/2}^{L/2} \frac{\partial u}{\partial x} dx = u(L/2,t) - u(-L/2,t) = 2u_0 \sin \omega t. \tag{5.12}$$

The constitutive equation describing the electric displacement with piezoelectric e_{31} constant of a piezoelectric k_{31} resonator is

$$D_3(t) = e_{31} \frac{\partial u}{\partial x} + \varepsilon_{33}^{x_1} \varepsilon_0 E_3(t). \tag{5.13}$$

- For the electrical boundary condition of zero electric potential case (i.e., short circuit), the external field is equal to zero. Therefore, $D_3(t) = e_{31}(\partial u/\partial x)$, and $\dot{D}_3 = e_{31}(\partial^2 u/\partial x \partial t)$. Now, the current can be written as

$$i(t) = \int_{A_e} \dot{D}_3 \, dA_e = b \int_{-L/2}^{L/2} \dot{D}_3 dx, \text{ and } i_0 = 2e_{31} u_0 b \omega_A = 2e_{31} b v_0. \tag{5.14}$$

The last transformation uses vibration velocity, v_0, at the plate edge ($x = \pm L/2$), given by $\omega u_0 = v_0$ for sinusoidal time varying displacement. Thus, the *force factor* (A_{31}), defined as the ratio between short-circuit current and edge vibration velocity, can then be written as

$$A_{31} = \frac{i_0}{v_0} = 2e_{31} b = 2b \frac{d_{31}}{s_{11}^E}. \tag{5.15}$$

- For open-circuit conditions, $D_3 = 0$, so the constitutive equation described in Equation 5.13 can be written as

$$E_3(x,t) = -\frac{e_{31}}{\varepsilon_{33}^{x_1} \varepsilon_0} \frac{\partial u}{\partial x}. \tag{5.16}$$

Assuming the variation of strain in thickness is negligible, the electric field across the thickness is uniform. Therefore, $E_3(x,t) = -V(t)/a$. Integrating across the length of the resonator

$$\int_{-L/2}^{L/2} E_3(x,t) dx = -\int_{-L/2}^{L/2} V(t)/a \, dx = \frac{e_{31}}{\varepsilon_{33}^{x_1} \varepsilon_0} \int_{-L/2}^{L/2} \frac{\partial u}{\partial x} dx.$$

This equation can be rewritten using Equation 5.12, assuming free natural vibration at the antiresonance frequency in open circuit conditions,

$$\frac{LV_0}{a} = \frac{e_{31}}{\varepsilon_{33}^{x_1}\varepsilon_0}2u_0, \text{ and } V_0 = \frac{2ae_{31}}{L\varepsilon_{33}^{x_1}\varepsilon_0}u_0.$$

Thus, the *voltage factor* (B_{31}), the ratio between open circuit voltage and displacement, can be written as

$$B_{31} = \frac{V_0}{u_0} = \frac{2a}{L}\frac{e_{31}}{\varepsilon_{33}^{x_1}\varepsilon_0} = \frac{2a}{L}\frac{g_{31}}{s_{11}^D} = \frac{2a}{L}h_{31}. \tag{5.17}$$

By applying the burst mode at resonance (short circuit) or antiresonance (open circuit) conditions, the force factor (A_{31}) or the voltage factor (B_{31}) can directly be obtained from the ratio between short-circuit current and edge vibration velocity, or from the ratio between open-circuit voltage and displacement. The mechanical quality factors (or elastic loss factors) and the real properties of the material can also be measured. For a damped linear system oscillating at its natural frequency, the quality factor can be described using the relative rate of decay of vibration amplitude. In general

$$Q_m = \frac{2\pi f}{2\ln(v_1/v_2)/(t_2 - t_1)} \tag{5.18}$$

This equation is valid for both resonance and antiresonance modes. At resonance, the current is proportional to the vibration velocity; therefore, its decay can be used. Similarly, the voltage decay can be used at antiresonance to determine the quality factor at antiresonance, which is supported by the above formula derivation. Note here that since we measure the mechanical quality factor change as a function of vibration amplitude, the time interval ($t_2 - t_1$) should be small enough in comparison with the measuring time period.

The first resonance frequency in the k_{31} resonator corresponds to s_{11}^E according to the equation

$$s_{11}^E = \frac{1}{(2L f_A)^2 \rho}. \tag{5.19}$$

By utilizing the measurement of the *force factor*, the piezoelectric charge coefficient can be computed as

$$d_{31} = A_{31}s_{11}^E/2b. \tag{5.20}$$

A more common approach to calculate this coefficient, frequently used in electrical resonance spectroscopy, is as follows: the *off-resonance permittivity* and *resonance elastic compliance* can be used to separate the piezoelectric charge coefficient, d_{31}, from the coupling coefficient measured from the relative frequency difference between the resonance and antiresonance $[(k_{31}^2/1 - k_{31}^2) = -(\pi/2)(\omega_A/\omega_R)/\tan((\pi/2)(\omega_A/\omega_R))]$, which can be expressed mathematically as follows:

$$d_{31} = k_{31}\sqrt{s_{11}^E\varepsilon_{33}^X\varepsilon_0}. \tag{5.21}$$

However, this approach assumes that ε_{33}^X does not change under the resonance condition. The calculation of d_{31} using the force factor does not make this assumption, so it is expected to be more accurate. Can we measure the permittivity at the resonance frequency? Or is the permittivity at an off-resonance frequency maintained even at the resonance frequency? This is a long-term frustration many researchers are facing for determining the piezoelectric constant from the admittance spectrum resonance technique.

5.2.3.2.2 Permittivity at the Resonance Frequency Range

By using the piezoelectric stress coefficient (e_{31}) calculated at the resonance (from the force factor) and the converse piezoelectric constant (h_{31}) calculated at the antiresonance, the longitudinally clamped permittivity, ε_{33}^{x}, can be calculated in resonance and antiresonance conditions directly. Then, ε_{33}^{X} can be calculated using k_{31}^2. Permittivity has never been calculated directly under the resonance condition according to the author's knowledge. Though Takahashi et al. reported permittivity under resonance conditions, they assumed that only the motional capacitance changes and the clamped capacitance in the resonance does not change.[11] Shekhani verified that this assumption is not correct.[13] From Equations 5.15 and 5.17, following equation can be derived:

$$\varepsilon_0\varepsilon_{33}^{X}\left(1-k_{31}^2\right)=\varepsilon_0\varepsilon_{33}^{x}=\frac{e_{31}}{h_{31}}=\frac{A_{31}}{B_{31}}\frac{a}{Lb} \tag{5.22}$$

Note that we still assume that the permittivities at the resonance and antiresonance are the same in this small frequency difference (i.e., resonance frequency range).

Recently, Daneshpajooh et al. extended Equation 5.22 by introducing the dielectric, elastic, and piezoelectric loss factors, and obtained the following relations:[14]

$$A_{31}=\frac{i_0}{v_0}=2e_{31}^{*}b=2b\frac{d_{31}^{*}}{s_{11}^{E*}}=2b\frac{d_{31}}{s_{11}^{E}}(1+j(\tan\phi_{11}'-\tan\theta_{31}')) \tag{5.23}$$

$$B_{31}=\frac{V_0}{u_0}=\frac{2a}{L}\frac{e_{31}^{*}}{\varepsilon_{33}^{x1*}\varepsilon_0}=\frac{2a}{L}\frac{d_{31}}{s_{11}^{E}\varepsilon_{33}^{x1}\varepsilon_0}(1+j(\tan\phi_{11}'-\tan\theta_{31}'+\tan\delta_{33}) \tag{5.24}$$

$$\varepsilon_0\varepsilon_{33}^{X}\left(1-k_{31}^2\right)=\varepsilon_0\varepsilon_{33}^{x1}=\mathrm{Re}\left\{\frac{A_{31}}{B_{31}}\right\}\frac{a}{bL} \tag{5.25a}$$

$$\tan\delta=\mathrm{Im}\left\{\frac{A_{31}}{B_{31}}\right\}\times\frac{bL}{a}\varepsilon_0\varepsilon_{33}^{x1} \tag{5.25b}$$

We should observe the phase lags between "current-vibration velocity" and "voltage-vibration displacement" for determining tan δ. The procedure to obtain the permittivity and dielectric loss tangent is summarized below, under the supposition that they are almost the same around the resonance frequency:

1. In the short-circuit measurement, the resonance frequency, ω_R; the ratio (current/vibration velocity) (A_{31}); and the phase lag of the current from the vibration velocity are obtained for a certain vibration velocity point.
2. In the open-circuit measurement, the antiresonance frequency, ω_A; the ratio (voltage/vibration amplitude) (B_{31}); and the phase lag for the voltage from the vibration amplitude are obtained for a certain vibration velocity point.
3. The ratio of A_{31} over B_{31} provides the longitudinally clamped permittivity, ε_{33}^{x1}, at that certain vibration velocity.

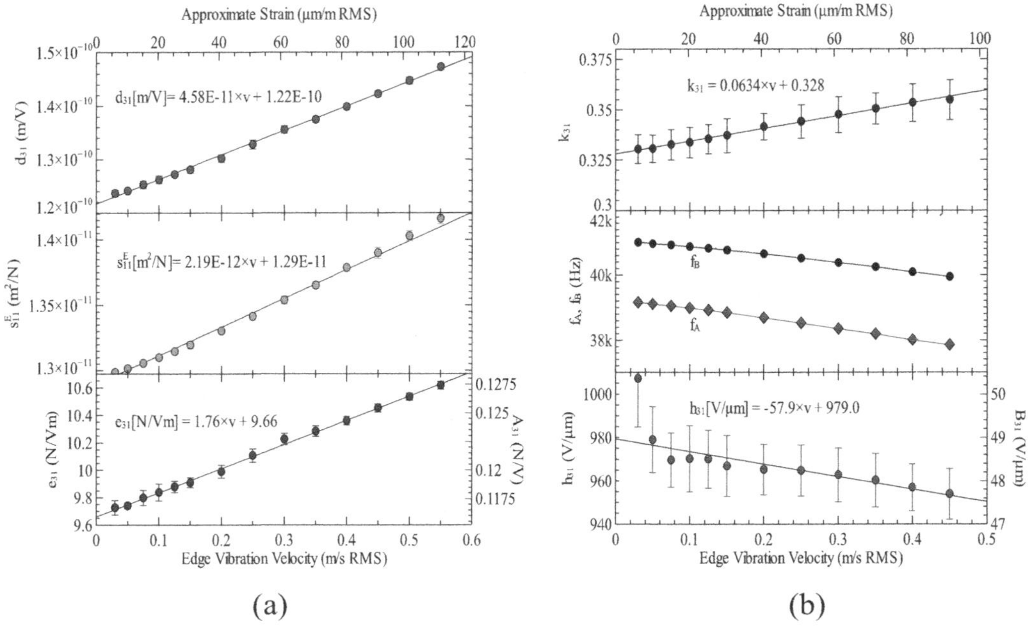

FIGURE 5.15 (a) Resonance characterization (force factor) of soft PZT, PIC 184 k_{31} and (b) antiresonance analysis (voltage factor) and electromechanical coupling factor. (From H. Shekhani et al.: *J. Am. Ceram. Soc.*, 1–10, 2016.)

4. From ω_R and ω_A, we obtain the electromechanical coupling factor, k_{31}, with the equation:

$$\frac{k_{31}^2}{1-k_{31}^2}=\frac{\pi}{2}\frac{\omega_A}{\omega_R}\tan\left[\frac{\pi(\omega_A - \omega_R)}{2\omega_R}\right]$$

5. The stress-free permittivity, ε_{33}^X, can be obtained from $\varepsilon_{33}^X\left(1-k_{31}^2\right)=\varepsilon_{33}^{x1}$.
6. The extensive dielectric loss, tan δ, is obtained from the difference between the phase lag, A_{31} and B_{31}, from Equation 5.25b.

The experimental results are shown in Figures 5.15 and 5.16.[13] Regarding resonance characterization, the current and vibration data were used to calculate the *force factor* and the piezoelectric stress constant. Using the resonance frequency, the compliance was calculated, and hence the piezoelectric charge coefficient could be calculated as well. The resonance characterization for soft PZT, PIC 184 is shown in Figure 5.15a, where the properties (e_{31}, s_{11}^E, d_{31}) change linearly with vibration velocity. By utilizing the displacement and open-circuit voltage at antiresonance, the voltage factor (B_{31}) and converse piezoelectric coefficient (h_{31}) were calculated (see Figure 5.15b). The coupling factor can also be calculated using the relative difference between the resonance and antiresonance frequencies. The coupling factor increases with the vibration velocity; h_{31} decreases with increasing vibration velocity, contrary to the behavior of the other properties. It has a much smaller dependence on vibration velocity than the other properties, namely those determined at resonance.

Using the decay of vibration at resonance and antiresonance, the mechanical quality factors were calculated. Each data point used amplitude data from two vibration measurements; therefore, the scale was readjusted as an average of the vibration velocity. Figure 5.16a shows the results; a log-log plot was used in able to more easily distinguish and compare the trends between the two compositions. PIC 144 (hard PZT) shows stable characteristics of the quality factor, until about 150 mm/s RMS,

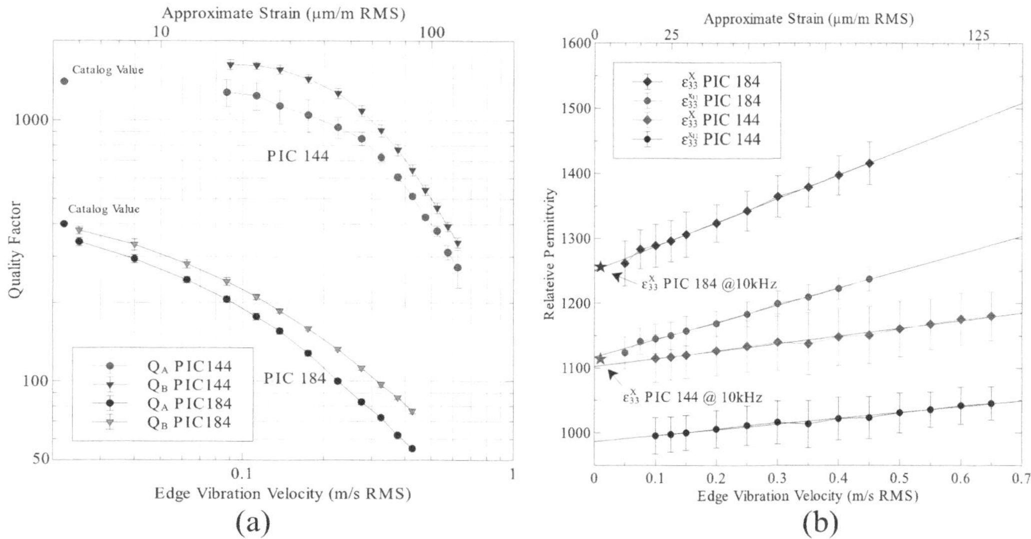

FIGURE 5.16 (a) Change in quality factors and (b) change in dielectric permittivity with vibration velocity for PIC 144 (hard PZT) and PIC 184 (soft PZT). (From H. Shekhani et al.: *J. Am. Ceram. Soc.*, 1–10, 2016.)

after which a sharp degradation in the quality factors occurred. PIC 184 (soft), however, showed an immediate decrease in its quality factors. Q_B was larger than Q_A for both materials.

Traditional methods cannot measure the permittivity under resonance conditions. This is because the large vibration does not allow the dielectric response to exhibit a unique and distinct feature that can be characterized in order to compute the permittivity. This is also true for the dielectric loss. Therefore, researchers have used one of the two approaches to estimate the permittivity in high power conditions. The first approach is to assume the permittivity measured at off-resonance applies to resonance conditions. This approach is problematic because the stress conditions and frequency are different from the resonance condition, and therefore the property is expected to change, similarly to other properties. The other approach is to assume a perturbation of the off-resonance frequency using the variation in the motional capacitance, which is proportional to d_{31}^2/s_{11}^E.

Using the *force factor* and the *voltage factor*, the permittivity under constant strain, $\varepsilon_{33}^{x_1}$, can be calculated directly (Equation 5.22). Using the coupling factor, k_{31}, ε_{33}^X can also be calculated. The permittivity vs. vibration velocity can be seen in Figure 5.16b. The off-resonance permittivity measured for the samples is in good agreement with the low vibration velocity permittivity measured through the burst technique. The off-resonance permittivity is represented as a star symbol. It is interesting that the clamped permittivity changes with increasing vibration velocity, in addition to the free permittivity. The vibration velocity dependence of the clamped permittivity may be originated from the AC stress dependence of the permittivity. The permittivity of PIC 184 is larger than that of PIC 144, and this is expected because PIC 184 is a soft PZT with larger off-resonance permittivity. From the low vibration state to the high one, the permittivity of both compositions increases. However, the increase in PIC 184 is larger, demonstrating that its properties have a larger dependence on vibration conditions. Unlike the expectation, the result shown in this study demonstrates that a majority of the change seen in the free permittivity can actually be attributed to the clamped permittivity change.

Figure 5.17 shows the vibration velocity dependence of three loss factors: (a) piezoelectric constant (d_{31}) and piezoelectric loss ($\tan\theta'_{31}$), (b) elastic compliance $\left(s_{11}^E\right)$ and elastic loss ($\tan\phi'_{11}$), (c), (d) free/clamped dielectric constants $\left(\varepsilon_{33}^E,\ \varepsilon_{33}^{LC}\right)$ and intensive/extensive dielectric loss $\left(\tan\delta'_{33}, \tan\delta_{33}\right)$, respectively.[14] All piezoelectric, elastic, and dielectric losses increase with vibration velocity, as do

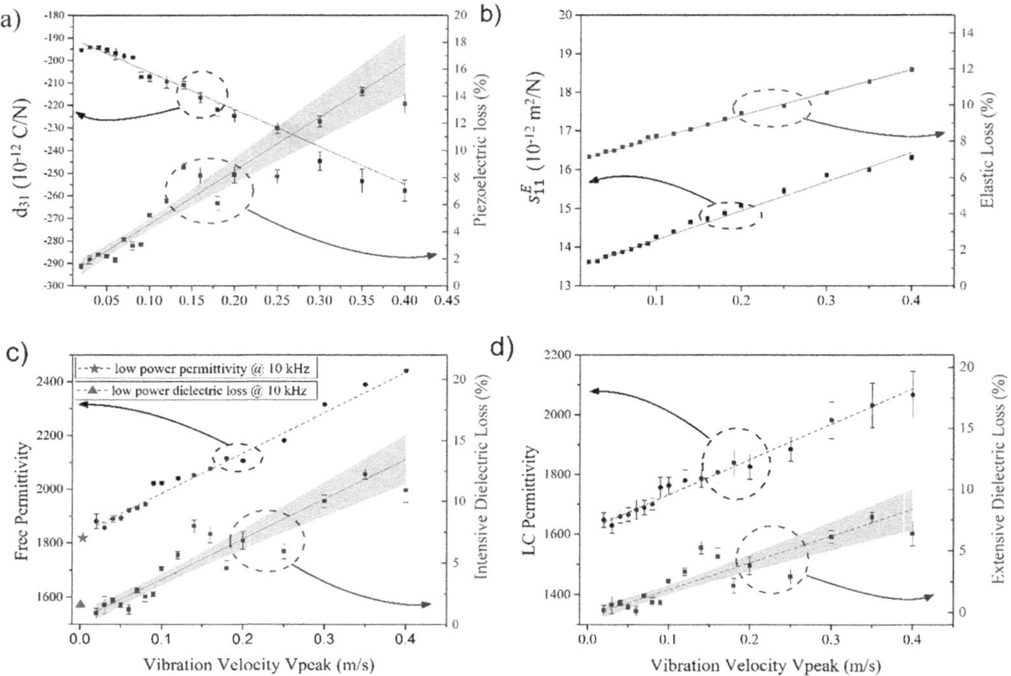

FIGURE 5.17 Vibration level dependence of (a) piezoelectric constant (d_{31}) and piezoelectric loss ($\tan\theta'_{31}$), (b) elastic compliance (s^E_{11}) and elastic loss ($\tan\phi'_{11}$), and (c,d) free/clamped dielectric constants (ε^E_{33}, ε^{LC}_{33}) and intensive/extensive dielectric loss ($\tan\delta'_{33}$, $\tan\delta_{33}$), respectively. (From H. Daneshpajooh et al.: *J. Amer. Ceram. Soc.*, 101, 1940–1948, 2018. doi:10.1111/jace.15338)

$|d_{31}|$, s^E_{11}, ε^X_{33}, and ε^{LC}_{33}. Note again that even the clamped permittivity and extensive dielectric loss change significantly with the vibration velocity!

5.3 FEEDBACK CONTROL

We will consider in this section how to design a microdisplacement control unit functioning as a *servo (mechanical feedback) system*. An example of use in a precision lathe machine, which can cut a rod with an accuracy of ±10 nm, still seems to be the most accurate cutting machine.[15] The positioning system is composed of (1) a multilayer electrostrictive actuator, (2) a potentiometer position sensor (magnetoresistive type), and (3) a feedback circuit, as shown in Figure 5.18. The schematic feedback system for this compact positioner is pictured in Figure 5.19. We begin by reviewing the fundamental principles of the *Laplace transform* and then determine the *transfer function* for this actuator system. We will conclude our examination of the system by evaluating its stability in terms of steady-state deviations induced by external disturbances. The reader is referred to the texts of Amemiya[16] and Davis[17] for a more advanced treatment of feedback systems.

5.3.1 The Laplace Transform

Let us first review an important mathematical tool, the *Laplace transform*. The Laplace transform is generally employed for treating the *transient response* to a pulse input. The *Fourier transform* is preferred for cases where a continuous sinusoidal input is applied, such as for resonance-type actuators, which is used in Section 5.5.

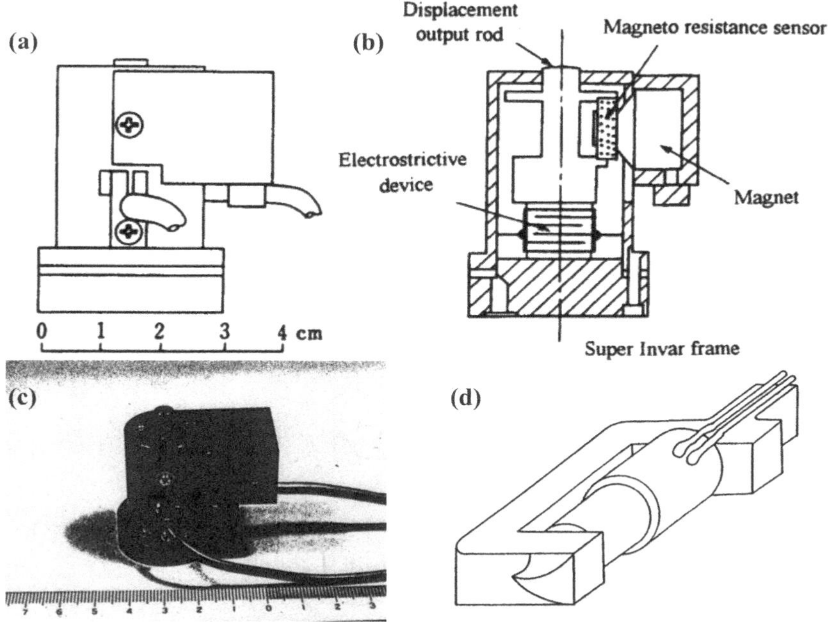

FIGURE 5.18 A smart ceramic actuator used as a micropositioner: (a) an external view, (b) a cross-sectional view, (c) photo, and (d) the actuator installed in a lathe machine cutting edge.

We consider a function, $u(t)$, that is defined for $t \geq 0$ ($u(t) = 0$ for $t < 0$), and satisfies $|u(t)| \leq k\, e^{\delta t}$ for all δ not less than a certain positive real number, δ_o. When these conditions are satisfied, $e^{-st}\, u(t)$ is absolutely integrable for $\mathrm{Re}(s) \geq \delta_o$. We define the Laplace transform:

$$U(s) = L[u(t)] = \int_0^{\infty} e^{-st} u(t)\, dt \tag{5.26}$$

The inverse Laplace transform is represented as $L^{-1}[U(s)]$. Applications of the useful theorems for the Laplace transform that are listed below reduce the work of solving certain differential equations by reducing them to simpler algebraic forms. The procedure is applied as follows:

1. Transform the differential equation to the s-domain by means of the appropriate Laplace transform.

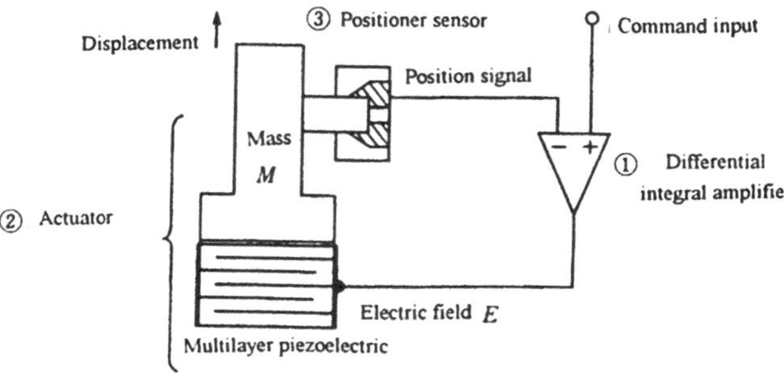

FIGURE 5.19 Schematic representation of the feedback system for the smart micropositioner.

2. Manipulate the transformed algebraic equation and solve for the output variable.
3. Obtain the inverse Laplace transform from Table 5.1.

Useful Theorems for the Laplace Transform:

a. *Linearity:*

$$L[au_1(t) + bu_2(t)] = aU_1(s) + bU_2(s) \tag{5.27}$$

$$L^{-1}[aU_1(s) + bU_2(s)] = au_1(t) + bu_2(t) \tag{5.28}$$

b. *Differentiation with respect to t:*

$$L\left[\frac{du(t)}{dt}\right] = sU(s) - u(0) \tag{5.29}$$

$$L\left[\frac{d^n u(t)}{dt^n}\right] = s^n U(s) - \sum s^{n-k} u^{k-1}(0) \tag{5.30}$$

c. *Integration:*

$$L\left[\int u(t)\,dt\right] = U(s)/s + (1/s)\left[\int u(t)\,dt\right]_{t=0} \tag{5.31}$$

d. *Scaling formula:*

$$L[u(t/a)] = aU(sa) \quad (a > 0) \tag{5.32}$$

e. *Shift formula with respect to t:* $u(t - k) = 0$ *for* $t < k$ (k: positive real number). The $u(t)$ curve shifts by k along the positive t-axis.

$$L[u(t - k)] = e^{-ks}U(s) \tag{5.33}$$

f. *Differentiation with respect to an independent parameter:*

$$L\left[\frac{\partial u(t,x)}{\partial x}\right] = \frac{\partial U(s,x)}{\partial x} \tag{5.34}$$

g. *Initial and final values*

$$\lim_{t\to 0}[u(t)] = \lim_{|s|\to\infty}[sU(s)] \tag{5.35}$$

$$\lim_{t\to\infty}[u(t)] = \lim_{|s|\to 0}[sU(s)] \tag{5.36}$$

EXAMPLE PROBLEM 5.1

Compute the Laplace transform of the Heaviside function (*step function*):

$$1(t) = 0 \quad \text{when } t < 0 \text{ and } 1(t) = 1 \text{ when } t \geq 0.$$

Solution

$$L[1(t)] = \int_0^\infty e^{-st}1(t)\,dt = \int_0^\infty e^{-st}dt = -(1/s)e^{-st}\Big|_0^\infty = 1/s \tag{P5.1.1}$$

TABLE 5.1
Some Common Forms of the Laplace Transform

	$H(t)$	$G(s)$
1	$1(t)$: Heaviside Step function $1(t) = 1,\ t > 0;\ 1(t) = 0,\ t < 0$	$1/s$
2	$\delta(t)$: Dirac Impulse function $\delta(t) = \infty,\ t = 0;\ \delta(t) = 0,\ t \neq 0$	1
3	$t^n/n!$ (*n*: positive integer)	$1/s^{n+1}$
4	e^{-at} (*a*: *complex*)	$1/(s+a)$
5	cos(at)	$s/(s^2 + a^2)$
6	sin(at)	$a/(s^2 + a^2)$
7	cosh(at)	$s/(s^2 - a^2)$
8	sinh(at)	$a/(s^2 - a^2)$
9	$e^{-bt}\cos(at)\quad a^2 > 0$	$\dfrac{s+b}{(s+b)^2 + a^2}$
10	$e^{-bt}\sin(at)\quad a^2 > 0$	$\dfrac{a}{(s+b)^2 + a^2}$
11		$\dfrac{1}{s}(e^{-as} - e^{-bs})$
12	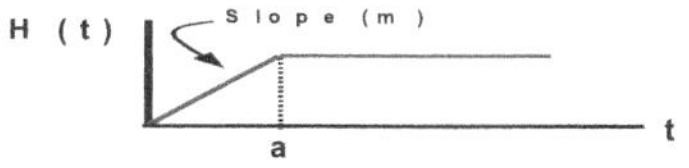	$\dfrac{m}{s^2}(1 - e^{-as})$
13	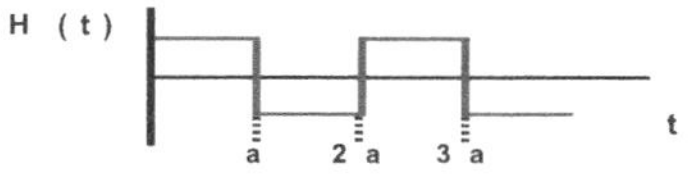	$\dfrac{1}{s}\tanh\left(\dfrac{as}{2}\right)$
14	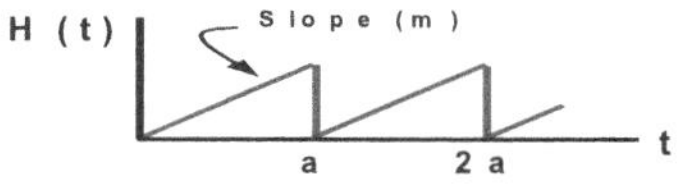	$\dfrac{m}{s^2} - \dfrac{ma}{2s}\left[\coth\left(\dfrac{as}{2}\right) - 1\right]$

Source: C. V. Newcomb and I. Flinn: *Electronics Lett.* 18, 442, 1982.

EXAMPLE PROBLEM 5.2

Using the result from the previous problem, $L[1(t)] = 1/s$, obtain the Laplace transform for a pulse function defined by the following:

$$P(t) = 0 \quad \text{when } t < a \text{ and } t > b \text{ (here, } 0 < a < b)$$

$$P(t) = 1 \quad \text{when } a < t < b$$

SOLUTION

$P(t)$ is obtained by superimposing the two step functions, $1(t - a)$ and $-1(t - b)$. Using the shift formula, we obtain the Laplace transform of $P(t)$:

$$L[P(t)] = e^{-as}(1/s) - e^{-bs}(1/s) = (1/s)(e^{-as} - e^{-bs}) \tag{P5.2.1}$$

5.3.2 The Transfer Function

5.3.2.1 Transfer Function of a Piezoelectric Actuator

First, let us consider a *transfer function* of a piezoelectric actuator. The displacement, $u(t)$, is induced in a piezoelectric actuator by an applied electric field, E, in terms of the simple model depicted in

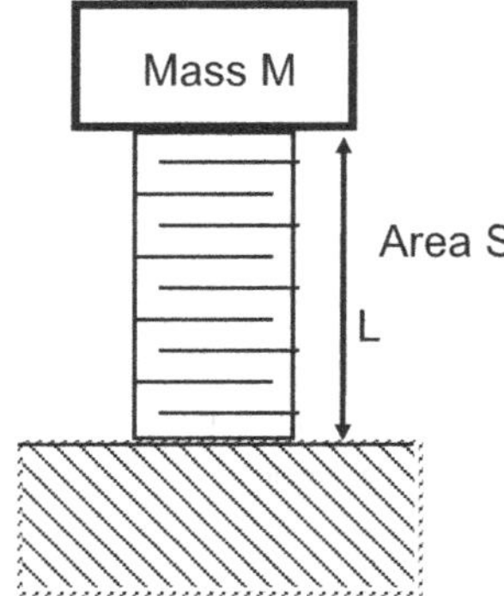

FIGURE 5.20 A simple actuator and mass model.

Figure 5.20. Let us assume a mass, M, is applied to the piezoelectric actuator spring, which has an elastic stiffness, c; piezoelectric strain coefficient, d; a length, L; and a cross-sectional area, A. The oscillation of the mass is sustained by the piezoelectric force (product of strain $d \cdot E$ and stiffness c), which is generated by the applied electric field and described by (Acd)E. The dynamic equation is written as follows:

$$M\frac{d^2u}{dt^2}+\zeta\frac{du}{dt}+\left(\frac{ac}{L}\right)u=\left(Acd\right)E \tag{5.37}$$

where ζ represents the *damping effect*, which originates from the material's loss.

Using the Laplace transformation, Equation 5.37 becomes:

$$Ms^2U+\zeta sU+\left(\frac{Ac}{L}\right)U=Acd\tilde{E} \tag{5.38}$$

Here, U and $\tilde{E}$ are the Laplace transforms of u and E, respectively. We may now define the following:

$$U(s)=G_2(s)\tilde{E}(s) \tag{5.39}$$

where the function $G_2(s)$ is given by:

$$G_2(s)=Acd/(Ms^2+\zeta s+Ac/L) \tag{5.40}$$

This function, which essentially relates the input, $\tilde{E}(s)$, to the output, $U(s)$, is called the *transfer function*. When the denominator of the transfer function includes the s^2 term, the transfer function is called the *second-order transfer function*, popularly analyzed in many applications. Try Example Problem 5.3 to become familiar with the response from the second-order transfer function.

EXAMPLE PROBLEM 5.3

The transfer function of a piezoelectric actuator is given by:

$$U(s)=G(s)\tilde{E}(s), \tag{P5.3.1}$$

$$G(s) = Acd/(Ms^2 + \zeta s + Ac/L) \tag{P5.3.2}$$

Calculate the displacement response to an impulse voltage, $\delta(t)$.

SOLUTION

In the case of an impulse input, $\tilde{E}(s) = 1$, *the output displacement, U(s), is provided directly by the transfer function, G(s):*

$$U(s) = (Acd/M\omega)[\omega/((s + \zeta/2M)^2 + \omega^2)] \tag{P5.3.3}$$

where:

$$\omega^2 = (Ac/ML) - (\zeta^2/4M^2) \tag{P5.3.4}$$

If the relationship can be expressed in the form of one of the common Laplace transforms found in Table 5.1, we can obtain the solution easily without using complex mathematics. Fortunately, we are able to take this simpler approach here (see #10 in Table 5.1), and the displacement response is found to be of the form:

$$u(t) = (Acd/M\omega)\exp[-(\zeta/2M)t]\sin(\omega t) \tag{P.5.3.5}$$

As illustrated in Figure 5.21 (arbitrary scale), we see that the sinusoidal vibration is damped according to the magnitude of ζ and that the resonance frequency is also affected by the damping factor, ζ, as expected from Equation P5.3.4.

5.3.2.2 Transfer Functions of a Position Sensor/Differential Amplifier

Next, referring to Figure 5.19, we consider the response of the feedback system, the block diagram of which can be translated into Figure 5.22a. The piezoelectric actuator is represented by the *second-order transfer function*, as we described in the previous section. We consider that the magnetoresistive position sensor used in this system generates a signal voltage, e_o, proportional to the input displacement u (i.e., *proportional amplifier*). The transfer function, G_3, for the sensor is simply:

$$G_3(s) = K_2 \tag{5.41}$$

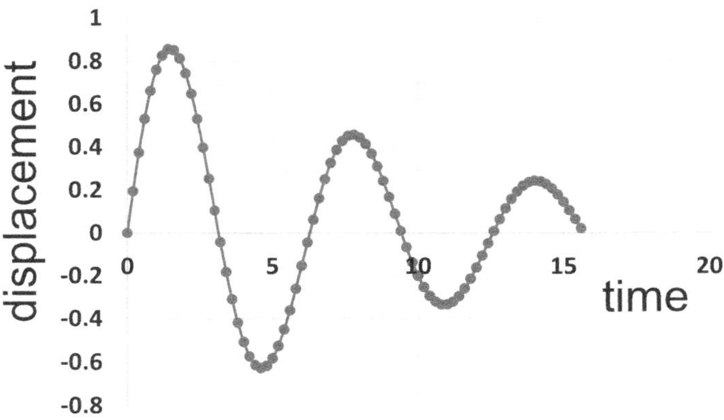

FIGURE 5.21 Response in second-order transfer function.

Finally, we consider the differential amplifier of the system. The command voltage, e_i, is at the input (say, move 0.1 μm!), and the differential voltage (e_i − e_o) from the position signal voltage, e_0, is generated (i.e., negative feedback). When the command voltage, e_i, is larger than the position signal, e_o, the voltage difference ($e_i - e_o$) becomes positive and after being amplified is applied to the piezoelectric actuator so as to increase the displacement. If $e_i < e_o$, the applied electric field is decreased. We assume the use of an *integral amplifier* in this example system, which is described by:

$$E = K_1 \int (e_i - e_o)dt \tag{5.42}$$

Thus,

$$\tilde{E} = G_1(s)\,(\tilde{e}_i - \tilde{e}_o) \tag{5.43}$$

$$G_1\,(s) = K_1/s \tag{5.44}$$

There are three types of amplifiers commonly used in a system such as this one: *proportional*, *integral*, and *derivative* (*PID controller*) with transfer functions K_1, K_1/s, and K_1s, respectively. Any one or a combination of these may be used in a given system. In general, system stability will increase with increasing integral components, but only at the expense of the response speed. On the other hand, the responsivity of the system is enhanced significantly with increasing derivative components, but, in this case, there is often some loss in stability, and overshoot phenomena.

5.3.2.3 Block Diagram

Let us describe this servo system in terms of the block diagram pictured in Figure 5.22a using the three transfer functions just obtained above. It should be noted at this point that an amplifier of rather low output impedance (less than 1 Ω) is typically employed a system of this kind. This is necessary because the actuator possesses a relatively large electrostatic capacitance, which can lead to a significant delay in system response if the amplifier output impedance is too high. (Note that the cut-off frequency is determined by the time constant, *RC*.)

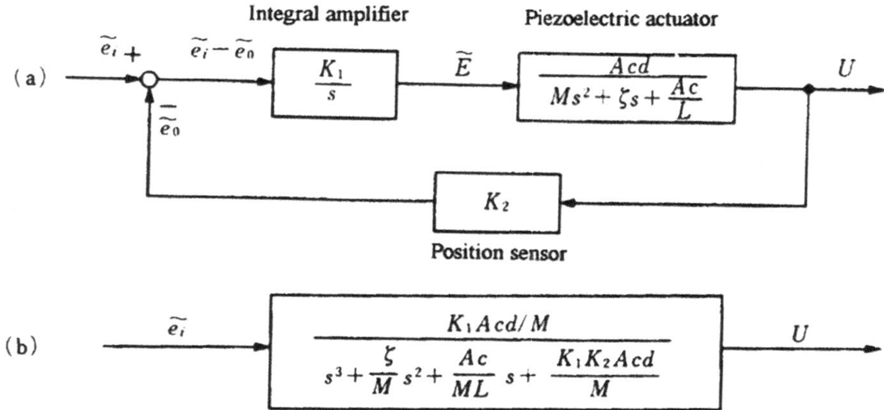

FIGURE 5.22 Block diagrams for a piezoelectric actuator servo system representing the function of: (a) the individual system elements and (b) the entire system.

There are generally two ways to unify the block diagram with a feedback loop: open and closed-loop transfer functions. The *open-loop transfer function* is defined as the "product of three components" transfer function (by disconnecting $-\tilde{e}_o$):

$$W_o = G_1G_2G_3 = \left[\frac{K_1K_2}{s}\right]\frac{Acd}{(Ms^2 + \zeta s + (Ac/L))} \tag{5.45}$$

To the contrary, the *closed-loop transfer function* is calculated by completing the feedback function:

$$W_c = G_1G_2 - G_1G_2(G_1G_2G_3) + G_1G_2(G_1G_2G_3)^2 - G_1G_2(G_1G_2G_3)^3 + \cdots \tag{5.46}$$

The first term provides the direct output from the input command signal, $\tilde{e}_i$, multiplied by integral amplifier, G_1, and piezoelectric actuator, G_2; the second term provides the one-cycle output, given by the negative open-loop transfer function (i.e., negative feedback) multiplied again by G_1G_2; similarly, the third term shows the two-cycle output, that is, multiplied another open-loop transfer function $G_1G_2G_3$; and so on. Knowing a general relation, $1 - r + r^2 - r^3 + r^4 - \cdots = 1/(1 + r)$, Equation 5.46 can be calculated as

$$\begin{aligned} W_c &= \frac{G_1G_2}{(1 + G_1G_2G_3)} \\ &= \frac{(K_1Acd/M)}{[s^3 + (\zeta/M)s^2 + (Ac/ML)s + (K_1K_2Acd/M)]} \end{aligned} \tag{5.47}$$

This function applies to the entire system as depicted in Figure 5.22b, which is the third-order transfer function (the s^3 term included).

5.3.3 Criterion for System Stability

In an ideal situation, where the input command voltage, e_i, changes stepwise as illustrated in Figure 5.23a, it is expected that the displacement, u, will also change in a stepwise fashion. However, in real situations, the actual displacement that occurs is quite different from the ideal response. A divergent case is depicted in Figure 5.23d, which is an undesirable response from a practical point of view and, in general, is referred to as *unstable control*. Even if a more converging response is achieved by the system, this in itself does not guarantee that the difference between the actual displacement voltage, e_0, and the input command voltage, e_i, will be exactly zero after a sufficient time lapse. This displacement deviation, which is called a *steady-state error*, is an important issue in control engineering. Simply reducing the steady-state error to a minimum is not necessarily the objective, however, as too small a displacement deviation can sometimes cause too long a decay time for the transient vibration, which is also not good control. So, we see that the decay time for the transient vibration is another key factor in evaluating a particular drive/control method.

5.3.3.1 Characteristic Equation

When the command input, e_i, is a *unit step*, the displacement response of the actuator system directly becomes the closed-loop transfer function and is described by:

$$U = W_ce_i = \frac{(K_1Acd/sM)}{[s^3 + (\zeta/M)s^2 + (Ac/ML)s + (K_1K_2Acd/M)]} \tag{5.48}$$

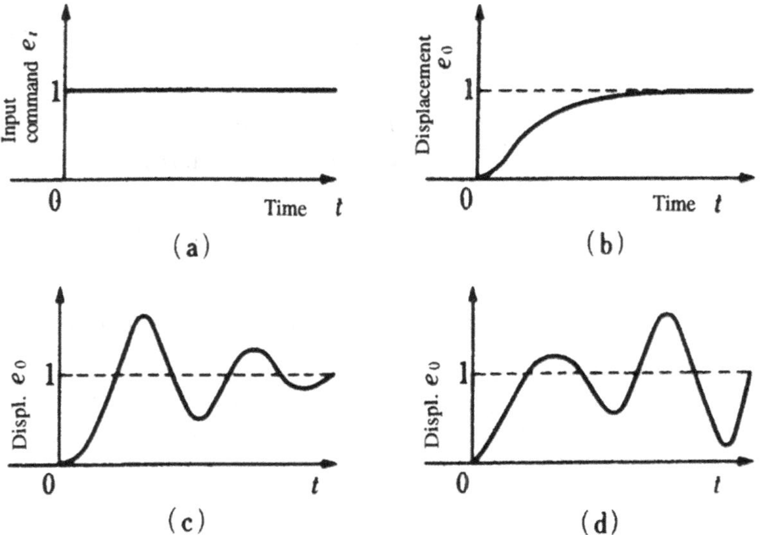

FIGURE 5.23 Various possible output voltages, e_o, for a step input, e_i.

We may simplify this expression for a practical case by assuming that: $(K_1Acd/M) = 4$, $(\zeta/M) = 4$, $(Ac/ML) = 6$ and $(K_1K_2Acd/M) = 4$, which now yields:

$$U = 4/s(s^3 + 4s^2 + 6s + 4) \tag{5.49a}$$

Further manipulation of this equation leads to the following:

$$U = \frac{4}{s(s+2)(s+1+j)(s+1-j)} = \left[\frac{1}{s} - \left(\frac{1}{s+2}\right) + \left(\frac{j}{s+1-j}\right) - \left(\frac{j}{s+1+j}\right)\right] \tag{5.49b}$$

When we make use of inverse Laplace transform #4 from Table 5.1, it leads to the following equation for the displacement, $u(t)$:

$$\begin{aligned} u(t) &= [1 - e^{-2t} + je^{-(1-j)t} - je^{-(1+j)t}]1(t) \\ &= [1 - e^{-2t} - 2e^{-t}\sin(t)]1(t) \end{aligned} \tag{5.50}$$

With increasing time, the second and third terms approach zero and $u(t)$ becomes closer to the step function; therefore, this system is considered stable. When the denominator of the closed-loop transfer function is set equal to zero, the resulting equation is called the *characteristic equation*. The system is judged to be stable, when the *real parts of the characteristic roots are all negative*. However, this judgment does not provide a further idea on the degree of stability.

5.3.3.2 The Nyquist Criterion for Stability

The stability of the system can be characterized in terms of the so-called *Nyquist criterion*, which can provide the level of stability. The Nyquist method utilizes the open-loop transfer function, W_o:

$$W_o = G_1G_2G_3 = \frac{(K_1K_2Acd/M)}{s[s^2 + (\zeta/M)s + (Ac/ML)]} \tag{5.51}$$

Since the mathematical verification of this equation is somewhat beyond the scope of this text, we will primarily focus our attention on applying the criterion to the system. By substituting $s = j\omega$ (where ω is the angular frequency) into Equation 5.51, the locus of W_o for ω increasing from 0 to $+\infty$ is plotted on the complex plane and the resulting curves constitute what is referred to as a *Nyquist diagram.* A set of such curves shown in Figure 5.24. Considering the tendency of these curves in the direction of ω 0 to $+\infty$ (see $K_2 = 3$ curve in Figure 5.24), we can say that *the system is stable if the point* $(-1 + j\,0)$ *is to the left of the locus.* When this point is to the right of the locus, the system is considered unstable, and when the Nyquist locus passes through this point, the system is exactly on the *critical limit of stability.*

Let us consider again the case associated with Equation 5.51, for which we assumed $(K_1Acd/M) = 4$, $(\zeta/M) = 4$, $(Ac/ML) = 6$. The general open-loop transfer function for this system may be written as:

$$W_o = 4K_2/j\omega(-\omega^2 + 4j\omega + 6) \tag{5.52}$$

The curves depicted in Figure 5.24 correspond to this general case for various values of K_2. We see in this situation that if the gain of the displacement sensor K_2 is adjusted to 6, the locus passes precisely through the point $(-1 + j\,0)$, exactly at the critical limit of stability. If the gain is reduced to half this value, the point $(-1 + j\,0)$ is to the left of the locus (with increasing frequency) and a stable state exists. When the gain is increased by a factor of 1.5, $K_2 = 9$, however, the point $(-1 + j\,0)$ is to the right side of the locus, resulting in an unstable state.

5.3.4 Steady-State Error

When the command voltage, e_i, is changed (say, cut 0.1 μm more), the sensor voltage, e_0, which corresponds to the actual position after a sufficient time lapse (Equation 5.48), will also change such that:

$$\tilde{e}_i - \tilde{e}_o = \left(1 - K_2W_c\right)\tilde{e}_i$$

$$= \frac{\tilde{e}_i s[s^2 + (\zeta/M)s + (Ac/ML)]}{[s^3 + (\zeta/M)s^2 + (Ac/ML)s + (K_1K_2Acd/M)]} \tag{5.53}$$

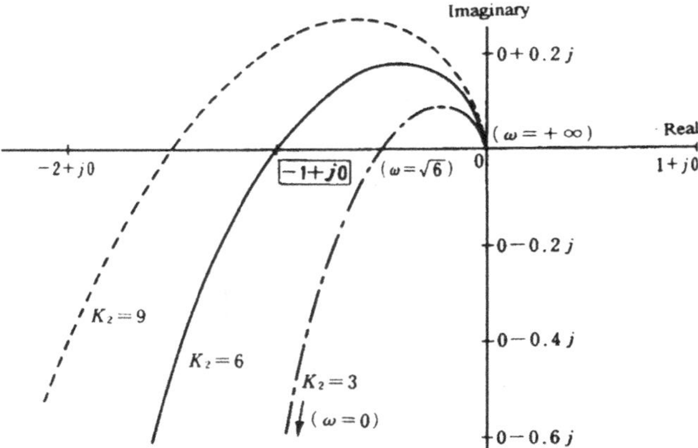

FIGURE 5.24 A Nyquist diagram for evaluating the stability of a piezo-actuator system.

When the input is a constant step voltage, [i.e., $e_i = 1(t)$], we also know that:

$$\lim_{t\to\infty}(e_i - e_o) = \lim_{s\to 0}[s(\tilde{e}_i - \tilde{e}_o)] = 0 \tag{5.54}$$

Thus, the steady-state error of the position is zero. Considering now an input that is a constantly increasing voltage, such that, $e_i = t$, $\tilde{e}_i = 1/s^2$, we find that

$$\lim_{t\to\infty}(e_i - e_o) = \lim_{s\to 0}[s(\tilde{e}_i - \tilde{e}_o)] = \frac{1}{K_1 K_2 dL} \tag{5.55}$$

and the steady-state error is no longer equal to zero. If the position sensor gain, K_2, is increased, the steady-state deviation decreases (i.e., response speed is increased). As we have already discussed, however, the system stability also deteriorates with increasing K_2; thus, we need to establish some kind of a compromise between the steady-state error and the system stability.

The data presented in Figure 5.25 show the experimental data of the position change for our compact actuator system under a step command input (say, "cut 0.1 μm" in Figure 5.25a). We see in Figure 5.25b how the actuator's position gradually drifts without feedback control due to the effects of thermal expansion. The displacement is more effectively regulated when feedback control is employed. Proper choice of the damping constant is also an important factor here in achieving the most stable response. We see in Figure 5.25d the effects of choosing a damping constant, ζ, that

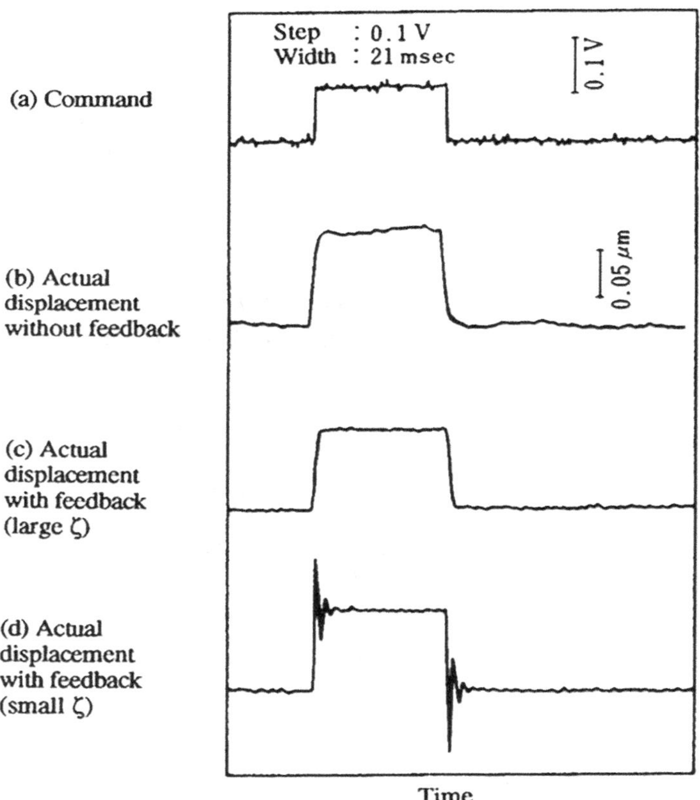

FIGURE 5.25 Step response of the piezoelectric actuator. (a) Pulse command signal, (b) without feedback, (c) with feedback (large damping), (d) with feedback (small damping).

is too small (no damper was attached in the experiment); a clear *overshoot* is observed under these conditions at the rise and fall edges of the output signal (the work will be overcut or damaged!). When the damping constant is increased by means of a *mechanical damper* such as a plate spring in our experiment, the overshoot is eliminated, as depicted in Figure 5.25c. It is important to keep in mind, however, in selecting an appropriate ζ that excessively high values will significantly decrease the responsivity of the system.

5.3.5 Advantages of Feedback Control

Adding a feedback control system that satisfies the Nyquist stability generally provides distinct benefits, namely: (1) a linear relationship between input and output, (2) an output response with a flat frequency dependence, and (3) minimization of external disturbance effects. These benefits are described in this section.

5.3.5.1 Linear Relation between Input and Output

Let us consider the nonlinear relation depicted in Figure 5.26a between an input voltage, V, as might be applied to an electrostrictive actuator, and the corresponding output response, Y, which in this case would be the position sensor voltage, such that:

$$Y = F(V) = KV^2 \tag{5.56}$$

for an electrostrictive (strain $\propto E^2$) actuator system. Let us add a feedback loop, as shown in Figure 5.26b. We will assume that K and β are constants, and that X, V, and Y are variable signals at the points of the system indicated on the block diagram. This system will be in equilibrium for a certain input $X = X_o$ (i.e., a command voltage corresponding to a desired position), and corresponding $V = V_o$, $Y = Y_o$. In this state, the following equations must be satisfied:

$$V_o = K(X_o - \beta Y_o) \tag{5.57}$$

$$Y_o = F(V_o) \tag{5.58}$$

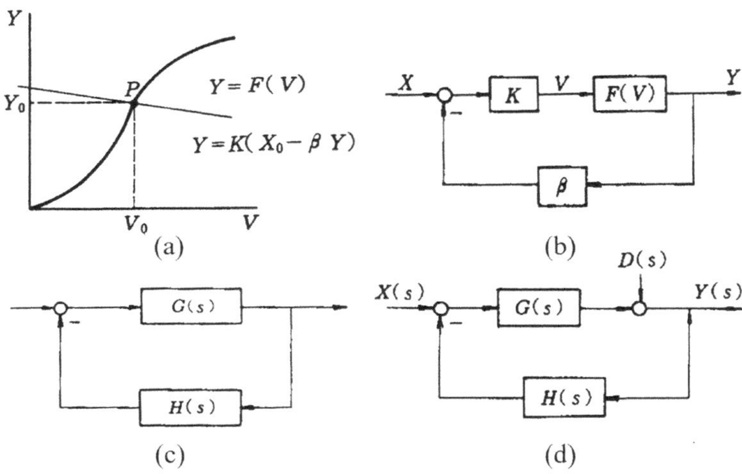

FIGURE 5.26 (a) Nonlinear relationship between Y (output voltage) and V (input voltage) (e.g., electrostrictor). (b) Feedback system with a nonlinear actuator. (c) Feedback system to explain flat frequency dependence. (d) Feedback system to explain the external disturbance.

The point (V_o, Y_o) is identified as point P in Figure 5.26a. Let's consider a small change in X by an incremental amount, x. The corresponding increments of V and Y are thus given by:

$$v = K(x - \beta y) \tag{5.59}$$

$$y = F'(V_o)v \tag{5.60}$$

Substituting and solving for y, we obtain:

$$y = \left[\frac{KF'(V_o)}{1 + \beta KF'(V_o)}\right]x \tag{5.61}$$

If the condition:

$$\beta KF'(V_o) \gg 1 \tag{5.62}$$

is satisfied, then

$$y = (1/\beta)x. \tag{5.63}$$

This means that *a linear relation is obtained between the input x and output y* in the feedback control system. Figure 5.27a shows the experimental result on quasistatic displacement characteristics of the actuators, where the open-loop performance ($a - 1$) exhibits electrostrictive displacement (parabolic curve with voltage) with slight hysteresis, and with the feedback control ($a - 2$), a perfect linear relation (zero hysteresis) between the displacement and command signal is obtained.

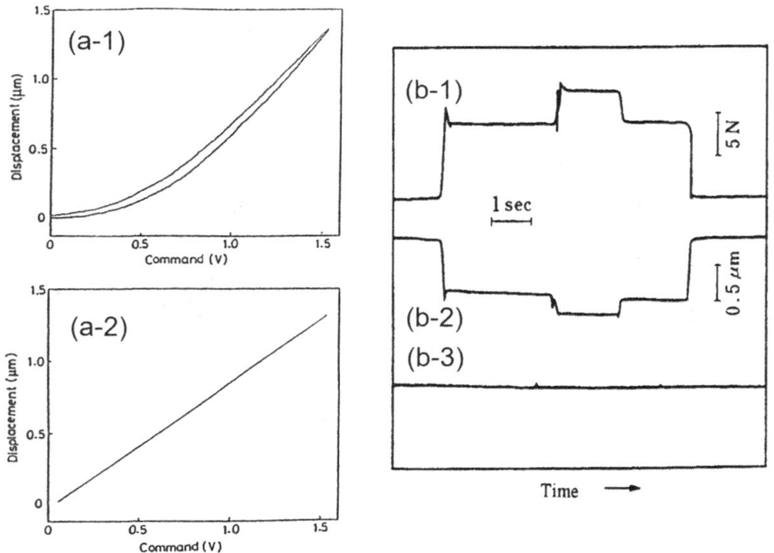

FIGURE 5.27 (a) Quasistatic displacement characteristics of the actuators: (a-1) open-loop, and (a-2) feedback-control. (b) Suppression of an external disturbance in a precision lathe machine using feedback: (b-1) the mechanical load disturbance, (b-2) the position change of the piezoelectric actuator without feedback, and (b-3) the position change of the actuator with feedback.

5.3.5.2 Output Response with a Flat Frequency Dependence

Let us now consider a forward transfer function, $G(s)$, and a feedback transfer function, $H(s)$, as depicted in the block diagram of Figure 5.26c. If we assume this feedback system is stable, its *closed-loop frequency response function, $M(j\omega)$*, is given by:

$$M(j\omega) = \frac{G(j\omega)}{[1 + G(j\omega)H(j\omega)]} \tag{5.64}$$

If the condition

$$|G(j\omega)H(j\omega)| \gg 1 \tag{5.65}$$

is satisfied, then

$$M(j\omega) = 1/H(j\omega). \tag{5.66}$$

In other words, in the frequency range where the gain of the open-loop transfer function, $|G(j\omega)H(j\omega)|$, is sufficiently larger than 1, the closed-loop frequency response function, $M(j\omega)$, depends only on the feedback transfer function, $H(s)$. If $H(j\omega) = \beta$ (constant), $M(j\omega)$ shows a constant characteristic with a gain $(1/\beta)$ and a phase angle of zero in this frequency range. The flat frequency dependence is realized only when the feedback gain is quite high.

On the other hand, if the condition

$$|G(j\omega)H(j\omega)| \ll 1 \tag{5.67}$$

is satisfied, then

$$M(j\omega) = G(j\omega). \tag{5.68}$$

and for the frequency range in which $|G(j\omega)\,H(j\omega)| \approx 1$, $M(j\omega)$ is obtained directly from Equation 5.64. A resonance is occasionally observed around this frequency (near the gain crossing frequency, ω_{cg}).

5.3.5.3 Minimization of External Disturbance Effects

We may now add an external disturbance (i.e., a deviation in position due to *thermal dilatation* or *mechanical noise vibration*) to the system and its associated transfer function, $G(s)$. This is represented as $D(s)$ in the block diagram of Figure 5.26d. The output, $Y(s)$, for this system will be described by:

$$Y(s) = \left[\frac{1}{[1 + G(s)H(s)]}\right](G(s)X(s) + D(s)) \tag{5.69}$$

The output deviation, $\Delta Y(s)$, due to the external disturbance, $D(s)$, is given by:

$$\Delta Y(s) = \frac{D(s)}{[1 + G(s)H(s)]} \tag{5.70}$$

Let us consider as an example a step disturbance, $d \cdot 1(t)$ (d: constant, $1(t)$: unit step function). The saturated output deviation $\Delta y(\infty)$ can be calculated using the Laplace theorem (g) in Section 5.3.1:

$$\Delta y(\infty) = \lim_{s\to 0}[s\Delta Y(s)] = \lim_{s\to 0}\left[\frac{s}{[1+G(s)H(s)]}(d/s)\right]$$
$$= \frac{d}{[1+G(0)H(0)]} \tag{5.71}$$

The external disturbance can be significantly diminished by a factor of $1/[1 + G(0)H(0)]$, as compared to the case where no feedback is employed. The open-loop transfer function under DC conditions, $G(0)H(0)$, should be sufficiently larger than 1 in order to minimize the output deviation.

If the external disturbance has a frequency spectrum $D(j\omega)$, the output deviation spectrum is:

$$\Delta Y(j\omega) = \frac{D(j\omega)}{[1 + G(j\omega)H(j\omega)]} \tag{5.72}$$

When the condition $|G(j\omega)H(j\omega)| \gg 1$ is satisfied in the frequency range corresponding to the external disturbance, the feedback control is effective for diminishing the external disturbance. The actuator response in a precision lathe machine to a mechanical step load appears in Figure 5.27b. Such a disturbance is generated in the lathe machine when the cutting edge is occasionally caught in a "sticky" material such as aluminum. The remarkable improvement from the 1-μm displacement deviation that occurs without feedback (Figure 5.27b-2) to only 0.002-μm maximum deviation with feedback (Figure 5.27b-3) is clearly demonstrated by these data.

EXAMPLE PROBLEM 5.4

Describe the *Bode diagram* for the standard second-order system with

$$G(j\omega) = \frac{1}{[(-\omega^2 T^2) + 2\zeta j\omega T + 1]} \tag{P5.4.1}$$

Solution

The Bode diagram is a representation of the transfer function (amplitude and phase) as a function of frequency on a logarithmic scale.

First, let us consider approximate curves for the low- and high-frequency regions.

- For $\omega \to 0$, $G(j\omega) \to 1$. Thus, gain $|G(j\omega)| = 1$, so that in decibels:

$$\text{dB} = 20\log_{10}(1) = 0 \tag{P5.4.2}$$

  0 dB/decade, that is, flat frequency dependence. Regarding the phase, the real number corresponds to 0°. Gain and phase Bode plots are shown in Figure 5.28.

- For $\omega \to \infty$, $G(j\omega) \to 1/(-\omega^2 T^2)$
  Gain:

$$|G(j\omega)| = \left|\frac{1}{(-\omega^2 T^2)}\right| \tag{P5.4.3a}$$

  so that in decibels:

$$\text{dB} = -20\log_{10}(\omega T)^2 = -40\log_{10}(\omega T) \tag{P5.4.3b}$$

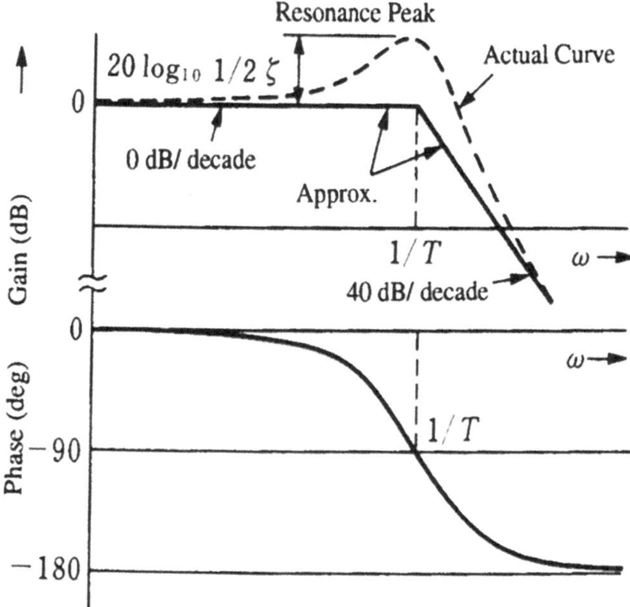

FIGURE 5.28 The Bode diagram for a standard second-order system.

40 dB down/decade with frequency. Regarding the phase, a negative real number corresponds to −180° as indicated by the gain and phase curves appearing in Figure 5.28.

The low- and high-frequency portions of the gain curve can be approximated with two straight lines, as shown in Figure 5.28. The high-frequency portion of the curve is approximated with a straight line with a negative slope of 40 dB/decade (or 12 dB/octave).

- Resonance range: we will now consider the deviation from these two lines around the bend-point frequency, which is obtained from the relation $\omega T = 1$. Substituting ($\omega T = 1$) in Equation P5.4.1 yields:

$$G(j\omega) = \frac{1}{(2\zeta j\omega T)} = \frac{1}{(2\zeta j)} \tag{P5.4.4}$$

so that the gain and phase become 1/(2ζ) and −90°, respectively. The constant ζ corresponds to the damping constant. If $\zeta = 0$, an infinite amplitude will occur at the bend-point frequency (i.e., the resonance frequency). When ζ is large (>1/2), however, the resonance peak will disappear, and monotonous decrease in amplitude is observed. It is notable that $(1/\zeta) \propto$ mechanical quality factor Q_m.

Finally, we will consider the displacement curve under a cyclic electric field in a piezoelectric actuator system. In a low frequency range, the displacement $\Delta L = d\ E\ L$ with zero phase lag, where the piezoelectric constant, d, is d_{33} or d_{31}, depending on the setting. In a piezo-multilayer actuator's case, it is d_{33}. ΔL exhibits a linear line with a slope proportional to the piezoelectric constant, depicted in Figure 5.29a (this is for the d_{31} case, where $d_{31} < 0$ and thus negative slope). In a high frequency range, the ΔL amplitude decreases dramatically as 40 dB down/decade. Because the phase is −180°, ΔL exhibits a linear line with a positive slope (opposite the piezoelectric constant sign) (Figure 5.29c). It is most intriguing at the resonance frequency, where ΔL is amplified significantly by a factor of the mechanical quality factor $(8/\pi^2)Q_m$ (e.g., 1000 times in a hard PZT), and the displacement curve does not exhibit a linear line but an elliptic curve because of 90° phase lag (clockwise rotation due to negative d_{31}). Refer to Figure 5.29b.

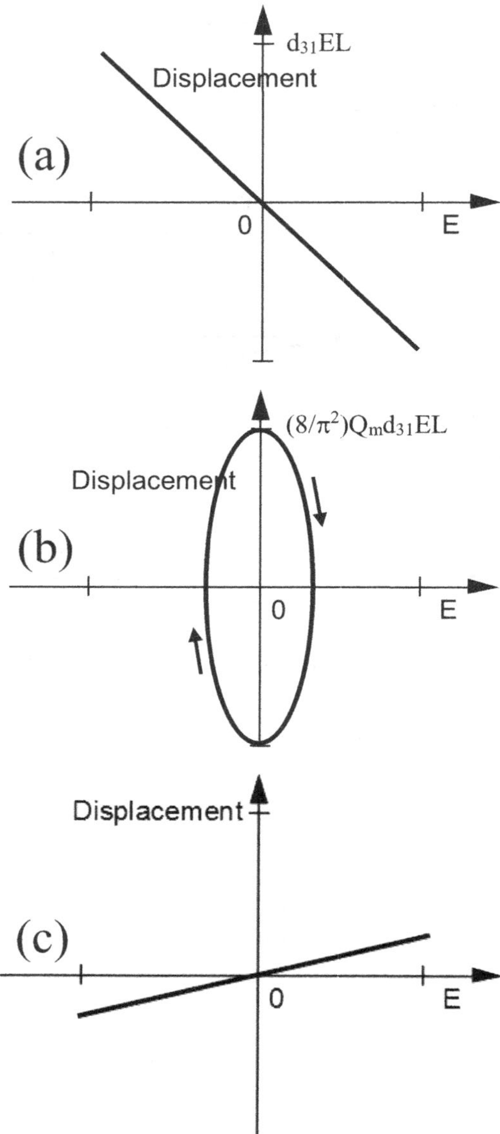

FIGURE 5.29 Displacement vs. electric field.

EXAMPLE PROBLEM 5.5

Figure 5.30 shows a mechanical system composed of a mass, a spring, and a damper. Derive the total transfer function for the input force $F(t)$ and the output displacement $x(t)$ from a system block diagram viewpoint composed of each component and velocity feedback. The transfer function should be:

$$G(s) = \frac{1}{s^2 m + s\zeta + c} \tag{P5.5.1}$$

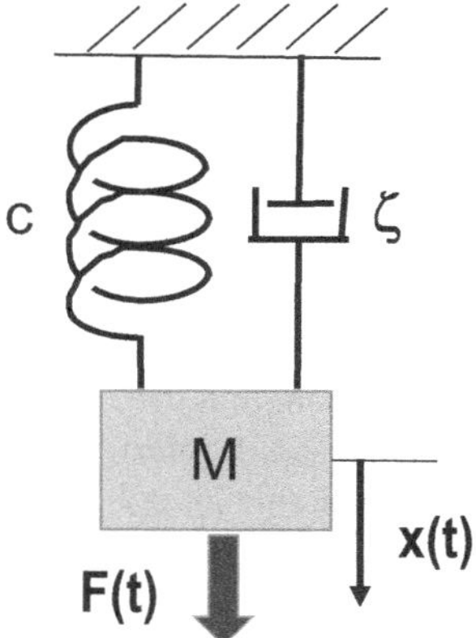

FIGURE 5.30 Spring-mass-damper system.

SOLUTION

The force, $F(t)$, acting on mass, M, has to be balanced with the mass inertial force, the damping force, and the spring force. Introducing the acceleration, a ($= d^2x/dt^2$), and velocity, v ($= dx/dt$), we can obtain the following relation:

$$F(t) = m \cdot a + \zeta \cdot v + c \cdot x \tag{P5.5.2}$$

Figure 5.31 shows the block diagram representation of each component and velocity feedback. Different from Figure 5.22, this block diagram includes two "positive" feedback branches, for adding velocity ($\zeta \cdot v$) and displacement ($c \cdot x$). You can notice that two ($1/s$)s (i.e., integration operation) to calculate "acceleration to velocity" and "velocity to displacement," successively. Thus, Equation P5.5.2 can be transformed (Laplace transformation) to:

$$\widetilde{F(s)} = (ms^2 + \zeta s + c)\widetilde{x(s)} \tag{P5.5.3}$$

Accordingly, total transfer function is obtained as:

$$\frac{\widetilde{x(s)}}{\widetilde{F(s)}} = \frac{1}{(ms^2 + \zeta s + c)} \tag{P5.5.4}$$

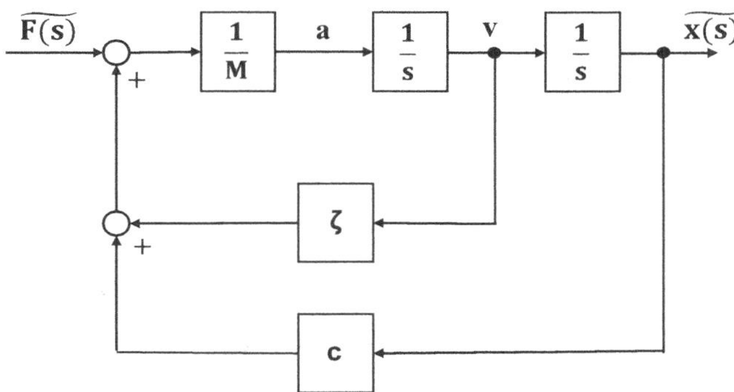

FIGURE 5.31 Block diagram of spring-mass-damper system.

5.3.6 Polarization Control Method

An interesting alternative technique, called *polarization control*, is used to minimize the effect of strain hysteresis in piezoelectric actuators without using a complex feedback (or servo) system. The strain curves for a PZT-based material plotted as a function of applied voltage (proportional to the electric field) and charge (proportional to the polarization) appear in Figure 5.32. Note that the strain curve representing the polarization dependence does not exhibit hysteresis, while the curve representing the electric field dependence does.[18] (The reasons for this have already been discussed in Section 2.6.3.) We see by these trends, then, that the strain hysteresis that occurs under a voltage control system will be significantly suppressed if polarization control is used instead.

An example of a current supply circuit suitable for polarization control is pictured in Figure 5.33.[18] A current proportional to the input voltage, V_{in}, is supplied to the piezoelectric actuator in this circuit. Another noteworthy feature of this current supply is its very high output impedance as compared to the very small output impedance of the conventional voltage supply.

The polarization control technique is effective only for pulse and AC drive methods. When used for pseudo-DC driving, the resistance of the actuator must be extremely large. Otherwise, charge leakage in the ceramic actuator prevents stable control for quasistatic positioning. A stacked actuator, for example, composed of 92 layers of PZT ceramic discs (each with a diameter of 20 mm and thickness of 0.5 mm) operated by polarization control is able to maintain its charge only for about 30 minutes.[18]

5.4 PULSE DRIVE

When a pulsed electric field is applied to a piezoelectric actuator, a mechanical vibration is excited, the characteristics of which depend on the pulse profile. Displacement overshoot and ringing are

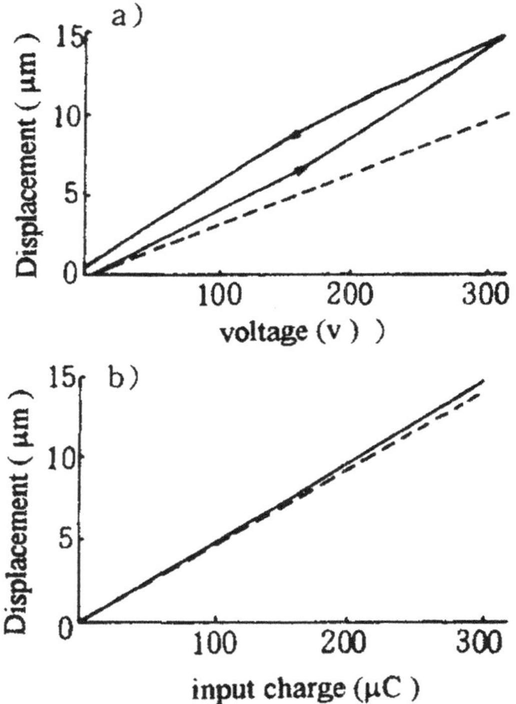

FIGURE 5.32 Strain curves for a PZT plotted as a function of: (a) applied voltage and (b) charge (proportional to the polarization). (From C. V. Newcomb and I. Flinn: *Electronics Lett.* 18, 442, 1982.)

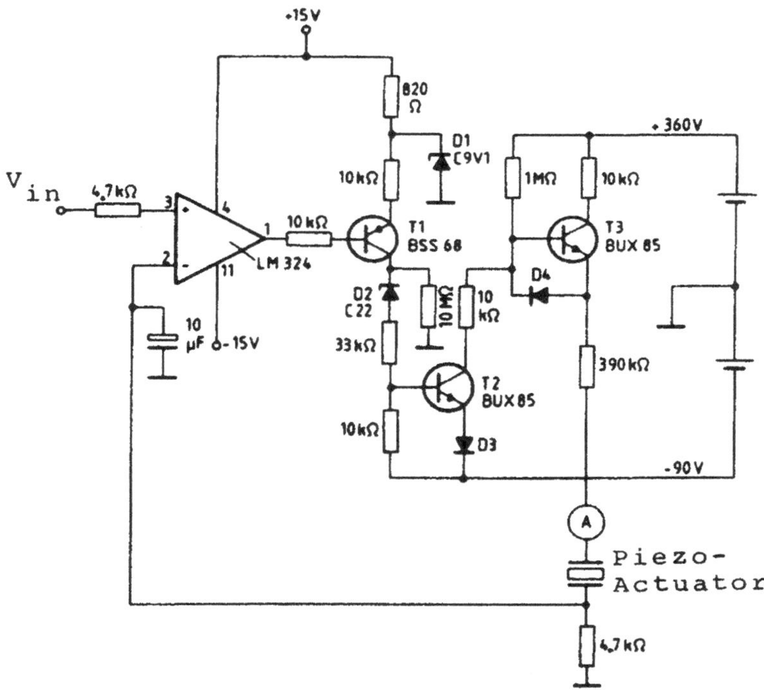

FIGURE 5.33 An example of a drive circuit for polarization control.

frequently observed. Thus, quick and precise positioning is difficult to achieve at the same time. Moreover, the pulse drive can lead to the destruction of the actuator due to the large tensile stress associated with overshoot. We will examine more closely the transient response of a piezoelectric device driven by a pulsed electric field in this section. Since we need the basics on the piezoelectric constitutive equations and dynamic vibration equations in a continuum medium, reviewing Section 2.8 is required prior to studying this section.

5.4.1 The Piezoelectric Equations and Vibration Modes

Only the key points are summarized in this section on the piezoelectric constitutive equations and dynamic vibration modes, which are also used in Section 5.5, "Resonance Drive."

5.4.1.1 Piezoelectric Constitutive Equations

When the applied electric field, E, and the stress, X, are small, the strain, x, and the electric displacement, D, induced in a piezoelectric can be represented by the following equations:

$$x_i = s_{ij}^E X_j + d_{mi} E_m \tag{5.73}$$

$$D_m = d_{mi} X_i + \varepsilon_0 \varepsilon_{mk}^X E_k \tag{5.74}$$

where ($i, j = 1, 2, \ldots, 6$; $m, k = 1, 2, 3$). These are the *piezoelectric constitutive equations*. There are 21 independent s_{ij}^E coefficients, 18 d_{mi} coefficients, and 6 ε_{mk}^X coefficients for the lowest symmetry "trigonal" crystal. When considering polycrystalline ceramic specimens, the poling direction is typically designated as the z-axis. A poled ceramic is isotropic with respect to this z-axis and has a

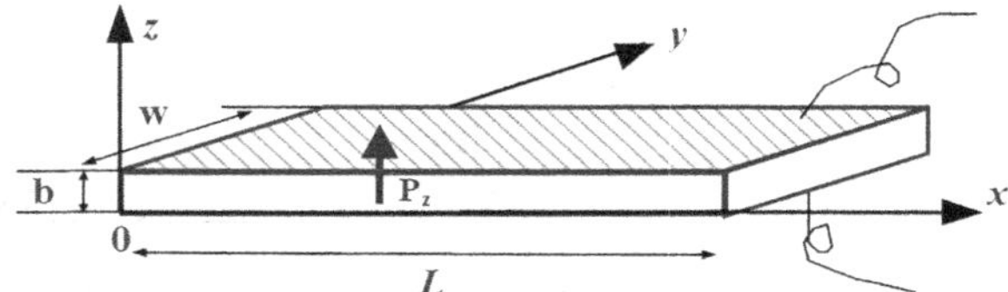

FIGURE 5.34 Longitudinal vibration k_{31} mode of a rectangular piezoelectric plate.

Curie group designation $C_{\infty v}$ (∞m). There are 10 nonzero matrix elements (s_{11}^E, s_{12}^E, s_{13}^E, s_{33}^E, s_{44}^E d_{31}, d_{33}, d_{15}, ε_{11}^X, and ε_{33}^X) that apply in this case.

5.4.1.2 Longitudinal Vibration Mode via Transverse Piezoelectric Effect (k_{31} Mode)

Let us consider a longitudinal mechanical vibration in a simple piezoelectric ceramic plate via the transverse piezoelectric effect, d_{31}, with thickness b, width w, and length L ($b \ll w \ll L$), pictured in Figure 5.34 (and as case (a) in Table 5.2). When the polarization is in the z direction and the x-y planes are the planes of the electrodes, the extensional vibration along the x direction is represented by the following dynamic equation:

$$\rho \frac{\partial^2 u}{\partial t^2} = F = \frac{\partial X_{11}}{\partial x} + \frac{\partial X_{12}}{\partial y} + \frac{\partial X_{13}}{\partial z} \tag{5.75}$$

where u is the displacement in the x direction of a small volume element in the ceramic plate, ρ is the density of the piezoelectric material, and X_{ij} are stresses (only the force along the x-direction is our target). The relations among the stress, electric field (only E_z exists, because $E_x = E_y = 0$ due to the electrodes on the top and bottom), and induced strains are described by the following set of equations:

$$\begin{aligned}
x_1 &= s_{11}^E X_1 + s_{12}^E X_2 + s_{13}^E X_3 + d_{31} E_z \\
x_2 &= s_{12}^E X_1 + s_{11}^E X_2 + s_{13}^E X_3 + d_{31} E_z \\
x_3 &= s_{13}^E X_1 + s_{13}^E X_2 + s_{33}^E X_3 + d_{33} E_z \\
x_4 &= s_{44}^E X_4 \\
x_5 &= s_{44}^E X_5 \\
x_6 &= 2\left(s_{11}^E - s_{12}^E\right) X_6
\end{aligned} \tag{5.76}$$

It is important to note at this point that when an AC electric field of increasing frequency is applied to this piezoelectric plate, length, width, and thickness extensional resonance vibrations are excited. If we consider a typical PZT plate with dimensions 100 × 10 × 1 mm, these resonance frequencies correspond roughly to 10, 100, and 1 MHz, respectively. We will consider here the fundamental mode for this configuration, the length extensional mode. When the frequency of the applied field is well below 10 kHz, the induced displacement follows the AC field cycle, and the displacement magnitude is given by $d_{31}E_3L$. As we approach the fundamental resonance frequency, a delay in the length displacement with respect to the applied field begins to develop, and the amplitude of the displacement becomes enhanced. At frequencies above 10 kHz, the length displacement no longer follows the applied field, and the amplitude of the displacement is significantly reduced.

TABLE 5.2
Characteristics of Various Resonators with Different Modes of Vibration

	Factor	Boundary Conditions	Resonator Shape	Definition
a	k_{31}	$X_1 \neq 0, X_2 = X_3 = 0$ $x_1 \neq 0, x_2 \neq 0, x_3 \neq 0$	3, 1	$\frac{d_{31}}{\sqrt{s_{11}^E \varepsilon_0 \varepsilon_{33}^X}}$
b	k_{33}	$X_1 = X_2 = 0, X_3 \neq 0$ $x_1 = x_2 \neq 0, x_3 \neq 0$	3 Fundamental Mode	$\frac{d_{33}}{\sqrt{s_{33}^E \varepsilon_0 \varepsilon_{33}^X}}$
c	k_p	$X_1 = X_2 \neq 0, X_3 = 0$ $x_1 = x_2 \neq 0, x_3 \neq 0$	3 Fundamental Mode	$k_{31}\sqrt{\frac{2}{1-\sigma}}$
d	k_t	$X_1 = X_2 \neq 0, X_3 \neq 0$ $x_1 = x_2 = 0, x_3 \neq 0$	3 Thickness Mode	$k_{33}\sqrt{\frac{\varepsilon_0 \varepsilon_{33}^x}{c_{33}^D}}$
e	k_p'	$X_1 = X_2 \neq 0, X_3 \neq 0$ $x_1 = x_2 \neq 0, x_3 = 0$	3 Radial Mode	$\frac{k_p - Ak_{33}}{\sqrt{1-A^2}\sqrt{1-k_{33}^2}}$
f	k_{31}'	$X_1 \neq 0, X_2 \neq 0, X_3 = 0$ $x_1 \neq 0, x_2 = 0, x_3 \neq 0$	3, 2, 1 Width Mode	$\frac{k_{31}}{\sqrt{1-k_{31}^2}}\sqrt{\frac{1+\sigma}{1-\sigma}}$
g	k_{31}''	$X_1 \neq 0, X_2 = 0, X_3 \neq 0$ $x_1 \neq 0, x_2 \neq 0, x_3 = 0$	3, 2, 1 Width Mode	$\frac{k_{31} - Bk_{33}}{\sqrt{1-k_{33}^2}}$
h	k_{31}'''	$X_1 \neq 0, X_2 \neq 0, X_3 \neq 0$ $x_1 \neq 0, x_2 = 0, x_3 = 0$	1, 2, 3 Thickness Mode	$\sqrt{\frac{\frac{(k_p - A\,k_{33})^2}{1-A^2} - (k_{31} - Bk_{33})^2}{1 - k_{33}^2 - (k_{31} - Bk_{33})^2}}$
i	k_{33}'	$X_1 \neq 0, X_2 = 0, X_3 \neq 0$ $x_1 = 0, x_2 \neq 0, x_3 \neq 0$	3, 1 Width Mode	$\frac{k_{33} - Bk_{31}}{\sqrt{(1-B^2)(1-k_{31}^2)}}$
j	$k_{24} = k_{24}$	$X_1 = X_2 = X_3 = 0, X_4 \neq 0$ $x_1 = x_2 = x_3 = 0, x_4 \neq 0$		$\frac{d_{31}}{\sqrt{\varepsilon_0 \varepsilon_{11}^X s_{44}^E}}$

$$\text{Here: } A = \frac{\sqrt{2}s_{13}^E}{\sqrt{s_{33}^E(s_{11}^E + s_{12}^E)}}, B = \frac{s_{13}^E}{\sqrt{s_{11}^E s_{33}^E}}$$

When a very long, thin plate is driven in the vicinity of this fundamental resonance, X_2 and X_3 may be considered zero throughout the plate. Since shear stress will not be generated by the applied electric field, E_z, only the following single equation applies:

$$X_1 = \frac{x_1}{s_{11}^E} - \left(\frac{d_{31}}{s_{11}^E}\right)E_z \quad (5.77)$$

Substituting Equation 5.77 into Equation 5.75, and assuming that $x_1 = \partial u/\partial x$ and $\partial E_z/\partial x = 0$ (since each electrode is at the same potential), we obtain the following dynamic equation:

$$\rho\frac{\partial^2 u}{\partial t^2} = \frac{1}{s_{11}^E}\frac{\partial^2 u}{\partial x^2} \quad (5.78)$$

5.4.1.3 Longitudinal Vibration Mode via Longitudinal Piezoelectric Effect (k_{33} Mode)

Let us consider next the longitudinal vibration mode via d_{33}. When the resonator is long in the z direction and the electrodes are deposited on each end of the rod, as depicted for case (b) in Table 5.2, the following conditions are satisfied:

$$X_1 = X_2 = X_4 = X_5 = X_6 = 0 \text{ and } X_3 \neq 0. \tag{5.79}$$

So that, from

$$X_3 = (x_3 - d_{33}E_z)/s_{33}^E \tag{5.80}$$

is the necessary equation in this configuration. Assuming a local displacement v in the z direction, the dynamic equation is given by:

$$\rho \frac{\partial^2 v}{\partial t^2} = \frac{1}{s_{33}^E}\left[\frac{\partial^2 v}{\partial z^2} - d_{33}\frac{\partial E_z}{\partial z}\right] \tag{5.81}$$

Different from Equation 5.78, the $\partial E_z/\partial z$ term cannot be removed. The electrical condition for the longitudinal k_{33} vibration is not $\partial E_z/\partial z = 0$, but rather $\partial D_z/\partial z = 0$, because the charge induced by the stress distribution cannot flow without side electrode along the z-direction. From Equation 5.74 and Equation 5.80

$$\varepsilon_0\varepsilon_{33}^X \frac{\partial E_Z}{\partial z} + \frac{d_{33}}{s_{33}^E}\left[\left(\frac{\partial^2 v}{\partial z^2}\right) - d_{33}\left(\frac{\partial E_Z}{\partial z}\right)\right] = 0,$$

or

$$\varepsilon_0\varepsilon_{33}^X\left(1 - k_{33}^2\right)\left(\frac{\partial E_Z}{\partial z}\right) = -\frac{d_{33}}{s_{33}^E}\left(\frac{\partial^2 v}{\partial z^2}\right) \tag{5.82}$$

Thus, introducing the electromechanical coupling factor, k_{33},

$$k_{33} = \frac{d_{33}}{\sqrt{s_{33}^E\varepsilon_0\varepsilon_{33}^X}} \tag{5.83}$$

we obtain the dynamic equation:

$$\rho\frac{\partial^2 v}{\partial t^2} = \frac{1}{s_{33}^D}\frac{\partial^2 v}{\partial z^2}\left(s_{33}^D = (1 - k_{33}^3)s_{33}^E\right) \tag{5.84}$$

Compared with the sound velocity $v = 1/\sqrt{\rho s_{11}^E}$ in Equation 5.78 in the surface electrode (E-constant) sample along the vibration direction, the nonelectrode k_{33} (D-constant) sample exhibits $v = 1/\sqrt{\rho s_{33}^D}$, which is faster (elastically stiffened) than that in the E-constant condition.

5.4.1.4 Other Vibration Modes

The dynamic equations for other vibration modes can be obtained in a similar fashion using the elastic boundary conditions and electromechanical coupling factors in Table 5.2.

EXAMPLE PROBLEM 5.6

Calculate the electromechanical coupling factor, k_{ij}, of piezoelectric ceramic vibrators: (a) longitudinal length extension mode (//E) k_{33}, and (b) planar extension mode of the circular plate, k_p (see Figure 5.35):

HINT

We will start with the energy expression:

$$U = \left(\frac{1}{2}\right) s_{ij}^E X_j X_i + 2\left(\frac{1}{2}\right) d_{mi} E_m X_i + \left(\frac{1}{2}\right) \varepsilon_0 \varepsilon_{mk}^X E_k E_m \\ = U_{MM} + 2U_{ME} + U_{EE} \tag{P5.6.1}$$

and make use of the following definition of the electromechanical coupling factor:

$$k = \frac{U_{ME}}{\sqrt{U_{MM} U_{EE}}} = \frac{d}{\sqrt{\varepsilon_0 \varepsilon^X s^E}} \tag{P5.6.2}$$

SOLUTION

a. The piezoelectric equations corresponding to the longitudinal 1D extension are:

$$x_3 = s_{33}^E X_3 + d_{33} E_3$$
$$D_3 = d_{33} X_3 + \varepsilon_0 \varepsilon_{33}^X E_3$$

Taking the necessary physical parameters, d, s, and ε, in Equation P5.6.2,

$$k_{33} = \frac{d_{33}}{\sqrt{s_{33}^E \varepsilon_0 \varepsilon_{33}^X}} \tag{P5.6.3}$$

b. The piezoelectric equations corresponding to the planar 2D extension are:

$$x_1 = s_{11}^E X_1 + s_{12}^E X_2 + d_{31} E_3$$
$$x_2 = s_{12}^E X_1 + s_{22}^E X_2 + d_{32} E_3$$
$$D_3 = d_{31} X_1 + d_{32} X_2 + \varepsilon_0 \varepsilon_{33}^X E_3$$

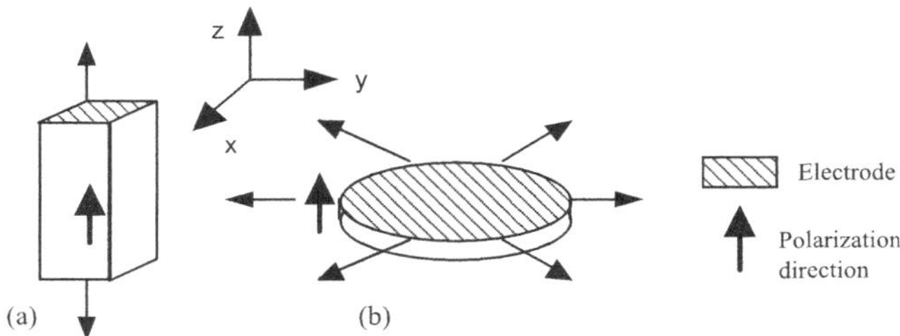

FIGURE 5.35 Two vibration modes of a piezoelectric device: (a) longitudinal length extension and (b) planar extension.

Assuming axial symmetry, $s_{11}^E = s_{22}^E$, $d_{31} = d_{32}$, and $X_1 = X_2$ $(= X_p)$ allows us to further simplify these equations into two:

$$x_1 + x_2 = 2\left(s_{11}^E + s_{12}^E\right)X_p + 2d_{31}E$$

$$D_3 = 2d_{31}X_p + \varepsilon_0\varepsilon_{33}^X E_3$$

Taking into account the physical parameters, d, s, and ε, in Equation P5.6.2,

$$k_p = \frac{2d_{31}}{\sqrt{2\left(s_{11}^E + s_{12}^E\right)\varepsilon_0\varepsilon_{33}^X}} = \frac{d_{31}\sqrt{2/(1-\sigma)}}{\sqrt{s_{11}^E\varepsilon_0\varepsilon_{33}^X}} = k_{31}\sqrt{2/(1-\sigma)} \tag{P5.6.4}$$

where σ is Poisson's ratio given by:

$$\sigma = -\frac{s_{12}^E}{s_{11}^E} \tag{P5.6.5}$$

5.4.2 Consideration of the Loss

When we consider the mechanical loss of the piezoelectric material, which is viscoelastic and proportional to the strain, the dynamic Equations 5.54 and 5.59 must be modified.[19] A damping term, $\zeta \cdot (\partial u/\partial t)$, is introduced as follows. The transverse piezo-effect vibration mode is now described by:

$$\rho\frac{\partial^2 u}{\partial t^2} = \frac{1}{s_{11}^E}\left[\frac{\partial^2(u + \varsigma(\partial u/\partial t))}{\partial x^2}\right], \tag{5.85}$$

and the longitudinal piezo-effect vibration mode by:

$$\rho\frac{\partial^2 v}{\partial t^2} = \frac{1}{s_{33}^D}\left[\frac{\partial^2(v + \varsigma(\partial v/\partial t))}{\partial z^2}\right] \tag{5.86}$$

Since Equations 5.78 and Equations 5.84 (and also Equations 5.85 and 5.86) have similar dynamic forms except for E constant (k_{31}) or D constant (k_{33}) conditions, we will consider just the k_{31} vibration mode in the following section as an example on how both cases might be treated.

5.4.3 Pulse Drive on the k_{31} Mode Specimen

5.4.3.1 General Solution for Longitudinal Vibration k_{31} Mode

Let us solve Equation 5.78 of Figure 5.34 using the Laplace transform. Denoting the Laplace transforms of $u(t,x)$ and $E_z(t)$ as $U(s,x)$ and $\widetilde{E(s)}$, respectively (x: coordinate along plate length), Equation 5.78 is transformed by Theorem (f) to:

$$\rho s_{11}^E s^2 U(s,x) = \frac{\partial^2 U(s,x)}{\partial x^2} \tag{5.87}$$

We will assume the following *initial conditions*:

$$\frac{\partial[u(0,x)]}{\partial t} = 0 \tag{5.88}$$

We may also make use of the fact that:

$$\rho s_{11}^E = \frac{1}{v^2} \tag{5.89}$$

where v is the speed of sound in the piezoelectric ceramic. To obtain a general solution of Equation 5.87 in terms of space coordinate x, we assume:

$$U(s,x) = Ae^{(sx/v)} + Be^{-(sx/v)} \tag{5.90}$$

The constants A and B can be determined by applying the *boundary conditions* $X_1 = 0$ at $x = 0$ and L:

$$X_1 = \frac{(x_1 - d_{31} E_z)}{s_{11}^E} = 0 \tag{5.91}$$

We also utilize the fact on the strain that:

$$L[x_1] = (\partial U/\partial x) = A(s/v)e^{(sx/v)} - B(s/v)e^{-(sx/v)} \tag{5.92}$$

in conjunction with the boundary conditions at $x = 0$ and L of Equation 5.91 to yield:

$$A(s/v) - B(s/v) = d_{31}\tilde{E} \tag{5.93a}$$

$$A(s/v)e^{(sl/v)} - B(s/v)e^{-(sl/v)} = d_{31}\tilde{E} \tag{5.93b}$$

Thus, we obtain

$$A = \frac{d_{31}\tilde{E}(1-e^{-sL/v})}{(s/v)(e^{sL/v} - e^{-sL/v})} \tag{5.94}$$

$$B = \frac{d_{31}\tilde{E}(1-e^{sL/v})}{(s/v)(e^{sL/v} - e^{-sL/v})} \tag{5.95}$$

and, consequently, Equations 5.90 and 5.92 become:

$$U(s,x) = \frac{d_{31}\tilde{E}(v/s)[e^{-s(L-x)/v} + e^{-s(L+x)/v} - e^{-sx/v} - e^{-s(2L-x)/v}]}{(1-e^{-2sL/v})} \tag{5.96}$$

$$L[x_1] = \frac{d_{31}\tilde{E}[e^{-s(L-x)/v} + e^{-s(L+x)/v} - e^{-sx/v} - e^{-s(2L-x)/v}]}{(1-e^{-2sL/v})} \tag{5.97}$$

The inverse Laplace transforms of Equations 5.96 and 5.97 now provide the displacement, $u(t, x)$, and strain, $x_1(t, x)$. Making use of the expansion series

$$1/(1 - e^{-2sl/v}) = 1 + e^{-2sl/v} + e^{-4sl/v} + e^{-6sl/v} \tag{5.98}$$

the strain, $x_1(t, x)$, can now be obtained by shifting the $d_{31}E_z(t)$ curves with respect to t according to Theorem (*e*). We may also consider that since $u(t, L/2) = 0$ [from $U(s, L/2) = 0$] and $u(t, 0) = -u(t, L)$ [from $U(s, 0) = -U(s, L)$], the total displacement of the plate device ΔL becomes equal to $2u(t, L)$. We finally arrive at the following:

$$U(s,L) = \frac{d_{31}\tilde{E}(v/s)(1-e^{-sL/v})}{(1+e^{-sL/v})} = d_{31}\tilde{E}(v/s)[\tanh(sL/2v)] \tag{5.99}$$

5.4.3.2 Displacement Response to a Step Voltage

We consider first a particular input of *Heaviside step* electric field $E(t) = E_0 H(t)$. Since the Laplace transform of the step function is given by $(1/s)$, Equation 5.99 can be expressed by

$$\begin{aligned} U(s, x = L) &= d_{31}E_0(v/s^2)(1 - e^{-(sL/v)})/(1 + e^{-(sL/v)}) \\ &= d_{31}E_0(v/s^2)(1 - 2e^{-(sL/v)} + 2e^{-(2sL/v)} - 2e^{-3sL/v} + 2e^{-4sL/v}\ldots.) \end{aligned} \tag{5.100}$$

Note that the base function of $U(s, L)$, $1/s^2$, gives the base function of $u(t, L)$ in terms of t (i.e., αt). The inverse Laplace transform of Equation 5.100 yields (by superposing the e^{-sk} terms):

$$\begin{aligned} &u(t,L) = d_{31}E_0v \cdot t && 0 < t < L/v \\ &u(t,L) = d_{31}E_0v[t - 2(t - L/v)] && L/v < t < 2L/v \\ &u(t,L) = d_{31}E_0v[t - 2(t - L/v) + 2(t - 2L/v)] && 2L/v < t < 3L/v \\ &u(t,L) = d_{31}E_0v[t - 2(t - L/v) + 2(t - 2L/v) - 2(t - 3L/v)] && 3L/v < t < 4L/v \\ &\ldots\ldots\ldots \end{aligned} \tag{5.101}$$

The transient displacement, ΔL ($= 2 \cdot u(t, L)$), produced by the step voltage is pictured in Figure 5.36 (since d_{31} is usually negative, Figure 5.36 is for $E_0 < 0$). The resonance period of this piezoelectric plate corresponds to $(2L/v)$, and the time interval in Eq. 5.101 is every (T/2). It is worth noting that the displacement changes linearly, not sinusoidally. On the contrary, the displacement of a discrete-component system changes sinusoidally, as demonstrated in Example Problem 5.7. This transient response difference comes from the vibration medium difference: continuum or discrete mechanical medium. In either case, continuous ringing occurs under a step input when the loss is neglected.

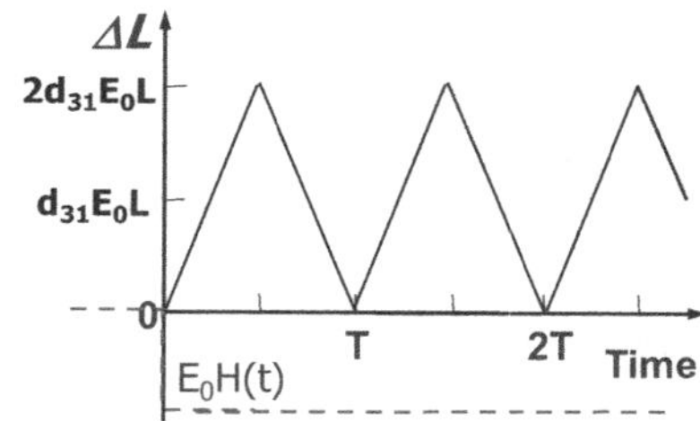

FIGURE 5.36 Triangular displacement response to a Heaviside step function voltage in a continuum piezoelectric plate (d_{31} mode).

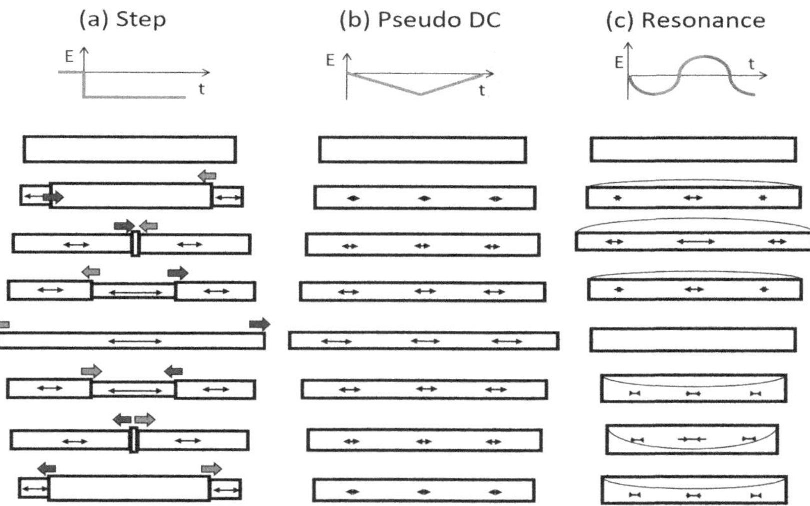

FIGURE 5.37 Strain distribution on a piezoplate during the deformation process under voltage: (a) step, (b) pseudo-DC triangle, (c) resonance sine.

The strain distribution on a piezoelectric plate is also intriguing for this step excitation case. From Equation 5.97, $L(x_1)$ is directly proportional to $\tilde{E}$; that is, the strain distribution, $x_1(x)$, follows exactly to $E(t)$, the Heaviside step function. The strain, x_1, at a certain point, x, becomes $d_{31}E_0$ suddenly from zero with a certain time lag depending on its coordinate, x. Thus, as illustrated in Figure 5.37a, the strained portion starts from both ends ($x = 0$ and L) of the piezoplates at the time of step voltage applied. These two symmetrical boundaries/walls between the strained and strain-free portion (analogous to a shock-wave) propagate with a piezo-material's sound velocity, v, oppositely each other, crossing over at the plate center, then generating the doubly strained part in the center area (i.e., $2d_{31}E_0$). Thus, when the wall reaches the plate end, the plate length becomes the maximum ($\Delta L = 2d_{31}E_0\, L$), and we can understand the reason 100% overshoot occurs under a step voltage applied. After that, the wall bounces back in the opposite direction, and the plate starts to shrink linearly according to the shrinkage of the strained portion. The linear displacement change originates from the constant wall velocity (which is equal to the piezomaterial's sound velocity). This triangular vibration ringing will continue for a long time if the loss is small. Figures 5.37b and 5.37c show the strain distribution on a piezoplate under a pseudo-DC triangular voltage and under a sinusoidal voltage at its resonance frequency, respectively, as references. The strain is uniformly generated in the plate under a pseudo-CD condition (no stress appears in the plate), while the strain and stress distribute sinusoidally with the maximum at the center part (i.e., nodal line) under the resonance drive.

EXAMPLE PROBLEM 5.7

The dynamic equations for a mechanical or electrical system composed of discrete components: mass and spring (damper is neglected for simplicity), or inductor, capacitor, and resistor, shown in Figures 5.38a and 5.38b, are expressed by

$$M(d^2u/dt^2) + cu = F(t), \tag{P5.7.1}$$

(u: displacement of a mass, M: mass, c: spring constant, F: external force)

$$L(d^2Q/dt^2) + R(dQ/dt) + (1/C)Q = V(t). \tag{P5.7.2}$$

(Q: charge, L: inductance, C: capacitance, R: resistance, V: voltage applied)

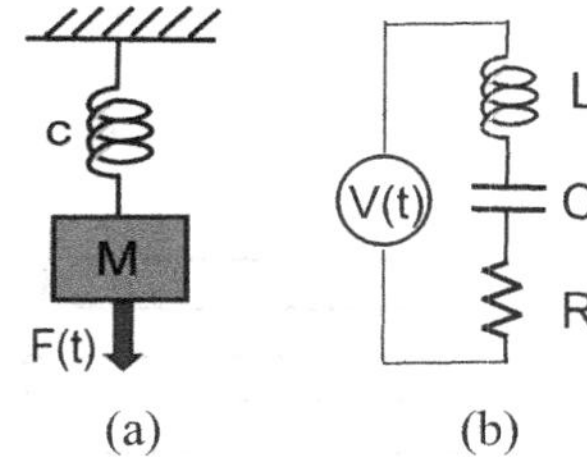

FIGURE 5.38 Discrete mechanical (a) and electrical (b) component systems.

Consider the transient response, $u(t)$ and $Q(t)$, in the case of the Heaviside step function of the force and voltage.

SOLUTION

Because these two second derivative equations are basically the same, we will consider only the mechanical system. Since the force is a Heaviside step function $F(t) = F_0\,H(t)$, the Laplace transform of the force is given by F_0/s. Equation P5.7.1 can be expressed by taking Laplace transformation as:

$$M[s^2 + (c/M)]U = F_0/s, \tag{P5.7.3}$$

where U is the Laplace transform of the displacement, u. U can be calculated by

$$\begin{aligned} U &= F_0/Ms\left(s^2+\omega_0^2\right) \quad \left[\omega_0^2 = c/M\right] \\ &= \left(F_0/M\omega_0^2\right)\{1/s-(1/2)[1/(s + j\omega_0)+1/(s-j\omega_0)]\} \end{aligned} \tag{P5.7.4}$$

Accordingly, using the inverse Laplace transformation, we can obtain:

$$u(t) = (F_0/c)[H(t) - \cos\omega_0 t]. \tag{P5.7.5}$$

As the result of Equation P5.7.5 is illustrated in Figure 5.39, the step force excites a sinusoidal displacement vibration of a discrete component of mass M, superposed on a step bias position. Here the time scale, T, is equal to the resonance period, $2\pi/\omega_0$ $(= 2\pi\sqrt{M/c})$. When we neglect the damping factor, the vibration ringing will continue forever. Note 100% overshoot for every vibration. When we consider the damping factor, ζ, the ringing will diminish gradually with the exponential envelope curve of the time constant, τ, inversely proportional to the damping factor. Refer to Example Problem 5.3.

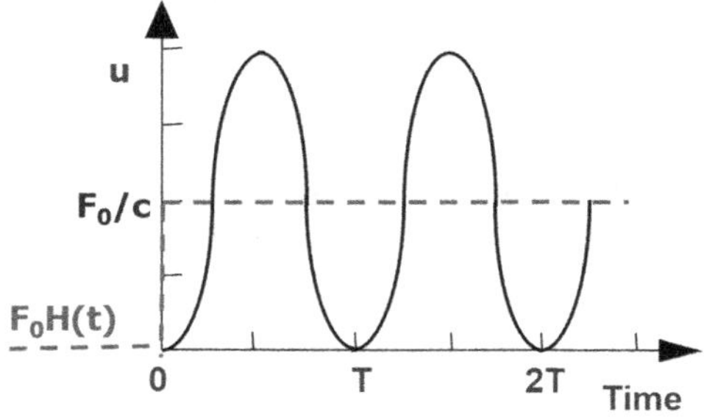

FIGURE 5.39 Sinusoidal displacement response to Heaviside step function force in a discrete component system.

5.4.3.3 Displacement Response to Pulse Drive

Let us consider next the response to a rectangular pulse voltage such as that pictured in the top left-hand corner of Figure 5.40a.[20] We begin by substituting

$$\tilde{E} = (E_0/s)(1 - e^{-(n\,sL)/v}) \tag{5.102}$$

into Equation 5.99, which allows us to obtain the displacement ΔL for $n = 1$, 2, and 3. The quantity n is a time scale based on half of the resonance period (= $T/2$) of the piezoelectric plate.

For $n = 1$,

$$\begin{aligned} U(s, x = L) &= d_{31}E_0(v/s^2)\,(1 - e^{-(sL/v)})^2/(1 + e^{-(sL/v)}) \\ &= d_{31}E_0(v/s^2)(1 - 3e^{-sL/v} + 4e^{-2sL/v} - 4e^{-3sL/v} + \cdots). \end{aligned} \tag{5.103}$$

Similar to the step case, the base function of $U(s, L)$, $1/s^2$, gives the base function of $u(t, l)$ in terms of t. The inverse Laplace transform of Equation 5.103 yields:

$$\begin{aligned} &u(t,L) = d_{31}E_0vt && 0 < t < L/v \\ &u(t,L) = d_{31}E_0v[t - 3(t - L/v)] && L/v < t < 2L/v \\ &u(t,L) = d_{31}E_0v[t - 3(t - L/v) + 4(t - 2L/v)] && 2L/v < t < 3L/v \\ &\cdots\cdots\cdots \end{aligned} \tag{5.104}$$

The transient displacement, ΔL, produced by the rectangular pulse voltage is pictured in Figure 5.40a for $n = 1$. The resonance period of this piezoelectric plate corresponds to $(2L/v)$. Notice how continuous ringing occurs under this condition.

For $n = 2$, since $\tilde{E} = (E_0/s)(1 - e^{-(2Ls)/v})$ includes the denominator of Equation 5.99,

$$U(s,L) = d_{31}E_0(v/s^2)(1 - 2e^{-sL/v} + e^{-2sL/v}) \tag{5.105}$$

Thus,

$$\begin{aligned} &u(t,L) = d_{31}E_0vt && 0 < t < L/v \\ &u(t,L) = d_{31}E_0v[t - 2(t - L/v)] && L/v < t < 2L/v \\ &u(t,L) = d_{31}E_0v[t - 2(t - L/v) + (t - 2L/v)] = 0 && 2L/v < t < 3L/v \end{aligned} \tag{5.106}$$

In this case, the displacement, ΔL, occurs in a single pulse and does not exhibit ringing as depicted in Figure 5.40a, bottom left. Remember again that the applied field $\tilde{E}$ should include the denominator term $(1 + e^{-sL/v})$ to realize finite expansion terms, leading to a complete suppression of vibrational ringing.

For $n = 3$, $U(s, L)$ is again expanded as an infinite series:

$$\begin{aligned} U(s, x = L) &= d_{31}E_0(v/s^2)\,(1 - e^{-(3sL/v)})(1 - e^{-(sL/v)})/(1 + e^{-(sL/v)}) \\ &= d_{31}E_0(v/s^2)(1 - 2e^{-sL/v} + 2e^{-2sL/v} - 3e^{-3sL/v} + 4e^{-4sL/v} - 4e^{-5sL/v} \ldots). \end{aligned} \tag{5.107}$$

The displacement response for this case is pictured in Figure 5.40a, bottom right. Note the displacement slope (plate edge vibration velocity) has twice the difference among the field applied period and zero field.

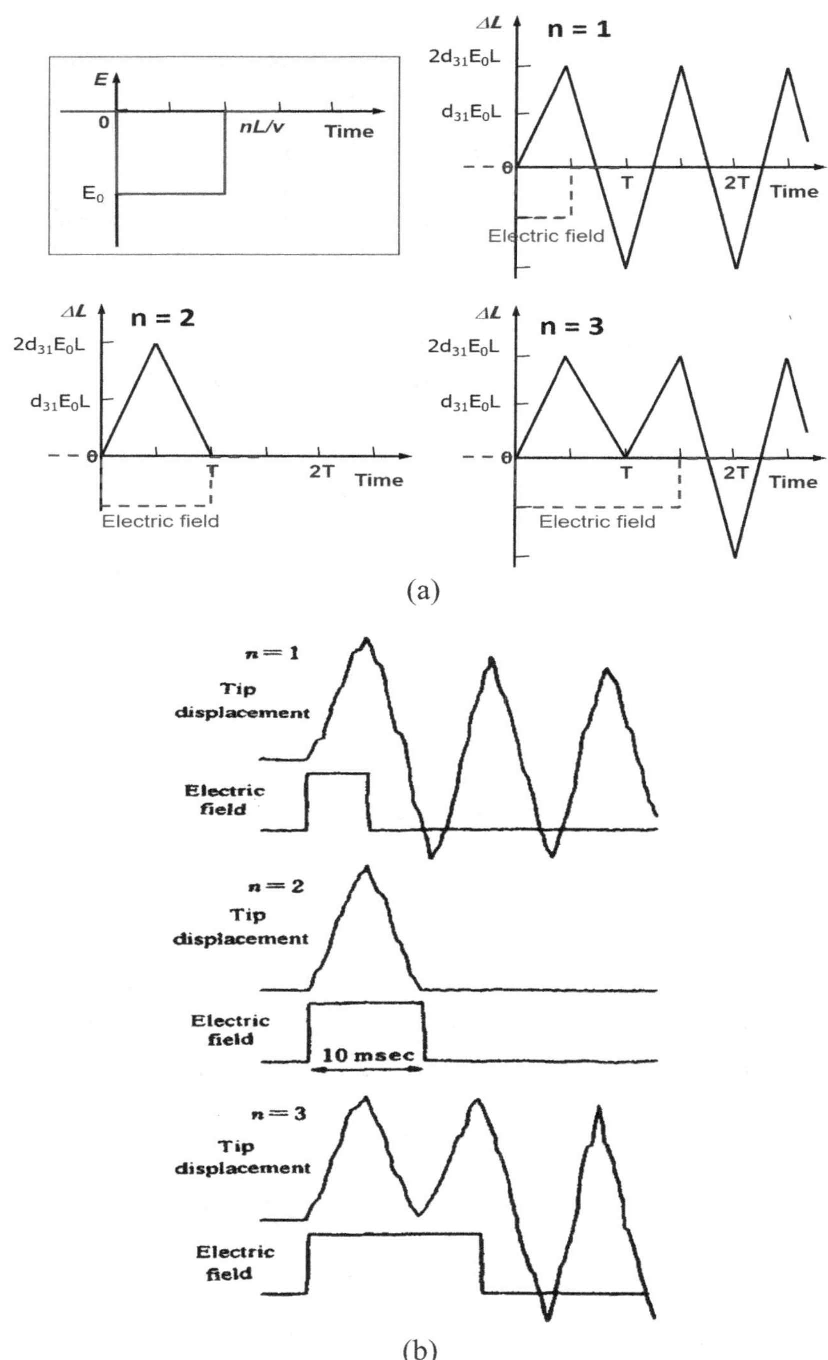

FIGURE 5.40 (a) Transient displacement ΔL produced by a rectangular pulse voltage. [Note that the time interval, $T = (2l/v)$, corresponds to the resonance period of the piezoelectric plate.] (b) Measurement on a bimorph tip displacement produced by a pulse voltage. (From Sugiyama and K. Uchino: *Proc. 6th IEEE Int'l Symp. Appl. Ferroelectrics*, p. 637, 1986.)

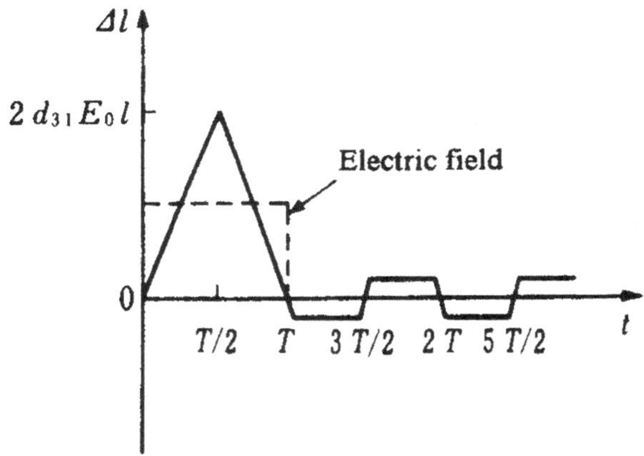

FIGURE 5.41 Transient displacement, ΔL, produced by a rectangular pulse voltage ($n = 1.9$).

Figure 5.40b shows the measurement data collected on a high mechanical quality factor ($Q_m = 1000$) PZT bimorph tip displacement produced by a pulse voltage, measured with an eddy-current type noncontact sensor. Ten msec corresponds to the resonance period of this bimorph. Notice that the ringing is completely eliminated in the case of $n = 2$; the pulse width is adjusted exactly to the resonance period, and that the displacement curve is linear or a sequence of triangulars.

How precisely does the pulse width need to be adjusted? The calculated transient vibration for $n = 1.9$ appears in Figure 5.41. Notice the small amount of ringing that occurs after the main pulse. The actual choice of n, then, will depend on the amount of ringing that can be tolerated for a given application.

5.4.3.4 Displacement Response to Pseudo-Step Drive

How about the displacement response of a rectangular plate to *pseudo-step voltage* as illustrated in the top left of Figure 5.42a? The Laplace transform is provided by the subtraction of a straight line, $1/s^2$, with the time difference of (nL/v):

$$\tilde{E} = (E_0 v / nLs^2)(1 - e^{-nsL/v}). \tag{5.108}$$

Substituting Equation 5.108 into Equation 5.99, we will repeat calculations similar to the above for obtaining the displacement, ΔL, for a time scale $n = 1$, 2, and 3 (the time unit is *half of the resonance period,* (L/v). The difference from Section 5.4.3.3 is in the base function of $U(s, L)$, $1/s^3$, which gives the base function of $u(t, L)$ as $t^2/2$ (parabolic curve).

For $n = 1$,

$$\begin{aligned} U(s,L) &= \frac{(d_{31}E_o/L)(v^2/s^3)(e^{-sL/v})^2}{(1+e^{-sL/v})} \\ &= (d_{31}E_o/L)(v^2/s^3)[1 - 3e^{-sL/v} + 4e^{-2sL/v} - 4e^{-3sL/v} + \cdots] \end{aligned} \tag{5.109}$$

Notice that the base function of $U(s, L)$, $1/s^3$, will lead to a base function of $u(t, L)$ in the form $t^2/2$ (parabolic curve), such that:

$$\begin{aligned} &u(t,L) = (d_{31}E_o v^2/2L)t^2 && 0 < t < L/v \\ &u(t,L) = (d_{31}E_o v^2/2L)[t^2 - 3(t - L/v)^2] && L/v < t < 2L/v \\ &u(t,L) = (d_{31}E_o v^2/2L)[t^2 - 3(t - L/v)^2 + 4(t - 2L/v)^2] && 2L/v < t < 3L/v \\ &\cdots\cdots\cdots \end{aligned} \tag{5.110}$$

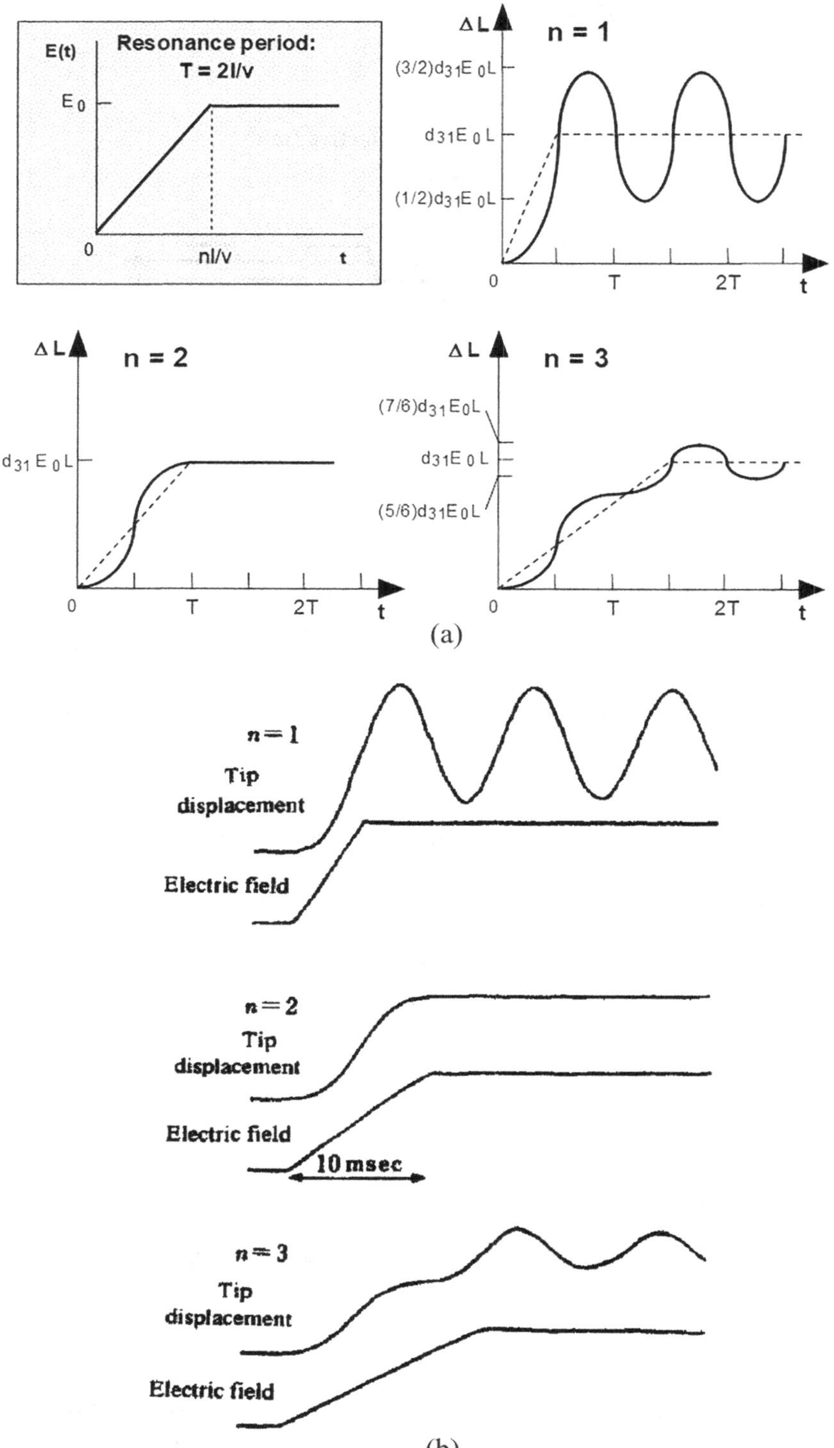

FIGURE 5.42 (a) Transient displacement, ΔL, induced in a rectangular plate for a pseudo-step voltage. The time scale, n, is based on 1/2 of the resonance period, T. (b) Measurement on a bimorph tip displacement produced by a pseudo-step voltage.

The transient displacement for an actuator driven by the pseudo-step voltage pictured in Figure 5.42a, top right, is seen to exhibit continuous ringing. Notice that this curve is actually a sequence of parabolic curves. It is *not* sinusoidal!

For $n = 2$,

$$U(s,L) = (d_{31}E_o/2L)(v^2/s^3)[1 - 2e^{-sL/v} + e^{-2sL/v}] \tag{5.111}$$

Thus,

$$\begin{aligned} &u(t,L) = (d_{31}E_o v^2/4L)t^2 && 0 < t < L/v \\ &u(t,L) = (d_{31}E_o v^2/4L)[t^2 - 2(t - L/v)^2] && L/v < t < 2L/v \\ &u(t,L) = (d_{31}E_o v^2/4L)[t^2 - 2(t - L/v)^2 + (t - 2L/v)^2] = (d_{31}E_o L/2) && 2L/v < t \end{aligned} \tag{5.112}$$

Neither overshoot nor ringing is apparent in the response for this case (i.e., the rise time is set exactly to the resonance period) represented in Figure 5.42a, bottom left. *When the applied field,* $\tilde{E}$, *includes the term* $(1 + e^{-sL/v})$, *the expansion series terminates in finite terms, leading to a complete suppression of mechanical ringing.*

For $n = 3$, $U(s, L)$ is again expanded as an infinite series:

$$U(s,L) = (d_{31}E_o/3L)(v^2/s^3)[1 - 2e^{-sL/v} + 2e^{-2sL/v} - 3e^{-3sL/v} + 4e^{-4sL/v} - 4e^{-5sL/v}\ldots] \tag{5.113}$$

The displacement response for this condition is represented by the curve appearing in Figure 5.42a, bottom right.

As shown in Figure 5.42a, ΔL does not exhibit overshoot or ringing for $n = 2$. But, for $n = 1$ and 3, the overshoot and ringing follow continuously. Note again that all the curves are composed of parabolic curves (*not* sinusoidal!) and that the heights of the overshoot are 1/2 and 1/6 of $d_{31}E_0L$, respectively, for $n = 1$ and 3.

You can also understand that the strain in the sample is generated linearly (not suddenly) with time, since the electric field changes thus. Figure 5.42b shows the measurement data collected on a PZT bimorph tip displacement produced by a pseudo-step voltage. Ten msec corresponds to the resonance period of this bimorph.

This derivation process suggests an empirical process on how to suppress the overshoot and/or ringing in a piezoelectric actuator system:

1. By applying a relatively steep rising voltage to the actuator, we can obtain the resonance period first from the time period between the overshoot peak and the successive peak point.
2. By adjusting the voltage rise time exactly to the resonance period next, we can eliminate the overshoot and/or ringing of the system vibration.

In other words, without using a mechanical damper (which loses energy), we can diminish vibration overshoot or ringing just by adjusting the pulse width or rise time of the applied voltage, which does not lose energy, in fact. This procedure can be adopted to the cutting edge/tool position change with a piezoelectric actuator in lathe, milling, or other cutting machines.

5.4.3.5 Consideration of the Loss in Transient Response

Recall that in order to properly account for loss effects, Equation 5.85 must be used for the transverse vibration rather than Equation 5.78. The displacement, $U(s, L)$, can then be obtained by making the following substitution in Equation 5.99:

$$s \to \frac{s}{\sqrt{1+\varsigma s}} \tag{5.114}$$

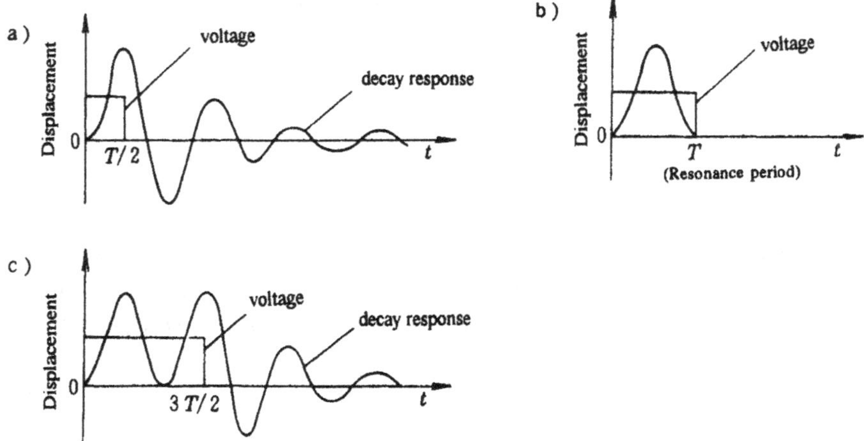

FIGURE 5.43 Displacement response of a lossy piezoelectric actuator under a rectangular pulse voltage. (From S. Smiley: US-Patent, No.3614486, 1971.)

This solution has not yet been obtained in an explicit form. Approximate solutions for the piezoelectric resonance state have been determined by Ogawa.[19]

Experimental results appear in Figure 5.43.[21] The displacement, ΔL, produced by a rectangular pulse voltage observed in this investigation is similar to the results shown in Figure 5.40, except for the vibrational damping. With the loss inclusion, the triangular shape is rounded and the amplitude is significantly damped. However, we still see that when the pulse width of this rectangular voltage is adjusted to the piezoelectric resonance period, T, or integral multiples of it, the vibrational ringing is eliminated.

5.4.4 Pulse Width Modulation Method

The *pulse width modulation* (PWM) method is useful to control a slow object by a quick actuator. We demonstrated controlling an oil pressure servo valve (see Section 8.5 for details).[22] The principle of the PWM method is illustrated in Figures 5.44a–c. A rectangular wave (RW) is generated, with a pulse width determined by the intersection of the *standard saw wave* (SW, 1.25 kHz) and the command voltage wave (CW), and then amplified and applied to a PZT bimorph flapper. When the bimorph flapper is driven by a rectangular wave ($\pm V_{max}$) of 1.25 kHz, the oil pressure outside the nozzle maintains a constant value (as if the flapper were located at some average position) proportional to the *duty factor* (the ratio of the period during which a pulse is sustained over the repetition period), because the response of the oil flow is much slower (about 50 Hz) than the piezo-actuator. The oil pressure outside the nozzle, and consequently the spool position, can be controlled in a linear fashion by the command voltage, thus diminishing the nonlinearity and hysteresis associated with the piezoelectric response of the PZT (see Figure 5.44d).[22]

Engineering control theory guarantees that the time-averaged PWM output is proportional to the command input if the following conditions are satisfied:

1. The ratio of the carrier frequency, f_c, to the input command frequency, f_e, should be greater than 7.
2. The carrier frequency must be high enough compared to the system response to eliminate all ripple in the system output.

As an example, let's consider an oil pressure valve like the one just described, which has a carrier wave frequency of 1250 Hz. Then, the command frequency must not exceed 1250/7 = 180 Hz.

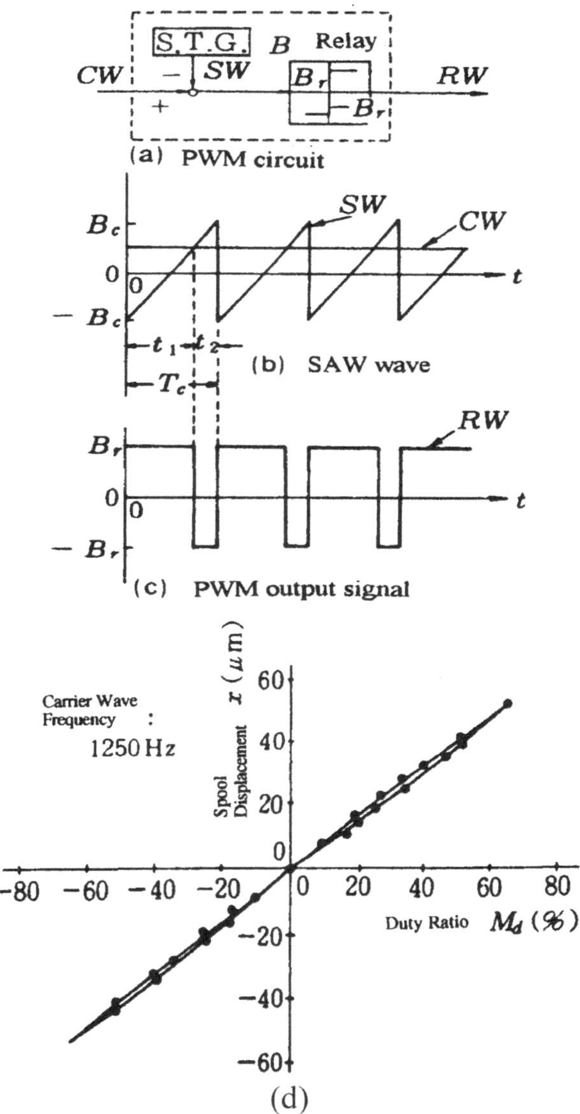

FIGURE 5.44 Pulse width modulation (PWM) method: (a) PWM circuit, (b) sawtooth carrier wave, (c) PWM output signal, (d) performance of the oil pressure servo valve.

Where the carrier frequency is much higher than the oil flow response (50 Hz) of the valve, the second condition is also satisfied.

The PWM is very attractive from an engineering control point of view, in particular for eliminating the hysteresis problems that will occur in an open-loop method. However, in general, special care must be taken when utilizing a bimorph actuator. First, the lifetime of the bimorph is limited to 10^8–10^9 cycles when it is driven with a bang-bang mode at maximum voltage. Delamination of the bonding layer may occur after only 10 hours when it is operated continuously at the 1-kHz carrier frequency. Second, since the bimorph motion is associated with a large hysteresis, it tends to generate a considerable amount of heat, accelerating the piezoelectric aging effect. A multilayer actuator with a lifetime longer than 10^{11} cycles is a possible solution for this application.

5.5 RESONANCE DRIVE

When an alternating electric field is applied to a piezoelectric ceramic, mechanical vibration is excited, and if the drive frequency is adjusted to the mechanical resonance frequency of the device, a large resonant strain is generated. This phenomenon is called *piezoelectric resonance* and is very useful for applications such as energy trap devices and actuators. We will consider in this section the steady-state response of such a device to a sinusoidally varying electric field.

5.5.1 Piezoelectric Resonance: Reconsideration

Let us reconsider a piezoelectric rectangular plate in which the transverse piezoelectric effect excites the length extensional mode, as depicted in Figure 5.34. We will apply a sinusoidal electric field ($E_z = E_o e^{j\omega t}$) to the plate and consider the *standing longitudinal vibration mode* based on the transverse piezoelectric effect, d_{31}, for $t \gg 0$. The following expression for the strain coefficient, $x_1(t, x)$, can be obtained by means of Equation 5.97, using the replacement of $s = j\omega$:

$$\begin{aligned} x_1(t,x) &= d_{31}\,E_o\,e^{j\omega t}\left[\frac{(e^{-j\omega(L-x)/v} - e^{-j\omega(L+x)/v} + e^{-j\omega x/v} - e^{-j\omega(2L-x)/v})}{(1-e^{-2j\omega L/v})}\right] \\ &= d_{31}E_o e^{j\omega t}\left[\frac{\sin[\omega(L-x)/v] + \sin[\omega x/v]}{\sin[\omega L/v]}\right] \\ &= d_{31}E_Z\left(\frac{\cos[(\omega(L-2x)/2v)]}{\cos(\omega L/2v)}\right). \qquad \left(v = 1/\sqrt{\rho s_{11}^E}\right) \end{aligned} \tag{5.115}$$

The piezoelectric actuator is an electronic component from the viewpoint of the driving power supply; thus, its *electrical impedance* [=*v*(applied)/*i*(induced)] plays an important role. The induced current is just the rate change of the electric displacement, D_z, which can be calculated by making use of the definition for D given by Equation 5.74:

$$i = w\left[\frac{\partial}{\partial t}\left(\int_0^l D_z dx\right)\right] = j\omega w\int_0^l\left[\left(\varepsilon_0\varepsilon_{33}^X - \left(\frac{d_{31}^2}{s_{11}^E}\right)\right)E_z + \left(\frac{d_{31}x_1}{s_{11}^E}\right)\right]dx \tag{5.116}$$

Substituting the expression given by Equation 5.115 for x_1 in this last equation and this expression for current into the defining equation for the admittance, Y, yields:

$$Y = \left(\frac{j\omega w L}{b}\right)\varepsilon_0\varepsilon_{33}^{LC}\left[1 + \left(\frac{d_{31}^2}{\varepsilon_0\varepsilon_{33}^{LC}s_{11}^E}\right)\left(\frac{\tan(\omega L/2v)}{(\omega L/2v)}\right)\right] \tag{5.117}$$

This describes the admittance of a mechanically free (unclamped) sample. Here, the width of the plate is w, its length L, and its thickness b. The quantity ε_{33}^{LC} is called the *longitudinally clamped dielectric constant*, which is given by:

$$\varepsilon_0\,\varepsilon_{33}^{LC} = \varepsilon_0\,\varepsilon_{33}^X - \left(d_{31}^2/s_{11}^E\right) = \varepsilon_0\varepsilon_{33}^X\left(1 - k_{31}^2\right) \tag{5.118}$$

The value ε_{33}^{LC} (1D-clamped permittivity) is close to ε_{33}^{X}, but not exactly equal to it theoretically, because ε_{33}^{X} is defined under the 3D-clamped condition.

- The *resonance state* is defined when the admittance becomes infinite (or the impedance becomes zero). It is described by Equation 5.117 when: $\tan(\omega l/2v) = \infty$ and $\omega L/2v = \pi/2$. The resonance frequency, f_R, can thus be provided as:

$$f_R = \frac{v}{2L} = \frac{1}{2L\sqrt{\rho s_{11}^E}} \tag{5.119}$$

- The *antiresonance state* is realized when the admittance becomes zero (and the impedance infinite). Under these conditions:

$$\left(\frac{\omega_A L}{2v}\right)\cot\left(\frac{\omega_A L}{2v}\right) = \frac{-d_{31}^2}{\varepsilon_0\varepsilon_{33}^{LC} s_{11}^E} = \frac{-k_{31}^2}{\left(1-k_{31}^2\right)} \tag{5.120}$$

where ω_A is the angular antiresonance frequency and, according to Table 5.2, the electromechanical coupling factor, k_{31}, is given by:

$$k_{31} = \frac{d_{31}}{\sqrt{s_{11}^E\ \varepsilon_0\varepsilon_{33}^X}} \tag{5.121}$$

EXAMPLE PROBLEM 5.8

When we apply an electric field on a k_{31} piezoplate (length: L), the total displacement ΔL can be obtained as:

$$U(s,L) = \frac{d_{31}\tilde{E}(v/s)(1-e^{-sL/v})}{(1+e^{-sL/v})} = d_{31}\tilde{E}(v/s)[\tanh(sL/2v)] \tag{P5.8.1}$$

When we start applying *cosine voltage* ($E_0 \cos \omega_0 t$, ω_0: resonance frequency) on this piezoplate, verify that the total displacement ΔL increases with the electric voltage cycle and finally reaches a saturated value, $(8/\pi^2)Q_m d_{31}E_z L$. Refer to Equation 3.54. How many cycles are required to realize the saturated value?

SOLUTION

Let us consider the response to a cosine voltage ($E_0 \cos \omega_0 t$). The Laplace transform is given by

$$\tilde{E}(s) = E_0 s/\left(s^2 + \omega_0^2\right) \tag{P5.8.2}$$

where $\omega_0 = (\pi v/L)$ (or $f_r = v/2L$) is the fundamental resonance frequency. The base function of $U(s, L)$ becomes $\left[1/\left(s^2 + \omega_0^2\right)\right]$, leading to the base function of $u(t, L)$ as $\sin \omega_0 t$. Thus,

$$\begin{aligned}
&u(t,L) = (d_{31}E_0 v)\sin \omega_0 t && 0 < t < L/v\\
&u(t,L) = (d_{31}E_0 v)[\sin \omega_0 t - 2\sin \omega_0(t - L/v)] && L/v < t < 2L/v\\
&u(t,L) = (d_{31}E_0 v)[\sin \omega_0 t - 2\sin \omega_0(t - L/v) + 2\sin\omega_0(t - 2L/v)] && 2L/v < t < 3L/v
\end{aligned} \tag{P5.8.3}$$

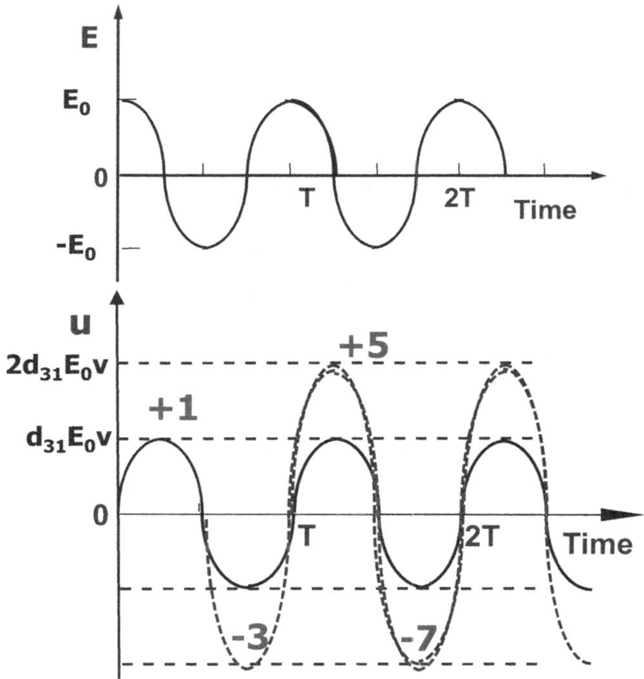

FIGURE 5.45 Displacement amplification mechanism under cosine voltage drive at the resonance frequency.

Figure 5.45 illustrates the displacement amplification mechanism under steady cosine electric field drive at the fundamental resonance frequency. After the original $\sin \omega_0 t$ curve, $-2\sin \omega_0(t - L/v)$ and $+ 2\sin \omega_0(t - 2L/v)$ are superposed every $T/2$ period, so that the vibration amplitude increases $+1$, -3, $+5$, -7, and so on; that is, $+4(d_{31}E_0v)$ for every resonance period cycle. The displacement increases linearly at the initial stage. However, due to the elastic loss, each sine term will decay the amplitude by a factor of $e^{-t/\tau}$. Thus, after reasonably many cycles, the displacement will be given in proportion to $4 \times (1 + e^{-T/\tau} + e^{-2T/\tau} + e^{-3T/\tau} + e^{-4T/\tau} + \cdots) = 4/(1 - e^{T/\tau}) = 4\tau/T$. Taking into account the relation: $Q_m = (1/2)\omega_0\ \tau$, the above saturated value becomes $4\tau/T = 4Q_m/\pi$. In conclusion, the number of cycles $\sim Q_m$ is required to obtain the saturated displacement amplitude, and the saturated amplitude is $(8/\pi^2)Q_m d_{31}E_zL$ theoretically for the fundamental resonance frequency.

5.5.2 Equivalent Circuits for Piezoelectric Vibrators

5.5.2.1 Equivalency between Mechanical and Electrical Systems

The dynamic equation for a mechanical system composed of a mass, a spring, and a damper illustrated in Table 5.3, left, is expressed by

$$M(d^2u/dt^2) + \zeta(du/dt) + c\,u = F(t),$$

or

$$M(dv/dt) + \zeta v + c\int_0^t v\,dt = F(t) \qquad (5.122)$$

where u is the displacement of a mass, M; v is the velocity ($= du/dt$); c the spring constant; ζ the damping constant of a dash-pot; and F the external force.

TABLE 5.3
Equivalency between a Mechanical and an Electrical System, Composed of *M, c, ζ; L, C, R*

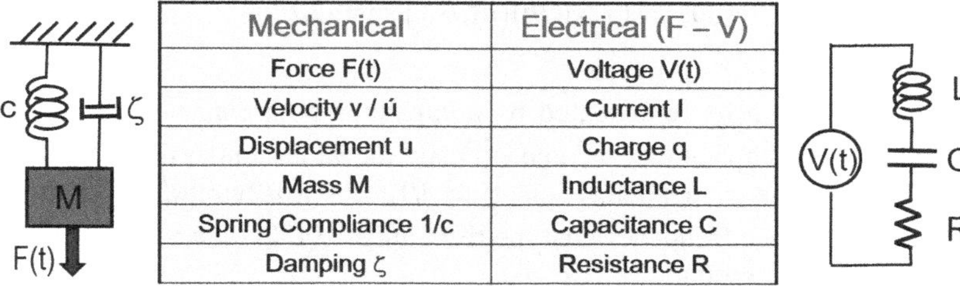

Mechanical	Electrical (F – V)
Force F(t)	Voltage V(t)
Velocity v / u̇	Current I
Displacement u	Charge q
Mass M	Inductance L
Spring Compliance 1/c	Capacitance C
Damping ζ	Resistance R

On the other hand, the dynamic equation for an electrical circuit composed of an inductance, L; a capacitance, C; and a resistance, R, illustrated in Table 5.3, right, is expressed by

$$L(d^2Q/dt^2) + R(dQ/dt) + (1/C)Q = V(t),$$

or

$$L(dI/dt) + RI + (1/C)\int_0^t I\,dt = V(t) \tag{5.123}$$

where Q is charge, I is the current ($= dQ/dt$), and V is the external voltage. Taking into account the equation similarity, the engineer introduces an equivalent circuit; that is, considering a mechanical system using an equivalent electrical circuit is intuitively simpler for an electrical engineer. In contrast, considering an electrical circuit using an equivalent mechanical system is intuitively simpler for a mechanical engineer. The equivalency between these two systems is summarized in Table 5.3. Basically, mass, M; spring compliance, $1/c$; and damping constant, ζ, correspond to inductance, L; capacitance, C; and resistance, R, respectively.

5.5.2.2 Equivalent Circuit (Loss-Free) of the k_{31} Mode

The equivalent circuit (EC) is a widely used tool that can greatly simplify the process of design and analysis of piezoelectric devices, in which the circuit, in a standard form,[23] can only graphically characterize the mechanical loss by applying a resistor (and sometimes dielectric loss). Damjanovic therefore introduced an additional branch into the standard circuit, which is used to present the influence of the piezoelectric loss.[24] However, concise and decoupled formulas of losses that can be used for the measurements of losses in piezoelectric material as a user-friendly method have not been derived yet.

We consider first the equivalent circuit (loss-free) for k_{31} mode, shown in Figure 5.34. We start from the admittance equation, already discussed:

$$Y = (j\omega wL/b)\varepsilon_0\varepsilon_{33}^{LC}[1 + \left(d_{31}^2/\varepsilon_0\varepsilon_{33}^{LC}s_{11}^E\right)(\tan(\omega L/2v)/(\omega L/2v)] \tag{5.124}$$

where v is the sound velocity of the piezoelectric given by

$$v = 1/\sqrt{\rho s_{11}^E} \tag{5.125}$$

By splitting Y into the damped admittance, Y_d, and the motional part, Y_m,

$$Y_d = (j\omega wL/b)\varepsilon_0\varepsilon_{33}^{LC} \tag{5.126}$$

$$Y_m = (j2vw/b)\left(d_{31}^2/s_{11}^E\right)(\tan(\omega L/2v) \tag{5.127}$$

The damped branch can be represented by a capacitor with capacitance $(wL/b)\,\varepsilon_0\varepsilon_{33}^{LC}$ (C_d in Figure 5.46). The damped capacitance, C_d and ε_{33}^{LC} , are the values under the 1D-clamped condition. For the motional branch (mechanical vibration), since $1/Y_m = -j(b/2vw)\left(s_{11}^E/d_{31}^2\right)(\cot(\omega L/2v)$ is close to zero around the resonance frequency (A-type resonance) $\omega_A L/2v = n\,\pi/2$ ($n = 1, 3, 5, \ldots$), taking the Taylor expansion series around $\omega_{A,n}$, we get

$$1/Y_m \approx j(b/2vw)\left(s_{11}^E/d_{31}^2\right)(\Delta\omega L/2v)). \tag{5.128}$$

where $\Delta\omega = \omega - \omega_{A,n}$. The suffix n indicates the nth high-order harmonics, taking the value, 1, 3, 5, and so on. Now, we convert the mass contribution to L and elastic compliance to C, then create L_n and C_n series connections, as shown in Figure 5.46. The impedance of this LC circuit around the resonance, ω_A, is provided by

$$1/Y_n = j\omega L_n + 1/j\omega C_n \approx j\left(L_n + 1/\omega_{A,n}^2 C_n\right)\Delta\omega \tag{5.129}$$

Since the resonance is realized when $L_n = 1/\omega_{A,n}^2 C_n$, by comparing Equations 5.128 and 5.129, we can obtain the following two equations that express L, C by the transducer physical parameters:

$$L_n = \frac{\left(bLs_{11}^E/4v^2wd_{31}^2\right)}{2} = (\rho/8)(Lb/w)\left(s_{11}^{E2}/d_{31}^2\right) \tag{5.130}$$

$$\begin{aligned} C_n &= 1/\omega_{A,n}^2 L_n = (L/n\pi v)^2(8/\rho)(w/Lb)\left(d_{31}^2/s_{11}^{E2}\right) \\ &= (8/n^2\pi^2)(Lw/b)\left(d_{31}^2/s_{11}^{E2}\right)s_{11}^E \end{aligned} \tag{5.131}$$

$$\omega_{A,n} = n\pi/L\sqrt{\rho s_{11}^E} \tag{5.132}$$

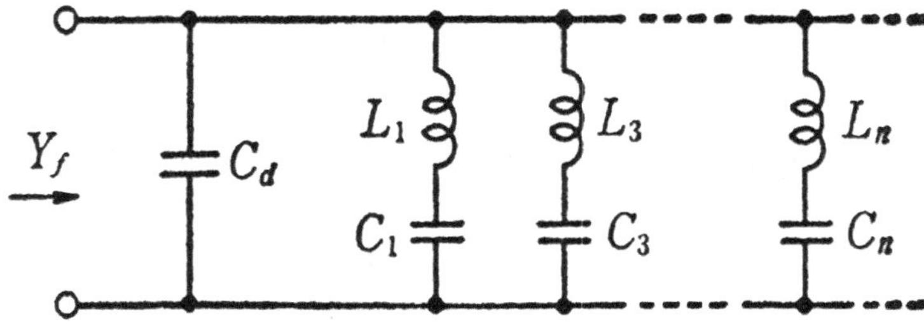

FIGURE 5.46 Equivalent circuit for k_{31} mode (loss-free).

Note initially that L_n is a constant, irrelevant to n (all harmonics have the same L), and directly proportional to the density ρ. C_n is proportional to $s_{11}{}^E$ ($d_{31}/s_{11}{}^E$) corresponds to the force factor introduced in the four-terminal equivalent circuit in Section 5.5.2.3.3. The total motional capacitance, $\sum_n C_n$, is calculated as follows, using the important relation $\Sigma[1/(2m-1)^2] = (\pi^2/8)$:

$$\sum_n C_n = \sum_n \frac{1}{n^2}\left(\frac{8}{\pi^2}\right)\left(\frac{Lw}{b}\right)\left(\frac{d_{31}^2}{s_{11}^E}\right) = \left(\frac{Lw}{b}\right)\left(\frac{d_{31}^2}{s_{11}^E}\right) = k_{31}^2 C_0 \tag{5.133}$$

Therefore, we can understand that the total capacitance, $C_0 = (wL/b)\varepsilon_0\varepsilon_{33}^X$, is split into the damped capacitance, $C_d = \left(1-k_{31}^2\right)C_0$, and the total motional capacitance, $k_{31}^2 C_0$, which is reasonable from an energy conservation viewpoint.

5.5.2.3 Equivalent Circuit (with Losses) of k_{31} Mode

We start from Hamilton's Principle, a powerful tool for "mechanics" problem solving, which can transform a physical system model to *variational problem* solving. We integrate loss factors directly into Hamilton's Principle for a piezoelectric k_{31} plate (Figure 5.34).[25,26] By skipping the detailed derivation process, we obtain the following admittance expression:

$$\begin{aligned} Y^* &= j\omega\cdot\frac{lw}{b}\cdot\left(\varepsilon_0\varepsilon_{33}^{X'} - \mathrm{Re}\left[\frac{\left(d_{31}^*\right)^2}{s_{11}^{E*}}\right]\right) + \omega\cdot\frac{lw}{b}\cdot\left(\varepsilon_0\varepsilon_{33}^{X''} + \mathrm{Im}\left[\frac{\left(d_{31}^*\right)^2}{s_{11}^{E*}}\right]\right) \\ &+ j\omega\cdot\frac{8lw}{b\pi^2}\cdot\mathrm{Re}\left[\frac{\left(d_{31}^*\right)^2}{s_{11}^{E*}}\right]\cdot\frac{\pi^2/l^2\rho s_{11}^{E*}}{\left(\pi^2/l^2\rho s_{11}^{E*}\right)-\omega^2} + j\omega\cdot\frac{8lw}{b\pi^2}\cdot\left(j\mathrm{Im}\left[\frac{\left(d_{31}^*\right)^2}{s_{11}^{E*}}\right]\right)\cdot\frac{\pi^2/l^2\rho s_{11}^{E*}}{\left(\pi^2/l^2\rho s_{11}^{E*}\right)-\omega^2} \end{aligned} \tag{5.134}$$

Among the above four terms in Equation 5.134, the first and second terms correspond to the damped capacitance and its 1D-clamped dielectric loss [i.e., close to the extensive dielectric loss (tan δ)], respectively, while the third and fourth terms correspond to the motional capacitance and the losses combined with intensive elastic and piezoelectric losses.

On the other hand, by introducing the complex parameters into the admittance formula, Equation 5.125, around the resonance frequency[27]: $\varepsilon_3^{X*} = \varepsilon_3^X(1-j\tan\delta_{33}')$, $s_{11}^{E*} = s_{11}^E(1-j\tan\phi_{11}')$, and $d_{31}^* = d_{31}(1-j\tan\theta_{31}')$:

$$\begin{aligned} Y &= Y_d + Y_m \\ &= j\omega C_d(1-j\tan\delta_{33}) \\ &\quad + j\omega C_d K_{31}^2[(1-j(2\tan\theta_{31}'-\tan\phi_{11}')]\left[(\tan\left(\omega L/2v_{11}^{E*}\right)/\left(\omega L/2v_{11}^{E*}\right)\right]. \end{aligned} \tag{5.135}$$

It is not difficult to verify the equivalency between Equations 5.134 and Equations 5.135.

5.5.2.3.1 Equivalent Circuit in Institute of Electrical and Electronics Engineers Standard

Figure 5.47 shows an IEEE Standard equivalent circuit with only one elastic loss (tan ϕ'). In addition to Equations 5.130 through 5.132, the circuit analysis provides the following R and Q (electrical quality factor, which corresponds to the mechanical quality factor in the piezoplate):

$$Q = \frac{\sqrt{L_A/C_A}}{R_A} \tag{5.136}$$

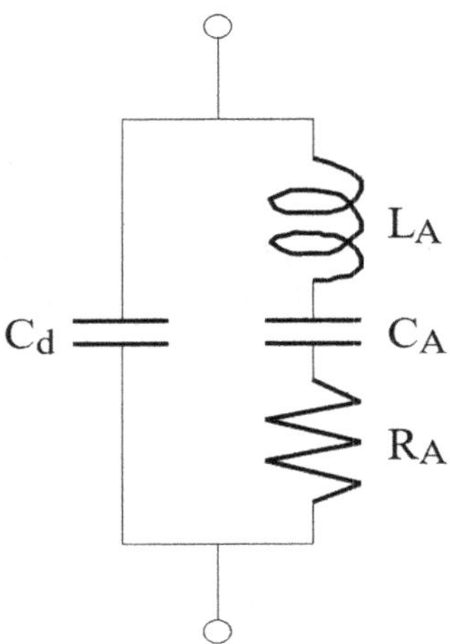

FIGURE 5.47 Equivalent circuit for k_{31} mode (IEEE).

PSpice is popular circuit analysis software for automatically maximizing the performance of circuits. EMA is distributing a free-download OrCAD Capture, this schematic design solution. How to use it is introduced in Chapter 6, "Computer Simulation of Piezoelectric Devices."

Figure 5.48a shows the equivalent circuit for the k_{31} mode of the IEEE type. L, C, and R values were calculated for PZT4 with 40 × 6 × 1 mm^3, and (b) plots the simulation results on the currents under 1 V_{ac}, that is, admittance magnitude and phase spectra. First, the damped admittance shows a slight increase with the frequency ($j\omega C_d$) with +90° phase in a full frequency range. Second, the motional admittance shows a peak at the resonance frequency, where the phase changes from +90° (i.e., capacitive) to –90° (i.e., inductive). In other words, the phase is exactly zero at the resonance (i.e., resistive characteristic). The admittance magnitude decreases above the resonance frequency with a rate of –40 dB down in a *Bode plot.* Third, by adding the above two, the total admittance is obtained. The admittance magnitude shows two peaks, maximum and minimum, which correspond to the resonance and antiresonance points, respectively. You can find that the peak sharpness (i.e., the mechanical quality factor) is the same for both peaks, because only one loss is included in the equivalent circuit. The antiresonance frequency is obtained at the intersect of the damped and motional admittance curves. Because of the phase difference between the damped (+90°) and motional (–90°) admittance, the phase is exactly zero at the antiresonance and changes to +90° above the antiresonance frequency. Remember that the phase is –90° (i.e., inductive) at a frequency between the resonance and antiresonance frequencies, which is discussed further in Section 5.5.7.

5.5.2.3.2 Equivalent Circuit with Three Losses

Damjanovic[24] introduced a motional branch to describe the third term in Equation 5.134, which contains a motional resistor, a motional capacitor, and a motional inductor. Meanwhile, an additional branch is also injected into the classical circuit[23] to pictorially express the last term in Equation 5.134 to present the influence of the piezoelectric loss, where the new resistance, capacitance, and inductance are all proportional to corresponding motional elements with the proportionality constant being

$$j\mathrm{Im}\left[\frac{\left(d_{31}^{*}\right)^{2}}{s_{11}^{E*}}\right] / \mathrm{Re}\left[\frac{\left(d_{31}^{*}\right)^{2}}{s_{11}^{E*}}\right].$$

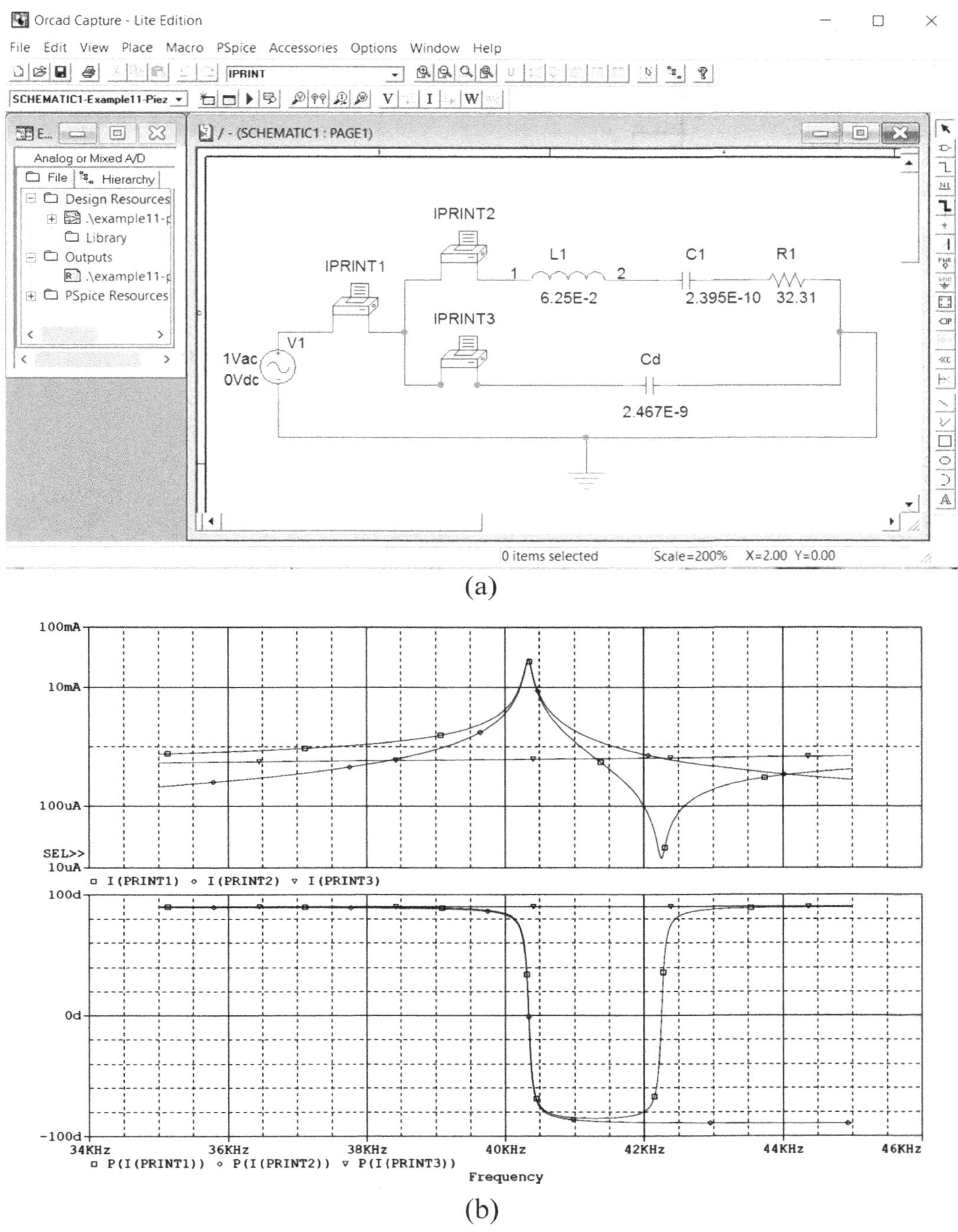

(a)

(b)

FIGURE 5.48 PSpice simulation of the IEEE-type k_{31} mode. (a) Equivalent circuit for k_{31} mode. L, C, and R values were calculated for PZT4 with $40 \times 6 \times 1$ mm^3. (b) Simulation results on admittance magnitude and phase spectra.

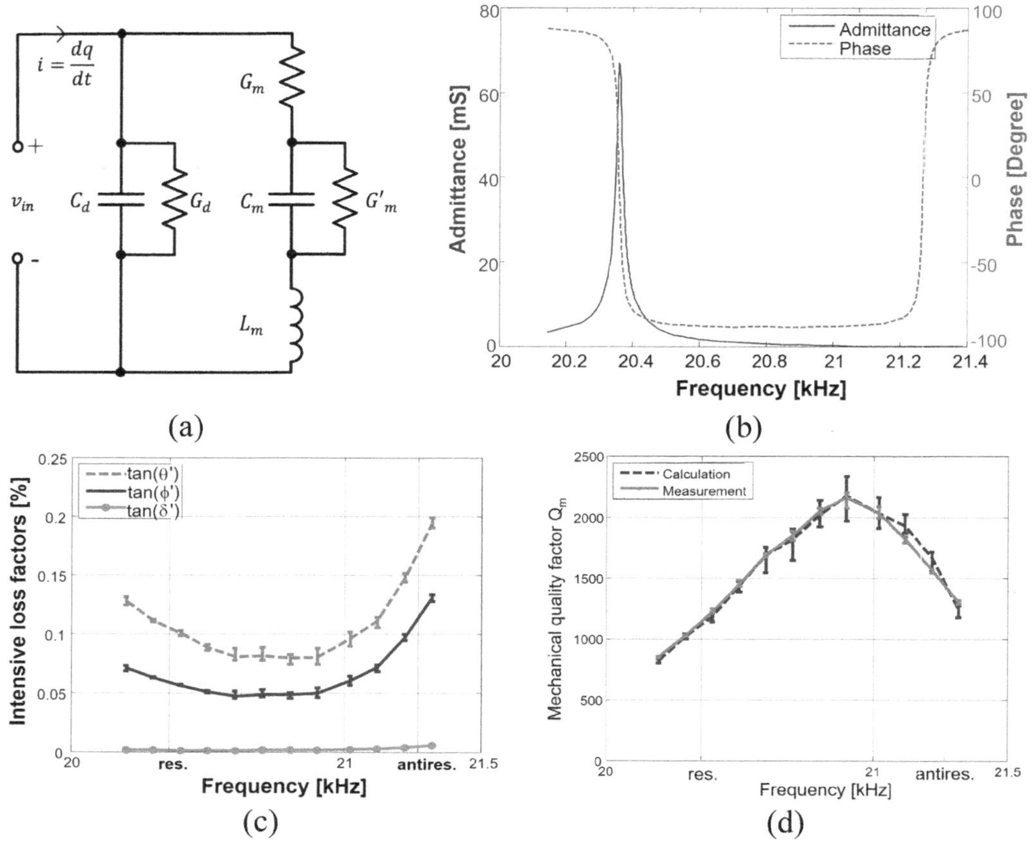

FIGURE 5.49 (a) Equivalent circuit proposed with three intensive loss factors, (b) admittance spectrum to be used in the simulation, (c) frequency spectra of intensive loss factors (i.e., dielectric, elastic, and piezoelectric losses) obtained from the admittance spectrum (b) fitting, and (d) calculated mechanical quality factor, Q_m, as a function frequency around the resonance and antiresonance frequencies.

Shi et al. proposed a more concise *equivalent circuit*, shown in Figure 5.49a with three losses.[25] Compared with the IEEE Standard EC with only one elastic loss or Damjanovic's EC with a full set of L, C, R, only one additional electrical element, G_m', is introduced into the classical circuit.[25] The new coupling conductance can reflect the coupling effect between the elastic and piezoelectric loss. This EC also can be mathematically expressed as:

$$Y^* = G_d + j\omega C_d + \frac{G_m' + j\omega C_m}{\left(1 + G_m'/G_m - \omega^2 L_m C_m\right) + j\left(\omega C_m/G_m + \omega L_m G_m'\right)} \tag{5.137}$$

The parameters of the EC can therefore be obtained by comparing Equation 5.134 with Equation 5.137 as new expressions of three intensive loss factors:

$$\tan\phi' = \frac{\omega C_m}{G_m} \tag{5.138a}$$

$$\tan\theta' = \tan(\phi' - \beta') \tag{5.138b}$$

$$\tan\delta' = k_{31}^2 \tan(2\theta' - \phi') + \frac{G_d}{\omega C_d} \tag{5.138c}$$

where the phase delay, $\tan\beta' = (\omega C_m / G_m') - \sqrt{(\omega C_m / G_m')^2 + 1}$, denotes the disparity between the piezoelectric and elastic components. The value of β' generally stays negative or approaches zero (when $G_m' \to 0$), which implies that the piezoelectric loss is persistently larger than or equal to the elastic component (in PZTs). The significance of the piezoelectric loss has therefore been verified theoretically from the equivalent circuit viewpoint.

Using the experimental data in Figure 5.49(b), almost frequency-independent circuit parameters as $C_d = 3.2\text{nF}$, $C_m = 0.29\text{nF}$, $L_m = 210\,\text{mH}$, and $G_d = 0$ (extensive dielectric loss, tan δ, is small), and frequency-dependent parameters (G_m and G_m') can be obtained. By manipulating Equations 5.138a through 5.138c, we can determine intensive dielectric, elastic, and piezoelectric losses as a function of frequency, as shown in Figure 5.49c.

The mechanical quality factor, Q_m, is always applied to evaluate the effect of losses. When arriving at steady state, it can be expressed by:

$$Q_m = 2\pi \cdot \frac{\text{energy stored/cycle}}{\text{energy lost/cycle}} \tag{5.139}$$

The denominator is supposed to compensate the dissipation, w_{loss}; that is, $\int_V w_{loss} dV = (\pi/2)\left|v_3^*\right|\left|q_3^*\right|\cos\varphi$, where the phase difference between current and input voltage, φ, ranges within $[-\pi/2, \pi/2]$. Meanwhile, the reactive portion of the input energy returns to the amplifier and is neither used nor dissipated. Furthermore, the maximum stored and kinetic energies also reach equilibrium in an electric cycle. With definitions of energy items and appropriate substitutions, Q_m can be calculated as:[26]

$$Q_m = \frac{\omega_a^2 - \omega_r^2}{\cos\varphi} \cdot \frac{\omega^2}{\left|\omega^2 - (\omega_r^*)^2\right|\left|\omega^2 - (\omega_a^*)^2\right|} \tag{5.140}$$

As ω^2 approaches ω_r^2 or ω_a^2, the phase difference will approach zero. Therefore, for low k_{31}^2 materials, with substituting Equation 5.140, mechanical quality factors at the resonance and antiresonance frequencies can be calculated as:

$$Q_A = \frac{1}{\tan\phi'} = \frac{1}{R_m}\sqrt{\frac{L_m}{C_m}} \tag{5.141}$$

$$Q_B = \frac{1}{\tan\phi' + \dfrac{8K_{31}^2}{\pi^2}[\tan\phi' + \tan\delta' - 2\tan\theta']} \tag{5.142}$$

Here, the effective coupling coefficient $K_{31}^2 = k_{31}^2 / 1 - k_{31}^2$. Equations 5.141 and Equations 5.142 obtained from a new equivalent circuit are basically the same as we derived analytically in Equation 5.1. Hence, the calculation of Q_m at these special frequencies has been verified by the well-accepted conclusion. Not only at these frequencies, Equation 5.140 also infers an advanced calculation method of Q_m for a wide bandwidth. Figure 5.49d shows the frequency spectrum of the mechanical quality factor, Q_m, calculated from Equation 5.140. You can clearly find that (1) the Q_B at antiresonance is larger than Q_A at resonance (this is true in PZTs), and (2) the maximum Q_m (i.e., the highest efficiency) can be obtained at a frequency between the resonance and antiresonance frequencies. This frequency can theoretically be obtained by taking the first derivative of Equation 5.140 in terms of ω to be equal to zero, which suggests the best operating frequency of the transducer to realize the maximum efficiency. Applications will be discussed in Section 5.7.2.

5.5.2.3.3 Four-Terminal Equivalent Circuit

Though the above two-terminal EC is useful for the basic no-load piezoelectric samples, we need to extend it to four- and six-terminal EC models in order to consider the load effect for practical transducer/actuator applications. Uchino proposed a new *four-terminal equivalent circuit* for a k_{31}-mode plate, including elastic, dielectric, and piezoelectric losses (Figure 5.50a), which can handle symmetrical external mechanical losses. The four-terminal EC includes an ideal transformer with a voltage step-up ratio Φ to connect the electric (damped capacitance) and mechanical (motional capacitance) branches, where $\Phi = 2wd_{31}/s_{11}^E$, called *force factor.* New capacitance, l and c_1, are related to L_1 and C_1 in the two-terminal EC given in Equations 5.130 and Equations 5.131:

$$l = \Phi^2 L;\ c_1 = C_1/\Phi^2 \qquad (5.143)$$

Regarding the three losses, as shown in Figure 5.50a in a PZT plate, in addition to the IEEE standard "elastic" loss, r_1, and "dielectric" loss, R_d, we introduce the coupling loss r_{cpl} in the force factor

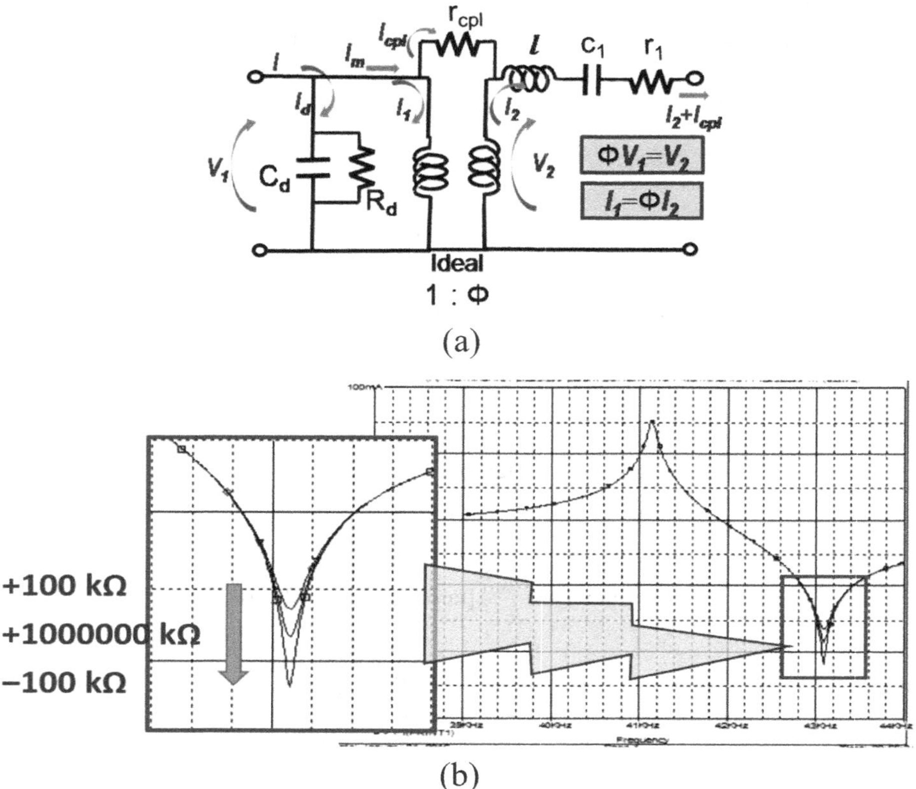

FIGURE 5.50 (a) Four-terminal (two-port) equivalent circuit for a k_{31} plate, including elastic, dielectric, and piezoelectric losses. r_1, R_d, and r_{cpl} correspond to these three losses. PSpice simulation results for three values of r_{cpl}.

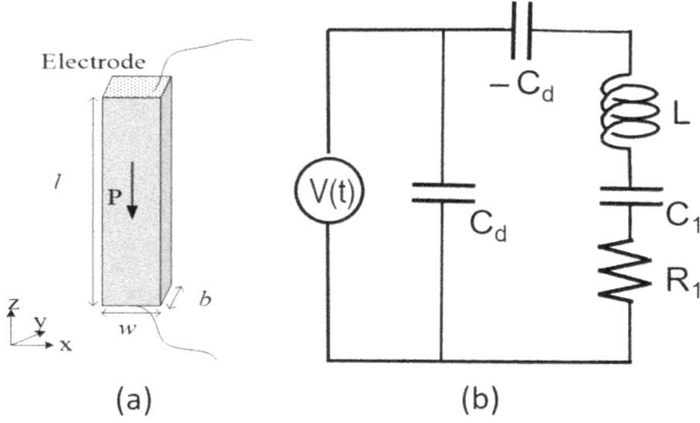

FIGURE 5.51 (a) k_{33}-mode piezoceramic rod; (b) equivalent circuit for k_{33} mode.

$\Phi = 2wd_{31}/s_{11}^E$ as proportional to (tan ϕ'– tan θ'), which can be either positive or negative, depending on the tan θ' magnitude. Figure 5.50b shows the PSpice software simulation results for three values of r_{cpl}. (1) The resonance, Q_A, does not change with changing r_{cpl}. (2) When $r_{cpl} = 100$ kΩ (i.e., tan $\theta' \approx 0$), $Q_A > Q_B$. (3) When $r_{cpl} = 1$ GΩ (i.e., tan ϕ'–tan $\theta' \approx 0$), $Q_A = Q_B$. (4) When $r_{cpl} = -100$ kΩ (i.e., tan ϕ' – tan $\theta' < 0$), $Q_A < Q_B$. The PZT results fit the case of $r_{cpl} = -100$ kΩ (i.e., tan ϕ' – tan $\theta' < 0$).

Dong et al. constructed a six-terminal equivalent circuit with three losses, which can handle asymmetric external loads for a k_{31}-mode plate[28] and Langevin transducer by integrating the head and tail mass loads,[29] then estimating the optimum (i.e., minimum required input electrical energy) driving frequency at which we can drive the transducer, as demonstrated with the highest efficiency. Section 5.7.2.3 discusses this issue in detail.

5.5.2.4 Equivalent Circuit of k_{33} Mode

Remember that k_{33} mode is governed by the sound velocity, v^D, not by v^E, the antiresonance is the primary mechanical resonance given by $f = v^D/2L$, and the resonance is the subsidiary mode originating from the electromechanical coupling factor, k_{33}. Therefore, its equivalent circuit includes so-called negative capacitance, $-C_d$ (exactly the same absolute value of the damped capacitance in the electric branch) in the motional branch, as shown in Figure 5.51. The closed-circuit impedance (corresponding to the antiresonance mode) should be the minimum, and the pure mechanical resonance status, the damped capacitance, should be compensated by this negative capacitance $-C_d$, while at the resonance, the open-circuit impedance should be the maximum, and the effective motional capacitance in the motional branch is provided by $1/[(1/C_1)+(1/-C_d)]$, which provides $s_{33}^D\left(= s_{33}^E\left(1-k_{33}^2\right)\right)$ rather than s_{33}^E (i.e., origin of C_1). In comparison with Equations 5.130, Equations 5.131, and Equations 5.136 in k_{31} mode, the components, L, C, and R can be denoted as:

$$\begin{aligned} C_n &= (8/n^2\pi^2)(wb/L)\left(d_{33}^2/s_{33}^{E2}\right)s_{33}^E\left(1-k_{33}^2\right) \\ L_n &= (\rho/8)(L^3/wb)\left(s_{33}^{E2}/d_{33}^2\right) \\ R_1 &= (L_1/C_1)^{1/2}/Q \end{aligned} \tag{5.144}$$

5.6 POSITION/FORCE SENSORS

Solid-state actuators are often used in conjunction with displacement sensors or force sensors in micropositioning systems. We will examine some of the most commonly used position and force sensors in this section.

TABLE 5.4
A Summary of Micro-Displacement Measurement Techniques

Principle	Method	Sensitivity	Response
Resistance	**Strain Gauge**		
	Metal	$\Delta l/l = 10^{-6}$	10 MHz
	Semiconductor	$\Delta l/l = 10^{-8}$	10 MHz
	Potentiometer		
	Contact	$\Delta l = 10^{-1}$ μm	100 kHz
	Non-Contact	$\Delta l = 10^{-3}$ μm	1 kHz
Electromagnetic Induction	**Differential Transformer**		
	AC	$\Delta l = 10^{-2}$ μm	100 Hz
	DC	$\Delta l = 10^{-1}$ μm	100 Hz
	Eddy Current	$\Delta l = 10^{-1}$ μm	50 kHz
Optical	**Optical Lever**	$\Delta l = 10^{-1}$ μm	1 kHz
	Optical Fiber	$\Delta l = 10^{-2}$ μm	100 kHz

5.6.1 Position Sensors

We will consider first various methods for measuring microscale displacements or strain. Precise displacement detection at this level is an important part of many drive/control systems, in particular servo displacement systems. Remarkable advances have been made with microscale position sensors, and currently attainable displacement resolutions are $\Delta l = 10^{-10}$ m (1 Å) using DC methods and $\Delta l = 10^{-13}$ m (10^{-3} Å) using AC methods. The techniques are classified into two general categories: electrical (resistance, electromagnetic induction, and capacitance) and optical (optical lever, optical grid, interferometric, and optical sensor) methods.[30] A summary of the sensitivity and response of these methods appears in Table 5.4 after the brief descriptions of each that follow.

5.6.1.1 Resistance Methods

Metal wire *strain gauges*, which are often used to measure the piezoelectric strain ($\Delta l/l$), typically have a resolution of 10^{-6}. The basic structure of this device is pictured schematically in Figure 5.52a. The strain gauge is usually attached to the sample with an appropriate resin. The resistance of the gauge wire changes when the sample deforms. A *Wheatstone bridge* is used to precisely measure the resistance change, from which the strain can then be determined. The change in resistance is actually due to two effects: the change in specimen size and the *piezoresistive effect*. In a metal wire gauge, the magnitudes of these two effects are comparable, while for a gauge made from a semiconductor, the piezoresistive effect is greater by 2 orders of magnitude than the size effect. The frequency response of the strain gauge is generally quite broad. Measurements at frequencies as high as 10 MHz are possible.

Contact potentiometers are similar to helical variable resistors. The contact position can be monitored according to the resistance. There are two general types of potentiometer sensors: the *angle detector* and the *linear position detector*. The resolution of a potentiometer device is limited by the thickness of the resistive helical wire, which is typically about 10 μm. The resolution is quite a bit better, however, for a newer variety, which makes use of a continuous conductive polymer. This type has a resolution to 0.1 μm and a response of 100 kHz.[31] *Noncontact potentiometers* operate through the *magnetoresistive effect*, by which certain semiconducting materials exhibit a large resistance change with the application of a magnetic field. The basic structure of a typical noncontact potentiometer, composed of a magnetoresistive sensor and a tiny permanent magnet, is pictured schematically in Figure 5.52d. This structure tends to have a long lifetime because noncontact eliminates problems of wear. It also offers an excellent resolution of 10^{-3} μm.

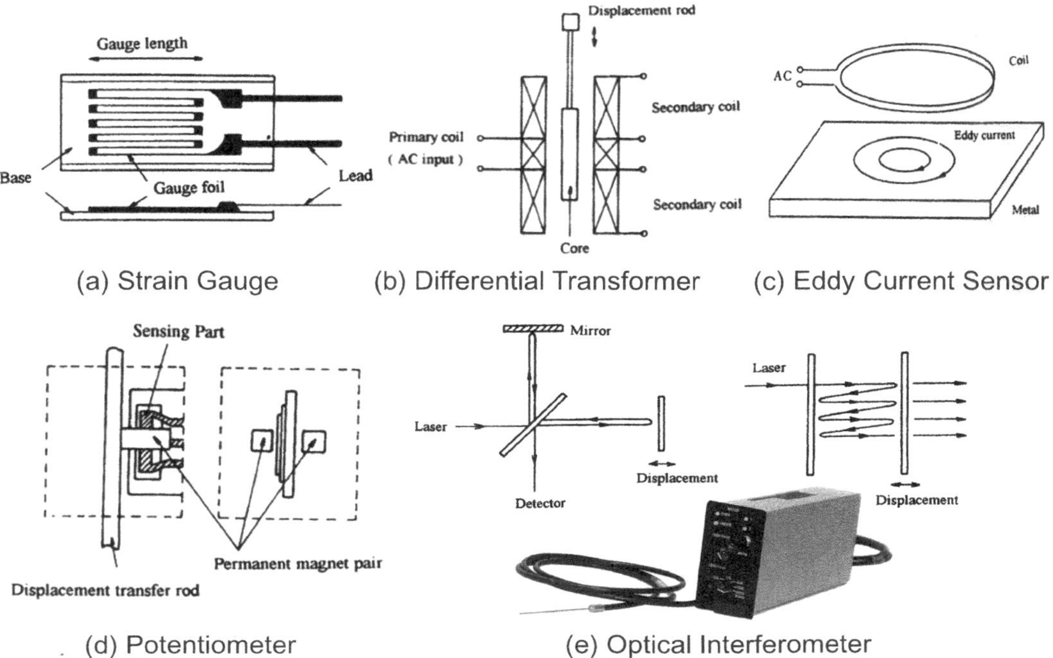

FIGURE 5.52 Popular displacement/strain sensors. (a) Strain gauge, (b) differential transformer, (c) eddy current sensor, (d) magnetoresistive potentiometer, and (e) optical interferometers.

5.6.1.2 Electromagnetic Induction Methods

A schematic representation of a *differential transformer* is pictured in Figure 5.52b. It is composed of two identical electromagnetic coils and a magnetic core, to which the monitor rod is attached. Any change in the position of the magnetic core will result in a change in the mutual inductance of the two coils. The difference in the voltages induced in the two coils is electronically processed (i.e., subtraction) and the resulting signal translated into the corresponding displacement. This is why the device is called a "differential" transformer. The typical resolution for this sensor is 1 μm, but resolutions on the order of 10^{-2} μm can also be achieved when a lock-in amplifier is incorporated into the system. The response speed is limited to about 100 Hz by both the frequency of the AC input voltage and the mechanical resonance, which is determined by the mass of the core.

Another method uses the *eddy current* in a noncontact device, as illustrated in Figure 5.52c. When a metal plate approaches a coil conducting an AC current, an eddy current (with a ring-shaped flow) is induced in the metal plate. The magnetic flux from the eddy current interacts with that of the coil, effectively decreasing the inductance of the coil. There are two types of eddy current sensors: the single-coil sensor, in which the single coil detects the change in inductance, and the triple-coil sensor, in which two secondary coils detect the eddy current generated by a primary coil. A small metal plate (e.g., aluminum foil) is attached to the sample. Since the mechanical load is small, the resonance of the sample can also be measured. The limits of resolution and response for this method are 0.1 μm and 50 kHz, respectively.

5.6.1.3 Capacitance Methods

The capacitance of an air-filled parallel-plate capacitor changes linearly with the distance between the two plates. A very sensitive dilatometer, based on this principle and utilizing a very precise capacitance bridge, can achieve AC displacement resolution of up to 10^{-3} Å.[32]

5.6.1.4 Optical Methods

The *optical lever*, which essentially consists of a laser source and a mirror, has been another popular apparatus for measuring the small displacements produced by such effects as thermal expansion. The sample displacement is transformed to mirror rotation, which is detected by the deflection of the reflected laser beam. The deflection can be measured by means as simple as a ruler on a screen or as sophisticated as a linear optical sensor, which produces an electrical signal directly proportional to the beam position. The resolution is ultimately determined by the laser beam path length, and the overall response of the system depends on the mechanical resonance of the mirror mechanism.

Interferometers can be used in conjunction with an optical fiber sensor or on an optical bench. When two or more monochromatic, coherent light beams with the same wavelength are superposed, an interference or "fringe" pattern is formed, which depends on the phase difference and the optical path difference between the "arms" of the interferometric system. Double-beam systems, such as the *Michelson* configuration depicted in Figure 5.52e, left, and multiple beam systems, such as the *Fabrey-Perot* configuration depicted in Figure 5.52e, right, can be employed for detecting microscale displacements.[33,34] When the optical path length difference between the two beams of a system like the Michelson interferometer changes, the resulting fringe shift will occur in integral multiples of ($\lambda/2$). Precise measurement of the changes in light intensity makes it possible to detect displacements of less than 1 nm by this method. Figure 5.52e, bottom, shows an optical fiber–type Polytec Doppler vibrometer for measuring vibration velocity directly.

5.6.2 Stress Sensors

Piezoelectric ceramics activated through the direct piezoelectric effect can function as stress and acceleration sensors. One such piezoelectric stress sensor designed by Kistler consists of a stack of quartz crystal plates (extensional and shear types) and can detect three-dimensional stresses.[35] Similar multilayer PZT structures have been proposed for use as three-dimensional actuators and sensors by PI Ceramics of Germany, which can also be applied for a 3-D force sensor.[36]

A *cylindrical gyroscope* produced by Tokin of Japan is depicted in Figure 5.53.[37] The cylinder has six divided electrodes, one pair of which is used to excite the fundamental bending mode of vibration, while the other two pairs are used to detect the acceleration. When rotational acceleration occurs about the axis of the gyro, the voltage generated on the electrodes is modulated by the *Coriolis force*. By subtracting the signals generated between the two pairs of sensor electrodes, a voltage that is directly proportional to the acceleration can be obtained.

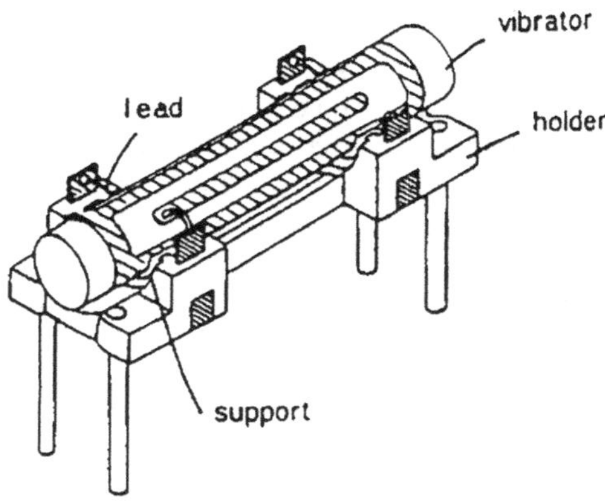

FIGURE 5.53 Piezoelectric cylindrical gyroscope (Tokin, Japan).

The *converse electrostrictive effect*, which essentially reflects the stress dependence of the dielectric constant, is also used in the design of stress sensors.[38] The *bimorph structure*, which utilizes the difference between the static capacitances of two laminated dielectric ceramic plates, provides superior stress sensitivity and temperature stability. The change in capacitance of the top and bottom plates will be opposite in sign for a uniaxial stress and of the same sign when a deviation in temperature occurs. The response speed is limited by the capacitance measurement frequency to about 1 kHz. Unlike piezoelectric sensors, electrostrictive sensors are effective in the low frequency range, especially for pseudo-DC applications.

EXAMPLE PROBLEM 5.9

An accelerometer is fabricated using a piezoelectric ceramic disk (piezoelectric voltage constant: g, thickness: h, and area: S) and a block mass (mass: M, and area: S) bonded together, as schematically shown in Figure 5.54. This accelerometer is installed on a base (such as vehicle chassis) excited by $D_0 \sin \omega t$ (D_0: the maximum displacement). Calculate the peak output voltage of the piezosensor.

Solution

Displacement $D = D_0 \sin \omega t$ provides the acceleration

$$d^2D/dt^2 = -\omega^2 D_0 \sin \omega t. \qquad \text{(P5.9.1)}$$

The stress applied to the piezodisk is given by

$$X = M(d^2D/dt^2)/S = -(\omega^2 D_0 M/S) \sin \omega t, \qquad \text{(P5.9.2)}$$

and the electric field generated is

$$E = gX = -(\omega^2 D_0 Mg/S) \sin \omega t. \qquad \text{(P5.9.3)}$$

Thus, the peak output voltage is provided by

$$V_0 = h\, E_{max} = \omega^2\, D_0\, M\, g\, h/S. \qquad \text{(P5.9.4)}$$

EXAMPLE PROBLEM 5.10

There is a strain gauge made of a thin metal string, as shown in Figure 5.52a (wire radius: r, total length: L, resistivity: ρ). Answer the following questions successively to understand the principle of the strain gauge by neglecting the piezoresistive effect.

a. Provide the initial resistance of this gauge wire.
b. When the strain, x, is applied along the length of this gauge wire, calculate the resistance change, which is proportional to the strain, x. Assume Poisson's ratio, σ, for the radial shrinkage. You may use a relation $1/(1-y)^2 = 1 + 2y$ (for $y \ll 1$).
c. Explain briefly what the piezoresistance effect is. There are two kinds: intrinsic and geometrical effect.

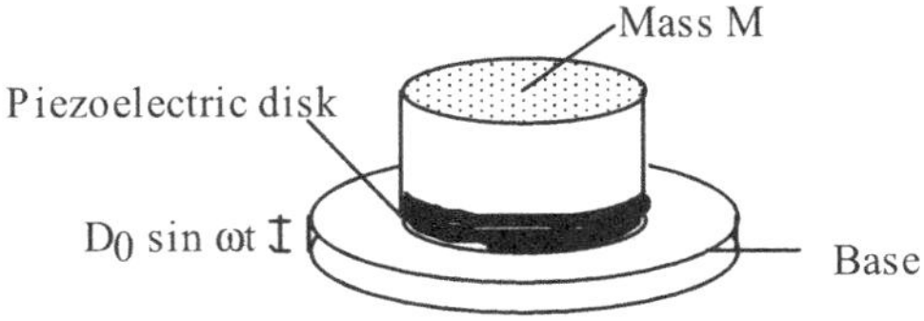

FIGURE 5.54 Principle of an accelerometer with a piezoelectric disk and a block mass.

SOLUTION

a.

$$\text{Resistance } R = \rho(L/\pi r^2) \quad \text{(P5.10.1)}$$

b. Taking into account Poisson's ratio, $L \rightarrow L(1 + x)$ and $r \rightarrow r(1-\sigma x)$. Thus,

$$\begin{aligned}\Delta R &= \rho[L(1 + x)]/\pi[r(1 - \sigma x)]^2\text{-}\rho(L/\pi r^2) \\ &= \rho(L/\pi r^2)(1 + 2\sigma)x,\end{aligned} \quad \text{(P5.10.2)}$$

by using a relation $1/(1 - y)^2 = 1 + 2y$ (for $y \ll 1$).
Resistance change is proportional to the strain x.

c. The *piezoresistive effect* describes the changing electrical resistance of a material due to applied mechanical stress. There are intrinsic and extrinsic (geometrical) effects in this effect. The latter was discussed above (b), which is the major contributor in metal wires. The former effect of semiconductor materials can be several orders of magnitudes larger than the geometrical effect and is present in materials like polycrystalline, amorphous silicon, and single crystal silicon. Hence, semiconductor strain gauges with a very high coefficient of sensitivity can be built. The resistance of *n*-conducting silicon mainly changes due to a shift of the three different conducting valley pairs. The shifting causes a redistribution of the carriers between valleys with different mobilities. This results in varying mobilities dependent on the direction of current flow.

5.7 POWER SUPPLY/DRIVE SCHEME

5.7.1 Power Supply Specifications

Prior to detailing the power supply and drive system scheme, let us consider how to decide the specifications of a required piezoelectric actuator drive system with the following Example Problem 5.11.

EXAMPLE PROBLEM 5.11

When Uchino et al. invented co-fired multilayer piezoelectric actuators in the late 1970s, the impedance measurement for characterizing the ML was the initial problem, because the conventional impedance analyzer could not supply sufficient voltage (max only 30 V) or current (max less than 0.2 A) with the output impedance 50 Ω. Solve this problem by determining the specifications for a ceramic actuator drive system with the following characteristics:

a. The multilayer actuator has 100 ceramic layers, each 100 μm thick with an area of 5 × 5 mm². The relative permittivity is 10,000. Calculate the capacitance of the actuator.
b. Assuming it has a density $\rho = 7.9$ g/cm³ and elastic compliance $s_{33}^D = 13 \times 10^{-12}\, \text{m}^2/N$, calculate the resonance frequency of the actuator. The ML length should be 10 mm in total. The electrode weight load may be ignored for this calculation.
c. Determine the current required, if 60 V is to be applied to the actuator as quickly as possible.
d. The cut-off frequency (1/RC) must be higher than the ML mechanical resonance frequency. Determine the required output impedance of the power supply.

SOLUTION

a.

$$\begin{aligned}C &= n\varepsilon_o\varepsilon(A/t) \\ &= (100)(8.854 \times 10^{-12}\text{F/m})(10{,}000)[(5 \times 5 \times 10^{-6}\text{m}^2)/(100 \times 10^{-6}\text{m})] \\ \rightarrow C &= 2.21 \times 10^{-6}\ (\text{F})\end{aligned} \quad \text{(P5.11.1)}$$

Note: Multilayer actuators have a capacitance higher than 1 μF.

b. The resonance frequency for the thickness vibration with $L = 10$ mm (neglecting the coupling with width vibrations) is given by:

$$f_R = \frac{1}{2L\sqrt{\rho s_{33}^D}} \tag{P5.11.2}$$

$$= \frac{1}{2[100][100\times10^{-6}(\text{m})]\sqrt{[7.9\times10^3(\text{kg/m}^3)][13\times10^{-12}(\text{m}^2/\text{N})]}}$$

$$\rightarrow f_R = 156\ (\text{kHz})$$

Note: The response speed of the power supply must be greater than the actuator's resonance frequency.

c. The relationship between the actuator voltage and the charging current is given by:

$$I = Q/\tau_R = CVf_R \tag{P5.11.3}$$

$$= [2.21\times10^{-6}(\text{F})][60(\text{V})][156\times10^3(\text{Hz})]$$

$$\rightarrow I = 21\ (\text{A})$$

Note: Ideally, a significant current is required from the power supply, even if just for a relatively short period (6 μsec). The *apparent power* is estimated to be [60 (V) × 21 (A)]. So, we see that more than 1 (kW) is needed for the resonance drive. A power of 12 (W) is needed for the 2 (kHz) drive in a dot matrix printer. The *real power* consideration is made in Section 5.7.2.

d. Assuming: $\omega_R = 2\ \pi\ f_R = 1/RC$:

$$R = 1/[2\pi f_R C] \tag{P5.11.4}$$

$$= 1/[2\pi[156\times10^3(\text{Hz})][2.21\times10^{-6}(\text{F})]]$$

$$\rightarrow R = 0.46\ [\Omega]$$

The output impedance of the power supply should be less than 1 Ω.

The power supply specifications for the ML resonance measurement should be:

Maximum Voltage: 200 (V), Maximum Current: 10 (A), Frequency Range: 0–500 (kHz), Output Impedance: <1 (Ω).

You may use a conventional impedance analyzer, on which you should recognize that the maximum voltage is only 30 V and maximum current less than 0.2 A with output impedance 50 Ω. Therefore, your measured admittance value on the ML device at its resonance is one order of magnitude smaller than the expected value, because the peak current cannot be supplied from the analyzer power system. Thus, through NF Corporation, Japan, Uchino's team initially developed and commercialized new power supplies for driving ML ultrasonic motors with the specs: 300 V/10 A and output impedance less than 0.2 Ω, which have practically accelerated progress in piezoelectric actuators, in particular in high-power applications.

5.7.2 Drive/Control Schemes of Piezoelectric Actuators

There are three basic schemes in driving/controlling piezoelectric actuators/transducers, as illustrated in the beginning of this chapter, Figure 5.1: (1) off-resonance, low frequency, (2) resonance (or antiresonance) frequency, and (3) an intermediate frequency between the resonance and antiresonance frequencies. When the operating frequency is much lower than the resonance one, this is considered "off-resonance" drive and the electric characteristic is purely "capacitive," with an admittance phase lag of 90°. When the operating frequency is resonance or antiresonance, the characteristic become "resistive," with a phase lag of 0°. In order to induce the same level of vibration velocity, low voltage and high current or high voltage and low current is required at the resonance or antiresonance drive, respectively. Note that both resonance and antiresonance are the mechanical resonance. We also introduce an operating frequency at an intermediate frequency between the resonance and antiresonance in the inductive region in order to minimize the required input drive power. On the contrary, the pulse drive includes a wide range of frequencies (pseudo-DC to harmonic resonance frequencies), which exhibit linear or parabolic (nonsinusoidal) total displacement in addition to the overshoot and vibration ringing. We review each classification in the following.

5.7.2.1 Off-Resonance (Capacitive) Drive

Most of the conventional linear and switching power systems have been developed for driving primarily resistive loads such as electromagnetic motors. Figure 5.55 illustrates energy flow charts for a switching power system in driving a resistive load (top) and a capacitive load (bottom) with practical numbers. Ninety percent of the input 160 W can be spent in the resistive load (90% efficiency), while only 2% can be used in a capacitive load (2% efficiency). Ninety percent of the input power is spent out in the power supply (mostly as heat). We had better consider a much better driving scheme for solid-state capacitive components such as multilayer piezo-actuators. When dealing with capacitive components, it is important to consider "energy" flow ("real" power $V \cdot I \cos \varphi$) rather than "apparent" power flow ($V \cdot I$). The key to escaping from this electric impedance mismatching is to insert an reactive (or inductive) component in the driving system in series to the capacitive device. Knowing that the conventional coil inductor kills the size and weight (and Joule loss) in the power supply significantly, Knowels et al. at Qortek, PA, introduced a negative capacitance in the switching power system, as illustrated in Figure 5.56.[39] The negative capacitance $-C_d/IT$, where C_d is the *damped capacitance* of the piezoelectric device, is inserted in series with the R-C circuit. Though the usage of power is still 2% (in the case of small electromechanical coupling factor, k, such as 20%), the remaining 98% will be recovered without losing much as heat in the

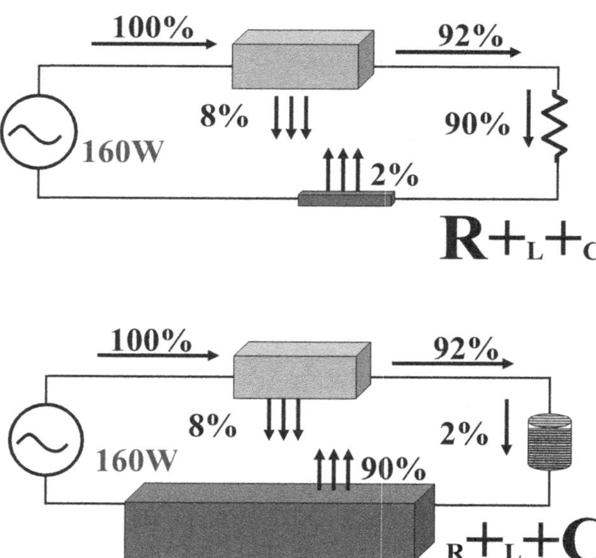

FIGURE 5.55 Efficiency of a switching power system in driving a resistive load 90% or a capacitive load 2%.

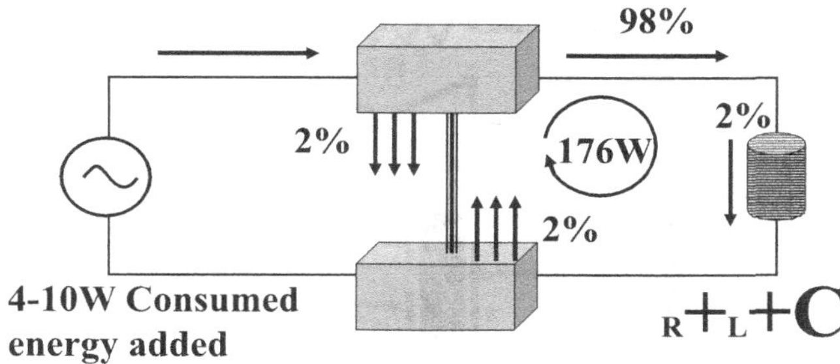

FIGURE 5.56 Energy flow of a switching power system with a negative capacitance in driving a capacitive load.[39]

power system because of the negative capacitance, leading to a high efficiency, 98%. Note that since the actually consumed electric energy level is just 4–10 W, depending on the mechanically consumed energy in the piezo-actuator (in addition to the energy loss in the power circuit), the total size/weight of the power system is significantly smaller than that of conventional ones.

5.7.2.2 Resonance/Antiresonance (Resistive) Drive

When the operating frequency is the resonance or antiresonance, the characteristic become resistive with a phase lag of 0°. Thus, conventional power supplies are possibly used, as long as the voltage and current specifications are satisfied. However, in order to induce the same level of vibration velocity (i.e., mechanically converted energy), low voltage and high current or high voltage and low current is required at the resonance or antiresonance drive, respectively. Note that both resonance and antiresonance are the mechanical resonance.

Remember also that the mechanical quality factor at the resonance, Q_A, is smaller than that at the antiresonance, Q_B, in PZTs (see Figure 5.5). The frequency dependence of the electromechanical

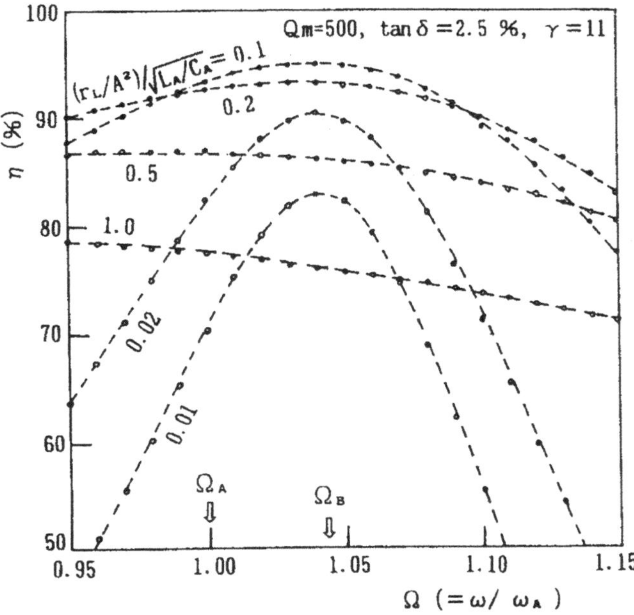

FIGURE 5.57 Frequency dependence of the electromechanical conversion efficiency of a longitudinally vibrating PZT ceramic bar transducer under various loads. ($\Omega = \omega/\omega_A$; A: resonance; B: antiresonance).

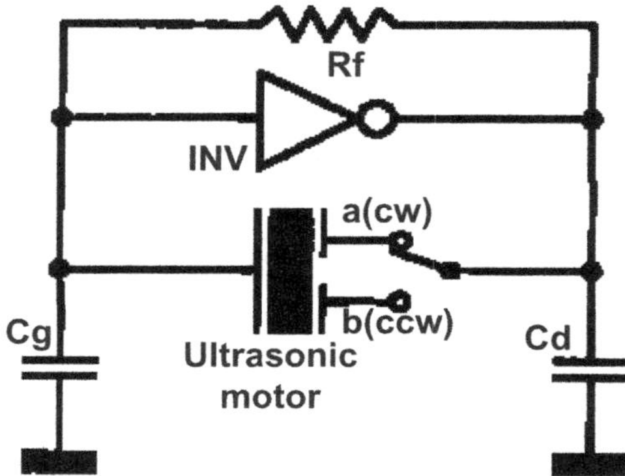

FIGURE 5.58 Self-oscillating ultrasonic motor.

conversion efficiency for this device is shown in Figure 5.57 for various applied loads, simulated from the equivalent circuit in Figure 5.6.[41] The efficiency exhibits the maximum at the antiresonance frequency when we include two loss factors (dielectric and elastic). The analysis, including piezoelectric loss, is discussed in Section 5.7.2.3. The difference between the resonance (A-type) and antiresonance (B-type) frequencies is also highlighted in this graph. When the load is not large, a significant variation in the efficiency with frequency is observed. As the load increases, the efficiency curve becomes flatter. When we consider the driving conditions that apply for each state, that is, a constant electric field, E, for the resonance mode and a constant electric displacement, D, for the antiresonance mode, the lower loss in the antiresonance mode makes sense. Recall that the strain hysteresis is significantly less when the strain is considered as a function of the electric displacement (or polarization) as compared with its electric field dependence. Moreover, antiresonance operation requires a low driving current and a high driving voltage, in contrast to the high current and low voltage required for resonance mode operation, thus allowing for the use of a conventional, inexpensive, low-current power supply.

Recently, compact ultrasonic motors much less than 5 mm have widely been installed in electronic devices such as mobile cameras, phones, and aerial drones, where the power supplies should also be miniaturized. Thus, self-oscillating circuits are popularly utilized. One example is shown in Figure 5.58, where the circuit oscillation frequency is adjusted exactly to the resonance of the ultrasonic motor.

5.7.2.3 Power Minimization (Reactive) Drive

As we already discussed in Sections 5.2.2.4, Real Electric Power Method and 5.5.2.3, "Equivalent Circuit (with Losses)," the antiresonance drive is much more efficient than the resonance drive in PZT-based piezoelectric ceramic devices. Further, a certain frequency between the resonance and antiresonance exhibits the best real power efficiency for generating the same output mechanical vibration velocity.

Yuan et al. proposed an innovative driving scheme of a Langevin piezoelectric transducer[40] under its reactive frequency range, which takes advantage of the maximum efficiency frequency between the resonance and antiresonance. In this approach, first, a constant vibration velocity measurement system is used to find the optimum driving frequency, which is defined as the point where the real input electric power ($V \cdot I \cos \varphi$) is the lowest for a given output mechanical vibration level. Figure 5.59 shows the resonance frequency 40.07 kHz and antiresonance frequency 42.05 kHz, while the lowest input power (i.e., maximum mechanical quality factor) is 41.27 kHz. The transducer has a *reactive behavior* at the optimum frequency (the phase lag is very close to −90°). Next in this approach, an equivalent circuit of the transducer based on the Butterworth-Van Dyke (BVD) model (C_d and L, C, R motional branch) is established (Figure 5.60, right), whose parameters are used to

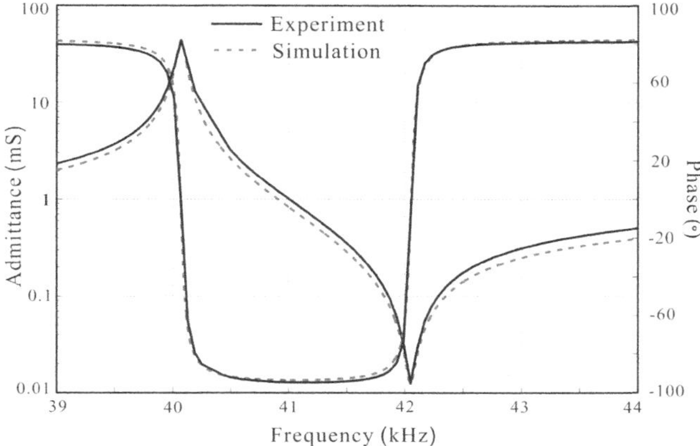

FIGURE 5.59 Admittance and corresponding phase comparisons between simulation and experimental result.

design *Class E inverter* driving circuits. Using MATLAB, a Class E inverter is precisely designed (Figure 5.60, left) to drive the transducer at the resonance frequency (resistive). Then, an impedance converter (basically capacitance C_{cir}) is added (slightly modified on the Class E inverter) when driving at the optimum frequency (reactive). Figure 5.61 summarizes the required electric power (*real power* $V \cdot I \cos \varphi$) for generating the same vibration velocity (measured at the edge of the Langevin head mass) at both driving frequencies, resonance (40.07 kHz) and the optimum condition (41.27 kHz), and the ratio of ($P_{optimum}/P_{resonance}$). The required power for the optimum frequency driving method is reduced by 39% compared with the resonance frequency driving method, and smaller heat generation is also revealed according to the experiments. Note that our driving scheme becomes more attractive with increasing the driving power level (i.e., the above ratio decreases with an increase in vibration velocity).

5.7.3 Fundamental Circuit Components

5.7.3.1 Switching Regulator

Let us start by reviewing the *Power Metal-Oxide-Semiconductor Field Effect* Transistor (MOSFET), which is a transistor that uses an electric field to control the electrical behavior of the device. The symbol of the MOSFET is shown in Figure 5.62a, and its drain current–drain-source voltage characteristics are plotted in (b), which corresponds to 500 V, 50 A Class Power MOSFET. Under keeping the drain-source voltage (V_{DS}) at 5 V, the drain current can be significantly changed from

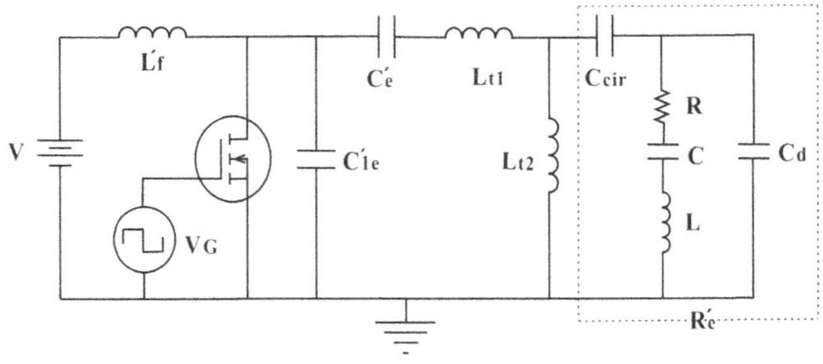

FIGURE 5.60 Class E inverter with impedance converter. (From T. Yuan et al.: *Sensors & Actuators A: Physical*, 261, A219, 2017.)

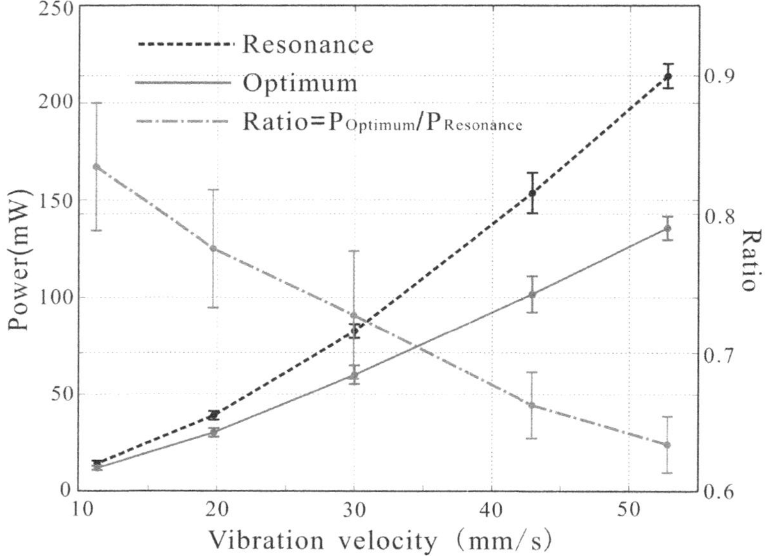

FIGURE 5.61 Power comparison of two driving methods and ratio of powers in different vibration velocities. (From T. Yuan et al.: *Sensors & Actuators A: Physical*, 261, A219, 2017.)

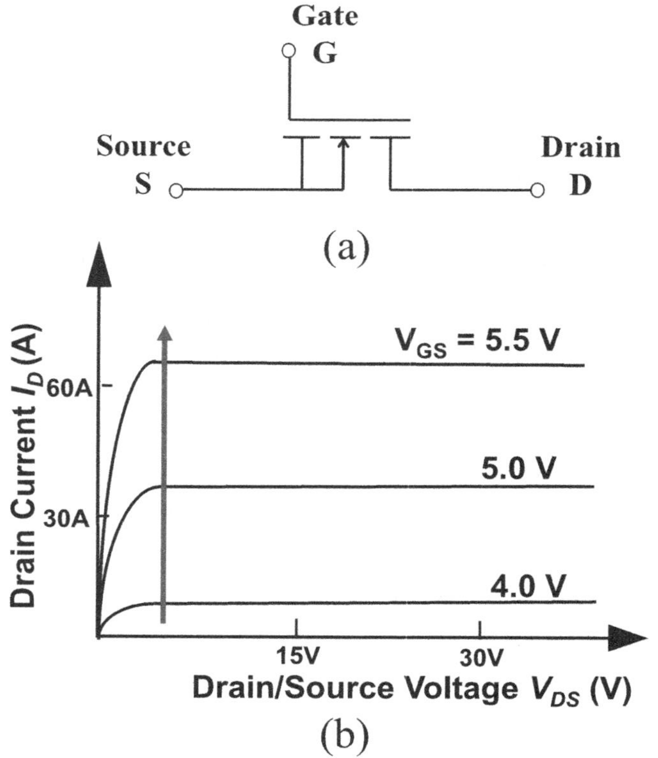

FIGURE 5.62 (a) MOSFET, (b) its output characteristics.

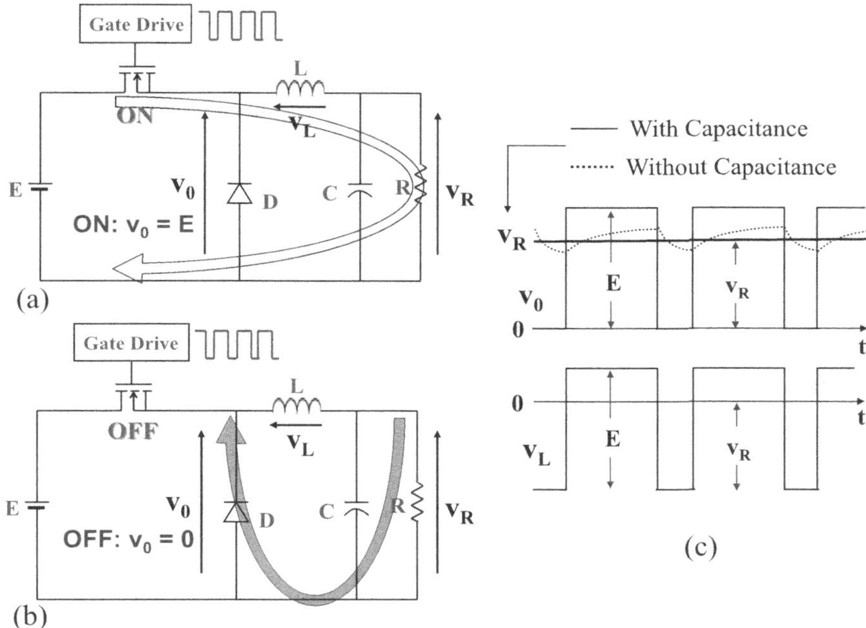

FIGURE 5.63 Switching regulator: (a) on, and (b) off operation. (c) Voltage wave forms of v_0, v_L, and v_R of the step-down buck chopper.

0 to 60 A by changing the gate-source voltage (V_{GS}) from 0 to 5.5 V. Thus, the current on-off switch is obtained by controlling the gate-source voltage with an on-off signal.

On the basis of the current switching effect of the MOSFET, we can design *switching regulators*, which are also called step-down buck choppers. The switching regulator is occasionally used as a step-down voltage (step-up current) converter (popularly DC-DC converter) without using an electromagnetic transformer, which kills the size and weight in a power system. Figure 5.63 shows the simplest switching regulator composed of a MOSFET, inductor L, diode, and resistive load R (capacitor C is occasionally added to stabilize the ripple wave form) under a DC constant voltage (battery). (a) and (b) illustrate the operations for the MOSFET on and off stages, respectively. When the gate-source voltage with a rectangular wave form [reasonably high carrier frequency with the *duty ratio* $d = (T_{on} / (T_{on} + T_{off}))$] is applied, the MOSFET behaves as an on/off switch. During the *on* stage, E – FET – L – R is the current flow route, so that $v_0 = E$, but v_R does not reach E so quickly because of the inductor (which needs to accumulate the electrical energy first). On the other hand, during the *off* state, $v_0 = 0$, but v_R does not reach 0 so quickly; moreover, because the inductor will release the electrical energy or generate the *reverse electromotive force*, the current still continues to flow in the route L – R – D now. Note first that the average voltage of a rectangular wave form (0 – E) with the duty ratio (d) is estimated as $d \cdot E$. Voltage wave forms of v_0, v_L, and v_R of the step-down buck chopper are illustrated as a function of time in Figure 5.63c, where you can find first v_0 as exactly the "similar" rectangular wave form to the MOSFET gate voltage with 0 to E voltage height (unipolar). On the contrary, v_L shows the same wave form, but a negative bias voltage equal to v_R. Since v_R is obtained as the subtraction $v_0 - v_L$, we can conclude that v_R is almost constant around the average voltage $d \cdot E$. More details on v_R behavior are described in Figure 5.63c. Without using a capacitance, v_R exhibits a ripple mode of an exponential curve with time constant $\tau = (L/R)$ in a L-R circuit. In order to minimize the ripple level, the carrier time period (inverse of the carrier frequency) should be chosen to be much less than the circuit time constant (L/R) first. An additional smoothing capacitance, C, helps more with realizing almost constant output voltage, $d \cdot E$.

5.7.3.2 On-Off Signal Generator

You will learn in this subsection how to synthesize the desired rectangular signal with a certain duty ratio. Figure 5.64 illustrates the principle of pulse width modulation based on a triangular carrier wave. When we use a triangular carrier wave, v_C, a certain input signal level, v_S, is easily converted into an on/off signal with a certain duty ratio by subtracting these two voltage values (Figure 5.64c). The subtraction operational amplifier is called the "comparator" (Figure 5.64a), and its practical device example, a low-power complementary metal–oxide–semiconductor (CMOS) clocked comparator, is shown in Figure 5.64b.

So far, we have demonstrated only a monopolar drive (0 to E V) switching regulator. However, we occasionally use a bipolar drive ($-E$ to $+E$ V) power supply, in particular in DC to AC converters. The key is to use a bridge circuit, illustrated in Figure 5.65a. In order to obtain "positive" E, we control Tr_1^+ (on and off) by keeping Tr_2^- (on), while to obtain negative E, we control Tr_2^- (on and off) by keeping Tr_1^+ (on). Figure 5.65b shows a DC-DC voltage converter from a small signal voltage, v_S, to amplified voltage, v_{ave}, utilizing ± triangular carrier. Note here that when we denote the duty ratio as d, the average output voltage $v_{ave} = k \cdot E$, where $k = 2d - 1$. Figure 5.65c shows the principle of AC voltage pulse width modulation. We will start from a triangular carrier signal, v_C, with $\pm$ v_C. We now consider two sine input signals, v_{Sa} and v_{Sb} [Figure 5.65c(1)], each of which generates a pulse-width modulated wave shown in v_a or v_b at the terminal a or b [Figure 5.65c(2) and (3)], respectively. Since the final output voltage, v_0, is provided by the subtraction $v_a - v_b$, we obtain the pulse-width modulated ± signals [Figure 5.65c(4)].

5.7.3.3 Piezoelectric Transformer

Because conventional inductive coil transformers kill the size/weight of analog power systems significantly, a piezoelectric transformer is one of the alternative components. Operating principles and some application examples are introduced.

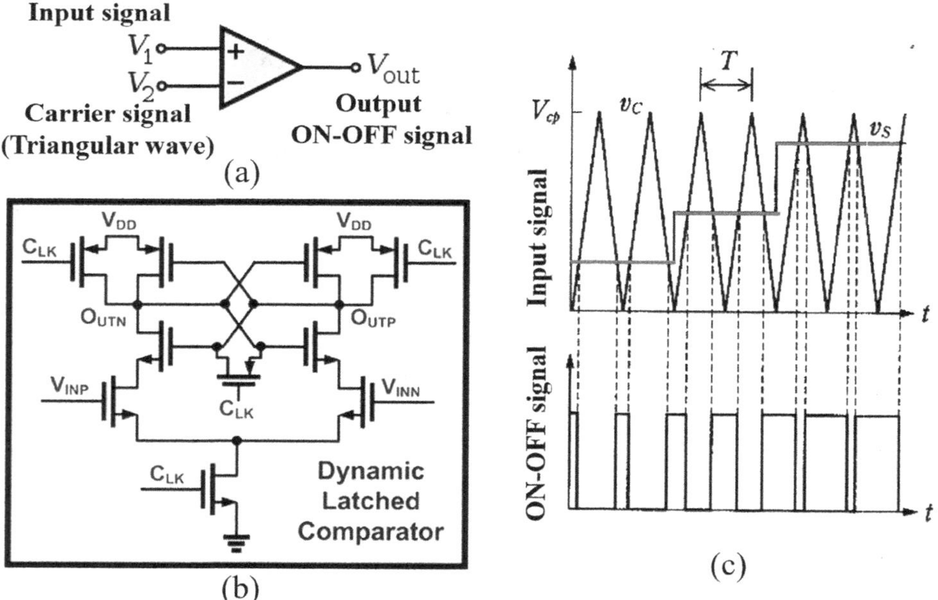

FIGURE 5.64 (a) Comparator, (b) low-power CMOS-clocked comparator, and (c) duty ratio realization with a triangular carrier signal.

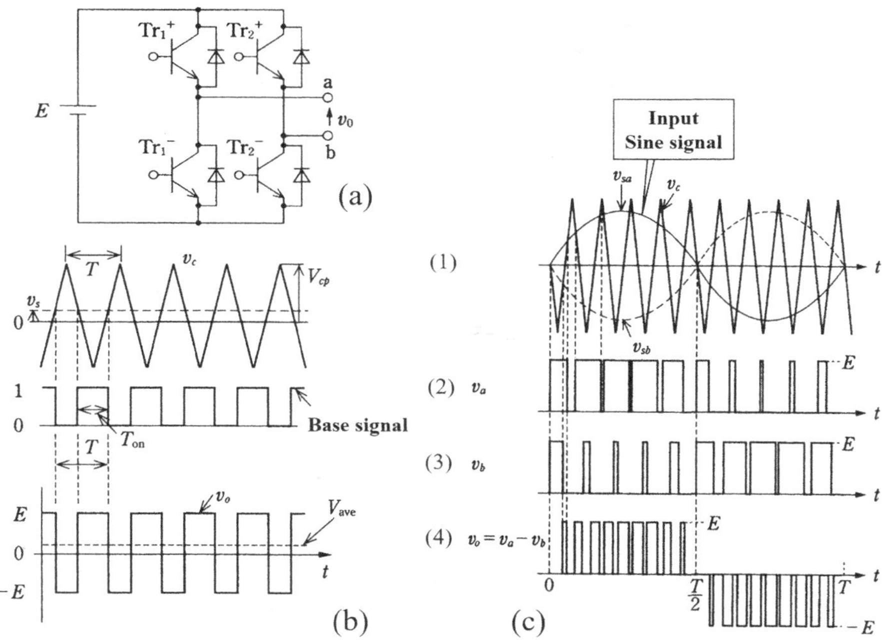

FIGURE 5.65 (a) Bridge circuit, (b) DC voltage converter, and (c) principles of AC voltage pulse width modulation.

5.7.3.3.1 Operating Principles

One of the bulkiest and most expensive components in solid-state actuator systems is the power supply with an electromagnetic transformer. Electromagnetic transformer losses occurring through the skin effect, thin wire loss, and core loss all increase dramatically as the size of the transformer is reduced. Therefore, it is difficult to realize miniature low-profile electromagnetic transformers with high efficiency. Piezoelectric transformers are an attractive alternative for such systems due to their high efficiency, small size, and lack of electromagnetic noise. They are highly suitable as miniaturized power inverter components, which might find application in lighting up the cold cathode fluorescent lamp behind a color liquid crystal display or in generating the high voltage needed for air cleaners.

The original design to step up or step down an input AC voltage using the converse and direct piezoelectric effects of ceramic materials was proposed by Rosen.[42] This type of transformer operates by exciting a piezoelectric element like the one pictured in Figure 5.66a at its mechanical resonance frequency. An electrical input is applied to one part of the piezoelectric element (at the top left electrode), which produces the fundamental mechanical resonance. This mechanical vibration is then converted back into an electrical voltage at the other end (right edge electrode) of the piezoelectric plate. Since the electric field level is similar among the input and output parts due to similar stress distribution in the input and output parts, the voltage step-up ratio (r) without load (open-circuit conditions) is primarily given by the electrode gap ratio:

$$r \propto k_{31}k_{33}\, Q_m\, (l/t), \tag{5.145}$$

where l and t are the electrode gap distances for the input and output portions of the transformer, respectively (Figure 5.66a). Note from this relationship how the length to thickness ratio, the electromechanical coupling factors, and/or the mechanical quality factor, Q_m, are the primary means

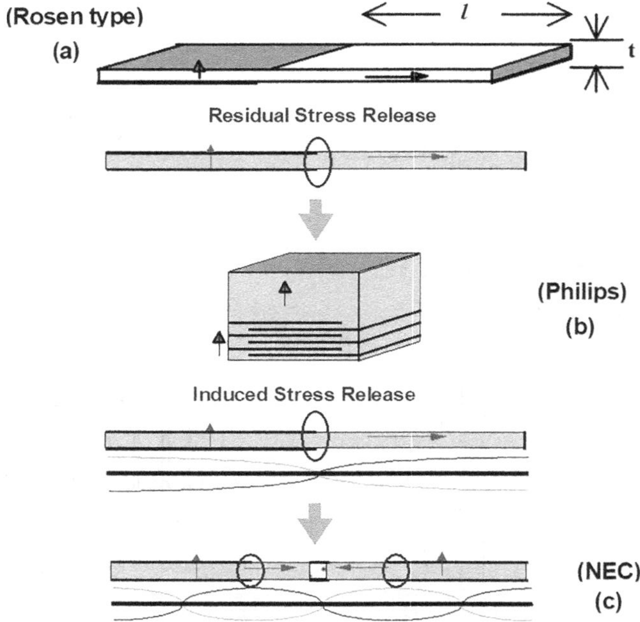

FIGURE 5.66 Piezoelectric transformer designs: (a) Rosen design, (From C. A. Rosen: "Ceramic Transformers and Filters," *Proc. Electronic Component Symp.*, p. 205–211, 1956.) (b) multilayer design by NEC, (From NEC: "Thickness Mode Piezoelectric Transformer," US Patent No. 5,118,982, 1992.), and (c) third resonant mode type developed by NEC. (From S. Kawashima et al.: *IEEE Int'l Ultrasonic Symp. Proc.*, Nov., 1994.)

of increasing the step-up ratio. This transformer was utilized on a trial basis in some color televisions during the 1970s.

In spite of its many attractive features, the original Rosen transformer design had a serious reliability problem. Mechanical failure tends to occur at the center of the device where the residual stress from the poling process is most highly concentrated. This happens also to be coincident with the nodal point of the vibration where the highest induced stress occurs. Two recently developed transformers pictured in Figure 5.66b and c are designed to avoid this problem and are commercially produced for use as backlight inverters in liquid crystal displays. Both of the newer designs make use of more mechanically tough ceramic materials. The NEC and Philips Components transformer shown in Figure 5.66b further alleviates the problem by using a multilayer structure to avoid the development of residual poling stress in the device.[43] Another NEC design pictured in Figure 5.66c makes use of an alternative electrode configuration to excite a third resonance excitation (longitudinal) in the rectangular plate to further redistribute the stress concentrations in a more favorable manner.[44]

5.7.3.3.2 Power Supply Applications for Piezoelectric Actuators

A variety of new methods for driving piezoelectric actuators have emerged in recent years. A typical system is illustrated in Figure 5.67.[45] If we tune the transformer's frequency to precisely match to the resonance frequency of an ultrasonic motor, it can be used as a driver for the system, thus creating a transformer-integrated motor.[46] Such a configuration might, for example, take the form of a ring transformer having the same dimensions as the ultrasonic motor.

A transformer coupled with a rectifier can be used to drive a multilayer or bimorph piezoelectric actuator.[45] A compact drive system designed for a piezoelectric vibration control device for a

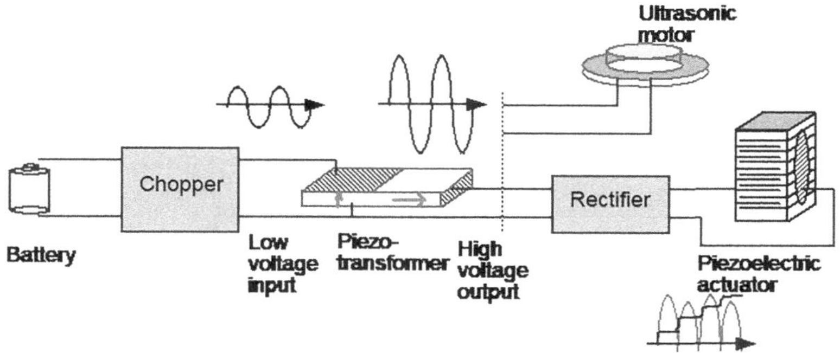

FIGURE 5.67 Typical piezoelectric actuator drive system (AC or pseudo-DC) with a piezoelectric transformer. (From K. Uchino et al.: Proc. 3rd Asian Mtg. on Ferroelectrics, Dec., 2000.)

helicopter is shown schematically in Figure 5.68. It includes a multilayer piezoelectric actuator to suppress the vibration and a piezoelectric transformer to drive it.

Two kinds of power supplies utilizing piezoelectric transformers powered by a $24V_{DC}$ helicopter battery have been developed. One serves as a high-voltage DC power supply (100–1000 V, 90 W) for driving a piezoelectric actuator, and the other is a DC adapter ($\pm 15V_{DC}$, 0.1–0.5 W) for driving the supporting circuitry. Large and small multistacked piezotransformer elements, both with an insulating glass layer between the input and output parts to ensure a completely floating condition, were used for the high voltage supply and the adapter, respectively. An actuator manufactured by Tokin Corporation with dimensions of $(10 \times 10 \times 20)$ mm^3 and capable of generating a 16-μm displacement under the maximum operating voltage of 100 V was used in this experiment.

Chopped 24V AC voltage is applied to the piezoelectric transformer with a step-up ratio about 10. This high-voltage AC signal is converted to high-voltage DC (300 V) through a rectifier, and the charge is collected on a capacitor (at point A in Figure 5.68). A power amplifier is used to regulate the voltage that is finally applied to the actuator. A *Class-D switching amplifier*, rather than a conventional switching or linear amplifier, was adopted because it allows more precise control of the amplitude and frequency of the drive signal and a higher actuator response rate due to the chopped DC input voltage. The signal from a pulse width modulation driving circuit is applied to

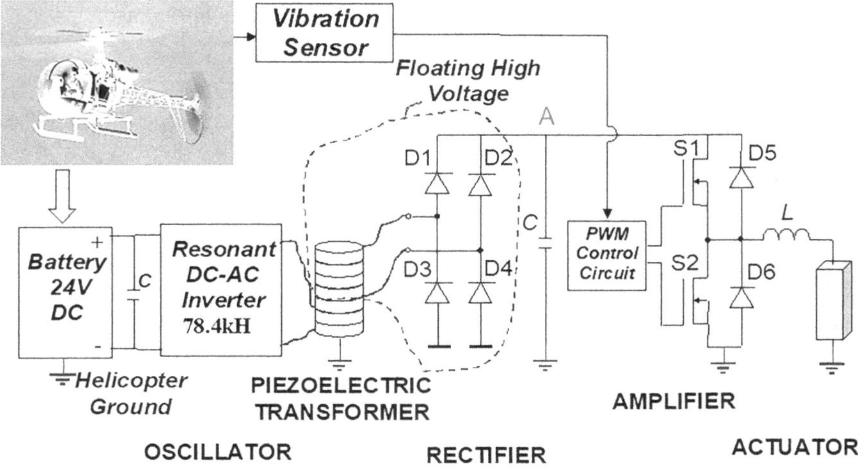

FIGURE 5.68 Compact drive system for piezoelectric actuator control. (From A. Vazquez Carazo and K. Uchino: *J. Electroceramics* 7, 197–210, 2001.)

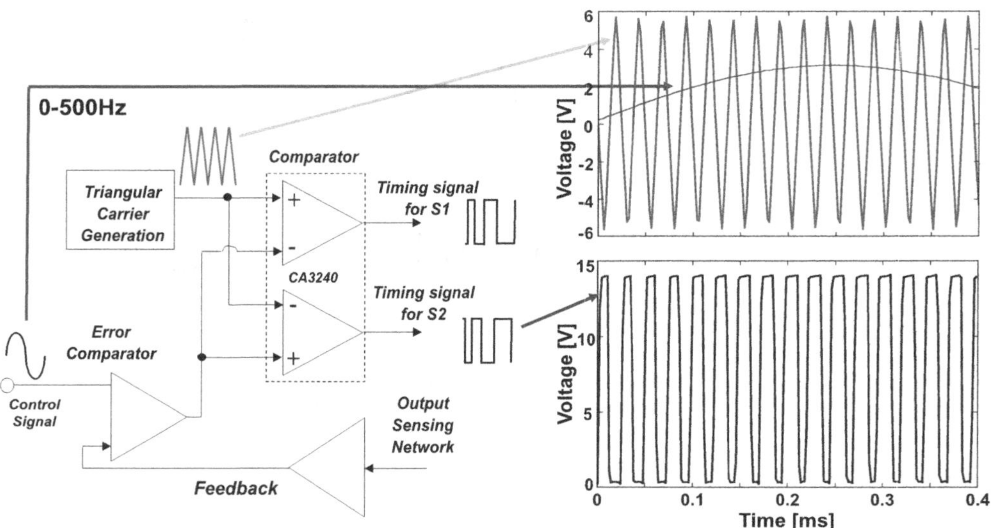

FIGURE 5.69 Pulse width modulation circuit for driving the power amplifier. (From A. Vazquez Carazo and K. Uchino: *J. Electroceramics* 7, 197–210, 2001.)

the two power MOSFETs of the half bridge, as shown in Figure 5.69, to chop the constant 300V_{DC} voltage from the piezoelectric transformer and to maintain the desired level of amplitude and frequency control.

The output voltage is applied to the piezoelectric actuator through a filtering inductance of 100 mH. The PWM carrier frequency is maintained at 40 kHz, which is below the mechanical resonance frequency of the piezoelectric actuator (~60 kHz). The displacement curves of an actuator driven by this newly developed power amplifier are shown in Figure 5.70, right. The displacement was directly measured with an eddy current sensor. As seen in this figure, a displacement of ±1.5 μm was controlled with an applied voltage of ±20 V. This drive system can be operated at frequencies up to 500 Hz, which is sufficient to maintain active vibration control on a helicopter.

In summary, the piezoelectric transformer can be used as part of the drive circuitry for a piezoelectric actuator. It has the advantages over conventional electromagnetic transformers of being lightweight, compact in size, highly efficient, and free of electromagnetic noise.

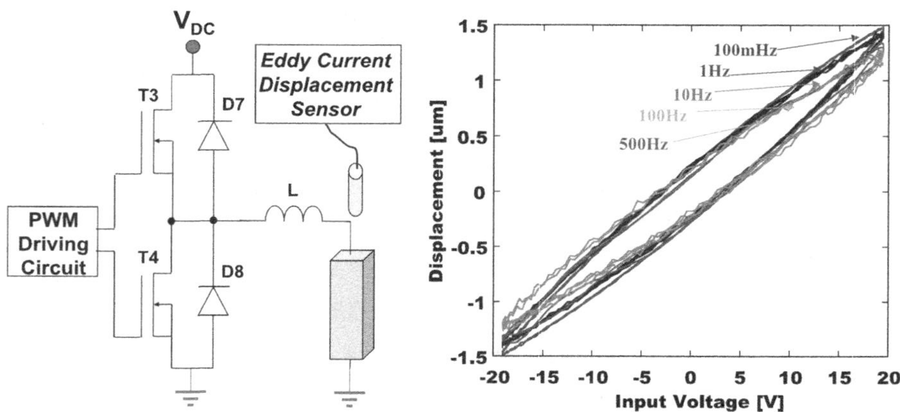

FIGURE 5.70 Displacement curves of a piezoelectric actuator driven by the newly developed power amplifier. (From A. Vazquez Carazo and K. Uchino: *J. Electroceramics* 7, 197–210, 2001.)

CHAPTER ESSENTIALS

1. *Piezoelectric characterization methods:*
   a. Admittance/impedance spectra
      1. Under constant voltage
      2. Under constant current
      3. Under constant vibration velocity
   b. Burst mode
      1. Short-circuit measurement
      2. Open-circuit measurement
   c. Precise input electric energy (including phase lag)
2. *Classification of ceramic actuators:*

Actuator Type	Drive	Device	Material Type
Rigid Displacement	Servo	Servo-Displacement Transducer	Electrostrictive
	On/Off	Pulse Drive Motor	Soft Piezoelectric
Resonant Displacement	AC	Ultrasonic Motor	Hard Piezoelectric

3. *Actuator control*:

• Open loop	On/off drive
	Pulse width modulation
	Polarization control
• Closed loop	Servo drive

4. *Pulse width modulation criteria*:
   a. The ratio of the carrier frequency, f_c, to the input command frequency, f_e, should be greater than 7.
   b. The carrier frequency, f_c, must be high enough (compared to the system response) to eliminate ripple in the system output.
5. *The Laplace transform*: A powerful tool for transient analysis has the general form:

$$U(s) = \int_0^\infty e^{-st}\, u(t)\, dt$$

   Among the theorems that apply to the Laplace transform, the following are especially useful:
   a. Differentiation with respect to t:

$$L\left[\frac{du(t)}{dt}\right] = sU(s) - u(0)$$

$$L\left[\frac{d^n u(t)}{dt^n}\right] = s^n U(s) - \sum s^{n-k} u^{k-1}(0)$$

   b. Shift formula with respect to t: $u(t - k) = 0$ for $t < k$
      (k: positive real number). This represents the $u(t)$ curve shifted by k along the positive t axis.

$$L[u(t-k)] = e^{-ks}U(s)$$

c. Differentiation with respect to an independent parameter:

$$L\left[\frac{\partial u(t,x)}{\partial x}\right] = \frac{\partial U(s,x)}{\partial x}$$

d. Initial and final values

$$\lim_{t \to 0}[u(t)] = \lim_{|s| \to \infty}[sU(s)]$$

$$\lim_{t \to \infty}[u(t)] = \lim_{|s| \to 0}[sU(s)]$$

6. *The transfer function*: $G(s) = U(s)/\tilde{E}(s)$ [where: $\tilde{E}(s)$: the input, $U(s)$: the output]. $G(s)$ can be obtained by inputting a unit impulse function.
7. *The transfer function of a piezoelectric actuator with a mass of M*:

$$U(s) = G(s)\widetilde{E(s)},$$

$$G(s) = [A\ c\ d/(M\ s^2 + \zeta s + Ac/l)]$$

8. *The Nyquist criterion of stability*: When the point $(-1 + j0)$ is to the left of the Nyquist diagram (increasing ω), the system is stable.
9. *Pulse drive*: The pulse width or rise time must be adjusted to exactly match the resonance period of the actuator system to eliminate vibrational ringing.
10. *Resonance drive*: A high Q_m material is essential in order to optimize the vibration amplitude and to suppress heat generation.
11. *Measuring techniques for micro displacements*:
    Resistance method: Strain gauge, potentiometer
    Electromagnetic induction methods: Differential transformer, eddy-current type
    Optical methods: Optical lever, optical fiber
12. *Power supply for piezoelectric devices*:
    a. Off-resonance (capacitive) drive
    b. Resonance/antiresonance drive
    c. Minimum energy (reactive) drive
13. *Switching regulators, piezoelectric transformers*: Are promising alternatives to conventional electromagnetic transformers to be used in the drive circuitry for solid-state actuators.

CHECK POINT

1. The electric field vs. strain relation exhibits a hysteresis curve in a PZT specimen during a cyclic measurement process. Is it a clockwise or counterclockwise trace?
2. (T/F) When we measure the admittance spectrum on a k_{31}-type piezoelectric plate specimen, the admittance minimum point corresponds to the resonance point. True or False?
3. (T/F) When we measure the admittance spectrum on a k_{33}-type magnetostrictive rod specimen, the admittance minimum point corresponds to the resonance point. True or False?

4. (T/F) The voltage supply has a small output impedance, while the current supply has a large output impedance. True or False?
5. (T/F) The transfer function of an actuator, $G(s)$ [defined by $U(s) = G(s)\cdot\tilde{E}(s)$], can be obtained from the displacement $u(t)$ measurement under a Heaviside step electric field. True or False?
6. When we measure the admittance spectrum on a k_{31}-type piezoelectric plate, the phase changes from +90° to −90° around the resonance point with an increase in drive frequency. What is the phase lag at the resonance point?
7. When we measure the admittance spectrum on a k_{31}-type piezoelectric plate, the phase changes from +90° to −90° around the resonance point with an increase in drive frequency. What do you call the frequencies that provide the phase ±45°?
8. Using the Laplace transformation definition: $U(s) = L[u(t)] = \int_0^\infty e^{-st}u(t)dt$, calculate $U(s)$ for the Heaviside step function $u(t) = 1(t)$ [$u(t) = 0$ $(t < 0)$; $u(t) = 1$ $(t > 0)$].
9. Derive the function $u(t)$ for the Laplace transform: $U(s) = (m/s^2)e^{-as}$. Note a kink on the curve.
10. Three distinct merits of the servo (feedback) control system include (1) an output response with a flat frequency dependence, and (2) minimization of external disturbance effects. What is the third merit?
11. Describe the resonance frequency, f, for the following electrical circuit:

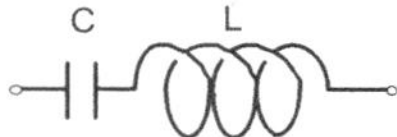

12. (T/F) When a piezoelectric actuator is driven by a rectangular pulse voltage, the mechanical ringing is completely suppressed when the pulse width is adjusted exactly to half of the resonance period of the sample. True or False?
13. When a piezoelectric actuator is driven by a Heaviside step voltage, the vibration displacement overshoot is excited (by neglecting mechanical loss). What is the maximum overshoot range (percentage), in comparison with the normal (pseudostatic) operation? 10%, 16.7%, 33%, 50%, 100%, or 200% larger than the normal displacement?
14. (T/F) Negative capacitance is utilized to drive a "capacitive" actuator device efficiently. True or False?
15. What is the full name of PWM?

CHAPTER PROBLEMS

5.1 Using a rectangular piezoelectric plate and the transverse d_{31} mode, consider the design of a flight actuator.
   a. Assuming a rectangular negative pulse ($-E_o$) is applied to the plate, which is installed normally and rigidly fixed at one end, verify that the velocity of sound, v, at the other end is given by $[2|d_{31}|E_o v]$, which is independent of the length.
   b. Suppose this velocity is acquired by a small steel ball of mass M, with no loss in energy. Calculate the maximum height the steel ball will attain when the impulse is applied entirely in the upward direction.

5.2. Let us consider open-loop and closed-loop transfer functions of the form:

$$W_o = 4/s(s^2 + 4s + 6),$$

$$W_c = 4/(s^3 + 4s^2 + 6s + 4),$$

for a ceramic actuator system.

a. Using the Nyquist criterion, check the stability of this feedback control system.
b. Describe the Bode diagram for $W_c(j\omega)$ using MATLAB software, and discuss the frequency dependence.

5.3. Explain how the type of servo mechanism can be determined from the Bode diagram shown below:

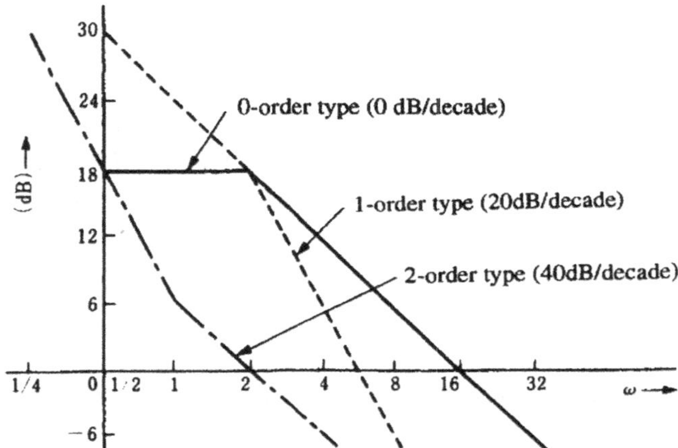

Hint: The slope of the low-frequency ($\omega \to 0$) portion of the Bode diagram identifies the type of the servo mechanism. When the open-loop transfer function has a pole at the origin, the Bode diagram will have a slope of –20 dB/decade in the low-frequency region.

Slope (dB/decade)	Type of Servo Mechanism
0	Type 0
−20	Type 1
−40	Type 2

5.4 Using a rectangular piezoceramic plate (length: l, width: w, and thickness: b, poled along the thickness), we can find the following parameters: k_{31}, d_{31}, and Q_m. Explain the fundamental principles for both the resonance and pulse drive methods.

The density, ρ, and dielectric constant, ε_{33}^{X}, of the ceramic must be known prior to the following experiments.

a. Using an impedance analyzer, the admittance for a mechanically free sample (that is, one that is supported at the nodal point at the center of the plate) is measured as a function of the drive frequency, f, and the admittance curve shown below is obtained. Explain how to determine the k_{31}, d_{31}, and Q_m values from these data. Also verify that the following approximate equation can be used for a low-coupling piezoelectric material:

$$k_{31}^2/\left(1-k_{31}^2\right)=(\pi^2/4)(\Delta f/f_R), \text{ where } (\Delta f = f_A - f_R).$$

b. Using a pulse drive technique, the transient displacement change is measured as a function of time, and the displacement curve pictured below is obtained. Explain how to determine the k_{31}, d_{31}, and Q_m values from these data. Use the relationship: $Q_m = (1/2)\omega_o T$.

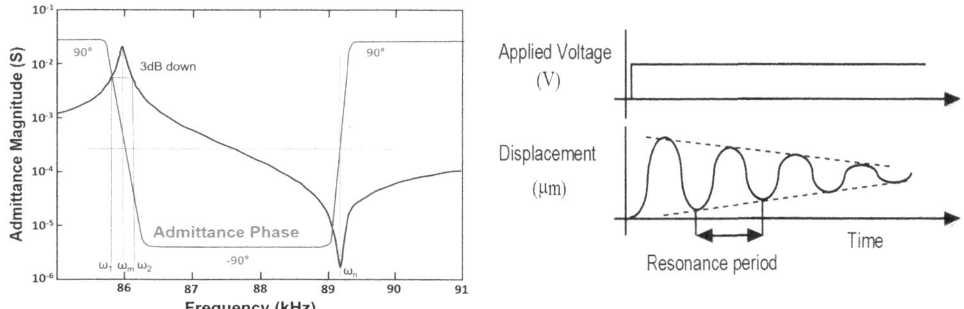

5.5 Knowing the mechanical system (mass, spring, and damper) and the electric circuit (inductance, capacitance, and resistance) equivalency, as shown in Figure (a) below, generate the electrical equivalent circuit corresponding to the mechanical system described in Figure (b).

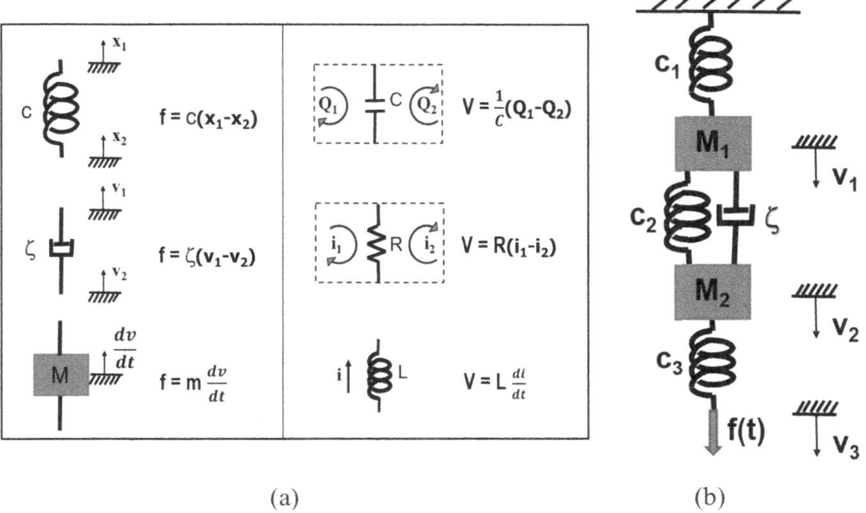

(a) (b)

5.6 Using Mason's equivalent circuits for two length expander bars, surface and end electroded, as shown below, calculate the step-up voltage ratio for a Rosen-type transformer under an open-circuit condition.

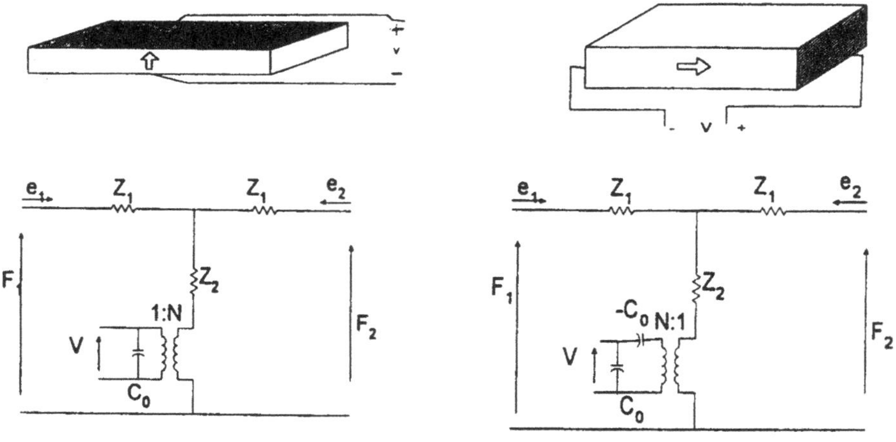

REFERENCES

1. ANSI/IEEE Std 176-1987, *IEEE Standard on Piezoelectricity*, The Institute of Electrical and Electronics Engineers, New York, 1987.
2. J. Zheng, S. Takahashi, S. Yoshikawa, K. Uchino and J. W. C. de Vries: *J. Amer. Ceram. Soc.*, 79, 3193–3198, 1996.
3. K. Uchino, J. Zheng, A. Joshi, Y. H. Chen, S. Yoshikawa, S. Hirose, S. Takahashi and J. W. C. de Vries: *J. Electroceramics*, 2, 33–40, 1998.
4. K. Uchino, H. Negishi and T. Hirose: *Japan. J. Appl. Phys.*, 28(Suppl. 28), 47–49. (Proceedings of FMA-7, Kyoto), 1989.
5. S. Hirose, M. Aoyagi, Y. Tomikawa, S. Takahashi and K. Uchino: *Proc. Ultrasonics Int'l '95*, Edinburgh, pp. 184–187, 1995.
6. Y. Zhuang, S. O. Ural, A. Rajapurkar, S. Tuncdemir, A. Amin, and K. Uchino: *Japan. J. Appl. Phys.*, 48, 041401, 2009.
7. Y. Zhuang, S. O. Ural, S. Tuncdemir, A. Amin, and K. Uchino: *Japan. J. Appl. Phys.*, 49, 021503, 2010.
8. S. O. Ural, S. Tuncdemir, Y. Zhuang and K. Uchino: *Japan. J. Appl. Phys.*, 48, 056509, 2009.
9. H. N. Shekhani, K. Uchino: *J. Euro. Ceram. Soc.* 35(2), 541–041544, 2014.
10. M. Umeda, K. Nakamura and S. Ueha: *Jpn. J. Appl. Phys.*, 38, 3327–3330, 1999.
11. S. Takahashi, Y. Sasaki, M. Umeda, K. Nakamura, and S. Ueha: *MRS Proceedings*. 604, 15, 1999.
12. M. Umeda, K. Nakamura, and S. Ueha: *Japan. J. Appl. Phys.*, 38, 5581, 1999.
13. H. Shekhani, T. Scholehwar, E. Hennig, and K. Uchino: *J. Am. Ceram. Soc.*, 100(3), 998–1011, 2017. doi: 10.1111/jace.14580
14. H. Daneshpajooh, H. Shekhani, M. Choi, and K. Uchino: *J. Amer. Ceram. Soc.*, 101, 1940–1948, 2018. doi: 10.1111/jace.15338
15. K. Uchino: *J. Industrial Education Soc. Jpn.*, 40(2), 28, 1992.
16. Y. Amemiya Edit: *Introduction to Mechatronics*, Japan Industrial Tech. Center, Tokyo, 1984.
17. S. A. Davis: *Feedback and Control Systems*, Simon & Schuster, Inc., New York, 1974.
18. C. V. Newcomb and I. Flinn: *Electronics Lett.* 18, 442, 1982.
19. T. Ogawa: *Crystal Physical Engineering*, Shoka-bo Pub., Tokyo, 1976.
20. S. Sugiyama and K. Uchino: *Proc. 6th IEEE Int'l Symp. Appl. Ferroelectrics*, p. 637, 1986.
21. S. Smiley: US-Patent, No.3614486, 1971.
22. Ikebe and Nakata: *J. Oil and Air Pressure* 3, 78, 1972.
23. W. P. Mason: *Proc. I.R.E.*, 23, 1252, 1935.
24. D. Damjanovic: *Ferroelectrics*, 110, 129–135, 1990.
25. W. Shi, H. N. Shekhani, H. Zhao, J. Ma, Y. Yao and K. Uchino: *J. Electroceram.* 35, 1–10, 2015.
26. W. Shi, H. Zhao, J. Ma, Y. Yao, and K. Uchino: *Japan. J. Appl. Phys.* 54, 101501, 2015. http://dx.doi.org/10.7567/JJAP.54.101501
27. K. Uchino: *Ferroelectric Devices*, 2nd edn, CRC Press, Boca Raton, FL, 2010.
28. X. Dong, M. Majzoubi, M. Choi, Y. Ma, M. Hu, L. Jin, Z. Xu and K. Uchino: *Sens. Actuators A. Phys.*, A256, 77–83, 2017. doi.org/10.1016/j.sna.2016.12.026.
29. X. Dong, T. Yuan, M. Hu, H. Shekhani, Y. Maida, T. Tou, and K. Uchino: *Rev. Sci. Instruments* 87, 105003, 2016. doi: 10.1063/1.4963920
30. P. H. Sydenham: *J. Phys. E: Sci. Inst.* 5, 721, 1972.
31. Midori-Sokki: *Potentiometer*, Production Catalog, Japan.
32. K. Uchino and L. E. Cross: *Ferroelectrics* 27, 35, 1980.
33. Tanida, Gomi and Nomura: *Abstract 44th Jpn. Appl. Phys. Mtg., Fall*, p. 27, 1983.
34. Evick Engineering: *Photonic Sensor*, Production Catalog.
35. *Kistler, Stress Sensor*: Production Catalog, Switzerland.
36. A. Bauer and F. Moller: *Proc. 4th Int'l Conf. New Actuators*, Germany, p. 128, 1994.
37. Tokin: *Gyroscope*, Production Catalog, Japan.
38. K. Uchino, S. Nomura, L. E. Cross, S. J. Jang and R. E. Newham: *Jpn. J. Appl. Phys.* 20, L367, 1981; K. Uchino: Proc. Study Committee on Barium Titanate, XXXI-171-1067, 1983.
39. http://www.qortek.com/en/products/piezo-drivers/polydrive-low-cost-lab-driver/
40. T. Yuan, X. Dong, H. Shekhani, C. Li, Y. Maida, T. Tou and K. Uchino: *Sensors & Actuators A: Physical*, 261, A219, 2017.
41. S. Hirose, S. Takahashi, K. Uchino, M. Aoyagi and Y. Tomikawa: *Proc. MRS '94 Fall Mtg.*, Vol. 360, p. 15, 1995.
42. C. A. Rosen: "Ceramic Transformers and Filters," *Proc. Electronic Component Symp.*, p. 205–101211, 1956.

43. NEC: "Thickness Mode Piezoelectric Transformer," US Patent No. 5,118,982, 1992.
44. S. Kawashima, O. Ohnishi, H. Hakamata, S. Tagami, A. Fukuoka, T. Inoue and S. Hirose: *IEEE Int'l Ultrasonic Symp. Proc.*, Nov., 1994.
45. K. Uchino, B. Koc, P. Laoratanakul and A. Vazquez Carazo: *Proc. 3rd Asian Mtg. on Ferroelectrics*, Dec., 2000.
46. S. Manuspiya, P. Laoratanakul and K. Uchino: *Ultrasonics*, 41(2), 83–87, 2003.
47. A. Vazquez Carazo and K. Uchino: *J. Electroceramics* 7, 197–210, 2001.

6 Computer Simulation of Piezoelectric Devices

The finite-element method (FEM) ATILA and circuit simulation software PSpice and their applications to piezoelectric devices are introduced in this chapter.

ATILA is a finite-element software package specifically developed for the analysis of two- or three-dimensional structures that contain various materials' data on piezoelectric and magnetostrictive crystals and ceramics, developed by Institut Supérieur de l'Electronique et du Numérique (ISEN), France, and distributed by Micromechatronics Inc., PA, USA. A limited educational version of ATILA Light is provided as a courtesy of Micromechatronics Inc., which can be installed onto the reader's personal laptop or desktop computer in learning the simulation process. Access: http://mmech.com/atila-fem/atila-downloads

PSpice is popular circuit analysis software for automatically maximizing the performance of circuits. EMA is distributing a free-download OrCAD Capture PSpice schematic design solution. The reader can access the download of OrCAD Light at http://www.orcad.com/resources/download-orcad-lite

6.1 ATILA FINITE-ELEMENT METHOD SOFTWARE CODE

We have learned so far the physical/analytical approach to designing solid-state actuators. However, with the increase in the complexity in the device design, it is rather difficult to create a simple model to simulate its actual operation. This is why the finite-element method is introduced.

6.1.1 Finite-Element Method Fundamentals

6.1.1.1 Domain and Finite Elements

Consider the piezoelectric specimen, denoted as *domain* Ω, within which the displacement field, $\boldsymbol{u}$, and electric potential field, ϕ, are to be determined. The $\boldsymbol{u}$ and ϕ fields satisfy a set of differential equations that represent the physics of the continuum problem considered. Boundary conditions are usually imposed on the domain's *boundary*, Γ, to complete the definition of the problem. The finite-element method is an approximation technique for finding solution functions.[1] The method consists of subdividing the domain Ω into subdomains, or *finite elements*, as illustrated in Figure 6.1. These finite elements are interconnected at a finite number of points, or *nodes*, along their peripheries. The ensemble of finite elements defines the *problem mesh*. Note that because the subdivision of Ω into finite elements is arbitrary, there is not a unique mesh for a given problem.

Within each finite element, the displacement and electric potential fields are uniquely defined by the values they assume at the element nodes. This is achieved by a process of *interpolation* or *weighing* in which *shape functions* are associated with the element. By combining, or *assembling*, these local definitions throughout the whole mesh, we obtain a trial function for Ω that depends only on the nodal values of $\boldsymbol{u}$ and ϕ and that is "piecewise" defined over all the interconnected elementary domains. Unlike the domain Ω, these elementary domains may have a simple geometric shape and homogeneous composition.

We will show in the following sections how this trial function is evaluated in terms of the variation principle to produce a system of linear equations whose unknowns are the nodal values of $\boldsymbol{u}$ and ϕ.[2]

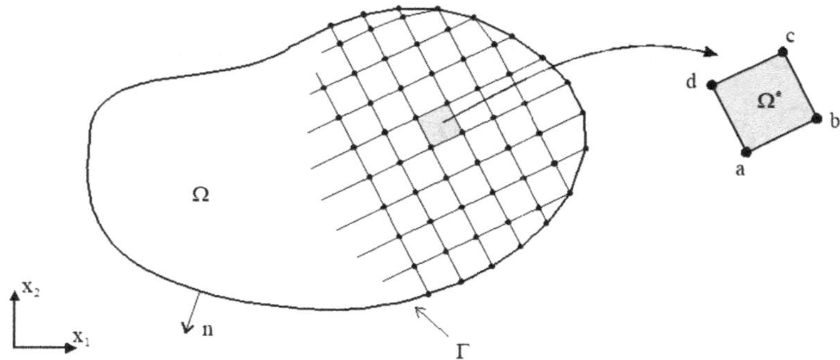

FIGURE 6.1 Discretization of the domain Ω.

6.1.1.2 Defining the Equations for the Problem

6.1.1.2.1 The Constitutive and Equilibrium Equations

The constitutive relations for piezoelectric media may be derived in terms of their associated thermodynamic potentials.[3,4] Assuming the strain, x, and electric field, E, are independent variables, the basic constitutive equations of state for the converse and direct piezoelectric effects are written:

$$\begin{cases} X_{ij} = c_{ijkl}^{E} x_{kl} - e_{kij} E_k \\ D_i = e_{ikl} x_{kl} + \chi_{ij}^{x} E_j \end{cases} \tag{6.1}$$

The quantities c^E (elastic stiffness at constant electric field), e (piezoelectric stress coefficients), and χ^x (dielectric susceptibility at constant strain) are assumed to be constant, which is reasonable for piezoelectric materials subjected to small deformations and moderate electric fields. Furthermore, no distinction is made between isothermal and adiabatic constants.

On the domain, Ω, and its boundary, Γ (where the normal is directed outward from the domain), the fundamental dynamic relation must be verified:

$$\rho \frac{\partial^2 u_i}{\partial t^2} = \frac{\partial X_{ij}}{\partial r_j} \tag{6.2}$$

where $\boldsymbol{u}$ is the displacement vector, ρ the mass density of the material, t the time, X the stress tensor, and $\boldsymbol{r} = < r_1\, r_2\, r_3 >$ a unit vector in the Cartesian coordinate system (Cartesian coordinates x_1, x_2, and x_3 are used in the analytical calculations in other chapters and Figure 6.1).

When no macroscopic charges are present in the medium, Gauss's theorem applies for the electric displacement vector, $\boldsymbol{D}$:

$$\frac{\partial D_i}{\partial r_i} = 0 \tag{6.3}$$

Considering small deformations, the strain tensor, $\boldsymbol{x}$, is written as:

$$x_{kl} = \frac{1}{2}\left(\frac{\partial u_k}{\partial r_l} + \frac{\partial u_l}{\partial r_k} \right) \tag{6.4}$$

Assuming electrostatic conditions, the electrostatic potential, ϕ, is related to the electric field, $\boldsymbol{E}$, by

$$\boldsymbol{E} = -\mathrm{grad}\phi \tag{6.5}$$

or, equivalently

$$E_i = -\frac{\partial \phi}{\partial r_i} \tag{6.6}$$

Using Equations 6.2, 6.3, and 6.6 in combination with Equation 6.1 yields:

$$\begin{cases} -\rho\omega^2 u_i = \dfrac{\partial}{\partial r_j}\left(c_{ijkl}^E x_{kl} - e_{kij} E_k\right) \\ \dfrac{\partial}{\partial r_j}\left(e_{ikl} x_{kl} + \chi_{ij}^x E_j\right) = 0 \end{cases} \tag{6.7}$$

6.1.1.2.2 *Boundary Conditions*

Mechanical and electrical boundary conditions complete the definition of the problem. The mechanical conditions are as follows:

- The *Dirichlet condition* on the displacement field, $\boldsymbol{u}$, is given by:

$$u_i = u_i^o \tag{6.8}$$

  where $\boldsymbol{u}^o$ is a known vector. For convenience, we name the ensemble of surface elements subjected to this condition S_u.
- The *Neumann condition* on the stress field, $\boldsymbol{X}$, is given by:

$$X_{ij} \cdot n_j = f_i^o \tag{6.9}$$

  where $\boldsymbol{n}$ is the vector normal to Γ, directed outward, and $\boldsymbol{f}^o$ is a known vector. For convenience, we name the ensemble of surface elements subjected to this condition S_X.

The electrical conditions are as follows:

- The conditions for the excitation of the electric field between those surfaces of the piezoelectric material that are not covered with an electrode and are, therefore, free of surface charges is given by:

$$D_i \cdot n_i = 0 \tag{6.10}$$

  where $\boldsymbol{n}$ is the vector normal to the surface. For convenience, we name the ensemble of surface elements subjected to this condition S_σ. Note that with the condition in Equation 6.9, we assume that the electric field outside Ω is negligible, which is easily verified for piezoelectric ceramics.
- When considering the conditions for the potential and excitation of the electric field between those surfaces of the piezoelectric material that are covered with electrodes, we assume that there are p electrodes in the system. The potential on the whole surface of the pth electrode is:

$$\phi = \phi_p \tag{6.11}$$

The charge on that electrode is:

$$-\iint_{S_p} D_i n_i dS_p = Q_p \tag{6.12}$$

In some cases, the potential is used, and in others it is the charge. In the former case, ϕ_p is known and Equation 6.11 is used to determine Q_p. In the latter case, Q_p is known and Equation 6.10 is used to determine ϕ. Finally, in order to define the origin of the potentials, it is necessary to impose the condition that the potential at one of the electrodes be zero ($\phi_o = 0$).

6.1.1.3 The Variational Principle

The *variational principle* identifies a scalar quantity, Π, typically named the *functional*, which is defined by an integral expression involving the unknown function, $\boldsymbol{w}$, and its derivatives over the domain, Ω, and its boundary, Γ. The solution to the continuum problem is a function, $\boldsymbol{w}$, such that

$$\delta\Pi = 0 \tag{6.13}$$

Π is said to be stationary with respect to small changes in $\boldsymbol{w}$, $\delta\boldsymbol{w}$ (this is analogous to energy minimization by changing variables slightly). When the variational principle is applied, the solution can be approximated in an integral form that is suitable for finite-element analysis. In general, the matrices derived from the variational principle are always symmetric.

Equation 6.7 and the boundary conditions expressed by Equations 6.8 through 6.12 allow us to define the so-called *Euler equations* to which the variational principle is applied such that a functional of the following form is defined that is stationary with respect to small variations in $\boldsymbol{w}$.

$$\begin{aligned}\Pi = &\iiint_{\Omega} \frac{1}{2}\left(x_{ij}c_{ijkl}^{E}x_{kl} - \rho\omega^2 u_i^2\right)d\Omega - \iint_{S_u}(u_i - u_i^o)n_j\left(c_{ijkl}^{E}x_{kl} - e_{kij}E_k\right)dS_u \\ &- \iiint_{\Omega} \frac{1}{2}\left(2x_{kl}e_{ikl}E_i + E_i\chi_{ij}^{x}E_j\right)d\Omega - \iint_{S_X} f_i u_i dS_X \\ &- \sum_{p=0}^{M}\iint_{S_p}(\phi - \phi_p)n_i\left(e_{ikl}x_{kl} + \chi_{ij}^{x}E_j\right)dS_p + \sum_{p=0}^{p}\phi_p Q_p\end{aligned} \tag{6.14}$$

Note that the first term of this expression for Π represents the *Lagrangian* of the mechanical state. Three mechanical forces, forces/stresses on a body element, surface point force, and surface traction, are taken into account in general on a three-dimensional body, as illustrated in Figure 6.2.

Satisfying the stationary condition for Π implies that all the conditions described by Equations 6.7 through 6.12 are satisfied.

6.1.2 Application of Finite-Element Method

6.1.2.1 Discretization of the Domain

The domain, Ω, is divided into subdomains, Ω^e, or finite elements (Figure 6.1), such that:

$$\Omega = \sum_{e}\Omega^e \tag{6.15}$$

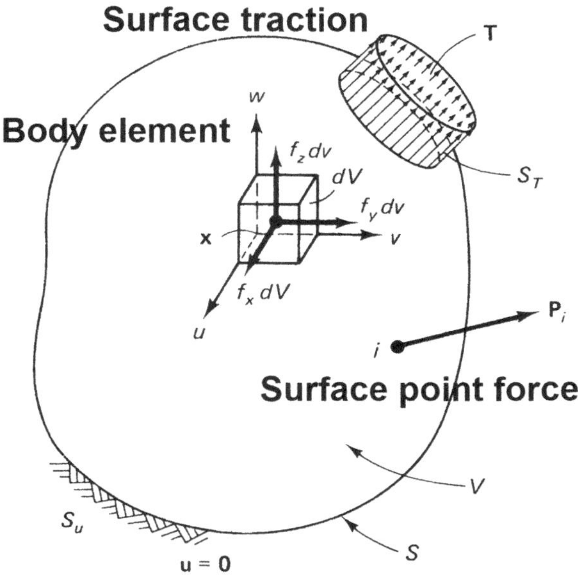

FIGURE 6.2 Mechanical forces on a 3D body.

Common finite elements available for discretizing the domain are shown in Figure 6.3. As a result of the discretization, the functional Π can then be written as:

$$\Pi = \sum_e \Pi^e \tag{6.16}$$

Note that the term Π^e contains only volume integral terms if the element Ω^e is inside the domain Ω. Elements having a boundary coincident with Γ will have a term Π^e that contains both volume and surface integrals.

6.1.2.2 Shape Functions

The finite-element method, being an approximation process, will result in the determination of an approximate solution of the form

$$\boldsymbol{w} \approx \hat{\boldsymbol{w}} = \sum \boldsymbol{N}_i \boldsymbol{a}_i \tag{6.17}$$

where the N_is are *shape functions* prescribed in terms of independent variables (such as coordinates) and the a_is are nodal parameters, known or unknown. The shape functions must guarantee the continuity of the geometry between elements. Moreover, to ensure convergence, it is necessary that the shape functions be at least C^m continuous if derivatives of the mth degree exist in the integral form. This condition is automatically met if the shape functions are polynomials complete to the mth order.

The construction of shape functions for an element, Ω^e, defined by n nodes, usually requires that if N_i is the shape function for node i, then $N_i = 1$ at node i, and $N_i = 0$ at the other nodes. Also, for any point, p, in Ω^e we must have

$$\sum_1^n N_i(\boldsymbol{p}) = 1 \tag{6.18}$$

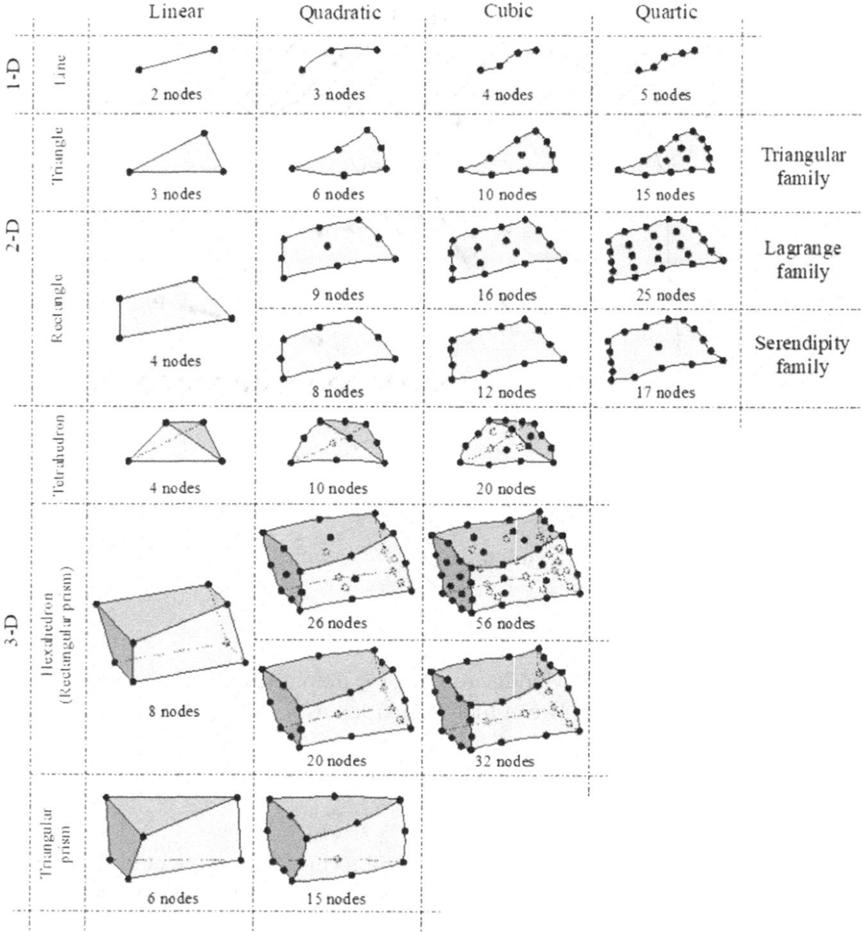

FIGURE 6.3 Common finite elements.

Polynomials are commonly used to construct shape functions. For instance, a Lagrange polynomial

$$N_i(\xi) = \prod_{\substack{k=1 \\ k \neq i}}^{n} \frac{\xi_k - \xi}{\xi_k - \xi_i} \tag{6.19}$$

can be used at node i of a one-dimensional element containing n nodes (see Figure 6.4a). It verifies the following conditions:

$$\begin{cases} N_i(\xi_i) = 1 \\ N_i(\xi_{k \neq 1}) = 0 \\ \sum_{1}^{n} N_i(\xi) = 1 \end{cases} \tag{6.20}$$

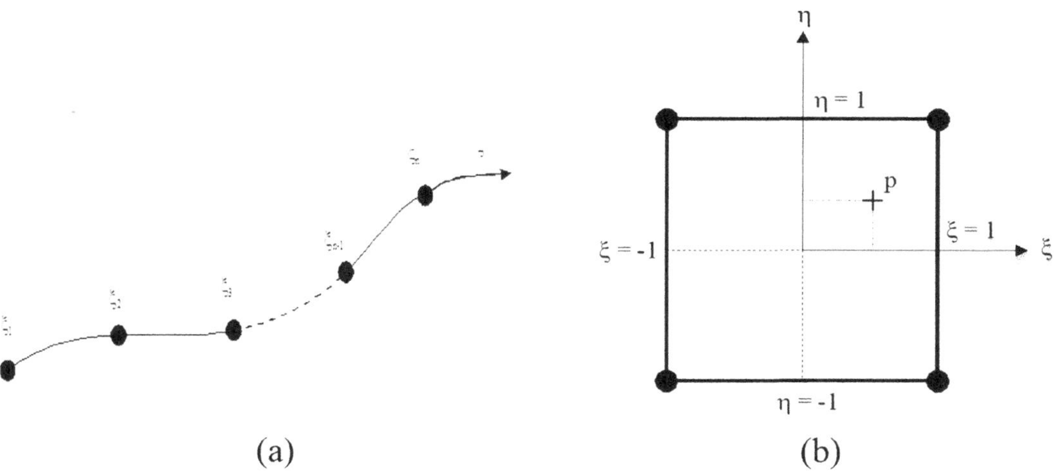

FIGURE 6.4 (a) Generalized n-node linear element. (b) Example of a four-node quadrilateral element.

The simplest 1D shape functions are linear fitting. Refer to Figure 6.5. Top-left shows an unknown function to be approximated. When we use two *linear shape functions*:

$$N_1 = \frac{1}{2}(1-\xi)$$
$$N_2 = \frac{1}{2}(1+\xi) \tag{6.21}$$

we can fit the function as shown at top-right in a linear fashion. However, this linear fitting will require a rather fine mesh size in order to obtain a reasonable agreement with the unknown curved function. Thus, the recent software integrates quadratic fitting (see quadratic column in Figure 6.3), where we use three *parabolic shape functions*, as shown in Figure 6.6:

$$N_1 = -\frac{1}{2}\xi(1-\xi)$$
$$N_2 = +\frac{1}{2}\xi(1+\xi) \tag{6.22}$$
$$N_3 = \frac{1}{2}(1-\xi)(1+\xi)$$

You can easily imagine that this parabolic fitting (in Figure 6.6, top right) seems to be much smoother than the linear fitting. Without increasing the node number, "quadratic" simulation provides better agreement with the initial unknown function by spending a shorter calculation period.

Regarding two-dimensional problems, let us expand the logic to a four-node rectangular element of Figure 6.4b; we can write the four shape functions as:

$$N(\xi,\eta) = N_i(\xi)N_j(\eta) \tag{6.23}$$

where i and j indicate the row and column of the node in the element and $i, j = 1,2$. The conditions described by Equation 6.20 can be verified for these functions. Finally, the position of any point p with coordinates (ξ, η) in the element is given by:

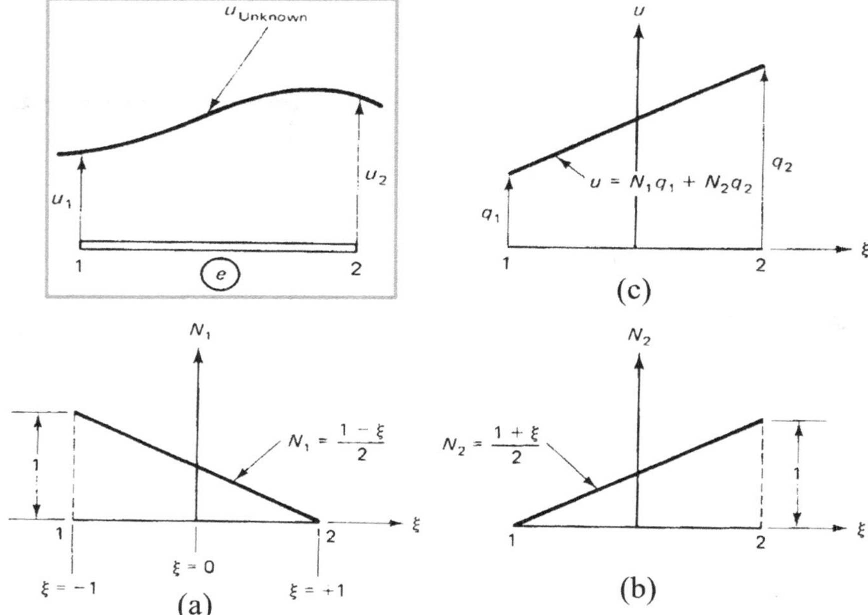

FIGURE 6.5 1D linear shape function fitting process.

$$\begin{bmatrix} \xi \\ \eta \end{bmatrix} = \left\langle \left(\frac{1}{4}\right)[(-1)+\xi][(-1)+\eta],\ \left(-\frac{1}{4}\right)[(1)+\xi][(-1)+\eta],\right.$$

$$\left. \left(-\frac{1}{4}\right)[(-1)+\xi][(1)+\eta],\ \left(\frac{1}{4}\right)[(1)+\xi][(1)+\eta] \right\rangle \begin{bmatrix} -1 & -1 \\ 1 & -1 \\ -1 & 1 \\ 1 & 1 \end{bmatrix} \tag{6.24}$$

This fitting is the combination of four flat planes (i.e., 2D linear fitting). In most cases, the same shape functions are used to describe the element geometry and to represent the solution, $\hat{W}$. It may not be difficult for the reader to extend the shape functions to a "quadratic" fashion.

6.1.2.3 Parent Elements

In order to better represent the actual geometry, it is generally useful to use curvilinear finite elements for the discretization of Ω. These elements are then mapped into *parent finite elements* (Figure 6.7) in order to facilitate the computation of Π^e. An isoparametric representation is commonly used to perform the mapping of the actual elements into the reference elements. Consider a volume element, Ω^e, of the domain, Ω, defined by n nodes. The position vector, $\boldsymbol{R}$, of a point, p, of Ω^e can be written as a function of parameters ξ, η, and ζ:

$$\xi \rightarrow R = R(\xi) \tag{6.25}$$

which is the same as:

$$R = \begin{Bmatrix} x \\ y \\ z \end{Bmatrix} = \begin{Bmatrix} x(\xi,\eta,\zeta) \\ y(\xi,\eta,\zeta) \\ z(\xi,\eta,\zeta) \end{Bmatrix} \tag{6.26}$$

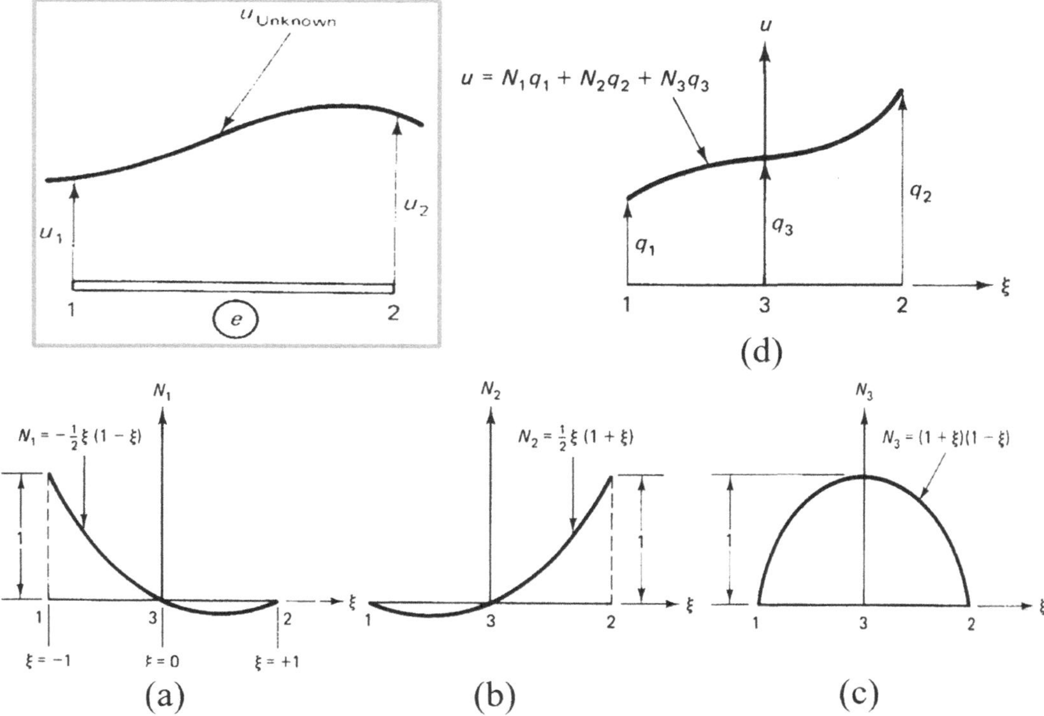

FIGURE 6.6 1D quadratic shape function fitting process.

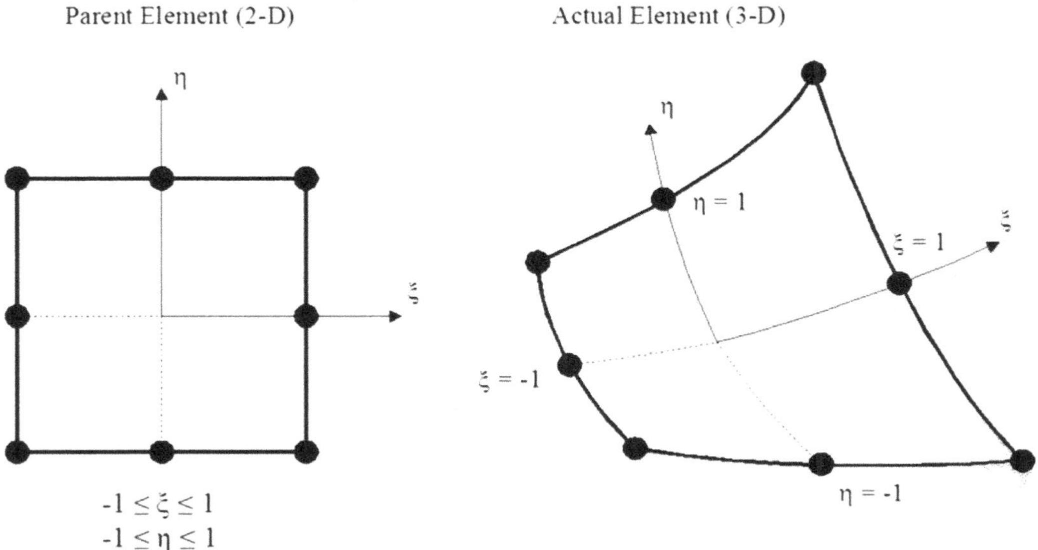

FIGURE 6.7 An example of a parent element for an eight-node quadrilateral element.

The finite-element representation can be written in the following form:

$$\begin{aligned} x &= \boldsymbol{N}(\xi,\eta,\zeta)\,\boldsymbol{x_n} \\ y &= \boldsymbol{N}(\xi,\eta,\zeta)\,\boldsymbol{y_n} \\ z &= \boldsymbol{N}(\xi,\eta,\zeta)\,\boldsymbol{z_n} \end{aligned} \tag{6.27}$$

where:

$$\begin{aligned} \boldsymbol{x_n} &= \langle x_1 \quad x_2 \quad \ldots \quad x_n \rangle \\ \boldsymbol{y_n} &= \langle y_1 \quad y_2 \quad \ldots \quad y_n \rangle \\ \boldsymbol{z_n} &= \langle z_1 \quad z_2 \quad \ldots \quad z_n \rangle \\ \boldsymbol{N} &= \langle N_1 \quad N_2 \quad \ldots \quad N_n \rangle \end{aligned} \tag{6.28}$$

The scalars x_i, y_i, and z_i represent the Cartesian coordinates at node i, and N_i is the shape function at node i. A differential element at point p is thus defined by:

$$d\boldsymbol{R} = \boldsymbol{F}_\xi\, d\xi \tag{6.29}$$

and

$$\frac{\partial}{\partial \xi} = \boldsymbol{F}_\xi^X \frac{\partial}{\partial \boldsymbol{R}} = \boldsymbol{J}\frac{\partial}{\partial \boldsymbol{R}} \tag{6.30}$$

where:

$$\boldsymbol{F}_\xi = \begin{bmatrix} x_{,\xi} & x_{,\eta} & x_{,\zeta} \\ y_{,\xi} & y_{,\eta} & y_{,\zeta} \\ z_{,\xi} & z_{,\eta} & z_{,\zeta} \end{bmatrix} = [\boldsymbol{a}_1\ \boldsymbol{a}_2\ \boldsymbol{a}_3] \tag{6.31}$$

and

$$d\boldsymbol{R} = \boldsymbol{a}_1 d\xi + \boldsymbol{a}_2 d\eta + \boldsymbol{a}_3 d\zeta \tag{6.32}$$

Vectors $\boldsymbol{a}_1$, $\boldsymbol{a}_2$, and $\boldsymbol{a}_3$ are the base vectors associated with the parametric space, $\boldsymbol{J}$ is the Jacobian matrix of the transformation, and $\boldsymbol{F}_\xi$ is the transformation matrix from the parametric space to the Cartesian space.

From Equations 6.29 and 6.30, we obtain:

$$d\xi = \boldsymbol{F}_\xi^{-1} d\boldsymbol{R} \tag{6.33}$$

and

$$\frac{\partial}{\partial \boldsymbol{R}} = \boldsymbol{J}^{-1}\frac{\partial}{\partial \xi} = \boldsymbol{j}\frac{\partial}{\partial \xi} \tag{6.34}$$

A volume element is defined by

$$dV = \left|(d\boldsymbol{x} \wedge d\boldsymbol{y}) \cdot d\boldsymbol{z}\right| \tag{6.35}$$

which, in Cartesian space, is $dV = dx\,dy\,dz$, and in parametric space:

$$dV = \left|(a_1 d\xi \wedge \boldsymbol{a}_2 d\eta) \cdot \boldsymbol{a}_3 d\zeta\right| \tag{6.36}$$

which is the same as

$$dV = Jd\xi d\eta d\zeta \qquad \text{where: } J = \left|\det \boldsymbol{J}\right| \tag{6.37}$$

Therefore, the quantity Π^e can be expressed in parametric space as

$$\int_{\Omega^e} (\ldots) dx\,dy\,dz = \int_{\Omega^{parent}} (\ldots) J\,d\xi\,d\eta\,d\zeta \tag{6.38}$$

which allows the computation of each element to be performed on the parent element rather than on the real element.

6.1.2.4 Discretization of the Variational Form

We now write the solution functions for the piezoelectric problem in terms of the shape functions. For each element, Ω^e, defined by n nodes, the electric field is obtained from Equation 6.6 and can be written in the form:

$$\boldsymbol{E} = -\boldsymbol{B}_\phi^e \Phi \tag{6.39}$$

where $\boldsymbol{\Phi}$ is the vector associated with the nodal values of the electrostatic potential:

$$\boldsymbol{B}_\phi^e = \left[\boldsymbol{B}_{\phi 1}^e \; \boldsymbol{B}_{\phi 2}^e \cdots \boldsymbol{B}_{\phi n}^e\right] \tag{6.40}$$

and

$$\boldsymbol{B}_{\phi i}^e = \begin{bmatrix} \dfrac{\partial N_i^e}{\partial x} \\ \dfrac{\partial N_i^e}{\partial y} \\ \dfrac{\partial N_i^e}{\partial z} \end{bmatrix} \tag{6.41}$$

The terms $\boldsymbol{B}_{\phi i}^e$ are the first spatial derivatives of the shape functions. Similarly, for each Ω^e, the strain tensor defined by Equation 6.4 becomes

$$x = -\boldsymbol{B}_u^e \boldsymbol{U} \tag{6.42}$$

where $\boldsymbol{U}$ is the vector of the nodal values of the displacement:

$$\boldsymbol{B}_u^e = \left[\boldsymbol{B}_{u1}^e \ \boldsymbol{B}_{u2}^e \cdots \boldsymbol{B}_{un}^e\right] \tag{6.43}$$

and

$$\boldsymbol{B}_{ui}^e = \begin{bmatrix} \frac{\partial N_i^e}{\partial x} & 0 & 0 \\ 0 & \frac{\partial N_i^e}{\partial y} & 0 \\ 0 & 0 & \frac{\partial N_i^e}{\partial z} \\ 0 & \frac{\partial N_i^e}{\partial z} & \frac{\partial N_i^e}{\partial y} \\ \frac{\partial N_i^e}{\partial z} & 0 & \frac{\partial N_i^e}{\partial x} \\ \frac{\partial N_i^e}{\partial y} & \frac{\partial N_i^e}{\partial x} & 0 \end{bmatrix} \tag{6.44}$$

Consequently, Equation 6.1 becomes:

$$\begin{cases} \boldsymbol{X} = \boldsymbol{c}^E \boldsymbol{B}_u^e \boldsymbol{U} + \boldsymbol{e}^x \boldsymbol{B}_\phi^e \Phi \\ \boldsymbol{D} = \boldsymbol{e} B_u^e U + \chi^x B_\phi^e \Phi \end{cases} \tag{6.45}$$

Finally, we can rewrite the functional Π^e on the element, as

$$\begin{aligned} \Pi^e = & \frac{1}{2}\iiint_{\Omega^e} U^{e^X}\left(B_u^{e^X} c^E B_u^e - \rho\omega^2 N^{e^X} N^e\right) U^e d\Omega^e \\ & + \iiint_{\Omega^e} U^{e^X} B_u^{e^X} e B_\phi^e \Phi^e d\Omega^e - \frac{1}{2}\iiint_{\Omega^e} \Phi^{e^X} B_\phi^{e^X} \chi^x B_\phi^e \Phi^e d\Omega^e \\ & - \iint_{S_X^e} U^{e^X} N^{e^X} f \, dS_X^e + \sum_{p=0}^{p} \phi_p Q_p \end{aligned} \tag{6.46}$$

After integrating the shape function matrices and their derivatives, we can write:

$$\Pi^e = \frac{1}{2} U^{e^X}\left(K_{uu}^e - \omega^2 M^e\right) U^e + U^{e^X} K_{u\phi}^e \Phi^e + \frac{1}{2}\phi^{e^X} K_{\phi\phi}^e \Phi^e - U^{e^X} F^e + \sum_{p=0}^{p} \phi_p Q_p \tag{6.47}$$

where

$$K_{uu}^{e} = \iiint\limits_{\Omega^{e}} B_{u}^{e^{X}} c^{E} B_{u}^{e} d\Omega^{e}$$
$$M^{e} = \iiint\limits_{\Omega^{e}} \rho\omega^{2} N^{e^{X}} N^{e} d\Omega^{e}$$
$$K_{\phi u}^{e} = \iiint\limits_{\Omega^{e}} B_{u}^{e^{X}} e B_{\phi}^{e} d\Omega^{e} \tag{6.48}$$
$$K_{\phi\phi}^{e} = -\iiint\limits_{\Omega^{e}} B_{\phi}^{e^{X}} \chi^{x} B_{\phi}^{e} d\Omega^{e}$$
$$F^{e} = \iint\limits_{S_{X}^{e}} N^{e^{X}} f dS_{X}^{e}$$

and K_{uu}^{e}, $K_{\phi u}^{e}$, and $K_{\phi\phi}^{e}$ are the elastic, piezoelectric, and dielectric susceptibility matrices, and M^{e} is the consistent mass matrix.

6.1.2.5 Assembly

The matrices in (6.47) must be rearranged for the whole domain, Ω, by a process called assembly. From this process, we obtain the following matrices:

$$K_{uu} = \sum_{e} K_{uu}^{e}$$
$$M = \sum_{e} M^{e}$$
$$K_{\phi u} = \sum_{e} K_{\phi u}^{e} \tag{6.49}$$
$$K_{\phi\phi} = \sum_{e} K_{\phi\phi}^{e}$$
$$F = \sum_{e} F^{e}$$

The application of the variational principle implies the minimization of the functional Π with respect to variations of the nodal values, $\boldsymbol{U}$ and $\boldsymbol{\Phi}$. Therefore:

$$\frac{\partial \Pi}{\partial u_{i}} = 0 \quad \forall\, i \tag{6.50}$$

and

$$\frac{\partial \Pi}{\partial \phi_{j}} = 0 \quad \forall\, j \tag{6.51}$$

Making use of Equations 6.47 and 6.16, and applying the stationary condition to Π, we obtain:

$$\begin{bmatrix} \boldsymbol{K}_{uu} - \omega^2 \boldsymbol{M} & \boldsymbol{K}_{u\phi} \\ \boldsymbol{K}_{u\phi}^X & \boldsymbol{K}_{\phi\phi} \end{bmatrix} \begin{bmatrix} \boldsymbol{U} \\ \Phi \end{bmatrix} = \begin{bmatrix} \boldsymbol{F} \\ -\boldsymbol{Q} \end{bmatrix} \tag{6.52}$$

The vector for the nodal charges, $\boldsymbol{Q}$, is such that for all nodes, i, that belong to an electrode, p, with potential ϕ_p, the sum of the charges, Q_i, is equal to ϕ_p. For all other nodes, j, that do not belong to an electrode, $Q_j = 0$.

6.1.2.6 Computation

Specific integration, diagonalization, and elimination techniques are employed to solve the system Equation 6.52 on a computer. A full description of these techniques is a topic that extends well beyond the scope of this text and thus will not be presented here. The matrix Equation 6.52 may be adapted for a variety of different analyses, such as the static, modal, harmonic, and transient types:

- Static Analysis: Slow operation at an off-resonance frequency.

$$\begin{bmatrix} \boldsymbol{K}_{uu} & \boldsymbol{K}_{u\phi} \\ \boldsymbol{K}_{u\phi}{}^X & \boldsymbol{K}_{\phi\phi} \end{bmatrix} \begin{bmatrix} \boldsymbol{U} \\ \Phi \end{bmatrix} = \begin{bmatrix} \boldsymbol{F} \\ -\boldsymbol{Q} \end{bmatrix} \tag{6.53}$$

- Modal Analysis: Only the resonance/antiresonance modes are analyzed.

$$\begin{bmatrix} \boldsymbol{K}_{uu} - \omega^2 \boldsymbol{M} & \boldsymbol{K}_{u\phi} \\ \boldsymbol{K}_{u\phi}{}^X & \boldsymbol{K}_{\phi\phi} \end{bmatrix} \begin{bmatrix} \boldsymbol{U} \\ \Phi \end{bmatrix} = \begin{bmatrix} 0 \\ -\boldsymbol{Q} \end{bmatrix} \tag{6.54}$$

- Harmonic Analysis: To obtain admittance/impedance frequency spectrum.

$$\begin{bmatrix} \boldsymbol{K}_{uu} - \omega^2 \boldsymbol{M} & \boldsymbol{K}_{u\phi} \\ \boldsymbol{K}_{u\phi}{}^X & \boldsymbol{K}_{\phi\phi} \end{bmatrix} \begin{bmatrix} \boldsymbol{U} \\ \Phi \end{bmatrix} = \begin{bmatrix} \boldsymbol{F} \\ -\boldsymbol{Q} \end{bmatrix} \tag{6.55}$$

- Transient Analysis: Transient response upon a pulse voltage.

$$\begin{bmatrix} \boldsymbol{M} & 0 \\ 0 & 0 \end{bmatrix} \begin{bmatrix} \ddot{U} \\ \ddot{\Phi} \end{bmatrix} + \frac{1}{\omega_o} \begin{bmatrix} \boldsymbol{K}'_{uu} & \boldsymbol{K}'_{u\phi} \\ \boldsymbol{K}'_{u\phi}{}^X & \boldsymbol{K}'_{\phi\phi} \end{bmatrix} \begin{bmatrix} \dot{U} \\ \dot{\Phi} \end{bmatrix} + \begin{bmatrix} \boldsymbol{K}_{uu} & \boldsymbol{K}_{u\phi} \\ \boldsymbol{K}_{u\phi}{}^X & \boldsymbol{K}_{\phi\phi} \end{bmatrix} \begin{bmatrix} \boldsymbol{U} \\ \Phi \end{bmatrix} = \begin{bmatrix} \boldsymbol{F} \\ -\boldsymbol{Q} \end{bmatrix} \tag{6.56}$$

Each of these cases requires specific conditioning and computation techniques. Some of the superior features in ATILA include:

- Materials data are already integrated, including various PZTs, quartz, Terfenol, and so on. You can easily modify the physical parameters for your special material.
- Three losses (elastic, dielectric, and piezoelectric losses; elastic, magnetic, and piezomagnetic losses) can be installed even for a low-symmetry triclinic crystal, which is the key to obtaining less than ± 2% error in the admittance peak values, and the difference among the Q values at the resonance and antiresonance modes.

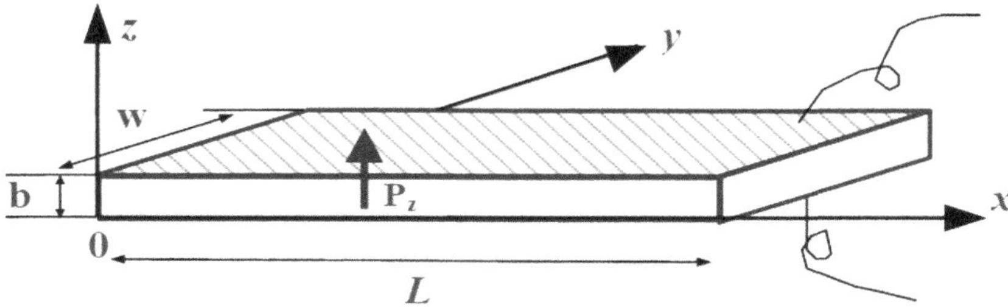

FIGURE 6.8 Longitudinal vibration through the transverse piezoelectric effect (d_{31}) in a rectangular plate ($L \gg w \gg b$).

6.1.3 ATILA Simulation Examples

The reader can learn how to use ATILA FEM software code from a textbook, *FEM and Micromechatronics with ATILA Software*, authored by Kenji Uchino, from CRC Press.[5] Four simulation examples are introduced in this section: (1) k_{31} resonance/antiresonance modes, (2) stress concentration in a multilayer actuator, (3) metal tube motor, and (4) piezoelectric transformer. More advanced simulation case studies can be found in Reference (6).

6.1.3.1 k_{31} Resonance/Antiresonance Modes

Let us consider the simplest k_{31}-type rectangular piezoelectric plate, as shown in Figure 6.8. The resonance and antiresonance states are both mechanical resonance states with amplified strain/displacement states, but they are very different from the driving viewpoints. The mode difference is described by the following intuitive model. In a high electromechanical coupling material with k almost equal to 1, the resonance or antiresonance states appear for $\tan(\omega L/2v) = \infty$ or 0 [i.e., $\omega L/2v = (m-1/2)\Pi$ or $m\Pi$ (m: integer)], respectively. The strain amplitude x_1 distribution for each state (calculated using Equation 2.155) is illustrated in Figure 6.9. In the resonance state, the strain distribution is basically sinusoidal with the maximum at the center of plate ($x = L/2$) (see the numerator). When ω is close to ω_R, $(\omega_R L/2v) = \Pi/2$, leading to the denominator $\cos(\omega_R L/2v) \rightarrow 0$. Significant strain magnification is obtained. It is worth noting that the stress, X_1, is zero at the plate ends ($x = 0$ and L), but the strain, x_1, is not zero, but is equal to $d_{31}E_Z$. According to this large strain amplitude, large capacitance changes (called *motional capacitance*) are induced, and under a constant applied voltage, the current can easily flow into the device (i.e., admittance Y is infinite). On the contrary, at antiresonance, the strain induced in the device compensates completely (because extension and compression are compensated), resulting in no motional capacitance change, and the current cannot flow easily into the sample (i.e., admittance Y is zero). Thus, for a high-k material, the first antiresonance frequency, f_A, should be twice as large as the first resonance frequency, f_R. There

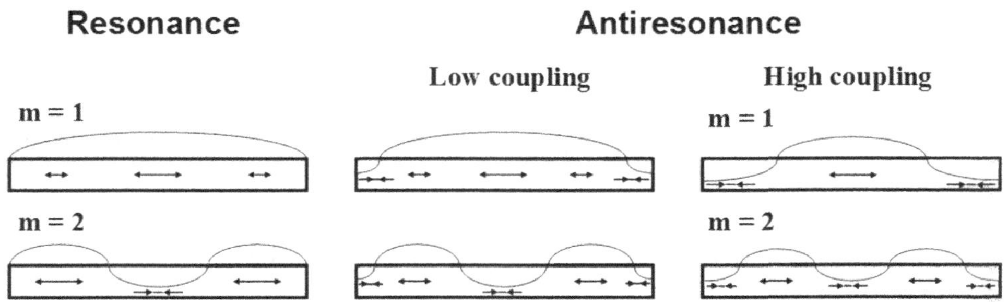

FIGURE 6.9 Strain distribution in the resonant or antiresonant state for a k_{31}-type piezoelectric plate.

is only one vibration node at the plate center for the resonance (top left in Figure 6.9), and there are an additional two nodes at both plate ends for the first antiresonance (top right in Figure 6.9). The reason is from the antiresonance drive, that is, high voltage/low current (minimum power) drive due to the high impedance. The converse piezo-effect strain under E directly via d_{31} (uniform strain in the sample) superposes on the mechanical resonance strain distribution (distributed strain with nodes in the sample), two strains of which have exactly the same level, theoretically, at the antiresonance for $k_{31} \approx 1$.

In a typical case, where $k_{31} = 0.3$, the antiresonance state varies from the previously mentioned mode and becomes closer to the resonance mode (top center in Figure 6.9). Low-coupling material exhibits an antiresonance mode where the capacitance change due to the size change (*motional capacitance*) is compensated completely by the current required to charge up the static capacitance (called *damped capacitance*). Thus, the antiresonance frequency, f_A, will approach the resonance frequency, f_R.

ATILA simulation results are introduced for hypothetical PZT samples ($L = 40$, $w = 6$, $b = 1$ mm) with $k_{31} = 0$, 0.35 to 0.96 under various intensive elastic loss, tan ϕ_{11}'. Figures 6.10a and b show the simulated admittance spectra and the displacement distribution profiles, respectively. Since ATILA integrates three intensive elastic, dielectric, and piezoelectric losses, we can recognize the difference of Q_A and Q_B first (i.e., $Q_A < Q_B$). Second, with increasing the tan ϕ_{11}', the admittance peak value decreases significantly. Third, with increasing the electromechanical coupling factor, k_{31}, the antiresonance frequency increases (by keeping the resonance frequency), approaching $f_B/f_A \approx 2$. Accordingly, the maximum displacement point at the antiresonance frequency moves from the near-plate-end to the 1/4 inside the plate length. All simulation results are consistent with the discussion above.

6.1.3.2 Stress Concentration in a Multilayer Actuator

As shown in Figure 6.11a, a multilayer piezoelectric actuator is composed of active and inactive areas, corresponding to the electrode-overlapped and -nonoverlapped portions. Accordingly, concentration of electric field and stress is expected around the internal electrode edges. Figures 6.11b and 6.12 are the 2D calculation results with ATILA under pseudo-DC drive in an eight-layered multilayer actuator in terms of the potential distribution and stress concentration around the internal electrode edges. The maximum tensile stress should be lower than the fracture strength of this PZT ceramic. In other words, knowing the fracture strength of the piezoceramics, we can evaluate the maximum voltage applicable on this ML actuator.

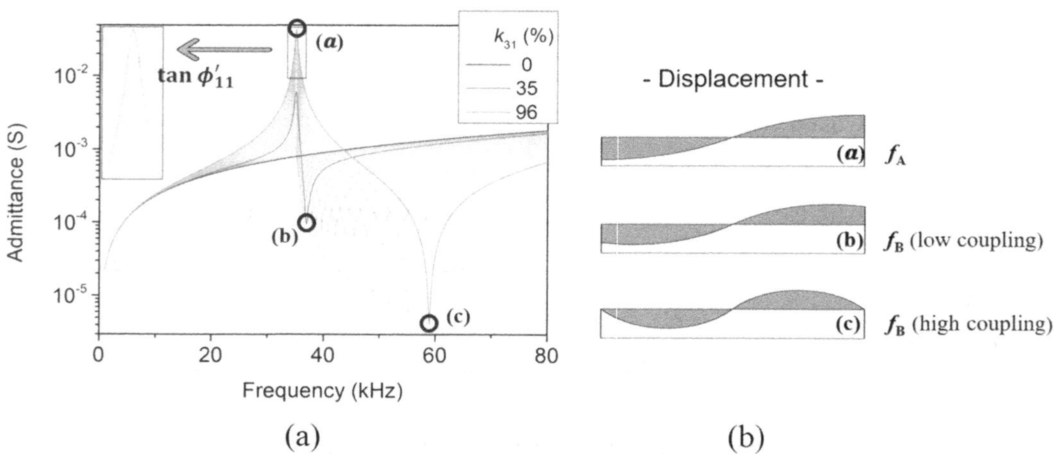

FIGURE 6.10 (a) Simulated admittance spectra for various k_{31} and intensive elastic loss, tan ϕ_{11}'; (b) displacement distribution profiles at the resonance and antiresonance modes.

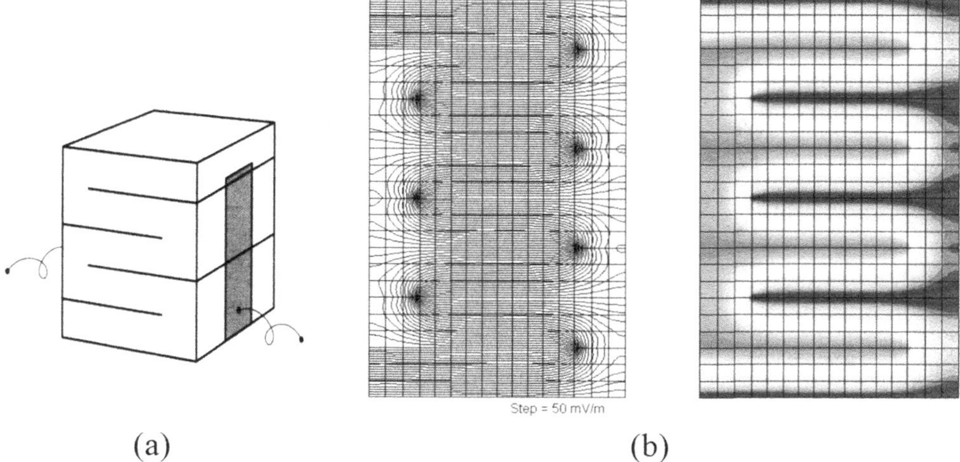

FIGURE 6.11 (a) Multilayer piezoelectric actuator with an interdigital electrode pattern; (b) potential distribution around the internal electrode edges in an eight-layered multilayer actuator (2D simulation results).

6.1.3.3 Metal Tube Motor

A metal tube motor is composed of two PZT rectangular plates bonded on a metal tube, as depicted in Figure 6.13.[7] Due to this asymmetric configuration of the two PZT plates, the resonance frequencies of the two orthogonal bending modes along x' and y' deviate slightly. Thus, driving the motor at the intermediate frequency exhibits a superposed vibration, like a hula-hoop mode, because of 90° phase lag between the above two split modes. Exciting either the X or Y plate generates a counterclockwise or clockwise wobbling motion of the metal tube, respectively. (see the ATILA calculation in Figure 6.14; time sequence is from top to bottom).

The ATILA software can calculate the stator vibration under mechanically free or forced condition, but in order to calculate the motor characteristics (speed, torque etc.) we need to combine another software package that treats the surface friction model.

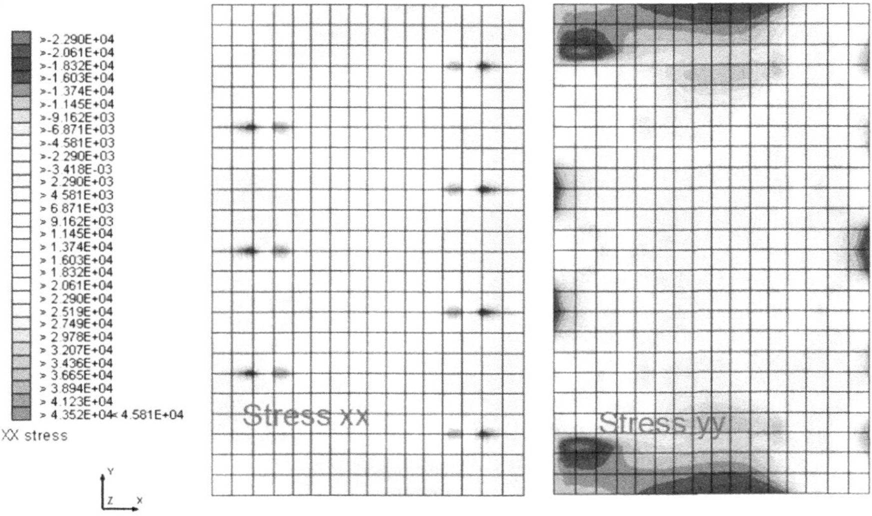

FIGURE 6.12 Stress concentration (XX and YY) around the internal electrode edges in a ML piezo-actuator (2D simulation results).

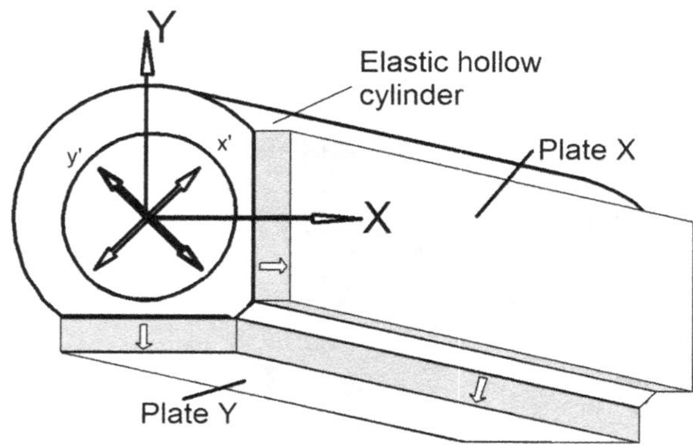

FIGURE 6.13 Structure of a metal tube motor. Two PZT plates are bonded asymmetrically on a metal tube.

6.1.3.4 Piezoelectric Transformer

Disk-type transformers have advantages over the rectangular plate Rosen types because of the usage of k_p instead of k_{31} (in general, $k_p > k_{31}$). One of the disk types is shown in Figure 6.15 with crescent curved electrodes.[8] Different from the conventional Rosen type, the circular shape (k_p) enhances the energy conversion rate and voltage step-up ratio. Also, the curved electrode contour excites local shear mode (k_{15}), which further enhances the voltage step-up ratio

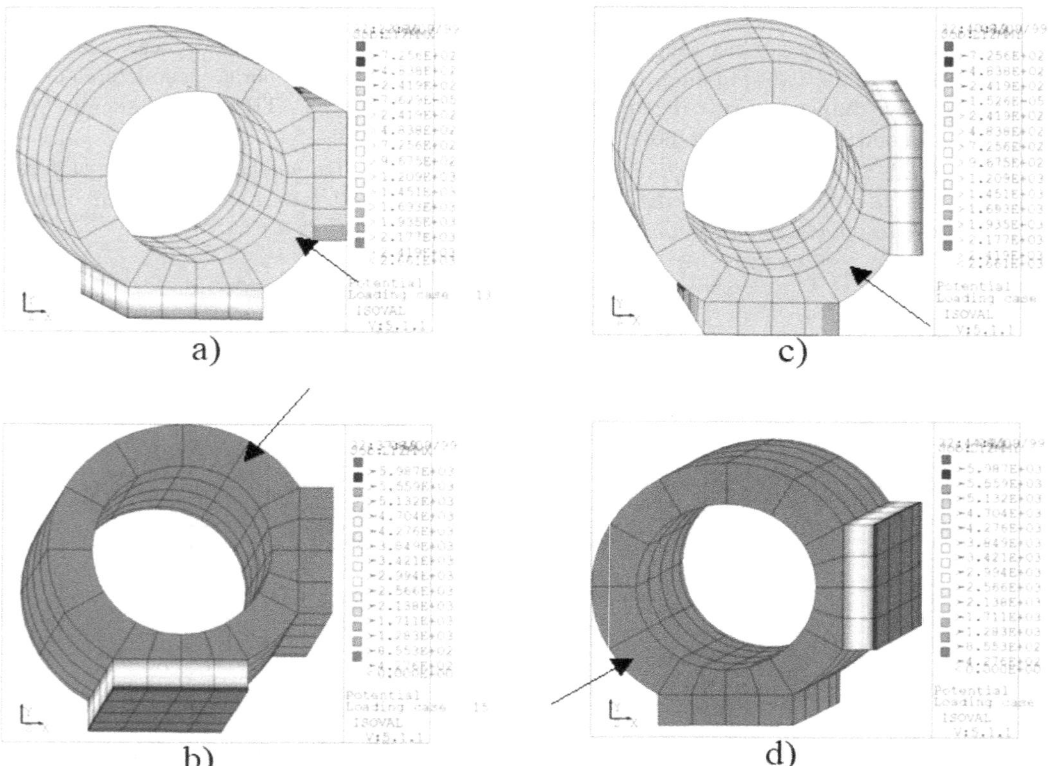

FIGURE 6.14 (a, b) Top view of orthogonal bending mode shapes when plate Y was excited (time sequence: a → b). (c, d) Bending mode shapes when plate X was excited (time sequence: c → d). X or Y plate excitation generates clockwise or counterclockwise rotation.

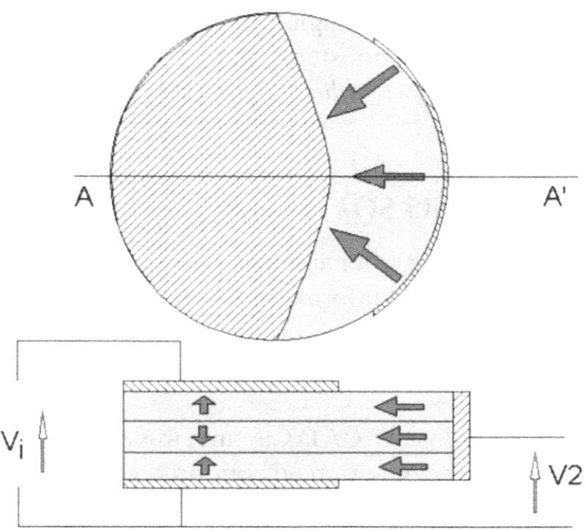

FIGURE 6.15 Disk shape piezoelectric transformer with crescent curved electrodes.

more than 300 (note again that $k_{15} > k_p$). Furthermore, usage of curved electrodes reduces the stress concentration, which can be calculated by ATILA. Figure 6.16 demonstrates higher-order vibration modes and the calculated potential distribution. A 25-mm diameter, 1-mm-thick PZT disk can generate 10 W or higher[8] so that this three-layer transformer can be applied to a small laptop computer AC/DC adaptor. We demonstrated lighting up the 4 W cold cathode fluorescent

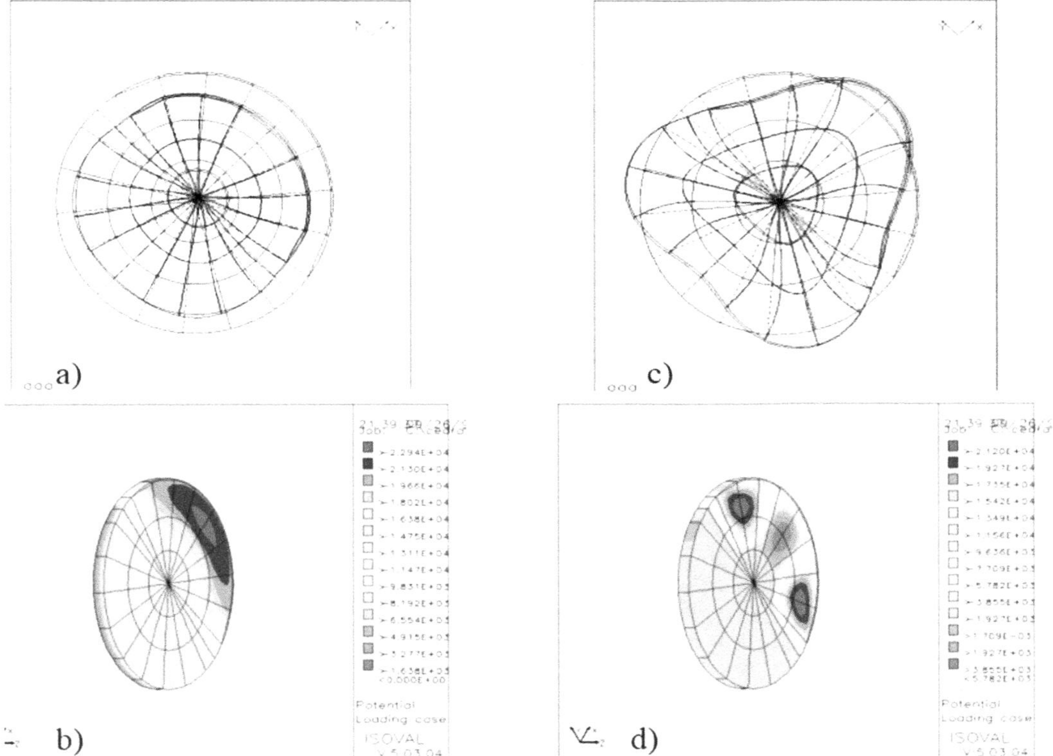

FIGURE 6.16 The first and third vibration modes [(a) and (c)] of a crescent electrode-type disk transformer, and their corresponding voltage distribution calculation by ATILA [(b) and (d)]. No load (open circuit).

lamp (CCFL) directly by 500 V using a single disk sample (25 mm in diameter and 1 mm in thickness) without using an additional booster coil.

ATILA can simulate the step-up/down voltage ratio even under a certain electrical load (L, C, and R) condition.

6.2 PSPICE CIRCUIT ANALYSIS SOFTWARE

PSpice is popular circuit analysis software for automatically maximizing the performance of circuits. Though PSpice cannot provide localized information such as strain/displacement distribution, such as in Figure 6.10b, it can provide information on the equivalent circuit of the piezo device, such as impedance spectra in Figure 6.10a, which is useful for designing the component, as well as for system optimization.

EMA is distributing a free-download OrCAD Capture, this schematic design solution. The reader can access the download site at: http://www.orcad.com/products/orcad-lite-overview?gclid=COaXitWJp9ECFcxKDQodCGMB0w

6.2.1 k_{31}-Type Piezoplate Simulation with Institute of Electrical and Electronics Engineers Equivalent Circuit

Corresponding to Section 6.1.3.1, we consider the same k_{31}-type piezoplate simulation with PSpice, first. Figure 6.17 shows an IEEE Standard equivalent circuit with only one elastic loss (tan ϕ'). In addition to following Equations 6.57–6.59,

$$L_n = \left(bLs_{11}^E/4v^2wd_{31}^2\right)/2 = (\rho/8)(Lb/w)\left(s_{11}^{E2}/d_{31}^2\right) \tag{6.57}$$

$$\begin{aligned} C_n &= 1/\omega_r^2 L_n = (L/n\pi v)^2(8/\rho)(w/Lb)\left(d_{31}^2/s_{11}^{E2}\right) \\ &= (8/n^2\pi^2)(Lw/b)\left(d_{31}^2/s_{11}^{E2}\right)s_{11}^E \end{aligned} \tag{6.58}$$

$$\omega_{\mathrm{A,n}} = \mathrm{n}\pi/\mathrm{L}\sqrt{\rho s_{11}^E} \tag{6.59}$$

the circuit analysis provides the following R and Q (electrical quality factor, which corresponds to the mechanical quality factor in the piezoplate):

$$Q = \sqrt{L_A/C_A}/R_A \tag{6.60}$$

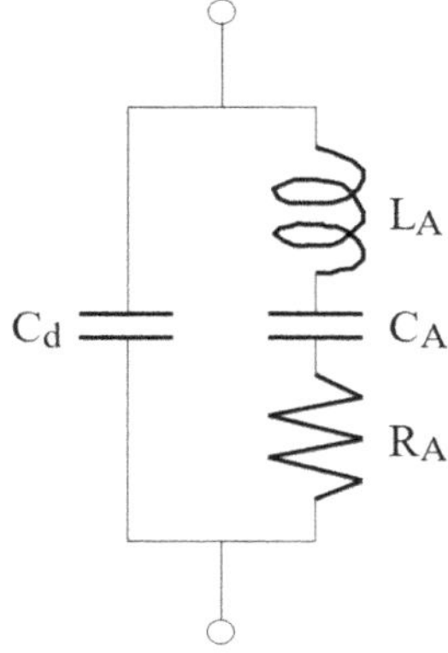

FIGURE 6.17 An IEEE Standard equivalent circuit for a k_{31}-type piezoplate.

Knowing the material's constants; density, ρ; elastic compliance, s_{11}^E; piezoelectric constant, d_{31}; dielectric constant, ε_{33}; and sample size, L, w, and b, we can calculate the above equivalent circuit components, L_A, C_A, and R_A. In order to calculate the damped capacitance, C_d, we use the following relation:

$$C_d = \varepsilon_0 \varepsilon_{33} \frac{Lw}{b}\left(1 - k_{31}^2\right), \tag{6.61}$$

where $k_{31}^2 = d_{31}^2 / \varepsilon_0 \varepsilon_{33} s_{11}^E$.

Figure 6.18 shows the PSpice simulation process for the IEEE-type k_{31} mode. (a) shows an equivalent circuit for k_{31} mode. L, C, and R values were calculated for PZT4 ($Q_m = 500$) with

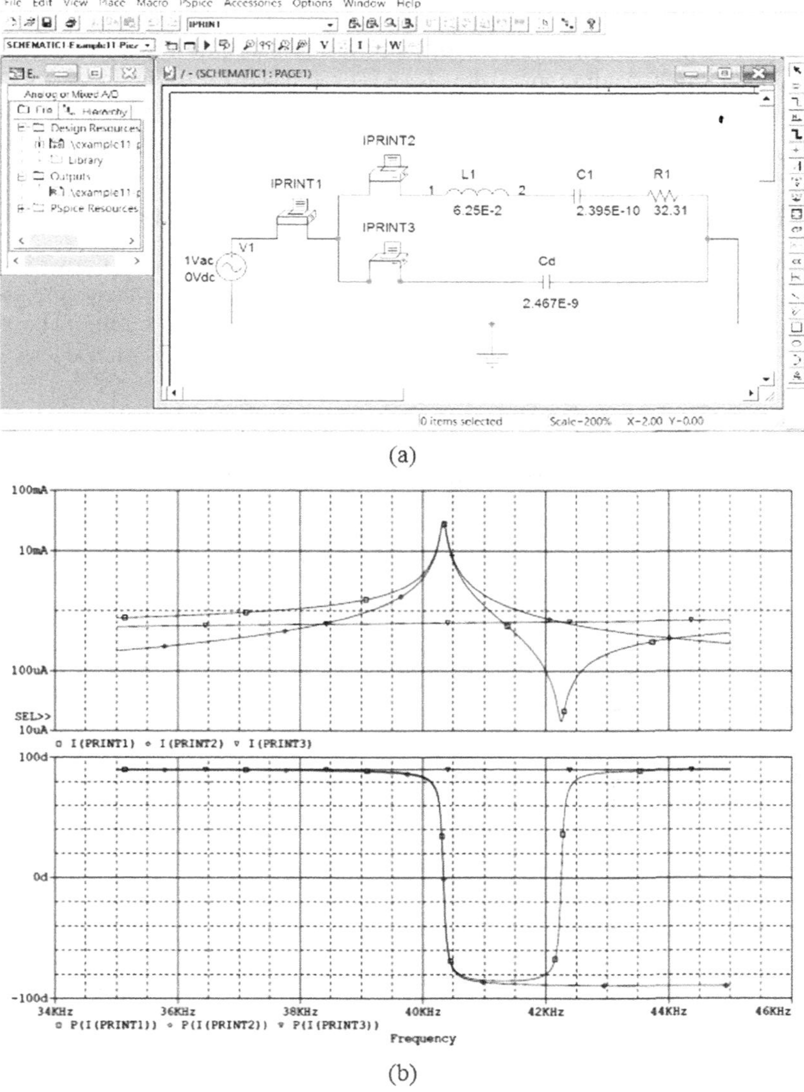

FIGURE 6.18 PSpice simulation of the IEEE-type k_{31} mode. (a) Equivalent circuit for k_{31} mode. L, C, and R values were calculated for PZT4 with $40 \times 6 \times 1$ mm^3. (b) Simulation results on admittance magnitude and phase spectra.

$40 \times 6 \times 1$ mm³, according to Equations 6.57–6.61. Figure 6.18b plots the simulation results on the currents under 1 V_{ac} (constant voltage drive), that is, admittance magnitude and phase spectra. IPRINT1 (current measurement), IPRINT2, and IPRINT3 are the measure of the total admittance (□ line), motional admittance (O line), and damped admittance (∇ line), respectively. First, the damped admittance shows a slight increase with the frequency (i.e., $j\omega C_d$) with +90° phase in a full frequency range. Second, the motional admittance shows a peak at the resonance frequency, where the phase changes from +90° (i.e., capacitive) to –90° (i.e., inductive). In other words, the phase is exactly zero at the resonance. The admittance magnitude decreases above the resonance frequency with a rate of –40 dB down in the Bode plot. Third, by adding the above two, the total admittance is obtained. The admittance magnitude shows two peaks, maximum and minimum, that correspond to the resonance and antiresonance points, respectively. You can find that the peak sharpness (i.e., the mechanical quality factor) is the same for both peaks, because only one loss is included in the equivalent circuit. The antiresonance frequency is obtained at the intersect of the damped and motional admittance curves. Because of the phase difference between the damped (+90°) and motional (–90°) admittance, the phase is exactly zero at the antiresonance and changes to +90° above the antiresonance frequency. Remember that the phase is –90° (i.e., inductive) at a frequency between the resonance and antiresonance frequencies.

6.2.2 k_{31} Piezoplate Equivalent Circuit Simulation with Three Losses

Shi et al. proposed a more concise equivalent circuit, shown in Figure 6.19a with three losses.[9] Compared to the IEEE Standard EC in Figure 6.17 with only one elastic loss, an additional two electrical elements, G_d and G'_m, are introduced into the classical circuit. The new coupling conductance can reflect the coupling effect between the elastic and piezoelectric loss. This EC also can be mathematically expressed as:

$$Y^* = G_d + j\omega C_d + \frac{G'_m + j\omega C_m}{(1 + G'_m/G_m - \omega^2 L_m C_m) + j(\omega C_m/G_m + \omega L_m G'_m)} \tag{6.62}$$

The parameters in this EC can be obtained as expressions of three loss factors:

$$\tan\phi' = \omega C_m/G_m \tag{6.63a}$$

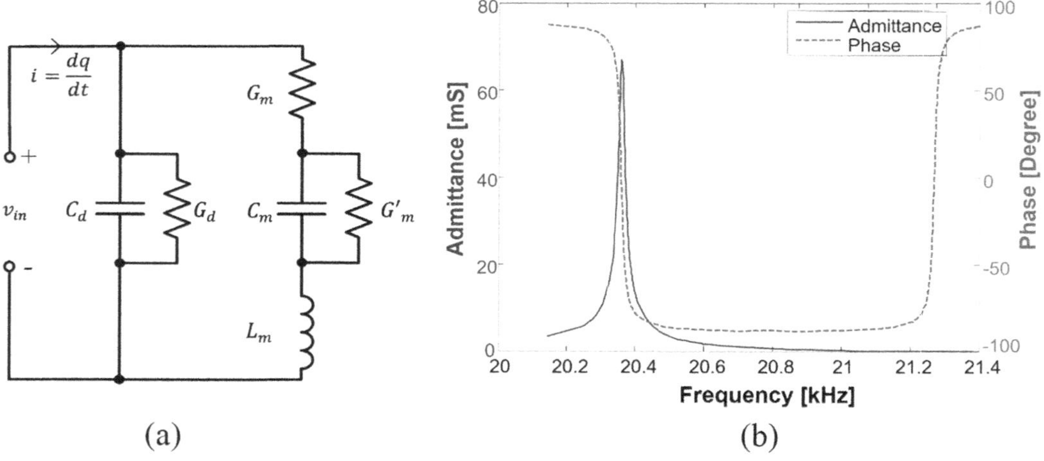

FIGURE 6.19 (a) A new equivalent circuit for a k_{31}-type piezoplate with three intensive loss factors, and (b) admittance spectrum simulated.

$$\tan\theta' = \tan(\phi' - \beta') \quad (6.63b)$$

$$\tan\delta' = k_{31}^2 \tan(2\theta' - \phi') + \frac{G_d}{\omega C_d} \quad (6.63c)$$

where the phase delay $\tan\beta' = \frac{\omega C_m}{G'_m} - \sqrt{\left(\frac{\omega C_m}{G'_m}\right)^2 + 1}$ denotes the disparity between the piezoelectric and elastic components. The value of β generally stays negative or approaches zero (when $G'_m \to 0$), which implies that the piezoelectric loss is persistently larger than or equal to the elastic component (in PZTs). The significance of the piezoelectric loss has therefore been verified theoretically from the equivalent circuit viewpoint. Figure 6.19b shows an admittance spectrum simulated for a PZT sample with integrating three intensive losses.

Uchino proposed a new *four-terminal equivalent circuit* for a k_{31}-mode plate, including elastic, dielectric, and piezoelectric losses (Figure 6.20a), which can handle symmetrical external mechanical loads. The four-terminal EC includes an ideal transformer with a voltage step-up ratio, Φ, to connect the electric (damped capacitance) and mechanical (motional capacitance) branches, where $\Phi = 2wd_{31}/s_{11}^E$, called the *force factor*. New capacitances, l and c_1, are related to L_1 and C_1 in the two-terminal EC given in Equations 6.57 and 6.58:

$$l = \Phi^2 L;\ c_1 = C_1/\Phi^2 \quad (6.64)$$

Regarding the three losses, as shown in Figure 6.20a in a PZT plate, in addition to the IEEE standard "elastic" loss, r_1, and "dielectric" loss, R_d, we introduce the coupling loss in the force factor ($\Phi = 2wd_{31}/s_{11}^E$) r_{cpl} as proportional to ($\tan\phi' - \tan\theta'$), which can be either positive or negative,

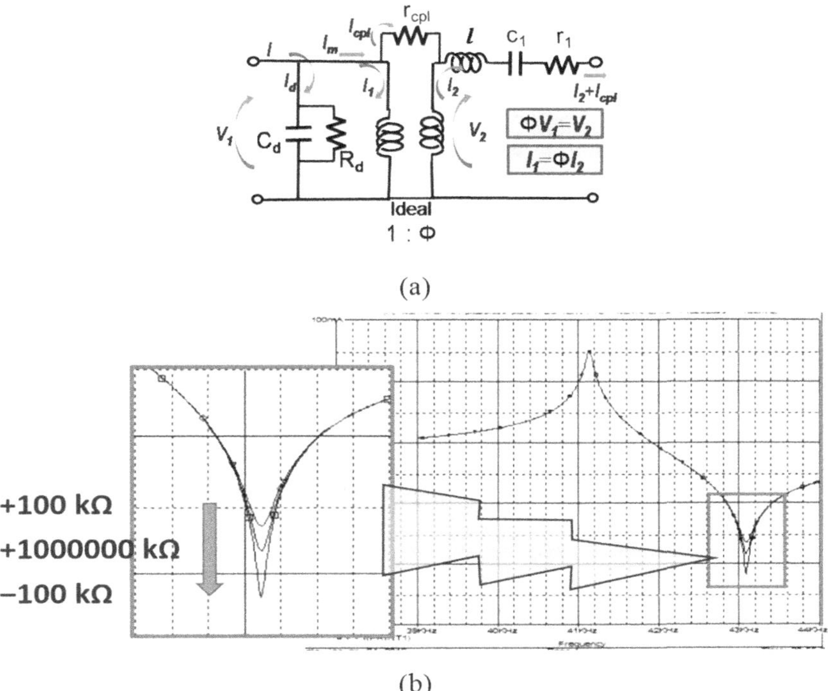

FIGURE 6.20 Four-terminal (two-port) equivalent circuit for a k_{31} plate, including elastic, dielectric, and piezoelectric losses. r_1, R_d, and r_{cpl} correspond to these three losses.

depending on the $\tan\theta'$ magnitude. Figure 6.20b shows the PSpice software simulation results for three values of r_{cpl}. (1) The resonance, Q_A, does not change with changing r_{cpl}. (2) When $r_{cpl} = 100$ kΩ (i.e., $\tan\theta' \approx 0$), $Q_A > Q_B$. (3) When $r_{cpl} = 1$ GΩ (i.e., $\tan\phi' - \tan\theta' \approx 0$), $Q_A = Q_B$. (4) When $r_{cpl} = -100$ kΩ (i.e., $\tan\phi' - \tan\theta' < 0$), $Q_A < Q_B$. It is notable that PSpice can handle negative resistance in the equivalent circuit. Since $Q_A < Q_B$ in PZTs, r_{cpl} should be negative, that is to say, intensive piezoelectric loss, $\tan\theta'$, should be larger than the elastic loss, $\tan\phi'$, in PZTs, as we have already discussed in Chapter 5.

Dong et al. constructed a six-terminal equivalent circuit with three losses, which can handle asymmetric external loads for a k_{31}-mode plate[10] and Langevin transducer by integrating the head and tail mass loads,[11] then estimating the optimum (i.e.,.minimum required input electrical energy) driving frequency at which we can drive the transducer, as demonstrated with the highest efficiency.

6.3 SUMMARY

The finite-element method and its application in smart transducer systems were introduced in the first part of this chapter. Section 6.1.1 described the fundamentals of finite-element analysis and defined the equations for the problem, and Section 6.1.2, "Application of Finite-Element Method" described meshing. Section 6.1.3, "ATILA Simulation Examples," introduced four cases: (1) k_{31} resonance/antiresonance modes, (2) stress concentration in a multilayer actuator, (3) metal tube motor, and (4) piezoelectric transformer. In the latter part, Section 6.2 handled PSpice software, which is useful to piezoelectric equivalent circuit analysis. How the admittance spectrum changes according to the physical parameters, as well as three dielectric, elastic, and piezoelectric loss factors, were simulated with popular equivalent circuits.

CHAPTER ESSENTIALS

ATILA Finite-Element Analysis

1. Within each finite element, the displacement and electric potential fields are uniquely defined by the values they assume at the element nodes (*degrees of freedom*). This definition is obtained by *interpolation*, or utilizing *shape functions* associated with the element. By combining, or *assembling*, these local definitions throughout the whole mesh, we obtain a trial function for Ω that depends only on the nodal values of $\boldsymbol{u}$ and ϕ and that is "piecewise" defined over all the interconnected elementary domains.
2. Mechanical and electrical boundary conditions complete the definition of the problem. The Dirichlet conditions apply to the displacement field, $\boldsymbol{u}$, and the Neumann conditions to the stress field, $\boldsymbol{X}$.

   Note that with this condition, we assume that the electric field outside Ω is negligible, which is easily verified for piezoelectric ceramics.
3. Using the system of equations and the defined boundary, we obtain the functional:

$$\Pi = \iiint_{\Omega} \frac{1}{2}\left(x_{ij}c_{ijkl}^{E}x_{kl} - \rho\omega^2 u_i^2\right)d\Omega - \iint_{S_u}\left(u_i - u_i^o\right)n_j\left(c_{ijkl}^{E}x_{kl} - e_{kij}E_k\right)dS_u$$

$$-\iiint_{\Omega} \frac{1}{2}\left(2x_{kl}e_{ikl}E_i + E_i\chi_{ij}^{x}E_j\right)d\Omega - \iint_{S_X} f_i u_i dS_X$$

$$-\sum_{p=0}^{M}\iint_{S_p}(\phi - \phi_p)n_i(e_{ikl}x_{kl} + \chi_{ij}^{x}E_j)dS_p + \sum_{p=0}^{P}\phi_p Q_p$$

4. The domain, Ω, is divided into subdomains, Ω^e, or finite elements (*discretization*), and the functional Π can be written as:

$$\Pi^e = \frac{1}{2} U^{e^X} (K^e_{uu} - \omega^2 M^e) U^e + U^{e^X} K^e_{u\phi} \Phi^e + \frac{1}{2} \phi^{e^X} K^e_{\phi\phi} \Phi^e - U^{e^X} F^e + \sum_{p=0}^{P} \phi_p Q_p$$

where

$$K^e_{uu} = \iiint_{\Omega^e} B^{e^X}_u c^E B^e_u d\Omega^e$$

$$M^e = \iiint_{\Omega^e} \rho\omega^2 N^{e^X} N^e d\Omega^e$$

$$K^e_{\phi u} = \iiint_{\Omega^e} B^{e^X}_u e B^e_\phi d\Omega^e$$

$$K^e_{\phi\phi} = -\iiint_{\Omega^e} B^{e^X}_\phi \chi^x B^e_\phi d\Omega^e$$

$$F^e = \iint_{S^e_X} N^{e^X} f dS^e_X$$

and $K^e_{uu}, K^e_{\phi u}$, and $K^e_{\phi\phi}$ are the elastic, piezoelectric, and dielectric susceptibility matrices, and M^e is the consistent mass matrix.

5. By applying the stationary condition to Π, the solution is derived from

$$\begin{bmatrix} \boldsymbol{K_{uu}} - \omega^2 \boldsymbol{M} & \boldsymbol{K_{u\phi}} \\ \boldsymbol{K^X_{u\phi}} & \boldsymbol{K_{\phi\phi}} \end{bmatrix} \begin{bmatrix} \boldsymbol{U} \\ \Phi \end{bmatrix} = \begin{bmatrix} \boldsymbol{F} \\ -\boldsymbol{Q} \end{bmatrix}$$

The matrix (6.52) may be adapted for a variety of different analyses, such as the static, modal, harmonic, and transient types.

PSpice Equivalent Circuit Simulation

1. PSpice software is useful for analyzing the equivalent circuit and optimizing the device design from the admittance/impedance electrical driving viewpoint.
2. The equivalent circuit with three intensive losses can analyze the loss contributions to the electromechanical performances; the maximum efficiency driving frequency can be identified between the resonance and antiresonance frequencies.

CHECK POINT

1. (T/F) In the ATILA 2D simulation with "plane stress" condition (on the *x-y* plane), stress along the *z*-axis should be zero. True or False?
2. (F/T) In the ATILA 2D simulation with "plane strain" condition (on the *x-y* plane), stress along the *z*-axis should be zero. True or False?
3. In the ATILA simulation, Analysis in Problem Data includes STATIC, MODAL, MODAL RESANTIRES, HARMONIC, and TRANSIENT. What additional information can we get

with Modal Resantires in comparison with Modal Analysis? Provide two additional pieces of informative data.

4. (T/F) In the ATILA simulation, the Utilities-Copy dialog box is useful to create "volume" from "surface." True or False?
5. In the ATILA simulation software, "Piezoelectric Materials" includes PZT4, PZT5A, PZT5AH, and PZT8. Which should we choose for a hard PZT application such as ultrasonic motors?

CHAPTER PROBLEMS

6.1 Simulate the admittance spectrum with PSpice software for a hypothetical high-k_{31} (97.5%) plate sample. Use an equivalent circuit shown in the right figure for the high-k PZT: Because the antiresonance frequency is around double the resonance frequency, the second harmonic mode ($n = 3$) should be calculated to get a better admittance spectrum. Also, because of the high k, the remaining capacitance ($C_{rem} = C_0\text{-}C_d\text{-}C_1\text{-}C_3$) of much higher-order harmonics should be integrated. Use the following physical parameters for the high-k PZT to calculate the necessary C, L, and R values for the equivalent circuit.

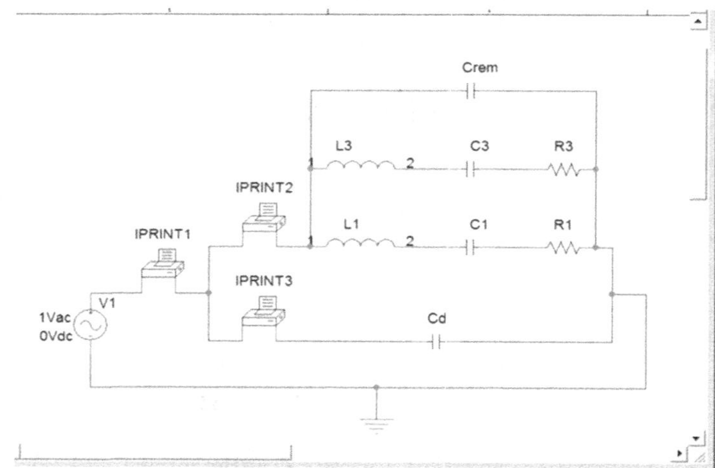

L	W	t	Density	s11E	d31	E33T/E0	E33T	DELTA M	Qm
4.00E-02	6.00E-03	1.00E-03	7500	1.23E-11	3.67E-10	1300	1.15E-08	0.002	500

LW/t	C0
Lt/W	Cd
d31^2/s11E^2	C1
k31^2	L1
fr	R1
x1/n^2	C3
x1	L3
xn	R3
C0-Cd-C1-C3	Crem

REFERENCES

1. O. C. Zienkiewicz: *The Finite Element Method*, 3rd expanded and revised ed., McGraw-Hill Book Company (UK) Limited, 1977. ISBN 0-07-084072-5.
2. J-N. Decarpigny: Application de la Méthode des Eléments Finis à l'Etude de Transducteurs Piézoélectriques, *Doctoral Thesis*, ISEN, Lille, France, 1984.

3. T. Ikeda: *Fundamentals of Piezoelectricity*, Oxford University Press, 1990. ISBN: 0-19-856339-6.
4. B. Dubus, J-C. Debus and J. Coutte: "Modélisation de Matériaux Piézoélectriques et Electrostrictifs par la Méthode des Eléments Finis", *Revue Européenne des Eléments Finis.*, 8(5–6), 581–606, 1999.
5. K. Uchino: *FEM and Micromechatronics with ATILA Software*, CRC Press, Boca Raton, FL, 2008. ISBN: 978-1-4200-5878-9
6. K. Uchino and J.-C. Debus, Editors, *Applications of ATILA FEM Software to Smart Materials—Case Studies in Designing Devices*, Woodhead Pub., Cambridge, UK, 2013. ISBN: 978-0-85709-065-2
7. B. Koc, S. Cagatay and K. Uchino: "A piezoelectric motor using two orthogonal bending modes of a hollow cylinder," *IEEE Ultrasonic, Ferroelectric, Frequency Control Trans.*, 49(4), 495–500, 2002.
8. B. Koc, S. Alkoy and K. Uchino: "A circular piezoelectric transformer with crescent shape input electrodes," *Proceedings of IEEE Ultrasonic Symposium*, Lake Tahoe, Nevada, October 17–21, 1999.
9. W. Shi, H. N. Shekhani, H. Zhao, J. Ma, Y. Yao and K. Uchino: *J. Electroceram.*, 35, 1–10, 2015.
10. X. Dong, M. Majzoubi, M. Choi, Y. Ma, M. Hu, L. Jin, Z. Xu and K. Uchino: "A new equivalent circuit for piezoelectrics with three losses and external loads,", *Sens. Actuators: A. Phys.*, A256, 77–83, 2017. doi: org/10.1016/j.sna.2016.12.026
11. X. Dong, T. Yuan, M. Hu, H. Shekhani, Y. Maida, T. Tou and K. Uchino: "Driving frequency optimization of a piezoelectric transducer and the power supply development," *Rev. Sci. Instrum.*, 87, 105003, 2016. doi: 10.1063/1.4963920

7 Piezoelectric Energy-Harvesting Systems

Energy recovery from wasted or unused power has been a topic of discussion for a long time. In recent years, industrial and academic research units have focused on harvesting energy from mechanical vibrations using piezoelectric transducers. These efforts have provided research guidelines and have brought to light the problems and limitations of implementing piezoelectric transducers. There are three major phases/steps associated with piezoelectric energy harvesting: (i) mechanical-mechanical energy transfer, including mechanical stability of the piezoelectric transducer under large stresses and mechanical impedance matching; (ii) mechanical-electrical energy transduction, related to the electromechanical coupling factor in the composite transducer structure; and (iii) electrical-electrical energy transfer, including electrical impedance matching, such as a DC/DC converter to accumulate the energy into a rechargeable battery. This chapter starts from the historical background of piezoelectric energy harvesting, followed by a brief review of recent research trends by pointing out several misconceptions by the researchers. The main parts deal with step-by-step detailed energy flow analysis in energy-harvesting systems with typical piezoelectrics, lead zirconate titanate (PZTs), in order to provide comprehensive strategies on how to improve the efficiency of the harvesting system. We also introduce a hybrid energy-harvesting system from magnetic field noise using a composite of piezoelectric and magnetostricive materials.

7.1 BACKGROUND

7.1.1 Necessity of Piezoelectric Energy Harvesting

The twenty-first century is called "The Century of Environmental Management." Though most of the major countries rely on fossil fuel and nuclear power plants at present, renewable energy development has become important for compensating for energy deficiency. Energy recovery from wasted or unused power has been a topic of discussion for a long period. Unused power exists in various electromagnetic and mechanical forms such as ambient electromagnetic noise around high power cables, noise vibrations, water flow, wind, human motion, and shock waves. In recent years, industrial and academic research units have focused their attention on harvesting energy from vibrations using piezoelectric transducers. These efforts have provided initial research guidelines and have brought to light the problems and limitations of implementing piezoelectric transducers.[1]

7.1.2 From Passive Damping to Energy Harvesting

Historically, Uchino's group at Sophia University, Japan, started the research on passive vibration damping using piezoelectric materials in the 1980s. Figures 7.1a and b show the results for damping vibration generated in a bimorph transducer.[1] A resistor-shunt was used with the bimorph, and the vibration damping was measured by changing the external resistance. As shown in Figure 7.1b, the quickest damping was observed with a 6.6 kΩ resistor, which is almost the same value as the electrical impedance of the bimorph (i.e., $R = 1/\omega C$). In these results, mechanical vibration energy was converted to electric energy, which was dissipated through the resistor as Joule heat, effectively leading to the quickest damping. In addition to the resistive shunt, capacitive, inductive, and switch shunts have been successively studied, aiming at adaptive vibration control after our studies.[2]

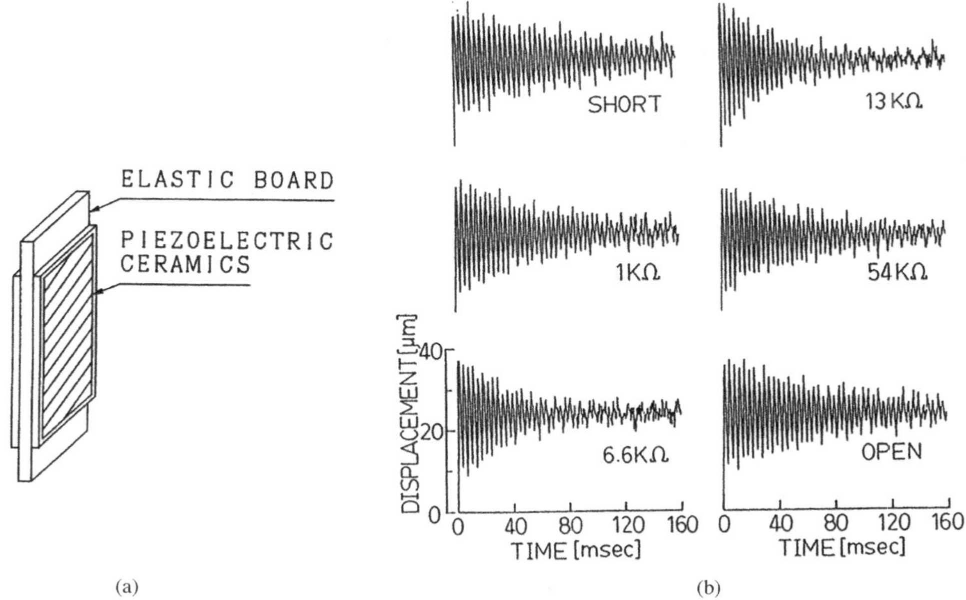

FIGURE 7.1 Vibration damping change associated with external resistance change. (a) Bimorph transducer for this measurement. (b) Damped vibration with external resistor. (From K. Uchino: *Ferroelectric Devices*, 2nd Ed., CRC Press, Boca Raton, FL, 2010.)

However, after the late 1990s, we decided to save and store this generated electrical energy into a rechargeable battery instead of dissipating it as Joule heat. This research target change is schematically illustrated in Figure 7.2, which fits to the renewable energy boom with good timing. Note that the maximum energy-harvesting condition corresponds to the largest vibration damping: "If you run after two hares (vibration suppression and energy harvesting), you will catch *both*," rather than "you will catch *neither*" (the original proverb), which seems to be the best development strategy! The detailed development process is described in Section 7.1.4.

7.1.3 Recent Research Trends

With respect to the marketing category of piezoelectric devices, a category of "actuators and piezogenerators" increased in these years; in particular, the "ecology and energy harvesting" share increased dramatically from almost zero to 7% of the total revenue (annual around $40 billion) during only these several years.[3]

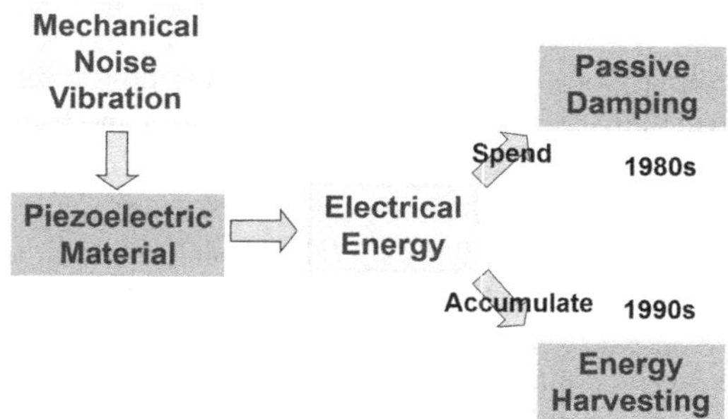

FIGURE 7.2 Piezoelectric passive vibration damping versus piezoelectric energy harvesting.

We summarize in this subsection the research trends, primarily after the 2000s, with reference to a book, Reference [4]. First, as one of the pioneers in piezoelectric energy harvesting, Uchino feels a sort of frustration regarding 90% of the recent research papers on the following points:

1. Though the electromechanical coupling factor k is the smallest (i.e., the energy conversion rate from the input mechanical to electric energy is the lowest) among various device configurations, the majority of researchers primarily use the "unimorph" design. Why?
2. Though the typical noise vibration is in a much lower frequency range, researchers measure the amplified resonance response (even at a frequency higher than 1 kHz) and report this unrealistically harvested electric energy. Why?
3. Though the harvested energy is lower than 1 mW, which is lower than the required electric energy to operate a typical energy-harvesting electric circuit with a DC/DC converter (typically around 2–3 mW), researchers report the result as an energy "harvesting" system. Does this situation mean actually energy "losing"? Why?
4. Few papers have reported complete energy flow or exact efficiency from the input mechanical noise energy to the final electric energy in a rechargeable battery via the piezoelectric transducer step by step. Why?

Interestingly, the unanimous answer from these researchers to my question, "Why?" is, "Because the previous researchers did so!"

7.1.3.1 Mechanical Engineers' Approach

Wallaschek et al.[5] and Smithmaitrie[6] reported comprehensive studies on bimorph transducer design optimization using a mechanical equivalent system with mass, spring, and dashpot in the former, and using a membrane/beam theory in the latter, including various references studied previously. The former indicated that the effective electromechanical coupling factor k can be increased by reducing the thickness ratio (PZT thickness/metal shim thickness), while in the latter paper, the voltage generation in the piezoelectric membrane shows the maximum around the resonance frequency (50 Hz). One of the significant problems of mechanical engineers is neglect of the piezoelectric ceramics PZT's loss and performance limitations. As we will discuss in Section 7.3, the maximum handling energy level in the current PZTs is only 10–30 W/cm^3. Thus, reducing the PZT thickness in order to increase the electromechanical coupling factor k dramatically reduces the handling energy level (note that the energy conversion amount is given by the product of input energy and k^2). We should understand that though we try to increase the handling electric energy level with the mechanical vibration input, additional input energy will convert to just heat generation due to hysteresis with increasing the handling power; that is, the PZT becomes a ceramic heater. Knowing the material's limitation is a primary essential prior to extending theoretical mechanical analysis.

7.1.3.1.1 Machinery Vibration (Resonance Usage)

Beeby and Zhu discussed commercially available energy-harvesting products in their article.[7] Products by Midé Technology (U.S.), Adaptivenergy (U.S.), Arveni (France), Advanced Cerametrics (U.S.), and Prepetuum (U.K.) were compared. These products are based on relatively large piezo-bimorph/unimorph designs (40–200 cm^3) in order to set the resonance frequency around 50 Hz with a narrow bandwidth less than 10 Hz. Power 1–20 mW is obtained under the acceleration 0.5–2 G. Most of them can be utilized to operate sensors of various machinery health-monitoring systems (the machine should be vibrating constantly at 50 Hz) and to transmit the wireless signals of the monitored data. This energy level is not sufficient for general energy storage applications.

Arms et al.[8] developed a smart system powered by piezoelectrics in a helicopter. The integrated structural health monitoring and reporting (SHMR) includes strain gauges, accelerometers, and load/torque cells. Data were stored in a central location. Wireless sensors were capable of logging 50,000 samples/sec, consuming 9 mA and 3 V DC (27 mW). The energy was supplied by piezoelectric

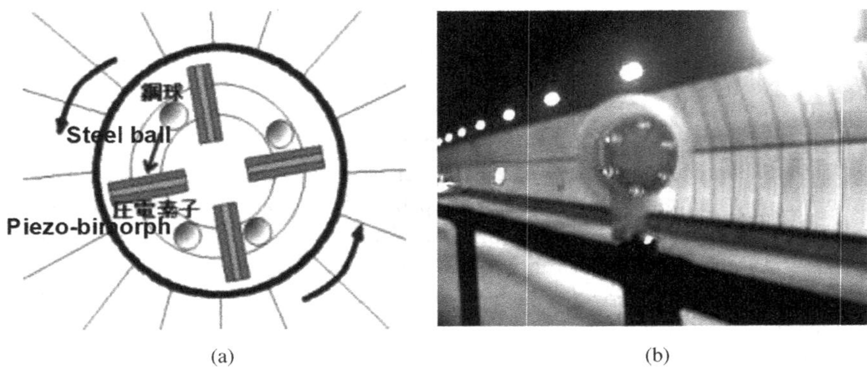

(a) (b)

FIGURE 7.3 Piezoelectric windmill structure (a) for driving an LED traffic light array system (b). (Courtesy NEC-Tokin.)

microgenerators capable of harvesting vibration energy from the helicopter gearbox. Microgenerators (similar to the dimensions of multiple American quarter coins) working under a constant blade rotation condition can generate 37 mW, more than enough to feed the sensors.

7.1.3.1.2 Human Motion

NEC-Tokin Corporation, Japan, had already commercialized two piezoelectric energy-harvesting products in the early 2000s by utilizing an impact-based mechanism.[9] First, finger-snapping action on a piezo-bimorph generates the electric energy to illuminate LED lamps equipped at the front of a key holder so that the user can easily find a keyhole on a house or car door. Second, "piezoelectric windmills" were produced and aligned in the northern part of Japan along the major highway roadsides in tunnels so that the wind generated by running vehicles or humans rotates the windmills, then the rotation of mills generates steel ball motion, as shown in Figure 7.3a, which hits piezo-bimorphs, leading to electric energy generation for operating an LED traffic light array system that can navigate vehicle drivers smoothly along the road curb (Figure 7.3b).

Another successful product (a million seller) is the Lightning Switch commercialized by Face Electronics (PulseSwitch Systems), VA, which is a remote switch for room lights, using a unimorph (Thunder) piezoelectric component. In addition to the living convenience, Lightning Switch (Figure 7.4) can reduce housing construction costs dramatically due to a significant reduction of copper electric wire and wire-aligning labor.[10] Harvesting energy from shoes was also reported using the Thunder component by Face Electronics.[11]

Renaud et al.[12] reported a similar impact-based energy harvester, which consists of two piezo-bimorphs and a movable elastic rod. The total volume of the device, 25 cm^3, could generate an output

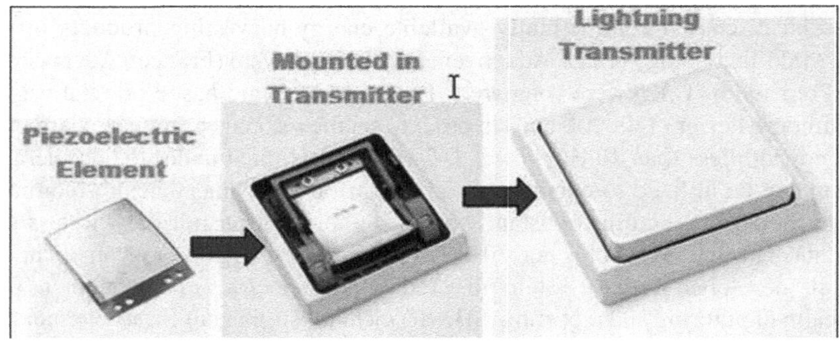

FIGURE 7.4 Lightning Switch with piezoelectric Thunder actuator. (Courtesy Face Electronics.)

power of 60 μW under a motion of 10 Hz with 10 cm linear motion amplitude (which seems to be too quick for human action).

Since human action is very slow and rather random, resonance-type devices are not suitable in general. Thus, there seem to be two choices: (1) usage of impact-based snap action, or (2) off-resonance, but carefully matching the acoustic/mechanical impedance with soft human tissue and soft piezoelectric devices.

7.1.3.2 Electrical Engineers' Approach

Because the electric energy obtained by piezoelectric energy-harvesting systems exhibits high voltage/low current (high impedance), we need to convert the electric impedance to a sufficiently low level (~50 Ω) in order to charge up a rechargeable battery smoothly or drive an ohmic-typical portable device.

Guyomar et al.[13] reviewed studies on "switch" shunt circuits combined with cyclic piezoelectric energy-harvesting systems. They initially compared two cases: a switching device is connected in parallel with a piezoelectric element (parallel synchronized switch harvesting on inductor; parallel SSHI) or in series between the active material and harvesting stage (series SSHI). The discussion was held only on a monochromatic excitation, which narrows the application of this technique.

As you will learn in Section 7.4.1, because of the randomness of the mechanical noise frequency, our strategy is basically to accumulate the original wide-frequency electric energy in a capacitor as DC charge; then, the electric impedance is to be changed using a DC/DC converter in order to facilitate charging a battery.

7.1.3.3 Micro-Electro-Mechanical Systems Engineers' Approach

Piezoelectric energy harvesting has been given a significant boost in the MEMS or NEMS area with a new term, "nano-harvesting." Lopes and Kholkin's paper[14] is a good review to reference. Designs for piezoelectric microharvesting are currently limited only to the unimorph types: (1) one-end clamp cantilever and (2) both-end clamp membrane. Because of the combination of this low electromechanical coupling k design ($k \leq 10\%$) and very thin (low-volume) piezoelectric film, it is difficult to expect high power or efficiency theoretically. 1–10 μW harvesting energy is reported at a resonance frequency higher than 1 kHz, which is 3 orders of magnitude smaller than the practically usable level (10 mW). Note that 1 mW is the minimum required for sending a wireless signal or sucking a drop of blood from a blood vessel.

A scientifically interesting topic is the usage of nanopiezoelectric fiber (i.e., single crystal). Chang et al.[15] reported direct-write, piezoelectric polymeric nanogenerators based on organic nonfibers with higher energy conversion efficiency. These nanofibers are made of poly-vinylidene di-fluoride with high flexibility, minimizing resistance to external mechanical motion in low frequencies. PVDF exhibits reasonable piezoelectric and soft mechanical properties, suitable to medical applications for realizing mechanical impedance matching soft human tissues. Chang et al. utilized an electrospinning process with a strong electric field (>10^7 V/m) and stretching forces from the naturally aligned dipoles in a nanofiber crystal such that the nonpolar R phase was transformed into the polar β phase, determining the polarity of the electrospun nanofiber. They reported an energy conversion rate of 12% as average, which corresponds to the electromechanical factor $k = 35\%$, which is 3 times higher than that of commercial PVDF films.

Another nanowire application was found in ZnO by Wang et al.[16] High energy can be produced by making an array of ZnO flexible nanowires where each individual wire can produce electricity. Further improvement was achieved by using a zig-zag configuration on the top electrode surface. Though they reported an energy-harvesting density of 2.7 mW/cm^3, 5 times higher than PZT cantilever types, the most serious practical problem is how to sum up the harvesting electric output from each nanofiber in phase without cancelling each other.

7.1.3.4 Military Application: Programmable Air-Burst Munition

Until the 1960s, the development of weapons of mass destruction (WMDs) was the primary focus, including nuclear bombs and chemical weapons. However, based on the global trend for "*Jus in Bello* (Justice in War)," environmentally friendly "green" weapons became the mainstream in the twenty-first century, that is, minimally destructive weapons with a pinpoint target, such as laser guns and rail guns. In this direction, programmable air-bust munitions (PABMs) were developed starting in 2003. After the World Trade Center was attacked by Al Qaeda on September 11th, 2001, the U.S. military started the "revenge" war against Afghanistan. U.S. troops initially destroyed all the buildings by bombs, which dramatically increased the war cost for restructuring new buildings. Thus, the U.S. Army changed the war strategy: without collapsing the building, it aimed just to kill the Al Qaeda soldiers inside by using a sort of micromissile. A micromissile passes through the window by making a small hole in the window glass and explodes in the air 3–5 m (programmable!) inside the window, so that the building structure damage is minimized. For this purpose, each bullet needs to have installed a microprocessor chip, which navigates the bullet to a certain programmed point. ATK Integrated Weapon Systems, AZ, started to produce button-battery–operated programmable air-burst munitions first.[17] Though they worked beautifully for an initial couple of months, due to severe weather conditions (it is incredibly hot in the daytime in Afghanistan), most of the batteries wasted out in 3 months. No soldier is willing to open a dangerous bullet to exchange a battery!

Under these circumstances, the Army Research Office asked Micromechatronics Inc., State College, PA (ICAT/PSU spin-off company), to develop a compact electric energy source to be embedded in the PABM bullet. The initial shooting impact can be converted to electricity via a piezoelectric device, which should fulfill energy to be spent in the microprocessor for 2–5 seconds (1–2 km distance) during maneuvering the bullet. Micromechatronics Inc. developed an energy source for 25-mm caliber "Programmable Ammunition." Instead of a battery, a multilayer PZT piezo-actuator is used for generating electric energy under the shot's mechanical impact to activate the operational amplifiers that ignite the burst according to the command program (Figure 7.5). This may be one of the highest-revenue products with piezoelectric energy harvesting. The development key is the multilayer usage for realizing a high electromechanical coupling factor, k, and high capacitance, low electrical impedance. Try Chapter Problem 7.1 to learn the development process of this device.

Since many of the current research strategies are not satisfactory enough with my development philosophy, the author introduces his personal development principles in the following sections.

7.1.4 Piezoelectric Energy-Harvesting Principles

7.1.4.1 Piezoelectric Constitutive Equations

In order to simplify the analysis, we assume the linear *constitutive piezoelectric equations* in the relationship between the extensive strain, x, electric displacement, D (or almost the same as polarization P in ferroelectrics), and the intensive stress, X, and electric field, E:

$$x = s^E X + d\,E, \tag{7.1}$$

$$D = d\,X + \varepsilon_0 \varepsilon^X E, \tag{7.2}$$

where s^E and ε^X are elastic compliance and relative permittivity (ε_0: vacuum permittivity), and d is the piezoelectric constant. Note that the direct piezoelectric effect (Equation 7.2) and the converse piezoelectric effect (Equation 7.1) keep the same d constant.

7.1.4.2 Piezoelectric Figures of Merit: Review

In order to refresh the reader's memory, we review the five figures of merit in piezoelectrics from the energy-harvesting viewpoint.

FIGURE 7.5 25-mm programmable air-burst munition. (Courtesy Micro-mechatronics, Inc.)

7.1.4.2.1 Piezoelectric Strain Constant, d

A figure of merit for actuator applications:

$$x = d\,E. \tag{7.3}$$

7.1.4.2.2 Piezoelectric Voltage Constant, g

A figure of merit for sensor applications:

$$E = g\,X. \tag{7.4}$$

$$g = d/\varepsilon_0\varepsilon^X. \quad (\varepsilon^X\text{: relative permittivity}) \tag{7.5}$$

7.1.4.2.3 Electromechanical Coupling Factor, k

The terms *electromechanical coupling factor*, *energy transmission coefficient*, and *efficiency* are sometimes confused, though all the terms are related to input and output energy conversion.[18]

a. The electromechanical coupling factor, k

$$k^2 = (\text{Stored mechanical energy/Input electrical energy}) \tag{7.6}$$

or

$$k^2 = (\text{Stored electrical energy/Input mechanical energy}). \tag{7.7}$$

Let us calculate Equation 7.7, when an external stress, X, is applied to a piezoelectric material. See Figure 7.6a first, when the top and bottom electrodes are short-circuited (i.e., E constant condition). Since the input mechanical energy is (1/2) $s^E X^2$ per unit volume and the stored electrical energy per unit volume under zero external electrical impedance (i.e., short-circuit) is given by (1/2) $D^2/\varepsilon_0\varepsilon^X$ = (1/2) $(d\,X)^2/\varepsilon_0\varepsilon^X$ (D [or P] is obtained by integrating the current i), k^2 can be calculated as

$$\begin{aligned} k^2 &= [(1/2)(d\,X)^2 / \varepsilon_0\varepsilon^X]/[(1/2)\,s^E X^2] \\ &= d^2 / \varepsilon_0\varepsilon^X \cdot s^E. \end{aligned} \tag{7.8}$$

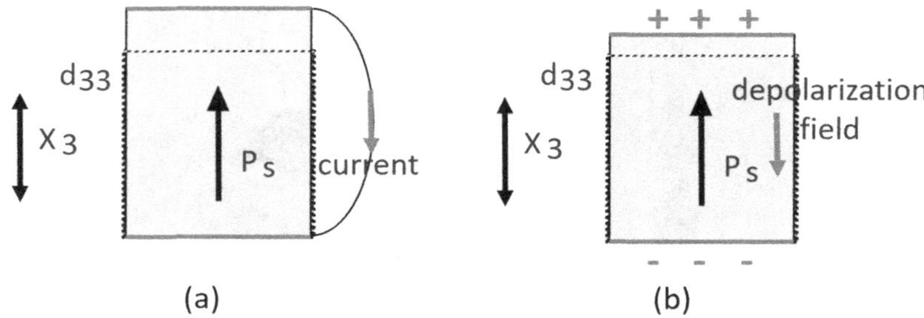

FIGURE 7.6 Calculation models of electromechanical coupling factor, k, for (a) short-circuit and (b) open-circuit conditions.

On the contrary, when the top and bottom electrodes are open circuited in Figure 7.6b, the depolarization field, E, is induced because of the piezoelectrically induced charge in order to satisfy $D = 0$ (i.e., D constant condition) in Equation 7.2:

$$E = -(d/\varepsilon_0\varepsilon^X)\,X. \tag{7.9}$$

Note that this depolarization field is valid only for a short time period (less than several minutes) by neglecting the charge drift. Thus, from Equation 7.1, the strain induced under the open-circuit condition should be smaller than that under the short circuit, as calculated:

$$x = s^E X + d\,E = s^E X - (d^2/\varepsilon_0\varepsilon^X)\,X = s^E\,[1 - (d^2/\varepsilon_0\varepsilon^X s^E)]\,X = s^E\,(1 - k^2)X.$$

If we denote $x = s^D X$ under the open circuit, we can obtain the following important equation:

$$s^D = s^E\,(1 - k^2).\ (k\text{: electromechanical coupling factor}) \tag{7.10}$$

Taking into account input mechanical energy (1/2) $s^D X^2$ and output electrical energy (1/2) $\varepsilon_0\varepsilon^X E^2$, the apparent electromechanical coupling factor, k', in the open-circuit condition can be defined as

$$k' = (1/2)\,\varepsilon_0\varepsilon^X E^2/(1/2)\,s^D X^2 = d^2/\varepsilon_0\varepsilon^X s^D = k^2/(1 - k^2).$$

It is worth noting that depending on the electrical load (i.e., impedance), input mechanical energy differs significantly even if we keep the same stress X level, because the elastic compliance of the piezoelectric material changes with the electrical constraint condition. Accordingly, the importance of the energy transmission coefficient also comes out in the piezo-energy harvesting system.

b. The energy transmission coefficient, λ_{max}. Not all the mechanically stored energy can actually be used under the electrical drive, and the actual work done in the piezoelectric actuator depends on the mechanical load. With zero mechanical load or a complete clamp (no strain), no output work is done. Vice versa, not all the electrically converted energy can actually be spent in the energy-harvesting case. With zero impedance (short circuit) or infinite impedance (open circuit), no electrical work can be obtained. Thus, the energy transmission coefficient is defined by

$$\lambda_{max} = (\text{Output mechanical energy/Input electrical energy})_{max} \tag{7.11}$$

or equivalently,

$$\lambda_{max} = (\text{Output electrical energy/Input mechanical energy})_{max} \tag{7.12}$$

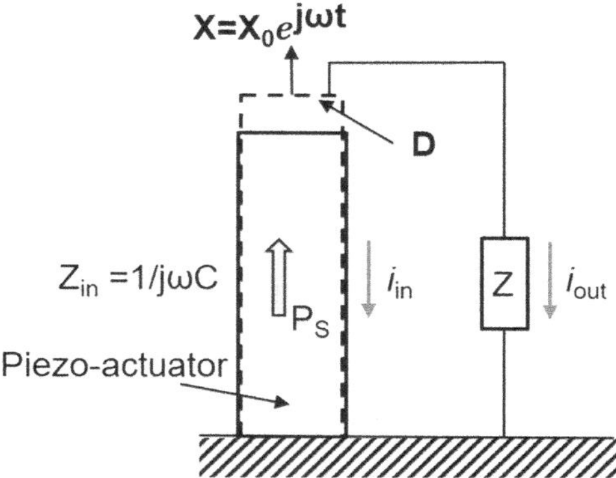

FIGURE 7.7 Electric energy-harvesting model under the external electrical impedance, Z, on a piezoelectric actuator.

The difference of the above from Equations (7.6) and (7.7) is "stored" or "output/spent." The energy transmission coefficient can be obtained as

$$\begin{aligned}\lambda_{max} &= \left[(1/k) - \sqrt{(1/k^2)-1}\right]^2 \\ &= \left[(1/k) + \sqrt{(1/k^2)-1}\right]^{-2}.\end{aligned} \tag{7.13}$$

Since the detailed derivation process under the electric field drive for actuator application is introduced in a textbook, Reference [18], only the key points are described here for energy harvesting under the mechanical stress drive, using Figure 7.7, which shows an electric energy-harvesting model under the external electrical impedance Z on a piezoelectric actuator. Figure 7.8 summarizes the calculation processes of the input mechanical and output electric energy under various impedance Zs: (a) the stress vs. electric displacement relation, (b) stress vs. strain relation to calculate the input mechanical energy, and (c) electric displacement vs. electric field to calculate the output electric energy.

When we assume sinusoidal input stress $X = X_0 e^{j\omega t}$ and output electric displacement $D = dX_0 e^{j\omega t}$, we can derive the following current and voltage relationships from Figure 7.7, knowing the piezo-actuator impedance $1/j\omega C$ under an off-resonance frequency:

$$\begin{aligned} i = \frac{\partial D}{\partial t} &= i_{in} + i_{out} = j\omega d X_0; \\ Z_{in} i_{in} &= Z i_{out}; \\ i_{out}(1 + j\omega CZ) &= j\omega d X_0. \end{aligned}$$

Thus, we can obtain the output electric energy as

$$|P| = \frac{1}{2} Z i_{out}^2 = \frac{1}{2} Z \frac{(\omega d X_0)^2}{(1 + (\omega CZ)^2)}. \tag{7.14}$$

Figure 7.9 shows the electric load dependence of the output electric energy, which concludes that the maximum electric energy $|P| = (1/4)(\omega d^2 X_0^2 / C)$ can be obtained at $Z = 1/\omega C$. In other words, the "stored" electric energy can be spent at maximum when the external load impedance matches exactly the internal impedance.

However, we need to be aware that since the input mechanical energy is changed (even if we keep the stress/force constant) due to the elastic compliance change with the external

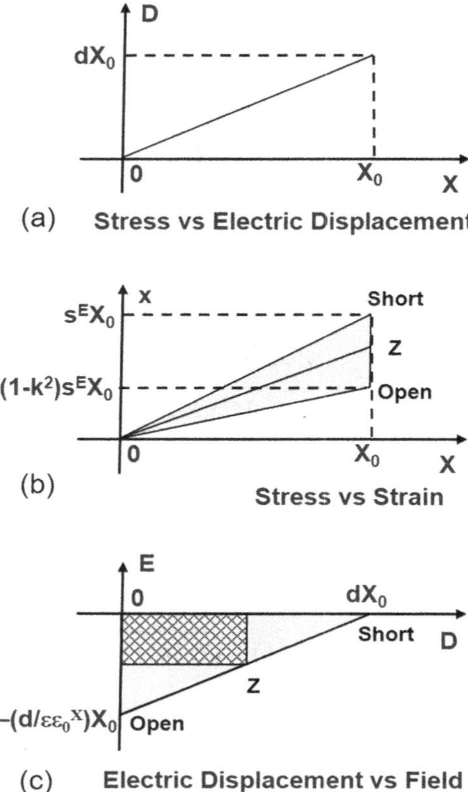

FIGURE 7.8 Calculation models of the input mechanical and output electric energy.

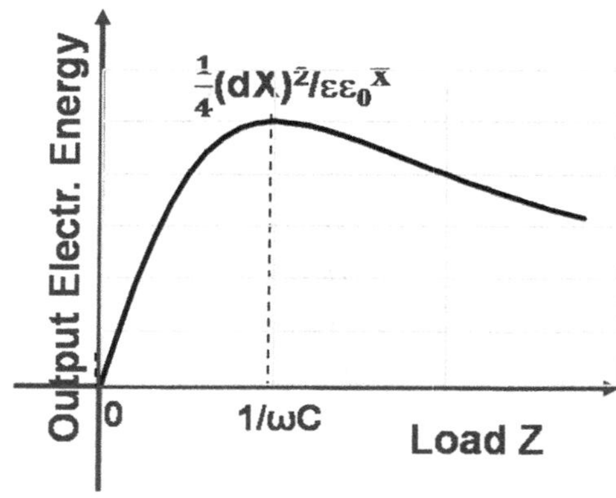

FIGURE 7.9 Output electric energy dependence on the external electrical load, Z, in a piezoelectric energy-harvesting system.

electrical impedance, the condition for realizing the maximum transmission coefficient is slightly off from the electrical impedance matching point. We can also notice that from Equation 7.13,

$$k^2/4 < \lambda_{max} < k^2/2, \tag{7.15}$$

depending on the k value. For a small k (<0.3), $\lambda_{max} \approx k^2/4$, and for a large k (~ 0.7), $\lambda_{max} \approx k^2/3 \sim k^2/2$. When k is unrealistically high (~ 0.95), λ_{max} approaches 1.

c. The efficiency, η

$$\eta = \text{(Output mechanical energy)/(Consumed electrical energy)} \tag{7.16}$$

or

$$\eta = \text{(Output electrical energy)/(Consumed mechanical energy).} \tag{7.17}$$

The difference from the efficiency definition from Equations 7.11 and 7.12 is "input" energy and "consumed" energy in the denominators. In a work cycle (e.g., a mechanical stress cycle), the input mechanical energy is transformed partially into electrical energy and the remaining is stored as mechanical energy (mechanical energy in a piezoelectric spring) in an actuator. In this way, the ineffective mechanical energy can be returned to the vibration source, leading to near 100% efficiency if the loss is small. Typical values of dielectric loss in PZT are about 1%–3%.

Figure 7.10 illustrates the difference between electromechanical coupling factor, k, energy transmission coefficient, λ_{max}, and efficiency, η, for a piezoelectric material with $k_{33} = 70\%$. Note that the efficiency is not related to the electromechanical coupling factor, k, as exemplified by a quartz crystal device ($k_t = 9\%$) with 99.99% efficiency.

7.1.4.2.4 Mechanical Quality Factor, Q_M

The mechanical quality factor, Q_M, is a parameter that characterizes the sharpness of the electromechanical resonance spectrum. When the motional admittance, Y_m, is plotted around the resonance frequency, ω_0, the mechanical quality factor, Q_M, is defined with respect to the full width ($2\Delta\omega$) at $Y_m/\sqrt{2}$ as:

$$Q_M = \omega_0/2\Delta\omega. \tag{7.18}$$

Also note that Q_M^{-1} is equal to the intensive mechanical loss (tan ϕ'). When we define a complex elastic compliance, $s^E = s^{E'} - j\, s^{E''}$, the mechanical loss tangent is provided by $\tan\phi' = s^{E''}/s^{E'}$. The Q_M value is very important in evaluating the magnitude of the resonant displacement and strain. The vibration amplitude at an off-resonance frequency ($dE{\cdot}L$, L: length of the sample) is amplified by a factor proportional to Q_M at the resonance frequency. For example, a longitudinally vibrating rectangular plate through the transverse piezoelectric effect, d_{31}, generates the maximum displacement given by $(8/\pi^2)\, Q_M\, d_{31}E\, L$. Refer to a textbook, Reference [1], for the details of the mechanical quality factors.

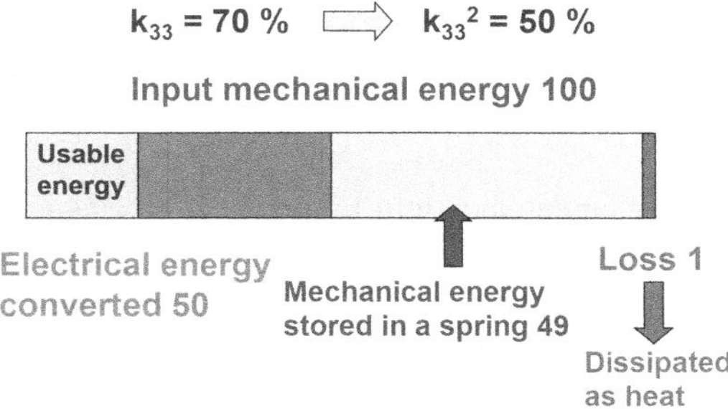

FIGURE 7.10 Schematic illustration of how much electric energy can be harvested from the electromechanical coupling factor's viewpoint.

7.1.4.2.5 Mechanical/Acoustic Impedance, Z

The mechanical impedance, Z, is a parameter used for evaluating the acoustic/sound wave energy transfer between two materials. It is defined, in general, by

$$Z^2 = (\text{pressure/volume velocity}). \tag{7.19}$$

In a solid material,

$$Z = \sqrt{\rho c}, \tag{7.20}$$

where ρ is the density and c is the elastic stiffness of the material.

The mechanical work done by one material on the other is evaluated by the product of the applied force, F, and the displacement, ΔL: $W = F \times \Delta L$. This corresponds to "Pushing a curtain," exemplified by the case when the acoustic wave is generated in water directly by a hard PZT transducer. Most of the acoustic energy generated in the PZT is reflected at the interface, and only a small portion of acoustic energy transfers into the water. On the other hand, if the material is very hard, the displacement will be very small, again leading to very small W. This corresponds to "Pushing a wall." Polymer piezoelectric PVDF cannot drive a hard steel part efficiently. Therefore, the *mechanical/acoustic impedance* must be adjusted to maximize the output mechanical power:

$$\sqrt{\rho_1 c_1} = \sqrt{\rho_2 c_2}, \tag{7.21}$$

where ρ is the density and c is the elastic stiffness, and the subscripts 1 and 2 denote the two materials. In practice, acoustic impedance matching layers (elastically intermediate materials between PZT and water, such as a polymer; more precisely, the acoustic impedance, Z, should be a *geometrical average* $\sqrt{Z_1 \cdot Z_2}$) are fabricated on the PZT transducer to optimize the transfer of mechanical energy to water.

In more advanced discussions, there are three kinds of impedances: specific acoustic impedance (pressure/particle speed), acoustic impedance (pressure/volume speed), and radiation impedance (force/speed). See Reference [19] for details.

7.1.4.3 Piezoelectric Passive Damper

The principle of the piezoelectric vibration damper is explained based on a piezoelectric ceramic single plate in Figure 7.11a,[20] as it was a product prior to the energy-harvesting devices. When an external impulse force is applied to the piezoplate, an electric charge is produced via direct piezoelectricity (Figure 7.11b). Accordingly, the vibration remaining after the removal of the external force induces an alternating voltage, which corresponds to the intensity of that vibration, across the terminals of the single plate. The electric charge produced is allowed to flow and is dissipated as Joule heat when a resistor is put between the terminals (see Figure 7.11c). As we discussed in Equation 7.14, when the external resistance is too large or small, the vibration intensity is not readily reduced, and we need to tune the resistance to match exactly the piezoplate impedance, that is, $1/j\omega C$, where ω is the cyclic frequency (i.e., the fundamental mechanical resonance of the piezoplate), and C is the piezoplate capacitance.

The bimorph piezoelectric element shown in Figure 7.1a, an elastic beam sandwiched with two sheets of piezoelectric ceramic plates, is a typical example of a combination of a vibration object and piezoelectric ceramics. The bimorph edge was hit by an impulse force, and the transient vibration displacement decay was monitored by an eddy current-type noncontact displacement sensor (SDP-2300 manufactured by Kaman). Figure 7.1b shows the measured displacement data, which vibrate at the bimorph resonance frequency (295 Hz), and Figure 7.12 shows the relationship between the damping time constant and an external resistance. It can be seen in the figure that the damping time constant was minimized in the vicinity of 8 kΩ, which is close to the impedance, $1/\omega C$.

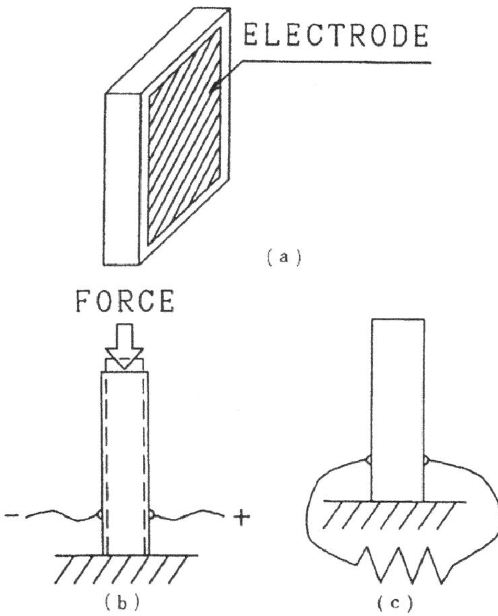

FIGURE 7.11 Piezoelectric mechanical damper. (a) Piezoelectric sample, (b) direct piezoelectric effect, and (c) electric energy dissipation via a resistance.

Let us evaluate the damping constant theoretically. The electric energy, U_E, generated can be expressed by using the electromechanical coupling factor, k, and the mechanical energy, U_M:

$$U_E = U_M \times k^2. \tag{7.22}$$

The piezoelectric damper transforms electric energy into heat energy when the resistor, R, is connected, and the transforming rate of the damper can be raised to a level of up to 50% when the electrical impedance is matched. Accordingly, the vibration energy is transformed at a rate of

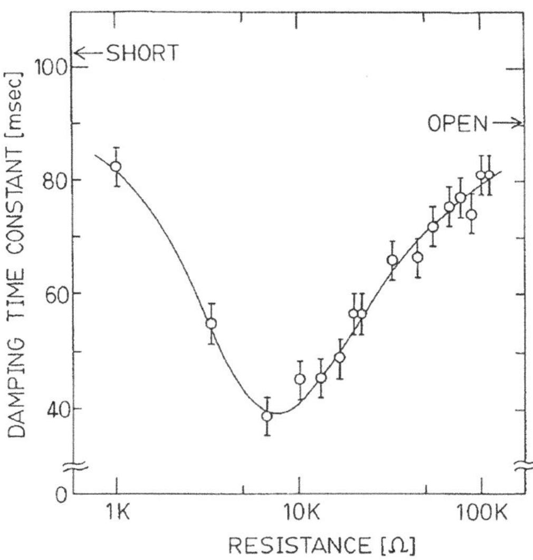

FIGURE 7.12 Relationship between the damping time constant and the external resistance. (From Uchino and T. Ishii: *J. Ceram. Soc. Japan.*, 96, 863, 1988.)

$(1 - k^2/2)$ times with energy vibration repeated, since $k^2/2$ multiplied by the amount of mechanical vibration energy is dissipated as heat energy. As the square of the amplitude is equivalent to the amount of energy, the amplitude decreases at a rate of $(1 - k^2/2)^{1/2}$ times with every vibration repeated. If the resonance period is taken to be T_0, the number of vibrations for t sec is $2t/T_0$. Consequently, the amplitude in t sec is $(1-k^2/2)^{t/T_0}$. If the residual vibration period is taken to be T_0, the damping in the amplitude of vibration is t sec and can be expressed as follows:

$$(1-k^2/2)^{t/T_0} = e^{-t/\tau}. \tag{7.23}$$

Thus, the following relationship for the time constant of the vibration damping is obtained.

$$\tau = -\frac{T_0}{\ln(1-k^2/2)} \tag{7.24}$$

Now, let us examine the time constant of the damping using the results for the bimorph in our study. Substitution in Equation 7.24 of $k = 0.28$ and $T_0 = 3.4$ msec produces $\tau = 85$ msec, which seems to be considerably larger than the value of approximately 40 msec obtained experimentally for τ. This is because the theoretical derivation Equation 7.24 was conducted under the assumption of a loss-free (high Q_M) bimorph. In practice, however, it originally involved mechanical loss, the time constant of which can be obtained as the damping time constant under a short-circuited condition, that is, $\tau_s = 102$ msec. The total vibration displacement can then be expressed as $e^{-t/\tau_{\text{total}}} = e^{-t/\tau_s} \times e^{-t/\tau}$. Accordingly,

$$\frac{1}{\tau_{\text{total}}} = \frac{1}{\tau_s} + \frac{1}{\tau} \tag{7.25}$$

Substitution in Equation 7.25 of $\tau_s = 102$ msec and $\tau = 85$ msec produces $\tau_{\text{total}} = 46$ msec. This conforms to the result shown in Figure 7.1b and agrees with the experiment.

Because we used a bimorph structure in this pioneering paper merely due to the simplest geometry, the author feels a sort of guilt for providing a strong influence on subsequent researchers on the bimorph design. "Because the previous researchers did so!" The bimorph is not an ideal piezodevice design at all, as we discuss in the following parts.

7.1.5 Three Phases in the Energy-Harvesting Process

In the 1990s, Uchino's group decided to collect the electric energy without just dissipating the energy via Joule heat, which was the beginning of piezoelectric energy harvesting: "Chasing two hares to obtain both." There are three major phases/steps associated with piezoelectric energy harvesting (see Figure 7.13): (i) *mechanical-mechanical energy transfer*, including mechanical stability of the piezoelectric transducer under large stresses, and mechanical impedance matching; (ii) *mechanical-electrical energy transduction*, related to the electromechanical coupling factor in the composite transducer structure; and (iii) *electrical-electrical energy transfer*, including electrical impedance matching. A suitable DC/DC converter is required to accumulate the electrical energy from a high impedance piezodevice into a rechargeable battery (low impedance).

The following sections mainly deal with detailed energy flow analysis in piezoelectric energy-harvesting systems with typical stiff "cymbals" (~100 mW) and flexible piezoelectric transducers smart material "macro fiber composites" (~1 mW) under cyclic mechanical load (off-resonance), in order to provide comprehensive strategies on how to improve the efficiency of the harvesting system. Energy transfer rates are practically evaluated for all three steps above. Our application target of the cymbal was set to hybrid vehicles with both an engine and an electromagnetic motor, reducing the engine vibration and harvesting electric energy to car batteries to increase the mileage, while the

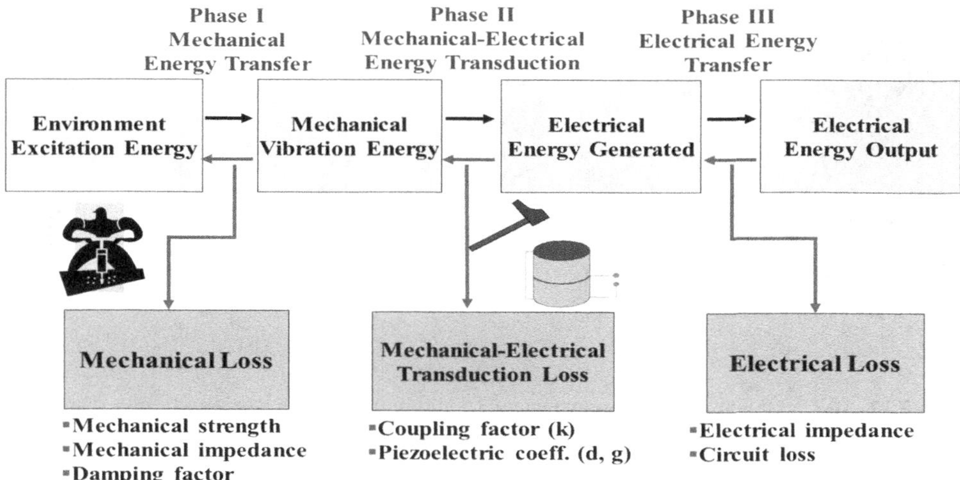

FIGURE 7.13 Three major phases associated with piezoelectric energy harvesting: (I) mechanical-mechanical energy transfer, (II) mechanical-electrical energy transduction, and (III) electrical-electrical energy transfer.

target of flexible piezocomposites was "wearable electric chargers" for portable electronic equipment such as mobile phones. The readers are requested to read this chapter contents, taking into account these practical applications.

We should also point out here that there is another research school of piezo-energy harvesting, that is, small energy harvesting (mW) for signal transfer applications, where efficiency is not a primary objective. This school usually treats an impulse/snap action load to generate instantaneous electric energy for transmitting signals for a short period (100 ms–10 s), without accumulating the electricity in a rechargeable battery. Successful products (million sellers) in the commercial market belong mostly to this category at present, including Lightning Switch[10] (remote switch for room lights using a unimorph piezoelectric component) by PulseSwitch Systems, VA, and the 25-mm-caliber Programmable Ammunition[17] (electricity generation with a multilayer piezo-actuator under shot impact) by ATK Integrated Weapon Systems, AZ, as introduced in Section 7.1.3.

7.2 MECHANICAL-TO-MECHANICAL ENERGY TRANSFER

First of all, wasted or unused mechanical energy (vibration source) should be transferred properly to an energy converter such as a piezoelectric device. Mechanical impedance matching is one of the important factors we have to take into account. The mechanical impedance of the material is defined by $Z = (\rho c)^{1/2}$, where ρ is the density and c is the elastic stiffness, or effective parameter values in a composite structure (see Equation 7.20). The receiving part of the mechanical energy in the piezodevice should be designed to match the mechanical/acoustic impedance with the vibration source. Otherwise, most of the vibration energy will be reflected at the interface between the vibration source and the harvesting piezoelectric device. Note a proverb, *pushing a curtain* and *pushing a wall* (useless task!), both cases of which will not transfer mechanical energy efficiently. Figures 7.14a and b exhibit two extreme examples we will treat in this article: (a) high energy harvesting from a "hard" machine, such as an engine, and (b) low energy harvesting from a "soft" machine, such as human motion. Not only mechanical impedance matching, but also mechanical strength and damping factor (i.e., loss) of the device are important.

The Cymbal transducer is a preferable device for the high-power purpose.[21] A Cymbal transducer consists of a piezoelectric ceramic disk and a pair of metal endcaps. The metal endcaps play an important role as displacement-direction convertor and displacement amplifier. (See Figure 7.15.)

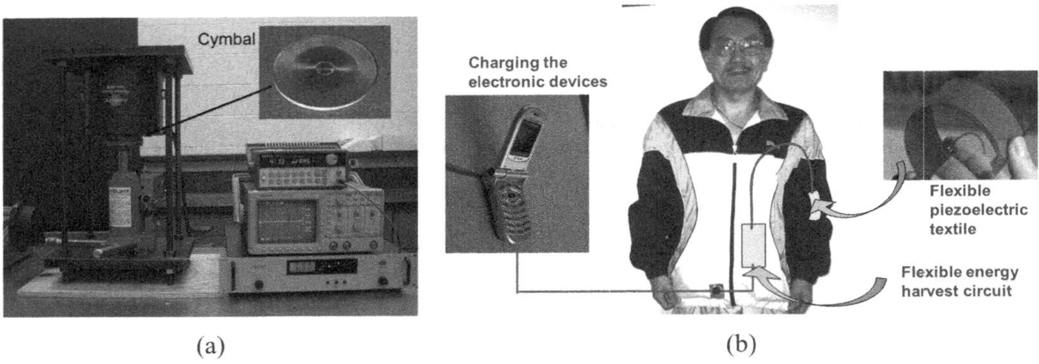

FIGURE 7.14 Two extreme vibration source examples: (a) high energy harvesting from a "hard" machine, such as an engine, and (b) low energy harvesting from a "soft" machine, such as human motion.

The Cymbal transducer has a relatively high coupling factor (k_{eff}) and a high stiffness in comparison with unimorphs/bimorphs, which makes it suitable for a high force mechanical source. Energy transfer or reflection rate from the hard electromagnetic shaker to a Cymbal was analyzed by changing the rigidity of the Cymbal endcap (i.e., by changing the endcap thickness, 0.3 and 0.4 mm). As shown in Figure 7.14a, a mechanical shaker (top part in Figure 7.14a) was employed to apply a maximum of 50 N force in a wide frequency range from pseudo-DC to 1 kHz (i.e., off-resonance usage). The force generated is proportional to the payload mass and the generated acceleration, which is controlled by the applied voltage. The acceleration of the payload was computed by performing the real-time differentiation of measured vibration velocity. The vibration velocity was measured using a Polytec Vibrometer (Tustin, CA). A payload mass (cylinder made from aluminum) of 85 g was used in the experiment. The Cymbal transducer was bonded at the top of the payload by using a silicone rubber sealer. A bias DC force was applied on the transducer using a home-built hydraulic system to avoid the separation problem, that is, bang-bang shock (the lower part in Figure 7.14a). The mechanical energy transferred to the Cymbal was evaluated from the Cymbal deformation and its effective stiffness. The mechanical-to-mechanical transfer rate of 46% for the 0.4-mm-thick sample was dramatically improved to 83%–87% by thinning the endcap down to 0.3 mm; that is, 0.4 mm seems to be too rigid for this shaker's elasticity, which will be discussed again in Table 7.3.

Note, however, that when we increased the bias force level up to 70 N by using a payload mass of 820 g, the sample with 0.3-mm-thick endcaps was damaged, and only the 0.4-mm sample could endure. Therefore, we should point out that the mechanical strength is another important factor, compromising with the efficiency.

To the contrary, the flexible transducer is a preferable device for "soft" application.[22] The macro fiber composite (MFC) is an actuator that offers reasonably high performance and flexibility in a cost-competitive manner (Smart Material Corp.) (Figure 7.16).[23] The MFC consists of rectangular

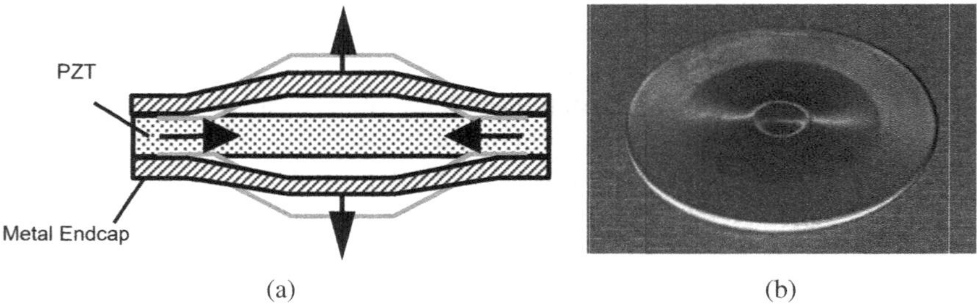

FIGURE 7.15 (a) Operation principle of the Cymbal, and (b) photo of the Cymbal.

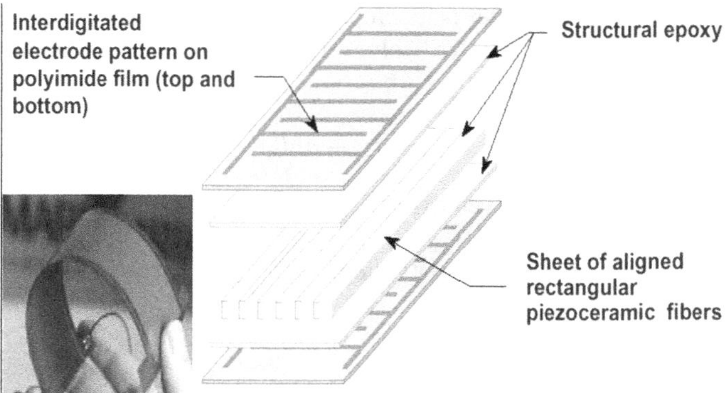

FIGURE 7.16 Macro fiber composite (MFC), Smart Material Corp. (From http://www.smart-material.com/Smart-choice.php?from=MFC.) (Courtesy Smart Materials.)

piezoceramic rods sandwiched between layers of adhesive and electroded polyimide film. This film contains interdigitated electrodes that transfer the applied voltage directly to and from the ribbon-shaped rods. This assembly enables in-plane poling, actuation, and sensing in a sealed, durable, ready-to-use package. Uchino et al. are developing intelligent clothing (IC) with a piezoelectric energy-harvesting system of flexible piezoelectric textiles,[22] aiming at a power source for charging up portable equipment such as cellular phones, health monitoring units, or medical drug delivery devices.

7.3 MECHANICAL-ELECTRICAL ENERGY TRANSDUCTION

7.3.1 Figure of Merit

Mechanical energy transferred to the transducer, such as a Cymbal and a flexible transducer, will be converted into electrical energy through the piezoelectric effect. Voltage induced in the transducer can be explained by Equations 7.26 and 7.27 for a bulk piezoelectric:

$$V = g \times \frac{F(\mathrm{N}) \times t(\mathrm{m})}{A(\mathrm{m}^2)} \tag{7.26}$$

$$g = \frac{d}{\varepsilon_0 \varepsilon_r} \tag{7.27}$$

Here, g, F, t, and A are piezoelectric voltage constant, applied force, thickness, and area of the piezomaterial, respectively. Since piezoelectric voltage constant g can be expressed by $g = d/\varepsilon_0\varepsilon_r$, the output electric power can be calculated as follows:

$$\begin{aligned} P &= \frac{1}{2} CV^2 \cdot f \\ &= \frac{1}{2} \cdot g_{33} \cdot d_{33} \cdot F^2 \cdot \frac{t}{A} \cdot f \end{aligned} \tag{7.28}$$

where C and f are the capacitance of the transducer and frequency of the vibration, respectively. Hence, output power can be evaluated by the product of g and d as figure of merit of the transducer:

$$P \propto g \cdot d \tag{7.29}$$

Note that when the piezocomponent is a composite or a hybrid structure, the above constants g and d should be replaced by the "effective" values. It is well known, for example, that the effective d coefficient for a cymbal is expressed by[24]

$$d_{eff} = d_{33} + A|d_{31}|\,, \tag{7.30}$$

where the amplification factor A is given by

$$A \propto (1/2)(\text{Cavity diameter/Cavity depth}). \tag{7.31}$$

7.3.2 Piezoelectric Material Selection

Popularly used piezoelectric materials are perovskite-type structures. Lead zirconate titanate-based ceramics have dominated the market for the last 60 years. We summarized the electromechanical properties of three commercially available piezoelectric PZT-based ceramics in Table 7.1: Hard APC 841, Soft APC 850 (American Piezo Ceramics, Mackeyville, PA), and High g D210 (Dong Il Technology, Korea). We can find the best material (D210) for an energy-harvesting application easily by comparing $g \cdot d$ product values.

7.3.3 Design Optimization

7.3.3.1 Cymbal

Cymbal design was optimized from the generated electrical energy level.[25] Figure 7.17 shows a rectification circuit with a full wave rectifier and a capacitor for storing the generated electrical energy of the Cymbal transducer in the case of off-resonance. Using this circuit, the output voltage and the power were initially measured across the resistive load directly without any amplification circuit to characterize the performance of different transducers. The maximum rectified voltage, V_{rec}, of a capacitor, C_{rec} (10 μF), was charged up to 248 V after saturation. Figure 7.18 shows the output electrical power from various Cymbal transducers under AC and DC mechanical loads as a function of external load resistance. Endcap thickness made of steel was changed from 0.3 to 0.5 mm. Prestress (DC bias load) to the Cymbal was 66 N, and applied AC (100 Hz) force was varied experimentally from 44 to 70 N. For a small force drive (40 and 55 N), the power level increased with decreasing the endcap thickness (from 0.5, then 0.4, and finally 0.3 mm). With increasing the force level up to 70 N, the maximum power of 53 mW was obtained at 400 kΩ with a 0.4-mm steel endcap, because the Cymbal sample with a 0.3-mm-thick endcap could not endure under this high force drive (i.e., the cavity depth collapsed). Note that the maximum power was obtained merely due to a large input mechanical vibration level. The 0.3-mm-thick samples were better from an energy efficiency viewpoint, which is discussed again in Table 7.3.

TABLE 7.1
Electromechanical Properties of the Commercial PZT Ceramics

Parameter	Hard (APC 841)	Soft (APC 850)	High g (D210)
ε_r	1350	1750	681
k_p (%)	60	63	58
d_{31}(10^{-12} C/N)	109	175	120
g_{31} (10^{-3} Vm/N)	10.5	12.4	20
Q_m	1400	80	89.7
T_c (deg.)	320	360	340
$g_{31} \cdot d_{31}$	1.14×10^{-12}	2.17×10^{-12}	2.40×10^{-12}

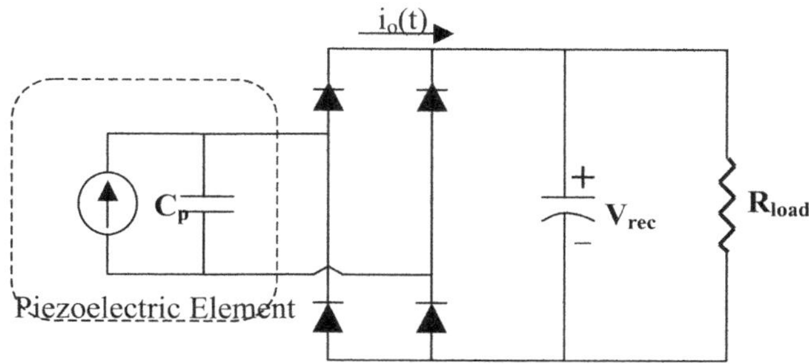

FIGURE 7.17 A full-bridge rectifier with a resistive load for piezoelectric energy harvesting.

In order to evaluate the mechanical-to-electrical converted energy in the Cymbal, the rectified voltage was used for charging the capacitor C_{rec} of 10 μF in open condition during the charging time (t), for which the capacitor was charged up to 200 V. Note that the measurement results in Table 7.3 include the energy Joule (not the power W, in this case) during a certain charging period.

7.3.3.2 Flexible Transducer

Bending mode is usually used for flexible transducers. However, bending vibration induces extension *and* compression stresses at the same time in the material, and induced electric energy cannot be obtained efficiently because of the partial cancellation. In order to avoid this problem, a unimorph design was adopted by bonding an elastic substrate to the piezoelectric plate.[26]

The d_{31}-mode type macro fiber composite (M8528 P2) from Smart Material Corporation is shown in Figure 7.19. The piezoceramic fibers in the MFC are cut by 350 μm width and 170 μm thickness from a piezoelectric wafer by computer-controlled dicing saw. The total dimensions of the MFC are 85 mm length, 28 mm width, and 0.3 mm thickness. Figure 7.20 shows the stress distribution of MFC when the mechanical force is applied. As shown in Figure 7.20a, without any substrate, the

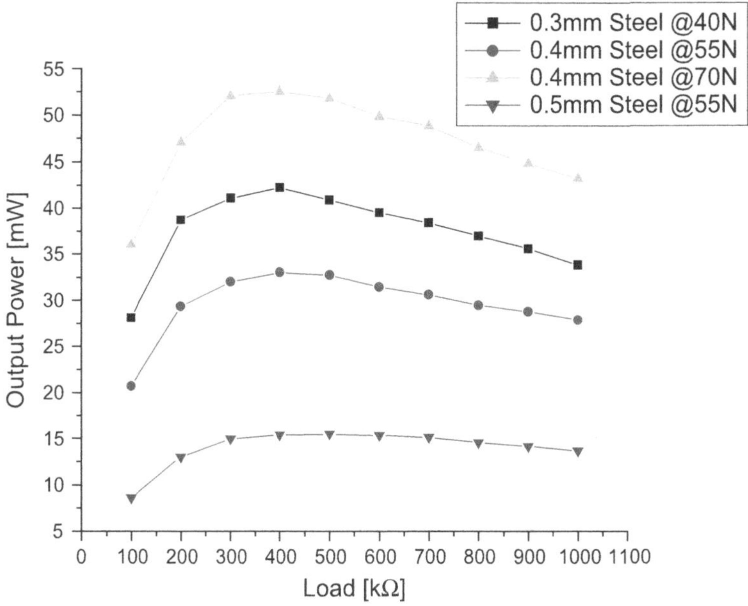

FIGURE 7.18 Change in output electrical power from the various Cymbal transducers under different 100 Hz AC mechanical load (shown as @ xx N under 66 N constant DC bias) with external electrical load resistance.

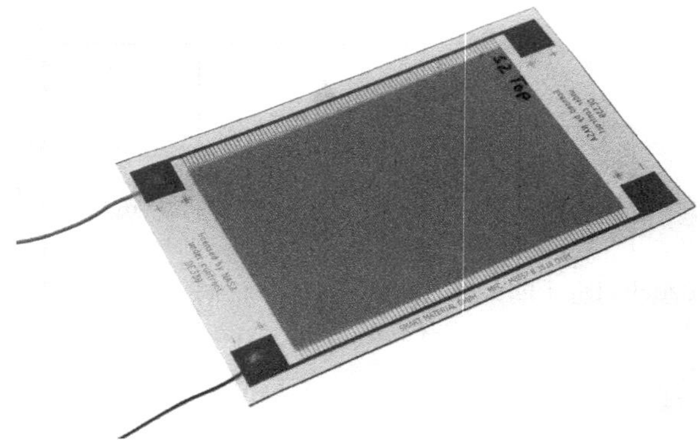

FIGURE 7.19 d_{31}-mode macro fiber composite (MFC), Smart Material Corporation. (Courtesy Smart Material Corp.)

extensive stress and compressive stress occurred on the top and the bottom of the MFC. In this case, the neutral line of the stress distribution is in the middle of the MFC. Because of this, the electrical output is almost cancelled and very small. However, in Figure 7.20b, where an additional substrate is bonded on the bottom of the MFC, the neutral line is shifted down in the substrate. Therefore, the electrical output is increased because the MFC has only a compressive stress or tensile stress in the whole volume. Though the uniformity of the compressive stress is increased with the elastic substrate thickness, the mechanical energy distribution to the MFC is decreased, leading to the optimum thickness for harvesting electrical energy. In addition, the thickness and material of the substrate should be carefully considered to keep the flexibility of the MFC.

The minimum thickness of the substrate was calculated by the FEM software code ATILA (distributed from Micromechatronics Inc., PA) depending on the material to shift the neutral line away from the MFC. Table 7.2 shows the neutral line position and displacement calculated.[26] Typical materials are considered to calculate the neutral line and flexural bending magnitude. For this calculation, the dimensions of the MFC are 85 × 28 × 0.17 mm, and the thickness of the substrate is 0.17 mm. Note that the signs of "+" and "−" indicate the distance from the adjacent line between the MFC and substrate. In the case of steel, brass, and copper, the neutral line is located in the substrate below the adjacent line. For the polymer, the neutral line is not changed from the center of the MFC, the same as an MFC without any substrate. The displacement in the table is a measure of the flexibility of the MFC with a substrate. This displacement is calculated under a fixed mechanical force condition. The aluminum substrate shows the best flexibility of any other metal substrates in Table 7.2. Therefore, aluminum was selected to be used for the substrate in the experiment, even though the neutral line is located slightly above the interface line.

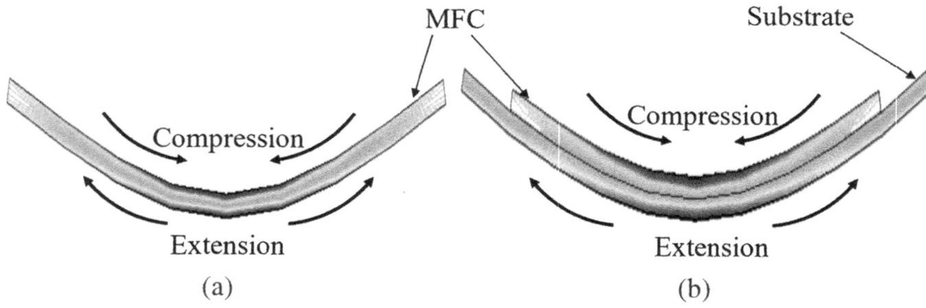

FIGURE 7.20 Stress distribution of the macro fiber composite in (a) single plate, and (b) with an elastic substrate.

TABLE 7.2
Neutral Line Position and Bending Displacement Calculated by FEM for Various Substrate Materials (t = 170 μm) under a Fixed Mechanical Force

	None	Steel	Brass	Copper	Aluminum	Polymer
Neutral Line	0.085	−0.03	−0.0025	−0.0125	0.005	0.0825
Displacement	901 μm	79 μm	114 μm	101 μm	129 μm	766 μm

Source: H.-W. Kim, K. Uchino and T. Daue: *Proc. CD 9th Japan Inter. SAMPE Symp. & Exhibit.*, SIT Session 05, November 29–December 2, 2005.

The method to apply the stress on the sample was based on the real situation. The mechanical shaker to make a vibration at a high frequency is not a close approach to piezoelectric energy harvesting with a small mechanical source such as human motion. The MFC was excited to generate a big bending motion by a small force at a frequency around 1–5 Hz. Figure 7.21 shows an experimental setup for measuring electrical output from a flexible piezocomponent under a fixed bending displacement. Note that the small mechanical force used in this experiment means the minimum force that can generate maximum strain in the flexible element without crack.

The voltage signal of the MFC is shown in Figure 7.22a.[26] The voltage of the MFC is considerably increased by bonding an aluminum substrate, as shown in Figure 7.22a. This signal was generated by a small mechanical force with a frequency of 5 Hz and monitored by an oscilloscope (Tektronix, TDS 420A). The addition of the aluminum substrate gave a lower flexibility to the MFC, but the output voltage from the MFC with small bending showed a much higher voltage signal. The output voltage signal from the MFC was passed through the rectifier and charged a capacitor, and successively discharged through a resistive load. The rectified voltage and output power are shown in Figure 7.22b. The generated electric power from a small mechanical force at 5 Hz was around 1.5 mW at 200 kΩ. This 200 kΩ corresponds roughly to the matched impedance value of $1/\omega C$ at 5 Hz.

7.3.4 Energy Flow Analysis

Table 7.3 summarizes the energy flow analysis on three types of Cymbal transducers: endcap thickness of 0.3 and 0.4 mm with and without bias force under various cyclic vibration levels and drive durations.[27,28] Note first that the mechanical-to-mechanical energy transfer rate is good for the

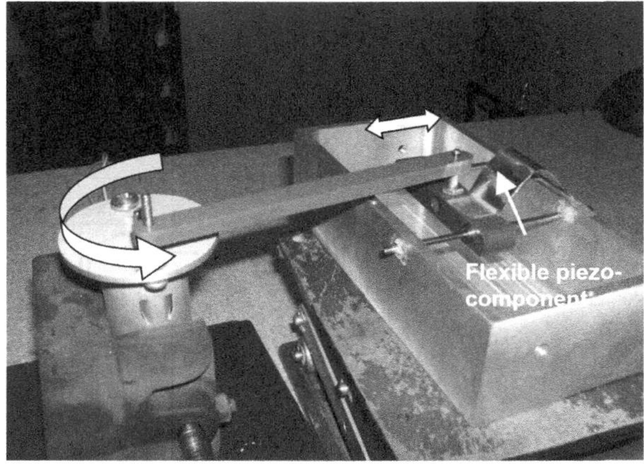

FIGURE 7.21 Experimental setup for measuring electrical output from a flexible piezocomponent.

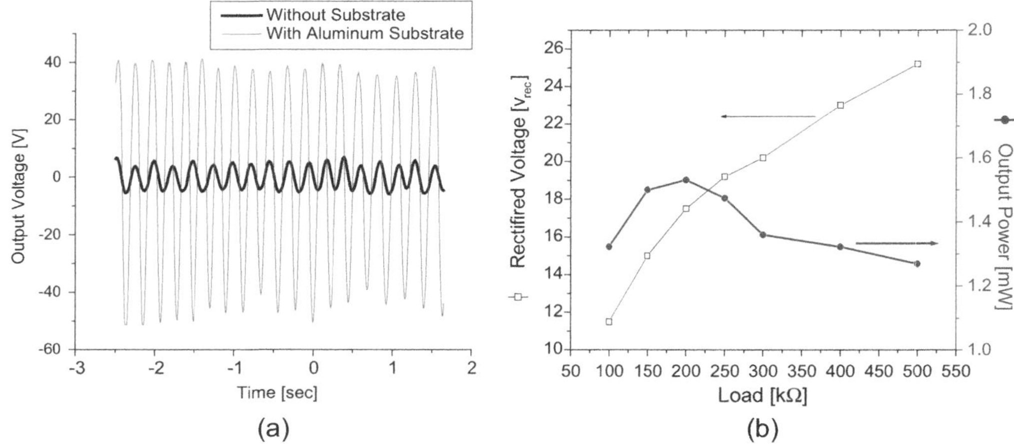

FIGURE 7.22 Output of the MFC at 5 Hz. (a) Output voltage signal of the MFC, and (b) rectified voltage and output power plotted as a function of resistive load.

0.3-mm-thick cymbals (83%–87%), while it is rather low for the 0.4-mm-thick cymbal (46%). This is related to the mechanical impedance matching; the 0.4-mm endcap seems to be too rigid (effective stiffness is too high) to match the vibration source shaker.

Second, the received mechanical energy to electrical energy transduction rate can be evaluated from the effective electromechanical coupling factor. Because the value of k_{eff} of the cymbal is around 25%–30%, the energy transduction rate k_{eff}^2 can be evaluated around 6.25%–9%, which

TABLE 7.3

Energy Flow Analysis on Three Types of Cymbal Transducers: Endcap Thickness of 0.3 and 0.4 mm with and without Bias Force under Various Cyclic Vibration Levels and Drive Durations

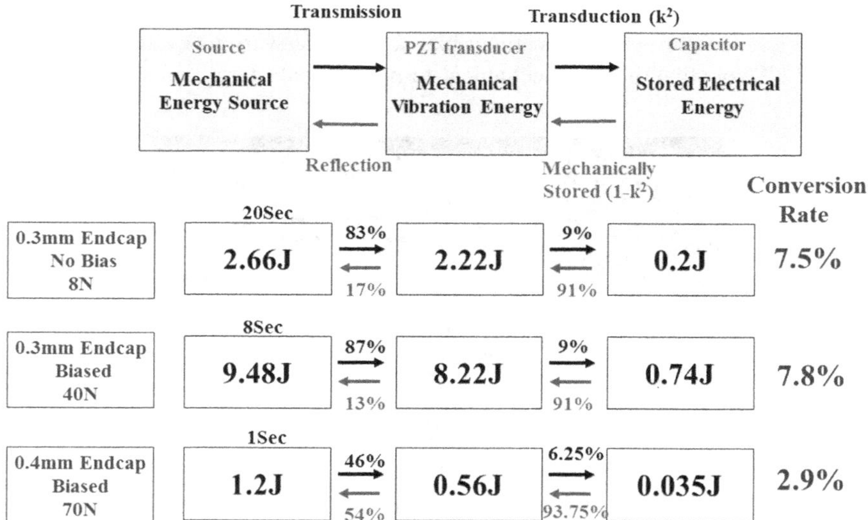

	Duration	Source	Transmission / Reflection	PZT transducer	Transduction / Mechanically Stored	Capacitor	Conversion Rate
0.3mm Endcap No Bias 8N	20Sec	2.66J	83% / 17%	2.22J	9% / 91%	0.2J	7.5%
0.3mm Endcap Biased 40N	8Sec	9.48J	87% / 13%	8.22J	9% / 91%	0.74J	7.8%
0.4mm Endcap Biased 70N	1Sec	1.2J	46% / 54%	0.56J	6.25% / 93.75%	0.035J	2.9%

Source: K. Uchino: *Proc. 5th Int'l Workshop on Piezoelectric Mater. Appl.*, State College, PA, October 6–10, 2008.

agrees very well with the experimental results in Table 7.3. Thus, we obtained the conversion rate from the vibration source energy to the stored electric energy in the cymbal transducer as 7.5%–7.8% for the 0.3-mm endcap cymbals, and 2.9% for the 0.4-mm endcap cymbal. The reduction of the conversion rate primarily originates from the mechanical/acoustic impedance mismatch.

7.4 ELECTRICAL-TO-ELECTRICAL ENERGY TRANSFER

Piezoelectric materials generally convert mechanical energy to electrical energy with relatively high voltage, which means output impedance is relatively high at an off-resonance frequency. On the other hand, energy storage devices such as a rechargeable battery have low input impedance (10–100 Ω). Thus, a large portion of the excited electrical energy is reflected back if we connect the battery immediately after the rectified voltage. In order to improve energy transfer efficiency, electrical impedance matching is required.

7.4.1 DC-DC Converter

As we discussed in Section 5.7.3, the basic principle of a switching regulator using a power MOSFET is the key to developing a suitable *DC-DC converter.* The output voltage, v_0, through a load, R, switches the maximum, E (some 100 s V), and minimum, 0, according to the gate/source voltage (rectangular voltage with the duty ratio of D), so that the average voltage is estimated by

$$v_0 = \frac{T_{on}}{T_{on} + T_{off}} E = dE. \quad (d\text{:duty ratio}) \tag{7.32}$$

If the loss of the electronic component is negligibly small, ideally input power = output power. Thus, using the duty ratio, d, the voltage step-up ratio and current step-down ratio should be d, leading to impedance change by the factor of d^2. If we take $d = 2\%$, the impedance can be reduced by 2500 theoretically.

Various converters can step down the voltage to adapt electrical impedance: forward converter, buck converter, buck-boost converter, flyback converter, and so on. These converters have low output impedance and low loss characteristics. A *DC-DC buck-converter*, shown in Figure 7.23, is the simplest topology using a single MOSFET, which allows transfer of 43 mW power out of 53 mW from the Cymbal (81% efficiency) in practice by converting the original impedance 300 kΩ down to 5 kΩ with a 2% duty cycle at a switching frequency of 1 kHz.[26,29] The impedance modification (60 times) differs from the above expectation (2500 times), which may be explained by the components' losses in the circuit.

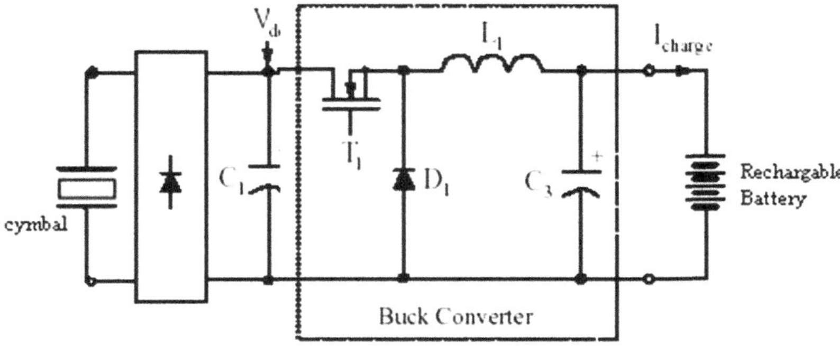

FIGURE 7.23 A DC-DC buck-converter designed to allow transfer of 43 mW power out of 53 mW from the Cymbal (81% efficiency) by converting the original impedance 300 kΩ down to 5 kΩ with a 2% duty cycle and at a switching frequency of 1 kHz.

7.4.2 Multilayered Cymbal

Since the buck-converter introduced in the previous section cannot reduce the output impedance sufficiently to match with the rechargeable battery (~50 Ω), we should further reduce the output impedance. The output impedance of the transducer can be changed by changing the transducer structure. Multilayered transducers have lower impedance.[27] Figure 7.24a shows a cross-sectional view of the multilayered Cymbal transducer. By increasing the number, n, of the layers, output impedance decreases by the factor of $1/n^2$, as you are familiar with. As shown in Figure 7.24b, the maximum output power shifts to the low resistance direction, which means output impedance can be controlled by the transducer structure. You can find that the matching impedance, 300 kΩ, in the single-layer cymbal is reduced down to 3 kΩ by using a 10-layer cymbal.

It is worth noting that the performance becomes much superior by combining the ML cymbal structure and the DC-DC converter. See Figure 7.25. When we compare the output power of the 10-layer cymbal with or without the DC-DC converter, using the converter reduces the maximum power level from 100 to 80 mW with a matching load shift from 5 to 2 kΩ. However, the output power around 50 Ω (matching to the rechargeable battery) differs significantly: 50 mW with the converter and less than 10 mW without the converter. The reason for this load-insensitive broadening effect is to be clarified in the future.

7.4.3 Usage of a Piezoelectric Transformer: Further Impedance Matching

Another unique circuit design is with a piezoelectric transformer.[30] The piezoelectric transformer used in the circuit has low output impedance, around 50 Ω (i.e., resistive at the transformer resonance frequency), and the efficiency of the piezoelectric transformer at its resonance can reach above 97%,

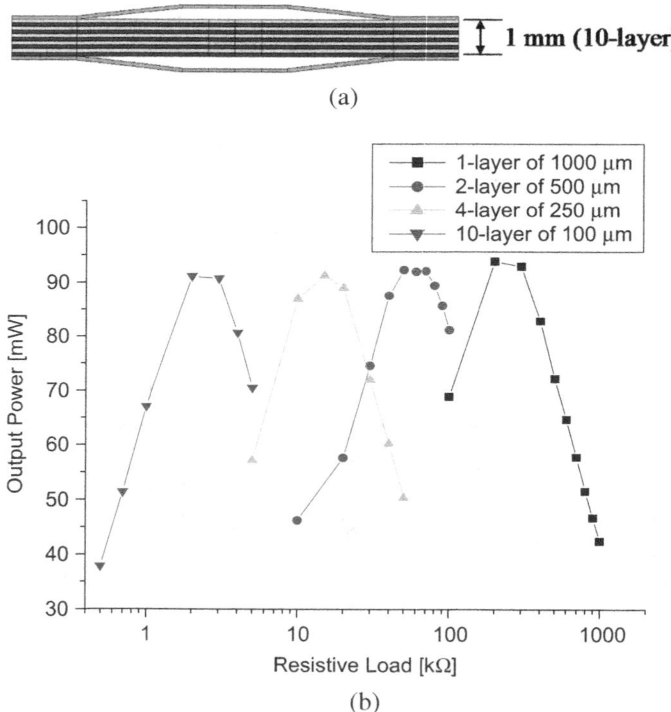

FIGURE 7.24 (a) Multilayered (ML) Cymbal transducer. (From H.-W. Kim, S. Priya, and K. Uchino: Japan. J. Appl. Phys., 45, 5836, 2006.) (b) Output power characteristics of the ML Cymbal transducers.

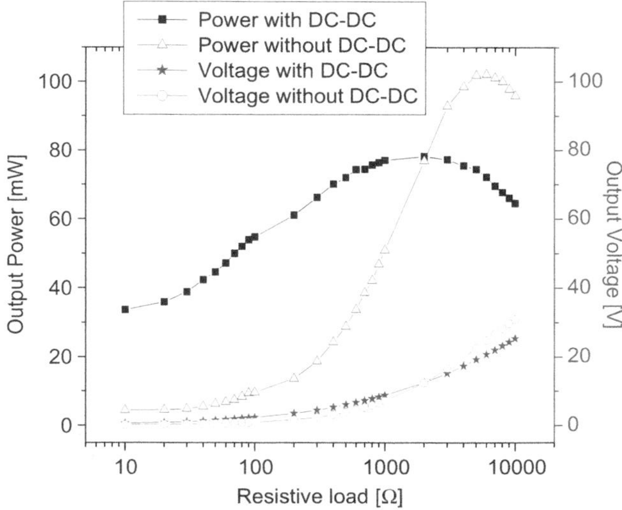

FIGURE 7.25 Load resistance dependence of the efficiency of the ML piezotransformer.

as shown in Figure 7.26. This low output impedance is suitable for impedance matching to energy storage devices. Figure 7.27a illustrates a block diagram of a piezoelectric energy-harvesting circuit with a piezoelectric transformer, and Figure 7.27b is an actual piezoelectric energy-harvesting circuit with a *ring-dot type multilayer transformer* (see the disk-shaped PZT transformer).[31] Input and output voltage signals of the piezotransformer and the final rectified voltage, which correspond to the three arrow parts in (a), are plotted in Figure 7.27c. The step-down voltage ratio seems to be 5.

7.5 TOTAL ENERGY FLOW CONSIDERATION

Let us summarize the energy flow/conversion from the input mechanical energy source to the output rechargeable battery. The sample was a cymbal with 0.3-mm-thick stainless steel endcaps, inserted below a 4-kg engine weight (40 N bias force). The electromagnetic shaker shook at 100 Hz for 8 seconds, and the accumulated energy during that time period was measured at each point. The energy levels at all states are summarized in Table 7.4.

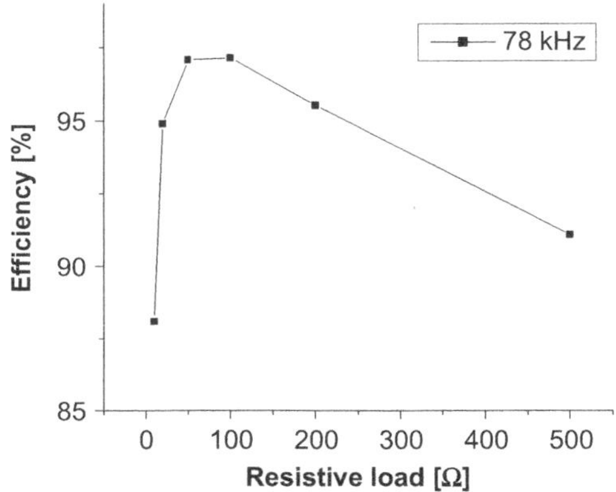

FIGURE 7.26 Load resistance dependence of the efficiency of the piezoelectric transformer. (From K. Uchino and A. Vazquez Carazo: *Proc. 11th Int'l Conf. New Actuators, A3.7*, p. 1, Bremen, Germany, June 9–11, 2008.)

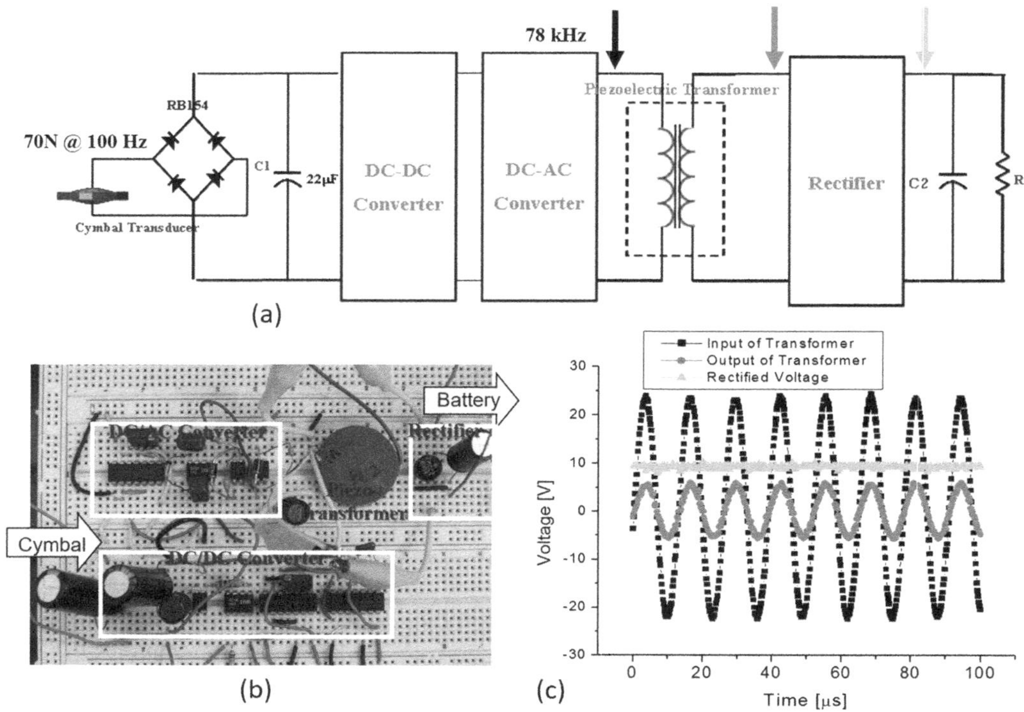

FIGURE 7.27 (a) Block diagram of a piezoelectric energy-harvesting circuit with a piezoelectric transformer. (b) Actual piezoenergy harvesting circuit with a ring-dot type multilayer transformer. (c) Input and output voltage signals of the piezotransformer and the final rectified voltage, which correspond to the three arrow parts in (a).

TABLE 7.4
Energy Flow/Conversion Analysis in the Cymbal Energy Harvesting Process

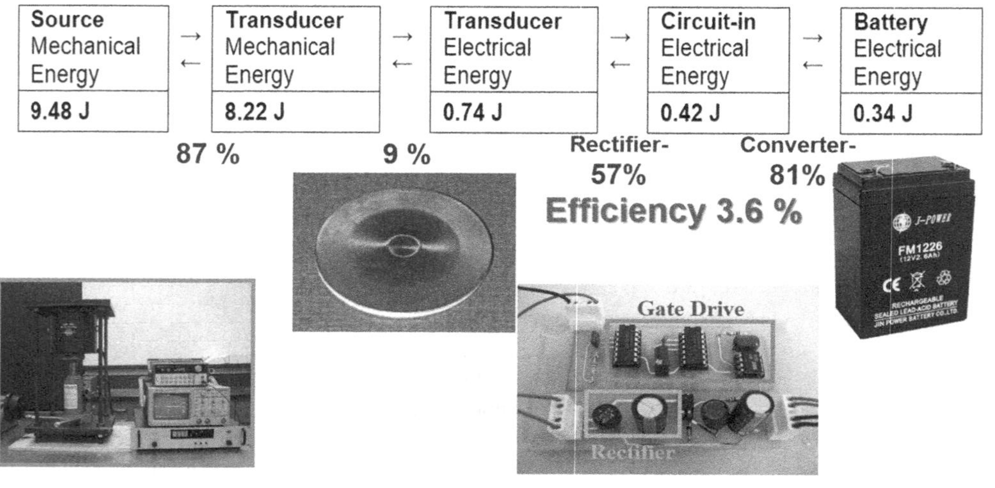

Let us discuss why the energy amount decreases with transmitting and transducing successively by numerical analyses:

1. *Source to Transducer:* Mechanical impedance mismatch. 8.22 J/9.48 J = 87% of the mechanical energy is transmitted from the source to the piezocymbal transducer. If the mechanical impedance is not seriously considered in the transducer (too mechanically soft or hard), the mechanical energy will not transfer efficiently, and a large amount will be reflected at the contact/interface point.
2. *Transduction in the Transducer:* Electromechanical coupling. The transduction rate is evaluated by k^2. Since k_{eff} = 0.25–0.30 in the cymbal, the energy conversion rate will be 9% maximum. The remaining portion will remain as the original mechanical vibration energy. 0.74 J/8.22 J = 9.0%. Since k_{eff} of the bimorph or unimorph is much smaller (10%–15%), the energy conversion rate is smaller than 2%. Using modes with higher k values is highly recommended.
3. *Transducer to Harvesting Circuit:* Electrical impedance matching. 0.42 J/0.74 J = 57% is related to the electrical impedance mismatch between the cymbal output and the circuit input. Unfortunately, half of the stored electrical energy in the piezodevice cannot be used effectively due to electrical rectification.
4. *Harvesting Circuit to Rechargeable Battery:* DC-DC converter. 0.34 J/0.42 J = 81%. This reduction partially originates from the energy consumption in the circuit and partially from the electrical impedance mismatch between the circuit output (still around 5 kΩ) and the rechargeable battery impedance (around 10–50 Ω). A MOSFET inevitably consumes 1 mW level during operation, leading to 2 mW as a DC-DC converter topology. Thus, a piezoelectric energy-harvesting component, which can generate less than 2 mW, is not an energy "harvesting" device, but a "losing" device in practice.

In summary, 0.34 J/9.48 J = 3.6% is the energy-harvesting rate from the vibration source to the storing battery in the current system. Taking into account the efficiency of popular amorphous silicon solar cells around 5%–9%, this prototype piezoelectric energy-harvesting system with 3.6% efficiency seems to be rather promising.

7.6 HYBRID ENERGY HARVESTING: MAGNETOELECTRIC DEVICES AND THE FUTURE

A sustainable society requires various sensors for monitoring hazardous environmental factors such as viruses and magnetic fields, which humans cannot sense biologically. Similar to nuclear radiation, magnetic irradiation cannot be easily felt by humans, but it may increase brain cancer probability. We cannot even purchase a magnetic field detector for a low frequency (50 or 60 Hz). Penn State, in collaboration with Seoul National University, Korea, developed a simple and handy magnetic noise sensor. Figure 7.28 shows a schematic structure of the device, in which a PZT disk is sandwiched by two Terfenol-D (magnetostrictor) disks. When a magnetic field is applied on this composite, Terfenol-D expands, which is mechanically transferred to PZT, leading to an electric

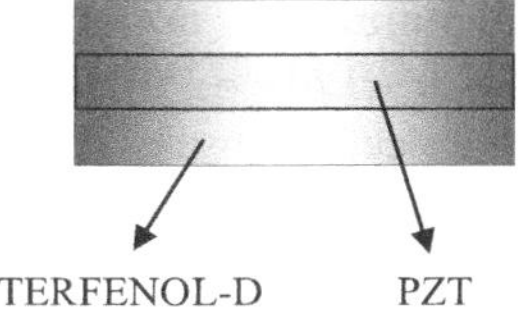

FIGURE 7.28 Magnetic noise sensor consisting of a laminated composite of a PZT and two Terfenol-D disks. (From J. Ryu, et al.: *Japan. J. Appl. Phys.*, 40, 4948, 2001.)

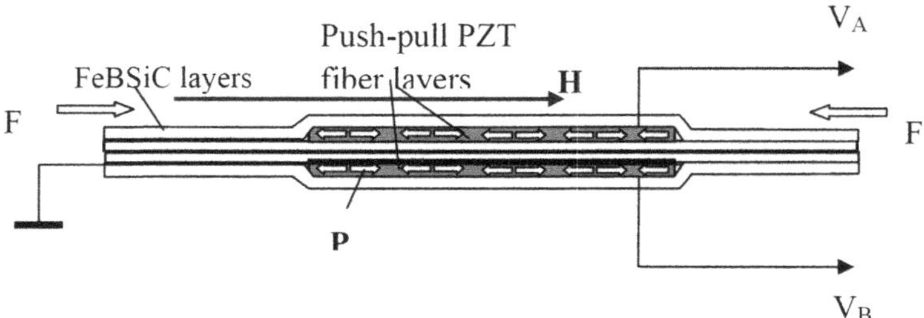

FIGURE 7.29 Hybrid energy-harvesting device with a lamination of PZT macro fiber composite and Metglass. (From J. Zhai: PhD Thesis, Virginia Tech, 2009.)

charge generation from the PZT. By monitoring the voltage generated in the PZT, we can detect the magnetic field. The key of this device is high effectiveness for a low frequency.[32] As long as the magnetostrictor-piezoelectric laminated composite can generate electric signal under magnetic noise, we can consider a hybrid energy-harvesting device on which either magnetic and/or mechanical noises can work.

A practical demonstration has been made by the Virginia Tech group.[33] Zhai et al. laminated Metglass (FeBSiC) and PZT macro fiber composite, as shown in Figure 7.29, to make a hybrid energy-harvesting device. Mechanical vibration generates the electrical energy via a bimorph-type "push-pull" PZT MFC, while the magnetic noise can generate electrical energy via an electromagnetic effect from the Metglass and MFC lamination. They demonstrated 2.5 V output voltage from the pure vibration, 2.5 V output voltage from the pure magnetic noise, and 5 V as a sum from a large industrial electromagnetic motor operated at 50 Hz.

Recently, Finkel et al.[34] demonstrated the above idea by using superior piezoelectric single crystal lead indium niobate–lead magnesium niobate–lead titanate (PIN-PMN-PT) and magnetostrictive single crystal Galfenol. An AC magnetic field ±250 G under a bias field of 250 G applied to the coupled device causes the magnetostrictive Galfenol element to expand, and the resulting stress forces a phase change in the relaxor ferroelectric PIN-PMN-PT single crystal. They have demonstrated high energy conversion (2 mW at the matching impedance 1 MΩ) in this magnetoelectric device by triggering the F_R-F_O transition in the single crystal by a small AC magnetic field in a broad frequency (off-resonance) range that is important for multidomain hybrid energy-harvesting devices.

For the future perspectives, there will be two development directions in piezoelectric energy-harvesting areas: (1) remote signal transmission (such as structure health monitoring) at the (mW) power level and (2) energy accumulation in rechargeable batteries at the (W) power level. Less than W energy harvesting does not attract clients in home appliance and automobile manufacturers. The former will use impulse/snap action mechanisms or low-frequency resonance methods, where efficiency is not a significant issue, but the minimum spec on the energy should be obtained, while the latter require dramatic research strategy changes. The researchers are requested to forget the current "biased" knowledge, or not to adopt a strategy "because the previous researchers did so." Remembering the fundamental principles explained in this chapter, challenge a totally unique design and/or idea in the future.

Though relatively large investments and research efforts are being put into MEMS/NEMS and "nanoharvesting" devices, a positive comment is not provided at the moment. Even for medical applications, obtained/reported energy level $pW \sim nW$ from one component (this level is called a "sensor," not "energy harvester," in practice) is a useless level, which originates from the inevitable small volume of the used piezoelectric material (i.e., thin films). Without discovering a genius idea of how to combine thousands of these nanodevices in parallel and synchronously in phase, the current efforts will be in vain. "Nanogrid" research to reach to a minimum 1 mW level by connecting thousands of nanoharvesting devices is highly encouraged, rather than merely the MEMS fabrication process from an academic viewpoint.

CHAPTER ESSENTIALS

1. Three phases/steps for piezoelectric energy harvesting are (1) mechanical impedance matching, (2) electromechanical transduction, and (3) electrical impedance matching. Systematic optimization for each phase is required in order to improve the total efficiency of the energy-harvesting system.
2. The Cymbal transducer is employed for energy harvesting from a high-power mechanical vibration, while the macro fiber composite is suitable for a small flexible energy vibration.
3. A buck-converter is used as a DC/DC converter for realizing better electrical impedance matching. Further improvement can be realized by using a highly capacitive multilayer design for a piezo-energy harvesting component. A piezoelectric transformer is a promising alternative component for the DC/DC converter.
4. As indicated in Table 7.4, the key to dramatic enhancement in the efficiency is to use a high k mode, such as k_{33}, k_t, or k_{15}, rather than flex-tensional modes. Figure 7.30 summarizes promising piezoelectric device designs for energy-harvesting applications. The hinge-lever mechanism is an ideal "mechanical transformer" without losing mechanical energy under an off-resonance condition, and it is easily manufactured using micromachining technologies, which may expand the tunability of the mechanical impedance by keeping a high electromechanical coupling factor, k, in the range of $k_{33} \approx 70\%$.
5. A hybrid energy-harvesting device that operates under either magnetic and/or mechanical noises was introduced by coupling magnetostrictive and piezoelectric materials.
6. Saving energy has become very important in recent years, and the analyses and results in this article can be applied to recover and store wasted or unused mechanical (and/or magnetic) energy efficiently.

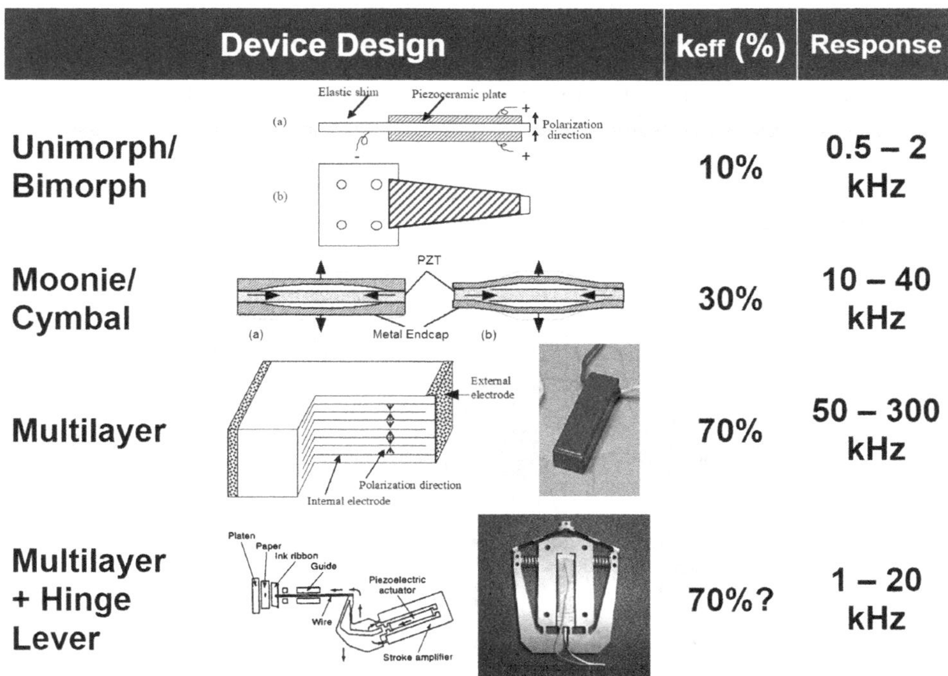

FIGURE 7.30 Promising piezoelectric device designs for energy-harvesting applications.

CHECK POINT

1. (T/F) When 1 J mechanical energy is input on a piezoelectric with an electromechanical coupling factor, k, we can expect k^2 J electrical energy converted in this piezomaterial. Thus, this piezoelectric can harvest up to k^2 J electrically to the outside load. True or False?
2. (T/F) Elastic compliances are the material constants in a piezoelectric. Thus, the resonance frequency is determined merely by the sample size, irrelevant to the external electrical load (L, C, or R). True or False?
3. There are three key factors to be considered for developing efficient piezoelectric energy-harvesting systems: (1) mechanical impedance matching, (2) electromechanical transduction, and (3) electrical impedance matching. Which factor is the primary problem for inefficiency in the following systems?
   - A piezoelectric energy-harvesting bimorph component is connected to a rechargeable battery immediately after a full rectification circuit.
   - A PZT thin film membrane on a Silicon wafer (MEMS) with a unimorph configuration.
   - A PVDF film bonded on a thick steel beam.
   - A PZT bulk disk attached directly on a human body, say, for energy harvesting for a pacemaker, and so on.
4. (T/F) When the loss of the electronic component is negligibly small, ideally input power = output power is expected in a back converter. If we take the duty ratio (on/off time period ratio) as 2%, the impedance can be reduced by 50 theoretically. True or False?

CHAPTER PROBLEMS

1. An energy source for 25-mm-caliber Programmable Ammunition was developed with a multilayer PZT piezo-actuator. Electric energy is generated under the shot's mechanical impact to activate the operational amplifiers, which ignite the burst according to the command program. According to the development flowchart below, design a suitable compact multilayer energy-harvesting component step by step.
   a. Knowing the general PABM caliber specs: weight = 100–200 g, acceleration $\approx 10^6$ m/s^2, muzzle velocity = 1–1.5 km/s, max effective range = 2–3 km, calculate the necessary energy consumption (in mJ unit) for a microprocessor. You may assume that its power consumption is 1 mW and a 6-sec flight period (maximum).
   b. Assuming the power of the microprocessor is 1 mW and its impedance is 1 MΩ, calculate the required current and voltage.

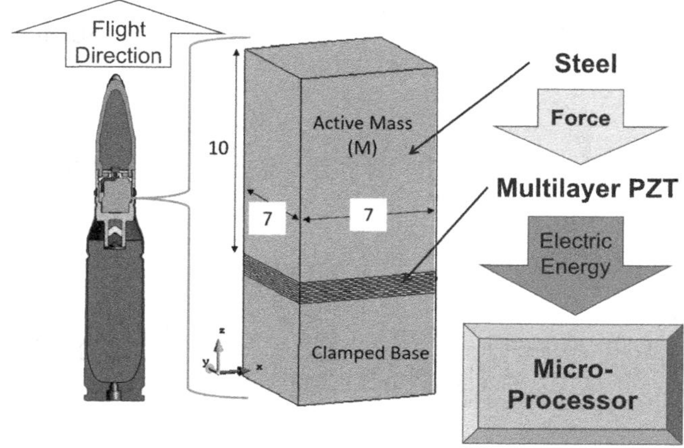

Model of a PABM energy harvesting system.

c. Because of the outer diameter of a bullet, 25 mm, the multilayer PZT total size is designed as 7 × 7 × 1 mm thick. Since the capacitance of each layer is obtained from $C = \varepsilon_0\varepsilon_{33}T \times (S/t)$, the total capacitance should be $C_{\text{total}} = \varepsilon_0\varepsilon_{33}^X \times (S/t_{\text{total}}) \times n^2$, where n is the layer number, keeping $t_{\text{total}} = 1$ mm. Supposing that we use a high-permittivity PZT-5H with $\varepsilon_{33}^X = 3400$, calculate the capacitance of the MLs for $n = 1$, 10, and 100, for example.

d. Supposing that the size of the steel mass on the multilayer PZT stack in this system is 7 × 7 × 10 mm, and the density is 7800 kg/m^3, calculate the mass, then the force and stress on the PZT, assuming the repulsive acceleration 5×10^6 m/s^2.

e. From the above-obtained stress on the ML PZT, calculate the generated charge, Q, and the energy, $U_{\text{gen}} = Q^2/C_{\text{total}}$. Note that U_{gen} is not related to the layer number, but merely to the total volume. Under a microprocessor's impedance, $Z_1 = 1$ MΩ, this energy must provide sufficient energy for a flight time period. So, the time constant of Z_1C_{total} should be longer than 6 seconds. From this condition, determine the layer number required for the PZT-5H case. You should obtain the number in the range from 10 to 100.

REFERENCES

1. K. Uchino: *Ferroelectric Devices*, 2nd Ed., CRC Press, Boca Raton, FL, 2010.
2. K. Uchino and K. Ohnishi: "Shock Preventing Apparatus", U.S. Patent No. 4,883,248.
3. Uchino estimated the revenue based on the resources, iRAP (Innovative Research & Products), Market Publishers, and IDTechEx.
4. N. Muensit (Ed.): *Energy Harvesting with Piezoelectric and Pyroelectric Materials*, Materials Science Foundations Vol. 72, Trans Tech Pub., Stafa-Zuerich, Switzerland, 2011.
5. J. Wallaschek, M. Neubauer and J. Twiefel: "Electromechanical Models for Energy Harvesting Systems", *Chapter 2 of Energy Harvesting with Piezoelectric and Pyroelectric Materials*, Materials Science Foundations Vol. 72, Trans Tech Pub., Stafa-Zuerich, Switzerland, 2011.
6. P. Smithmaitrie: "Vibration Theory and Design of Piezoelectric Energy Harvesting Structures", *Chapter 3 of Energy Harvesting with Piezoelectric and Pyroelectric Materials*, Materials Science Foundations Vol. 72, Trans Tech Pub., Stafa-Zuerich, Switzerland, 2011.
7. S. P. Beeby and D. Zhu: "Energy Harvesting Products and Forecast", *Chapter 9 of Energy Harvesting with Piezoelectric and Pyroelectric Materials*, Materials Science Foundations Vol. 72, Trans Tech Pub., Stafa-Zuerich, Switzerland, 2011.
8. S. W. Arms, C. P. Townsend, J. H. Galbreth, L. C. David and P. Nam: "Synchronized System for Wireless Sensing, RFID, Data Aggregation & Remote Reporting", 2009.
9. K. Uchino: "Piezoelectric Actuators 2004—Materials, Design, Drive/Control, Modeling and Applications", *Proc. 9th Int'l Conf. New Actuators, A1.0*, p. 38–48, Bremen, Germany, June 14–16, 2004.
10. K. Uchino: "Piezoelectric Actuators 2010—Piezoelectric Devices in the Sustainable Society", *Proc. 12th Int'l Conf. New Actuators, A3.0*, Bremen, Germany, June 14–16, 2010. http://www.lightningswitch.com/
11. H. Kim, Y. Tadesse and S. Priya: "Energy Harvesting Technologies", Eds. S. Priya and D. Inman, Springer Science+Business Media LLC, New York, 2009.
12. M. Renaud, P. Fiorini, R. van Schaijk and C. van Hoof: *Smart Mater. Struct.*, 18, 035001, 2009.
13. D. Guyomar, M. Lallart, N. Muensit and C. Lucat: "Conversion Enhancement for Energy Harvesting", *Chapter 5 of Energy Harvesting with Piezoelectric and Pyroelectric Materials*, Materials Science Foundations Vol. 72, Trans Tech Pub., Stafa-Zuerich, Switzerland, 2011.
14. R. P. Lopes and A. Kholkin: "Energy Harvesting for Smart Miniaturized Systems", *Chapter 6 of Energy Harvesting with Piezoelectric and Pyroelectric Materials*, Materials Science Foundations Vol. 72, Trans Tech Pub., Stafa-Zuerich, Switzerland, 2011.
15. C. Chang, V. H. Tran, J. Wang, Y.-K. Fuh and L. Lin: *Nano Lett.*, 10, 726, 2010.
16. Z. L. Wang and J. Song: *Science*, 321, 242, 2006.
17. http://www.atk.com/MediaCenter/mediacenter_videogallery.asp
18. T. Ikeda: *Fundamentals of Piezoelectric Materials Science*, Ohm Publishing Co., Tokyo, 1984.
19. L. E. Kinsler, A. R. Frey, A. B. Coppens and J. V. Sanders: *Fundamentals of Acoustics*, John Wiley & Sons, New York, 1982.

20. Uchino and T. Ishiix: *J. Ceram. Soc. Japan.*, 96, 863, 1988.
21. H.-W. Kim, A. Batra, S. Priya, K. Uchino, D. Markley, R. E. Newnham and H. F. Hofmann: *J. Appl. Phys.*, 43, 6178, 2004.
22. K. Uchino: *J. Japan. Soc. Appl. Electromag.*, 15(4), (20071210), 399, 2008.
23. http://www.smart-material.com/Smart-choice.php?from=MFC.
24. B. Koc, A. Dogan, J. F. Fernandez, R. E. Newnham and K. Uchino: *Japan. J. Appl. Phys.* 35, 4547, 1996.
25. H.-W. Kim, S. Priya, K. Uchino and R. E. Newnham: *J. Electroceramics*, 15, 27, 2005.
26. H.-W. Kim, K. Uchino and T. Daue: *Proc. CD 9th Japan Inter. SAMPE Symp. & Exhibit.*, SIT Session 05, November 29–December 2, 2005.
27. H.-W. Kim, S. Priya, and K. Uchino: *Japan. J. Appl. Phys.*, 45, 5836, 2006.
28. K. Uchino: *Proc. 5th Int'l Workshop on Piezoelectric Mater. Appl.*, State College, PA, October 6–10, 2008.
29. H.-W. Kim, S. Priya, H. Stephanau and K. Uchino: *IEEE Trans. UFFC*, 54, pp. 1851 2007.
30. K. Uchino and A. Vazquez Carazo: *Proc. 11th Int'l Conf. New Actuators, A3.7*, p. 1, Bremen, Germany, June 9–11, 2008.
31. H.-W. Kim, S. Priya, and K. Uchino: "A New Method of Impedance Adaptation for Piezoelectric Energy Harvesting Using a Cymbal Transducer," *Proc. 10th Int'l Conf. New Actuators, A5.4*, p. 189–192, Bremen, Germany, June 14–16, 2006.
32. J. Ryu, A. Vazquez Carazo, K. Uchino and H. E. Kim: *Japan. J. Appl. Phys.*, 40, 4948, 2001.
33. J. Zhai: PhD Thesis, Virginia Tech, 2009.
34. P. Finkel, R. P. Moyet, M. Wun-Fogle, J. Restorff, J. Kosior, M. Staruch, J. Stace and A. Amin: "Non-Resonant Magnetoelectric Energy Harvesting Utilizing Phase Transformation in Relaxor Ferroelectric Single Crystals", *Actuators*, 5, 2; doi: 10.3390/act501000, 2016.

8 Servo Displacement Transducer Applications

Piezoelectric/electrostrictive actuators are classified into three categories, based on the type of driving voltage applied to the device and the nature of the strain induced by the applied voltage: (1) servo displacement transducers, (2) pulse drive devices operated in a simple on/off switching mode, and (3) ultrasonic resonant devices. The fields of application of ceramic actuators are also classified into three categories: (1) positioning, (2) motor application, and (3) vibration suppression.

This chapter will focus on servo displacement transducers and applications. These devices are used for positioning and vibration suppression in optical control systems (such as those implementing deformable mirrors) and mechanical systems implementing such mechanical devices as microscope stages, linear motion guide mechanisms, and oil pressure servo valves. Many of the devices for these applications are made from lead magnesium niobate-based electrostrictive ceramics because the hysteresis in their strain response is small compared with that exhibited by normal piezoelectric materials.

8.1 DEFORMABLE MIRRORS

8.1.1 Monolithic Piezoelectric Deformable Mirror

Precise wave front control with as small a number of parameters as possible and compact construction is a common and basic requirement for *adaptive optical systems*. Continuous surface deformable mirrors, for example, tend to be more desirable than segmented mirrors in terms of controllability. The monolithic piezoelectric deformable mirror pictured in Figure 8.1a has been produced and is composed of a PZT bulk ceramic (the ceramic monolith) with electrodes arranged in a two-dimensional configuration.[1] Applying suitable voltages to the electrodes causes the mirror to deform, as if pushing the mirror by rods from the back side. One application of this mirror is demonstrated in Figure 8.1b in which the phase of the reflected light is effectively modulated by adjusting the contour of the mirror surface. A deformable mirror can also be used to correct the distortion that occurs in a telescope image due to atmospheric conditions, as illustrated in Figure 8.2. Various electrode configurations have been proposed to produce a variety of mirror contours.[2]

8.1.2 Multimorph Deformable Mirror

The monolithic piezoelectric deformable mirror requires many electrodes (350 elements) and the same number of electrical leads to the individual electrode elements. Furthermore, since each element does not function independently, a minicomputer is required for its control. A much simpler multimorph deformable mirror has been proposed, which can be more simply controlled by means of a microcomputer.[3,4]

In the case of the *two-dimensional multimorph deflector design*, a static deflection along the z-axis, $f_i(x, y)$, is generated by the voltage distribution on the ith layer, $V_i(x, y)$, which can be obtained by solving the following system of differential equations:

$$A\left[\frac{\partial^4 f_i(x,y)}{\partial x^4}+2\frac{\partial^4 f_i(x,y)}{\partial x^2 \partial y^2}+\frac{\partial^4 f_i(x,y)}{\partial y^4}\right]+B_i\left[d_{31}\frac{\partial^2 V_i(x,y)}{\partial x^2}+d_{32}\frac{\partial^2 V_i(x,y)}{\partial y^2}\right]=0 \tag{8.1}$$

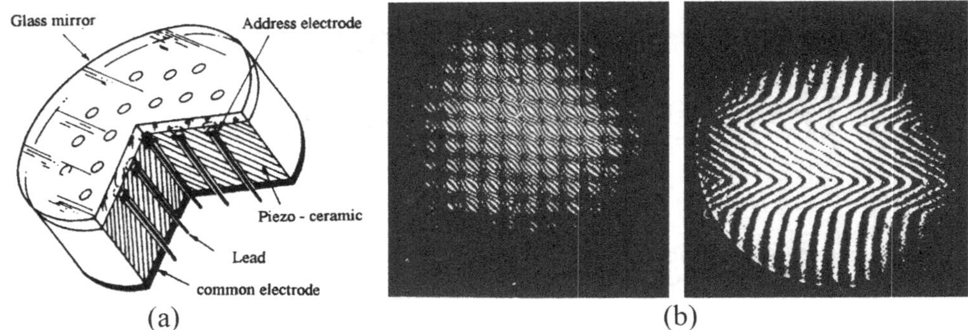

FIGURE 8.1 (a) A deformable mirror made from a piezoelectric monolith. (b) Application of the monolithic piezoelectric deformable mirror for phase modulation. (From J. W. Hardy et al.: *J. Opt. Soc. Amer.*, 67, 360, 1977.)

where $[i = 1, 2, \ldots.. I]$, A and B_i are constants, d_{31} and d_{32} are piezoelectric strain coefficients in a piezoelectric ($d_{31} = d_{32}$ in isotropic piezoelectric ceramics), and I is the total number of layers in the multimorph. This equation is derived specifically for piezoelectric layers, but may also be adapted to describe a device composed of electrostrictive layers, provided the nonlinear relationship between the electric field and the induced strain characteristic of electrostrictive materials is properly taken into account. When the conditions $d_{31} = d_{32}$ and $f_i(x, y) = 0$ under $V_i(x, y) = 0$ $(i = 1, 2, \ldots, I)$ are assumed, then Equation 8.1 reduces to

$$\nabla^2 f_i(x,y) + C_i V_i(x,y) = 0 \tag{8.2}$$

where C_i is a constant that is different for each layer and ∇^2 is the Laplacian operator. Thus, we see in general that the contour of the mirror surface can be changed by means of an appropriate electrode configuration and *applied voltage distribution* on each electroactive ceramic layer.

The contour of the mirror surface can be represented by the *Zernike aberration polynomials*, whereby an arbitrary surface contour modulation, *g(x, y)*, can be described as follows:

$$g(x,y) = C_r(x^2 + y^2) + C_c^1 x(x^2 + y^2) + C_c^2 y(x^2 + y^2) + \cdots \tag{8.3}$$

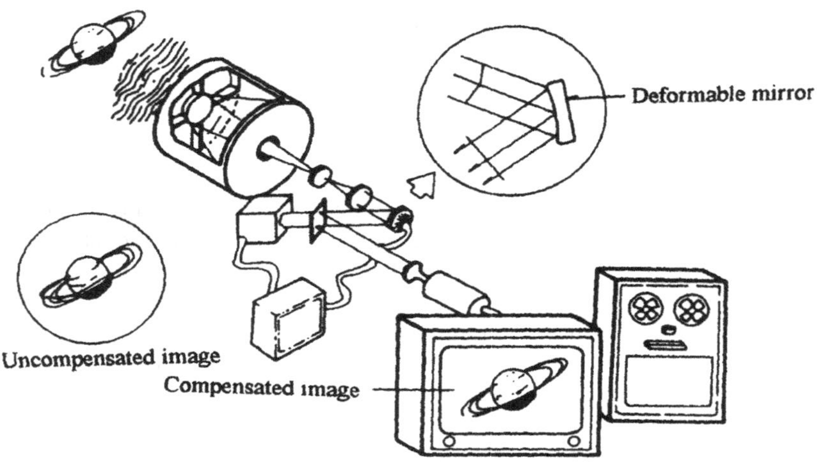

FIGURE 8.2 A telescope image correction system using a monolithic piezoelectric deformable mirror. (From J. W. Hardy et al.: *J. Opt. Soc. Amer.*, 67, 360, 1977.)

Notice that the Zernike polynomials are *orthogonal* and therefore completely independent of each other. The C_r and C_c terms represent *refocusing* and the *coma aberration*, respectively. As far as human vision is concerned, correction for aberrations, up to the second order in this series (representing astigmatism), is typically sought to provide an acceptably clear image. In more accurate computer imaging processes, high-order components are integrated.

The important parameters for designing the electrode configurations are as follows:

a. *The Effective Optical Area*: Typically, only the central portion of the mirror surface (amounting to about one-quarter of the total mirror surface) is used since the outer portions are difficult to deform when the boundary conditions for optimum operation are applied. The electrode covering the active area is divided into J identical elements (an 8×8 array in our study).
b. *The Deflection of Each Unit*: The deflection of a particular unit with the application of unit voltage is given by

$$u_{ij}(x,y) = \frac{16C_i}{\pi^4}\sum_{p,q}\left[\frac{1}{pq(P^2+Q^2)}\right]\Pi \tag{8.4}$$

   where:

$$\Pi \equiv \sin(P\pi\xi_j)\sin(Q\pi\eta_j)\sin(P\pi u/2)\sin(Q\pi v/2)\sin(P\pi x)\sin(Q\pi y)$$

$$P = (p/l_x) \text{ and } Q = (q/l_y)$$

   and the coordinates (ξ_j, η_j) identify the center of the jth electrode element. All elements have the same area ($\boldsymbol{u} \times \boldsymbol{v}$). The suffix i refers to the ith layer.
c. *The Voltage Distribution*: The level of voltage V_{ij}, $j = 1, 2, \ldots, J$ applied to each element to generate the deflections $f_i(x, y)$ is determined such that the mean-square error between the generated and the required deflections over the effective optical area is minimized.
d. *Electrode Shape*: Certain groups of elements are addressed with the same voltage to effectively establish a desired electrode shape and associated mirror contour.

As examples, let us consider the cases where we wish to refocus and to correct for the coma aberration. A uniform large-area electrode can provide a parabolic (or spherical) deformation with the desired focal length. In the case of coma correction, one could employ the electrode pattern shown in Figure 8.3, which consists of only six elements addressed by voltages applied in a fixed ratio. This study may be one of the first trials of the finite-element method on piezoelectric actuator developments.

Measurement of the deflection of a deformable mirror has been carried out using the holographic interferometric system pictured schematically in Figure 8.4.[3] The presence of the *hologram* in the system effectively eliminates effects of the initial deformation (due to manufacturing process) of the mirror from the final interferogram. Some experimental results characterizing mirror deformations established for the purpose of refocusing, coma correction, and both refocusing and coma correction are shown in Figure 8.5. Good agreement between the desired and generated contours is seen in each case. These results also demonstrate the effectiveness of achieving a superposition of deformations by means of an appropriate configuration of discrete electrodes. The deformable mirror examined in this study was found to have a linear response to sinusoidal input voltages with frequencies up to 500 Hz.

Interferograms from deformable mirrors incorporating PZT piezoelectric and PMN electrostrictive elements under a series of applied voltages are shown in Figure 8.6.[3] A distinct hysteresis for the PZT mirror is apparent, while no significant hysteretic effect is observed for the PMN mirror.

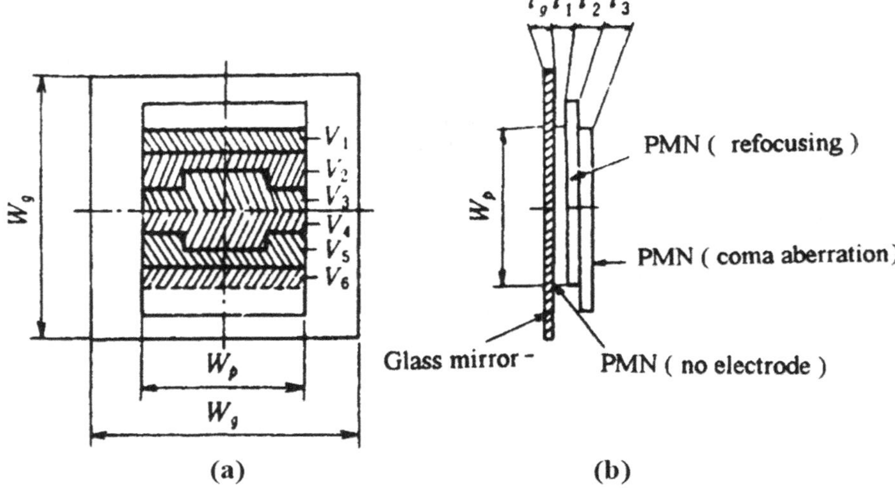

FIGURE 8.3 Two-dimensional multimorph deformable mirror designed for refocusing and coma aberration correction: (a) front view and (b) a cross-sectional view of the structure. (From T. Sato et al.: *Appl. Optics*, 21, 3669, 1982.)

8.1.3 Articulating Fold Mirror

An active mirror for use in space imaging systems called the articulating fold mirror has been developed.[5] The system is depicted schematically in Figure 8.7. Three articulating fold mirrors were incorporated into the optical train of the Jet Propulsion Laboratory's Wide Field and Planetary Camera-2, which was installed into the Hubble Space Telescope in 1993. Each articulating fold mirror utilizes six PMN electrostrictive multilayer actuators to precisely tip and tilt a mirror in

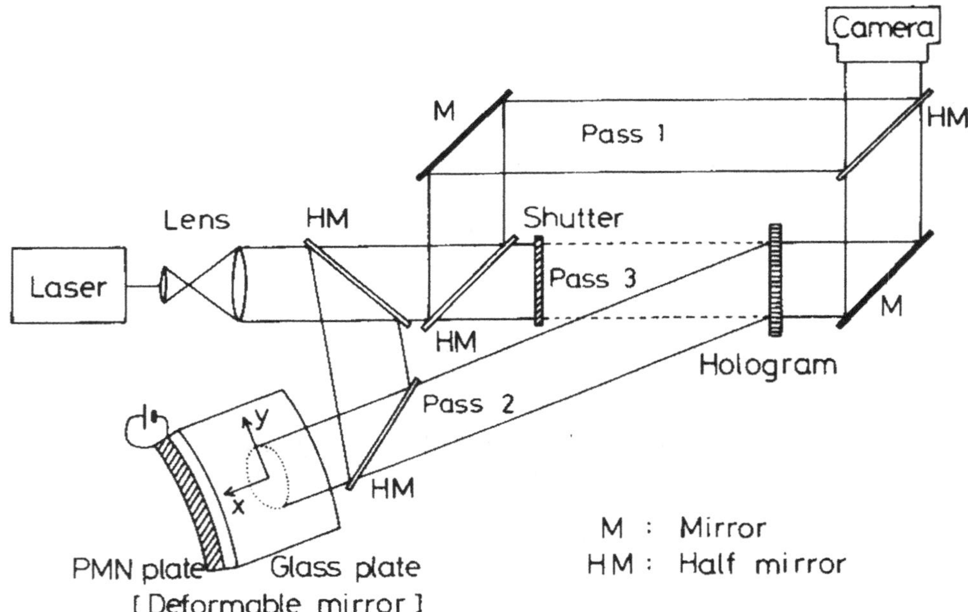

FIGURE 8.4 Optical system for the deformable mirror. (From K. Uchino et al.: *Appl. Optics*, 20, 3077, 1981.)

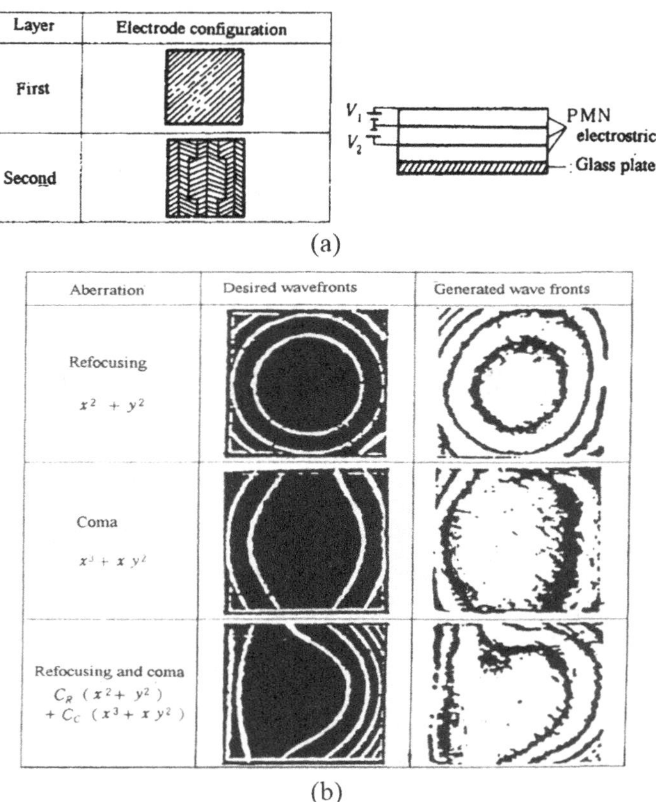

FIGURE 8.5 PMN-based multilayer deformable mirror: (a) diagrams of internal electrode configurations and the multilayer structure and (b) interferograms revealing the surface contours produced on the mirror. (From K. Uchino et al.: *Appl. Optics*, 20, 3077, 1981.)

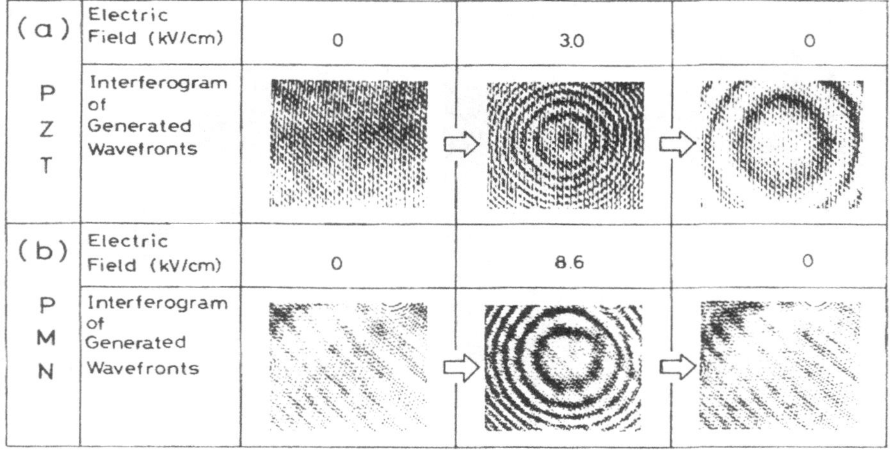

FIGURE 8.6 Interferograms from: (a) PZT and (b) PMN deformable mirrors. (From K. Uchino et al.: *Appl. Optics*, 20, 3077, 1981.)

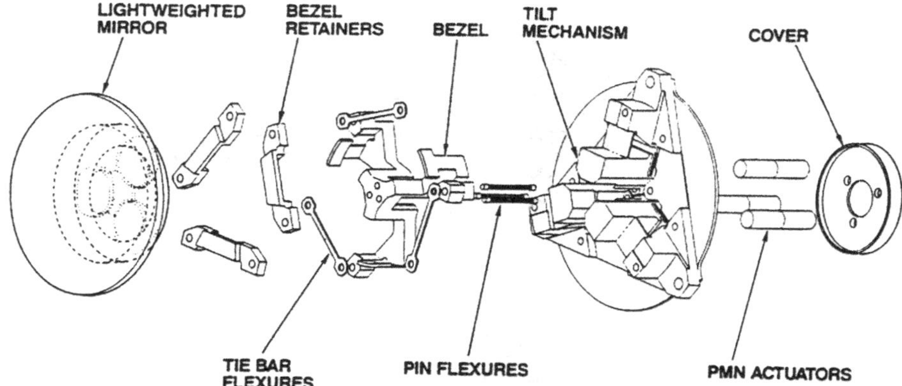

FIGURE 8.7 Schematic diagram of an articulating fold mirror utilizing PMN actuators. (From J. L. Fanson and M. A. *Ealey: Active and Adaptive Optical Components and Systems II, SPIE 1920*, Albuquerque, 1993.)

order to correct for the aberration in the Hubble telescope's primary mirror. Images of the core of M100, a spiral galaxy in the Virgo cluster, before and after the correction of the aberration appear in Figure 8.8a and b, respectively. The corrections have helped to restore the Hubble system to its original imaging specifications.

8.2 CAMERA LENS CONTROL

Optical microscopes exhibit very shallow "depth of focus"; that is, even for observing a object surface contour in a couple of mm, we need to readjust the lens position to get the best focus. In order to achieve the best focal point, there are two methods: (1) shifting the lens position or (2) deforming the lens position. In the former, we need to shift the objective lens position approximately 3 mm for shifting the focal plane by 4 mm, as is illustrated in Figure 8.9a, while in the latter, deforming the convex lens shape by only 25 μm is sufficient to achieve a 4-mm focal plane shift.

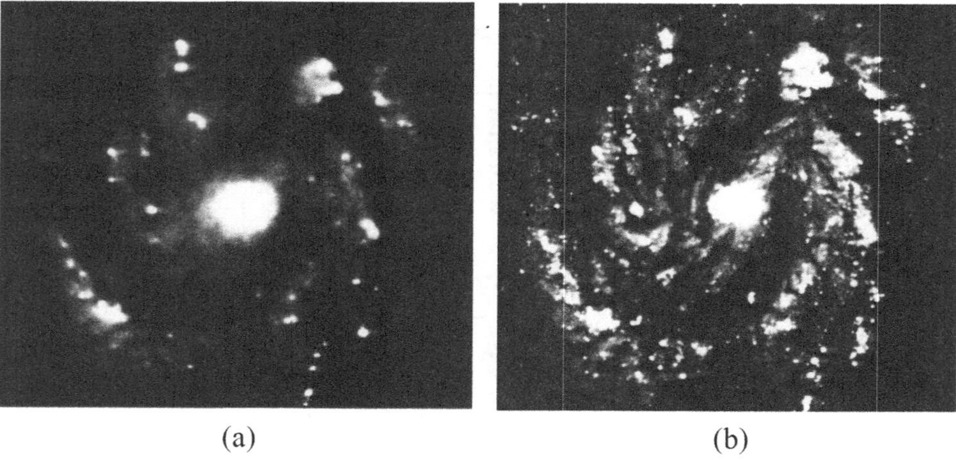

FIGURE 8.8 Hubble space telescope images of the M100 galaxy: (a) before and (b) after installation of articulating fold mirrors to correct for aberration in the system. (From J. L. Fanson and M. A. *Ealey: Active and Adaptive Optical Components and Systems II, SPIE 1920*, Albuquerque, 1993.)

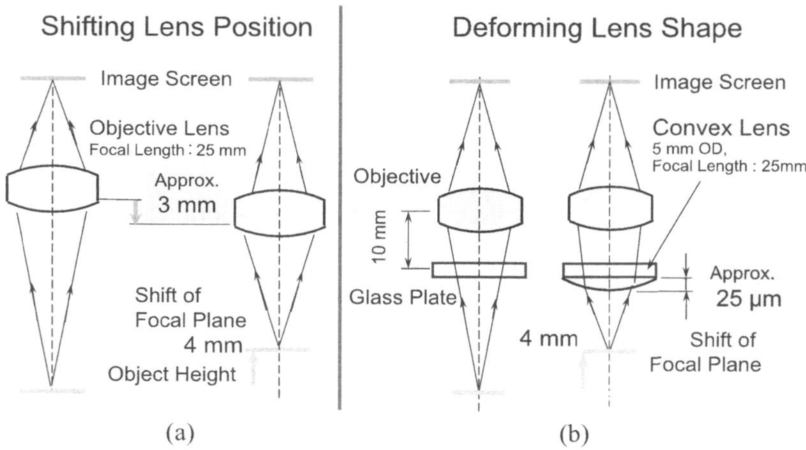

FIGURE 8.9 Two methods for changing the focal plane: (a) shifting lens position, and (b) deforming lens shape.

8.2.1 Helimorph

One Limited, U.K., developed a sophisticated helical and circular structure bimorph (Helimorph™), which is composed of a long and narrow-width bimorph wound helically, as shown in Figure 8.10a.[6] The circular diameter is 5–10 mm. By applying a reasonable voltage (5–10 V), the tip of the Helimorph moves up/down by a couple of sub-mm, which is sufficient for the lens position shift for focusing (not enough for "zoom"). As illustrated in Figure 8.10b, a lens is assembled in the center of the Helimorph capsule to ensure precise parallel up-down motion.

8.2.2 Deformable Lens

Denso Corporation, Japan, commercialized deformable lens systems for factory surveillance purposes. The operation manager in an assembly factory, say, automobile, needs to survey a 200-m-long assembly line at a glance on the TV monitor. If we need to manually change the focus during this surveillance process, it is rather time consuming. Denso demonstrated a deformable lens mechanism[7] for a microscope "all-distance-focus" application. Figure 8.11 shows a picture of a deformable lens device (a) and the lens-deforming principle with PZT bimorph actuators (b).[7] The deformation of the glass diaphragms with silicon oil optical medium is generated by multiple-layered PZT bimorph

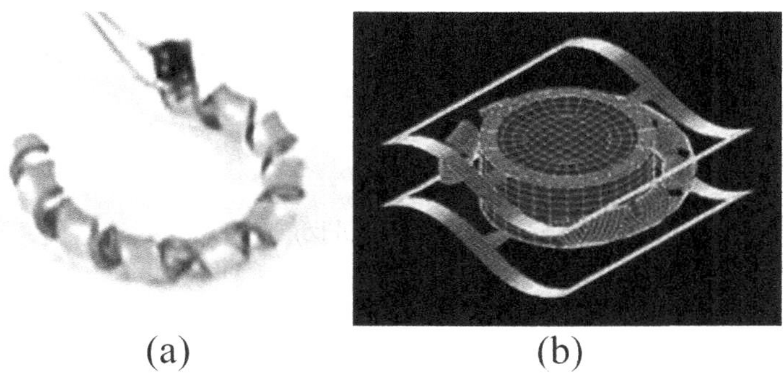

FIGURE 8.10 (a) Helimorph component (One Limited), and (b) lens assembly (up-down motion) with the Helimorph.

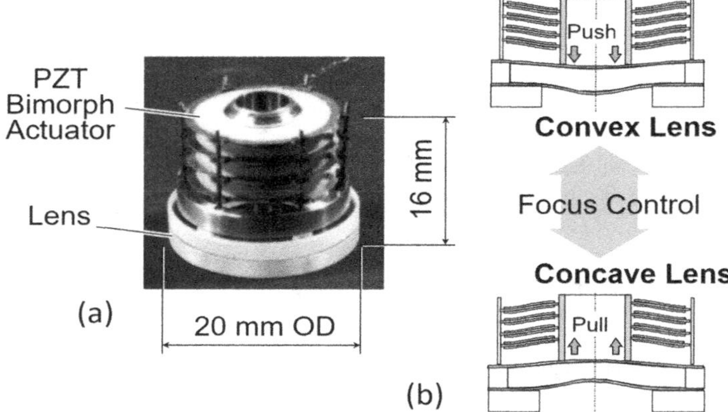

FIGURE 8.11 (a) Picture of a deformable lens device; (b) lens deforming principle with PZT bimorph actuators. (From T. Kaneko et al.: *Proc. of 35th ICAT Smart Actuator Symp., State College, PA*, April 18–19, 2002.)

actuators (Figure 8.11b). A multiple array is required to realize a large and quick enough stroke. The transparent fluid-embedded diaphragm behaves as a deformable lens by deforming into convex and concave shapes, according to the force direction. Figure 8.12a shows the principle of the "all-distance-focus" mechanism, where the deformable lens is operated under 60 Hz between the top-surface and bottom-surface focus points, quickly enough for a human to recognize as a "all-distance-focus" picture. Figure 8.12b demonstrates the deformable lens function in an optical microscope, which exhibits (a) low-point and (b) high-point focus. The principle and feasibility of the deformable lens are already confirmed. However, usage of a fluid in the cell phone presents a problem, namely the cellular phone's functioning temperature range (in high and freezing low temperatures).

The Institute of Materials Research and Engineering (IMRE, now A*STAR), Singapore, also fabricated "liquid lens."[8] Their design and the appearance are shown in Figure 8.13, where pressure

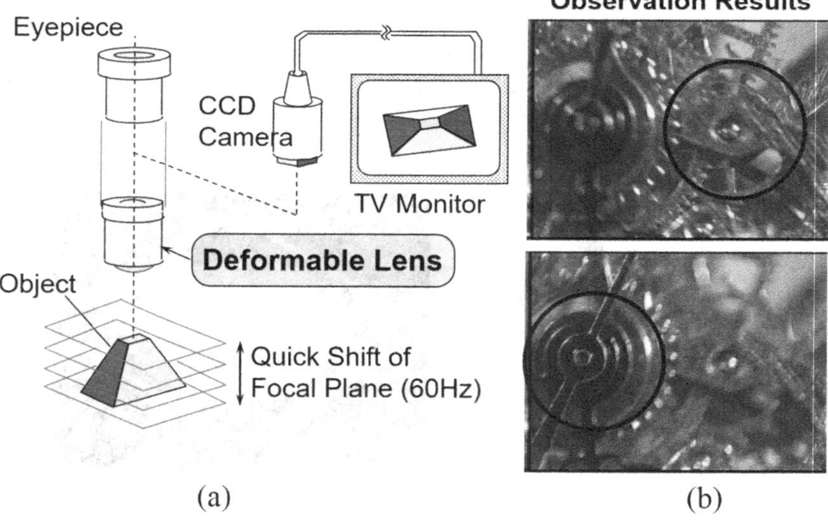

FIGURE 8.12 (a) Principle of the "all-distance-focus" mechanism; (b) Observation results, bottom surface-focus (top figure) and top-focus (bottom).

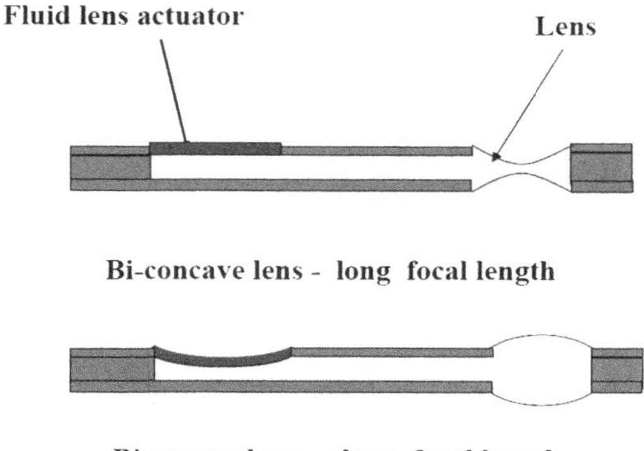

FIGURE 8.13 Basic design of the "liquid lens" developed by IMRE. (From Cover article: *Perspectives*, 7(3), 3rd Quarter, 2005.)

(by a finger) can change the liquid pressure, leading to the lens shape change. If we adopt a flex-tensional piezo-actuator (such as Cymbal and Bimorph), this mechanism becomes electrically driven.

8.3 MICROSCOPE STAGES

An adjustable sample stage for optical and electron microscopes has been developed that utilizes a monolithic hinge lever mechanism.[9] A schematic depiction of the stage and its components appears in Figure 8.14a. The displacement induced in the PZT multilayer actuator under an applied voltage of 1 kV is amplified by a factor of 30 by means of a two-stage lever mechanism, leading to an actual displacement of 30 μm at the center of the stage. The sample stage must be compact, especially when it is to be used in the electron microscope, and able to withstand vacuum conditions of about 10^{-9} mm Hg. In order to meet the latter condition, a suitable nonvolatile bonding resin should be used between the layers of the multilayer actuator, and electrode materials such as zinc and cadmium, which are highly volatile under these conditions, must be avoided.

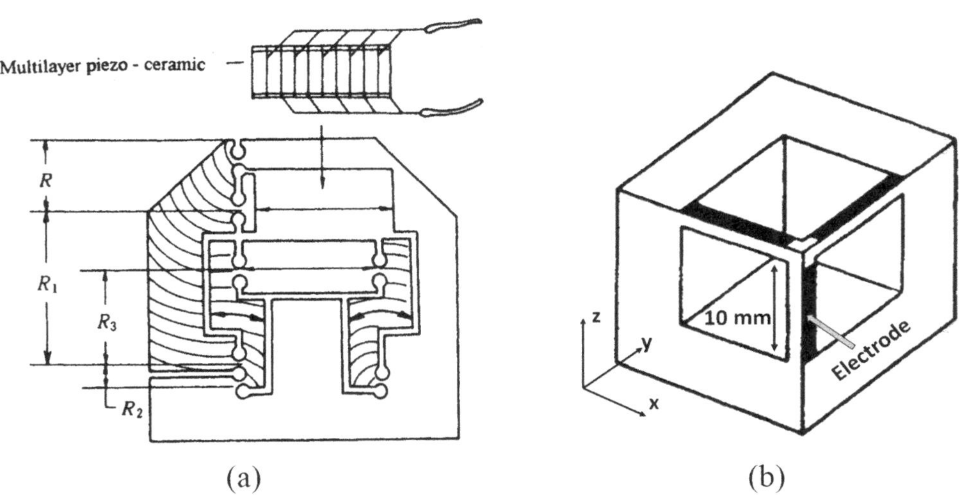

FIGURE 8.14 (a) Adjustable microscope stage using a monolithic hinge lever mechanism. (From K. Uchino: *Proc. Actuator 2008*, A5.0, June 9–11, Bremen, Germany; http://www.onelimited.com/ (as of March, 2008).) (b) Probe control tripod-type actuator for a scanning tunneling microscope. (From T. Kaneko et al.: *Proc. of 35th ICAT Smart Actuator Symp., State College, PA*, April 18–19, 2002.)

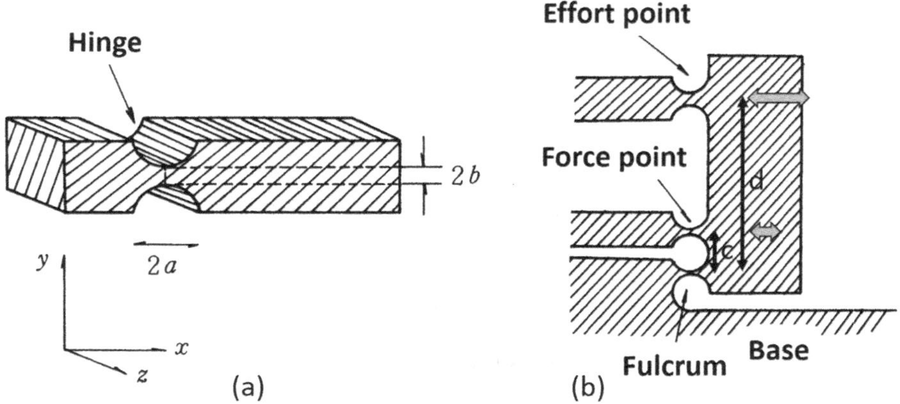

FIGURE 8.15 (a) A magnified structure at the hinge part; and (b) a lever principle schematic.

A probe scanning actuator for a scanning tunneling microscope (STM) has been proposed that makes use of a PZT tripod structure, as illustrated in Figure 8.14b.[10] One primary limitation with this design is the displacement hysteresis of the probe actuator, which directly affects the reproducibility and resolution of the STM image. A more reliable PMN-based probe actuator design that is virtually hysteresis free has been developed to overcome this limitation.[11]

EXAMPLE PROBLEM 8.1

One of the important parts in micro stages and MEMS structures is a *monolithic hinge lever mechanism* to amplify small displacement generated in a piezoelectric or other smart material. Figure 8.15 illustrates a magnified structure at the hinge part (a) and a lever principle schematic (b). Discuss the design principles of the mechanism in terms of the dimensions, a, b, c, and d.

Solution

When we use the lever mechanism as a displacement magnification mechanism of a multilayer piezo-actuator, the actuator is installed at the force point in Figure 8.15b or in Figure 8.14a. In this case, the effort point moves in proportion to the distance ratio *d/c*, according to the lever rule, when the hinge is ideally thin enough. However, with reducing the thickness *b* of the hinge part, longitudinal elastic strength dramatically decreases, and this lever rule is not maintained. From our experiences on the dynamic response, we recommend adjusting the hinge thickness to *2b* as the thickness at which the actual displacement amplification factor becomes half of the apparent amplification factor (*d/c*) (unpublished). This condition satisfies both generative force and response speed reasonably without sacrificing displacement significantly.

Figure 8.16 shows mechanical amplifier design improvement during dot-matrix printer development in the 1980s (Private communication from Dr. Takeshi Yano, former NEC General Manager), for the reader's reference: (a) one-stage amplifier, (b) two-stage amplifier, and (c) two-stage differential motion mechanical amplifier. Dynamic mechanical energy transfer efficiency from the actuator to the printing wire changed significantly from $<10\%$ (a), to $<25\%$ (b), and finally $>60\%$. The FEM computer software code is very helpful for modifying mechanical amplifier configurations (weight and the curvature of the arms are the key).

8.4 HIGH-PRECISION LINEAR MOTION DEVICES

Recent developments in precision diamond cutting and polishing have been remarkable. The high degree of precision for the flatness required of laser interferometric mirrors to adequately satisfy the

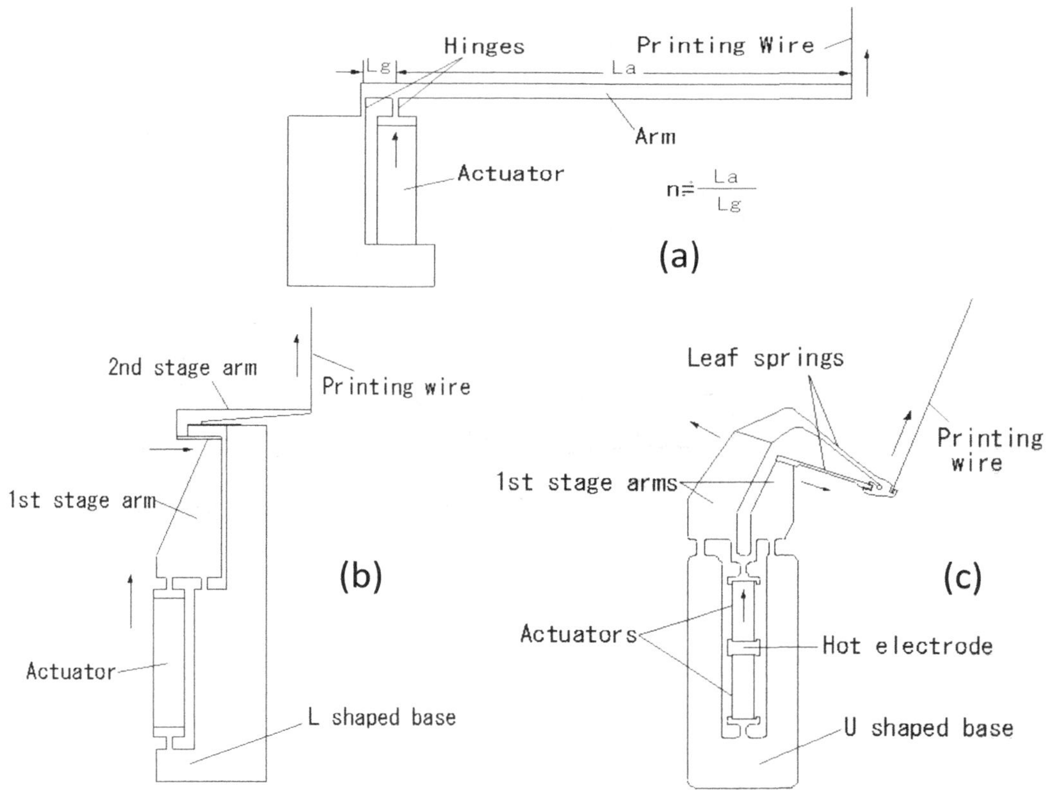

FIGURE 8.16 Mechanical amplifier improvement: (a) one-stage amplifier, (b) two-stage amplifier, and (c) two-stage differential motion mechanical amplifier.

requirements for most applications to date has been achieved with these machines. The requirements for newly developed applications, however, indicate an even higher precision in the flatness of the laser mirror surfaces (on the order of 0.05 μm), and thus more sophisticated precision linear motion guide mechanisms must be employed on the cutting apparatus. Conventional guide mechanisms, such as air bearings and roller-type bearings, produce residual noise vibration (ringing) due to the low friction contact between surfaces, which results in a slow response. An ultrahigh-precision linear guide system that makes use of PZT actuators has been developed in Hitachi Central Research Laboratories, Japan, with a linear motion tolerance of ± 0.06 μm/200 mm.[12] A similar mechanism can be employed to produce an ultraprecise *x-y* positioning stage.[13]

8.4.1 Ultrahigh-Precision Linear Motion Guide Mechanism

A block diagram of the linear guide mechanism is shown in Figure 8.17. The linear motion is driven by the transfer screw mechanism pictured in the lower left-hand side of the figure. As the motion occurs, the transverse motion and the pitching and rolling of the moving table relative to the master scale (made of fused quartz) installed on the base is monitored by the three differential transformers installed in the bottom of the movable table, as shown in the figure. The three PZT positioners situated at the edges of the table (two of which are shown in the figure) respond to signals processed by the feedback circuit so as to maintain a constant position.

The key component of this mechanism is the feedback circuit. The flatness of the master scale is initially measured as a function of position by a position encoder, and the data are recorded as

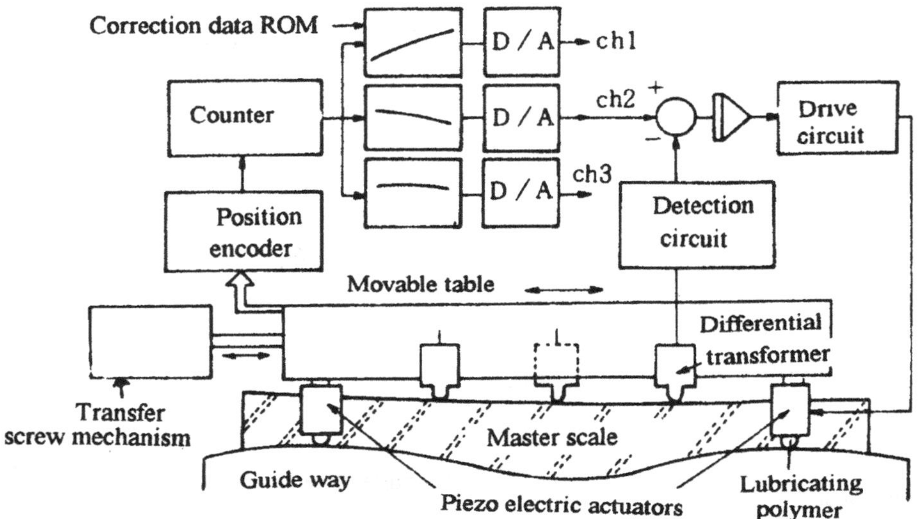

FIGURE 8.17 Block diagram of the ultrahigh-precision linear motion guide system by Hitachi. (From S. Moriyama and F. Uchida: *Proc. Jpn. Procision Eng. Soc.*, 201, 61(Spring), 1984.)

correction data in the *read-only memory* (ROM) of the system. The table position at a given time is measured and the correction signal corresponding to that particular table position is combined with the differential transformer signal. The resulting signal is processed and then fed back to the PZT actuators. This process is effective in producing a precision in the linear motion that actually exceeds that of the master scale. The flatness error of the master scale is depicted in Figure 8.18. The correction data collected by the three differential transformers are stored in the ROM of channels 1–3 as reference data for correction by the PZT actuators. Based on the correction data processed by the system, the master scale exhibits some "concavity," with a flatness error amounting to a 0.2-μm deviation. Even though we pay a high cost for expert polishing, this concavity cannot be avoided due to the flattening polish process.

Some additional features of this system are depicted in Figure 8.19. The movable table is guided on a slider medium, which for this system is a polymer lubricant, as shown in Figure 8.19a. The linear motion of the table is typically accompanied by a ± 2-μm transverse displacement. The linear drive system is composed of an electromagnetic rotary motor and a transfer screw mechanism. The PZT actuators installed between the moving table and the polymer slider have a cylindrical configuration (see Figure 8.19a) and can produce a displacement up to ± 3 μm depending on the magnitude of the applied voltage. The three differential transformers are installed beneath the table, as shown in

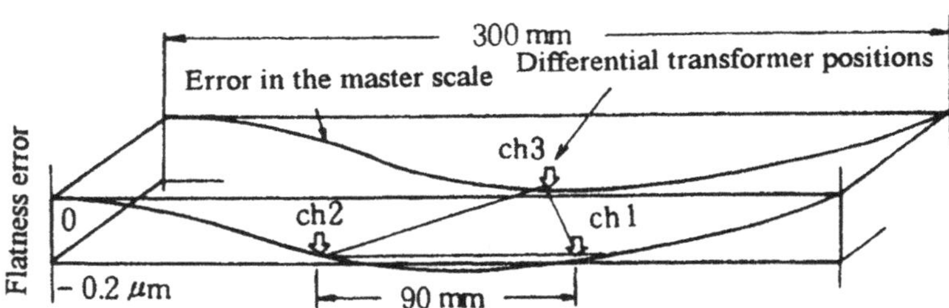

FIGURE 8.18 A depiction of the flatness error of the master scale for the ultrahigh-precision linear motion guide system. (From S. Moriyama and F. Uchida: *Proc. Jpn. Procision Eng. Soc.*, 201, 61(Spring), 1984.)

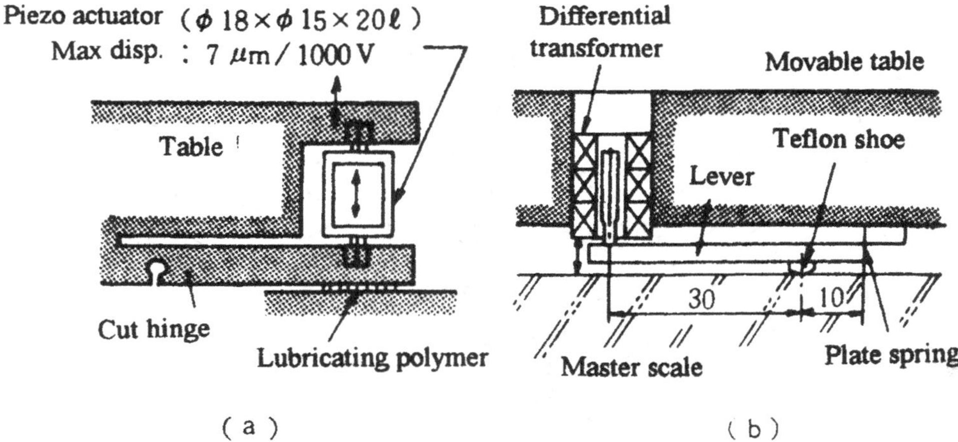

FIGURE 8.19 Details of the ultrahigh-precision linear motion guide system: placement of (a) a piezoelectric actuator and (b) a differential transformer in the system. (From S. Moriyama and F. Uchida: *Proc. Jpn. Procision Eng. Soc.*, 201, 61(Spring), 1984.)

Figure 8.19b. The transverse motion is monitored after being amplified by a factor of 4 by means of the lever mechanism.

The data shown in Figure 8.20 indicate the effectiveness of this high-precision linear guide system in maintaining a linear translation with a significantly reduced transverse displacement. The data appearing in Figure 8.20a represent the motion of the table without the servo correction. The $\pm$ 2-μm up-down motion is apparent in these data, which corresponds to the roughness of the guide way. The data in Figure 8.20b represent the motion of the table when the servo mechanism is coupled with the master scale. These data indicate that up-down displacement has been reduced to $\pm$ 0.15 μm, which corresponds to the surface roughness of the expensive master scale (see Figure 8.18). The data in

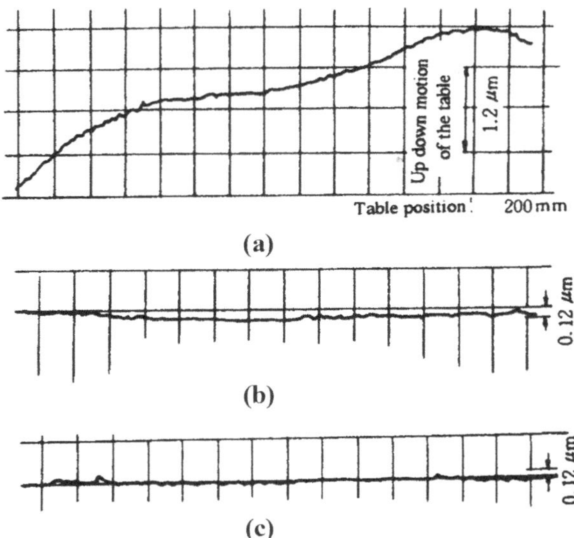

FIGURE 8.20 Reduction of the up-down displacement in the ultrahigh-precision linear guide: (a) without the servo mechanism, (b) with the servo mechanism (master-scale), and (c) with the servo mechanism and scale error correction applied. (From S. Moriyama and F. Uchida: *Proc. Jpn. Procision Eng. Soc.*, 201, 61(Spring), 1984.)

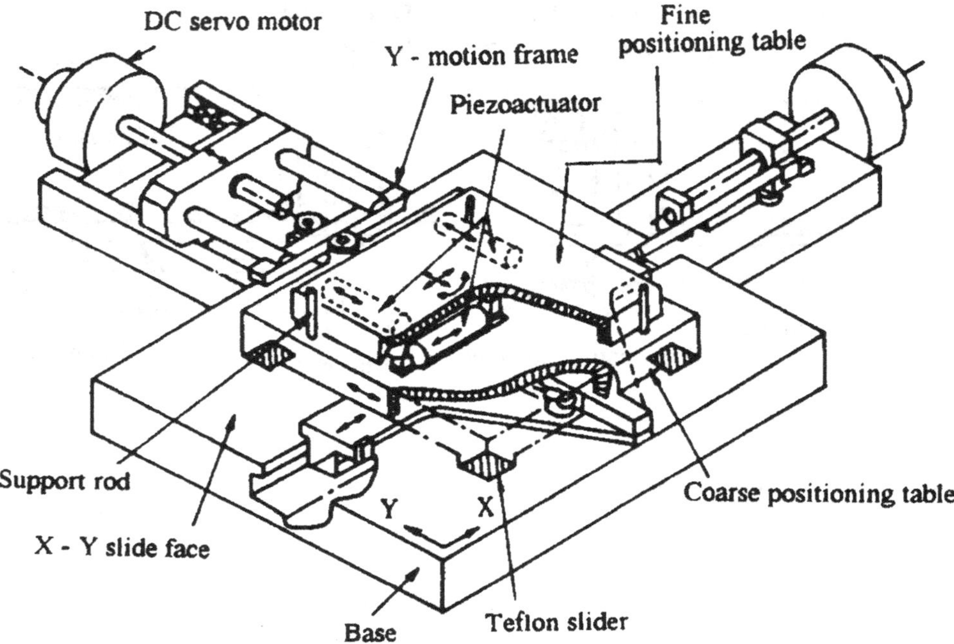

FIGURE 8.21 Ultrahigh-precision *x-y* positioning stage. (From S. Moriyama et al.: *Precision Machine*, 50, 718, 1984.)

Figure 8.20c represent the motion of the table when the servo mechanism is employed with additional correction of the master scale (reference) data. This results in an up-down displacement that has been further reduced to about ± 0.06 μm. Although we have focused here on the transverse displacement correction, it should be noted that this system is also effective in reducing the pitching motion from ± 0.8 second without the servo to ± 0.1 second when the feedback mechanism is in operation. It is important to note that the final displacement error can be made much less than the master scale error by using the additional correction via ROM data of the scale error in the feedback system. We do not need to spend much to purchase a very precise master scale, in other words.

8.4.2 Ultraprecise *x-y* Stage

An ultraprecise *x-y* positioning stage based on a similar mechanism has also been developed by Moriyama et al.[13] A schematic representation of the stage appears in Figure 8.21. Highly precise positioning, on the order of ± 0.05 μm accompanied by only a 1 μrad yaw, is achieved with this stage for manipulations over a relatively broad area of 120 × 120 mm. Moreover, movement of this stage can be as rapid as 10 mm/200 ms. The use of ultra-precise *x-y* stages such as this has become increasingly important in the photolithographic processing of very-large-scale integration (VLSI) semiconductor chips and micro-electro-mechanical systems.

8.5 HYDRAULIC SERVO VALVES

A classification of hydraulic servo systems with respect to power and response is shown in Figure 8.22.[14] We see from this layout that a system employing a combination of electric and oil pressure control (i.e., an electrohydraulic system) is required to achieve both high power and quick response. An even quicker response, on the order of 1 kHz, is desired for the electrohydraulic system. Though this is actually the maximum and generally least attainable speed, as indicated in Figure

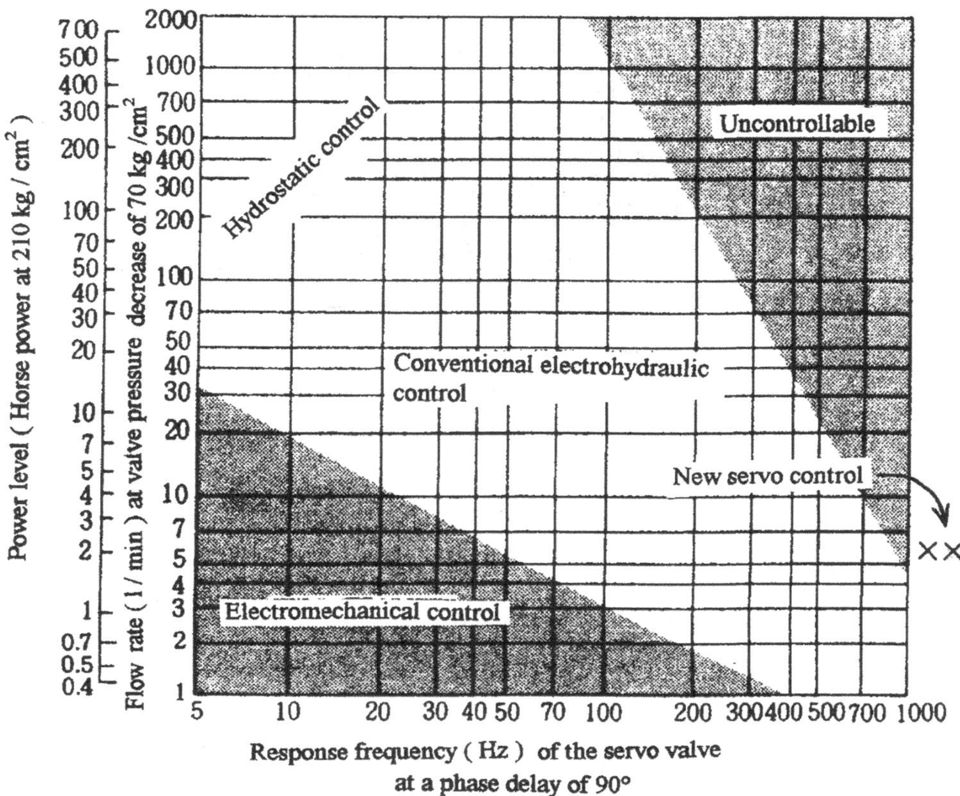

FIGURE 8.22 Classification of oil pressure servo valves with respect to power and response. (From H. Ohuchi et al.: *Proc. Fluid Control & Measurement*, Pergamon Press, Tokyo, p. 415, 1985.)

8.22, it has yet to be realized for any of the most commonly employed systems featured in this classification. Among the more recent developments for enhancing the performance of devices in this class is a simplified valve structure that makes use of a piezoelectric PZT flapper.[14,15] However, this design proves to be limited in its usefulness, due to the hysteretic response of the PZT. In order to effectively compensate for the hysteresis, the device is controlled by a *pulse width modulated drive* voltage, which causes the flapper to constantly vibrate in sympathy with the carrier signal, thus severely limiting the high-frequency response and the durability of the servo valve. A modification of this basic design, incorporating an electrostrictive PMN bimorph for the flapper instead of the hysteretic piezoelectric PZT, has been developed by our group that effectively overcomes these limitations.[16,17]

8.5.1 Oil Pressure Servo Valves with Ceramic Actuators

A schematic depiction of a two-stage four-way valve appears in Figure 8.23.[17] The first stage of the valve contains a PMN-based electrostrictive flapper, aiming at less displacement hysteresis, in comparison with piezoelectric actuators. The second-stage spool is the smallest currently available, with a diameter of only 4 mm and a nominal flow rate of 6 liter/min.

The electrostrictive composition (0.45)PMN-(0.36)PT-(0.19)BZN (PMN: lead magnesium niobate, PT: lead titanate, BZN: barium zinc niobate) is used because of its large displacement and small hysteresis. The flapper has a multimorph structure, in which two PMN thin plates are bonded on both

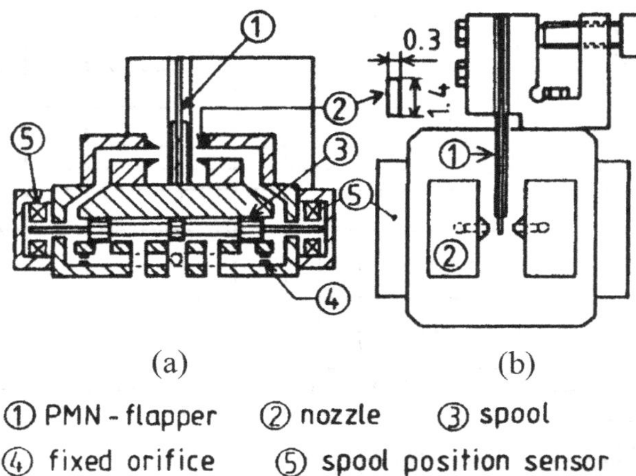

FIGURE 8.23 A schematic depiction of a two-stage four-way valve: (a) front view and (b) top view. (From H. Ohuchi et al.: *Proc. Fluid Control & Measurement*, Pergamon Press, Tokyo, p. 415, 1985.)

sides of a phosphor bronze shim, as shown in Figure 8.24a. The *multimorph* structure is used for its enhanced tip displacement, generative force, and response speed. The structure is addressed such that the top and bottom electrodes and the metal shim have a common ground potential, as shown in the figure, and a high voltage (designated by V_1 and V_2 in the figure) is applied on the electrodes between the two PMN plates on each side of the shim. The tip deflection exhibits a quadratic dependence on the applied voltage, as would be expected for an electrostrictive response. Thus, in order to obtain a linear relation, a *push-pull driving* method is adopted, in which the applied voltages are controlled such that

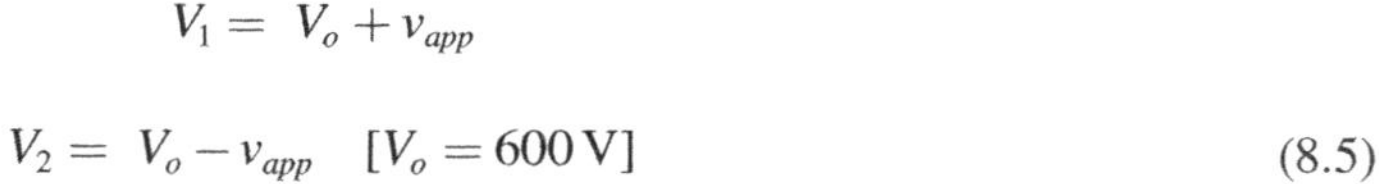

$$V_1 = V_o + v_{app}$$

$$V_2 = V_o - v_{app} \quad [V_o = 600\,\mathrm{V}] \tag{8.5}$$

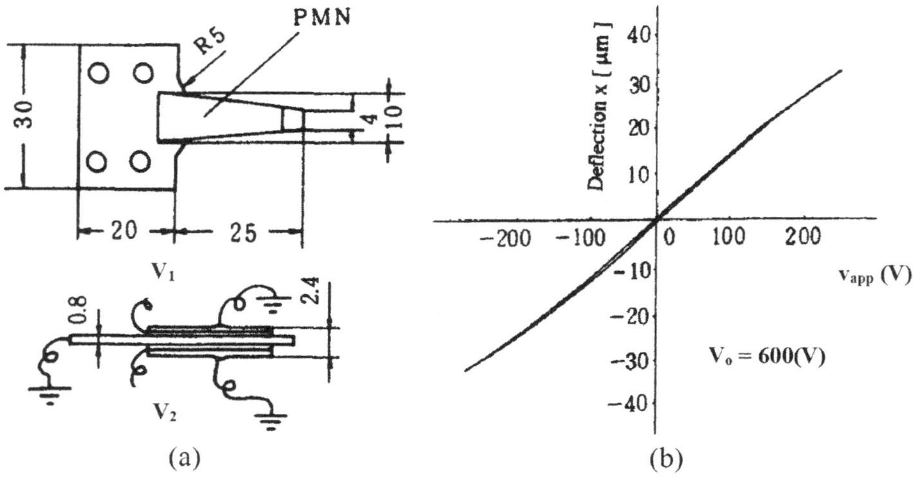

FIGURE 8.24 (a) Multimorph electrostrictive flapper. (b) Push-pull drive characteristics of the electrostrictive flapper. (From H. Ohuchi et al.: *Proc. Fluid Control & Measurement*, Pergamon Press, Tokyo, p. 415, 1985.)

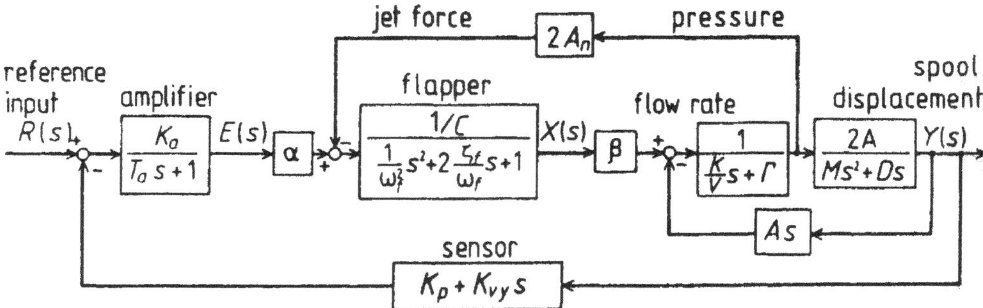

FIGURE 8.25 Block diagram of the oil pressure servo valve. (From H. Ohuchi et al.: *Proc. Oil and Air Pressure Soc.*, 3(Fall), 9, 1983; H. Ohuchi et al.: *Proc. Fluid Control & Measurement*, Pergamon Press, Tokyo, p. 415, 1985.)

where v_{app} is the applied signal voltage. A plot of v_{app} as a function of displacement x, as shown in Figure 8.24b, is seen to be nearly linear. Notice that the displacement hysteresis for the PMN-based ceramic is much smaller than what would occur for a PZT piezoelectric. The resonance frequency of this flapper in oil is about 2 kHz.

The conventional force feedback method makes use of a structure in which the flapper tip and the spool are connected by a spring, which limits the responsivity of the system. In order to overcome this limitation, an alternative electric feedback method, for which a signal associated with the spool position is used, has been developed.[16,17] A compact differential transformer (operated at 50 kHz) was employed to detect the spool position. This method has various merits, including a readily variable feedback gain and the option of a *speed feedback* mode of operation.

A block diagram of the oil pressure servo valve from reference input to spool displacement is shown in Figure 8.25. Several approximations are adopted for the frequency range below 1 kHz:

1. The power supply (amplifier) is represented by $[K_a/(T_a\, s + 1)]$ (i.e., integral amplifier).
2. The transfer function of the flapper has a second-order form (i.e., regular linear piezoelectric actuator).
3. The compressibility of the oil is neglected.
4. The visco-resistance between the spool and sleeve is also neglected.

It is assumed in the following analyses that the speed feedback mode of operation has not been employed. When the input-output relation for the system depicted in Figure 8.25 is expressed in a simple *second-order form*, the following transfer function is obtained:

$$\frac{Y(s)}{R(s)} = \left[\frac{1}{K_p}\right] \frac{1}{a_o s^2 + a_1 s + 1} \tag{8.6}$$

where

$$a_o = \frac{[1 + (2A_n\beta/C\Gamma)]M + 2[(2\zeta_f/\omega_f) + T_a](A^2/\Gamma)}{a_2}$$

$$a_1 = \frac{2(A^2/\Gamma)}{a_2}$$

$$a_2 = \frac{2A\beta\alpha K_a K_b}{C\Gamma}$$

EXAMPLE PROBLEM 8.2

Neglecting the higher-order terms and K_{vy} and D, derive Equation 8.6 from the block diagram shown in Figure 8.25.

Solution

There are four variables represented in Figure 8.25: $R(s)$, $E(s)$, $X(s)$, and $Y(s)$. We may write the following three equations for the three nodes indicated by open circles in the block diagram (left to right):

$$\left[\frac{T_a s+1}{K_a}\right]E(s) = R(s) - [K_p + K_{vy}s]Y(s) \tag{P8.2.1}$$

$$C\left[\frac{1}{\omega_f^2}s^2 + 2\frac{\zeta_f}{\omega_f}s+1\right]X(s) = \alpha E(s) - 2A_n\left[\frac{Ms^2+Ds}{2A}\right]Y(s) \tag{P8.2.2}$$

$$\left[\frac{K}{V}s+\Gamma\right]\left[\frac{Ms^2+Ds}{2A}\right]Y(s) = \beta X(s) - AsY(s) \tag{P8.2.3}$$

Solving Equation P8.2.1 for $E(s)$, substituting this expression into Equation P8.2.2, and solving for $X(s)$ yields:

$$X(s) = \frac{\left[\left(\frac{\alpha K_a}{T_a s+1}\right)R(s) - \left(\frac{\alpha K_a(K_p+K_{vy}s)}{T_a s+1}\right)Y(s) - \left(\frac{A_n(Ms^2+Ds)}{A}\right)Y(s)\right]}{C\left[\frac{1}{\omega_f^2}s^2 + \frac{2\zeta_F}{\omega_F}s+1\right]} \tag{P8.2.4}$$

Combining Equations P8.2.4 with P8.2.3 allows us to write:

$$\left[\frac{((K/v)s+\Gamma)(Ms^2+Ds)}{2A} + As\right]Y(s) = \beta X(s) \tag{P8.2.5}$$

where

$$\beta X(s) = \frac{\beta\left[\left(\frac{\alpha K_a}{T_a s+1}\right)R(s) - \left(\frac{\alpha K_a(K_p+K_{vy}s)}{T_a s+1}\right)Y(s) - \left(\frac{A_n(Ms^2+Ds)}{A}\right)Y(s)\right]}{C\left[\frac{1}{\omega_f^2}s^2 + \frac{2\zeta_F}{\omega_F}s+1\right]}$$

We can now obtain the inverse transfer function in the initial form:

$$\frac{R_s}{Y_s} = \Sigma(T_a s+1) + (K_p + K_{vy}s) \tag{P8.2.6}$$

where

$$\Sigma \equiv \left[\frac{C}{2\alpha\beta K_a A}\left(\frac{1}{\omega_f^2}s^2 + \frac{2\zeta_f}{\omega_f}s+1\right)\left[\left(\frac{K}{V}s+\Gamma\right)(Ms^2+Ds) + 2A^2s\right] + \frac{A_n(Ms^2+Ds)}{\alpha K_a A}\right]$$

Neglecting terms of order higher than s^2 results in the somewhat more simplified form of the inverse transverse function:

$$\frac{R(s)}{Y(s)} = K_p + sK_p\mathbf{\Phi} + s^2K_p\mathbf{\Psi} \tag{P8.2.7}$$

where

$$\mathbf{\Phi} \equiv \frac{K_{vy}}{K_p} + \frac{A_nD}{\alpha K_a A K_p} + \frac{C\Gamma}{2A\alpha\beta K_a K_p}\left(\frac{2A^2}{\Gamma} + D\right)$$

$$\mathbf{\Psi} \equiv [A_nM + A_nDT_a]\frac{1}{\alpha K_a A K_p} + \left[\frac{KD}{V} + \Gamma M\right]\frac{C}{2A\alpha\beta K_a K_p} + \left[\frac{C2\zeta_f}{\omega_f} + CT_a\right]\frac{(\Gamma D + 2A^2)}{2A\alpha\beta K_a K_p} + \frac{T_a A_n D}{\alpha K_a A K_p}$$

Neglecting the terms containing K_{vy} and D and further simplifying Equation P8.2.7 produces Equation 8.6.

The static spool displacement of the oil pressure servo valve utilizing the electronic feedback system is shown as a function of reference input in Figure 8.26a. The slight hysteresis and nonlinearity manifested in the response curve for the nonregulated valve shown in Figure 8.24b do not appear in the data recorded in Figure 8.26a due to the action of the feedback mechanism implemented in this system. The dynamic characteristic of the spool displacement is shown in Figure 8.26b. The 0 dB gain level was adjusted at 10 Hz. A slight peak in the gain curve occurs at 1800 Hz, and the 90° retardation frequency is about 1200 Hz. The maximum spool displacement at 1 kHz is ± 0.03 mm. In terms of the classification scheme appearing in Figure 8.22, this new servo system is placed in the position of the two xs indicating that it is the fastest servo valve currently available.

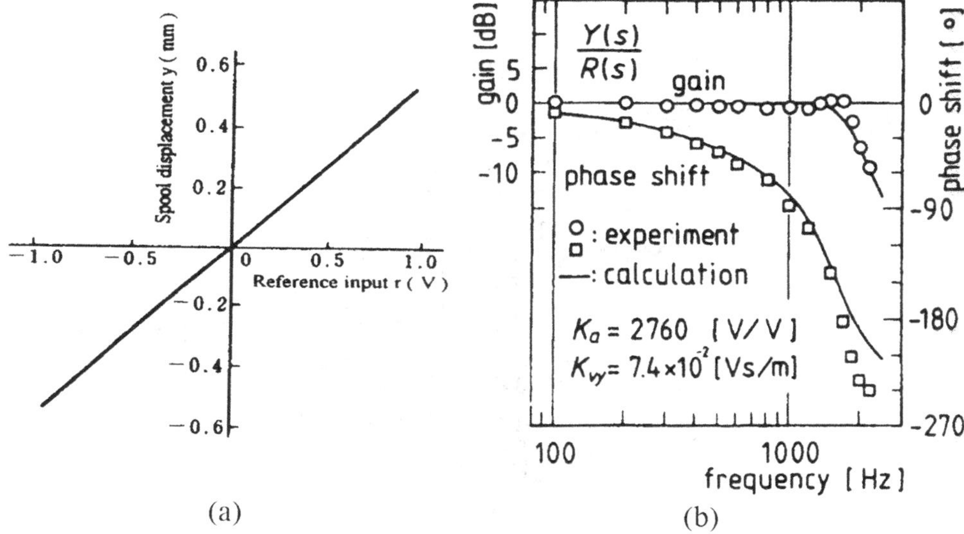

FIGURE 8.26 (a) Static spool displacement of the oil pressure servo valve utilizing the electronic feedback system plotted as a function of reference input. (b) Normalized frequency characteristics of the oil pressure servo valve utilizing the electronic feedback system.

8.5.2 Air Pressure Servo Valves

Similar nozzle-flapper mechanisms can be applied for air pressure control.[18] The main difference between the oil and air pressure control systems is the force applied to the flapper. The typical range of diameter for most nozzle-flapper mechanisms utilized in oil and air pressure control systems is $d_n = 0.3–0.8$ mm, and the maximum force on the flapper, f_{max}, is determined by the following:

$$f_{max} < (1/2)\pi d_n^2 p \tag{8.7}$$

where p is the nozzle pressure. The typical range of p for oil pressure systems is 70–210 kgf/cm^2, and for air pressure systems 3 kgf/cm^2. The nozzle-flapper distance is generally adjusted to less than $d_n/4$.

8.5.3 Direct Drive Spool Servo Valve

A direct drive spool servo valve has also been developed that uses two multilayer PMN-based actuators coupled with a sophisticated hinge lever mechanism, as shown in Figure 8.27.[17] This system exhibits an even quicker response with a 90° phase retardation frequency of about 1400 Hz.

8.6 VIBRATION AND NOISE SUPPRESSION SYSTEMS

8.6.1 Vibration Damping

The U.S. Army is interested in developing a rotor control system for helicopters. A bearingless rotor flexbeam with integrated piezoelectric strips is pictured in Figure 8.28a.[19] An active fiber composite ply has been designed to generate the twisting motion of the helicopter blades.[20] The fibers are aligned transversely, as shown in Figure 8.28b, in order to sense and actuate in-plane stresses and strains. Semicontinuous 130-μm-diameter PZT fibers are currently used. The matrix material is composed of B-staging epoxy and a fine PZT powder, which is used to increase the dielectric constant of the matrix so that a higher electric field can be established within the high-permittivity

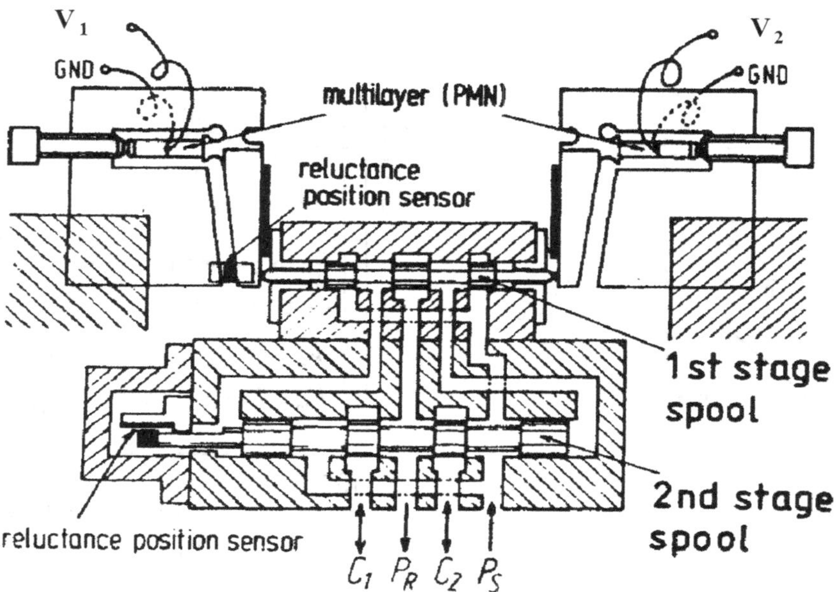

FIGURE 8.27 Schematic diagram of a direct drive spool servo valve incorporating two PMN-based multilayer actuators. (From H. Ohuchi et al.: *Proc. Fluid Control & Measurement*, Pergamon Press, Tokyo, p. 415, 1985.)

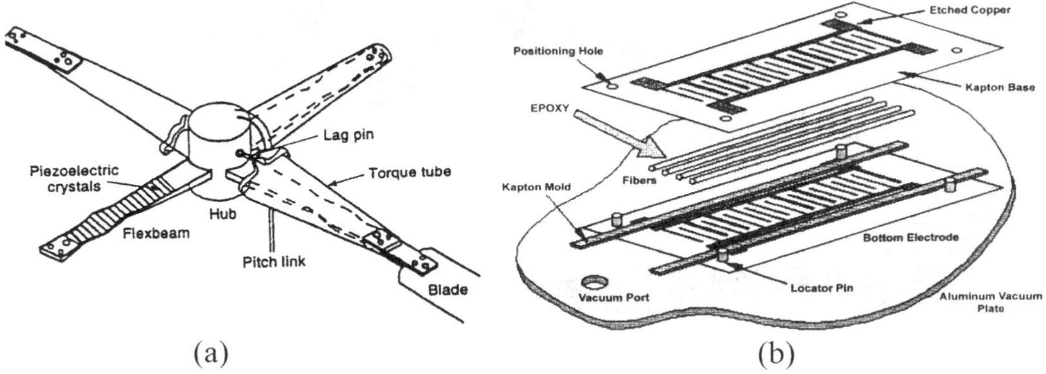

FIGURE 8.28 (a) Bearingless rotor flexbeam with integrated piezoelectric strips. (From N. W. Hagood: Introduction to AMSL/MIT, http://amsl.mit.edu/research.) (b) Active fiber composite ply designed for use in the bearingless rotor flexbeam. (From P. C. Chen and I. Chopra: *Smart Mater. Struct.*, 5, 35, 1996.)

fibers. Glass fibers, 9 μm in diameter, can also be included to further enhance the strength of the blade. Various types of PZT sandwiched beam structures have been investigated for this and similar flexbeam applications and for active vibration control.[21]

Shape memory alloy is used for various vibration damping applications, in particular for large machinery such as engines.[22] An aircraft for surveillance purposes such as taking sky-maps needs to be isolated completely from the propeller engine vibration. Figure 8.29 illustrates an engine floating mechanism on a plane wing with shape memory alloy wires. Low-resonance frequency vibrations at 5 and 9 Hz originating from the eigen frame vibration modes can be dramatically reduced by 20 dB down with the feedback; that is, extension or shrinkage of the shape memory wires is made in order to compensate for the frame resonance vibrations.

8.6.2 Noise Elimination

8.6.2.1 Acoustic Stealth and Sound Elimination

There are two ways to see an object acoustically: (1) sonar or (2) hydrophone. A sonar transducer generates and transmits an acoustic wave (at a specific frequency such as 30 kHz) first, then the reflected wave from the object is received and analyzed for visualizing the object image. In order to make an object (such as a submarine) transparent or stealthy under sonar surveillance, we should

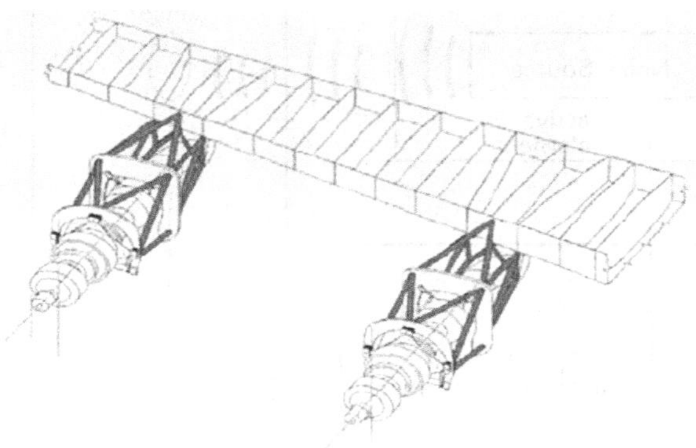

FIGURE 8.29 Engine float mechanism on a surveillance aircraft.

FIGURE 8.30 Engine float mechanism on a surveillance aircraft.

eliminate the wave reflection. Two ways are possible: (1) absorbing all transmitted waves not to reflect, or (2) generating a conjugate wave to superpose on the transmitted wave and to cancel the reflection. Cummer et al., Duke University, developed the 3D acoustic cloak,[23] which reroutes sound waves to create the impression that both the cloak and anything beneath it are not there (Figure 8.30). The geometry of the plastic sheets and placement of the holes interacts with sound waves and resonates with a particular frequency (30 kHz), so that the transmitted wave is absorbed, not to reflect or to make it transparent.

On the contrary, a hydrophone is primary a highly sensitive microphone to monitor the sound generated in/on the object (such as submarine propeller noise). In order to make the object undetectable, we should (1) stop noise generation in the object, or (2) generate a conjugate wave to superpose on the acoustic noise (such as propeller noise) not to be detected by hydrophone. A block diagram of a system for eliminating acoustic noise is schematically shown in Figure 8.31.[24] The acoustic noise is detected by a piezoelectric microphone, and the signal is fed back to a piezoelectric high-power speaker to generate a conjugate wave (identified as the "antinoise" signal in the figure), which effectively cancels the noise signal as shown.

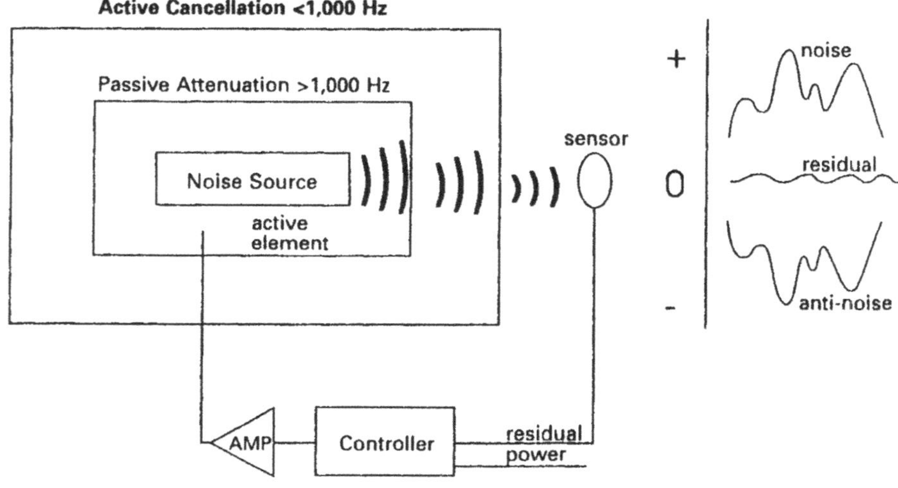

FIGURE 8.31 A block diagram of a system for eliminating acoustic noise. (From H. B. Strock: *Spectrum-Performance Mater.*, 51-1, Decision Resources, 1993.)

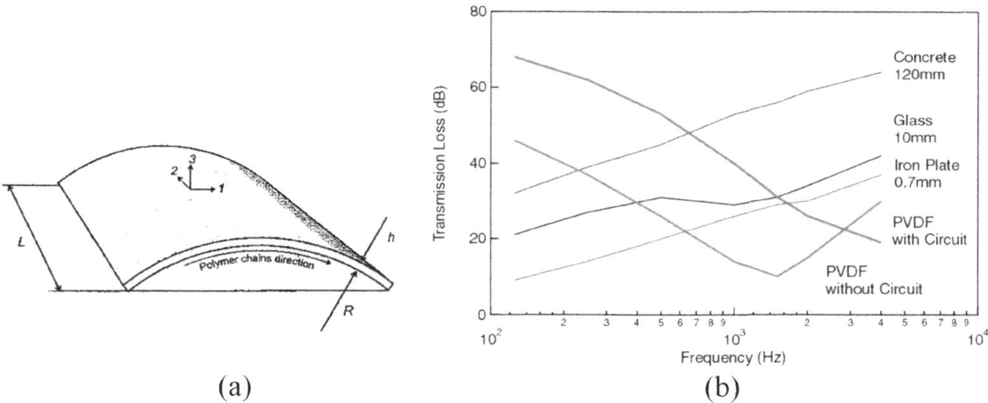

FIGURE 8.32 (a) Noise barrier with PVDF films; (b) Transmission loss verses frequency for noise barriers. (From E. Fukada et al.: *J. Materials Tech.*, 19(2), 83, 2004.)

8.6.2.2 Noise Barrier

Mechanical noise (vibration and sound noise) is also a hazardous environmental problem, nowadays. Automobile sounds should be banned in residential areas. Ishii and Uchino reported the first passive damping concept with PZT ceramics in early 1980s.[25] When a piezoelectric is adopted in a noise vibration system, the cyclic electric field is excited. If this electric energy is consumed via a suitable resistor as Joule heat, mechanical noise vibration is significantly suppressed (see Section 7.1.2). Kobayashi Institute of Physical Research, Japan, developed large curved PVDF panels for highway noise active cancellation application (that is, Noise Barrier™), as shown in Figure 8.32a. A 30~40-dB reduction of sound noise was achieved around a low frequency range (100–300 Hz) by combining a negative capacitance circuit. Figure 8.32b shows transmission loss of these PVDF barriers (with and without a negative capacitance circuit), compared with the regular concrete, glass, or iron walls.[26] Referring to Section 5.7.2.1, a negative capacitance is popularly used to enhance the effectiveness of the system for a specific frequency range (i.e., low-frequency noise [100–300 Hz] is a current environmental problem).

CHAPTER ESSENTIALS

1. The deformable mirror surface contour can be represented by the *Zernike aberration polynomials*, which can be used to optimize the design of the system.
   a. Each orthogonal term represents the function of individual devices independent of the other elements in the system.
   b. Higher-order corrections can be made by increasing the number of terms used.
   c. Error evaluation is also possible.
2. An ultrahigh-precision linear guide system with PZT actuators can exhibit motion accuracy better than the master scale in terms of the transverse, pitching, and rolling of the moving table. The key is to use the master scale error in the ROM, which is integrated additionally in the feedback control.
3. A servo (feedback) system stabilizes the actuator (nonlinear or hysteretic behaviors) for realizing (1) linearity in the input and output relation, (2) an output response with a flat frequency dependence, and (3) minimization of external disturbance effects.
4. There are two methods of vibration suppression: (1) passive and (2) active damping. Complete absorption of the transmitting wave, or "antinoise" generation for cancelling the transmitting wave.
5. A negative capacitance component is useful in a feedback control circuit to enhance system performance and keep high power efficiency.

CHECK POINT

1. Which is the most suitable for servo displacement transducer applications, electrostrictor, soft piezoelectric, or hard piezoelectric?
2. Fill in the blank (a) below to make the sentence correct: The three distinct benefits of a feedback control system include: (1) minimization of external disturbance effects, (2) an output response with a flat frequency dependence, and (3) (a).
3. How do we change the temperature for controlling a shape memory device, in addition to an external heater such as a hair dryer? Answer simply.
4. What is the key reason we use an "integral amplifier" in a "piezoelectric" oil-pressure servo valve control scheme? Answer simply.

CHAPTER PROBLEMS

8.1 Describe the merits of implementing the expansion series for designing an actuator device.

8.2 Is the following argument correct? Discuss this popular misconception.
"Given a servo system with a position sensor and an actuator, the precision of position control will not exceed the specified master scale precision."

8.3 What are the primary differences between negative capacitance and inductor components from an application viewpoint?

8.4 Write a brief definition for each of the following important micropositioning terms. Refer to the figure below, left side.
   a. Pitch
   b. Roll
   c. Yaw

8.5 The figure below, right side, shows a typical performance (gain and phase) Bode plot for a piezoelectric oil-pressure servo valve with an "integral" amplifier. If we use a regular "proportional" amplifier instead of "integral" type, which part of the plot will be significantly changed?

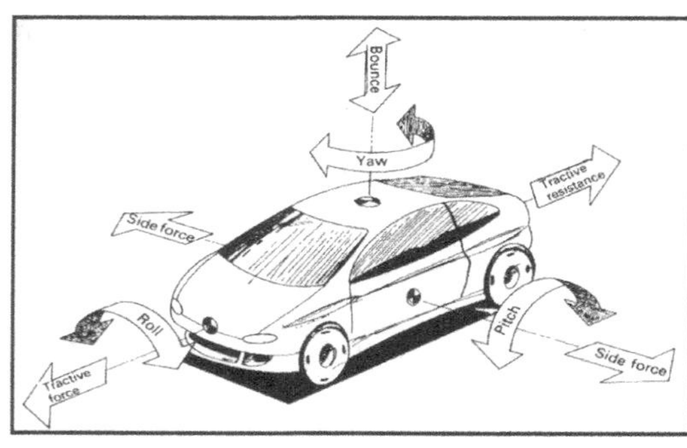

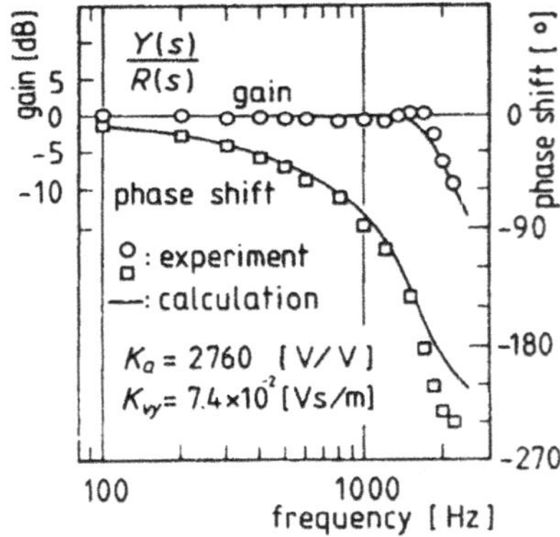

REFERENCES

1. J. W. Hardy, J. E. Lefebre and C. L. Koliopoulos: *J. Opt. Soc. Amer.*, 67, 360, 1977.
2. J. W. Hardy: *Proc. IEEE*, 66, p. 651, 1978.
3. K. Uchino, Y. Tsuchiya, S. Nomura, T. Sato, H. Ishikawa and O. Ikeda: *Appl. Optics*, 20, 3077, 1981.
4. T. Sato, H. Ishikawa, O. Ikeda, S. Nomura and K. Uchino: *Appl. Optics*, 21, 3669, 1982.
5. J. L. Fanson and M. A. Ealey: *Active and Adaptive Optical Components and Systems II, SPIE 1920*, Albuquerque, 1993.
6. K. Uchino: *Proc. Actuator 2008, A5.0*, June 9–11, Bremen, Germany; http://www.onelimited.com/ (as of March, 2008).

7. T. Kaneko, K. Tsuruta and N. Kawahara: *Proc. of 35th ICAT Smart Actuator Symp.*, State College, PA, April 18–19, 2002.
8. Cover article: *Perspectives*, 7(3), 3rd Quarter, 2005.
9. F. E. Scire and E. C. Teague: *Rev. Sci. Instruments*, 49, 1735, 1978.
10. S. Okayama, H. Bando, H. Tokumoto and K. Kajimura: *Jpn. J. Appl. Phys.*, 24(Suppl. 24-3), 152, 1985.
11. K. Uchino: *Ceramic Databook'88 (Chap. Ceramic Actuators)*, Inst. Industrial Manufacturing Technology, Tokyo, 1988.
12. S. Moriyama and F. Uchida: *Proc. Jpn. Procision Eng. Soc.*, 201, 61(Spring), 1984.
13. S. Moriyama, T. Harada and A. Takanashi: *Precision Machine*, 50, 718, 1984.
14. Ikebe and Nakada: *Proc. Jpn. Measurement and Automation Soc.*, 7–5, 480, 1971.
15. Ikebe and Nakada: *Oil and Air Pressure*, 3, 78, 1972.
16. H. Ohuchi, K. Nakano, K. Uchino, S. Nomura and H. Endo: *Proc. Oil and Air Pressure Soc.*, 3(Fall), 9, 1983.
17. H. Ohuchi, K. Nakano, K. Uchino, H. Endoh and H. Fukumoto: *Proc. Fluid Control & Measurement*, Pergamon Press, Tokyo, p. 415, 1985.
18. K. Yamamoto, A. Nomoto and Y. Ueda: *Proc. SICE*, 1408, p. 121, 1982.
19. F. K. Straub: *Smart Mater. Struct.*, 5, 1, 1996.
20. N. W. Hagood: Introduction to AMSL/MIT, http://amsl.mit.edu/research.
21. P. C. Chen and I. Chopra: *Smart Mater. Struct.*, 5, 35, 1996.
22. C. Mavroidis, C. Pfeiffer, and M. Mosley: "5.1 Conventional actuators, shape memory alloys, and electrorheological fluids." Ed. Y. Bar-Cohen, *Automation, Miniature Robotics and Sensors for Non-Destructive Testing and Evaluation*, American Society for Nondestructive Testing, Columbus, OH, 1999.
23. L. Zigoneanu, B. Popa, S. A. Cummer: *Nat. Mater.*, March 9, 2014. DOI: 10.1038/NMAT3901; https://pratt.duke.edu/about/news/acoustic-cloaking-device-hides-objects-sound
24. H. B. Strock: Spectrum- Performance Materials, 51-1, Decision Resources, 1993.
25. K. Uchino and T. Ishii: *J. Japan. Ceram. Soc.*, 96(8), 863–36867, 1988.
26. E. Fukada, M. Date and H. Kodama: *J. Materials Tech.*, 19(2), 83, 2004.

9 Pulse Drive Motor Applications

Pulse drive motors are used in imaging systems, inchworms and impulse drive motors, piezoelectric relays, automobile adaptive suspension systems, and inkjet and diesel injection valves. Low-permittivity actuator materials (soft PZTs) are essential for these applications, due to the requirement of small capacitance for realizing low current to the devices from the power amplifier used in these sudden operation systems. Vibration overshoot and ringing are critical drive/control issues for this class of devices; thus, effective means of suppressing these detrimental effects will also be considered in this chapter. Refer to Section 5.4. The pulse width or rise time must be adjusted to exactly match the resonance period of the actuator system to eliminate vibrational ringing.

9.1 IMAGING SYSTEM APPLICATIONS

The swing charge-coupled device (CCD) image sensor and swing pyroelectric sensor are introduced here as examples of how actuators can be used in imaging systems. Each offers unique advantages over the conventional devices previously used in these systems.

9.1.1 Swing Charge-Coupled Device Image Sensors

Charge-coupled devices are popular as solid-state image sensors and are employed widely in cameras and video cameras. Although the resolution and sensitivity have recently undergone considerable improvement, there are still problems in image resolution, especially in the horizontal direction.[1] One method for improving the horizontal resolution involves alternately swinging a CCD chip by a half picture unit cell (pixel) as the image data are sampled.[2]

A magnified view of an interline transfer-type CCD is shown in Figure 9.1a.[2] The sensing part of the optical image system is made up of photodiodes, and the remaining parts, which include vertical CCD registers and transfer lines and overflow drains, are shielded by an aluminum coating. The principle of the swing imaging method is illustrated schematically in Figure 9.1b. The CCD chip substrate is shifted alternately in the horizontal direction by half of the horizontal pixel pitch (P_H) with respect to the input image light. The timing of this swing motion is adjusted to correspond to the frame period, as shown in Figure 9.1b, bottom, so as to sample the image data at two spatial points (A and B). The image signals obtained in the A and B fields are combined and processed such that a two-fold improvement in the horizontal resolution is achieved. This method also leads to an effective sensing area twice that obtained by a conventional system. This can significantly reduce the appearance of *Moiré patterns* (an interference phenomenon) in the image.

The structure of the swing CCD sensor developed by Toshiba, Japan is shown in Figure 9.2.[2] The CCD substrate is translated by a pair of piezoelectric bimorphs supported at both ends. The both-end support was required to protect the bimorph from external noise vibrations. A special brace was invented to keep rigidity without decreasing the deflection of the bimorph. A drive voltage of only 15 V is required to generate an 11-μm displacement, which corresponds to half of the CCD pixel pitch, so that no high-voltage supply is required. In order to suppress the vibration overshoot and ringing that occurs when the system is driven by a step voltage, a trapezoidal waveform with a rise/fall time of 4 ms, which is almost equal to the resonance period, is used (see Figure 5.42).[2]

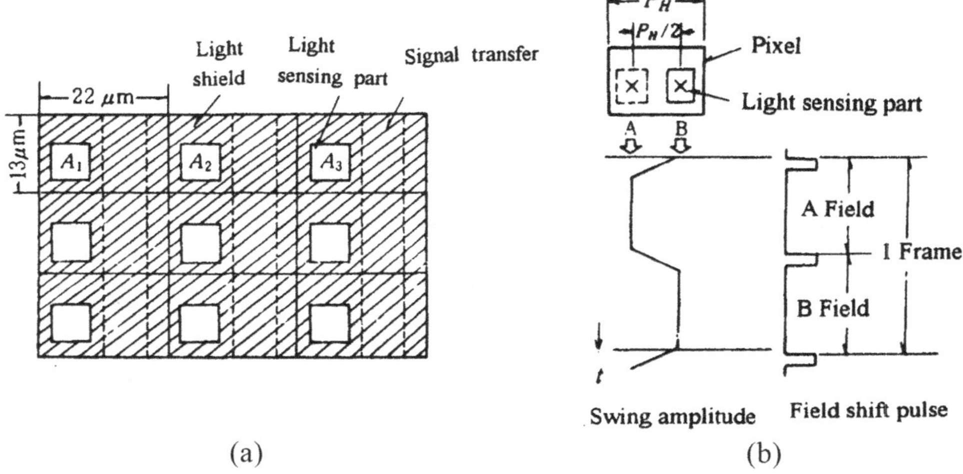

FIGURE 9.1 (a) Schematic of an interline transfer-type CCD. (b) Operation of the swing CCD image sensor: the CCD chip is shifted alternately in the horizontal direction by half of the horizontal pixel pitch (P_H) with a timing of the swing motion adjusted to the frame period. (From K. Yokoyama and C. Tanuma: Electronic Ceramics 15, Spring, p.45, 1984.)

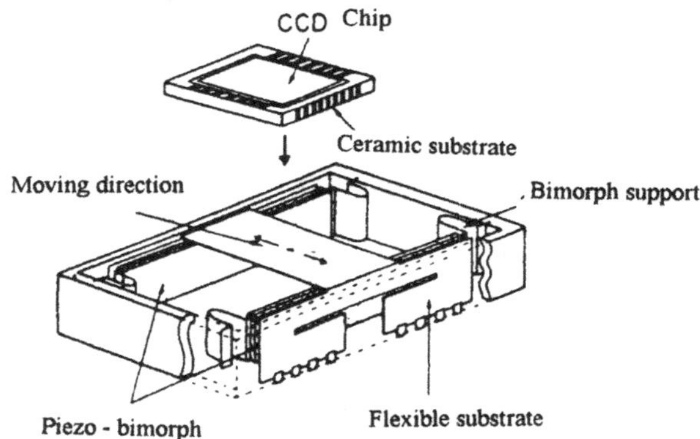

FIGURE 9.2 Structure of the swing CCD sensor developed by Toshiba. (From K. Yokoyama and C. Tanuma: Electronic Ceramics 15, Spring, p.45, 1984.)

9.1.2 Swing Pyroelectric Sensor

The piezoelectric bimorph can also be used in a similar manner with pyroelectric sensors. A large electromagnetic motor has conventionally been used for chopping infrared light to monitor the temperature. A tiny pyroelectric sensor that makes use of a piezoelectric bimorph light chopper developed by SANYO is shown in Figure 9.3.[3] The compact size, light weight, and low cost of the unit make it a suitable temperature sensor for microwave ovens. A trapezoidal voltage waveform at 10–20 Hz with a rise/fall time equal to the resonance period was used for the swing operation not to generate the displacement overshoot or ringing.

9.2 INCHWORM DEVICES

The motion produced by an inchworm device is similar to that produced by an ultrasonic motor in terms of the general speed and execution of the motion, but the mechanisms for their operation

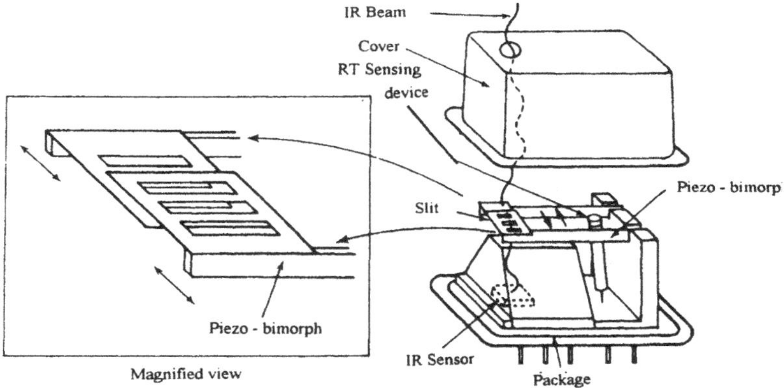

FIGURE 9.3 Structure of the swing pyroelectric sensor. (K. Shibata et al: *Jpn. J. Appl. Phys.*, 24(Suppl. 24–3), 181, 1985.)

are completely different. The inchworm is driven by a rectangular signal at a frequency below its resonance frequency, and its movement is intermittent and discrete.

9.2.1 Microangle Goniometer

A microangle goniometer consisting of a fixed inchworm and a turntable is pictured in Figure 9.4a.[4] The inchworm is composed of an electrostrictive unimorph and two electromagnets. Let us consider the operation of the inchworm with reference to the timing diagram shown in Figure 9.4b. The operation cycle is initiated with the movable electromagnet on and the fixed one off. The drive voltage is applied to the electrostrictive unimorph to displace the movable magnet. The angular position of the turntable is thereby advanced. The two electromagnets are then alternately switched on and off, the voltage on the unimorph is removed, and the movable magnet is returned to its initial position with the turntable kept in its displaced position. Repetition of this process drives the turntable sequentially, and the rotation angle increases. The rotation direction is reversed by exchanging the clamp timing of the two electromagnets.

The operation of the microgoniometer at 0.5 Hz is illustrated in Figure 9.5. The angular position as a function of time has a clear triangular form composed of discrete steps. The angular velocity

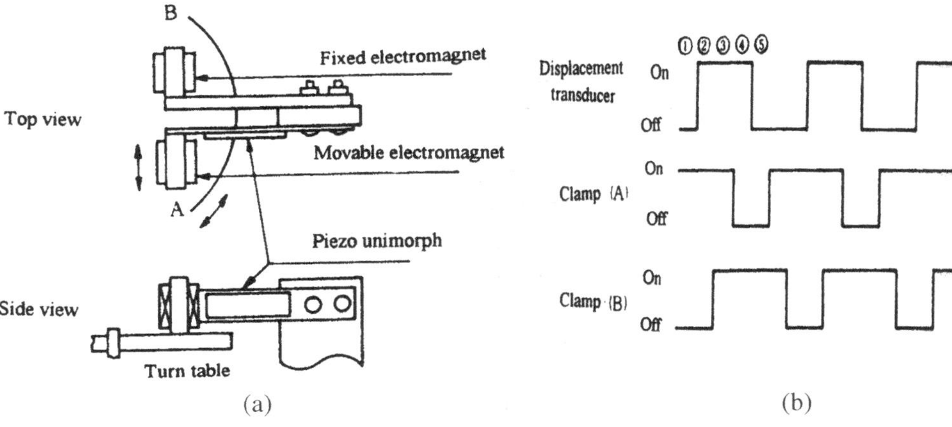

FIGURE 9.4 (a) Microangle goniometer consisting of a fixed inchworm and a turntable. (b) Timing diagram for the components of the inchworm. (From M. Aizawa et al: *Jpn. J. Appl. Phys.*, 22, 1925, 1983.)

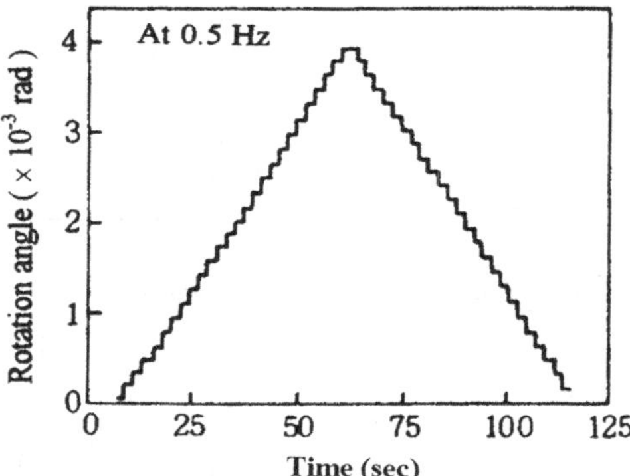

FIGURE 9.5 Response of the inchworm microangle goniometer system operated at 0.5 Hz. (From M. Aizawa et al: *Jpn. J. Appl. Phys.*, 22, 1925, 1983.)

of the turntable is controlled by the voltage applied to the unimorph, current passing through the electromagnets, and the drive frequency. The system produces a maximum angular speed of 10^{-2} rad/s and a maximum shift of 6×10^{-4} rad/step. A modified version of this inchworm microgoniometer, utilizing three PZT actuators for both the clamp and shifting operations, has also been proposed by a group at Hitachi.[5]

9.2.2 Linear Walking Machines

Sophisticated linear walking machines have been developed by two German companies. A linear drive inchworm, incorporating two d_{33} (longitudinal mode) and two d_{31} (transverse mode) multilayer actuators, by Philips is pictured in Figure 9.6.[6] Very precise positioning of less than 1 nm has been reported for this device. There are some problems with this design, however: in particular, the rather high level of audible noise in operation, significant heat generation when it is driven at high frequency, and high manufacturing costs due to the use of several multilayer piezoelectric actuators. A two-legged inchworm was manufactured by Physik Instrumente.[7] The motion of a pair of inchworm units, consisting of two multilayer actuators, are coupled such that there is a 90° phase difference between them in order to produce smoother, more continuous movement of the device.

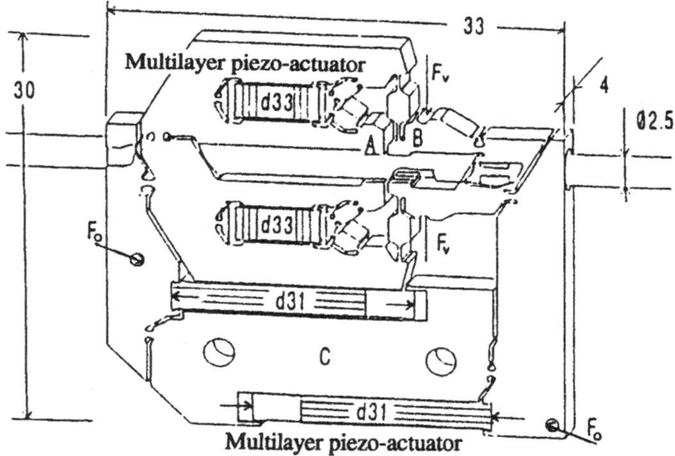

FIGURE 9.6 Linear drive inchworm with two d_{33} and two d_{31} MLs (Philips). (From M. P. Coster: *Proc. 4th Int'l Conf. on New Actuators, 2.6*, p. 144, Germany, 1994.)

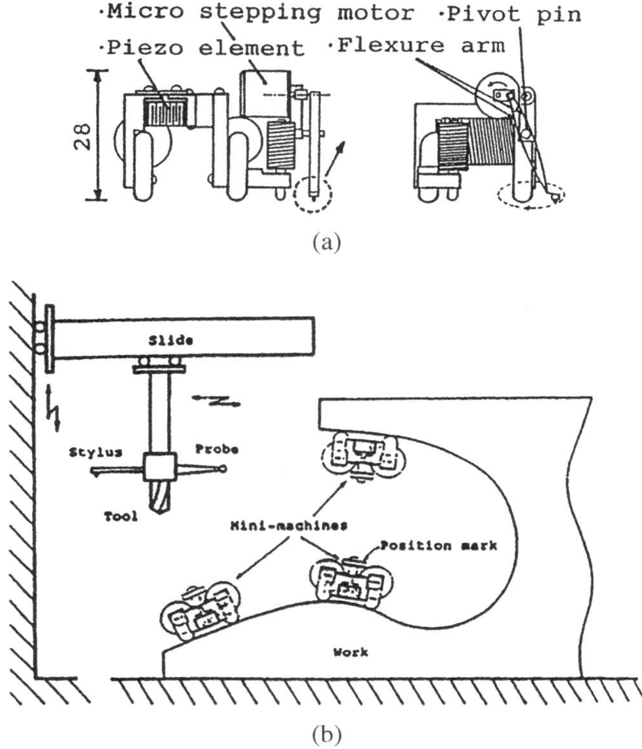

FIGURE 9.7 (a) Microwalking machining inchworm vehicles with piezo-actuators and electromagnets. (From N. Aoyama et al.: *Proc. J. Precision Engr. Soc. P.897*, 1992.) (b) Micromachining using multiple compact vehicles. The attractive force is magnetic. (From N. Aoyama et al.: *Proc. J. Precision Engr. Soc. P.897*, 1992.)

Microwalking vehicles incorporating an inchworm mechanism composed of two electromagnets and a multilayer piezoelectric actuator are pictured in Figure 9.7a.[8] When a diamond cutting edge is installed on this vehicle, it becomes a compact machining tool. The tool can move over any sloped surface provided the surface is magnetic, as illustrated in Figure 9.7b. Tasks ranging from microgroove cutting to microdust elimination can be achieved by attaching different tools to the vehicle.

9.3 IMPULSE DRIVE MOTORS

We now consider impulse and inertial motors. The principle of the "stick & slick" motion on a drive rod is illustrated in Figure 9.8. By applying a sawtooth-shaped voltage on a piezoelectric actuator, alternating slow expansion and quick shrinkage are excited on a drive friction rod (i.e., stator) under a drive frequency much lower than the resonance of a piezo-actuator-rod system. A moving object (i.e., slider) placed on the drive rod will "stick" on the rod due to friction during a slow expansion period, while it will "slip" during a quick shrinkage period, so that the slider moves from left to right in this case. When the voltage sawtooth shape is reversed, an opposite motion can be obtained. There are two types: (1) stator sawtooth displacement (Figure 9.8) or (2) slider sawtooth displacement, which are visualized in Figure 9.9a and b, respectively.

9.3.1 Stator Impulse Drive

Konica-Minolta, Japan, developed a smooth impact drive mechanism (SIDM) using a multilayer piezoelement.[9] The idea comes from the "stick & slick" condition of the ring object attached on a

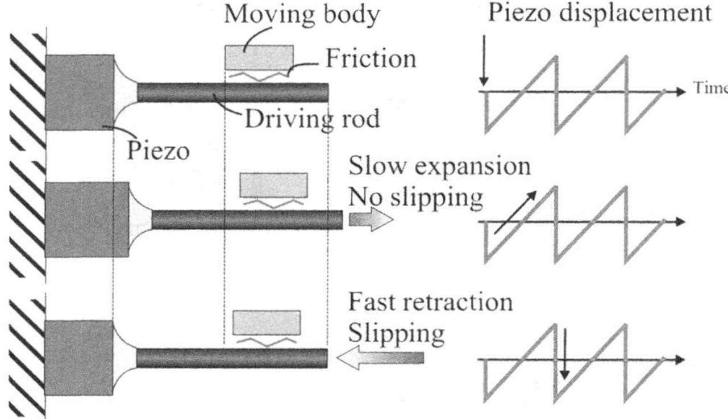

FIGURE 9.8 Principle of the "stick & slick" motion of the slider on a drive rod. (From Y. Okamoto et al: *Konica Minolta Tech. Report*, 1, 23, 2004.)

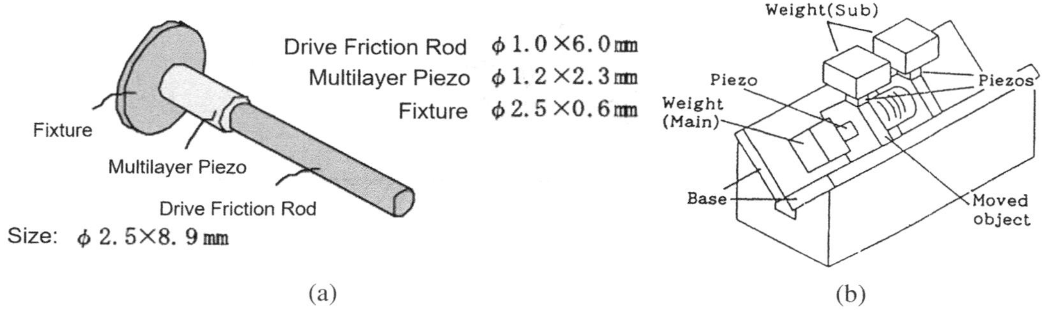

FIGURE 9.9 Impulse drive motors with "stick & slick" motion on the stator (a) (From Y. Okamoto et al: *Konica Minolta Tech. Report*, 1, 23, 2004.) and on the slider (b). (From T. Higuchi et al: *J. Precision Engr.*, 54(11), 2107, 1988.)

drive rod in Figure 9.9a. Figure 9.10a shows the camera module structure, including the SIDM and lenses. By applying a sawtooth-shaped voltage to a multilayer actuator, alternating slow expansion and quick shrinkage are excited on a drive friction rod (Figure 9.10b). A ring slider placed on the drive rod will "stick" on the rod due to friction during a slow expansion period, while it will "slide" during a quick shrinkage period, so that the slider moves from the bottom to the top. The lens is attached to this slider (Figure 9.10a), which moves up continuously, as shown in Figure 9.10b. When the voltage saw shape is reversed, the opposite motion can be obtained.

Piezo Tech, Korea, developed a similar SIDM motor, but using a bimorph instead of an ML actuator, that suppressed the cost significantly: the tiny ultrasonic linear actuator (TULA).[10] Though a flexural bimorph is used, the driving frequency is much higher than 40 kHz (ultrasonic range) due to the small size. We discuss the SIDM driven at its resonance frequency in Chapter 10.

9.3.2 Slider Impulse Drive

Impact mechanism walking machines have also been proposed by Higuchi et al.[11] The structure of such a walker is pictured in Figure 9.9b. The moving table is attached to a v-shaped guide rail. The weight table is moved by the counterforce of the main weight when a sudden expansion or contraction of the multilayer piezoelectric actuator is induced at an appropriate rate of the sawtooth voltage. Note that the clamping force of the table is provided by the frictional force, which is determined by the coefficient of friction between the surfaces and the weight of the table.

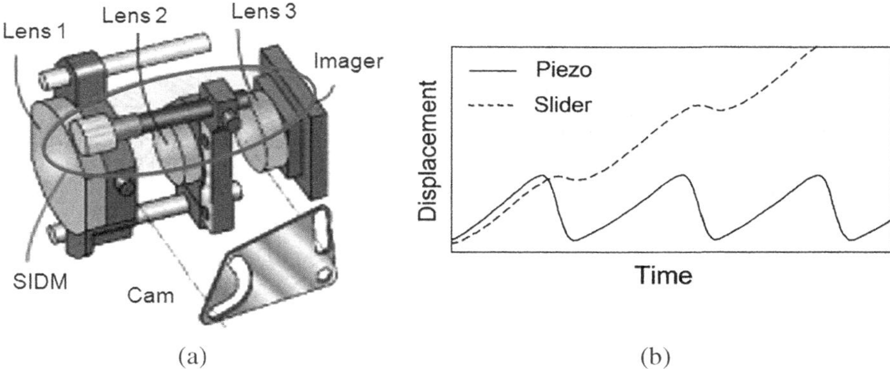

FIGURE 9.10 (a) Konica-Minolta SIDM camera module for an optical-zoom mechanism. (b) Displacements of the piezoelement (sawtooth type) and the slider (continuous). (From Y. Okamoto et al: *Konica Minolta Tech. Report*, 1, 23, 2004.)

9.4 PIEZOELECTRIC RELAYS

9.4.1 Piezoelectric Relays

The relay is a basic component commonly found in electrical equipment. Most current relays are mechanical switches driven by electromagnetic solenoids. The need for increasingly fast response and more compact size has stimulated the development of new types of relays. The piezoelectric relay is a promising alternative. The basic structure of a piezoelectric relay proposed by Piezo Electro Products, Inc. (PEPI) is pictured in Figure 9.11a.[12] The primary advantages of the piezo-relay over the electromagnetic type are: (1) low energy consumption, (2) no heat generation, (3) no generation of electromagnetic noise, (4) quick response, (5) compact size, and (6) suitability for integration. The structure, which utilizes the coupled action of a piezoelectric bimorph and a spring, exhibits mechanical bistability function and is readily integrated into relay arrays. Relay cards, 6 mm in thickness and with a density of several pieces per square centimeter, are commercially available.

Another relay incorporating a piezoelectric bimorph manufactured by OMRON Corporation is pictured in Figure 9.11b.[13] This design is two-fold smaller than design (a), and includes an especially sensitive snap-action switch with a leaf spring, which is activated with only several tens of micron displacement of the piezo-bimorph and leads to an overall response speed of less than 5 ms.

When a piezoelectric bimorph is used as the actuator for a relay of this type, the generative force is decreased with the deflection amplification, thus leading to a problem in secure contact. A snap-action spring or a permanent magnet is often used in conjunction with the bimorph to ensure proper contact by producing a system with *mechanical bistability*. Both the on and off states are mechanically stable, and no continuous application of electric field is required. This is the fundamental principle of *latching relays*. The piezoelectric bimorph functions in this case only to switch between the two states. The latching function is especially useful for emergency shutdown situations. The relay retains the initial state so that the system is restored to normal operation when restarted. Latching relays typically constitute only 10% of the total relay production, but the unit price is about 10 times higher than nonlatching varieties, leading to comparable revenue overall.

A multilayer piezoelectric actuator relay has also been introduced by NEC on a trial basis in order to improve the contact stress level.[14] The schematic diagram of this device pictured in Figure 9.11c illustrates the operation of the monolithic hinge lever mechanism. The reader can understand the three-stage displacement amplification with three arms, 1, 2, and 3, which correspond to roughly double, triple, and six-fold amplifications, respectively (40 times in total). Initial 10-μm displacement in the ML generates 0.4-mm displacement at the contact tip on the right side by keeping high contact force ~25 N.

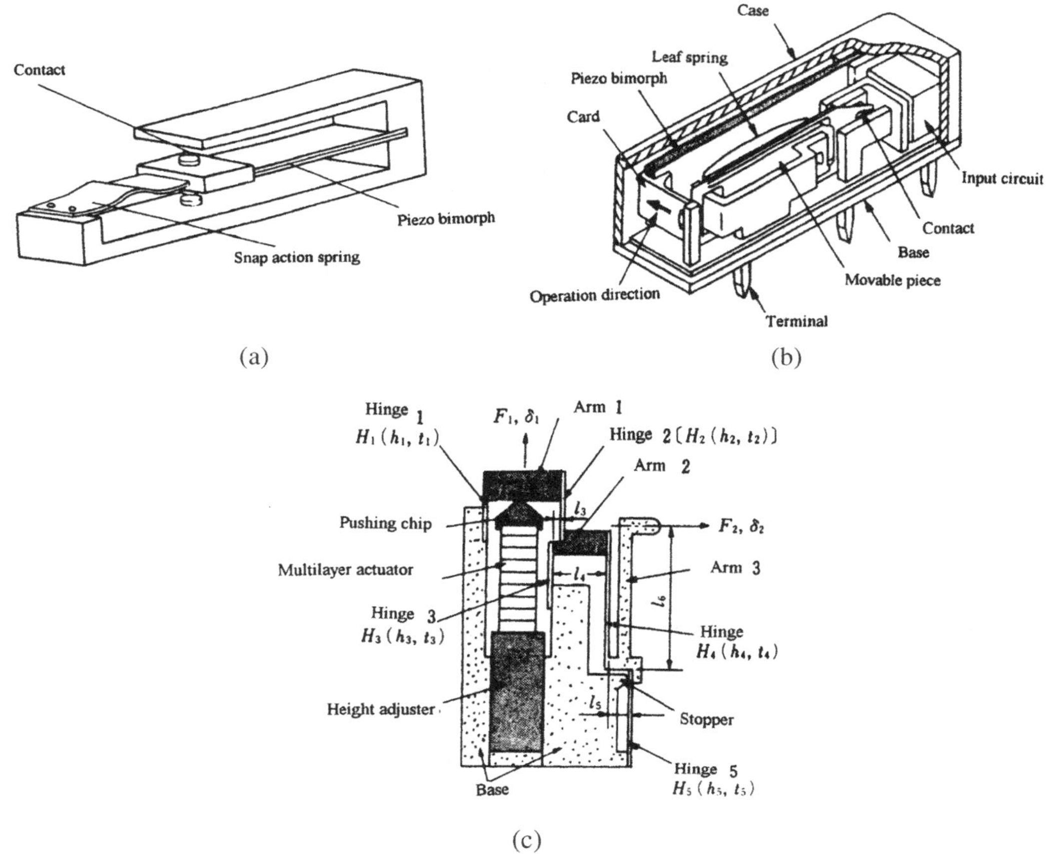

FIGURE 9.11 Piezoelectric relays: (a) by Piezo Electro Products, Inc. (PEPI) (From Piezo Electric Products Inc., Product Catalog.); (b) by OMRON (From Sato, Taniguchi and Ohba: *OMRON Technics*, (70), 52, 1983.); and (c) by NEC. (From S. Mitsuhashi et al: *Jpn. J. Appl. Phys*., 24(Suppl. 24–3), 190, 1985.)

9.4.2 Shape-Memory Ceramic Relays

A compact latching relay has been developed utilizing a *shape-memory ceramic* unimorph.[15] The structure of the relay is similar to OMRON's design pictured in Figure 9.11b, with the shape-memory ceramic unimorph (made from an antiferroelectric-ferroelectric phase transition material as described in Section 3.3.2) replacing the piezoelectric bimorph. Once the bending is induced in the unimorph by applying a pulse voltage, it can be maintained for more than a week. Recovery of the original shape is obtained by applying an inverse bias voltage. The static condition for switching is shown in Figure 9.12a. The dynamic responses of the shape-memory latching relay and a conventional electromagnetic coil relay are compared in Figure 9.12b and c.[15] Although some chattering is apparent in the responses of both relays, the shape-memory ceramic one exhibits much quicker action (1.8-ms rise) and lower energy consumption (7 mJ) than the electromagnetic relay. Use of a smart material in the design of a relay can thus be effective in reducing the number of components and the total size of the device.

9.4.3 Piezoelectric Micro-Electro-Mechanical System Relay

The Electronic Materials Research Laboratory at RWTH Aachen University developed piezoelectrically actuated MEMS switches used for reconfigurable microwave applications.[16]

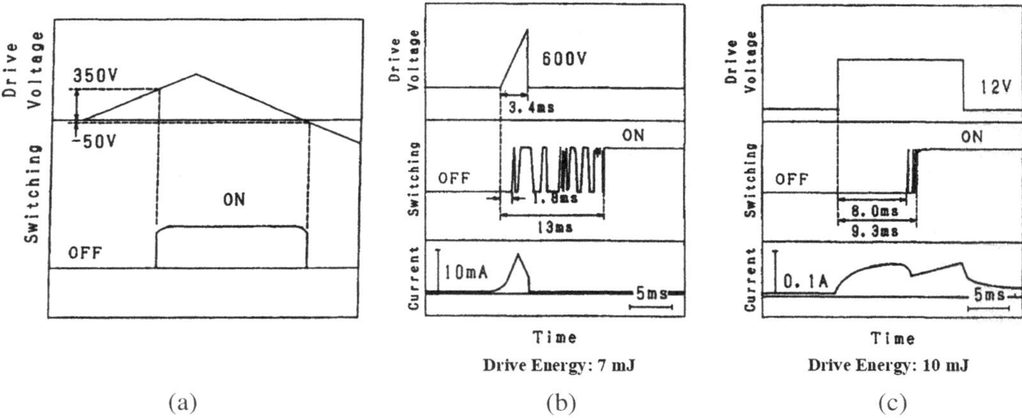

FIGURE 9.12 (a) Static condition for switching of a compact latching relay with a shape-memory ceramic. Dynamic response of two latching relays: (b) shape-memory ceramic relay and (c) electromagnetic coil relay.

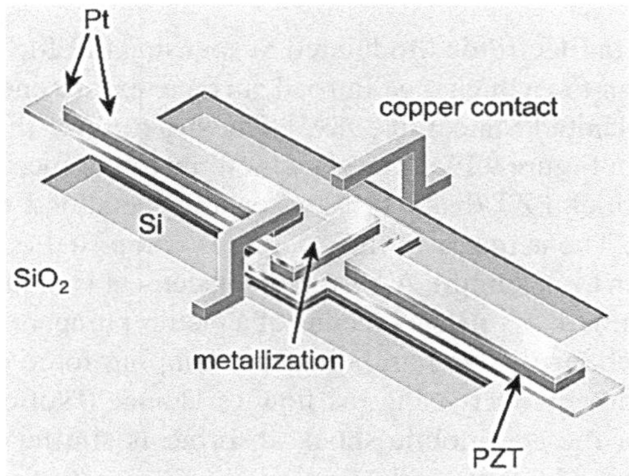

FIGURE 9.13 Piezoelectric MEMS microrelay with PZT unimorphs. (From http://www.emrl.de/r_a_2.html)

In comparison with conventional "electrostatic" MEMS switches, piezoelectric MEMS types require lower operation voltages (below 10 V) and do not suffer from striction of electrodes, which occasionally limits the lifetime. Figure 9.13 shows the design principle of a piezoelectric microrelay bridge structure. The voltage applied to the SiO_2/Pt/PZT/Pt unimorph stack generates a bridge bend up or down in order to interconnect or release the two contact bars. Such a mechanical contact design can handle relatively large signal powers in comparison with a semiconductor-type relay, and exhibits relatively high mechanical stability compared to other cantilever switch designs.

9.5 AUTOMOBILE ADAPTIVE SUSPENSION SYSTEM

In general, when a shock absorber of an automobile provides a higher damping force (referred to as a "hard" damper), the controllability and stability of the vehicle are improved. Driving comfort is diminished, however, because the road roughness is easily transferred to the passengers. The purpose of the electronically controlled shock absorber is to provide both controllability and comfort simultaneously. When normal, relatively smooth road conditions exist, the system is set to provide a low damping force ("soft" damping) so as to optimize passenger comfort, and when rougher terrain is encountered, the damping force is increased accordingly to improve the controllability while still maintaining a reasonable level of comfort. In order to effectively respond to variations in the road surface, a highly responsive sensor/actuator combination is required for the *shock-absorbing system.*

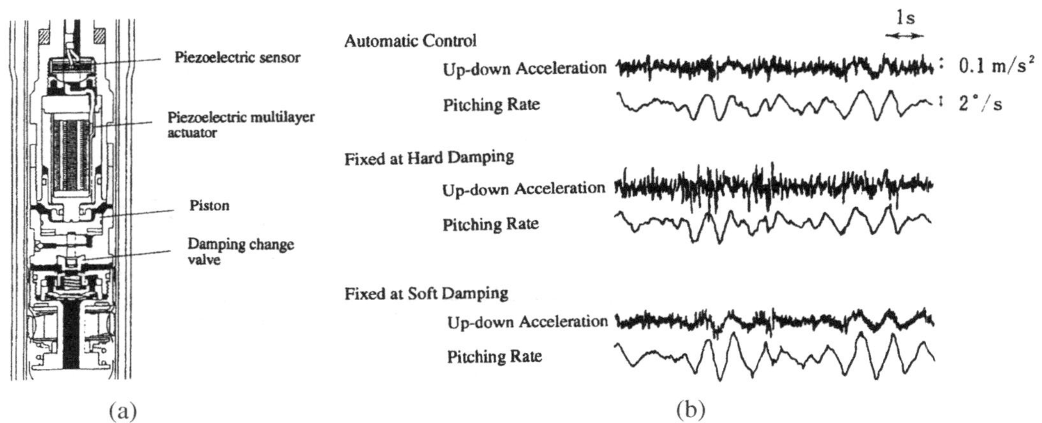

FIGURE 9.14 (a) Piezoelectric Toyota Electronic Modulated Suspension (TEMS) system. (b) Response of the adaptive TEMS system as compared to two conditions of fixed damping. (From Y. Yokoya: *Electronic Ceramics*, 22(111), 55, 1991.)

The piezoelectric Toyota Electronic Modulated Suspension (TEMS) system, which is designed to adapt its damping action to each change in road surface texture encountered, was installed on the Celcio (Lexus in the United States) in 1989.[17] The structure of this electronically controlled shock absorber is shown in Figure 9.14a. The sensor on the top of the shock absorber is composed of five layers of 0.5-mm-thick PZT disks. It has a sensing speed of 2 ms, and the road roughness resolution is about 2 mm. The actuator in this system is composed of 88 layers of 0.5-mm-thick PZT disks (about 50 mm in total length). A 50-μm displacement is induced with an applied voltage of 500 V, and this is magnified 40 times by means of a piston/plunger pin combination. The stroke of the piston pushes the change valve regulating the damping force down, thereby opening the bypass oil route and effectively decreasing the flow resistance ("softening" the damping effect). Note that the principle of the automobile shock absorber is similar to a syringe, where liquid flow rate/pressure is changed by the nozzle size; wide and narrow nozzles provide soft and hard damping, respectively. This adjustable damper is basically a binary (hard or soft) control system by switching the soft and hard damping status with the piezo-actuation every 2 seconds according to the road conditions.

The response of the adaptive system as compared to two conditions of fixed damping is shown in Figure 9.14b. The up-down acceleration and pitching rate (the sensor signals from the front and rear shock absorbers should be in phase and in 180° opposite, respectively) were monitored when the vehicle was driven on a rough road. When the TEMS system was used (top set of curves), the up-down acceleration was suppressed to a level comparable with that obtained with fixed soft damping, providing comfortable driving conditions. At the same time, the pitching rate was also suppressed to a level comparable with that obtained with fixed hard damping, allowing for better controllability.[17]

9.6 INKJET PRINTERS

The piezoelectric dot-matrix printer is the first mass-produced device by NEC, Japan, using multilayer piezoelectric actuators, and one of the most successful piezo-actuator products.[18–20] One of the intriguing technologies still to be noted is a flight actuation. The flight actuator is shown in Figure 9.15 for a printer head application.[21] The impulse actuation of the multilayer actuator can hit a steel ball like a pinball machine, which is connected with a spring. A large displacement amplification (mm-distance) can be realized with this impact drive. Elimination of mechanical ringing or double strike can be achieved with the proper adjustment of the applied voltage pulse width.

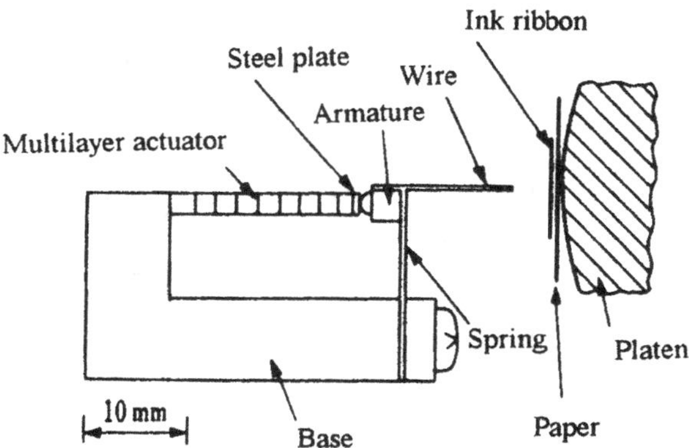

FIGURE 9.15 A flight actuator printer head. (From T. Ota et al: *Jpn. J. Appl. Phys.* 24(Suppl. 24–3), 193, 1985.)

The current generation of printers, following the impact dot-matrix types, includes inkjet and laser printers. These satisfy the need for higher image quality and faster printing.

9.6.1 Basic Design of the Piezoelectric Inkjet Printer Head

In this section, we examine the basic design of a piezoelectric inkjet printer head capable of printing intermediate grayscales.[22–24] A drop-on-demand type of piezoelectric printer head is shown in Figure 9.16. The head is composed of a piezoelectric ceramic cylinder, a micro one-way valve, and a flow resistance component, which together function as a tiny pump. This micropump responds over a broad range of frequencies and can produce a wide range of ink drop sizes.[22] The key component of this printer head is the microvalve. The valve and base patterns are successively plated onto a substrate using electroforming techniques. They are then removed from the substrate by a delamination process.

The area of the printed dot can be adjusted to four distinct levels with the application of an appropriate drive voltage pulse width. The smallest dot diameter possible for this print head is 80 μm. Grayscale printing is achieved by using dot-area modulation. The dot density for a 2×2 matrix created with five gray levels is 8 dots/mm (which corresponds to a pixel density of 4 pel/mm). The reflection density from the grayscale pattern printed with this print head, operated with a drive frequency of 10 kHz, exhibits a range of 17 intermediate grayscales. Another piezoelectric inkjet printer head incorporating a piezoelectric bimorph is shown in Figure 9.17a.[25]

9.6.2 Integrated Piezo-Segment Printer Head (Piezo Bimorph)

A high-speed, high-resolution color printer that makes use of integrated piezoelectric segments has been developed by Seiko-Epson.[26] The inkjet head structure incorporating the multilayer ceramic with hyperintegrated piezoelectric segments (MLChips) is pictured in Figure 9.17b. The inkjet is activated primarily by the bending mode of a piezoelectric unimorph, similar to Figure 9.17a. Layers of zirconia (to form the ink chambers), the vibration plate, and the PZT actuator plate are co-fired in the fabrication process. This produces a stable structure, which allows for adequate vibration isolation between ink chambers, as well as some retardation of mechanical aging effects.

An MLChips unit 10×10 mm in size is composed of two arrays of 48 ink chambers. High-precision nozzle fabrication is essential for proper ink drop generation and ink impact accuracy; thus, two advanced processes are employed: submicrometer-precision pressing and microchip water repellency treatment. The so-called face eject back contact structure of the printer head is depicted in Figure 9.18a. This laminated structure, which more readily allows for various

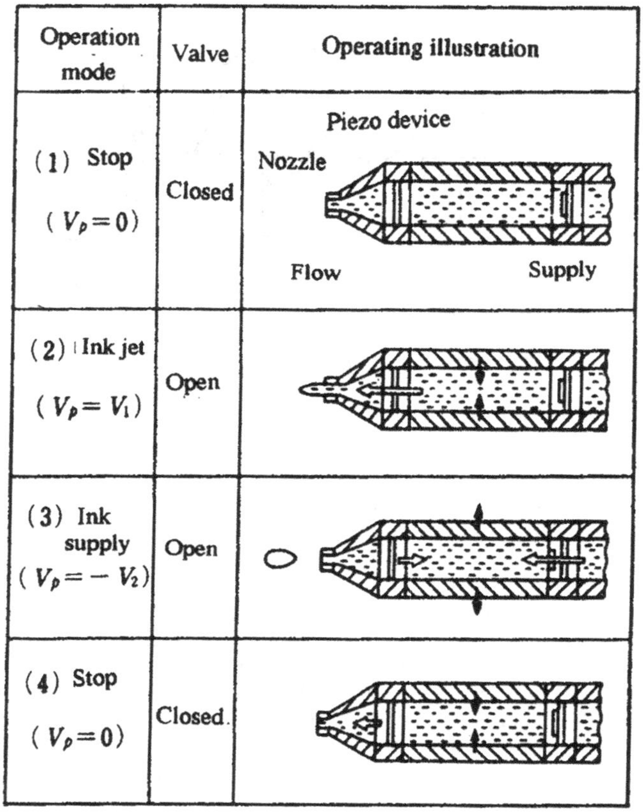

Operation mode	Valve	Operating illustration
(1) Stop ($V_p = 0$)	Closed	Piezo device Nozzle Flow Supply
(2) Ink jet ($V_p = V_1$)	Open	
(3) Ink supply ($V_p = -V_2$)	Open	
(4) Stop ($V_p = 0$)	Closed	

FIGURE 9.16 Operation of a drop-on-demand–type piezoelectric printer head. (From Tsuzuki, M et al: *Symp. Jpn. Electr. Commun. Soc.*, IE 83–59, 1983.)

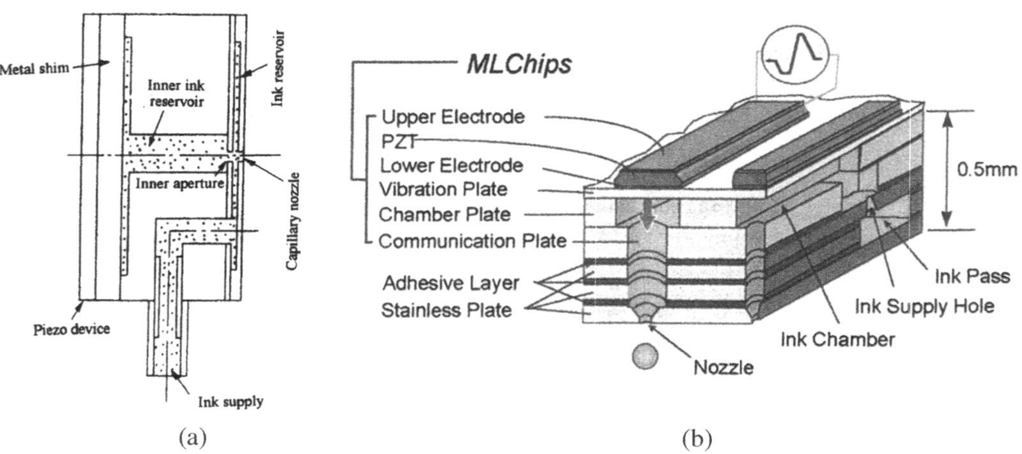

FIGURE 9.17 (a) Piezo-bimorph inkjet print head (From Nikkei Electronics, Oct. 3, p.72, 1979.); (b) inkjet head structure incorporating the multilayer ceramic with hyperintegrated piezoelectric segments (MLChips) developed by Seiko-Epson. (From N. Kurashima: *Proc. Machine Tech. Inst. Seminar*, MITI, Tsukuba, Japan, 1999.)

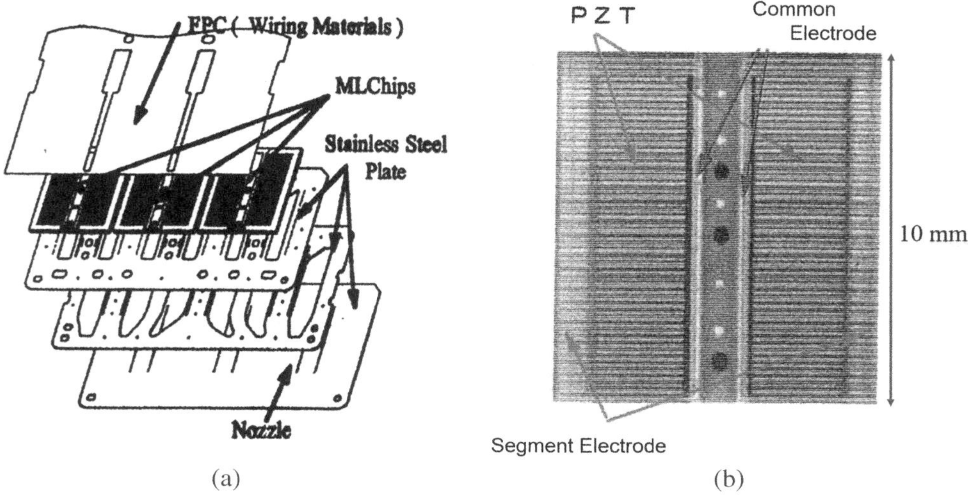

FIGURE 9.18 (a) The face eject back contact structure of the printer head incorporating MLChips by Seiko-Epson. (b) MLChips photo. (From N. Kurashima: *Proc. Machine Tech. Inst. Seminar*, MITI, Tsukuba, Japan, 1999.)

configurations of the inkjet nozzles in a 2-dimensional array, is well suited to mass production at low cost (Figure 9.18b).

High-speed meniscus control is achieved by implementing the pull-push-pull (PPP) method. The concept is illustrated in Figure 9.19a for normal-size dot generation. An intermediate voltage is applied to the PZT actuator in its initial state, as shown in Figure 9.19b. The first "pull" stage of the process occurs when the applied voltage is reduced slowly and the ink meniscus is smoothly drawn into the nozzle. The "push" occurs with a rapid step up to the highest voltage, thereby exciting a large vibration in the vibration plate that causes ink to be ejected from the nozzle. Finally, in the second "pull" stage, the voltage is reduced to an intermediate level at a rate that effectively damps the meniscus vibration quickly. Note that the Step (2) and (3) quick periods should be adjusted to the ink chamber resonance period so as not to generate unnecessary vibration ringing. This three-step meniscus control process provides precise and rapid ink drop formation and delivery to the page.

In order to achieve higher-definition imaging, Epson invented the multi-dot-size printing method by using two-stage dot generation, the operation of which is illustrated in Figure 9.20a. Compared with normal dot generation (Figure 9.19a), by introducing a much sharper initial "pull" process, a microdot can be generated in the former half of one cycle period. In the latter half of one cycle, a normal dot is generated successively. By suitably combining these two micro and normal dot generation processes, four levels of dot size, that is, "zero," "micro dot," "normal dot," or "micro + normal dot" can be obtained in one cycle. These ink droplet size changes are simulated and visualized in Figure 9.20b.

The piezoelectric printer head characteristics when commercialized were as follows:

1. Ink Chamber Density: 120 dpi
2. Number of Nozzles [N]: $48N \times 6$ colors
3. Response: 28.2 kHz
4. Minimum Ink Drop: 4 pl
5. High-Definition Mode: Multisized Dot Technology 4/7/11 pl
6. High-Speed Mode: MSDT 11/23/39 pl

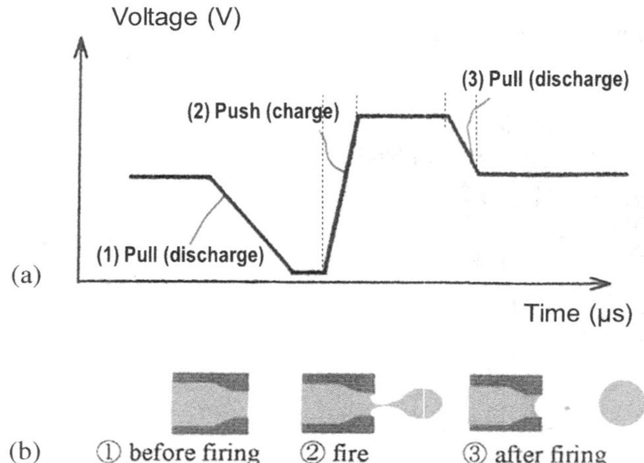

FIGURE 9.19 (a) The pull-push-pull meniscus control method. (From N. Kurashima: *Proc. Machine Tech. Inst. Seminar*, MITI, Tsukuba, Japan, 1999.) (b) Ink droplet shape change (simulation).

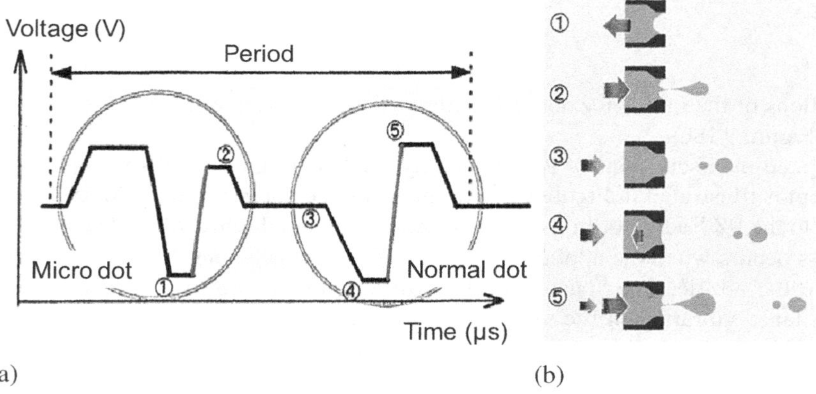

FIGURE 9.20 (a) The two-stage pull-push-pull meniscus control method to generate micro- and normal dots.) (b) Ink droplet shape change. (From N. Kurashima: *Proc. Machine Tech. Inst. Seminar*, MITI, Tsukuba, Japan, 1999.

9.6.3 Piezo Multilayer Inkjet

On the contrary, Canon developed the so-called "bubble jet" printer by using a local thermal boiling technique of ink, which shows slower printing speed than the piezoelectric bimorph inkjet, but finer resolution. Thus, in order to compete with the bubble-jet fine resolution, Seiko-Epson used the multilayer piezoelectric actuator with narrow width to replace a wide bimorph structure[26] (private communication with Akio Owatari, then general manager). Compare Figure 9.21 with Figure 9.17b. Though the price is higher than the bubble jet type, the new piezo ML jet exhibits much quicker printing speed with a higher level of resolution.

9.7 DIESEL PIEZO-INJECTION VALVE

Diesel engines are recommended rather than regular gasoline cars from an energy conservation and global warming viewpoint. Why? We need to consider the total energy of gasoline production; well-to-tank and tank-to-wheel. The energy efficiency, measured by the total energy required to

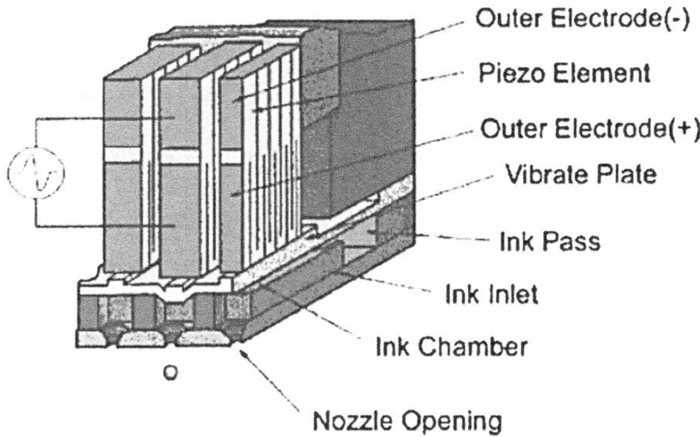

FIGURE 9.21 Piezo ML jet (Seiko-Epson). (From N. Kurashima: *Proc. Machine Tech. Inst. Seminar*, MITI, Tsukuba, Japan, 1999.) (Courtesy of Seiko-Epson.)

realize unit drive distance for a vehicle (MJ/km), is of course better for high-octane gasoline than diesel oil. However, since the electric energy required for purification is significant, gasoline is inferior to diesel.[27] However, as is well known, the conventional diesel engine generates toxic exhaust gases such as SO_x and NO_x. In order to solve this problem, injection valves should make a very fine mist of diesel fuel in order to burn it completely to minimize the generation of "particulate matter" (PM) dust as well as toxic gases. New diesel injection valves were developed by Siemens, Bosch, and Toyota with piezoelectric multilayer actuators (Bosch news release in 2004; Reference (28)).

In order to eliminate toxic SO_x and NO_x and increase diesel engine efficiency, high-pressure fuel and quick injection control are required by using a so-called "common rail system." Figure 9.22a shows the concept of how to create a fine diesel fuel mist. Two key issues: (1) high common-rail back pressure and (2) quick multiple injection are required. As shown in the injection timing chart of Figure 9.22b, five (pilot, pre, main, after, and post) injections are required during one engine cycle (typically 60 Hz). Thus, the actuator response should be a minimum 100 μs (or quicker), which the electromagnetic actuators cannot satisfy, but merely piezoelectric MLs. Note that the ML piezo-actuator is driven basically by an on-off voltage, or more precisely a trapezoidal voltage. The voltage rise and fall period should be adjusted exactly to the actuator system resonance period. Otherwise, the displacement overshoot and following ringing would create a significant problem for stable fuel injection. The piezoelectric actuator is a key component to increase the burning efficiency and minimize the toxic exhaust gas elements.

However, the highest reliability of the actuator component at an elevated temperature (150°C) for a long period (10 years) is required, which took a long research period.

Of course, just from the environmental viewpoint, electric cars with fuel cells or rechargeable batteries will be the best. However, because very high investment is required to increase the mileage rate, electric cars are still not popular. The short-range target may be, in the author's opinion, hybrid cars with a fuel cell and a diesel engine (not the present gasoline engine!).

9.7.1 Piezo-Actuator Material Development

Required actuator specifications for diesel injection valve applications are summarized as:

- Displacement: 50 μm/length 50 mm
    - Strain level: 0.14% at 2 kV/mm (effective $d_{33} \approx 700$ pm/V)
- Temperature stability: ± 5% for −40°C to 150°C
    - Curie temperature higher than 335°C

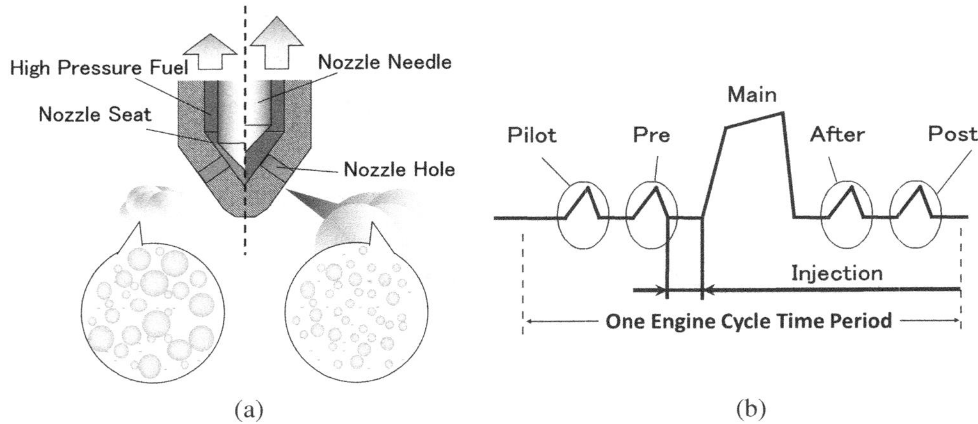

FIGURE 9.22 (a) Schematic illustration on how to create fine diesel fuel mist and (b) multiple injection timing chart for the fine mist generation.

- Life time: longer than 10^9 cycles
  - Under a severe drive condition: switching time = 50 μs
  - Electric field = 2 kV/mm
- Heat generation
- Manufacturing price
  - Tape-casting co-firing technique is required.

Though detailed compositions are a "trade secret" in each company, they are PZT-based with a Curie temperature around 300°C, the low-voltage $d_{33} \approx 350$ pm/V, mechanical quality factor ≈500. Future research directions for actuator materials will include: (1) lower-temperature sinterable PZTs, aiming at Cu internal electrode ink usage and manufacturing energy saving, and (2) Pb-free piezoelectric ceramics to overcome social regulations such as RoHS.

9.7.2 Multilayer Design

The internal electrode design is not a conventional "interdigital" type for the automobile applications, but still keeps the unelectroded part at the corner point of each electrode layer in order to make the external side electrode inexpensively. With increasing the actuator length up to 50 mm, these unelectroded inactive parts start to accumulate significant tensile stress under the electric field application. Figure 9.23a shows experimentally obtained results by Siemens.[29] With increasing the applied electric field on the long ML actuator, the first crack occurred at 960 V/mm at the length center of this actuator stack corner. Then, still an increase in the field, cracks 2 and 3 occurred at almost the same fields: 1130, 1140 V/mm, now at the length quarter of the stack. Since the length center part already released the stress concentration once the first crack happened, the second cracks happened again at the center of the segmented 1/2 length. More cracks can be expected to occur when we apply the electric field up to the maximum 2 kV/mm.

According to the above results, Denso Corporation engineers decided to laminate 2-mm-thick piezo ML units by using epoxy resin to create a 50-mm-long piezo-ML actuator, as shown in Figure 9.23b. Each 2-mm piezo unit consists of 80 μm PZT thin films (25 layers), which does not generate any cracks during operation up to 2 kV/mm. An 80-μm thickness was selected by the economical situation, not by the technological restriction. Refer to Chapter Problem 9.3. The resin mold design can reduce the unnecessary tensile stress accumulation from the stack actuators.

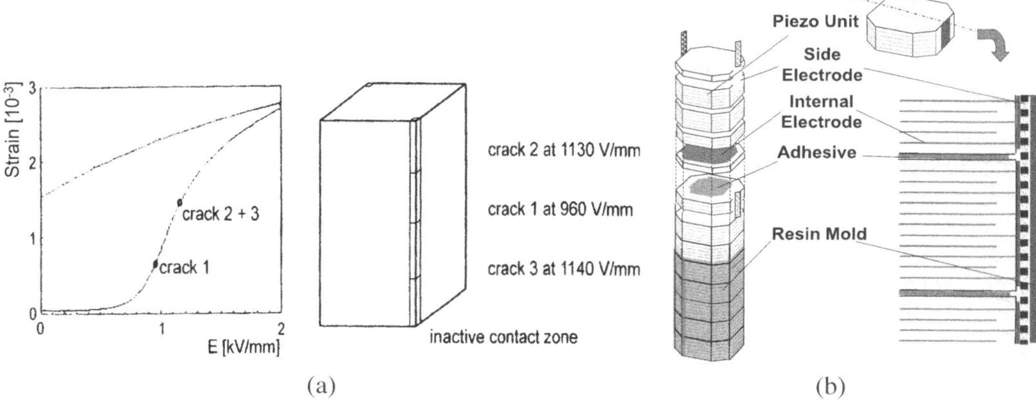

FIGURE 9.23 (a) Side crack initiation processes in a multilayer piezoelectric actuator. (From K. Reichmann: *107th Annual Meeting—Amer Ceram. Soc.*, April 10–13, Baltimore, MD, 2005.) (b) Multistacking of ML actuators for diesel injection valves. (From A. Fujii, *Proc. Smart Actuators/Sensors Study Committee*, JTTAS, Dec. 2, Tokyo, 2005.)

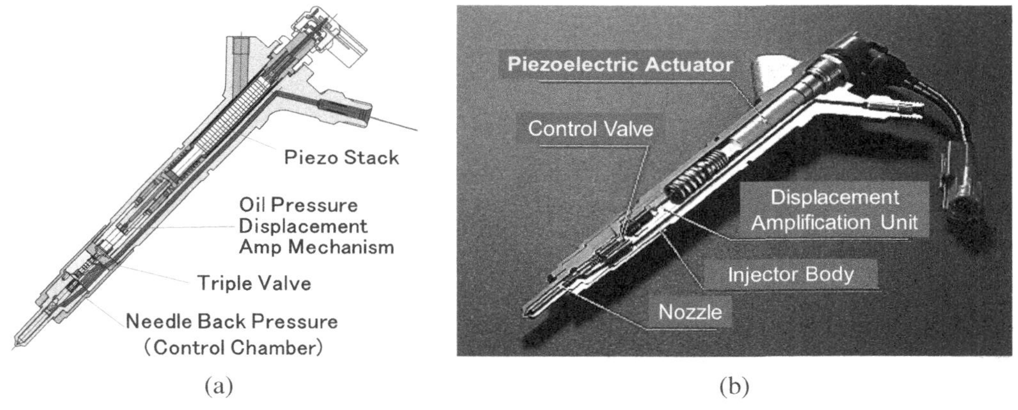

FIGURE 9.24 (a) Diesel injection valve structure with a piezo-ML actuator and an oil-pressure displacement amplifier and (b) diesel injection valve cross-section picture. (Courtesy by Denso Corporation, Japan.)

9.7.3 Diesel Injection Valve Assembly

Since the displacement 50 μm/length 50 mm from the piezo-stack actuator is not sufficient to move the injection needle, they coupled the oil pressure displacement amplification mechanism. Figure 9.24a shows the diesel injection valve structure with a piezo ML actuator and an oil-pressure displacement amplifier, and its cross-section picture is inserted in Figure 9.24b.[28] The oil-pressure mechanism can amplify the needle displacement up to the submillimeter level by keeping the responsivity.

CHAPTER ESSENTIALS

1. *Merits of the Multilayer Actuator*: as compared to the bimorph type:
   a. Longer lifetime
   b. Faster printing speed
   c. Higher impact force
   d. Higher efficiency

2. *Impulse Drive Motor*: Sawtooth-type voltage on the actuator generates a sawtooth-type displacement on the stator, leading to continuous motion of the slider via "stick & slip" action on the contact point.
3. *Latching Relays*: Typically constitute only 10% of the total relay production, but the unit price is about 10 times higher than nonlatching varieties, leading to comparable total revenue—highly functional and valuable devices!
4. *Shape-Memory Ceramic Latching*: Use of a smart material such as this in the design of a relay can be effective in reducing the number of components and the total size of the device.
5. *MEMS Mechanical Relays*: Higher current flow capability than semiconductor electronic relays.
6. *Automobile Adaptive Suspension System*: Switching the narrow and wide nozzle sizes for changing the oil friction, according to the road roughness, to provide both drive controllability and comfort.
7. *Advantages of the Piezoelectric Printer*: Compared to conventional electromagnetic types:
   a Low energy consumption
   b Low heat generation (electromagnetic solenoids generate significant heat!)
   c Fast printing speed
8. *Co-Fired Integrated Multilayer Printer Head Structure*: Allows for successful mass production of the ink-jet printer with high mechanical stability.
9. *Diesel injection valve application*: required the developments in (1) new material, (2) new ML design, and (3) new drive scheme.

CHECK POINT

1. There is a piezoelectric multilayer actuator with a length 10 mm and a cross section area 5×5 mm^2. We coupled a monolithic hinge mechanism to amplify the displacement up to 100 μm. How large a force is expected from the tip of this hinge mechanism? 1 N, 10 N, 100 N, 1 kN, or 10 kN?
2. (T/F) When a piezoelectric actuator is driven by a rectangular pulse voltage, the mechanical ringing is completely suppressed when the pulse width is adjusted exactly to half of the resonance period of the sample. True or False?
3. When a piezoelectric actuator is driven by a rectangular pulse voltage, the vibration displacement overshoot is excited (by neglecting mechanical loss). What is the maximum overshoot range (percentage) in comparison with the normal (pseudostatic) operation? 16.7%, 33%, 50%, 100%, or 200% larger?
4. In a diesel injection valve application, the ML piezo-actuator is driven basically by on-off, or more precisely a trapezoidal voltage. How should we adjust the voltage rise or fall time period to realize a stable injection operation?

CHAPTER PROBLEMS

9.1 Piezoelectric actuators can be employed in laser and ink-jet printers. Survey the recent literature on this issue.

9.2 List and briefly describe the major problems of using a piezoelectric ceramic in a pulse drive device (in terms of the tensile stress, heat generation, etc.).

9.3 The multilayer piezoelectric actuator for a diesel injection valve application consists of 80-μm PZT thin films (25 layers for each unit) at present, which is driven by 160 V (or electric field 2 kV/mm). Just from technological restrictions, 10–20-μm thickness is more popularly used, or even 2–3 μm is not difficult for ML capacitors. Why was 80 μm selected? Consider from a manufacturing cost viewpoint.

Hint

Refer to the figure below. When we use an Ag/Pd (30%) internal electrode for the ML actuators, the cost of the actuator is primarily determined by the number of internal electrodes; that is, the cost increases exponentially with reducing the drive voltage. On the contrary, the power supply costs more with increases in the voltage and current range. In order to minimize the system cost, we should consider the sum of the ML actuator cost and the power supply price, leading to a drive voltage around 160 V and a layer thickness of 80 μm.

REFERENCES

1. Y. Endo, Y. Egawa, N. Harada and G. Yoshida: *Proc. Jpn. Television Soc.*, 3–19, 1983.
2. K. Yokoyama and C. Tanuma: *Electronic Ceramics*, Spring, 15, p.45, 1984.
3. K. Shibata, K. Takeuchi, T. Tanaka, T. Yokoo, S. Nakano and Y. Kuwano: *Jpn. J. Appl. Phys.*, 24(Suppl. 24–3), 181, 1985.
4. M. Aizawa, K. Uchino and S. Nomura: *Jpn. J. Appl. Phys.*, 22, 1925, 1983.
5. Tojo and Sugihara: *Proc. Jpn. Precision Eng.*, p. 423, Fall, 1983.
6. M. P. Coster: *Proc. 4th Int'l Conf. on New Actuators, 2.6*, p. 144, Germany, 1994.
7. R. Gloess: *Proc. 4th Int'l Conf. on New Actuators, P26*, p. 190, Germany, 1994.
8. N. Aoyama et al.: *Proc. J. Precision Engr. Soc. P.897*, 1992.
9. Y. Okamoto, R. Yoshida and H. Sueyoshi, *Konica Minolta Tech. Report*, 1, 23, 2004.
10. http://www.piezo-tech.com/eng/product/ 2008.
11. T. Higuchi, M. Watanabe and K. Kudo: *J. Precision Engr.*, 54(11), 2107, 1988.
12. Piezo Electric Products Inc., Product Catalog.
13. Sato, Taniguchi and Ohba: *OMRON Technics*, (70), 52, 1983.
14. S. Mitsuhashi, K. Wakamatsu, Y. Aihara and N. Okihara: *Jpn. J. Appl. Phys.*, 24(Suppl. 24–3), 190, 1985.
15. A. Furuta, K. Y. Oh and K. Uchino: *Sensor and Mater.*, 3(4), 205, 1992.
16. http://www.emrl.de/r_a_2.html
17. Y. Yokoya: *Electronic Ceramics*, 22(111), 55, 1991.
18. K. Yano, T. Hamatsuki, I. Fukui and E. Sato: *Proc. Jpn. Electr. Commun. Soc.*, 1–157, Spring, 1984.
19. K. Yano, I. Fukui, E. Sato, O. Inui and Y. Miyazaki: *Proc. Jpn. Electr. Commun. Soc.*, 1–156, Spring, 1984.
20. K. Yano, T. Inoue, S. Takahashi and I. Fukui: *Proc. Jpn. Electr. Commun. Soc.*, 1–159, Spring, 1984.
21. T. Ota, T. Uchikawa and T. Mizutani: *Jpn. J. Appl. Phys.*, 24(Suppl. 24–3), 193, 1985.
22. M. Tsuzuki, M. Suga and H. Banno: *Symp. Jpn. Electr. Commun. Soc.*, IE 83–59, 1983.
23. M. Suga, M. Tsuzuki and H. Banno: *Symp. Jpn. Electr. Commun. Soc.*, EMC 84–46, 1984.
24. M. Suga and M. Tsuzuki: *Jpn. J. Appl. Phys.*, 23, 765, 1984.
25. Nikkei Electronics, Oct. 3, p.72, 1979.
26. N. Kurashima: *Proc. Machine Tech. Inst. Seminar*, MITI, Tsukuba, Japan, 1999.
27. www.marklines.com/ja/amreport/rep094_200208.jsp
28. A. Fujii, *Proc. Smart Actuators/Sensors Study Committee*, JTTAS, Dec. 2, Tokyo, 2005.
29. K. Reichmann: *107th Annual Meeting—Amer. Ceram. Soc.*, April 10–13, Baltimore, MD, 2005.

10 Ultrasonic Motor Applications

The ultrasonic motor is driven usually by a *sinusoidal AC voltage* at its *resonance frequency*, while the pulse drive motor described in the previous chapter is driven by a *rectangular or sawtooth waveform* at a frequency less than the resonance frequency. High power density ultrasonic devices are divided into three general categories. The first type is operated in a mode that produces a simple ultrasonic vibration for uses, such as the ultrasonic scalpel. The second type is designed to transmit energy to air or liquid for applications such as piezoelectric fans, piezoelectric pumps, and ultrasonic diagnostic systems. *Ultrasonic motors* constitute the final category, which will be the primary focus of this chapter. The ceramic piezoelectric materials typically used in the fabrication of high power ultrasonic motors are similar to the conventional piezoelectric PZT-based materials used for vibrator/resonator applications, but a high mechanical quality factor, Q_m, is generally required for the motors, as described in Section 5.1.

10.1 BACKGROUND OF ULTRASONIC MOTORS

Electromagnetic motors have actually existed for more than 100 years. While these motors still dominate industrial applications, further dramatic improvement of the current designs is limited. These improvements await the development of suitable new magnetic and/or superconducting materials. Efficient motors of this type smaller than 1 cm are difficult to produce, because an electromagnetic motor with a submillimeter rotor used in a wristwatch, for example, requires an electromagnet still 12 mm in length. Without using an electromagnet, the rotor does not rotate! Due to a very thin Cu wire in the electromagnet, the efficiency of these micromotors does not exceed 1% (i.e., most of the input electric energy is spent for heat generation via Joule heat). Refer to Figure 1.4. Therefore, piezoelectric ultrasonic motors, whose efficiency is insensitive to their size, have gained much attention for micromotor applications.

The basic structure of an ultrasonic motor is depicted in Figure 10.1. It consists essentially of a *high-frequency power supply* (i.e., ultrasonic range, inaudible to humans; typically 30–300 kHz is allowed in the industrial standard), a *stator*, and a *slider/rotor* piece. The stator is composed of a piezoelectric driver and an elastic vibrator. The slider has an elastic moving component with a *friction coat* on the side facing the stator. The elastic surface vibrates in an elliptical shape (i.e., two directional vibrations are superposed with a certain phase lag), so that the slider attached will move to one direction owing to "mechanical rectification." The rectification details are discussed in Section 10.8. The motor is driven by a compact and inexpensive drive circuit (sine or rectangular wave form), which, in conjunction with its simple design, makes it especially suitable for mass production.

The first practical ultrasonic motor was proposed by Barth of IBM in 1973 (Figure 10.2).[1] The rotor in this design is in contact with two horns placed on either side of it. When one of the horns is driven by its piezoelectric vibrator, the horn moves up/down at its resonance frequency. Since the horn tip contacts a rotor with a cant angle, the rotor is driven in one direction via the friction force. Rotation in the opposite direction occurs when the other horn is driven. Various mechanisms based on virtually the same principle have been subsequently proposed.[2,3] However, the motors of this design were ultimately not useful, due to changes in the vibration amplitude that tend to occur at higher temperatures (thermal expansion) and as a consequence of mechanical wear. In the 1980s, more precise and sophisticated positioners, which do not generate magnetic field noise, were required by integrated circuit manufacturers as the pattern density on chips continued to increase. This trend helped to accelerate the development of new ultrasonic motors. The merits and demerits of modern ultrasonic motors are summarized in Table 10.1.

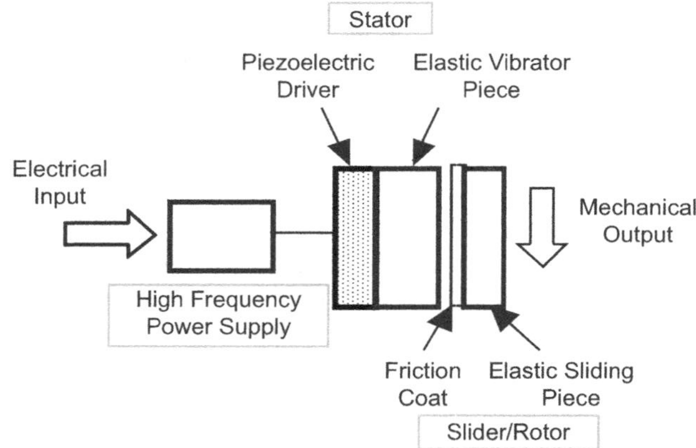

FIGURE 10.1 Basic structure of the ultrasonic motor.

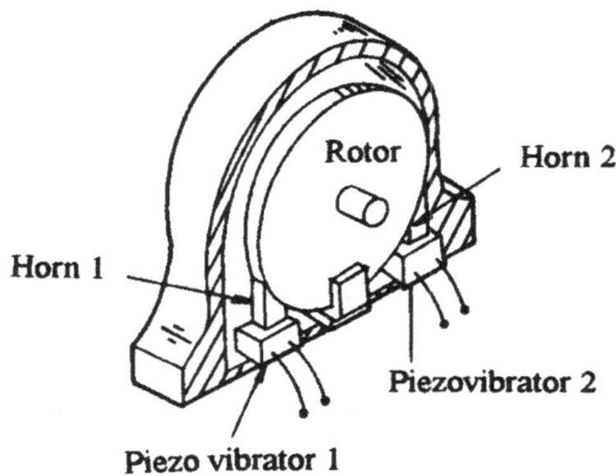

FIGURE 10.2 The first practical ultrasonic motor proposed by H. V. Barth of IBM in 1973. (From H. V. Barth: *IBM Technical Disclosure Bull.*, 16, 2263, 1973.)

TABLE 10.1
The Merits and Demerits of Ultrasonic Motors Compared with Electromagnetic Motors

Merits	Demerits
• Low speed and high torque	• High-frequency power supply required
• Direct drive	• Torque vs. speed droop
• Quick response	• Low durability (surface ware)
• Wide range of speeds	
• Hard brake with no backlash	
• Excellent controllability	
• Fine position resolution	
• High power-to-weight ratio	
• Quiet drive—Inaudible to human	
• Compact size and light weight	
• Simple structure and easy Production	
• No generation of electromagnetic radiation	
• Unaffected by external electric or magnetic fields	

EXAMPLE PROBLEM 10.1

Write expressions for the input current density, the induced strain, and the stored electric and elastic energies and any related quantities for a longitudinal piezoelectric actuator $k_{33}(d_{33}, s_{33}^E, \varepsilon_{33}, \text{length}: l)$ under (1) DC, (2) AC off-resonance, and (3) AC resonance drive conditions.

SOLUTION

1. DC Drive

Electric Field:	$E_3 = V/l$
Polarization:	$P_3 = \varepsilon_0 \varepsilon_{33} E_3$
Generative Strain:	$x_3 = d_{33} E_3$
Generative Stress:	$X_3 = x_3/s_{33}^E = (d_{33}/s_{33}^E)E_3$
Stored Electric Energy:	$U_{EE} = (P_3E_3)/2 = (\varepsilon_0\varepsilon_{33}E_3^2)/2$
Stored Piezoelectric Energy:	$U_{ME} = (x_3X_3)/2 = (d_{33}^2E_3^2)/2s_{33}^E$
Electromechanical Coupling Factor:	$k_{33}^2 = U_{ME}/U_{EE} = d_{33}^2/\varepsilon_0\varepsilon_{33}s_{33}^E$

2. AC Drive (off-resonance)

Electric Field:	$E_3 = [V_o\sin(\omega t)]/l$
Current Density:	$J_3 = Y_oE_3 = \omega\, \varepsilon_0\, \varepsilon_{33}\, E_3$
Vibration Strain:	$x_3 = d_{33}E_3$
Vibration Velocity:	$v = \omega\, d_{33}\, E_3$
Input Electric Energy Density:	$u_{EE} = J_3E_3 = \omega\varepsilon_0\varepsilon_{33}E_3^2$
Mechanical Energy Density:	$u_{ME} = \omega\left(d_{33}^2/s_{33}^E\right)E_3^2 = \omega k_{33}^2\varepsilon_0\varepsilon_{33}E_3^2$

3. AC Drive (resonance k_{33} mode)

Electric Field:	$E_3 = [V_o\sin(\omega_o t)]/l$
Current Density:	$J_3 = Y_m\, E_3 = \omega\alpha\, \varepsilon_0\, \varepsilon_{33}\, E_3$
Sound Velocity:	$v = 1/\sqrt{\rho s_{33}^D}$ [ρ: mass density; $s_{33}^D = s_{33}^E\left(1-k_{33}^2\right)$]
Resonance Frequency:	$f_o = v/2\, l$
Resonance Amplification:	$\alpha = Y_m/Y_o$ [$\propto Q_m$]
Vibration Strain:	$x_3 = \alpha d_{33}E_3$
Maximum Vibration Velocity:	$v_m = \omega_o\alpha d_{33}E_3\, l = \pi\, \alpha\, v\, d_{33}\, E_3$
Sound Pressure:	$p = \rho v(v_m) = \pi\alpha\rho v^2 d_{33}E_3 = \pi\alpha(d_{33}/s_{33}^E)E_3$
Input Electric Power:	$P = IV = J_3SE_3l = \omega_o\alpha\varepsilon_0\varepsilon_{33}E_3^2v_m$
Mechanical Energy Density:	$u_{ME} = \omega_o\alpha E_3 l(d_{33}^2/s_{33}^E)E_3S = \omega_o\alpha(d_{33}^2/s_{33}^E)E_3v_m$

Note: Some of these parameters are α times larger than those defined for the off-resonance mode. α is proportional to the mechanical quality factor, Q_m.

10.2 CLASSIFICATION OF ULTRASONIC MOTORS

There are two general types of ultrasonic motors: the *standing-wave type* and the *traveling wave type*.

The vibration induced in the motor is either in the form of a standing wave or a traveling wave. Recall that a *standing wave* is described by

$$u_s(x,t) = A\,(\cos(kx))\cdot(\cos(\omega t)) \quad (k\text{: wave vector, } \omega\text{: frequency}) \tag{10.1}$$

[node $\cos(kx)=0$ does not move with t], while the equation for a *traveling wave* is

$$u_t(x,t)=A\cos(kx-\omega t) \tag{10.2}$$

[node moves with time lapse t with the velocity ω/k].

Equation 10.2 can be rewritten by using a trigonometric relation as

$$u_t(x,t)=A\,(\cos(kx))\cdot(\cos(\omega t))+A\,(\cos(kx-\pi/2))\cdot(\cos(\omega t\text{-}\pi/2)) \tag{10.3}$$

Two terms are both composed of the product of space and time terms; that is to say, both terms are the standing waves. This leads to an important concept. *A traveling wave can be generated by superposing two standing waves with a phase difference of 90° between them in terms of space and time.* Taking into account the fact that a traveling wave is not stable in a finite size matter, but only a standing wave remains in it, this superposition is the only feasible way to generate a stable traveling wave in a structure of finite size and volume.

USMs can further be distinguished by their physical shape (rod, ring, cylinder, etc.), as depicted in Figure 10.3.

10.2.1 Standing-Wave Motor Principles

The standing-wave motor is sometimes referred to as a *vibratory coupler* or a *woodpecker motor.* A vibratory piece is connected to the piezoelectric driver and the tip portion executes an elliptical displacement, as shown in Figure 10.4.[4] Canting the x-axis of the vibratory piece slightly (angle θ)

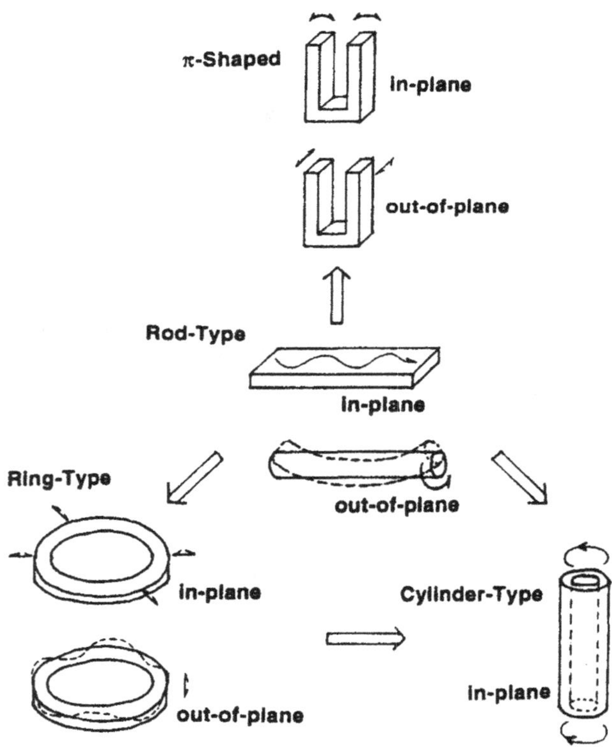

FIGURE 10.3 Various USM configurations.

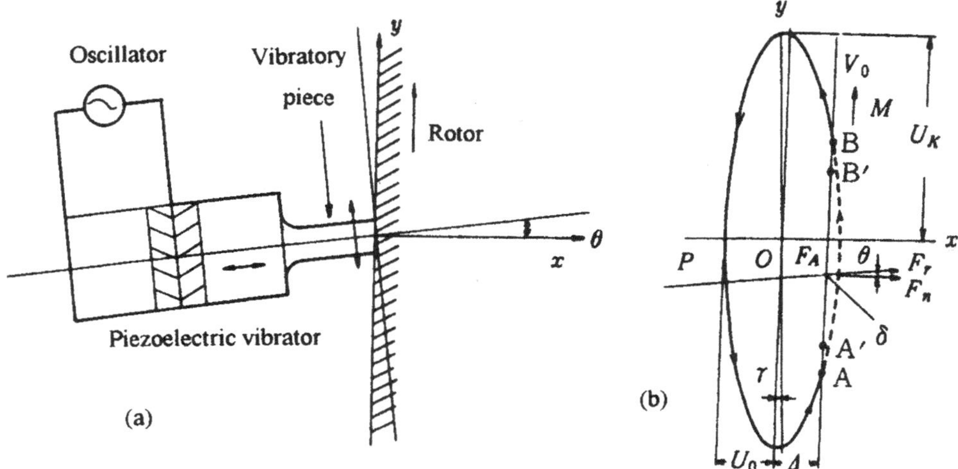

FIGURE 10.4 Vibratory coupler type of motor: (a) system configuration and (b) locus of vibrator tip locus. (From T. Sashida: *Oyo Butsuri*, 51, 713, 1982.)

from the normal to the rotor face is the key to inducing the y-direction vibration, as seen in Figure 10.4a, leading to the elliptic tip locus. When a displacement

$$u_x = u_o \sin(\omega t + \alpha) \tag{10.4}$$

is excited in the piezoelectric vibrator, the vibratory piece will bend due to its contact with the rotor (with cant angle θ), causing the tip of the vibrator to move along the rotor face between points A and B, as shown in Figure 10.4b. The tip then moves freely from point B to point A. When the resonance frequency of the vibratory piece, f_1, is adjusted to the resonance frequency of the piezovibrator, f_o, and the bending deformation is sufficiently small compared with the length of the piece, the displacement of the tip from point B to point A is described by

$$x(t) = u_o \sin(\omega t + \alpha) \text{ and } y(t) = u_1 \sin(\omega t + \beta) \tag{10.5}$$

which produces an *elliptical locus*. The torque is thus applied only as the vibrator tip moves from A to B. Though the driving force is intermittent, the rotation speed ripple is minimized due to the inertia of the rotor. The numerical analysis is conducted in Section 10.8.

The standing-wave motor has the advantages of low production cost and high efficiency (theoretically up to 98%). The disadvantages of this design include the *unidirectional drive* in general and some variability of the locus depending on temperature. Though the moving direction can be changed between the clockwise and counterclockwise rotations by a slight geometric asymmetry and/or drive frequency shift, the rotation speed differs somewhat between these two directions.

10.2.2 Traveling-Wave Motor Principle

The superposition of two standing waves with a 90° phase difference both in space and time produces a traveling wave in the type of motor that bears this name (also known as the *surface-wave* or *surfing* motor). A particle or point on the surface of the elastic body executes a displacement with an *elliptical locus* due to the coupling of longitudinal and transverse waves, as shown in Figure 10.5. This motor requires two vibration sources for generating the two standing waves so as to introduce the superposed traveling wave, leading to a demerit of low efficiency (not higher than 50% because

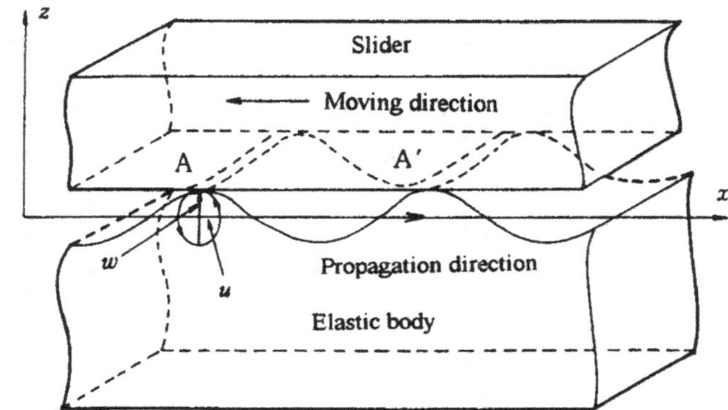

FIGURE 10.5 Traveling-wave-type motor.

two waves create one traveling wave). On the contrary, the rotation direction is easily changed by switching the phase difference from 90° to –90°.

10.2.3 Ultrasonic Motor Classification

The following categories are employed in this chapter for presenting and comparing the essential features of the most popular and widely used ultrasonic motors:

1. *Standing-Wave Motors*: In which *one major vibrational mode* is induced by means of one actuator component. Note that we include the vibratory coupler motor in this category even though it is technically not a purely standing-wave device, but rather exhibits a coupling of the standing-wave and bending-mode vibrations. The operation of the device can still be adequately described in terms of the conventional standing-wave model if appropriate allowances are made for the bending mode.
2. *Mixed-Mode Motors*: In which *two vibrational modes* are induced by means of two actuators. This type is, in a sense, a variation of the standing-wave motor, but in this case two standing waves are induced independently and combined.
3. *Traveling-Wave Motors*: In which *two vibrational modes with a phase difference of 90 degrees* between them are excited by two actuators and superposed to create a traveling wave.
4. *Mode Rotation Motors*: In which one vibrational mode is excited through the action of several actuators. This is, in a sense, a variation of the traveling-wave type, but in this case, multiple actuators are used to induce the individual vibrations with an appropriate phase difference (three-phase, four-phase, six-phase drive) to create a desired elliptic displacement when combined.

10.3 STANDING-WAVE MOTORS

10.3.1 Standing-Wave Rotary Motors

10.3.1.1 Vibratory Piece Motor (Shinsei)

The prototype structure of the rotary motor developed by Sashida is pictured in Figure 10.6a, and has four vibratory pieces (similar to Figure 10.4a), installed on the end of a cylindrical vibrator and pressed onto the rotor.[4] It can attain a rotation speed of 1500 rpm, a torque of 0.8 kgf-cm, and an output of 12 W (with 40% efficiency) from an electric input of 30 W applied at 35 kHz. This type of ultrasonic motor can achieve a much higher speed than is possible for the inchworm (described

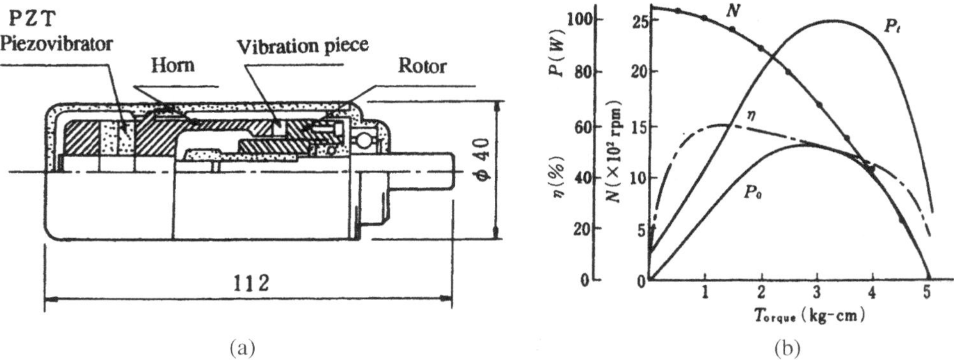

FIGURE 10.6 (a) Sashida's rotary motor structure. (b) Power-related characteristics of the rotary motor. (From T. Sashida: *Oyo Butsuri*, 51, 713, 1982.)

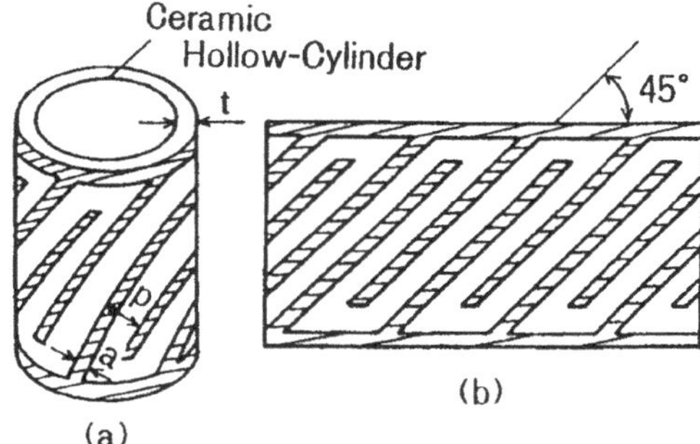

FIGURE 10.7 (a) Piezoelectric hollow cylinder torsional vibrator; (b) spread view of the interdigital electrode pattern. (From Y. Fuda and T. Yoshida: *Ferroelectrics*, 160, 323, 1994.)

in Chapter 9) because it is operated at a higher frequency corresponding to its resonance frequency where an amplified displacement occurs ($\Delta l = \alpha\ d\ E\ l$, $\alpha = 100$–500, which is proportional to the mechanical quality factor Q_m). The motor performances measured for this motor are plotted as a function of torque in Figure 10.6b. The rotation speed decreases monotonously from 2,600 rpm to 0 with increasing the torque from 0 to 5 kgf-cm (typical drooping characteristic). On the other hand, input and output power show the maximum around the load torque of 3 kgf-cm, and the maximum efficiency (P_o/P_i) 60% obtained is at $T = 1$ kgf-cm. The production of prototypes such as this one in the early 1980s motivated the surge of research that has occurred in the last three decades concerned with the development and application of ultrasonic motors.

10.3.1.2 Hollow Piezoceramic Cylinder Motor (Tokin)

Another variation on this basic design that makes use of a hollow piezoelectric ceramic cylinder as a torsional vibrator appears in Figure 10.7a.[5] An interdigital-type electrode pattern is printed at 45° angle on the cylinder surface, as shown in Figure 10.7b, so that torsional vibration can be induced in the structure. A rotor attached to the end of the cylinder is thus rotated in response to this vibration. The problem with ceramic cylinder motors is the weakness under the crash test for portable electronic equipment.

10.3.1.3 Metal Tube Motor

A simpler and more compact motor has been designed that makes use of a metal tube rather than a piezoelectric PZT, which enhances significantly the fracture toughness under the crash test and

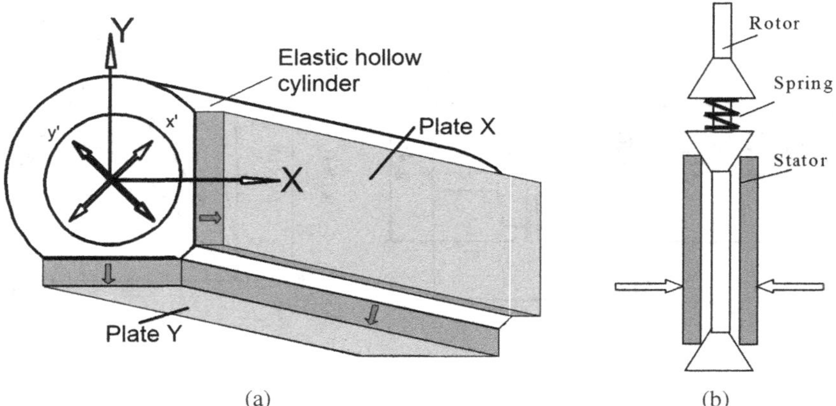

FIGURE 10.8 Metal tube rotary motor: (a) structure of the metal tube stator and (b) assembly of the entire motor. (From B. Koc et al.: *Proc. Actuator 2000 7th Int'l Conf. New Actuators*, pp. 242–2245, June 19–21, 2000.)

decreases the manufacturing cost.[6] The basic structure of the stator is shown in Figure 10.8a. Two rectangular poled PZT pieces are bonded onto the metal tube. When Plate X is driven, a bending vibration is excited along the x-axis. Due to the asymmetric mass (Plate Y) arrangement, another spurious bending vibration is induced along the y-axis that lags in phase the bending vibration established in the x direction. This results in an elliptical clockwise displacement. On the other hand, when Plate Y is driven, a counterclockwise wobble motion is excited. The rotor is a cylindrical rod with a pair of stainless ferrules separated with a spring, as shown in the diagram of the entire motor assembly appearing in Figure 10.8b. The spring force F creates the maximum friction force μF (μ is the friction constant 0.2~0.4), then the max torque $T = \mu F \times$ (inner radius). The cost for the materials and simple construction of this motor is quite low as compared with some of the more elaborate designs. Another advantage lies in the fact that only a single-phase power supply is required to drive the motor.

A metal tube ultrasonic motor of 1.8 mm in diameter and 4 mm in length is pictured in Figure 10.9a.[6] The rotor is a thin hollow tube spring, through which an optical fiber can pass. The motor performances of a metal tube motor 2.4 mm in diameter and 12 mm in length are plotted as a function of torque in Figure 10.9b. The motor was driven at 62.1 kHz in both rotation directions. A no-load speed of 1800 rpm and an output torque up to 1.8 mNm were obtained for rotation in both directions under an applied rms voltage of 80 V. The very high level of torque produced by this motor

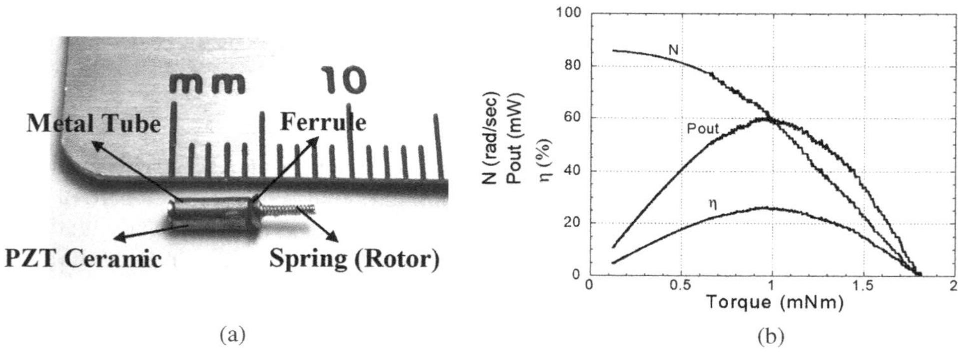

FIGURE 10.9 (a) Metal tube ultrasonic motor (diameter: 1.8 mm, length: 4 mm). (b) Motor performances of a metal tube motor plotted as a function of torque. (From B. Koc, J. F et al.: *Proc. Actuator 2000 7th Int'l Conf. New Actuators*, pp. 242–2245, June 19–21, 2000.) (diameter: 2.4 mm, length: 12 mm, operating frequency: 62.1 kHz).

is due to the dual stator configuration and the high contact force between the metal stator and rotors. The rather high maximum efficiency of about 28% for this relatively small motor is a noteworthy feature of the data presented in Figure 10.9b.

10.3.1.3.1 Camera Module Application

In collaboration with Penn State University, Samsung Electromechanics, Korea, developed a zoom and focus mechanism with two microrotary motors in 2003. Two metal tube motors 2.4 mm in diameter and 14 mm in length were installed to control zooming and focusing lenses independently in conjunction with screw mechanisms, as illustrated in Figure 10.10a.[7] A screw is rotated through a pulley, which then is transferred to the lens up-down motion. The square chip (3×3 mm^2) in Figure 10.10b is a high-frequency drive voltage supply.

Sunnytec Electronics, Taiwan, utilized the metal tube stator in a different configuration, as shown in Figure 10.10c.[8] They used the anti-node point (rod center) of the "wobbling" cylinder as a linear actuation by arranging the metal tube in an orthogonal way, so that the height along the lens motion can be dramatically reduced (i.e., flat) in comparison with the design in Figure 10.10a, where the stators are arranged in parallel to the lens movement.

10.3.1.3.2 Medical Application

The smallest metal tube motor (1.3 mm$\phi \times 3$ mm$L$) was installed on a medical catheter, and blood clot removal was successfully conducted by Mechatronics Inc., PA (unpublished). A drill larger than the motor was installed at the tip (Figure 10.11a) and a blood clot (generated in a pig body) was removed successfully (Figure 10.11b).

10.3.1.3.3 Microvehicle Application

A four-wheel vehicle ($7 \times 7 \times 7$ mm^3) produced at Penn State University is shown in Figure 10.12.[9] It is composed of two metal tube motors (4 mm in length), four wheels (active two wheels) and a mass weight (to give thrust via friction force mg·μ). The forward, backward, and sideways motions of the vehicle are controlled by a joystick remote-control device. Using rubber coat on the wheels increased traction force to climb up even on a human finger as shown in the figure.

10.3.2 Standing-Wave Linear Motors

10.3.2.1 π-Shaped Linear Motor

A *π-shaped linear motor* composed of a multilayer piezoelectric actuator and fork-shaped metallic legs is pictured in Figure 10.13a.[10] When the two legs are perfectly symmetric, pure symmetric

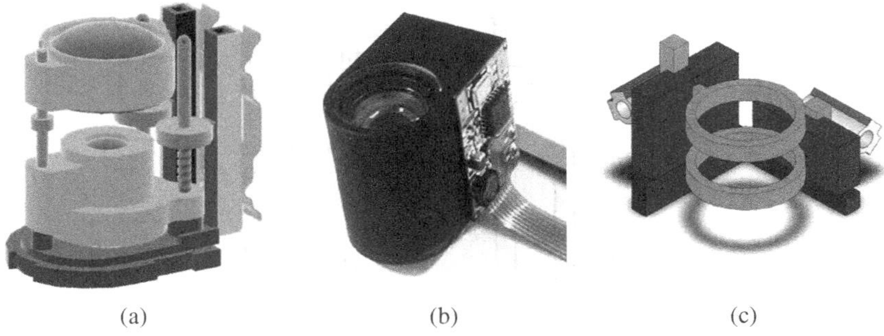

(a) (b) (c)

FIGURE 10.10 (a) Samsung camera auto zooming/focusing mechanism with two metal tube motors; (b) commercialized Samsung camera module. (From K. Uchino: Proc. *New Actuator 2004* Bremen, June14–16, p.127, 2004.) (c) Sunnytec design with two stators in an orthogonal arrangement. (From R. Lee: *Proc. Proc. Int'l Actuator Symp.*, ICAT/PSU, State College, PA, Oct., 2007.)

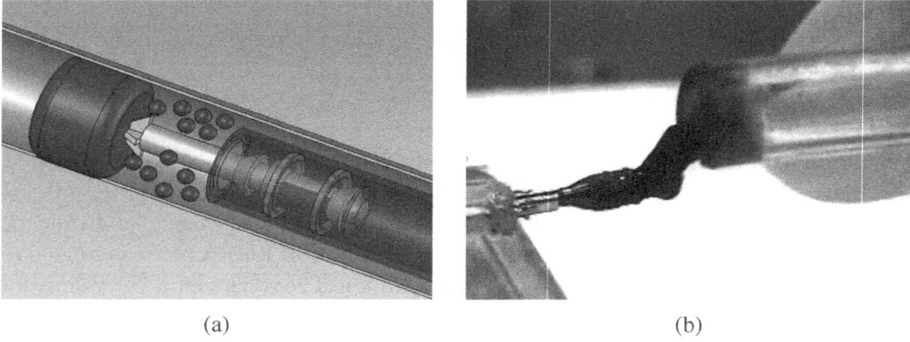

FIGURE 10.11 (a) Medical catheter drilling tool with a metal tube motor, and (b) blood clot removal test. (Courtesy by Micromechatronics Inc.)

FIGURE 10.12 Four-wheel vehicle ($7 \times 7 \times 7$ mm^3) with two metal tube motors. (From B. Akbiyik: Undergraduate Thesis, Electr. Engr., Kirikkale University, Turkey, 2001.)

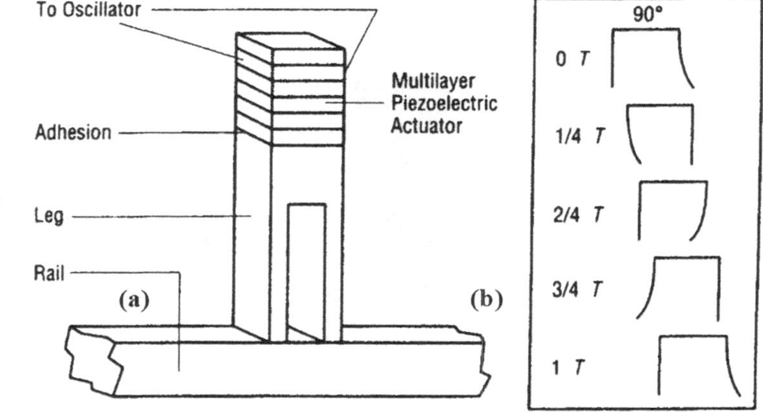

FIGURE 10.13 π-shaped linear ultrasonic (walking) motor: (a) basic structure and (b) motion sequence showing the walking action of the legs in operation. (From K. Uchino et al.: *Ferroelectrics*, 87, 331, 1988.)

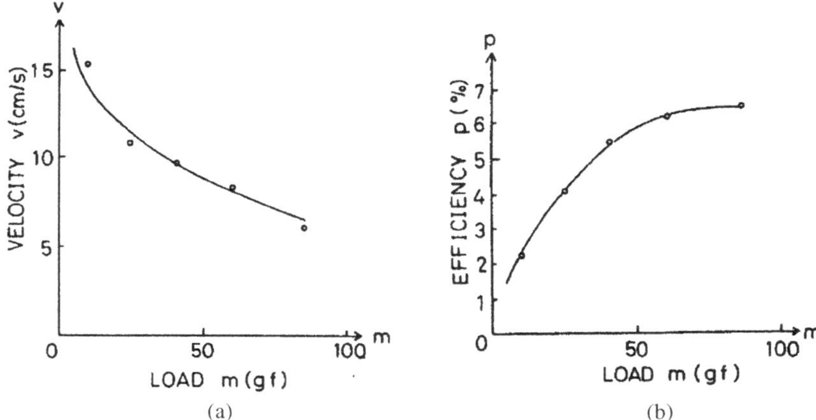

FIGURE 10.14 Motor performances of a π-shaped motor: (a) velocity and (b) efficiency as a function of load. (From K. Uchino et al.: *Ferroelectrics*, 87, 331, 1988.)

(open/closing) and antisymmetric vibrations will occur (i.e., no motion is expected). Our invention is *slight asymmetry* in these legs. Due to the slight difference in mechanical resonance between the two legs, there will be a phase difference of 90 or –90° (depending on the drive frequency) between the bending vibrations set up in the legs (Figure 10.13b). The walking slider moves similarly to the way a horse uses its fore and hind legs to trot. A motor with dimensions $20 \times 20 \times 5$ mm^3 exhibits a maximum speed of 15 cm/s and a maximum thrust of 0.2 kgf with an efficiency of 20% when driven by a 6 V, 98 kHz (actual power = 0.7 W) input signal as indicated in Figure 10.14. This motor has been employed in a precision printer stage for drawing large maps.

10.3.2.2 Poly-Vinylidene-Difluoride Walker

Two designs for PVDF bimorph walking machines are illustrated in Figure 10.15.[11] The devices were fabricated from two PVDF films, each with a thickness of 30 μm, bonded together and bent with a curvature of 1 cm^{-1}. The legs of the walker are of slightly different widths so that there should be a difference between their resonance frequencies. Accordingly, the leg movement required for both clockwise and counterclockwise rotations of the device can be achieved.

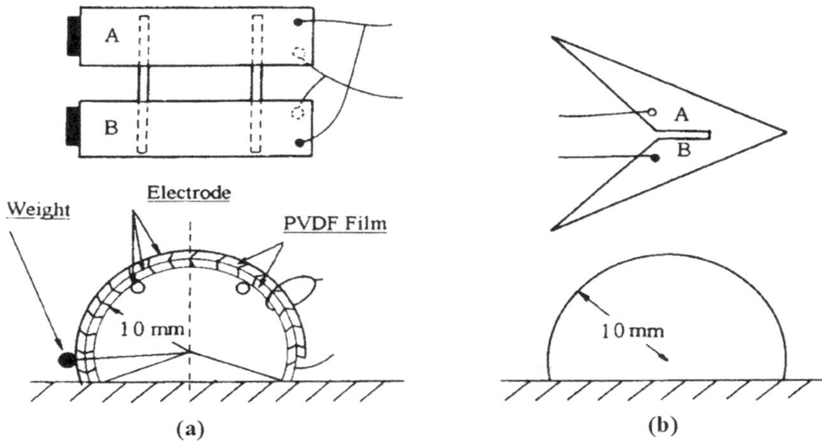

FIGURE 10.15 Two designs for PVDF bimorph walking machines: (a) arch and (b) triangular configurations. (From T. Hayashi: *Proc. Jpn. Electr. Commun. Soc.*, US84-8, p.25, June, 1984.)

10.4 MIXED-MODE MOTORS

10.4.1 "Kumada" Motor

A significantly improved rotary motor has been developed by Kumada at Hitachi Maxell that makes use of a torsional coupler instead of the vibratory pieces used in the conventional design.[12] The torsional coupler transforms the longitudinal vibration generated by the Langevin vibrator to a transverse (or torsional) vibration, producing an elliptical rotation of the tip. The motor shown in Figure 10.16a exhibits very high torque and efficiency due to the action of the torsional coupler and an enhanced contact force with the rotor. A motor 30 mm $\phi \times 60$ mm L in size with a 20°–30° cant angle between the leg and vibratory pieces in Figure 10.16b provides a torque as high as 13 kgf-cm (1.3 N·m) with an efficiency of 80%. These attractive values excited the USM research fever in the late 1980s. However, this type produces only *unidirectional* rotation. Note that the output rotation is quite smooth due to the inertia of the rotor, even though this motor is driven by an intermittent mechanical input (i.e., mechanical rectification).

10.4.2 "Windmill" Motors

The *windmill motor* pictured in Figure 10.17 has a flat and wide configuration that is driven by a metal-ceramic composite structure.[13] The motor is composed of four basic components: a stator, a rotor, a ball bearing, and the housing structure, as shown in Figure 10.17a. The piezoelectric part is simply a ring with a diameter of 3.0 mm (smallest-size case) electroded on its top and bottom surfaces that has been transversely poled (i.e., across its thickness). The metal rings are fabricated by *electric discharge machining* and have four arms placed 90° apart on the inner circumference of the ring, as shown in Figure 10.17b, left. The metal and piezoelectric rings are bonded together, but the arms remain free, allowing them to move like cantilever beams. The length and cross-sectional area of each arm are selected such that the resonance frequency of the secondary bending mode of the arms is close to the resonance frequency of the radial mode of the stator. The rotor is placed at the center of the stator and rotates when an electric field is applied with a frequency between the radial and bending resonance modes. The truncated cone shape at the rotor end ensures constant contact with the tips of the arms.

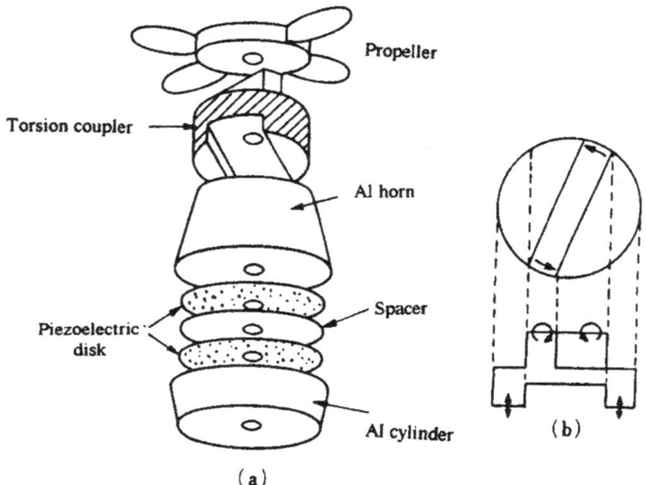

FIGURE 10.16 Mixed-mode ultrasonic motor with a torsional coupler: (a) structure of the entire motor and (b) motion of the torsional coupler. (From A. Kumada: *Jpn. J. Appl. Phys.*, 24(Suppl. 24–2), 739, 1985.)

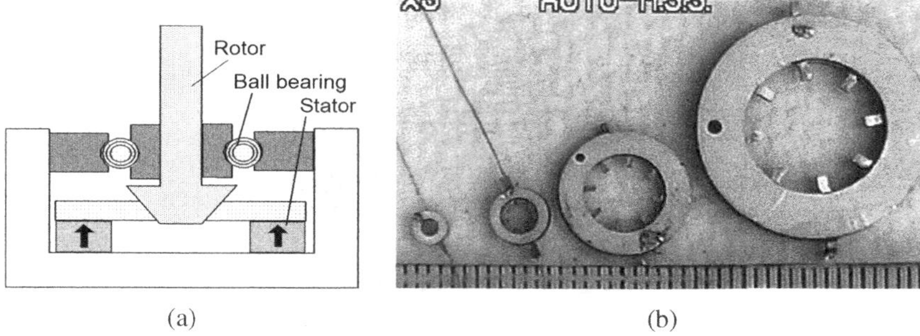

FIGURE 10.17 The "windmill" motor: (a) cross-sectional view and (b) various stators (diameter: 3–20 mm). (From B. Koc et al.: *IEEE Trans.-UFFC*, 47, 836, 2000.)

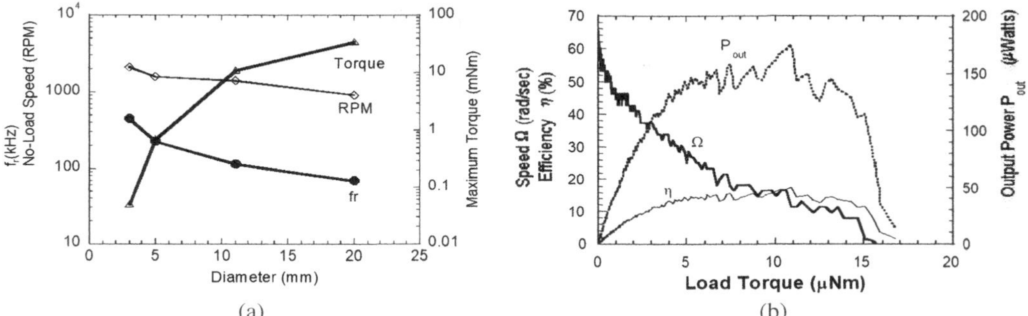

FIGURE 10.18 (a) Radial mode resonance frequency, no-load speed, and starting torque as a function of stator diameter (drive: 15.7 V, 160 kHz). (b) Speed, efficiency, and output power as a function of load torque for a 3-mm-diameter Windmill motor. (From B. Koc et al.: *IEEE Trans.-UFFC*, 47, 836, 2000.)

Rotation is produced by the alternate contraction and expansion of the stator. During the contraction portion of the cycle, the four arms at the center of the metal ring clamp the rotor and apply a tangential force to it. Since the radial mode frequency of the stator is close to the secondary bending mode frequency of the arms, the deformations add and the tips of the arms bend down. During the expansion portion of the cycle, the arms release the rotor at a different position. The resulting motion is similar to the action produced by a human hand as it grasps and twists an object.

The radial mode resonance frequency, no-load speed, and maximum torque are plotted as a function of motor size in Figure 10.18a. When a motor with a 5-mm diameter is driven at 160 kHz, a maximum speed of 2000 rpm and a maximum torque of 0.8 mNm are attained. The speed, efficiency, and output power of a 3-mm diameter motor are plotted as a function of load torque in Figure 10.18b. The starting torque of 17 μNm for this motor is one order of magnitude higher than that of a thin film motor of a similar size.

10.4.3 Dual-Vibration Coupler Motors

A dual-vibration coupler motor is introduced in Figure 10.19.[14] A torsional Langevin vibrator acts in conjunction with three multilayer actuators to generate larger transverse and longitudinal surface displacements of the stator, as well as to control their phase difference between them. The rotation direction is switched by changing the phase difference. Since it is quite difficult to

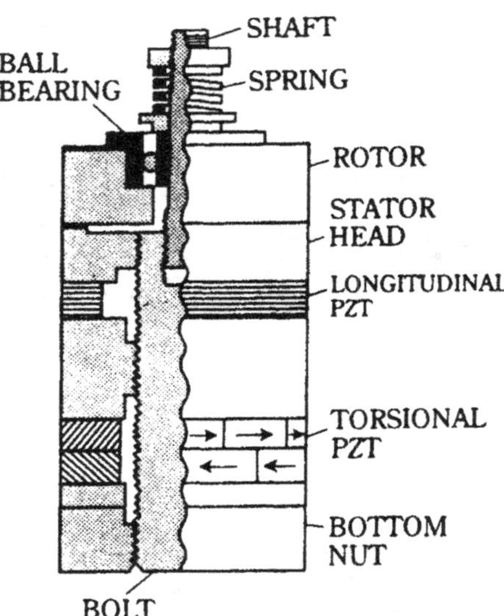

FIGURE 10.19 A dual-vibration (longitudinal and torsional) vibration coupler motor. (From K. Nakamura et al.: *Proc. Jpn. Acoustic Soc.*, No.11-18, p.917, Oct., 1993.)

match the torsional and longitudinal resonance frequencies, the motor is generally operated at a frequency corresponding to the torsional resonance frequency. In this dual actuator type, the elliptic locus on the stator head surface can be changed by changing the applied voltages on the torsional and longitudinal actuators independently, which corresponds to speed and torque control, respectively.

10.4.4 Piezoceramic Multilayer Ultrasonic Motor

Mitsui Sekka, Japan, developed a piezoceramic multilayer motor with its electrode pattern depicted in Figure 10.20a.[15] Longitudinal (L_1) and bending (B_2) modes are simultaneously excited in the device when the appropriate input signals are applied to the external electrodes. The center large external electrode is for generating L_1 mode, the voltages on the top and bottom electrodes are in opposite phases to each other for bending, and the right and left short electrodes should also be in opposite phases to each other to generate the second bending mode preferably. The resonance frequencies for both L_1 and B_2 modes should be adjusted to be the same. They put two friction coat legs under the bottom in order to trot on the rail. The speed and efficiency of the motor are shown as a function of load in Figure 10.20b. Though the efficiency is less than half in comparison with the π-shaped motor, the speed is in the same range. This motor design seems to be the first trial (1994) using the mass-production-capable ML preparation process.

Samsung Electromechanics, Korea, developed an ML piezo-actuator, as shown in Figure 10.21a, for a camera module application in 2006.[16] The multilayer is composed of three parts (top, bottom, and intermediate); the top and bottom generate the second bending mode, while the intermediate part contributes the first longitudinal vibration (similar to Mitsui Sekka's type). They put a friction contact point on the length side. The superposed motion is an elliptic vibration at the tip point of this ML chip with $5 \times 1.5 \times 1.5$ mm^3 (the right-hand-side tip in Figure 10.21a). An optical AF camera module with the size $10 \times 9.8 \times 5.6$ mm^3 is shown in Figure 10.21b, which has been installed in a Samsung phone, SCH-i718. Samsung Galaxy 6 or higher versions still use similar two ML motors (smaller that this) for a camera module (zoom and focus roles).

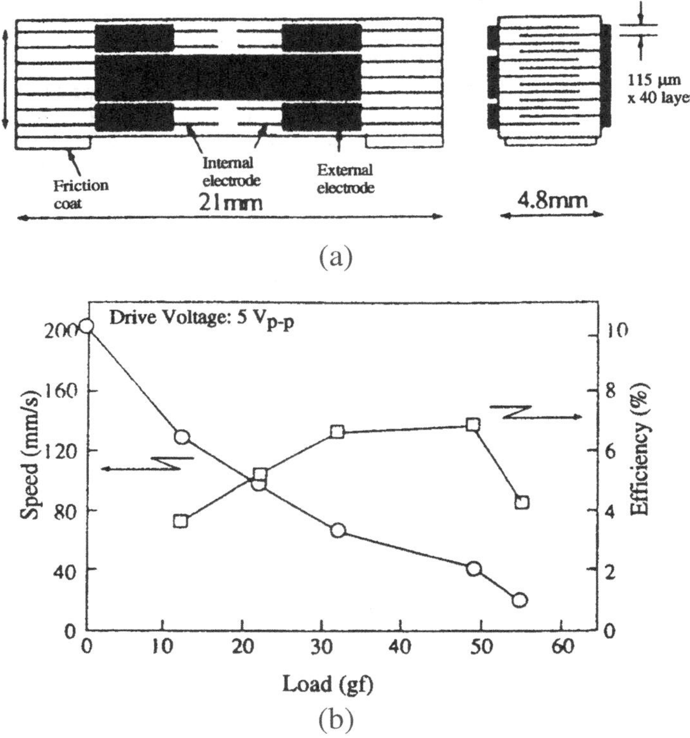

FIGURE 10.20 (a) Piezoceramic multilayer linear ultrasonic motor. (b) Speed and efficiency of the ceramic multilayer linear ultrasonic motor as a function of load. (From H. Saigo: *15th Symp. Ultrasonic Electronics (USE 94)*, No. PB-46, p.253, Nov. 1994.)

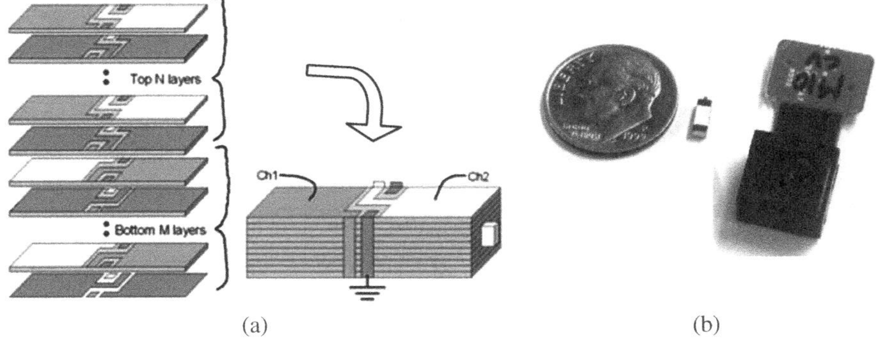

FIGURE 10.21 (a) Multilayer piezomotor by Samsung Electromechanics; (b) camera module with an ML linear USM. (From B. Koc et al.: *Proc. New Actuator 2006*, Bremen, June 14–16, p. 58, 2006.)

10.5 TRAVELING-WAVE MOTORS

10.5.1 Traveling-Wave Linear Motors

A linear motor driven by a traveling wave is illustrated in Figure 10.22a.[17,18] The two piezoelectric vibrators (Langevin transducer with a horn amplification mechanism) installed at both ends of a transmittance steel rod transmit and receive the traveling transverse wave on the rod (rather

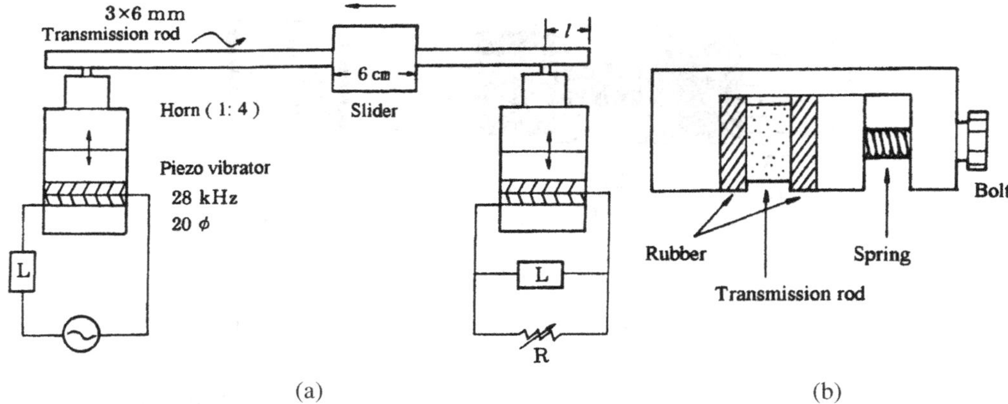

FIGURE 10.22 (a) Linear motor driven by a traveling wave. (b) Slider structure to clamp the transmission rod. (From Nikkei Mechanical, Feb. 28 issue, p.44, 1983; M. Kurosawa et al.: *J. Acoust. Soc. Amer.*, 77, 1431, 1985.)

than superposing sine and cosine waves), which is an antisymmetric fundamental *lambda wave*. Adjustment of the load resistance of the absorbing vibrator produces a nearly perfect traveling wave. Movement in the reverse direction is made by exchanging the transmitting and receiving roles of the piezoelectric vibrators.

Let us consider the numerical design of the motor system. The bending vibration transmitted via the rail rod is represented by the following differential equation:

$$\frac{\partial^2 w(x,t)}{\partial t^2} + \left(\frac{YI}{\rho A}\right)\frac{\partial^2 w(x,t)}{\partial x^2} = 0 \tag{10.6}$$

where $w(x,t)$ is the transversal displacement (see Figure 10.5), x is the coordinate along the rod axis, Y is the Young's modulus of the rod metal, A, its cross-sectional area, ρ its density, and I its moment of inertia. Assuming a general sinusoidal solution of the form in this rail:

$$w(x,t) = W(x)\ [A\ \sin(\omega t) + B\ \cos(\omega t)] \tag{10.7}$$

the wave transmission velocity, v, and the wavelength, λ, are given by

$$v = \left(\frac{YI}{\rho A}\right)^{1/4} \sqrt{\omega} \tag{10.8}$$

$$\lambda = 2\pi \left(\frac{YI}{\rho A}\right)^{1/4} \frac{1}{\sqrt{\omega}} \tag{10.9}$$

When the motor is driven in this manner, a wavelength as short as several millimeters can be readily produced by adjusting either the cross-sectional area, A, or the moment of inertia, I, of the rod, thereby ensuring adequate surface contact with the slider. A typical wavelength is about 26.8 mm.

The transmission efficiency is strongly affected by the contact position of the vibration sources on the rod. The periodic variation of the transmission efficiency with the position of the piezoelectric driver with respect to the free end of the rod (this length is labeled l in Figure 10.22a) is shown in

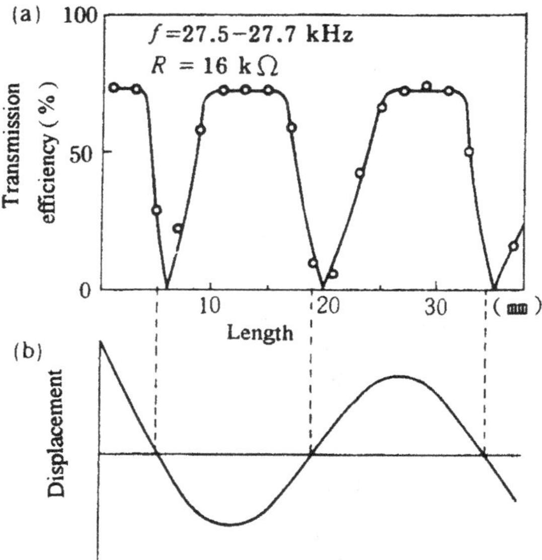

FIGURE 10.23 Characteristics of the linear motor as a function of position of the piezoelectric driver from the rod end: (a) the transmission efficiency and (b) the displacement. (From Nikkei Mechanical, Feb. 28 issue, p.44, 1983; M. Kurosawa et al.: *J. Acoust. Soc. Amer.*, 77, 1431, 1985.)

Figure 10.23. The optimum position for the vibration source is at a distance l from the end of the rod that corresponds to exactly one wavelength, 26.8 mm.

The structure of the slider clamped to the transmission rod with an appropriate force is shown in Figure 10.22b. The contact face is coated with rubber or a vinyl resin. The slider is a 60-mm-long clamp, where its length is selected to be approximately two wavelengths of the vibration. It is driven at a speed of 20 cm/s and produces a thrust of 5 kgf at 28 kHz. A serious problem with this type of linear motor lies in its low efficiency (around 3%) since the entire rod must be excited while only a relatively small portion of it is utilized for the output. The ring-type motors described in Section 10.5.2 generally have a much higher efficiency, because the stator and the rotor have the same length and the entire rod can be utilized.

EXAMPLE PROBLEM 10.2

At the tip of the Langevin transducer in Figure 10.22a, you see a horn (1:4). Describe the role of the horn in ultrasonic transducers.

Solution

A "horn" is an AC resonance displacement amplification mechanism. *It effectively produces an amplification of the resonance displacement that is inversely proportional to the cross-section area of the vibrator.* Three commonly used types are depicted in Figure 10.24. The *exponential cut horn* (Figure 10.24a) exhibits the highest energy transmission efficiency, but the fabrication procedure needed to produce a precisely cut horn of this shape is not simple and thus not cost efficient from a production point of view. The *linear taper horn* (Figure 10.24b) exhibits an intermediate efficiency and is somewhat easier to fabricate. The *step-contoured horn* (Figure 10.24c, which is actually used in Figure 10.22a) is the easiest to fabricate, but is the least efficient due to reflection of a portion of the vibration energy at the neck of the structure.

Figure 10.25 shows a $LiNbO_3$ surface acoustic wave motor.[19] Rayleigh waves are excited in two mutually perpendicular directions on the surface of a Y-cut $LiNbO_3$ crystal plate (127.8° rotation)

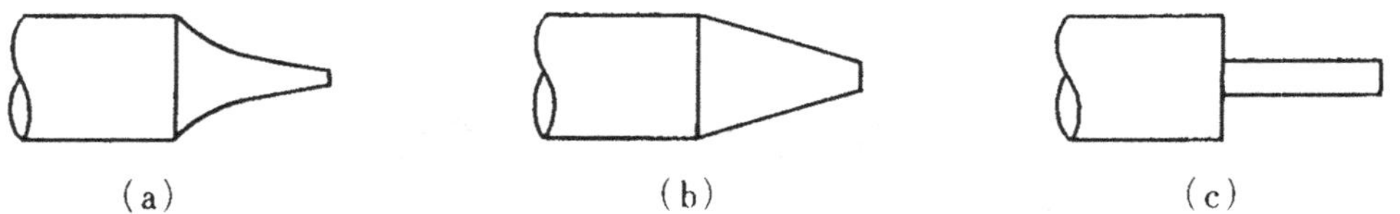

FIGURE 10.24 Three commonly used horn types: (a) exponential cut horn, (b) linear taper horn, and (c) step contoured horn.

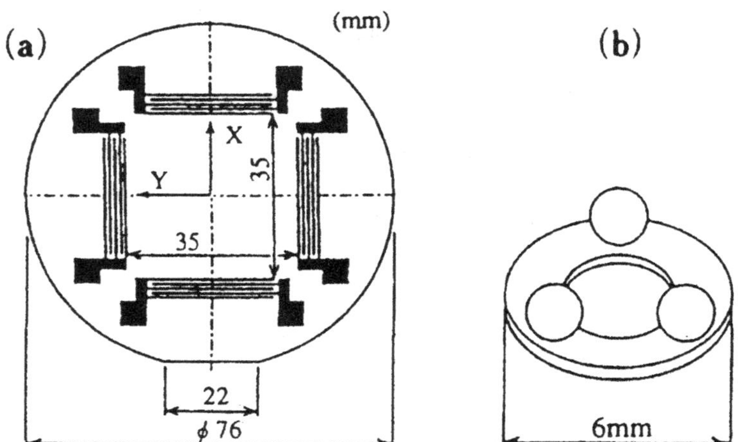

FIGURE 10.25 $LiNbO_3$ surface acoustic wave motor: (a) the stator structure and (b) the slider structure. (From M. Takahashi et al.: *Proc. 6th Symp. Electro-Magnetic Dynamics '94*, No. 940-26 II, D718, p.349, July, 1994)

via two pairs of interdigital surface electrodes arranged as shown in Figure 10.25a. The slider structure with three balls as legs appears in Figure 10.25b. The driving vibration amplitude and the wave velocity of the Rayleigh waves are adjusted to 6.1 nm and 22 cm/s, respectively, in both the x and y directions. It is important to note that even though the vibration amplitude is much smaller ($<1/10$) than the surface roughness of the $LiNbO_3$, the slider moves smoothly. The mechanisms for this have not yet been fully identified, but one possibility may be related to a local enhancement of the frictional force through the ball contact.

A rectangular plate dual-mode vibrator motor is shown in Figure 10.26.[20] The fundamental longitudinal (L_1) mode and eighth harmonic bending (B_8) mode vibrations, which have practically the same resonance frequency, are utilized in the operation of this device. When input signals with a phase difference of 90° (i.e., sine and cosine) are applied to the L- and B-mode electrodes, identical elliptical displacements are generated at both ends of the plate, causing the rollers in contact with these areas to rotate. Such a system is well suited for the task of conveying paper and cards.

Referring to Figure 10.3, a π shape is the topological equivalent to a rod type. The π-shaped device described in Section 10.3.2 can be modified to operate as a traveling wave motor in the slider, rather than generating a traveling wave on the rail.[21] Two multilayer actuators are installed at the two corners of the π-shaped frame with a cant angle of 45°, as shown in Figure 10.27. When input signals with 90° phase difference (i.e., sine and cosine) are applied to these two actuators, a sort of traveling wave moves from one leg to the other, leading to a "trotting" motion in these legs of the device.

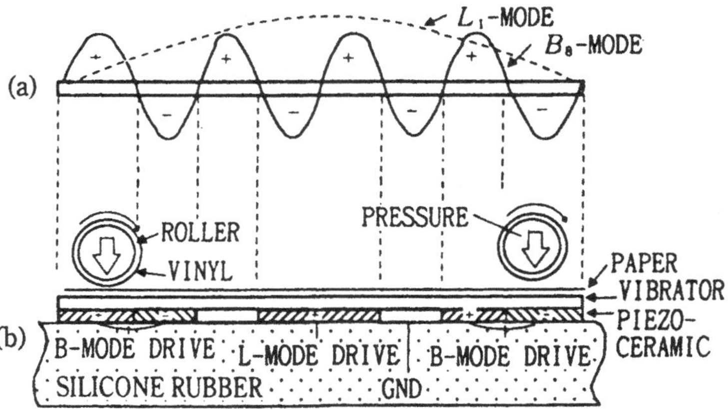

FIGURE 10.26 A rectangular plate dual-mode vibrator motor: (a) longitudinal L_1 and bending B_8 mode vibrations and (b) schematic diagram of the motor principle. (From Y. Tomikawa et al.: *Sonsors & Mater.*, 1, 359, 1989.)

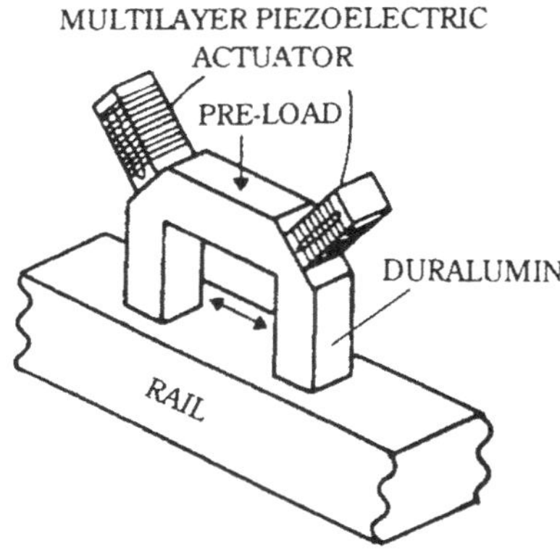

FIGURE 10.27 π-shaped ultrasonic linear motor of the traveling-wave type. (From K. Ohnishi et al.: *J. Acoust. Soc. Jpn.*, 47, 27, 1991.)

10.5.2 Traveling-Wave Rotary Motors

10.5.2.1 Basics of Traveling-Wave Rotary Motors

Refer to Figure 10.3. When the rod discussed in the previous section is folded to form a ring topologically, it can be operated as a rotary motor. Two types of ring motors are pictured in Figure 10.28: (a) bending mode and (b) extensional mode of vibration.[22] Although the basic operation principle for these motors is similar to that for the linear type, more sophisticated methods of piezoceramic poling must be applied and somewhat complex mechanical support structures are needed, as described later.

When a source of vibration is acting at a given position on a closed ring (either a circular or square) at the resonance frequency of the ring, only a standing wave is excited due to the interference of the two disturbances that proceed from the source around the ring in opposite directions. When more than one vibrator is applied, the superposition of the disturbances generated by the sources (two from each source) can produce a traveling wave around the ring.

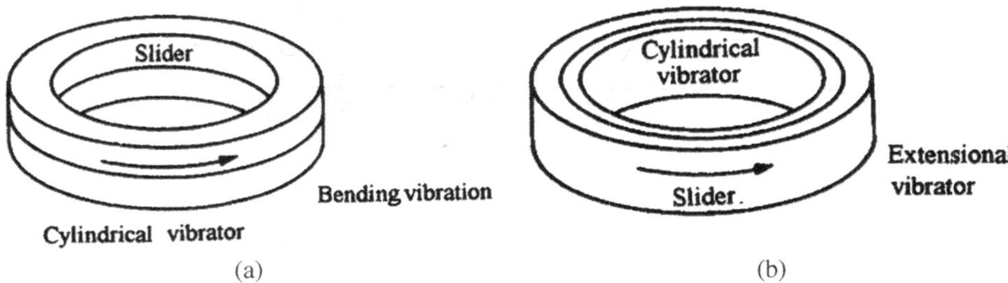

FIGURE 10.28 Two types of ring ultrasonic motors utilizing (a) a bending mode and (b) an extensional mode vibration. (From S. Ueha and M. Kuribayashi: *Ceramics*, 21(1), 9, 1986.)

Assuming a vibration of the form $A\cos(\omega t)$ is applied at position $\theta = 0$ on an elastic ring, the nth mode standing wave established in the ring is described by

$$u(\theta,t) = A\,(\cos(n\theta))\cdot(\cos(\omega t)); \tag{10.10}$$

that is, the product of a space term and time term. On the other hand, a general expression for a traveling wave is

$$u(\theta,t) = A\cos(n\theta - \omega t); \tag{10.11}$$

that is, the node ($u = 0$) is shifting with time. Alternatively, a traveling wave can be represented as the superposition of two standing waves in an equation of the form

$$u(\theta,t) = A(\cos(n\theta))\cdot(\cos(\omega t)) + A(\cos\,(n\theta - \pi/2))\cdot(\cos(\omega t - \pi/2)) \tag{10.12}$$

from which we see that: *a traveling wave can be generated by superposing two standing waves whose phases differ by 90° with respect to both position and time.* In more general terms, we can say that the superposition of standing waves with any constant phase difference except π can result in a traveling wave. In principle, excitation at only two positions on the ring is sufficient to generate a traveling wave. Commonly used vibration source configurations are illustrated in Figure 10.29. In practice, the largest number of vibration sources possible is preferred in order to increase the mechanical output, but not preferred from the cost viewpoint. When deciding upon the total number and placement of the vibration sources, the symmetry of the electrode pattern must be carefully considered.

The displacements produced in ring motors of the bending and extensional mode types are depicted in Figure 10.30.[22] The displacement locus becomes an elongated ellipse normal to the surface for the thin ring motor operated in the bending mode, while the locus is a flat ellipse independent of the thickness for the device operated in the extensional mode. The thrust direction on the top and bottom surfaces is the same for the bending-mode motor, but they are in opposite directions for the extensional-mode motor. The main advantage of the extensional mode over the bending mode is its higher speed due to its relatively high resonance frequency.

10.5.2.2 Sashida's "Surfing" Rotary Motors

The first successful "surfing" rotary motor was developed by Sashida.[23] The stator structure for this device and side view of the assembled motor are pictured in Figure 10.31. A traveling wave is induced in the elastic body by the thin piezoelectric ring to which it is bonded. A ring-shaped slider in contact with the rippled surface of the elastic ring can be driven in either the clockwise or counterclockwise direction depending on the phase of the input voltage (exchanging sine and cosine) applied to the piezoelectric. The motor's thin design and a large hole make it suitable for installation in a camera as

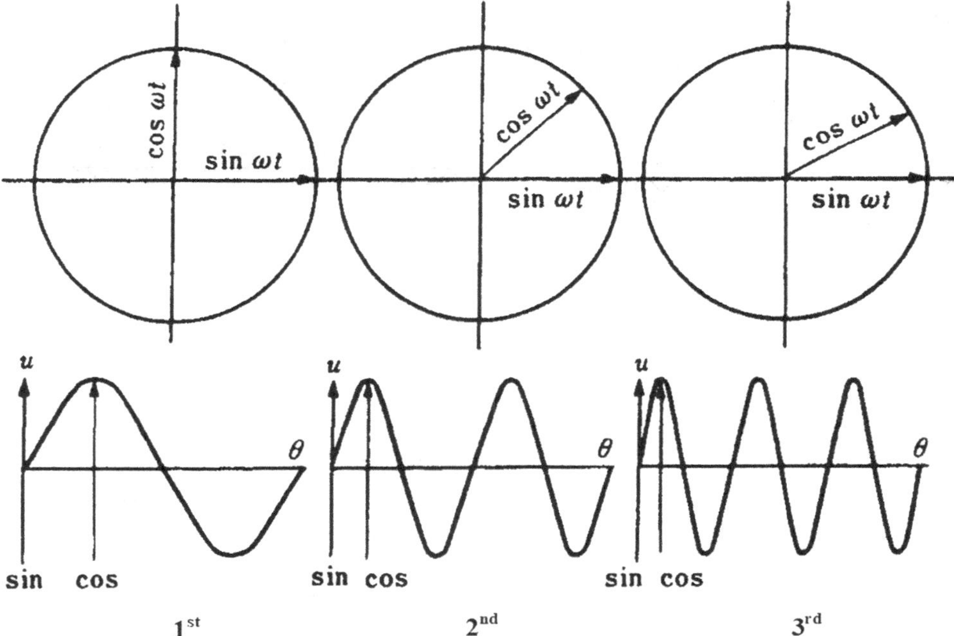

FIGURE 10.29 Commonly used vibration source configurations for generating a traveling wave in a ring motor. (From S. Ueha and M. Kuribayashi: *Ceramics*, 21(1), 9, 1986.)

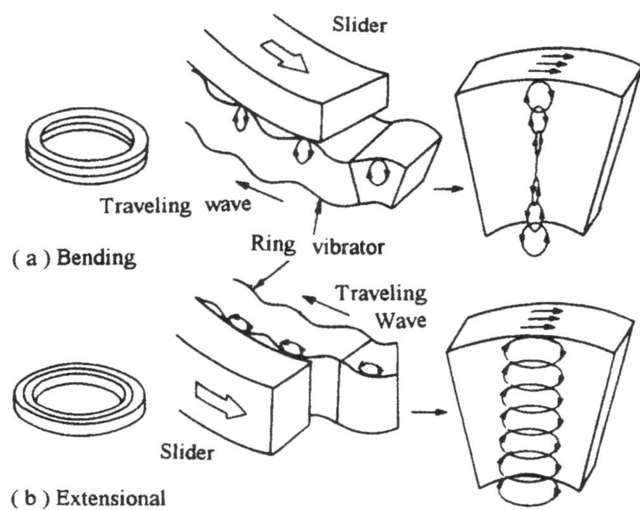

FIGURE 10.30 Ring motor displacements for: (a) the bending mode and (b) extensional mode types. (From S. Ueha and M. Kuribayashi: *Ceramics*, 21(1), 9, 1986.)

part of the automatic focusing system. Eighty percent of the exchange lenses in Canon's EOS camera series have already been replaced by the ultrasonic motor mechanism.

As shown in Figure 10.31a, the PZT piezoelectric ring is divided into 16 positively and negatively poled regions and two asymmetric electrode gap regions so as to generate the 9th-mode traveling wave at 44 kHz. The electrode configuration for this design is asymmetric and requires more complex lead connections than the simple basic structure described in Figure 10.29. Note that since

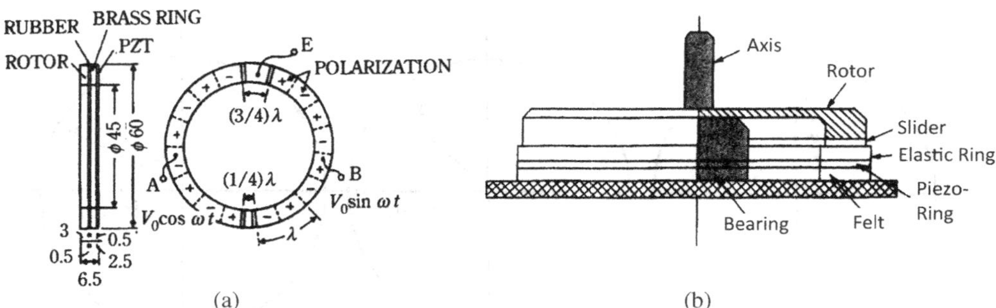

FIGURE 10.31 (a) Stator structure of Sashida's rotary motor. (From T. Sashida: *Mech. Automation of Jpn.*, 15(2), 31, 1983.) (b) Side view schematic of Sashida's rotary motor.

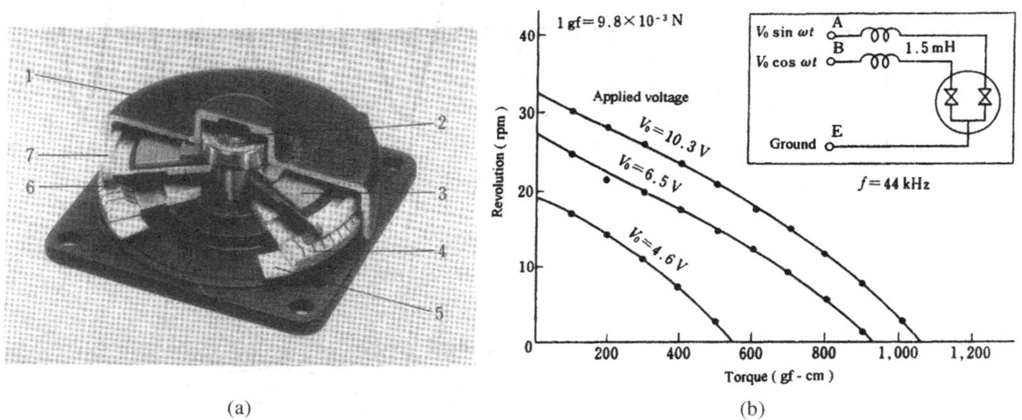

FIGURE 10.32 (a) Photograph of Sashida's surfing rotary motor (Shinsei Industry). (1. housing, 2. bearing, 3. plate spring, 4. stator, 5. PZT ring, 6. contact material, 7. rotor); (b) surfing motor speed plotted as a function of torque for several levels of applied voltage. (From T. Sashida: *Mech. Automation of Jpn.*, 15(2), 31, 1983.)

the lower-order resonance frequency for this size ring is in the audible range, they used a higher-order (9th) mode in order to keep the operating frequency in the range of 30–300 kHz (allowance by the Industrial Standard).

A side view schematic and a photograph of the motor appear in Figures 10.31b and 10.32a, respectively. It is composed of a 2.5-mm-thick brass ring with a 60-mm outer diameter and a 45-mm inner diameter bonded onto a 0.5-mm-thick PZT ceramic ring with divided electrodes printed on the back side. The front side has a full electrode for ground with the brass ring. The rotor is a polymer ring that has been coated with a hard rubber or polyurethane. The motor speed is plotted as a function of torque for several levels of applied voltage in Figure 10.32b.

The automatic focusing mechanism incorporating this rotary motor that was developed for the Canon camera is illustrated in Figure 10.33.[23] The placement of the Shinsei motor into the system is shown in Figure 10.33a. The compact size of the device allows it to fit nicely above and beneath the lens array. A more detailed view of the motor's contact with the lens components appears in Figure 10.33b. The stator actually has a ring of teeth (Figure 10.32a), which effectively amplify the transverse (and circular direction) elliptical displacement to increase the speed. This teeth groove practically helps with trapping metal wear dust during motor operation, so as not to damage the contact surface and to keep a smooth rotation. The lens position is shifted by means of the screw mechanism shown in Figure 10.33b.

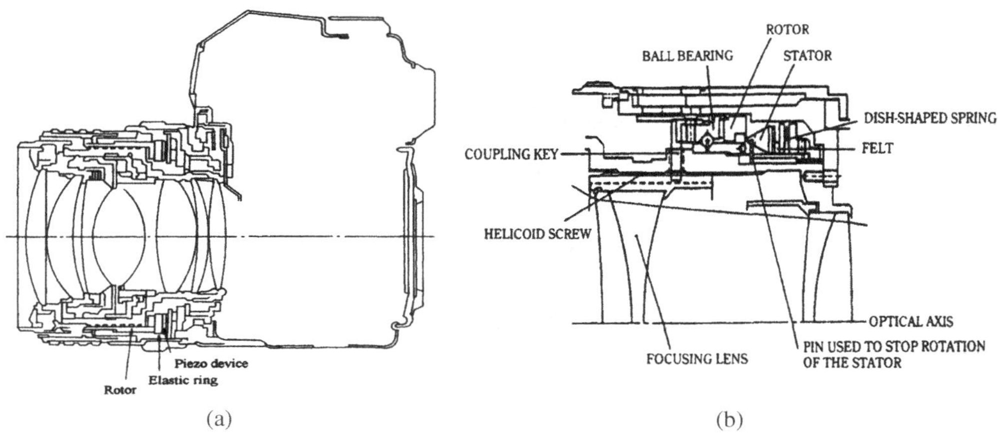

FIGURE 10.33 Canon automatic focusing mechanism: (a) placement of the motor in the system, and (b) detail showing the motor's contact with the lens components. (From T. Sashida: *Mech. Automation of Jpn.*, 15(2), 31, 1983.)

The advantages of this motor over a conventional electromagnetic motor are:

1. *Silent Drive*: Due to the ultrasonic operating frequency (i.e., inaudible) and the absence of a gear-box mechanism. This makes the system especially suitable for video cameras with microphones.
2. *Compact Size*: Allowing it to be incorporated in a space-efficient manner in systems such as the automatic focusing camera. Lenses can be set in the center part of the ring motor.
3. *High Torque, Low Speed*: Since the motor requires no gear mechanism, it can operate at higher torque and lower speed than its electromagnetic counterpart. No speed reduction mechanism with gears is required.
4. *Energy Saving*: This compact, friction-type motor unit is highly efficient in a compact size.

10.5.2.3 Other "Surfing" Traveling-Wave Motors

One important consideration in the design of these traveling-wave motors concerns the support of the stator. In a standing-wave motor, the nodal points or lines are stationary and thus serve as ideal contact points for the support structure. However, a traveling-wave device will not have this feature and care must be taken in selecting and making contact with the support points such that the bending vibration is not significantly suppressed. The stator for this motor is supported along a ring-axial direction with a felt interface to establish a "gentle" frictional contact as shown in Figures 10.31b and 10.33b. The repulsive rotation/motion of the stator is stopped by means of pins, which latch onto the stator teeth when engaged. An alternative means of support for a traveling-wave stator has been proposed by Matsushita Electric. It is essentially a nodal line support method applied to a higher-order vibration mode, as illustrated in Figure 10.34b.[24] Note the relative position of the nodal circle used for support with respect to the placement of the stator teeth on an adjacent antinodal circle, as shown in Figure 10.34a.

A much smaller version of this basic surfing motor design, measuring only 10 mm in diameter and 4.5 mm in thickness, has been produced by Seiko Instruments.[25] The structure of this miniature motor is pictured in Figure 10.35. When it is driven with a driving voltage of 3 V and a current of 60 mA, a no-load speed of 6000 rev/min with a torque of 0.1 mN·m is generated. The "Mussier-kun" toy (two-wheel vehicle) developed by Seiko Instruments is a tiny vehicle with dimensions $10 \times 10 \times 10$ mm^3 incorporating ultrasonic motors 8 mm in diameter. The toy is activated by illuminating an optical sensor in Mussier-kun's eye.[26]

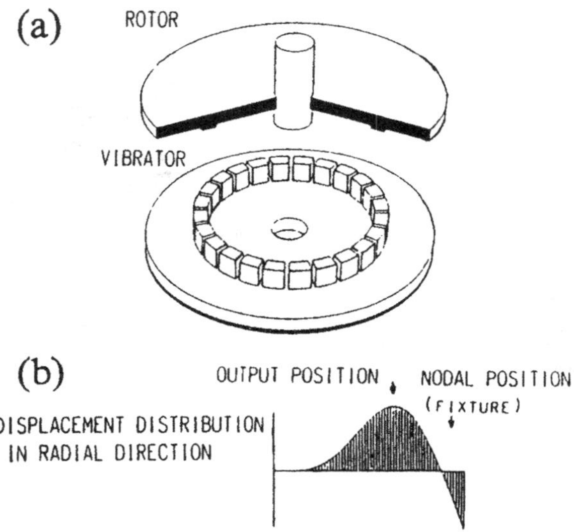

FIGURE 10.34 Support mechanism for the surfing rotary motor stator: (a) a ring of teeth is placed on an antinodal circle and (b) the support is placed on a nodal circle associated with a higher-order radial vibration mode. (From K. Ise: *J. Acoust. Soc. Jpn.*, 43, 184, 1987.)

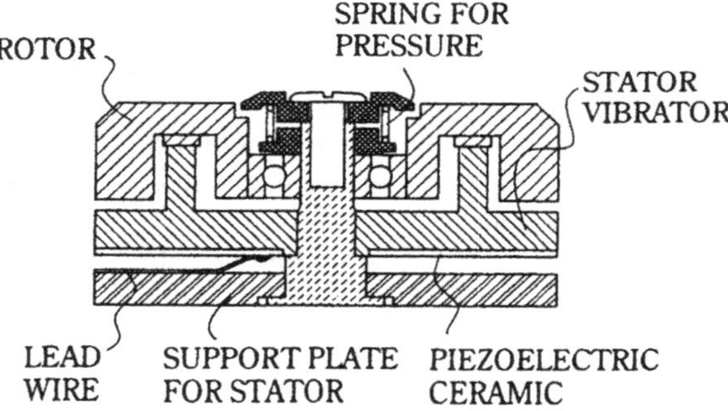

FIGURE 10.35 A miniature version of the surfing rotary motor produced by Seiko Instruments. (Diameter: 10 mm, Thickness: 4.5 mm). (From M. Kasuga et al.: *J. Soc. Precision Eng.*, 57, 63, 1991.)

Smaller motors similar to these have also been developed by Allied Signal to be used as mechanical switches for launching missiles.[27]

10.5.2.4 Disk and Rod Traveling-Wave Motors

Simple disk structures are preferable over unimorphs for motors because the bending action of the unimorph cannot generate sufficient mechanical power and its electromechanical coupling factor is generally less than 10%.[28,29] Therefore, instead of the unimorph structure (in the previous section), a simple disk is used for the motors here. The (1,1), (2,1), and (3,1) vibration modes of a simple disk, which are axial-asymmetric modes, are depicted in Figure 10.36. Excitation of these modes can produce a rotation of the outer circumference, resulting in "hula-hoop" action.

A similar principle with a rod design referred to as the "spinning plate" motor was developed by Tokin, Japan, as pictured in Figure 10.37.[30] A combination rotary/bending (i.e., "wobbling" mode) vibration is excited in a PZT rod (not a tube!) when sine and cosine voltages are applied to the

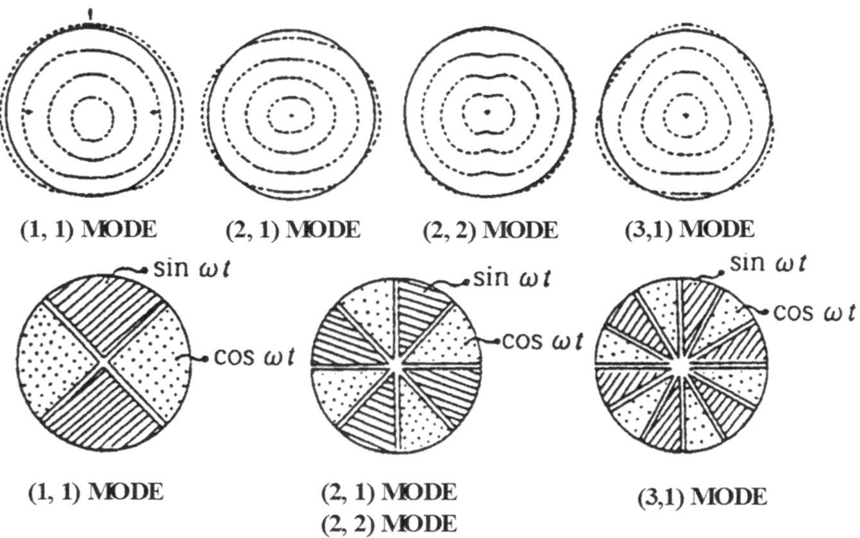

FIGURE 10.36 A disk-shaped "hula hoop" rotary motor. (From A. Kumada: *Ultrasonic Technology*, 1(2), 51, 1989; Y. Tomikawa and T. Takano: *Nikkei Mechanical*, (Suppl.), 194, 1990.)

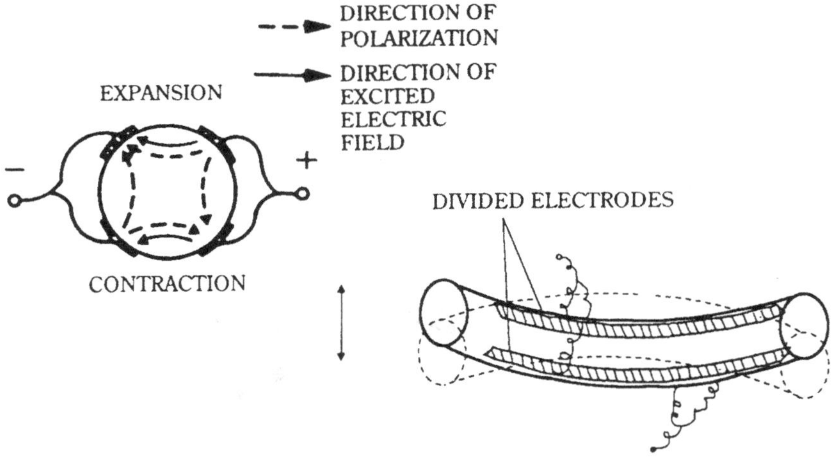

FIGURE 10.37 The "spinning plate" rotary motor. (From T. Yoshida: *Proc. 2nd Memorial Symp. Solid Actuators of Japan: Ultra-precise Positioning Techniques and Solid Actuators for Them*, p.1, 1989.)

divided electrodes. The inner surface of a cuplike rotor is brought in contact with the "spinning" rod to produce the rotation.

10.5.2.5 Δ-Shaped Motors

Penn State developed an intriguing 2 degrees of freedom Δ-shaped motor.[31] We demonstrate ultrasonic motors prepared by the low-temperature co-fired ceramic (LTCC) technique for 2D stage application with submicron control accuracy. This fabrication process allows inexpensive and onboard motors, which can be used not only for the precise positioning stage application but also for optical fiber alignment, providing an essential solution to the current optoelectronics industries' demand.

The original motor design was a Λ shape consisting of two piezoelectric rectangular bars joined at 90° angle (Λ shape). The design of the motor was modified and optimized by changing width, thickness, angle, and connection part between motor and LTCC dielectric ceramics, using the ATILA FEM (finite element method) software. To obtain 2-dimensional movement, we proposed a bimorph-type Δ-shaped motor, which has four input electrodes. Due to the problems of wire connection and angle change during sintering, the delta-shape was newly developed instead of the original Λ shape. A sandwiched delta-shaped motor was fabricated by co-firing with silver internal electrodes. See Figure 10.38a. Basically, double phases (sine and cosine voltages) on the right and left legs were used for the *x*, *y*, and diagonal axis movement (Figure 10.38b), while double phases (sine and cosine) on the top and bottom PZTs were used for the out-of-plane elliptic motion. Figures 10.38c and d show the co-fired Λ-shaped USMs and ultrasonic motor co-fired with LTCC for optical fiber alignment. The fabricated motor size was below 1 cm^2, and several tens of nanometer range movement in 2-dimensional space were observed by laser interferometers. Refer to Reference [32] for the optical fiber alignment capability.

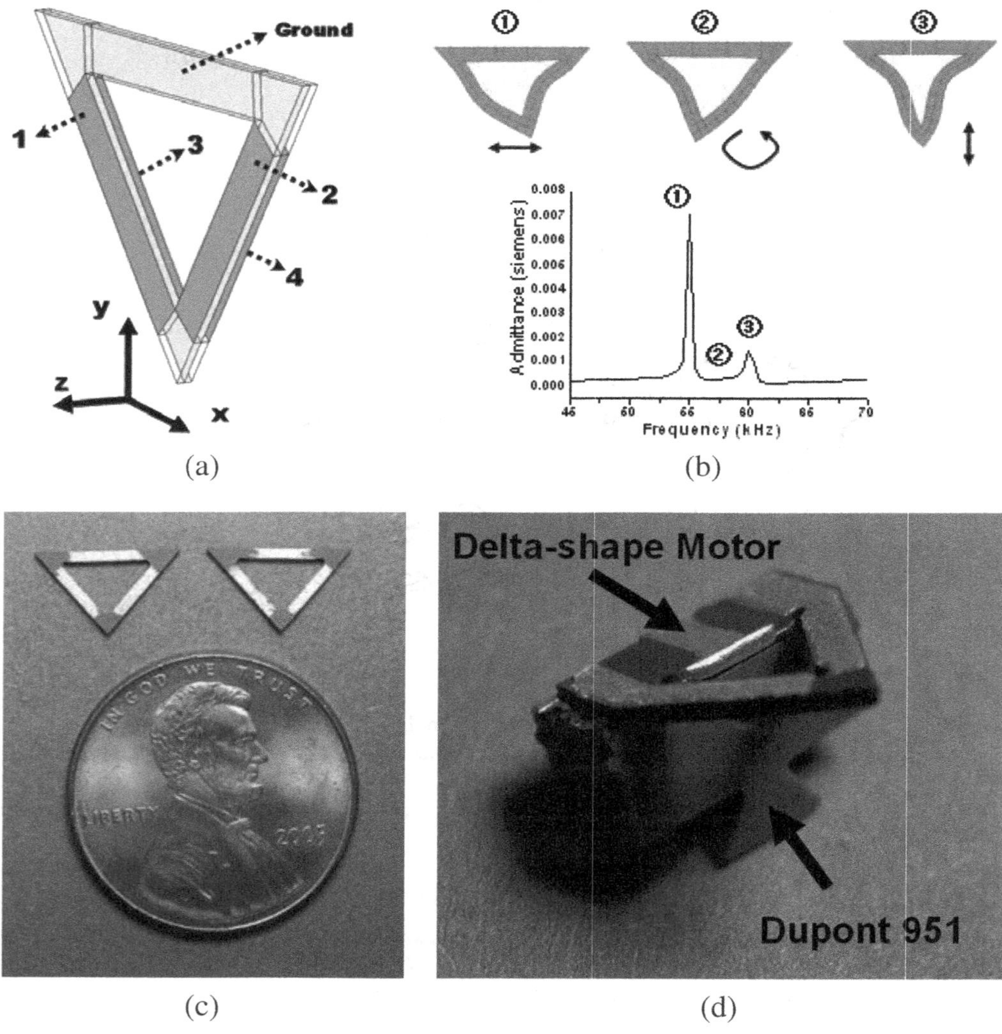

FIGURE 10.38 The Λ-shaped bimorph USM. (a) Four-arm structure; (b) in-plane mode simulation; (c) co-fired Λ-shaped USMs; (d) ultrasonic motor co-fired with LTCC for optical fiber alignment. (From S. -H. Park et al: *Proc. 10th Int'l Conf. on New Actuators*, p. 432, 14–16 June, Bremen, Germany, 2006.)

10.6 MODE ROTATION MOTORS

The PZT tube motor shown in Figure 10.39, which was developed jointly by Pennsylvania State University and the Institute of Materials Research and Engineering, Singapore, has a long, thin configuration.[33] It is similar to the Tokin motor[30] in appearance, but the mechanism for its operation is quite different. Four segmented electrodes are applied to the PZT tube (having an outer diameter of either 1.5 or 2.2 mm) and it is uniformly poled along the radial direction. A rotary bending mode of vibration is excited in the PZT cylinder when sine and cosine voltages are applied on the segmented electrodes as inserted in Figure 10.39: that is, equivalent to the four-phase (sine, cosine, –sine, and –cosine) drive. The vibrating PZT stator thus simultaneously drives the two rotors in contact with each end of the "wobbling" cylinder. The motion is analogous to that produced by the "spinning plate" motor operated by a four-phase voltage in this case.

A PZT tube motor of this type with inner and outer diameters of 1.5 and 2.2 mm ϕ, respectively, 7 mm in length and with a mass of 0.3 g generates a torque of 0.1 mNm and has a no-load speed of 1000–2000 rpm. The specifications for the PZT tube motor are compared with those of two other commercially available motors, which are widely used in mobile phones and wrist watches, in Table 10.2. A tube motor of this size generally has an efficiency of more than 20%, which is about one order of magnitude higher than those exhibited by electromagnetic motors. The power density of the PZT tube motor exceeds that of the Seiko motor by about one order of magnitude, which is due primarily to the utilization of the rotary bending mode of vibration in the operation of the device. The tube motor is highly suitable for a variety of precision micromechanical applications, such as intravascular medical microsurgery. The primary disadvantage of this design lies with the difficulty in manufacturing the delicate PZT tubes with the required degree of uniformity in wall thickness and cylinder symmetry, which leads to a rather high manufacturing cost. The metal tube motor is thus a less expensive alternative for these types of applications. Samsung's camera module utilized metal tube types, in practice.

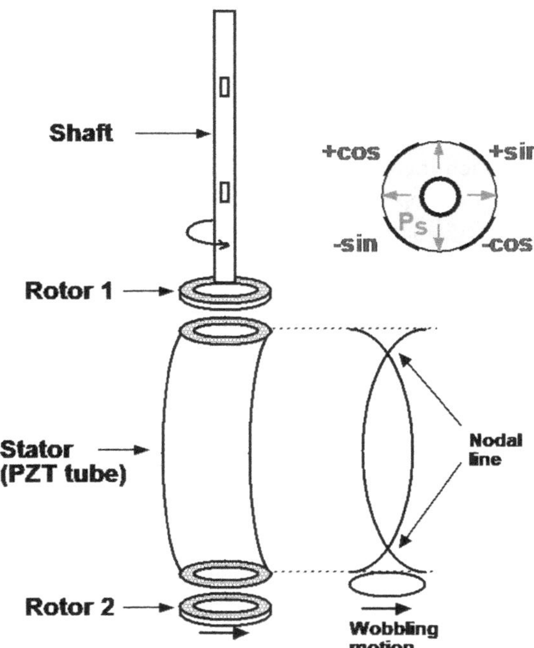

FIGURE 10.39 PZT tube motor developed by Penn State and IMRE, Singapore. (From S. Dong et al: *IEEE UFFC Trans.*, 50(4), 361–367, 2003.)

TABLE 10.2
Specifications for the PSU/IMRE Tube Motor and Two Other Commercially Available Motors

	Motorola Electromagnetic Micromotor	Sieko Ultrasonic Micromotor	PSU/IMRE Ultrasonic Micromotor
Outer Diameter (mm)	7	8	2.2
Length (mm)	16	4.5	8
Input Power (V)	1.5	1.5–3.5[a]	3–6[a]
(mA)	126	60–12	2–5
No-Load Speed (rpm)	5000	1200	1000–2000
Starting Torque (mN m)	0.075	0.05–0.1	0.1

[a] A booster circuit is required.

10.7 COMPARISON AMONG VARIOUS ULTRASONIC MOTORS AND THEIR SYSTEM INTEGRATION

10.7.1 Comparison Among Various Ultrasonic Motors

Standing-wave motors are generally among the least expensive, primarily because they require only a single vibration source. This type also offers an exceptionally high efficiency (as high as 98%, theoretically), but the motion in both the clockwise and counterclockwise directions is generally not well controlled. Traveling wave–type motors require more than one vibration source to generate the propagating wave, leading to a much lower efficiency (typically not more than 50%), but their rotational direction is generally far more easily controllable, just by switching sine and cosine voltages. The performance comparison of three motors representing the standing-wave type (Hitachi Maxell vibratory coupler), a hybrid type (Matsushita "compromise teeth vibrator"), and the traveling-wave type (Shinsei Industry) is summarized in Table 10.3.[34]

The traveling-wave motor such as Shinsei types does not lend itself well to miniaturization. A sufficient gap between adjacent electrodes is required to ensure proper insulation between them. During electrical poling, which involves alternating the polarity across adjacent poled regions in the ceramic, cracks readily develop in the electrode gap regions where residual stresses tend to concentrate. This problem ultimately precludes any further miniaturization of the traveling wave type of motor. Standing-wave motors, on the other hand, with their much simpler structures, are much better suited for miniaturization. They simply require a uniformly poled piezoelectric element, a few lead wires, and a single power supply.

Another problem associated with the traveling-wave motor concerns the support of the stator. In the case of a standing-wave motor, the nodal points or lines are generally supported, which causes a minimal effect on the resonant vibration. A traveling wave does not have stable nodal points or lines and therefore relatively elaborate support structures must be designed that will not significantly suppress the bending vibration. One of the simplest means is the method previously described that is depicted in Figure 10.31b (Shinsei motor), whereby the stator is supported along an axial direction with a layer of felt between the pieces to minimize the damping effects on the vibration.

In general, the following design concepts must be optimized in the development of new microscale ultrasonic motors:

1. *Simple structure* for which the number of components is as small as possible.
2. *Simple poling configuration* not to generate piezoceramic damage.
3. *Minimization of the number of vibration sources* and drive circuit components required for operation. (Standing-wave motors meet this requirement nicely.)

TABLE 10.3
Performance Comparison of Three Motors; Standing-Wave Type (a vibratory coupler by Hitachi Maxell), Hybrid Type ("compromise teeth vibrator" by Matsushita), and Traveling-Wave Type (by Shinsei Industry)

CHARACTERISTICS / TYPES	Rotation	Rotation Speed [rpm]	Rotation Torque [kgf·cm]	Efficiency [%]	Size	Analogy
Vibratory Coupler Type	Uni-Direction	600	13	80	Slim & Long	Euglena
Compromise Type	Reversible	600	1	45	⋮	Paramecium
Surface Wave Type	Reversible	600	0.5	30	Wide & Thin	Ameba

Source: K. Uchino: *Solid State Phys., Special Issue "Ferroelectrics"*, 23(8), 632, 1988.

10.7.2 System Integration of the Ultrasonic Motor

The current semiconductor chip manufacturing process requires handling 300-mm-diameter silicon wafers with positioning accuracy 1 nm. Thus, the manufacturing stage needs to have a minimum of this size, stage moving distance 300 mm, speed 300 mm/s, and accuracy 1 nm. Though an ultrasonic motor can provide this speed, the positioning accuracy is usually just sub-micron meter, because the operation principle is the use of vibration amplitude of sub-μm. In order to increase the positioning accuracy, the solution may be coupling the USM resonance mode with the off-resonance serve displacement mode in the same piezoelectric transducer.

Nanomotion, Israel, commercialized a stage (Figure 10.40) that is capable of successive operation mode shift from USM mode to servo displacement mode to reach 1-nm accuracy in 50-ms period.[35]

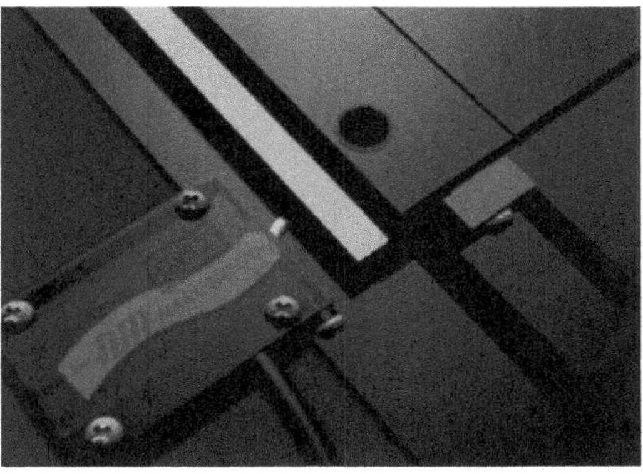

FIGURE 10.40 Nanomotion stage system with an USM. (From N. Karasikov: *Proc. 67th ICAT/JTTAS Joint Int'l Symposium on Smart Actuators*, State College, PA, Sept. 30 – Oct. 1, 2014.)

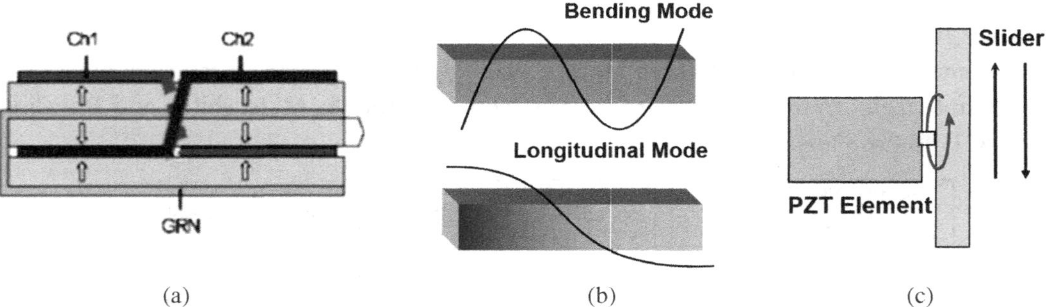

FIGURE 10.41 (a) PZT laminated motor developed by Nanomotion. (b) Second bending and first longitudinal modes are superposed (c) to generate an elliptic tip motion to move the slider. (From N. Karasikov: *Proc. 67th ICAT/JTTAS Joint Int'l Symposium on Smart Actuators*, State College, PA, Sept. 30 – Oct. 1, 2014.)

Figure 10.41a illustrates the PZT laminated motor developed by Nanomotion, where the top two layers with a bimorph design generate the second harmonic bending mode, while the bottom layer provides the fundamental longitudinal/extensional mode. This situation is depicted in Figure 10.41b. By superposing these two modes at the same resonance frequency, we can obtain an elliptic tip motion of the stator to move the slider, as shown in Figure 10.41b.[35] Note also that when DC voltage is applied on the top two layers, bimorph up/down DC displacement can be obtained. Similarly, the bottom layer generates left/right DC displacement.

Now we consider the stage positioning control of the motor in Figure 10.40. The initial quick motion is demonstrated in Figure 10.42a. The step motion with 400 nm is shown within only 10 ms. The second displacement serve drive is demonstrated in Figure 10.42b, where continuous triangular motion with ± 40 nm under a drive frequency of 1 Hz of 90 $V_{p\text{-}p}$ is demonstrated. The resolution of the feedback sensor (encoder) is 1 nm.

Another intriguing system integration is exemplified by Micron by Newscale Technologies, NY, which is handshake stabilization in microsurgery. Six metal tube USMs (Squiggle Motor™) are combined to make six-degree-of-freedom motion to assist the surgeon's handshake.[36]

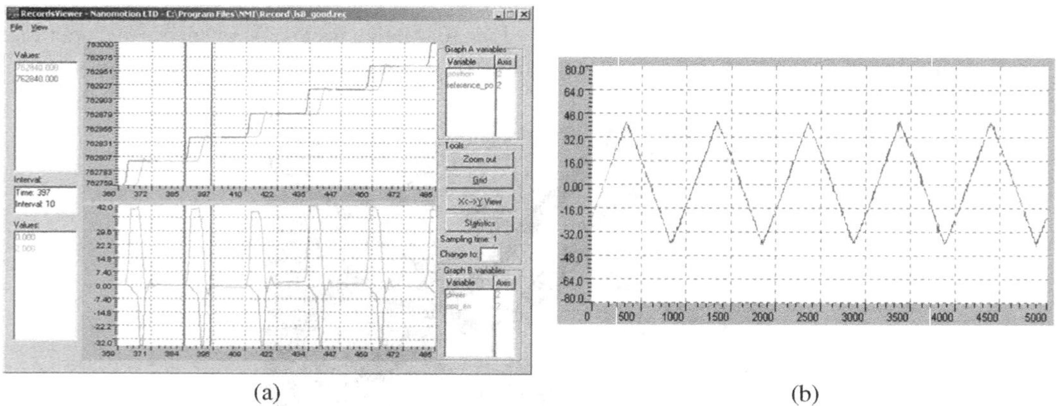

FIGURE 10.42 Nanomotion stage control scheme: (a) USM mode: step motion with 400 nm within only 10 ms. (b) DC mode: triangular motion with ± 40 nm under a drive frequency of 1 Hz of 90 $V_{p\text{-}p}$. The resolution of the feedback sensor (encoder) is 1 nm. (From N. Karasikov: *Proc. 67th ICAT/JTTAS Joint Int'l Symposium on Smart Actuators*, State College, PA, Sept. 30 – Oct. 1, 2014.)

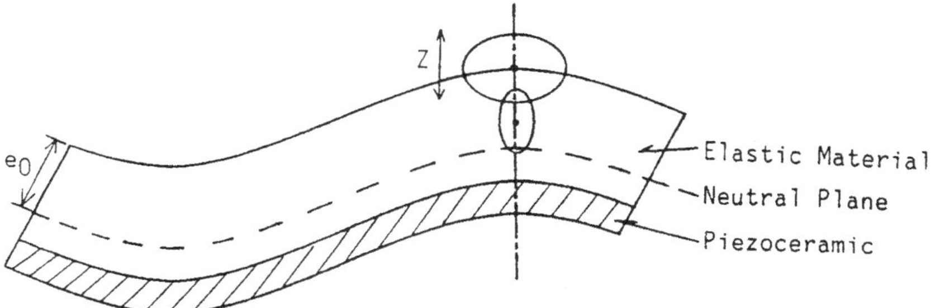

FIGURE 10.43 Stator displacement of a surface-wave-type motor.

10.8 CALCULATIONS FOR THE SPEED AND THRUST OF ULTRASONIC MOTORS

Calculations for the speed and thrust of an ultrasonic motor include parameters associated with both the type of motor and the contact conditions between the slider and the stator of the motor. The contact models typically employed represent the following conditions: (1) rigid slider and rigid stator, (2) compliant slider and rigid stator, and (3) compliant slider and compliant stator. Further, the intermittent drive of a vibratory-coupler (standing-wave) type motor and the continuous drive of a surface-wave (traveling-wave) type must be incorporated into the calculation.

10.8.1 Surface Wave Motor Calculations

The rigid slider and rigid stator contact model is employed in the analysis of surface wave motors. Of course, the stator is supposed not to be very "rigid," since the vibration wave is generated on the surface. That means the contact between the slider and stator will not change significantly the surface contour on both surfaces. The slider speed can be obtained from the horizontal velocity of the stator surface portion of the stator as it undergoes the displacement depicted in Figure 10.43. The phase velocity of the vibration, v_{sw}, is simply given by

$$v_{sw}=f\lambda \tag{10.13}$$

where f is the frequency and λ wavelength of the stator vibration. The speed of the slider, v, is described by

$$v=\frac{(4\pi^2)Ze_of}{\lambda} \tag{10.14}$$

where Z is the transverse vibration amplitude and e_o is the distance between the surface and the neutral plane (i.e., zero strain plane) as shown in Figure 10.43.

EXAMPLE PROBLEM 10.3

Calculate the phase velocity of the surface vibration, v_{sw}, and the slider speed, v, for a surface wave–type motor, using the following typical measured values: Bulk Vibration Velocity: 5050 (m/s), Vibration Frequency: 30 (kHz), Vibration Wavelength: $\lambda=28$ (mm), Distance e_o: 3.5 (mm), and Transverse Surface Vibration Amplitude: 1 (μm).

SOLUTION

Using the known values of vibration frequency and wavelength, the phase velocity of the surface vibration is calculated as follows:

$$v_{sw} = f\lambda = [30\times 10^{3}\ (\mathrm{Hz})]\ [28\times 10^{-3}\ (\mathrm{m})] \tag{P10.3.1}$$

$$\to\to\ v_{sw} = 850\ (\mathrm{m/s})$$

which is found to be much slower than the bulk vibration velocity (5 km/s).
The slider speed may be determined from the given data according to the following:

$$\begin{aligned} v &= \frac{(4\pi^2)Ze_o f}{\lambda} \\ &= \frac{(4\pi^2)[1\times 10^{-6}(\mathrm{m})][3.5\times 10^{-3}(\mathrm{m})][30\times 10^{3}(\mathrm{Hz})]}{[28\times 10^{-3}(\mathrm{m})]} \\ &\to\to v = 150(\mathrm{mm/s}) \end{aligned} \tag{P10.3.2}$$

which is considerably slower than the bulk vibration velocity!

10.8.2 VIBRATION COUPLER MOTOR CALCULATIONS

The compliant slider and rigid stator contact model is employed in the analysis of vibration coupler motors. The horizontal and vertical displacements of the rigid stator shown as a and b, respectively, in Figure 10.44, are described by:

$$a = a_o\ [\cos(\omega t)] \text{ and } b = b_o\ [\sin(\omega t)] \tag{10.15}$$

The horizontal speed of the stator may thus be expressed as:

$$v_h = \frac{\partial a}{\partial t} = -a_o\omega[\sin(\omega t)] \tag{10.16}$$

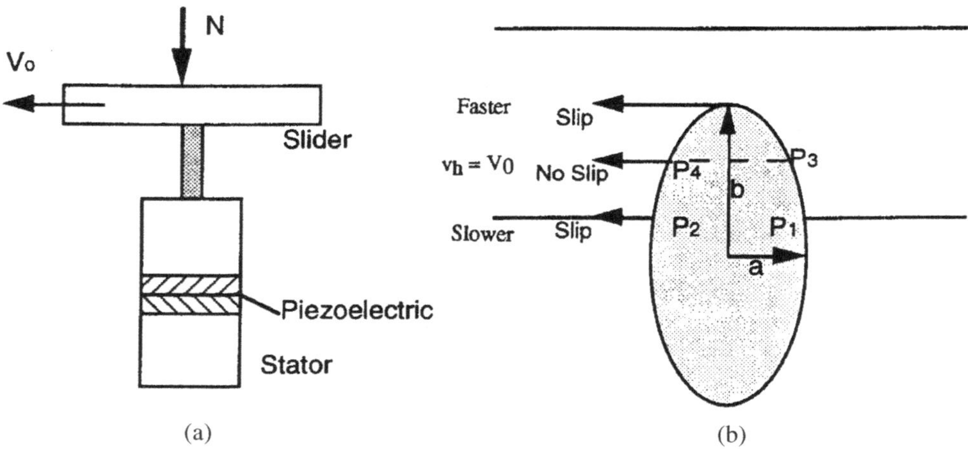

FIGURE 10.44 Vibration coupler motor: (a) schematic of the motor structure and (b) displacement of the rigid stator on a compliant slider.

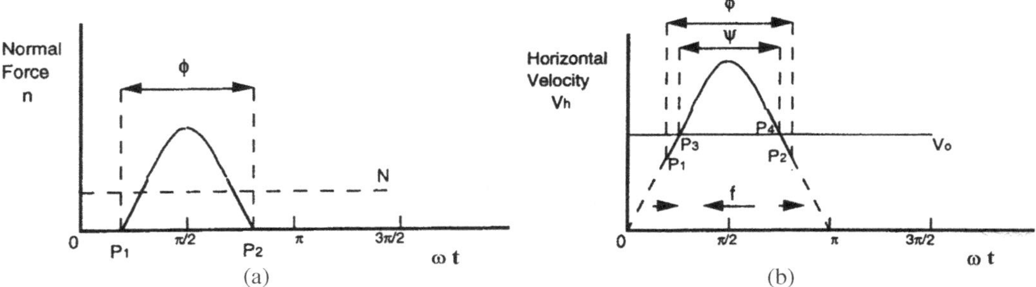

FIGURE 10.45 (a) Normal force acting on a vibratory coupler plotted as a function of phase. (b) Horizontal stator speed of a vibratory coupler motor plotted as a function of phase.

10.8.2.1 Normal Force Assumption

The normal force, n, is assumed as a function of phase, ωt, in Figure 10.45a. The curve shown is defined as:

Slider and Stator in Contact: $(\pi/2 - \phi/2) < \omega t < (\pi/2 + \phi/2)$

$$n = \beta[\sin(\omega t) - \cos(\phi/2)] \tag{10.17a}$$

Slider and Stator Not in Contact: $0 < \omega t < (\pi/2 - \phi/2)$ and $(\pi/2 - \phi/2) < \omega t < (2\pi)$

$$n = 0 \tag{10.17b}$$

where ϕ is a measure of the contact period (P_1 to P_2) as shown in Figure 10.44 and β is a parameter defined in the analysis that follows.

The integrated normal force, N, is given by:

$$N = \int n d(\omega t) = \left(\frac{1}{2\pi}\right) \int_{\pi/2-\phi/2}^{\pi/2+\phi/2} \beta[\sin(\omega t) - \cos(\phi/2)] d(\omega t)$$
$$N = \left(\frac{\beta}{\pi}\right)[\sin(\phi/2) - (\phi/2)\cos(\phi/2)] \tag{10.18}$$

from which we may define the parameter, β, as:

$$\beta = \frac{\pi N}{[\sin(\phi/2) - (\phi/2)\cos(\phi/2)]} \tag{10.19}$$

10.8.2.2 Slider Speed Assumption

The no-load slider speed corresponds to the average horizontal stator speed (v_h), and is identified as v_o in the plot of v_h as a function of phase (ωt) in Figure 10.45b. It is thus evaluated over the phase interval between P_1 and P_2 by:

$$v_o = \int v_h \, d(\omega t)$$
$$= \left(\frac{1}{\phi}\right) \int_{\pi/2-\phi/2}^{\pi/2+\phi/2} -a_o\omega[\sin(\omega t)] d(\omega t) \tag{10.20}$$
$$v_o = \frac{-a_o\omega[\sin(\phi/2)]}{(\phi/2)}$$

When the phase ψ conditions identified by P_3 and P_4 in Figure 10.45b exist, no slip is observed and

$$v_h = v_o = -a_o\omega[\cos(\psi/2)] \tag{10.21}$$

Comparison of Equations 10.20 and 10.21 indicates that:

$$\cos(\psi/2) = \frac{\sin(\phi/2)}{(\phi/2)} \tag{10.22}$$

10.8.2.3 Thrust Calculation Assumption

The frictional force, f, is defined by the kinetic coefficient of friction between the slider and the stator, μ_k, and under conditions where it acts as an accelerating force or a drag force, it is expressed as follows:

Accelerating Force: $(\pi/2 - \psi/2) < \omega t < (\pi/2 + \psi/2)$

$$f = \mu_k\, n \tag{10.23a}$$

Dragging Force: $(\pi/2 - \phi/2) < \omega t < (\pi/2 - \psi/2)$ and $(\pi/2 - \psi/2) < \omega t < (\pi/2 + \phi/2)$

$$f = -\mu_k\, n \tag{10.23b}$$

The maximum thrust, F, is given by the integrated frictional force as:

$$\begin{aligned} F &= \int f\, d(\omega t) \\ &= \left(\frac{1}{2\pi}\right)\int_{\pi/2-\phi/2}^{\pi/2+\phi/2} \mu_k\, \beta[\Gamma(\omega t)]d(\omega t) \\ &= \left(\frac{\mu_k\, \beta}{2\pi}\right)\left[\int_{\pi/2-\phi/2}^{\pi/2-\psi/2} [\Gamma(\omega t)]d(\omega t) - \int_{\pi/2-\psi/2}^{\pi/2+\psi/2} [\Gamma(\omega t)]d(\omega t) + \int_{\pi/2+\psi/2}^{\pi/2+\phi/2} [\Gamma(\omega t)]d(\omega t)\right] \end{aligned}$$

where $\Gamma(\omega t) \equiv [\sin(\omega t) - \cos(\phi/2)]$.

Integrating yields:

$$F = \mu_k\, N\left[1 - \left(\frac{[\sin(\psi/2) - (\psi/2)\cos(\phi/2)]}{[\sin(\phi/2) - (\phi/2)\cos(\phi/2)]}\right)\right] \tag{10.24}$$

- When $\psi < \phi \ll 1$:

$$F \approx \mu_k\, N[1 - 2(\psi/\phi)] \approx \mu_k\, N[1 - 2(1/\sqrt{3})] \tag{10.25}$$

- When $\phi = 0$ and $\psi = 0$, $v_o = [-a_o\omega]$ and $F = (-0.155)\mu_k N$. When $\phi = \pi$ and $\cos(\psi/2) = (2/\pi)$, $v_o = [-(2/\pi)a_o\omega]$ and $F = (-0.542)\mu_k N$. Thus, we understand that as the contact period, ϕ, is increased, the thrust, F, increases at the expense of the slider speed.

In summary, if the contact period, ϕ, under a certain normal force, N, can be determined experimentally or theoretically, the no-load speed, v_o, can be calculated from Equation 10.20 and the no-slip position angle, ψ, from Equation 10.22. The thrust, F, can then be obtained from Equation 10.24.

10.9 DESIGNING FLOW OF THE ULTRASONIC MOTOR

A practical designing process of an ultrasonic motor is demonstrated by using the π-shaped linear motors pictured in Figure 10.46. Multilayer and bimorph types are depicted in Figure 10.46a[37] and b. We follow the process outlined in the design flowchart presented in Figure 10.47.

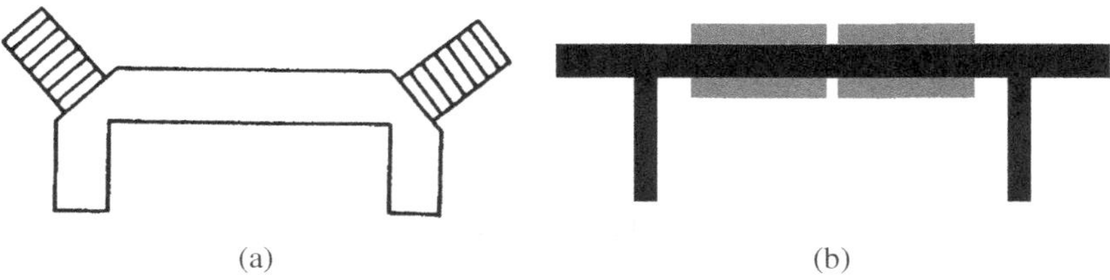

FIGURE 10.46 Two π-shaped linear motors: (a) a multilayer type (From K. Onishi: Ph.D. Thesis, Tokyo Institute of Technology, Japan, 1991.) and (b) a bimorph type.

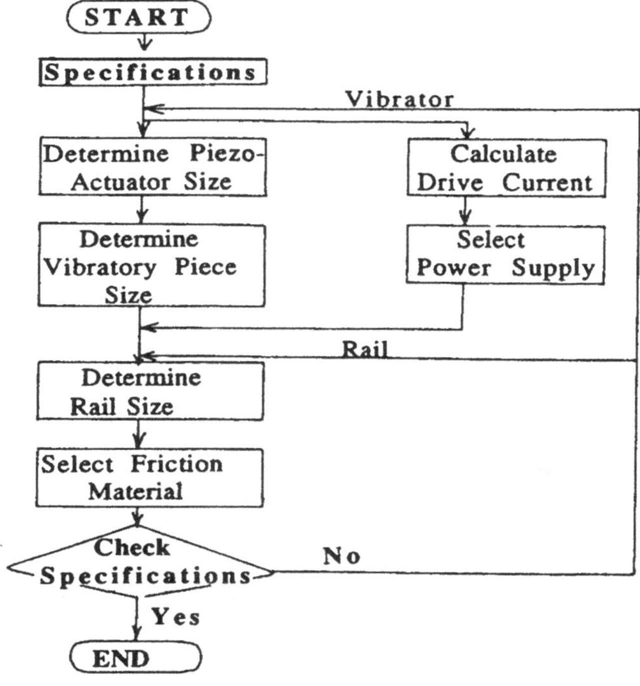

FIGURE 10.47 Flowchart for an ultrasonic motor design.

10.9.1 Defining the Specifications of the Motor

Let us assume that the customer has provided the following required specifications for a π-shaped linear motor.

10.9.2 Determining the Size of the Piezoelectric Actuator

The size of the piezoelectric actuator is determined largely by the resonance frequency at which it will be operated. The π-shaped multilayer motor requires an actuator that is operated in a thickness longitudinal mode of vibration. The resonance frequency for the desired vibrational mode is given by:

$$f_r = \left(\frac{1}{2l}\right)\sqrt{\frac{1}{\rho s_{33}^D}} \tag{10.26}$$

Assuming a mass density for the piezoceramic of $\rho = 7.8 \times 10^3$ (kg/m^3), a compliance of $s_{33}^D = 25\times10^{-12}$ (m^2/N), and a length $l = 9$ (mm), the resonance frequency for the thickness longitudinal mode will be:

$$f_r = \left(\frac{1}{2[9\times10^{-3}(\text{m})]}\right)\sqrt{\frac{1}{[7.8\times10^3(\text{kg/m}^3)][25\times10^{-12}(\text{m}^2/\text{N})}}$$

$$\rightarrow\rightarrow f_r = 1.25\times10^5(\text{Hz})$$

The resonance frequency of the piezoelectric actuator should actually be selected slightly higher than the drive frequency to allow for the presence of the vibratory piece, which will be installed later, due to additional mass. The size of the multilayer actuator is usually chosen from the mass-production components (i.e., off the shelf), and the above length is for the NEC/Tokin product $5\times5\times9$ (mm^3).

10.9.3 Determining the Size of the Vibratory Piece

10.9.3.1 Analytical Approach

The two legs and crossbar of the π-shaped structure are referred to as the "cantilevers" and the "free bar," respectively. The resonance frequencies for both types are described by a general equation of the form:

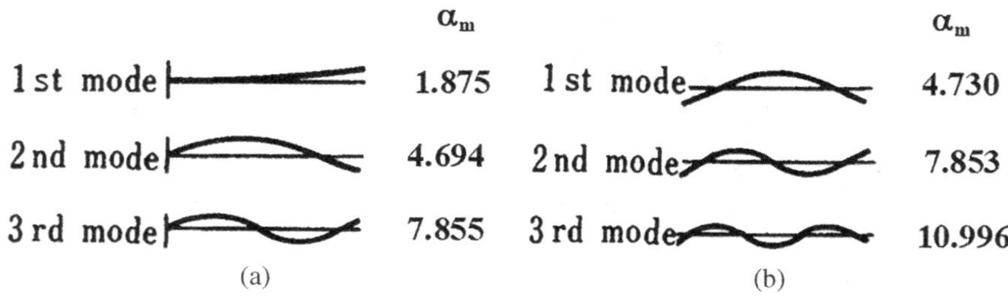

FIGURE 10.48 Values of the parameter α_m corresponding to the first three resonance modes for the elements of a π-shaped linear motor: (a) the cantilevers and (b) the free bar.

$$f = \left(\frac{\alpha_m^2}{2\sqrt{3}(2\pi)}\right)\left(\frac{t}{l^2}\right)\sqrt{\frac{Y}{\rho}} \tag{10.27}$$

where l is the length of the element, t is its thickness, Y is the Young's modulus, ρ is the mass density, and α_m is a parameter determined by the vibration mode. Values for α_m are summarized in Figure 10.48 for the first three resonance modes of the cantilever and free bar elements. The second mode resonance frequencies ($m = 2$) for the cantilevers (a) and the free bar (b) are thus defined as follows:

$$f_{2c} = 1.01\left(\frac{t}{l_c^2}\right)\sqrt{\frac{Y}{\rho}} \tag{10.28a}$$

$$f_{2b} = 2.83\left(\frac{t}{l_b^2}\right)\sqrt{\frac{Y}{\rho}} \tag{10.28b}$$

Using the mass density [$\rho = 2.76 \times 10^3$ (kg/m^3)] and Young's modulus [$Y = 7.03 \times 10^{10}$ (N/m^2)] for aluminum, and the dimensions determined by finite-element analysis indicated in Figure 10.49, the cantilever and free bar second mode resonance frequencies are found to be: $f_{2c} = 255$ (kHz) and $f_{2b} = 106$ (kHz).

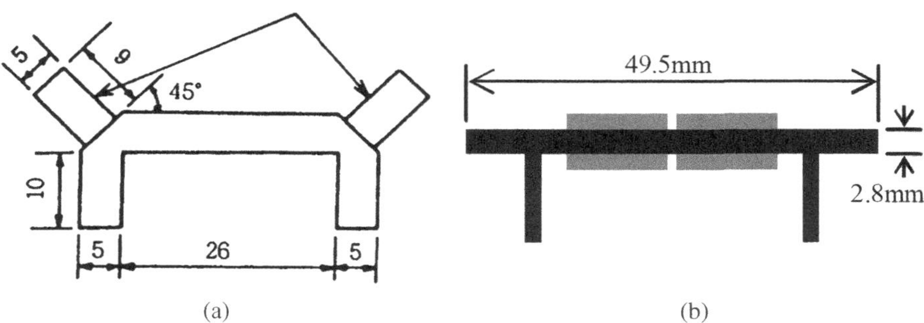

FIGURE 10.49 Dimensions of two π-shaped linear motors determined by both analytical approach and finite-element method (FEM): (a) the multilayer type (From K. Onishi: Ph.D. Thesis, Tokyo Institute of Technology, Japan, 1991.) and (b) the bimorph type.

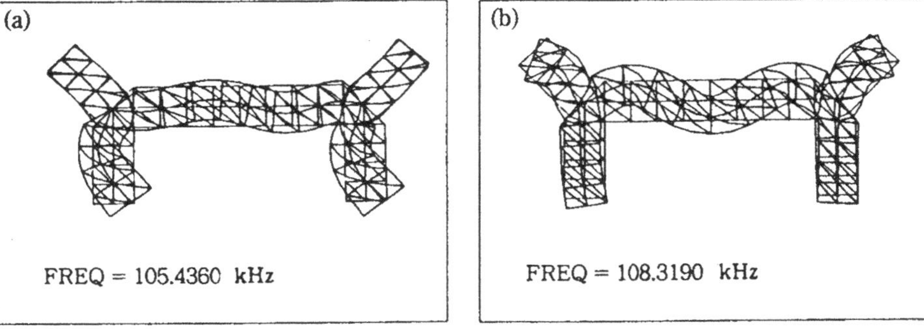

FIGURE 10.50 Some results of the free vibration finite element analysis on a bimorph-type π-shaped linear motor: (a) leg bending is excited at 105 kHz, and (b) transverse leg motion is excited at 108 kHz.

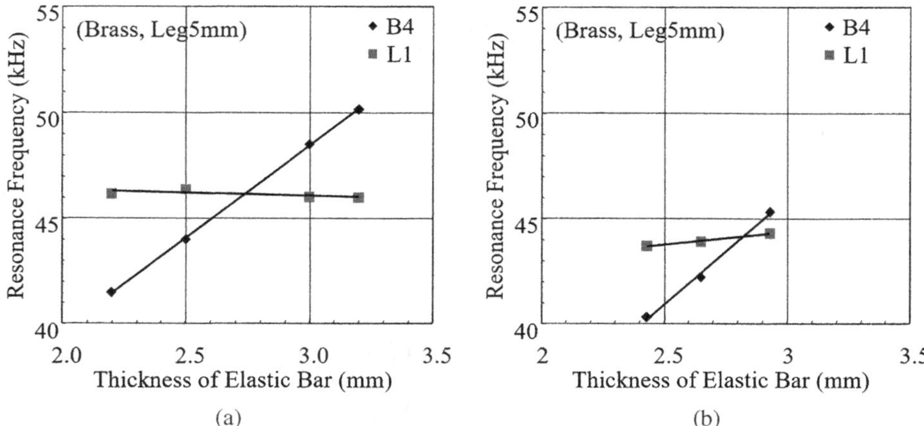

FIGURE 10.51 Resonance frequencies for the bending (*L*1) and transverse vibrational (*B*4) modes of the cantilever leg element on a bimorph π-shaped motor plotted as a function of an elastic bar thickness determined by: (a) finite-element analysis and (b) experiment.

10.9.3.2 Finite-Element Method Simulation

Free vibration analysis is performed using a finite-element method (ATILA) program on the structures. Some results are presented in Figure 10.50. In order to achieve high speed, both the longitudinal and flexural modes must be excited at the same drive frequency ideally. FEM simulated and experimentally determined resonance frequencies for the bending (*L*1) and transverse (up and down) vibrational (*B*4) modes of the cantilever leg element on a bimorph-type π-shaped linear motor are plotted as a function of thickness of the bar element in Figures 10.51a and b, respectively. The data obtained by finite analysis and the experimental results are in good agreement with a discrepancy of less than 4% for both resonance frequencies. It is important for the proper operation of this device that these two resonances be adjusted to the same frequency. We see from the data of Figure 10.51 that this condition is satisfied when the element has a thickness of 2.75–2.8 mm.

10.9.4 Determining the Rail Size

The rail size depends primarily on the motor size. Once this has been decided upon according to the preceding considerations, an appropriate rail configuration can be selected. In the case of the π-shaped motor discussed here, a rail with a cross-section of 5×5 mm^2 is used.

10.9.5 Selecting the Proper Drive Conditions

10.9.5.1 Impedance Spectrum Measurement

Resonance frequencies of ultrasonic devices are usually measured with an impedance analyzer. The impedance analyzer analyzes the output current and phase lag against the input voltage (i.e., constant voltage method); that is, two-terminal measurement. On the contrary, the bimorph-type π-shaped linear motor is driven with the sine and cosine voltage (four-terminal) for generating a superposed vibration, that is, a "horse trotting mode" with 90° phase lag between two legs. Though the ATILA FEM simulation code can calculate the two impedance spectra for these two inputs, it is not possible using a commercial impedance analyzer. Thus, we use a sort of trick to measure the symmetric modes (zero phase lag in voltage) and antisymmetric modes (180° phase lag in voltage; in practice, ML actuator's lead wires [red and black] are exchanged to connect to the impedance analyzer input terminal) separately. Two actuators are connected together with the same polarity for the symmetric

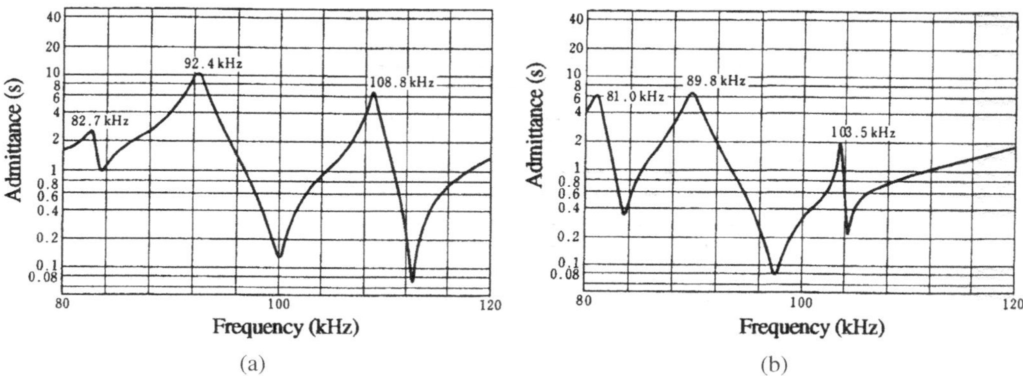

FIGURE 10.52 Admittance spectra for a π-shaped ultrasonic motor: (a) the symmetric and (b) and antisymmetric vibration modes.

modes (a 180° phase difference in displacement between the legs) and with the opposite polarity for the antisymmetric modes (a zero phase difference in displacement between the legs). Example spectra for the π-shaped motor are shown in Figure 10.52. 108.8 and 103.5 kHz in Figure 10.52a and b correspond to the symmetric (open/close of two legs) and antisymmetric (parallel wagging of two legs) modes, respectively. On the contrary, 105.4 kHz in Figure 10.50a corresponds to the "trotting" mode with 90° phase lag.

The motor is driven most efficiently over the frequency range that extends from a resonance mode that produces a large longitudinal displacement to one corresponding to a large bending displacement. The drive frequency should also fall between the frequencies from symmetric and antisymmetric leg motion, ideally such that there is either a 90° or a –90° phase difference between the legs conducive to the "trotting" motion of the device.

10.9.5.2 Stator Operation Test

Next, driven vibrational analysis is performed on the device. The trace of the leg movement when a sinusoidal drive is applied to one actuator and a cosine drive to the other, as shown in Figure 10.53, is determined. Both actuators are driven at 90 (kHz) with a force of 6 (N). The elliptical loci of the legs'

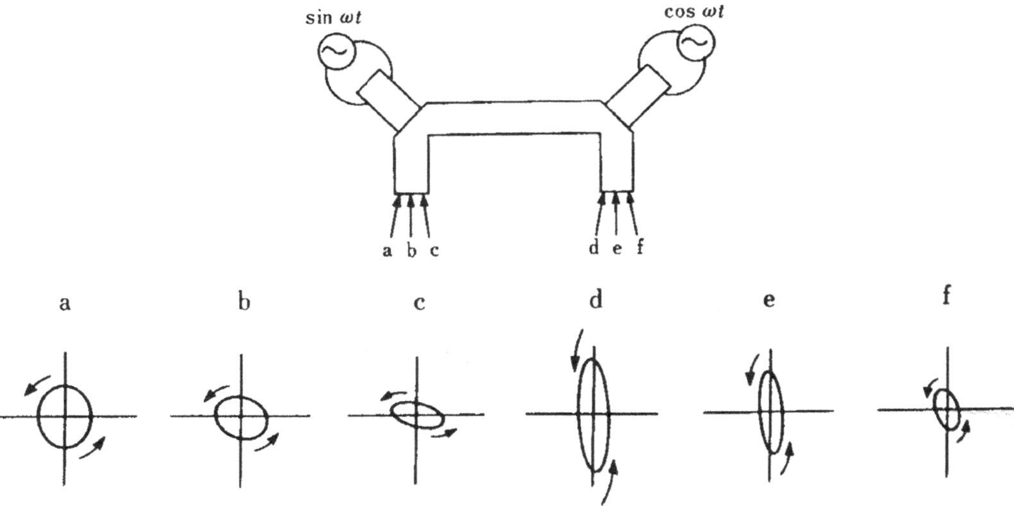

FIGURE 10.53 Driven vibrational analysis of the π-shaped linear motor.

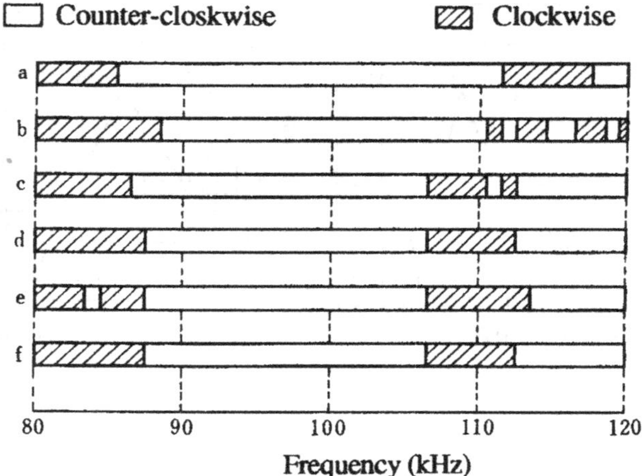

FIGURE 10.54 Rotation direction of the elliptical vibration of the two legs of a π-shaped ultrasonic linear motor as determined by the laser Doppler method. (a, b, c: first leg, d, e, f: second leg).

displacement for various positions on the legs are shown in Figure 10.53, bottom. The counterclockwise rotation of all points produces an efficient drive force against the rail (no drag force!).

Driven vibration analysis by FEM also supports the same counterclockwise elliptical vibration of both legs. The displacement of the legs can also be examined experimentally using a laser Doppler method. Typical data collected by this method for a π-shaped linear ultrasonic motor are summarized in Figure 10.54. We see from these data that there exists a frequency range over which the same rotation direction is produced in both legs. This agrees well with FEM predictions.

10.9.6 Checking the Motor Specifications

The dependence of the motor traveling direction on drive frequency for a π-shaped motor is shown in Figure 10.55, and the speed of the motor as a function of drive voltage and load are shown in Figure 10.56a and b, respectively. The maximum no-load speed and maximum load are approximately 20 cm/s and 200 gf (2 N), respectively. The traveling direction of the motor is reversed by exchanging the drive signals applied to the two multilayer actuators. Tests of the motor's action in both directions result in similar speed versus drive voltage curves. The measured parameters for the motor thus meet the original device specifications in Table 10.4 requested by the customer.

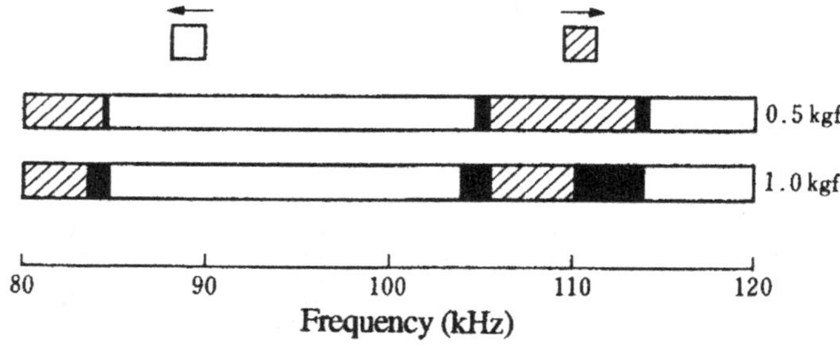

FIGURE 10.55 Dependence of the motor traveling direction on the drive frequency for the π-shaped ultrasonic linear motor.

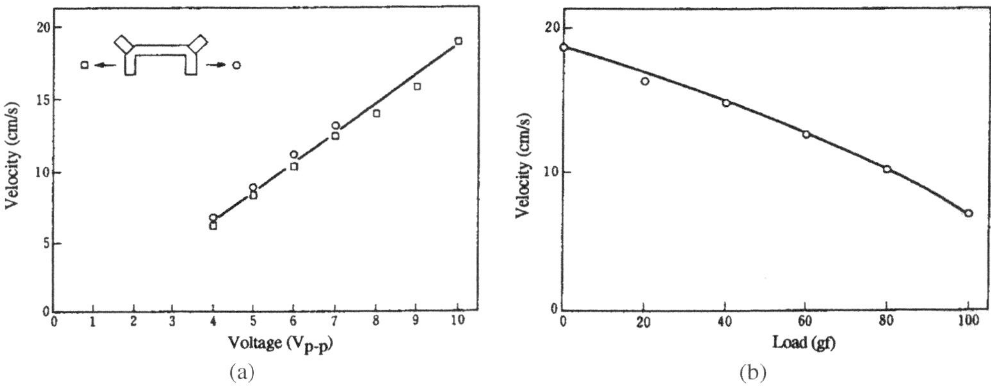

FIGURE 10.56 Speed of the π-shaped ultrasonic linear motor plotted as a function of: (a) drive voltage and (b) load.

TABLE 10.4
Desired Specifications for a π-Shaped Linear Motor

Thrust (gf)	200
Speed (cm/s)	20
Size (mm^3)	20 × 40 × 5
Drive Frequency (kHz)	90

10.10 RELIABILITY OF ULTRASONIC MOTORS

We have considered in this chapter the essential elements involved in the development of ultrasonic motors. In particular, we should examine the following key concepts:

1. **Basic Materials Development:** identifying materials with low loss that are able to sustain high vibration rates.
2. **Methods for Measuring High Field Electromechanical Couplings**
3. **Fundamental Ultrasonic Motor Design Considerations:**
   a. *Displacement Magnification Mechanisms*: such as the horn and hinge lever mechanisms.
   b. *Basic Ultrasonic Motor Types*: classified according to their mode of operation, such as the standing-wave type and the traveling-wave type.
   c. *Frictional Contact*: between the stator and moving parts of the motor.
4. **Drive and Control of the Ultrasonic Motor:**
   a. *High Frequency/High Power Supplies*
   b. *Resonance/Antiresonance Modes of Operation*

Important in the consideration of all these relevant areas is the matter of reliability. We consider here the reliability of an ultrasonic motor in terms of three critical issues: (1) *heat generation* within the device, (2) the *friction materials* used, and (3) the *drive/control techniques* employed.

10.10.1 Heat Generation

The primary factor affecting the reliability of an ultrasonic motor is heat generation within the device. The heat generated during operation can be great enough to cause the temperature to increase as high as 120°C and cause serious degradation in the performance of the motor due to depoling of the piezoelectric ceramic. A hard piezoelectric with a high mechanical quality factor, Q_m, is thus required to inhibit the excessive generation of heat and the detrimental effects of thermal depoling. Selection of an appropriate material is also influenced by the fact that the amplitude of the mechanical vibration at resonance is directly proportional to the magnitude of Q_m. Materials from the solid-solution system PZT-Pb(Mn,Sb)O_3 described in Section 3.5.3 are excellent choices in this regard, as motors in which they are contained have been found to operate at input/output power levels an order of magnitude higher than those incorporating conventionally used hard PZT materials. No significant heating was observed for these devices when operated at these higher power levels.[38,39]

10.10.2 Frictional Coating and Motor Lifetime

The friction coat in the USM as in Figure 10.1 is important for improving the USM's performance; (1) higher friction constant increases the torque level, and (2) lower wear enhances the motor lifetime. The efficiency of the Shinsei ultrasonic motor (Figure 10.31) incorporating various friction materials with respect to maximum output is shown in Figure 10.57.[40] The highest rank materials among those represented in the figure are those in the A group, PTFE (polytetrafluoroethylene, Teflon), PPS (Ryton), PBT (polybutylterephthalate), and PEEK (polyethylethyl ketone). Note that PTFE (Teflon) is used usually as a lubricating material ($\mu_k < 0.1$) under small stress (Figure 8.19), but is used as a high-friction material ($\mu_k \approx 0.4$) under large stress in USMs. Some of the more popular materials used

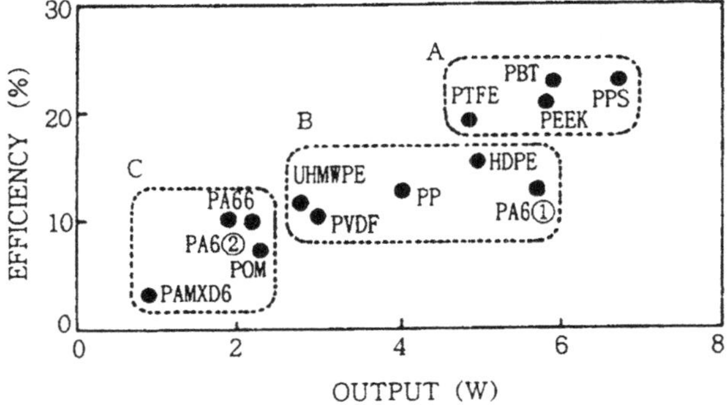

FIGURE 10.57 Efficiency of the Shinsei ultrasonic motor incorporating various friction materials with respect to maximum output. (From Y. Tada et al.: *Polymer Preprints*, 40, 4-17-23, 1408, 1991.)

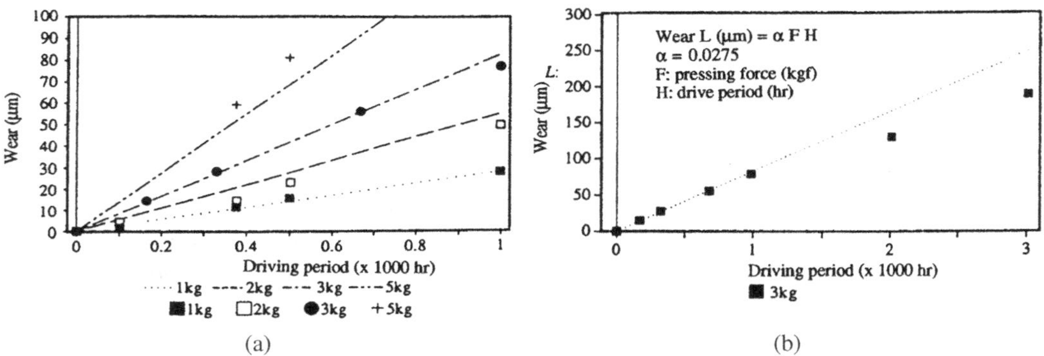

FIGURE 10.58 Wear as a function of driving period for a π-shaped ultrasonic motor incorporating CFRP friction material: (a) over a range of contact forces and (b) with a contact force of 3 (kgf). (From K. Ohnishi et al.: *SAW Device 150 Committee, Jpn. Acad. Promotion Inst., Abstract 36th Mtg.*, p.5, Aug., 1993.)

in commercially available motors include Econol (Sumitomo Chemical), carbon fiber–reinforced plastic (Japan Carbon), PPS (Sumitomo Bakelite), and polyimide ($\mu_k \approx 0.4$ and high wear resistance to be used in automobile brakes). The wear for the π-shaped ultrasonic motor incorporating CFRP friction material is plotted as a function of driving period in Figure 10.58. The data indicate that a 0.3-mm-thick coat of this material will last approximately 3000 hours.[41]

Until recently, the lifetime of an ultrasonic motor was largely determined by the durability of the friction material used in the device. The emergence of new, more durable friction materials has led to considerably longer lifetimes for some of the more recently developed motors. As one example, let us consider the performance of the Shinsei motor (USR 30).[40] The motor is driven continuously for 2000 hours such that there is alternately a clockwise rotation for 1 minute and then counterclockwise for the next minute, a rotation rate of 250 rpm, and a load of 0.5 kgf-cm (50 mN·m). The change in the rate of rotation after this drive period was less than 10%. A second test examines the deterioration of motor characteristics under intermittent drive conditions whereby the motor rotates clockwise for one revolution and then counterclockwise for the next revolution with no load applied. After 250 million revolutions under these conditions, no significant degradation in motor characteristics was observed. Considering the lifetime requirements for motors such as this used in a VCR, about 2000–3000 hours, we see that the lifetime of the ultrasonic motor will quite adequately meet the required specifications.

10.10.3 Drive and Control of the Ultrasonic Motor

Four principal methods for controlling an ultrasonic motor are summarized in Figure 10.59. The "voltage change" and "frequency change" methods have a lack of slow motion, while "phase change" in the drive voltage shows low efficiency. "Pulse width modulation" is the best of the four in terms of the high degree of controllability and efficiency that can be achieved.

Referring to Figure 5.5, the mechanical quality factor, Q_m, of a PZT ceramic sample driven at its fundamental resonance (A-type) is lower than that at antiresonance (B-type) frequency over the entire vibration velocity range, indicating that the same level of mechanical vibration can be achieved

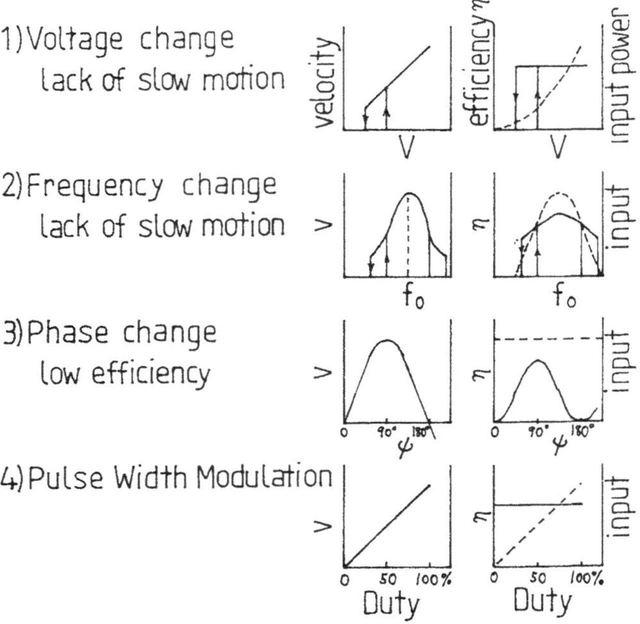

FIGURE 10.59 Four principal methods for controlling an ultrasonic motor.

under smaller input electric power in antiresonance mode than in resonance mode, with less heat generation. Thus, when the motor is driven at the antiresonant frequency, rather than at the resonance frequency, the load on the piezoelectric ceramic and the power supply can be reduced.[42] Moreover, when it is operated in antiresonance mode, the admittance is very low; therefore, it can be powered by a conventional inexpensive low current, high voltage source, in contrast to the high current, low voltage source required for driving the motor at resonance.

One final feature of ultrasonic motors should be highlighted before we close our discussion on this class of devices. They are generally characterized by low speed and high torque, in contrast to the high speed and low torque characteristic of conventional electromagnetic motors. Thus, ultrasonic motors do not require gear mechanisms, which allows for their very *quiet operation* and their *compact size*. The relative simplicity of their design is also favorable from a reliability point of view, as there are fewer components that will be subject to wear and failure over the lifetime of the motor.

10.11 RESONANCE IMPULSE MOTORS

In order to simplify the motor structure and make the manufacturing cost inexpensive in comparison with ultrasonic motors, an inertial motor has been investigated. The inertial motor was first commercialized by Konica-Minolta, Japan, called a smooth impact drive mechanism using an ML piezoelement.[43] The principle of the "stick & slick" motion on a drive rod is explained in Section 9.3. By applying a sawtooth-shaped voltage on a piezoelectric actuator, alternating slow expansion and quick shrinkage are excited on a drive friction rod. A slider placed on the drive rod will "stick" on the rod due to friction during a slow

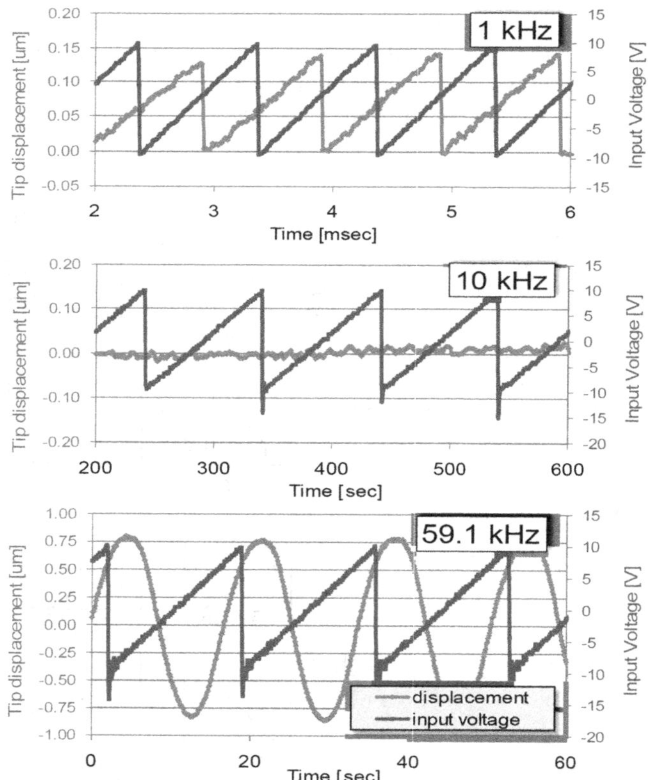

FIGURE 10.60 Frequency dependence of the output displacement on sawtooth-type input voltage. (From S. Tuncdemir et al: *Japan. J. Appl. Phys.*, 50, 027301, 2011.)

expansion period, while it will "slide" during a quick shrinkage period, so that the slider moves from the left to the right. When the voltage saw shape is reversed, the opposite motion can be obtained.

10.11.1 Problems in Smooth Impact Drive Mechanisms

The recent trend is to increase the drive frequency of the SIDM up to its mechanical resonance, aiming at improvement in speed and thrust. Tuncdemir et al. pointed out the problem in its drive voltage form: the saw-type voltage wave cannot generate saw-type displacement with approaching the resonance frequency. The displacement gradually diminishes around 10 kHz and becomes sinusoidal just around the resonance frequency (with significant amplification), as shown in Figure 10.60.[44]

10.11.2 Higher-Order Harmonics Combination

In order to utilize the "stick & slip" mechanism, sawtooth-type displacement is essential. Taking into account that the sawtooth wave can be expressed by a Fourier transform as

$$\begin{aligned} f(x) &= \frac{a_0}{2} + \sum_{n=1}^{\infty}[a_n \cos(nx) + b_n \sin(nx)] \\ &= 2\sum_{n=1}^{\infty}\frac{(-1)^{n+1}}{n}\sin(nx), \end{aligned} \tag{10.29}$$

Morita's group proposed the first, second, and third harmonic combination drive for exhibiting a sawtooth-type displacement mode.[45] Though this technique can be verified as feasible, the harmonic combination drive requires multiple voltage sources according to the number of higher-order harmonics to consider, leading to a significant increase in the total system cost.

10.11.3 Variable Duty-Ratio Rectangular Pulse Drive

In order to simplify the drive circuit and reduce the system cost, Tuncdemir et al. proposed using a rectangular voltage wave at the resonance frequency with variable duty ratio, as shown in Figure 10.61. A schematic illustration of the principle is shown in Figure 10.61a.[46–48] Using a "translational-rotary" multi-degree-of-freedom piezoelectric ultrasonic motor as pictured in Figure 10.62, we computer-simulated and measured the tip motion of the driving rod. The stator of this motor consists of four slanted piezoelectric plates bonded on a metal rod. Dual function output, which is observed on the ring-shaped slider, is controlled by single source excitation signal. The PZT ceramics are excited at the resonance frequency of first longitudinal mode (~59 kHz) for translational operation, or of first torsional mode (~34 kHz) for rotational output motion. Figures 10.61b and c show the ATILA computer simulation of the tip displacement and the measured tip displacement for various duty ratios of the rectangular wave voltage. Try Example Problem 10.4 to understand the detailed calculation process.

EXAMPLE PROBLEM 10.4

The detailed analytical solution and simulation of how we can obtain sawtooth displacement from the variable duty-ratio rectangular pulse drive are not conducted in the above main content. Calculate the processes we skipped above.

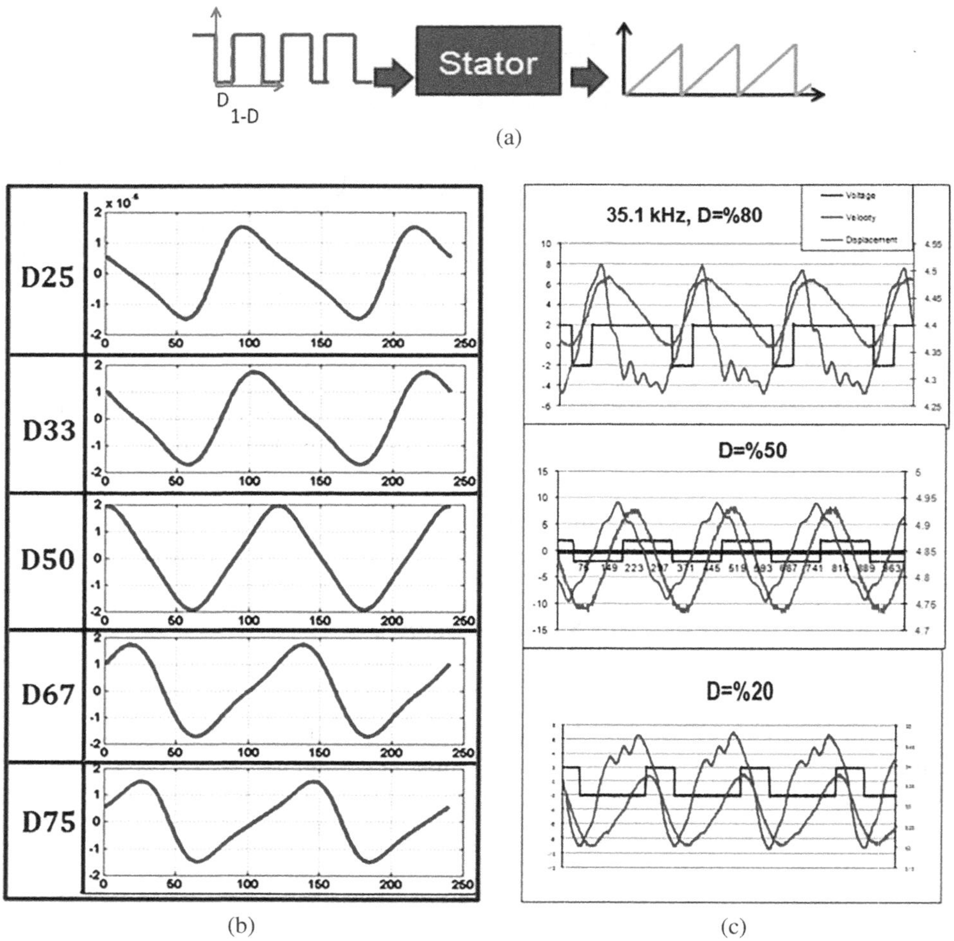

D
1-D
Stator

(a)

D25
D33
D50
D67
D75

(b)

35.1 kHz, D=%80
Voltage
Velocity
Displacement

D=%50

D=%20

(c)

FIGURE 10.61 Resonance-type inertial motor drive with a variable duty-ratio rectangular wave voltage. (a) Schematic illustration of the principle, (b) ATILA computer simulation of the tip displacement, and (c) measured tip displacement for various duty ratio of the rectangular wave voltage. (From S. Tuncdemir et al: *Japan. J. Appl. Phys.*, 50, 027301, 2011.)

FIGURE 10.62 Translational-rotary ultrasonic motor with four slanted PZT ceramic plates.

Solution

We proposed using a rectangular voltage wave at the resonance frequency with the variable duty ratio, as the principle is schematically shown in Figure 10.61a. The rectangular pulse waveform generator, one of the easiest and simplest electric circuits, is a promising driving method with its functionality to contain multiple orders of frequency components. It can be written in Fourier functions as:

$$V(t,D)=\sum_{n=1}^{\infty}[a_n\cos(n\omega t)+b_n\sin(n\omega t)]$$

$$a_n=\frac{2A}{n\pi}\sin(2n\pi D), b_n=\frac{2A}{n\pi}[1-\cos(2n\pi D)], \quad \text{(P10.4.1)}$$

where A is the voltage amplitude, ω is the fundamental resonant frequency, and D is the duty ratio of the pulse waveform. The output displacement of the stator driven by the pulse waveform is determined by four factors: (1) duty ratio of the pulsed waveform (D), (2) ratio of resonance frequencies (n), (3) ratio of vibration amplitude (Q_r), and (4) vibration phase difference (ϕ_e) of its fundamental and higher harmonic mode.[49] Note, however, that the mixing ratio of higher-order harmonics cannot be determined independently in the asymmetric rectangular voltage drive, different from the case of using independent multiple driving sources.

In the following analysis, only the harmonic modes closest to the fundamental mode (usually the second or third harmonic mode) are considered higher-order harmonics. Much higher harmonics are often highly damped, and their influence on the final output is insignificant (i.e., the voltage amplitude is proportional to $1/n$ in Equation 10.30). The contribution from the fundamental and higher harmonic modes can be tuned by changing the duty ratio, D, without using multiple power supplies (such as three power supplies for ω, 2ω, 3ω) to provide the necessary magnitudes of the fundamental and higher harmonic modes. Specifically, we consider the duty ratio, 25%, 33%, 50%, 67%, and 75%:

- When $D=1/3$ (33%) [or $D=2/3$ (67%)]

$$V(t)=\pm\frac{\sqrt{3}A}{\pi}\cos(\omega t)+\frac{3A}{\pi}\sin(\omega t)\mp\frac{\sqrt{3}A}{2\pi}\cos(2\omega t)+\frac{3A}{2\pi}\sin(2\omega t) \quad \text{(P10.4.2)}$$

Notice that at duty ratios of 33% and 67%, 3ω makes no contribution.

- When $D=1/4$ (25%) [or $D=3/4$ (75%)]

$$V(t)=\pm\frac{2A}{\pi}\cos(\omega t)+\frac{2A}{\pi}\sin(\omega t)+\frac{2A}{\pi}\sin(2\omega t)\mp\frac{2A}{3\pi}\cos(3\omega t)+\frac{2A}{3\pi}\sin(3\omega t) \quad \text{(P10.4.3)}$$

It can be seen from above equations that harmonic mechanical resonances at frequencies of 2ω and 3ω can both contribute to the modulation of fundamental vibration. However, from the simulation and experiment below, we can see that only mechanical resonance at 2ω contributes to the sawtooth-type displacement output.[49]

Knowing the tip displacement simulated value for a unit voltage ($-1\cdot\sin n\omega t$) of three frequencies, ω, 2ω, and 3ω, including the phase, taking into account the voltage weight-ratio in (a) for $D=1/3$ and 1/4, we can calculate the expected tip displacement. When we apply -1 V_{ac} (i.e., $V=-\sin(n\omega t)$) at the frequencies ω, 2ω, 3ω, the tip generates the displacements shown in Figure 10.63. Note that -1 V_{ac} is intentionally taken to make positive displacement under the $d_{31}<0$ condition. Taking into account the voltage amplitude weights for ω, 2ω, 3ω in the cases of $D=1/3$, 1/4, we obtain the superposed displacement curves as a function of time, as shown in Figure 10.61b. Figure 10.61c shows the measured tip displacement for different duty ratios of the rectangular wave voltage in the multi-DoF USM in Figure 10.62.

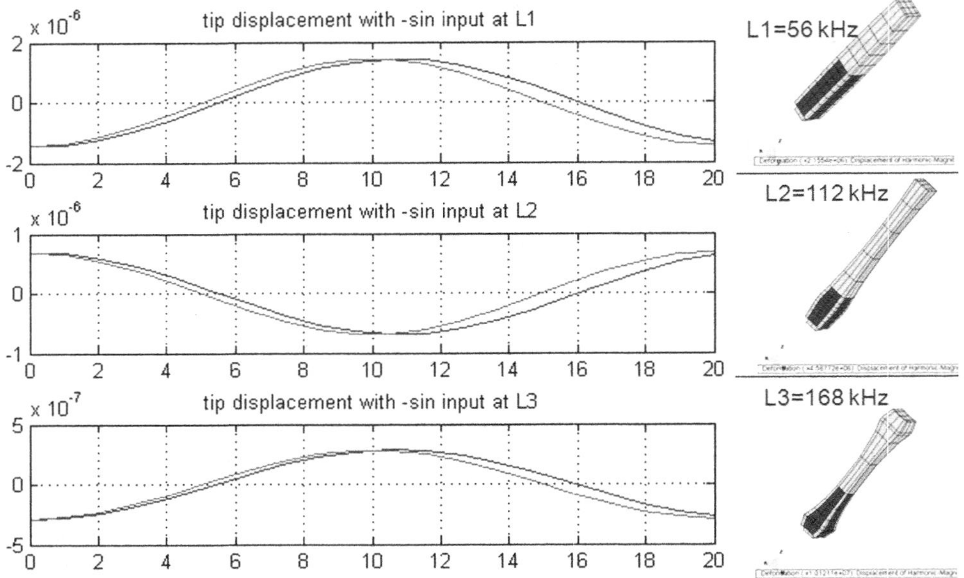

FIGURE 10.63 ATILA harmonic analysis results, showing the tip displacement of the stator at L1, L2, and L3 modes.

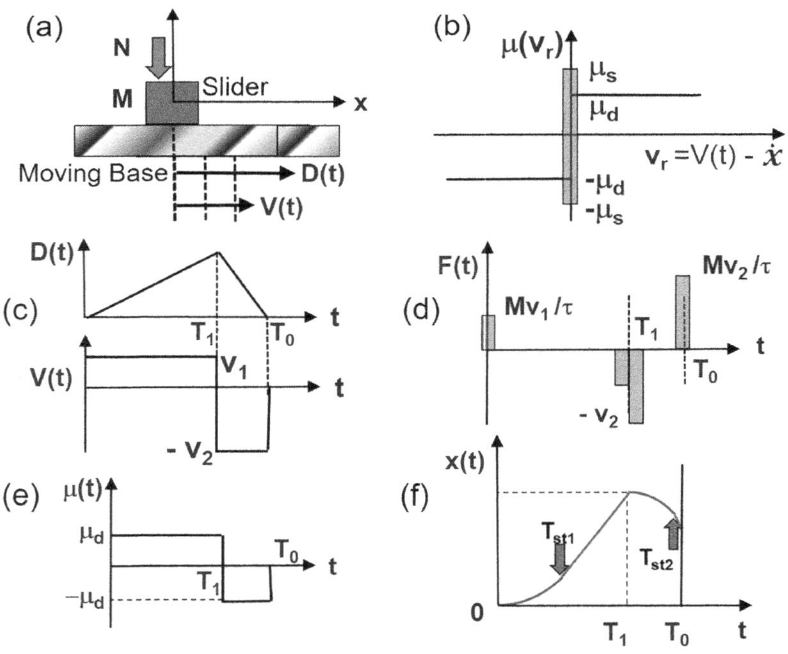

FIGURE 10.64 (a) Stick-slick motion model via friction model; (b) friction constant as a function of velocity; (c) sawtooth displacement and velocity vs. time; (d) impact force vs. time; (e) friction constant vs. time; and (f) slider position change with time. (From K. Uchino: *Proc. 6th Conf. Noise and Vibration Emerging Methods*, May 7–9, Ibiza, Spain, 2018.)

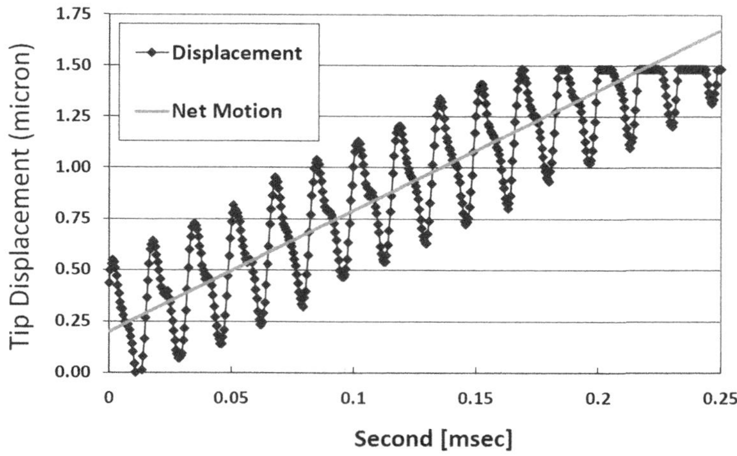

FIGURE 10.65 Practical slider motion of the inertial motor under a drive condition of 59 kHz and $D = 67\%$.

10.11.4 "Stick & Slip" Motion Model

We propose here a simple "stick-slick" motion analysis model. Figure 10.64a shows a model via friction model, where M is the slider mass, N normal force, and $D(t)$ and $V(t)$ the displacement and velocity of the moving base. The friction constant, μ, can be different for static and dynamic conditions, but we handle them as the same and constant as a function of velocity (Figure 10.64b). When the displacement and velocity change as an ideal sawtooth shape as a function of time (Figure 10.64c), impact force can be estimated as a function of time (Figure 10.64d). The initial impact force (Mv_1/τ) generates a constant acceleration slipping motion of the slider (thus, parabolic increase in displacement) until T_{st1}; then, the slider is stuck on the moving base with almost constant velocity. After T_1, the base moves backward very quickly with velocity v_2, which induces another slipping motion until T_{st2} (almost all quick motion period). The dragging friction is shown in Figure 10.64e. The slider is again stuck from T_{st2} to T_0, but this period is very short. The slider motion is schematically shown in Figure 10.64f by a combination of quadratic and linear lines.[50] Practical slider translational motion under a drive condition of 59 kHz and $D = 67\%$ is demonstrated in Figure 10.65.[47] In the test setup, the motor is held on the mobile ring element and the stator movement is monitored by a laser vibrometer. The linear average speed of the mobile element, superposed with zig-zag vibrational displacement, can be calculated as 5 mm/s under a condition of 4 mN blocking force, which is provided by the spring.

10.11.5 Drive Technique Summary

Figure 10.66 summarized the drive condition (frequency and duty ratio) for obtaining translational or rotary operation on the slanted-PZT-plate inertial motor (Figure 10.62). Thirty-four and 59 kHz denote the resonance frequencies of the first torsional and first longitudinal modes, respectively. Being

Duty Ratio / Frequency	Type and Direction of Slider Motion		
	33%	67%	Type
34 kHz	Clockwise	Counter	Rotation
59 kHz	Inward	Outward	Linear

FIGURE 10.66 Command matrix of dual function, bidirectional operation.

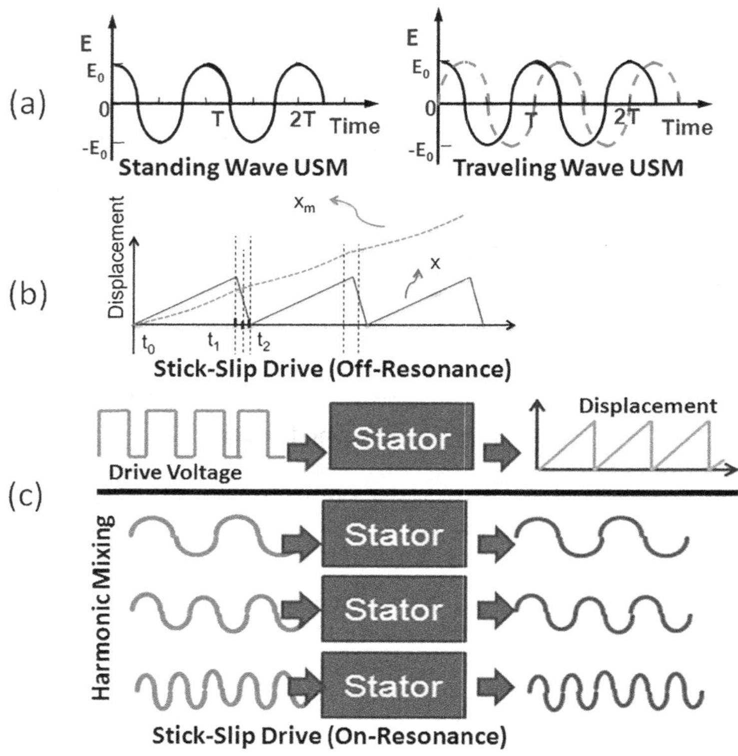

FIGURE 10.67 Drive techniques of piezomotors. (a) Standing- and traveling-wave ultrasonic motors; (b) inertial motor under off-resonance; (c) inertial motor under resonance: variable duty-ratio rectangular-wave and higher-harmonic combination. (From K. Uchino et al: *15th US-Japan Seminar on Dielectric and Piezoelectric Ceramics*, Nov., Kagoshima, Japan, 2011.)

able to command the bidirectional dual function motor with a single source drive by manipulating the frequency and duty ratio realizes the main objective of this research.

Before closing this section, we overview the traditional and advanced drive methods of piezomotors, including resonance ultrasonic motors, off-resonance stick-slick motors, and the compromised resonance stick-slick motors. Drive techniques of piezomotors are summarized in Figure 10.67 for (a) standing-wave (left) and traveling-wave (right) ultrasonic motors; (b) inertial motor under off-resonance (low frequency); and (c) inertial motor under resonance drive with a higher-harmonic combination voltage (bottom) or with a variable duty-ratio rectangular-wave voltage (top).[46]

A standing-wave-type ultrasonic motor is driven at the mechanical resonance frequency with one piezoelectric driving source, while a traveling-wave-type USM is driven by two vibration sources, whose phase differs 90° both in time and space. A mode-rotation type includes multiple vibration sources with phase difference of 120°, 90°, 60°, and so on, depending on the number of segmented electrodes. On the other hand, the inertial "stick-slick" motor is driven by a sawtooth-type voltage to generate a sawtooth-type displacement on a drive rod, and the frequency is typically much lower than the resonance. The inertial motor at the resonance frequency needs a special drive technique, because the sawtooth-type voltage cannot generate sawtooth-type displacement with approaching the resonance frequency; that is, the displacement becomes sinusoidal! A higher-harmonic-mode combination method and a rectangular-pulse method with variable duty ratio are currently investigated. This inertial drive USM is a compromised design between the conventional USM and SIDM, and also exhibits compromised superior performances: displacement amplification, simple motor and drive circuit configurations suitable for further miniaturization, and the possibility of multi-degree-of-freedom motion in one design (such as rotary and translation).

10.12 OTHER ULTRASONIC DEVICES

Current advances in biomedical research and ultraprecise surgical techniques require sophisticated tiny actuators for manipulation of such tools as optical fibers, catheters, and microscale surgical blades. This section introduces additional actuator-related ultrasonic devices, which may be useful for medical applications.

10.12.1 Ultrasonic Surgical Knife

The blades of conventional surgical scalpels tend to become coated with clotted blood during an operation, thus impeding the cutting of tissue. Ultrasonic vibration of the knife is an effective means of maintaining a clear blade for longer periods of time during the procedure.[51] Ultrasonic vibration in the vicinity of the cut also helps to accelerate the rate of blood clotting near the wound after the incision is made. A very thin ultrasonic tube-shaped knife has been proposed for fine procedures such as cataract surgery.[52] Figure 10.68 demonstrates the ultrasonic vibration benefits on medical surgery. Olympus, Japan, developed a hybrid surgical device (electric surgery knife superposed with ultrasonic vibration) for a laparoscopic liver resection purposes. They demonstrated the device in a porcine model (i.e., raw egg).[53] Figures 10.68a and c show a procedure for just an electric surgery knife without using ultrasonic vibration and a procedure for an electric surgery knife and ultrasonic vibration. We can see a beautiful cauterizing function with the ultrasonic vibration. Figure 10.68c shows a procedure for cornea incision. Micromechatronics Inc., PA, developed an ultrasonic blade

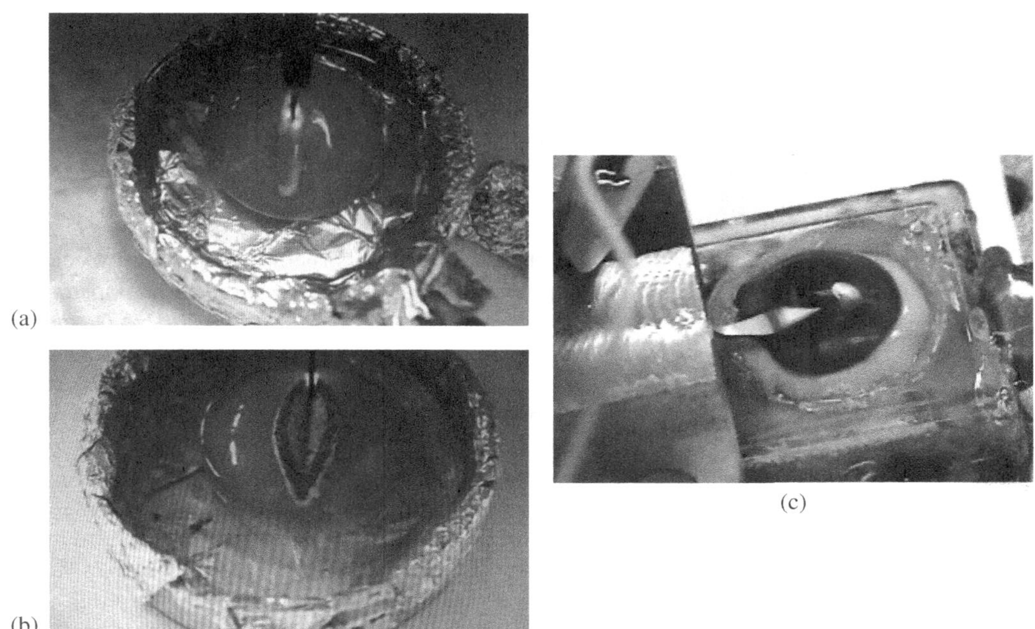

FIGURE 10.68 Electric surgery knife test: (a) procedure for just an electric surgery knife without using ultrasonic vibration and (b) procedure for electric surgery knife and ultrasonic vibration. (Courtesy by Olympus). (c) ultrasonic blade for keratotomy using an in-plane cut ultrasonic vibration: *in vitro* test with a pig eyeball. (Courtesy by Micromechatronics.)

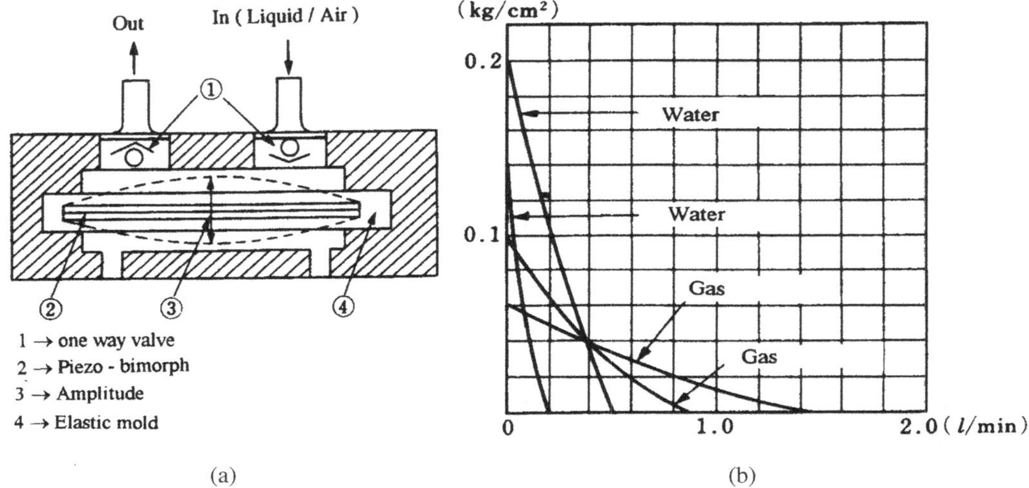

FIGURE 10.69 (a) Piezoelectric pump by Bimor, Inc. (From T. Narasaki: *Proc. 13th Intersoc. Energy Conversion Eng. Conf., Soc. Automotive Eng.*, p.2005, 1978.) (b) Flow rate of the piezoelectric pumps operated in air and water.

for keratotomy using an in-plane cut ultrasonic vibration (unpublished). This figure shows the *in vitro* test with a pig eyeball, where no eyeball distortion is observed, nor is vitreous humor leakage observed during insertion.

10.12.2 Piezoelectric Pump/Fan

An ultrasonic device designed to displace air is the piezoelectric fan, which has been used for cooling electronic circuits[54] and for suspending particulate substances.[55] Devices designed to displace liquids include piezoelectric pumps, ink jets, and evaporators.[56,57] Medical applications of these devices are highly anticipated because of the MRI facility compatibility (no magnetic interference).

10.12.2.1 Piezoelectric Pumps

Piezoelectric pumps are generally composed of a dual bimorph structure coupled to two one-way valves, as shown in Figure 10.69a, which depicts a pump produced by Bimor, Inc., Japan.[58] This device can be driven directly by the commercial line voltage (50/60 Hz and 100 V) because the bimorph resonance frequency has been tuned to approximately 50/60 Hz. The flow characteristics of piezoelectric pumps operated in air and water are shown in Figure 10.69b. These pumps are most extensively used for injection of intravenous drip in hospitals and for automatic seasoning equipment in hotels.

Can the reader create a pump without using one-way valves? An innovative piezoelectric pump is designed by coupling a straw with piezoelectric actuators, such as a cymbal. The new design, pumping principle, and performance of a miniaturized, ultrasonic, valveless pump are introduced.[59] The cymbal actuator has been shown to have larger displacement and generative force than bimorph actuators. The key is the high underwater pressure induced by the cymbal. Gosain et al. used a double-layered cymbal with the advantage that it can behave both as an "in-phase" extensional or an "out-of-phase" bending actuator. The setup promises a hassle-free, internal system that offers considerably smaller size and weight (Figure 10.70a), thereby facilitating the current trend of size reduction in the electronics industry. A simple way of understanding the principle is to analyze the

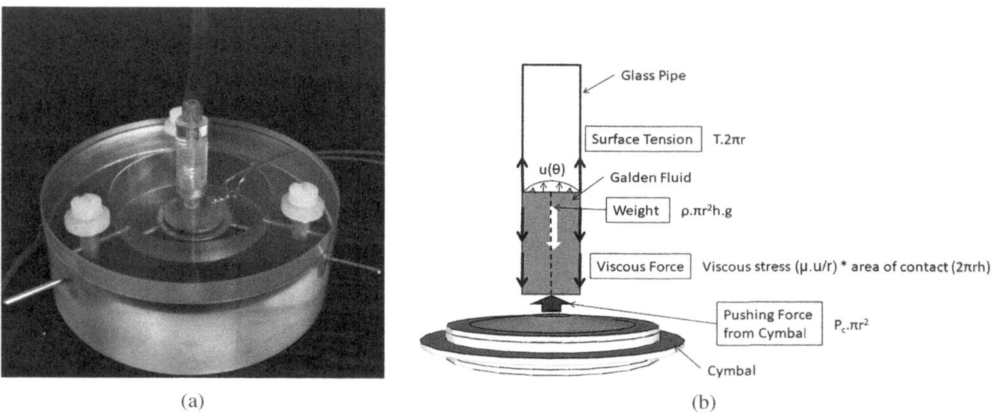

(a) (b)

FIGURE 10.70 (a) Photograph of a pump prototype. (b) Free-body diagram of cymbal and fluid in pipe system.

fluid mass rising in the pipe as a single entity or a "free body," at any time t inside the pipe and treating the whole setup as a Newtonian free-body diagram (Figure 10.70b). The forces acting on this fluid mass are (a) pushing acoustic force from the cymbal ($P_c \cdot \pi r^2$) supporting the flow; (b) surface tension ($T \cdot 2\pi r$) in the direction of flow; (c) the weight of the fluid mass ($\rho \cdot \pi r^2 \cdot h \cdot g$) against the direction of flow; and (d) viscous force [viscous stress ($\mu \cdot u/r$) × area of contact ($2\pi \cdot r \cdot h$)], where r and h are the tube radius and water height, as shown in Figure 10.70b. The net sum of these forces can then be equated to the inertial force which is given by $[\rho \cdot \pi r^2 h \cdot g] \cdot d^2h/dt^2$. Here, h is the height of the fluid column and is a function of time, r is the internal radius of the pipe, μ is the coefficient of kinematic viscosity, and ρ is the fluid density. As the period of oscillation of the fluid particles is orders of magnitude less than the period of oscillation of the cymbal, the acoustic force from the cymbal (P_c) can be simplified and assumed as a constant pressure source. In this case, the fluid velocity, u, can be taken as the first derivative of the fluid column height (dh/dt). On summing the net forces, we get,

$$P_c \cdot \pi r^2 + T \cdot 2\pi r - \left[\rho \cdot \pi r^2 h \cdot g + (2\pi \mu r) h \cdot \frac{dh}{dt}\right] = \left[\rho \cdot \pi r^2 h \cdot g\right] \cdot \frac{d^2 h}{dt^2} \tag{10.30}$$

Taking inspiration from current and past trends, balancing the trade-off between efficient performance and device size is essential. Therefore, a key feature of this pump is performance with respect to size (volume less than 1.5 cm^3). The pump was able to generate a maximum back pressure greater than 5.3 kPa and an approximate maximum flow rate of 10 ml/min.

10.12.2.2 Piezoelectric Fans

Some electronic components such as high-power transistors generate heat and therefore require a local compact cooling system that does not generate electromagnetic noise. A compact piezoelectric fan developed by Piezo Electric Products, Inc. is shown in Figure 10.71.[60] The device includes a pair of piezoelectric bimorphs that are driven out of phase in order to optimize the fan action. A flexible blade attached to a weight is installed at the tip of each bimorph. This configuration causes the bottom and top of the blade to have a phase difference of 90° when driven by the bimorph so that it executes a motion similar to a swimming fish. A piezoelectric fan unit with dimensions $4 \times 2 \times 1.2$ cm^3 is capable of decreasing the heat generated from a 30-W heater to several mW. Compact piezoelectric fans with a similar design are also fabricated by Misuzu-Erie and Nippon Denso[61] of Japan.

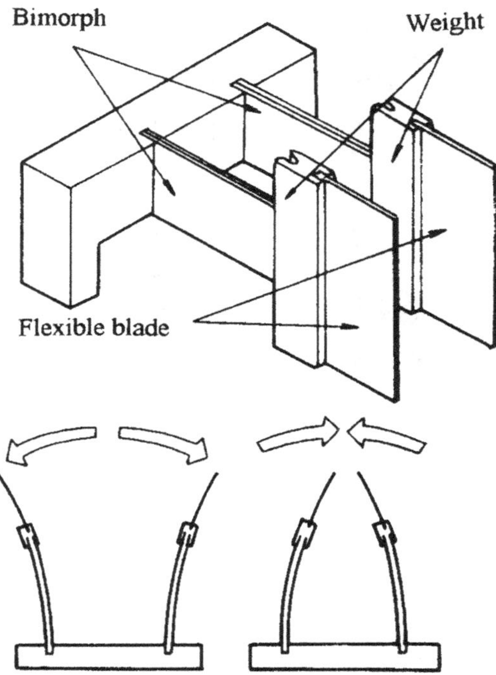

FIGURE 10.71 Compact piezoelectric fan by Piezo Electric Products, Inc. (PEPI). (From Piezo Electric Product Inc.: Products Catalog.)

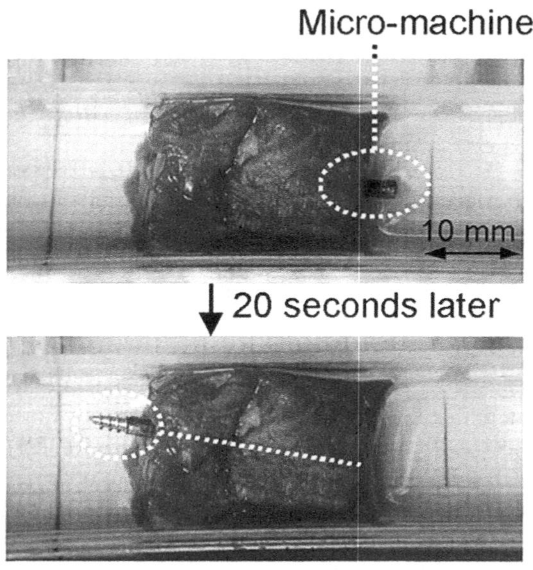

FIGURE 10.72 Magnetic screw motor. (From K. Arai: www.riec.tohoku.ac.jp/Lab/Arai 2001.)

10.12.2.3 Evaporators

The structure of piezoelectric evaporators is quite similar to that of the inkjet. The major differences between these two devices are related to the drive resonance and the maximum evaporation rate of the liquid. One significant application for piezoelectric evaporators has been in cool vapor humidifiers, which are preferred in homes with small children for safety reasons. A more recent application that has attracted considerable attention is their use as part of a system for evaporating heavy oil in diesel engines to increase the combustion efficiency of the fuel.

10.12.3 Magnetic Actuators

Motors activated by magnetic actuators have the capability of operating by remote control. A variety of prototype micromagnetic devices have been demonstrated that can fly, walk, swim, and drill.[62] One intriguing design has the shape of a tiny butterfly, which can actually fly between a pair of magnetic poles when it is magnetically driven at a frequency corresponding to the resonance frequency of the butterfly wings. Another interesting and potentially useful device is the magnetic screw motor pictured in Figure 10.72. The photos depict the device drilling through a piece of beef as it is driven by a cyclic magnetic field. The rate of rotation of the magnetic drill is thus synchronized with the frequency of the applied field and, in this demonstration, the rate of drilling occurred at approximately 1 mm/s. Note that there is no electric lead wire (i.e., remote control!).

CHAPTER ESSENTIALS

1. *Merits and Demerits of Ultrasonic Motors*:

Merits	Demerits
• Low Speed and High Torque	• High Frequency Power Supply Required
• Direct Drive	• Torque vs. Speed Droop
• Quick Response	• Low Durability
• Wide Range of Speeds	
• Hard Brake with No Backlash	
• Excellent Controllability	
• Fine Position Resolution	
• High Power-to-Weight Ratio	
• Quiet Drive	
• Compact Size and Light Weight	
• Simple Structure and Easy Production	
• No Generation of Electromagnetic Radiation	
• Unaffected by External Electric or Magnetic Fields	

2. *Classification of Ultrasonic Motors*:
   a. Standing-Wave Type
   b. Mixed-Mode Type
   c. Traveling-Wave Type
   d. Mode Rotation Type
3. *Principle of Operation for the Traveling-Wave Type*: A traveling wave can be generated by superposing two standing waves with a 90° phase difference between them.
4. *Contact Models for Speed/Thrust Calculations*:
   a. Rigid Slider and Rigid Stator
   b. Compliant Slider and Rigid Stator
   c. Compliant Slider and Compliant Stator
5. *Essential Elements Involved in the Development of Ultrasonic Motors*:
   a. Basic Materials Development
   b. Methods for Measuring High-Field Electromechanical Couplings

c. Fundamental Ultrasonic Motor Design Considerations:
   i. Displacement Magnification Mechanisms
   ii. Basic Ultrasonic Motor Types
   iii. Frictional Contact
d. Drive and Control of the Ultrasonic Motor:
   i. High-Frequency/High-Power Supplies
   ii. Resonance/Antiresonance Modes of Operation

6. Resonance Impulse Motor Drive:
   - Higher-harmonic combination voltage
   - Variable duty-ratio rectangular-wave voltage

   Superior performances include: displacement amplification, simple motor and drive circuit configurations suitable for further miniaturization, cost reduction, possibility of multi-degree-of-freedom motion in one design (such as rotary and translation).

CHECK POINT

1. Describe the resonance frequency, f, for the following electrical circuit:

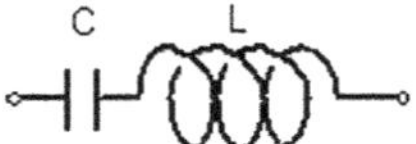

2. Fill in the blank (a) in the following equation:

$$\cos kx \cos \omega t + \cos (kx\text{-}\pi/2) \cos (\omega t\text{-}\pi/2) = \cos [\ (a)\ ]$$

3. The piezoelectric motor has the following four benefits in a compact shape (<30 W) compared with the traditional electromagnetic motors: (1) more suitable to miniaturization, (2) higher efficiency, (3) nonflammable. What is the remaining benefit? Answer simply.

4–9 To summarize the comparison between the normal standing-wave and traveling-wave type ultrasonic motors, fill in the following table (Do not consider elastic loss or a specially modified version):

Ultrasonic Motors	Standing-Wave Type	Traveling-Wave Type
Number of vibration sources	(4)	(5)
Controllability (uni- or bidirectional)	(6)	(7)
Efficiency (theoretical maximum)	(8)	(9)

CHAPTER PROBLEMS

10.1 Describe the general principle for generating a traveling wave on an elastic ring using a unimorph structure. Then, assuming a displacement produced by the second-order vibration mode of the form $u(\theta,t) = A\cos(2\theta - \omega t)$, describe two possible electrode and poling configurations that can be applied to the piezoelectric ceramic ring to produce the traveling wave.

10.2 Verify Equations 10.18 through 10.25 for the compliant slider and rigid stator model. Evaluate the average speed, v_o, and the maximum thrust, F, for the cases of $\phi = 0$ and $\phi = \pi$.

10.3 Survey the recent literature (i.e., from the last 5 years) and prepare a brief report on the research that has been done on piezoelectric ultrasonic motors during this period. Be sure to include the following in your report:

a. Include a list of papers (at least five) that report on ultrasonic motors.
b. Prepare a classification of the motors reported using the categories described in this chapter (i.e., standing-wave type or traveling-wave type, rotary or linear type, etc.).
c. Provide a brief explanation of the basic operating principles of the motors included in your report.
d. Summarize and briefly describe any intended applications for the motors included in your report.

REFERENCES

1. H. V. Barth: *IBM Technical Disclosure Bull.*, 16, 2263, 1973.
2. V. V. Lavrinenko, S. S. Vishnevski and I. K. Kartashev: *Izvestiya Vysshikh Uchebnykh Zavedenii, Radioelektronica*, 13, 57, 1976.
3. P. E. Vasiliev et al.: UK Patent Application GB 2020857 A, 1979.
4. T. Sashida: *Oyo Butsuri.*, 51, 713, 1982.
5. Y. Fuda and T. Yoshida: *Ferroelectrics*, 160, 323, 1994.
6. B. Koc, J. F. Tressler and K. Uchino: *Proc. Actuator 2000 7th Int'l Conf. New Actuators*, p.242–2245, June 19–21, 2000.
7. K. Uchino: *Proc. New Actuator 2004*, Bremen, June14–16, p.127, 2004.
8. R. Lee: *Proc. Proc. Int'l Actuator Symp.*, ICAT/PSU, State College, PA, Oct., 2007.
9. B. Akbiyik: Undergraduate Thesis, Electr. Engr., Kirikkale University, Turkey, 2001.
10. K. Uchino, K. Kato and M. Tohda: *Ferroelectrics*, 87, 331, 1988.
11. T. Hayashi: *Proc. Jpn. Electr. Commun. Soc.*, US84-8, p.25, June, 1984.
12. A. Kumada: *Jpn. J. Appl. Phys.*, 24(Suppl. 24–2), 739, 1985.
13. B. Koc, P. Bouchilloux and K. Uchino: *IEEE Trans.-UFFC*, 47, 836, 2000.
14. K. Nakamura, M. Kurosawa and S. Ueha: *Proc. Jpn. Acoustic Soc.*, No.11-18, p.917, Oct., 1993.
15. H. Saigo: *15th Symp. Ultrasonic Electronics (USE 94)*, No. PB-46, p.253, Nov. 1994.
16. B. Koc, J. Ryu, D. Lee, B. Kang and B. H. Kang: *Proc. New Actuator 2006*, Bremen, June 14–16, p. 58, 2006.
17. Nikkei Mechanical, Feb. 28 issue, p.44, 1983.
18. M. Kurosawa, S. Ueha and E. Mori: *J. Acoust. Soc. Amer.*, 77, 1431, 1985.
19. M. Takahashi, M. Kurosawa and T. Higuchi: *Proc. 6th Symp. Electro-Magnetic Dynamics '94*, No. 940-26 II, D718, p.349, July, 1994.
20. Y. Tomikawa, T. Nishituka, T. Ogasawara and T. Takano: *Sonsors & Mater.*, 1, 359, 1989.
21. K. Ohnishi, K. Naito, T. Nakazawa and K. Yamakoshi: *J. Acoust. Soc. Jpn.*, 47, 27, 1991.
22. S. Ueha and M. Kuribayashi: *Ceramics*, 21(1), 9, 1986.
23. T. Sashida: *Mech. Automation of Jpn.*, 15(2), 31, 1983.
24. K. Ise: *J. Acoust. Soc. Jpn.*, 43, 184, 1987.
25. M. Kasuga, T. Satoh, N. Tsukada, T. Yamazaki, F. Ogawa, M. Suzuki, I. Horikoshi and T. Itoh: *J. Soc. Precision Eng.*, 57, 63, 1991.
26. Seiko Instruments: Product Catalogue "Mussier," 1997.
27. J. Cummings and D. Stutts: *Amer. Ceram. Soc. Trans.* "Design for Manufacturability of Ceramic Components," p.147, 1994.
28. A. Kumada: *Ultrasonic Technology*, 1 (2), 51, 1989.
29. Y. Tomikawa and T. Takano: *Nikkei Mechanical*, (Suppl.,)194, 1990.
30. T. Yoshida: *Proc. 2nd Memorial Symp. Solid Actuators of Japan: Ultra-Precise Positioning Techniques and Solid Actuators for Them*, p.1, 1989.
31. S. -H. Park, Y. -D. Kim, J. Harris, S. Tuncdemir, R. Eitel, A. Baker, C. Randall and K. Uchino: *Proc. 10th Int'l Conf. on New Actuators*, p. 432, Bremen, Germany, 14–16 June, 2006.
32. S. -H. Park, S. J. Agraz, S. Tuncdemir, Y.-D. Kim, R. E. Eitel, A. Baker, C. A. Randall and K. Uchino: *Japan. J. Appl. Phys.*, 47(1), 313–318, 2008.
33. S. Dong, S. P. Lim, K. H. Lee, J. Zhang, L. C. Lim and K. Uchino: *IEEE UFFC Trans.*, 50(4), 361–367, 2003.
34. K. Uchino: *Solid State Phys., Special Issue "Ferroelectrics"*, 23(8), 632, 1988.
35. N. Karasikov: *Proc. 67th ICAT/JTTAS Joint Int'l Symposium on Smart Actuators*, State College, PA, Sept. 30–Oct. 1, 2014.

36. https://www.youtube.com/watch?v=Tlo5z6rz7S4
37. K. Onishi: *Ph.D. Thesis*, Tokyo Institute of Technology, Japan, 1991.
38. S. Takahashi, Y. Sasaki, S. Hirose and K. Uchino: *Proc. MRS '94 Fall Mtg.* Vol. 360, p.2–5, 1995.
39. Y. Gao, Y. H. Chen, J. Ryu, K. Uchino and D. Viehland: *Jpn. J. Appl. Phys.*, 40, 79–85, 2001.
40. Y. Tada, M. Ishikawa and N. Sagara: *Polymer Preprints* 40, 4-17-23, 1408, 1991.
41. K. Ohnishi et al.: *SAW Device 150 Committee, Jpn. Acad. Promotion Inst., Abstract 36th Mtg.*, p.5, Aug., 1993.
42. S. Hirose, S. Takahashi, K. Uchino, M. Aoyagi and Y. Tomikawa: *Proc. Mater. for Smart Systems, Mater. Res. Soc.*, Vol. 360, p.15, 1995.
43. Y. Okamoto, R. Yoshida and H. Sueyoshi: *Konica Minolta Tech. Report*, 1, 23, 2004.
44. S. Tuncdemir, S.O. Ural, B. Koc and K. Uchino: *Japan. J. Appl. Phys.*, 50, 027301, 2011.
45. T. Nishimura and T. Morita: *Proc. 12th Int'l Conf. New Actuators*, p. 181, Bremen, Germany, June 2010.
46. K. Uchino, S. Tuncdemir and Y. Bai: *15th US-Japan Seminar on Dielectric and Piezoelectric Ceramics*, Nov., Kagoshima, Japan, 2011.
47. S. Tuncdemir, S. O. Ural, B. Koc, and K. Uchino: *Proc. 12th Int'l Conf. New Actuators*, Bremen, Germany, June, 14–16, 2010.
48. S. Tuncdemir, Y. Bai and K. Uchino: *Proc. 13th Int'l Conf. New Actuators*, Germany, June, 2012.
49. Y. Ma, H. Shekhani, X. Yan, M. Choi, and K. Uchino: *Sensors and Actuators A*, 248, 29–37, 2016.
50. K. Uchino: *Proc. 6th Conf. Noise and Vibration Emerging Methods*, Ibiza, Spain, May 7–9, 2018.
51. Y. Tsuda, E. Mori and S. Ucha: *Jpn. J. Appl. Phys.*, 22(Suppl., 3), 1983.
52. Y. Kuwahara: *Cataract Surgery*, Igaku-Shoin, Tokyo, 1970.
53. https://journals.sagepub.com/doi/suppl/10.1177/1553350618812298
54. E. Kolm and H. Kolm: *Chemtech*, 3, 180, 1983.
55. H. Hatano, Y. Kanai, Y. Ikegami, T. Fujii and K. Saito: *Jpn. J. Appl. Phys.*, 21, 202, 1982.
56. R. Aoyagi and H. Shimizu: *Proc. Jpn. Acoust. Soc.*, p.251, 1981.
57. K. Asai and A. Takeuchi: *Jpn. J. Appl. Phys.*, 20(Suppl. 20–3), 169, 1981.
58. T. Narasaki: *Proc. 13th Intersoc. Energy Conversion Eng. Conf., Soc. Automotive Eng.*, p.2005, 1978.
59. R. Gosain, S. O. Ural, Y. Zhuang, S. Tuncdemir, A. Amin and K. Uchino: *Japan. J. of Appl. Phys.*, 49, 095201, 2010.
60. Piezo Electric Product Inc.: Products Catalog.
61. M. Yorinaga, D. Makino, K. Kawaguchi and M. Naito: *Jpn. J. Appl. Phys.*, 24(Suppl. 24–3), 203, 1985.
62. K. Arai: www.riec.tohoku.ac.jp/Lab/Arai 2001.

11 The Future of Solid State Actuators in Micromechatronic Systems

After 40 years of intensive research and development of solid-state actuators, the focus has gradually shifted to applications and commercialization. Automatic focusing mechanisms (Canon) in cameras, inkjet printers (Epson), and compact camera modules for mobile phones (Samsung Electromechanics), and multilayer piezoelectric actuators for diesel injection valves (TDK-EPCOS) have been mass-produced on a scale of 1–100 millions of pieces per month. Throughout this period of commercialization, especially over the last decade, new actuator designs and methods of drive and control have been developed to meet the requirements of the latest applications. At a somewhat slower pace, advances in device reliability and strength have likewise occurred, although some considerable work is yet to be done in further extending the lifetime of devices and ensuring consistent performance over that period. An overview of these recent trends in development is presented in this chapter, followed by a projected view of future trends and design concepts.

11.1 PIEZOELECTRIC DEVICE MARKET TRENDS

Actuator applications of piezoelectrics started in the late 1970s, and enormous investment was made in practical developments during the 1980s, aiming at consumer applications such as precision positioners with high strain materials, multilayer device design, and mass-fabrication processes for portable electronic devices, ultrasonic motors for microrobotics, and smart structures. After the slump due to the worldwide economic recession in the late 1990s, we are now facing a sort of renaissance of piezoelectric actuators according to the social environmental changes for sustainability (i.e., energy saving, biomedical areas) and crisis technologies. Figure 11.1 shows the piezoelectric device market estimated from multiple sources.[1] The current (2017) revenue is around US$40 billion (Figure 11.1a). Actuator/piezogenerator (energy harvesting) is the largest category, followed by transducer/sensor/accelerometer/piezotransformer. Then, the resonator/acoustic device/ultrasonic motor category follows (Figure 11.1b). Figure 11.2 shows a pie chart of the piezoelectric device application areas.[1] The most remarkable expansion can be found in the "biomedical" and "ecology & energy harvesting" areas in only these several years.

11.2 ENGINEERING HISTORY

11.2.1 History of Solid-State Actuators

Actuator applications of piezoelectrics started in the late 1970s, and enormous investment was made in practical developments during the 1980s, aiming at consumer applications such as precision positioners with high strain materials, multilayer device designing, and mass-fabrication processes for portable electronic devices, ultrasonic motors for microrobotics, and smart structures. After the slump due to the worldwide economic recession in the 1990s, we are now facing a sort of renaissance of piezoelectric actuators, according to the social environmental changes. The twenty-first century faces becoming a *sustainable society*. Global regulations are strongly called for in ecological and human health care issues, and government-initiated technology (i.e., *politico-engineering*) has

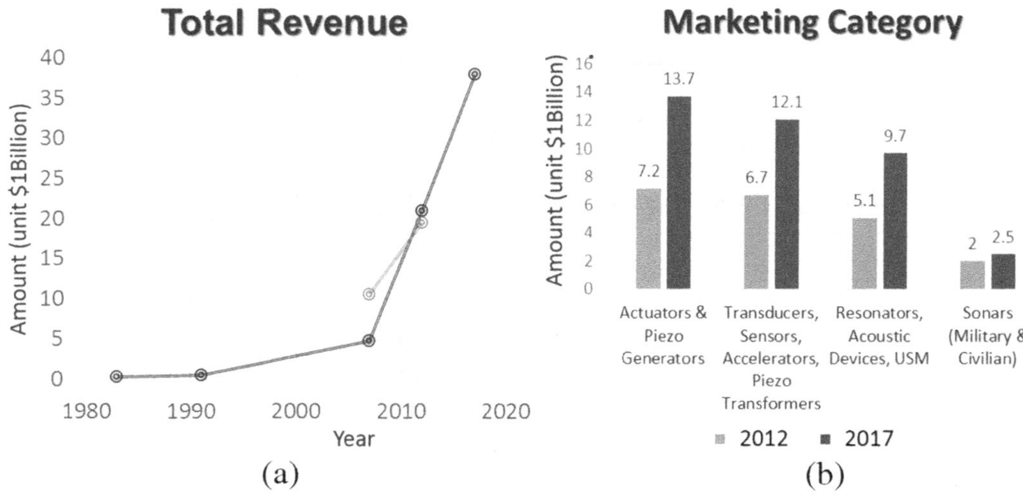

FIGURE 11.1 Piezoelectric device market trends: (a) worldwide piezodevice revenue change; (b) piezoelectric category share. (From Information Resources: *Multiple Year Market Research Data from iRAP (Innovative Research & Products)*, Market Publishers; IDTechEx.)

become essential.[2] Because of the significantly high energy efficiency of piezoelectrics in compact size (typically less than 30 W) in comparison with other actuators such as chemical engines and electromagnetic components, piezoelectric actuators have been refocused upon most recently in the sustainable society (i.e., a renaissance in piezoelectric actuators). This chapter reviews the recent advances in materials, designing concepts, and applications of piezoelectric actuators/transducers, as the reader's review purpose, then describes the future perspectives (the author's own) in this area.

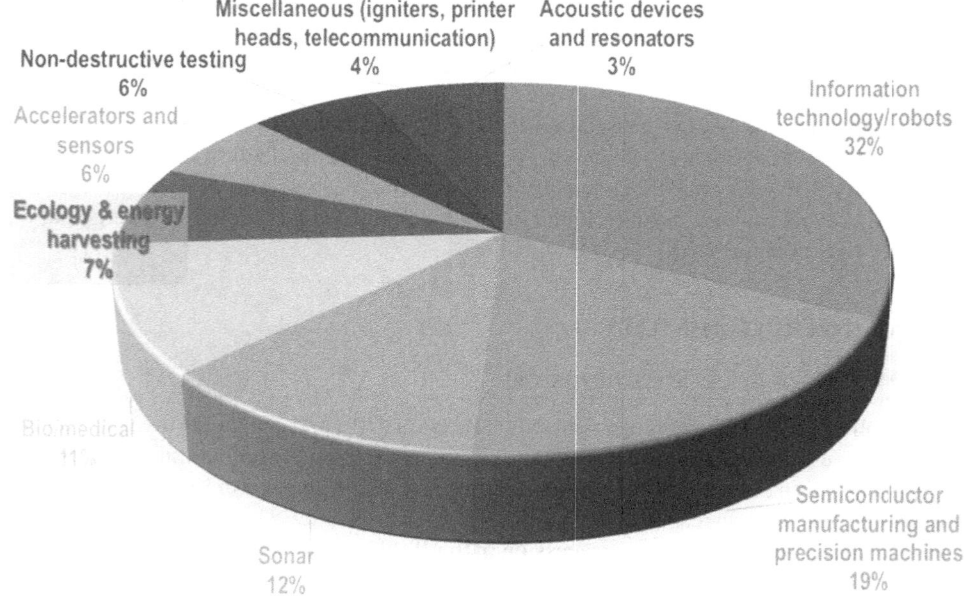

FIGURE 11.2 Piezoelectric device application areas as a pie chart. (From Information Resources: *Multiple Year Market Research Data from iRAP (Innovative Research & Products)*, Market Publishers; IDTechEx.)

11.2.2 History of Piezoelectricity

Piezoelectricity was discovered by Jacque and Pierre Curie in 1880 in a quartz single crystal. Though the shipwreck of the *Titanic* in 1912 motivated the application of piezoelectrics to sonar transducers to survey the undersea, we needed to wait for real investment until World War I (1914–18). Paul Langevin developed underwater transducers (so-called "Langevin Transducers") by sandwiching tiny quartz single crystals with metal plates in order to survey German U-boats (submarines). Another epoch-making discovery in piezoelectrics was a new ceramic, barium titanate, during World War II (1939–45). $BaTiO_3$ was first developed for creating high-capacitance materials for compact radar system applications, based on Tita-Con (the then-famous TiO_2-based condenser) independently by researchers in the United States, Japan, and Russia.[3–5] The discovery of piezoelectricity on $BaTiO_3$ after electric-poling treatment by Gray facilitated wide application of piezoceramics after WWII.[6] Rath, Germany, seems to have discovered BT earlier, according to the patent filing (UK Patent #445,495, 1936). Lead zirconate titanate with a monomorphic crystal structure (i.e., perovskite) was systematically studied[7] and piezoelectric performance was improved significantly in comparison with BT. PZT has been the most dominant piezoelectric material in the last 60 years.

11.2.3 Engineering Trends after World War II

After WWII, the country power (GDP) changed with the year for Japan, the United States (20 years ahead), and China (30 years behind) in a typical S-shape growth curve, as visualized in Figure 11.3. Let us review the product planning strategy taken historically by Japanese industries. In the 1960s, the four-Chinese-character slogan was "重厚長大 (*heavier thicker longer and larger*)"; that is, manufacturing heavier ships, thicker steel plates, constructing longer buildings, and larger power plants (dams) were the key strategies for recovering from the ruins of World War II (i.e., domestic politics). Refer to the initial rise of the growth curve in Figure 11.3. Though Japanese people became wealthy, subsidiary effects started to surface: that is, *industrial pollution*. Steel industries produced air pollution and "asthma" even for small kids. Traffic congestion generated severe acoustic noise even in suburban areas. One of the most advanced industries, nuclear power plants, leaked hazardous radioactive waste multiple times. The meltdown accident of Three Mile Island, Pennsylvania happened in 1979.

A completely opposite slogan started in the 1980s; that is, "軽薄短小 (*lighter thinner shorter and smaller*)." Printers and cameras became lighter in weight, thinner computers and TVs (flat panel) gained popularity, printing time and information transfer periods became shorter, and air conditioners and tape recorders (e.g., the Walkman by SONY) were smaller. Though serious industrial pollution diminished gradually during this period in proportion to the country power (high GDP per capita), different subsidiary effects started: (1) the greenhouse effect and *global warming* due to CO_2 gas

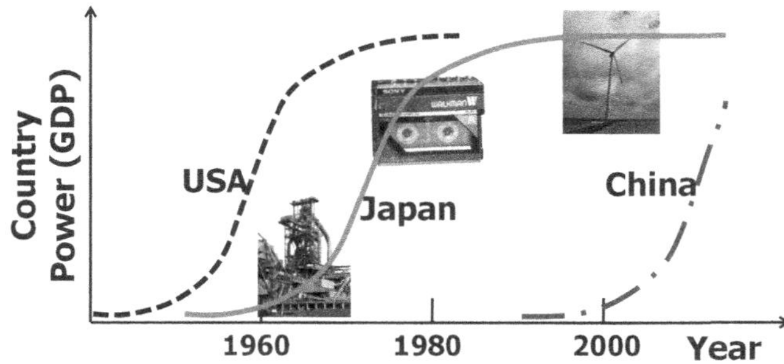

FIGURE 11.3 Country power (GDP) change with year for Japan, USA, and China visualized by typical S-shape growth curves. (From K. Uchino: "Politico-Engineering in Piezoelectric Devices," *Proc. 13th Int'l Conf. New Actuators*, Bremen, Germany, June 18–20, p. A1.0, 2012.)

generated by overproduced automobiles; (2) energy crisis due to overconsumption of energy and lack of fossil energy sources (oil), in addition to political mismanagement; and (3) population growth due to advanced medical technologies. A longer lifetime is welcomed by individuals (now the average age for Japanese females is approaching 87, the age for males is 81 already; the world's eldest). However, overpopulation of humans will create an imbalance against other animals/nature, and the senior population, in particular, causes societal and economic problems (pension, health insurance, work force, etc.).

When the twenty-first century began, as consequent results, environmental degradation, resource depletion, and food famine have become major problems. Global regulations (i.e., global regime) are strongly called for, and government-initiated technology, that is, *politico-engineering*, has become important again in order to overcome the regulations. On the other hand, multinational terrorist attacks have risen after the Cold War ending. The author proposed a new four-Chinese-character keyword for the era of politico-engineering, "協守減維 (cooperation protection reduction and continuation)."[2] Global coordination and international cooperation in standardization of internet systems and computer cables became essential to accelerate mutual communication. The Kyoto Protocol in December 1997 is an international agreement linked to the United Nations Framework Convention on Climate Change for reducing greenhouse gas emission (now the Paris Agreement).[8] Protection of the territory and environment from enemies or natural disaster and prevention of infectious disease spread are mandatory. Reduction of toxic materials such as lead, heavy metals, and dioxin, and of the use of resources and energy consumption is also key, and society's continuation, that is, status quo or sustainable society, is important to promote.

According to the above social atmosphere change, the research trends in micromechatronics have also been changing, and are discussed in the following.

11.3 RECENT RESEARCH TRENDS AND THE FUTURE PERSPECTIVES

We discuss five key trends in this section for providing future perspectives on research trends; "Performance to Reliability," "Hard to Soft," "Macro to Nano," "Homo to Hetero," and "Single to Multifunctional."

11.3.1 Performance to Reliability

11.3.1.1 Pb-Free Piezoelectric Ceramics

In 2006, the European Community started RoHS, which explicitly limits the usage of lead (Pb) in electronic equipments. Pb (lead)-free piezoceramics started to be developed after 1999, and are basically classified into three groups; $(Bi,Na)TiO_3$, $(Na,K)NbO_3$, and tungsten bronze (TB), most of which are revival materials after the 1970s. The share of the patents for bismuth compounds [bismuth layered type and $(Bi,Na)TiO_3$ type] exceeds 61%. This is because bismuth compounds are easily fabricated in comparison with other compounds. Honda Electronics, Japan, developed Langevin transducers using BNT-based ceramics for ultrasonic cleaner applications.[9] Their composition $0.82(Bi_{1/2}Na_{1/2})TiO_3 - 0.15BaTiO_3 - 0.03(Bi_{1/2}Na_{1/2})$-$(Mn_{1/3}Nb_{2/3})O_3$ exhibits $d_{33} = 110 \times 10^{-12}$C/N, which is only 1/3 of that of a hard PZT, but the electromechanical coupling factor $k_t = 0.41$ is larger because of much smaller permittivity ($\varepsilon = 500$) than that of PZT. Furthermore, the maximum vibration velocity of a rectangular plate (k_{31} mode) is close to 1 m/s (rms value), which is higher than that of hard PZTs.

$(Na,K)NbO_3$ systems exhibit the highest performance among the present Pb-free materials, because of the morphotropic phase boundary usage. Figure 11.4 shows the current best data reported by Toyota Central Research Lab, where strain curves for oriented and unoriented (K,Na,Li) $(Nb,Ta,Sb)O_3$ ceramics are shown.[10] Note that the maximum strain reaches up to 1500×10^{-6}, which is equivalent to the PZT strain. Drawbacks include their sintering difficulty and the necessity of a sophisticated preparation technique (topochemical method for preparing flaky raw powder).

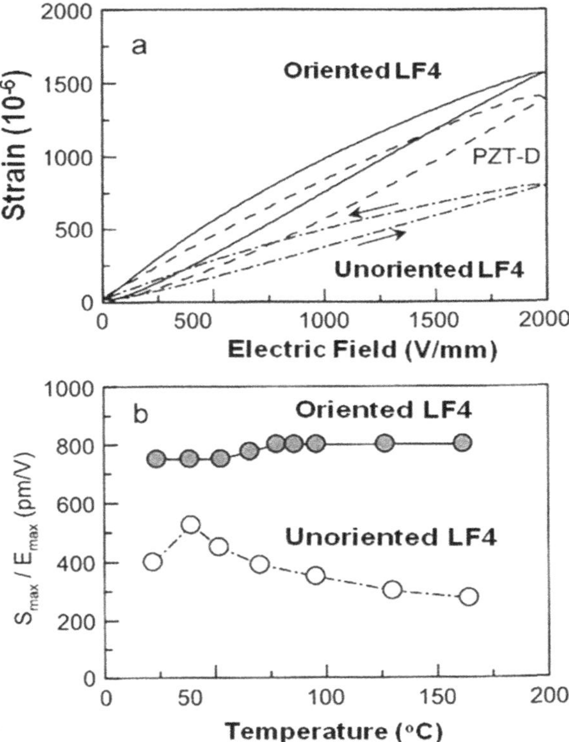

FIGURE 11.4 Strain curves for oriented and unoriented (K,Na,Li) (Nb,Ta,Sb)O_3 ceramics. (From Y. Saito: *Jpn. J. Appl. Phys.*, 35, 5168–73, 1996.)

Tungsten-bronze types are another alternative choice for resonance applications because of their high Curie temperature and low loss. Taking into account the general consumer attitude on the disposability of portable equipment, Taiyo Yuden, Japan, developed micro ultrasonic motors using non-Pb multilayer piezo-actuators.[11] Their composition is based on TB [$(Sr,Ca)_2NaNb_5O_{15}$] without heavy metal. The basic piezoelectric parameters in TB ($d_{33} = 55\sim80$ pC/N, $T_C = 300°C$) are not very attractive. However, once *c*-axis-oriented ceramics are prepared, the d_{33} is dramatically enhanced, up to 240 pC/N. Further, since the Young's modulus $Y_{33}^E = 140$GPa is more than twice of that of PZT, higher generative stress is expected, which is suitable to ultrasonic motor applications. Taiyo Yuden developed a sophisticated preparation technology for oriented ceramics with a multilayer configuration, that is, preparation under strong magnetic field, much simpler than the flaky powder preparation.

11.3.1.2 Biodegradable Polymers

The above Pb-free materials are nontoxic and disposable. Murata Manufacturing Co. is further seeking biodegradable devices using L-type poly-lactic acid. PLLA is made of vegetable corn-based composition.[12] Because it exhibits pure piezoelectric without pyroelectric effect, the stress sensitivity is sufficient for leaf-grip remote controllers (with Nintendo GameBoy), which do not need a very long lifetime.

11.3.1.3 Low-Loss Piezoelectrics

High-power piezoelectrics with low loss have become a central research topic from the energy-efficiency improvement viewpoint; that is to say, "real (strain magnitude) to imaginary performance (heat generation reduction)." Reducing hysteresis and increasing the mechanical quality factor to amplify the resonance displacement is the primary target from the transducer application viewpoint. We discussed a universal loss characterization methodology in smart materials, piezoelectrics, and magnetostrictors; namely, by accurately measuring the mechanical quality factors Q_A for the resonance and Q_B for the antiresonance in the admittance/impedance curve, we can derive physical losses in Section 5.2.[13,14]

There are three losses in piezoelectrics: dielectric tan δ, elastic tan ϕ, and piezoelectric tan θ, each of which is further categorized into intensive (observable) and extensive (material parameter) losses as defined by: $\varepsilon^{X*} = \varepsilon^{X}(1 - j\tan\delta')$, $S^{E*} = S^{E}(1 - j\tan\phi')$, $d^{*} = d(1 - j\tan\theta')$; $\kappa^{x*} = \kappa^{x}(1 + j\tan\delta)$, $C^{D*} = C^{D}(1 + j\tan\phi)$, $h^{*} = h(1 + j\tan\theta)$. Though previous researchers neglected the piezoelectric loss (tan θ), we pointed out that piezoelectric loss in PZTs has almost a comparable magnitude with dielectric and elastic losses, and is essential to explain the admittance/impedance spectrum.

High power characterization of Pb-free piezoelectric and PZT disk samples is shown in Figure 11.5, where the resonance, Q_A, and antiresonance, Q_B, are plotted as a function of vibration velocity. Compared with the maximum vibration velocity (defined by the velocity that generates a 20°C temperature rise on the sample) of 0.3 m/s (rms) in hard PZTs, Pb-free piezoelectrics can exhibit a maximum vibration velocity higher than 0.5 m/s, double the energy density of a transducer.

11.3.2 Hard to Soft

We are now facing the revival polymer era after the first in the 1980s because of their elastically soft superiority. Larger, thinner, lighter, and mechanically flexible human interfaces are the current necessity in portable electronic devices, leading to the development of elastically soft displays, electronic circuits, and speakers/microphones.

11.3.2.1 Elastomer Actuators

Dielectric elastomer actuators (nonpiezoelectric, nonferroelectric) are based on the deformation of a soft polymer that acts as a dielectric between highly compliant electrodes. This effect is dominated by the Maxwell's stresses imposed by the compliant electrodes. Extremely high strains at low frequencies have been reported by Pelrine et al.[15] In-plane strains of more than 100% and 200% were observed in silicone and acrylic elastomers, respectively.

11.3.2.2 Electrostrictive Polymers

Polyvinylidene difluoride-trifluoroethylene copolymer is a well-known piezoelectric, which has been popularly used in sensor applications such as keyboards. Zhang et al. reported that the field induced

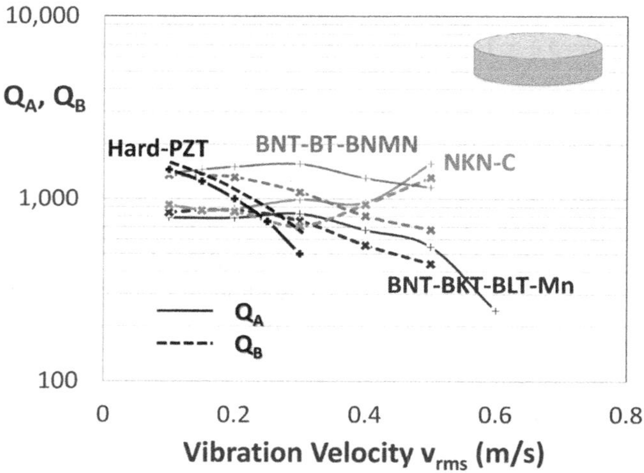

FIGURE 11.5 High-power characterization of Pb-free piezoelectrics and PZT. Maximum vibration velocity is larger for Pb-free materials.

strain level can be significantly enhanced up to 5% by using high-energy electron irradiation onto PVDF films, leading to an electrostrictive performance.[16]

11.3.2.3 Lead Zirconate Titanate Composites

Fuji Film news released their new bendable and foldable speakers using a 1:3 composite (PZT fine powder was mixed in a polymer film).[17] Superior acoustic performance seems to be promising for flat-speaker applications.

11.3.2.4 Large Strain Ceramics

$Pb(Zn_{1/3}Nb_{2/3})O_3$-$PbTiO_3$ (PZN-PT) or $Pb(Mg_{1/3}Nb_{2/3})O_3$-$PbTiO_3$ (PMN-PT) single crystals became focused due to the rubberlike-soft piezoceramic strain 25 years after the discovery. Since enhancement of induced strain level is a primary target, single crystals with a better capability to generate larger strains are being used in these days. In 1981, Kuwata et al. first reported an enormously large electromechanical coupling factor, $k_{33} = 92{\sim}95\%$, and piezoelectric constant $d_{33} = 1500$ pC/N in solid solution single crystals between relaxor and normal ferroelectrics, PZN-PT.[18,19] This discovery has been marked practically after more than 15 years when high-k materials have been given attention in medical acoustics. These data have been reconfirmed, and improved data were obtained recently, aiming at medical acoustic applications.[20,21] Strains as large as 1.7% can be induced practically for PZN-PT solid solution single crystals (Figure 11.6). It is notable that the highest values are observed for a rhombohedral composition only when the single crystal is poled along the perovskite [001] axis, not along the [111] spontaneous polarization axis.

11.3.3 Macro to Nano

In the micro (nano) electro-mechanical system (MEMS/NEMS) area, *piezoelectric-MEMS* is one of the miniaturization targets for integrating piezo-actuators in microscale devices, aiming at bio/medical applications for maintaining human health. PZT thin films are deposited on a silicon wafer, which is then micromachined to leave a membrane for fabricating microactuators and sensors, that is, microelectromechanical systems. Figure 11.7 illustrates a blood tester developed by Penn State

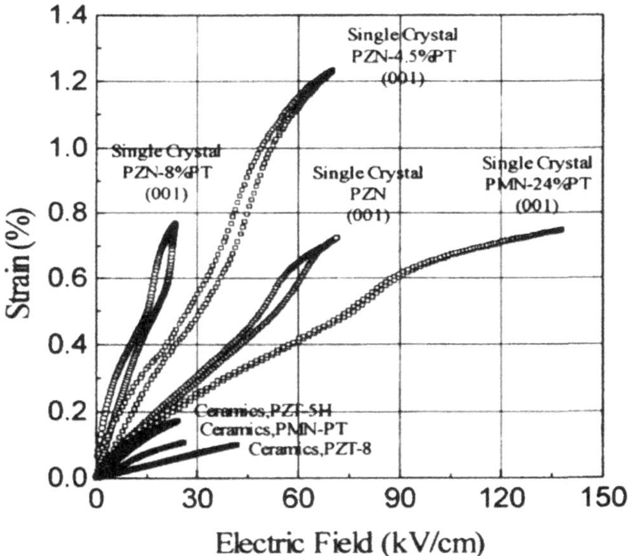

FIGURE 11.6 Field-induced strains in single crystals $Pb(Zn_{1/3}Nb_{2/3})O_3$-$PbTiO_3$ in comparison with PZT ceramics. (From S. E. Park and T. R. Shrout: *Mat. Res. Innovt.*, 1, 20, 1997.)

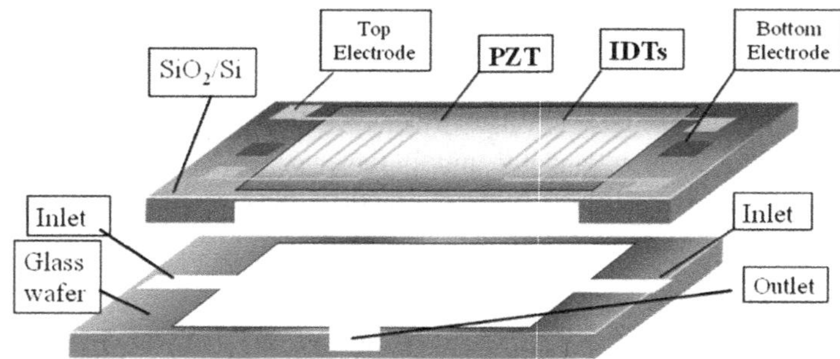

FIGURE 11.7 Structure of a PZT/silicon MEMS device, blood tester. (From S. Kalpat: Ph.D. Thesis, Penn State University, Fall, 2001.)

in collaboration with OMRON Corporation, Japan.[22] Applying voltage to two surface interdigital electrodes, the surface PZT film generates surface membrane waves, which soak up blood and the test chemical from the two inlets, then mix them in the center part, and send the mixture to the monitor part through the outlet. Finite element analysis (FEA) calculation was conducted to evaluate the flow rate of the liquid by changing the thickness of the PZT or the Si membrane, inlet and outlet nozzle size, and cavity thickness. However, care should be taken to choose rather thick (>10 micron) PZT films for generating reasonable mechanical energy density higher than 1 mW. Refer to [23] for updated piezoelectric MEMS studies.

11.3.4 Homo to Hetero

"Homo to hetero" structure change is also a recent research trend: stress gradient in terms of space in a dielectric material exhibits piezoelectric-equivalent sensing capability (i.e., *flexoelectricity*), while electric-field gradient in terms of space in a semiconductive piezoelectric can exhibit bimorph-equivalent flextensional deformation (*monomorph*).

Space gradient of stress or electric-field generates a direct or converse flexoelectric effect, expressed respectively by:

$$\boldsymbol{P}_l = \mu_{ijkl}(\partial \boldsymbol{x}_{ij} / \partial x_k), \tag{11.1}$$

$$\boldsymbol{X}_{ij} = \mu_{ijkl}(\partial \boldsymbol{E}_l / \partial x_k), \tag{11.2}$$

where $\boldsymbol{P}_l$, $\boldsymbol{E}_l$ are electric polarization, electric field; $\boldsymbol{X}_{ij}$, $\boldsymbol{x}_{ij}$ are elastic stress, strain; x_k is coordination in $\boldsymbol{x}_{ij}$ or $\boldsymbol{E}_l$; and μ_{ijkl} is denoted as a flexoelectric coefficient, which has a fourth-rank polar tensor symmetry, similar to the electrostrictive tensor.[24] This means that even a paraelectric material can generate charge under stress when the strain gradient is generated artificially in the material. Cross et al. demonstrated this "piezoelectric"-equivalent effect in various artificial designs, as shown in Figure 11.8.[25] A $Ba_{0.67}Sr_{0.33}TiO_3$ (BST) paraelectric composition sample with a trapezoid shape exhibited 10^{-7} C/m^2 of polarization under a strain gradient of 10^{-3}/m.

Conventional bimorph bending actuators are composed of two piezoelectric plates or two piezoelectrics and an elastic shim bonded together. The bonding layer, however, causes both an increase in hysteresis and a degradation of the displacement characteristics, as well as delamination problems. Furthermore, the fabrication process for such devices, which involves cutting, polishing, electroding, and bonding steps, is rather laborious and costly. Thus, a monolithic bending actuator (monomorph) that requires no bonding is a very attractive alternative structure.

Such a monomorph device can be produced from a single ceramic plate.[26] The operating principle is based on the combined action of a semiconductor contact phenomenon and the piezoelectric or

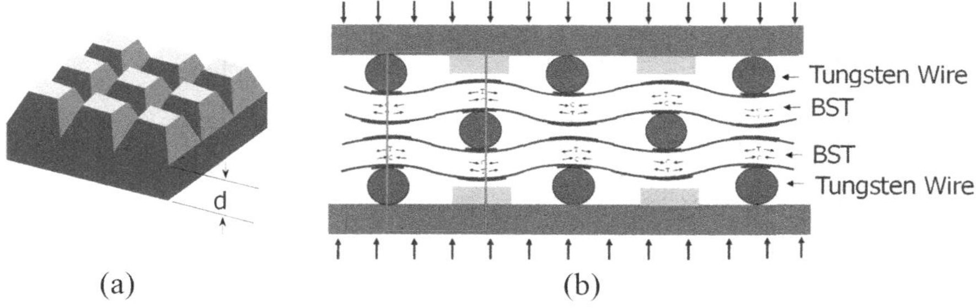

FIGURE 11.8 Space gradient of stress can be obtained in trapezoidal BST samples (a) or wire-inserted BST laminates (b).

electrostrictive effect. When metal electrodes are applied to both surfaces of a semiconductor plate and a voltage is applied as shown in Figure 11.9, the electric field is concentrated on one side (that is, the Schottky barrier), thereby generating a nonuniform field within the plate. When the piezoelectric (or electrostrictor) is slightly semiconductive, contraction along the surface occurs through the piezoelectric effect only on the side where the electric field is concentrated. The nonuniform field distribution generated in the ceramic causes an overall bending of the entire plate. The energy diagram of a modified structure including a very thin insulative layer is represented in Figure 11.10a.[27] The thin insulator layer increases the breakdown voltage. The rainbow actuator by Aura Ceramics is a modification of the basic semiconductive piezoelectric monomorph design, where half of the piezoelectric plate is reduced so as to make a thick semiconductive electrode that enhances the bending action.[28] The energy diagram for the rainbow device is shown in Figure 11.10b.[27]

11.3.5 Single to Multifunctional

Some new functions can be realized by coupling two effects. We developed magnetoelectric devices (i.e., voltage is generated by applying a magnetic field) by laminating magnetostrictive Terfenol-D and piezoelectric PZT materials, and demonstrated photostriction by coupling photovoltaic and piezoelectric effects in donor-doped PLZT.

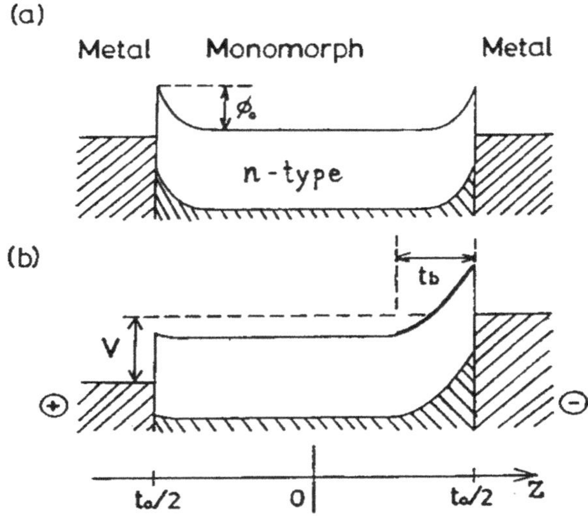

FIGURE 11.9 Schottky barrier generated at the interface between a semiconductive (n-type) piezoceramic and metal electrodes.

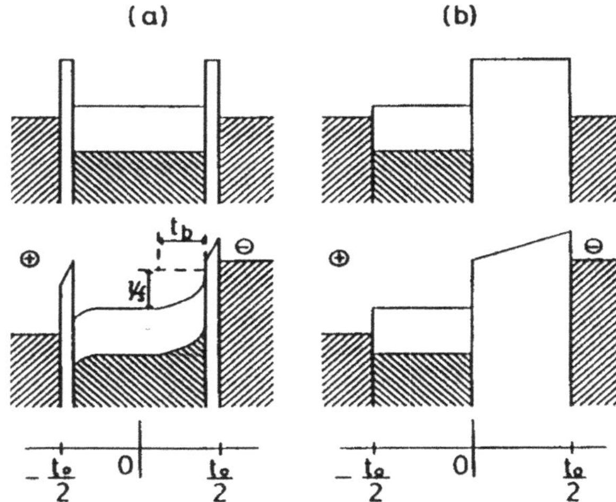

FIGURE 11.10 Energy diagrams for modified monomorph structures: (a) a device incorporating a very thin insulating layer and (b) the rainbow structure.

11.3.5.1 Magnetoelectric Effect

Similar to nuclear radiation, magnetic irradiation cannot be easily felt by humans. We cannot even purchase a magnetic field detector for a low frequency (50 or 60 Hz). Penn State, in collaboration with Seoul National University, developed a simple and handy magnetic noise sensor for these environmental monitoring purposes, for example, below a high-voltage power transmission line. Figure 11.11 shows a schematic structure of this device, in which a PZT disk is sandwiched by two Terfenol-D (magnetostrictor) disks.[29] When a magnetic field is applied on this composite, Terfenol will expand, which is mechanically transferred to PZT, leading to an electric charge generation from PZT. By monitoring the voltage generated in the PZT, we can detect the magnetic field. The key of this device is highly effective for a low frequency such as 50 Hz.[29]

11.3.5.2 Photostriction

A photostrictive actuator is a fine example of an intelligent material, incorporating "illumination sensing" and self production of "drive/control voltage" together with final "actuation." In certain ferroelectrics, a constant electromotive force is generated with exposure to light, and a photostrictive

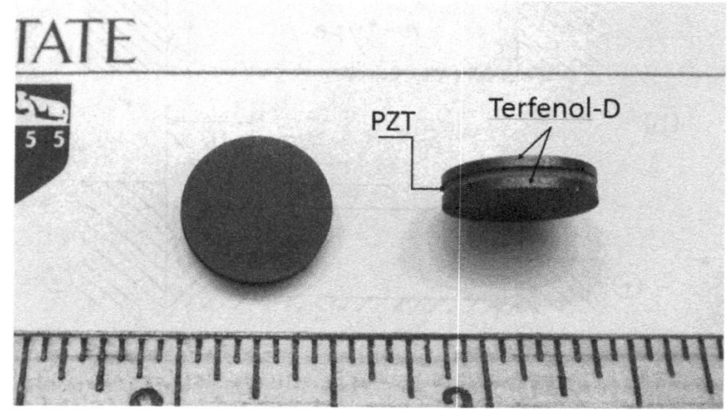

FIGURE 11.11 Magnetic noise sensor consisting of a laminated composite of a PZT and two Terfenol-D disks. (From J. Ryu et al.: *Jpn. J. Appl. Phys.*, 40, 4948–4951, 2001.)

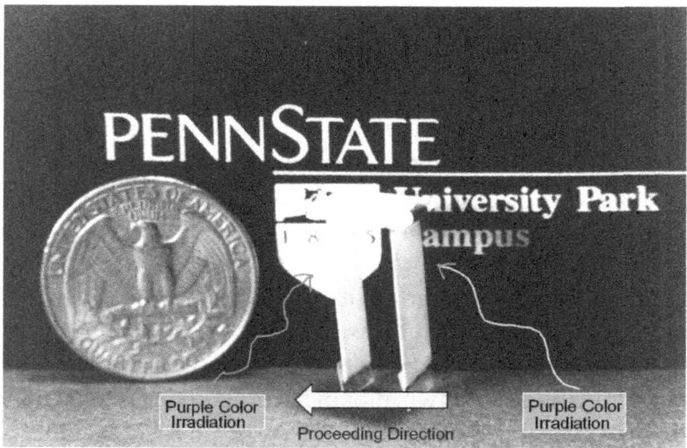

FIGURE 11.12 Photo-driven walking device using photostrictive PLZT bimorphs. (From K. Uchino: *J. Rob. Mech.*, 1(2), 124, 1989.)

strain results from the coupling of this bulk photovoltaic effect with inverse piezoelectricity. A bimorph unit has been made from PLZT 3/52/48 ceramic doped with slight addition of tungsten.[30] The remnant polarization of one PLZT layer is parallel to the plate and in the direction opposite to that of the other plate. When a violet light is irradiated to one side of the PLZT bimorph, a photovoltage of 1 kV/mm is generated, causing a bending motion. The tip displacement of a 20-mm bimorph with 0.4 mm in thickness was 150 μm, with a response time of 1 sec.

A photo-driven microwalking device, designed to begin moving with light illumination, has been developed.[31] As shown in Figure 11.12, it is simple in structure, having neither lead wires nor electric circuitry, with two bimorph legs fixed to a plastic board. When the legs are irradiated alternately with light, the device moves like an inchworm with a speed of 100 μm/min. In pursuit of thick film–type photostrictive actuators for space structure applications, in collaboration with the Jet Propulsion Laboratory, Penn State investigated the optimal range of sample thickness and surface roughness dependence of photostriction. Thirty-μm-thick PLZT films exhibit the maximum photovoltaic phenomenon.[32]

11.4 NEW APPLICATION DEVELOPMENT

11.4.1 Normal Technologies

One of the normal technologies, which are initiated politically, is sustainability technology. Sustainability technologies include:

- Power and energy (lack of oil, nuclear power plant, new energy harvesting)
- Rare material (rare-earth metal, Lithium)
- Food (rice, corn—biofuel)
- Toxic material
    - Restriction (heavy metal, Pb, dioxin)
    - Elimination/neutralization (mercury, asbestos)
    - Replacement material
- Environmental pollution
- Energy efficiency

In the application area, the global regime for ecological sustainability particularly accelerated new developments in ultrasonic disposal technology of hazardous materials, diesel injection valves for air pollution, and piezoelectric renewable energy harvesting systems.

11.4.1.1 Ultrasonic Disposal Technology

Ultrasonic lens cleaner is commonly used in homes. Industrial ultrasonic cleaners are widely utilized in the manufacturing lines of silicon wafers and liquid crystal glass substrates. Honda Electronics added another ultrasonic cleaner in conjunction with a washing machine produced by Sharp. (Private communication, Honda Electronics Company brochure) By using an *L-L* coupler horn to generate water cavitation, their machine can remove dirt on a shirt collar. It is noteworthy that we can reduce the amount of detergent (one of the major causes of river contamination) significantly by this technique.

By increasing the power level of water cavitation, we can make hazardous waste innocuous, because the cavitation (cyclic adiabatic compression) generates more than 3000°C locally for a short period. Hazardous wastes in underground or sewer water include dioxin, trichloroethylene, PCB, and environmental hormones.[33] As is well known, dioxin becomes another toxic material when it is burned at a low temperature, while it becomes innocuous only when burned at a high enough temperature.

11.4.1.2 Reduction of Contamination Gas

Diesel engines are recommended rather than regular gasoline cars from the energy conservation and global warming viewpoint. When we consider the total energy of gasoline production, both well-to-tank and tank-to-wheel, the energy efficiency, measured by the total energy required to realize unit drive distance for a vehicle (MJ/km), is of course better for high-octane gasoline than diesel oil. However, since the electric energy required for purification is significant, gasoline is inferior to diesel.[34] As is well known, the conventional diesel engine, however, generates toxic exhaust gases such as SO_x and NO_x. In order to solve this problem, new diesel injection valves were developed by Siemens, Bosch, and Toyota with piezoelectric multilayered actuators. Figure 11.13 shows such a common rail-type diesel injection valve with a ML piezo-actuator that produces high-pressure fuel and quick injection control. The highest reliability of these devices at an elevated temperature (150°C) for a long period (10 years) has been achieved.[35] The piezoelectric actuator is the key to increasing burning efficiency and minimizing toxic exhaust gases.

11.4.1.3 New Energy Harvesting Systems

One of the most recent research interests is piezoelectric energy harvesting. A cyclic electric field excited in a piezoelectric plate by environmental noise vibration is now accumulated into a rechargeable battery without consuming it as Joule heat. NEC-Tokin developed an LED traffic light array system driven by a piezoelectric windmill, which is operated by wind generation effectively

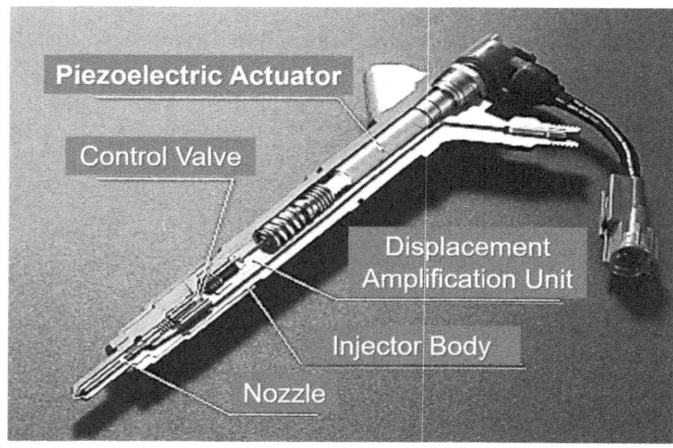

FIGURE 11.13 Common rail-type diesel injection valve with a piezoelectric multilayer actuator. (Courtesy Denso Corporation. From A. Fujii, *Proc. JTTAS Meeting on December* 2, 2005, Tokyo, 2005.)

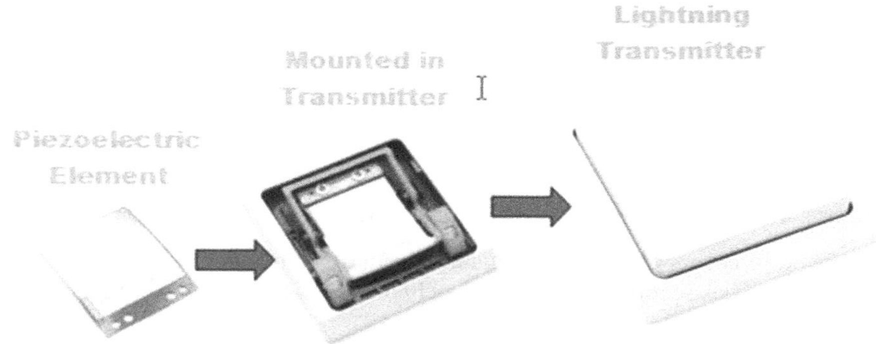

FIGURE 11.14 Lightning Switch with piezoelectric Thunder actuator. (Courtesy by Face Electronics.)

by passing automobiles. Successful products (million sellers) in the commercial market include Lightning Switch (a remote switch for room lights using a unimorph piezoelectric component) by PulseSwitch Systems, VA.[36] In addition to the living convenience, Lightning Switch (Figure 11.14) can reduce housing construction cost drastically due to a significant reduction of copper electric wire and aligning labor.

The Penn State group developed energy-harvesting piezoelectric devices based on a cymbal structure (29 mm ϕ, 1–2 mm thick), which can generate electric energy up to 100 mW under an automobile engine vibration.[37] By combining three cymbals in a rubber composite, a washerlike energy-harvesting sheet was developed for a hybrid car application, aiming at 1 W-level constant accumulation to a fuel cell. These efforts have provided the initial research guidelines and limitations of implementing the piezoelectric transducer. There are three major phases/steps associated with piezoelectric energy harvesting, as illustrated in Figure 11.15:[38] (1) mechanical-mechanical energy

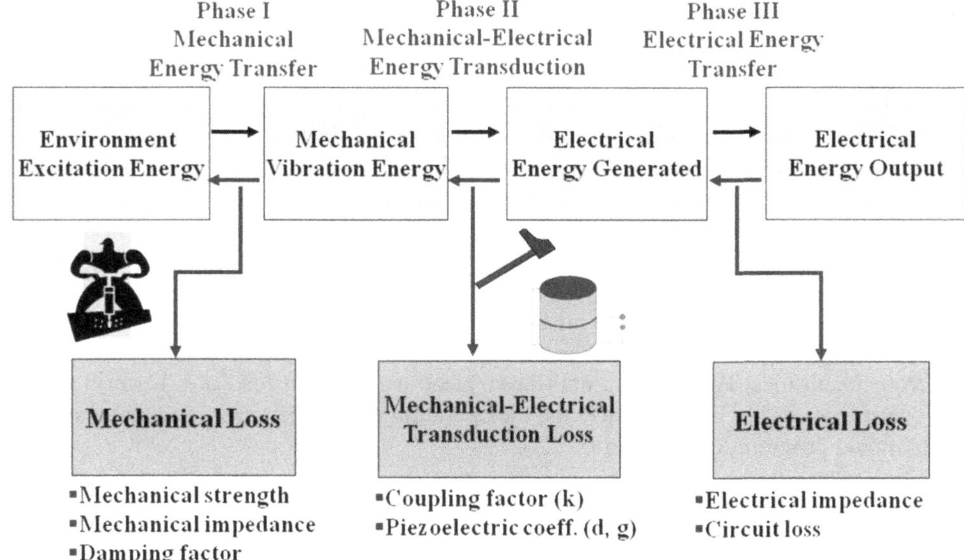

FIGURE 11.15 Three major phases associated with piezoelectric energy harvesting): (i) mechanical-mechanical energy transfer, (ii) mechanical-electrical energy transduction, and (iii) electrical-electrical energy transfer. (From K. Uchino: Energy Harvesting with Piezoelectric and Pyroelectric Materials. *Energy Flow Analysis in Piezoelectric Harvesting Systems*, Ed. N. Muensit, Partial Charge Chapter 4, Trans Tech Publications, Zurich, Switzerland, 2011.)

transfer, including mechanical stability of the piezoelectric transducer under large stresses, and mechanical impedance matching; (2) mechanical-electrical energy transduction, related to the electromechanical coupling factor in the composite transducer structure; and (3) electrical-electrical energy transfer, including electrical impedance matching, such as a DC/DC converter to accumulate the energy into a rechargeable battery.

11.4.2 Crisis Technology

Politico-engineering covers (1) legally regulated normal technologies such as sustainability, as discussed in the previous section, and (2) crisis technologies, which are further classified into five types of crises:[2]

- Natural disasters (earthquakes, tsunamis, tornadoes, hurricanes, lightning, etc.)
- Epidemic/infectious diseases (smallpox, polio, measles, and HIV)
- Enormous accident (Three Mile Island core meltdown accident, BP oil spill, etc.)
- Intentional accidents (acts of terrorism, criminal activity)
- Civil war, war, territorial aggression

As infectious or contagious disease involves some association with terrorist activities, those five are related to each other. In the United States, politicians were attacked with anthrax in 2001. In order to neutralize the biological attack, Pezeshk et al. at Penn State University developed a portable hypochlorous-acid disinfection device using a piezoelectric ultrasonic humidifier.[39] Hypochlorous acid is a strong disinfectant with no side effects on humans and would be ideal for disinfecting office and hospital buildings against viruses like SARS and anthrax. Coupled with the atomization of the acidic solution, much higher disinfection effects can be expected. This acid is not sold as a pure solution since it naturally disintegrates after a few hours. We designed a corrosion-resistant electrolytic cell to produce hypochlorous acid from brine. An ultrasonic piezoelectric atomizer was utilized to generate microdroplets of the diluted acid.

Regarding natural disasters, the research themes of urgent need in the actuator/sensor area include: (1) prediction technologies such as for earthquakes, tornadoes; (2) accurate monitoring and surveillance techniques; (3) technologies for gathering and managing crisis information and informing the public in a way not to bring about panic reaction; and (4) rescue technologies (autonomous unmanned underwater, aerial, land vehicles, robots, etc.).

Until 1960s, the development of weapons of mass destruction was the primary focus, including nuclear bombs and chemical weapons. However, based on the global trend for "*Jus in Bello* (Justice in War)," environmentally friendly "green" weapons became the mainstream in the twenty-first century, that is, minimally destructive weapons with a pinpoint target such as laser guns and rail guns. In this direction, programmable air-bust munitions were developed successfully starting in 2004. The 25-mm-caliber Programmable Ammunition by ATK Integrated Weapon Systems, AZ, and Micromechatronics, PA,[40] uses a multilayer piezo-actuator (instead of a battery) to generate electric energy under shot impact to activate the operational amplifiers that ignite a burst according to the command program (refer to Figure 7.5).

CHAPTER ESSENTIALS

1. Product Planning Strategy in the first 25 years of the twenty-first century: "politico-engineering" – Keywords: "cooperation, protection, reduction, and continuation (協守減維)"
2. Five key research trend changes in engineering:
   - "Performance to Reliability" (Pb-free piezoelectrics, biodegradable piezopolymer, low loss piezoelectric)
   - "Hard to Soft" (foldable piezopolymer film, PMN/PZN single crystals)

- "Macro to Nano" (piezo-MEMS)
- "Homo to Hetero" (flexoelectricity, monomorph)
- "Single to Multifunctional" (magnetoelectrics, photostriction)

3. Application areas: "politico-engineering"
   - The global regime for ecological sustainability particularly accelerated new developments in ultrasonic disposal technology of hazardous materials, diesel injection valves for air pollution, and piezoelectric renewable energy harvesting systems.
     - Power and energy (lack of oil, nuclear power plants, new energy harvesting)
     - Rare material (rare-earth metal, lithium)
     - Food (rice, corn—biofuel)
     - Toxic material
     - Restriction (heavy metals, Pb, dioxin)
     - Elimination/neutralization (mercury, asbestos)
     - Replacement material
     - Environmental pollution
     - Energy efficiency
   - Crisis technologies include:
     - Natural disasters (earthquakes, tsunamis, tornadoes, hurricanes, lightning, etc.)
     - Epidemic/infectious diseases (smallpox, polio, measles, and HIV)
     - Enormous accident (Three Mile Island core meltdown accident, BP oil spill, etc.)
     - Intentional accidents (acts of terrorism, criminal activity)
     - Civil war, war, territorial aggression

CHECK POINT

1. What are the four keywords to describe the twenty-first century "product planning strategy" in the "politico-engineering" period? Answer these four "– tions."
2. Provide a remaining recent research trend change during the last 10 years in actuators, in addition to the following four: "Hard to Soft," "Macro to Nano," "Homo to Hetero" and "Single to Multifunctional."
3. (T/F) High-power ultrasound energy can make dioxin into innocuous material without increasing temperature. True or False?
4. (T/F) "Sustainability" and "crisis" technologies are usually controlled by politicians. True or False?

CHAPTER PROBLEMS

11.1 Based on a literature survey from the past 5 years, discuss and summarize recent research related to the reliability of piezoelectric actuators. The report should deal with issues such as actuator design, low-hysteresis/high-power piezoelectrics, manufacturing reproducibility, and high temperature characteristics. Your survey should include at least five published papers.

   a. Classify the reliability issues addressed in the papers in terms of materials, design, and drive/control-related issues.
   b. Summarize the problems addressed in the papers and their possible solutions.
   c. Provide a complete list of the papers used for your report.

11.2 Create a new crisis technology device with using smart materials. A brief device development proposal is required. New crisis technologies with using smart materials are exemplified by:

   - Piezoelectrics—Nuclear power plant–monitoring sensors
   - Pyroelectrics—Missile target-chasing sensors

You may expand the coverage on crisis technologies including the following incidents:

- Natural disaster (earthquake, tsunami, tornado, typhoon, thunder)
- Infectious/contagious disease (smallpox, poliomyelitis, measles, HIV)
- Enormous accident (Three Mile Island nuclear power plant meltdown, BP deep-water oil flow)
- Intentional (terrorist/criminal) incident
- External and civil war/territorial invasion

REFERENCES

1. Information Resources: *Multiple Year Market Research Data from iRAP (Innovative Research & Products)*, Market Publishers; IDTechEx.
2. K. Uchino: "Politico-Engineering in Piezoelectric Devices," *Proc. 13th Int'l Conf. New Actuators*, Bremen, Germany, p. A1.0, June 18-20, 2012.
3. E. Wainer and N. Salomon: "High titania dielectrics," *Trans. Electrochem. Soc.*, 89, 1946.
4. T. Ogawa: "On barium titanate ceramics" [in Japanese], *Busseiron Kenkyu*, (6), 1–27, 1947.
5. B. M. Vul: "High and ultrahigh dielectric constant materials" [in Russian], *Electrichestvo*, (3), 1946.
6. B. Jaffe, W. Cook and H. Jaffe: *Piezoelectric Ceramics*, Academic Press, London, 1971.
7. E. Sawaguchi: "Ferroelectricity versus anti-ferroelectricity in the solid solutions of $PbZrO_3$ and $PbTiO_3$," *J. Phys. Soc. Japan*, 8, 615–629, 1953.
8. http://unfccc.int/kyoto_protocol/items/2830.php
9. T. Tou, Y. Hamaguchi, Y. Maida, H. Yamamori, K. Takahashi and Y. Terashima: *Jpn. J. Appl. Phys.*, 48, 2009. 07GM03
10. Y. Saito: *Jpn. J. Appl. Phys.*, 35, 5168–73, 1996.
11. Y. Doshida: Proc. 81st Smart Actuators/Sensors Study Committee, *JTTAS, December 11*, Tokyo, 2009.
12. http://www.murata.co.jp/corporate/ad/article/metamorphosis16/Application_note/
13. Y. Zhuang, S. O. Ural, S. Tuncdemir, A. Amin, and K. Uchino: *Jpn. J. Appl. Phys.*, 49, 021503, 2010.
14. K. Uchino, Y. Zhuang, and S. O. Ural: "Loss determination methodology for a piezoelectric ceramic: New phenomenological theory and experimental proposals," *J. Adv. Dielectrics*, 1(1), 17–31, 2011.
15. R. E. Pelrine, R. D. Kornbluh, Q. Pei, J. P. Joseph: "High-speed electrically actuated elastomers with strain greater than 100%," *Science*, 287, 836, 2000.
16. V. Bharti, H. S. Xu, G. Shanti, Q. M. Zhang and K. Liang: *J. Appl. Phys.*, 87, 452, 2000.
17. http://techon.nikkeibp.co.jp/english/NEWS_EN/20130201/263651/
18. J. Kuwata, K. Uchino and S. Nomura: "Phase transitions on the $Pb(Zn_{1/3}Nb_{2/3})O_3$-$PbTiO_3$ system," *Ferroelectrics*, 37, 579–582, 1981.
19. J. Kuwata, K. Uchino and S. Nomura: Japan. "Dielectric and piezoelectric properties of $0.91Pb(Zn_{1/3}Nb_{2/3})O_3$—$0.09PbTiO_3$ single crystals," *Jpn. J. Appl. Phys.*, 21, 1298–1302, 1982.
20. K. Yanagiwawa, H. Kanai and Y. Yamashita: *Jpn. J. Appl. Phys.*, 34, 536, 1995.
21. S. E. Park and T. R. Shrout: *Mat. Res. Innovt.*, 1, 20, 1997.
22. S. Kalpat: *Ph.D. Thesis*, Penn State Universtity, Fall, 2001.
23. S. Tadigadapa and K. Mateti: "Piezoelectric MEMS sensors: state-of-the-art and perspectives," *Meas. Sci. Technol.*, 20, 092001, 2009.
24. A. K. Tagantsev: *Phys. Rev. B*, 34, 5883, 1986.
25. L. E. Cross: *J. Mater. Sci.*, 41, 53, 2006.
26. K. Uchino, M. Yoshizaki, K. Kasai, H. Yamamura, N. Sakai and H. Asakura: *Jpn. J. Appl. Phys.*, 26, 1046, 1987.
27. K. Uchino, M. Yoshizaki and A. Nagao: *Jpn. J. Appl. Phys.*, 26(Suppl. 26-2), 201, 1987.
28. Aura Ceramics, Inc.: USA, Catalogue "Rainbow."
29. J. Ryu, A. Vazquez Carazo, K. Uchino and H. E. Kim: *Jpn. J. Appl. Phys.*, 40, 4948–4951, 2001.
30. K. Uchino: *Mat. Res. Innovat.*, 1, 163, 1997.
31. K. Uchino: *J. Rob. Mech.*, 1(2), 124, 1989.
32. P. Poosanaas, K. Tonooka and K. Uchino: *J. Mechatronics*, 10, 467–487, 2000.
33. I. Hua and M. R. Hoffmann, "Optimization of ultrasonic irradiation as an advanced oxidation technology," *Environ. Sci. Technol.*, 31(8), 2237–2243, 1997.
34. http://www.marklines.com/ja/amreport/rep094_200208.jsp
35. A. Fujii, *Proc. JTTAS Meeting on December 2, 2005*, Tokyo, 2005.

36. K. Uchino and T. Ishii, *J. Jpn. Ceram. Soc.*, 96(8), 863–867, 1988.
37. H. W. Kim, S. Priya, K. Uchino and R. E. Newnham, *J. Electroceramics*, 15, 27–34, 2005.
38. K. Uchino: Energy Harvesting with Piezoelectric and Pyroelectric Materials." *Energy Flow Analysis in Piezoelectric Harvesting Systems*, Ed. N. Muensit, Partial Charge Chapter 4, Trans Tech Publications, Zurich, Switzerland, 2011.
39. A. Pezeshk, Y. Gao and K. Uchino, "Ultrasonic piezoelectric hypochlorous acid humidifier for disinfection applications," *NSF EE REU Penn State Annual Research Journal*, II, ISBN 0-913260-04-5, 2004.
40. http://www.atk.com/MediaCenter/mediacenter_video gallery.asp

Index

G

H

I

J

K

L

M

N

Q

R

V

W

Y

Z

9781032240695